KEY OF ATOM COLORS USED IN MOLECULAR MODELS IN *HOLT CHEMISTRY*

Element	Color	Element	Color
Hydrogen, H		Silicon, Si	
Helium, He		Phosphorus, P	
Carbon, C		Sulfur, S	
Nitrogen, N		Chlorine, Cl	
Oxygen, O		Argon, Ar	
Fluorine, F		Iron, Fe	
Neon, Ne		Copper, Cu	
Sodium, Na (similar color used for all Group 1 metals)		Bromine, Br	
Magnesium, Mg (similar color used for all Group 2 metals)		Silver, Ag	
Aluminum, Al		Iodine, I	

COMMON IONS

Cation	Symbol	Cation	Symbol	Anion	Symbol	Anion	Symbol
Aluminum	Al^{3+}	Lead(II)	Pb^{2+}	Acetate	CH_3COO^-	Hydrogen sulfate	HSO_4^-
Ammonium	NH_4^+	Magnesium	Mg^{2+}	Bromide	Br^-	Hydroxide	OH^-
Arsenic(III)	As^{3+}	Mercury(I)	Hg_2^{2+}	Carbonate	CO_3^{2-}	Hypochlorite	ClO^-
Barium	Ba^{2+}	Mercury(II)	Hg^{2+}	Chlorate	ClO_3^-	Iodide	I^-
Calcium	Ca^{2+}	Nickel(II)	Ni^{2+}	Chloride	Cl^-	Nitrate	NO_3^-
Chromium(II)	Cr^{2+}	Potassium	K^+	Chlorite	ClO_2^-	Nitrite	NO_2^-
Chromium(III)	Cr^{3+}	Silver	Ag^+	Chromate	CrO_4^{2-}	Oxide	O_2^-
Cobalt(II)	Co^{2+}	Sodium	Na^+	Cyanide	CN^-	Perchlorate	ClO_4^-
Cobalt(III)	Co^{3+}	Strontium	Sr^{2+}	Dichromate	$Cr_2O_7^{2-}$	Permanganate	MnO_4^-
Copper(I)	Cu^+	Tin(II)	Sn^{2+}	Fluoride	F^-	Peroxide	O_2^{2-}
Copper(II)	Cu^{2+}	Tin(IV)	Sn^{4+}	Hexacyanoferrate(II)	$Fe(CN)_6^{4-}$	Phosphate	PO_4^{3-}
Hydronium	H_3O^+	Titanium(III)	Ti^{3+}	Hexacyanoferrate(III)	$Fe(CN)_6^{3-}$	Sulfate	SO_4^{2-}
Iron(II)	Fe^{2+}	Titanium(IV)	Ti^{4+}	Hydride	H^-	Sulfide	S^{2-}
Iron(III)	Fe^{3+}	Zinc	Zn^{2+}	Hydrogen carbonate	HCO_3^-	Sulfite	SO_3^{2-}

HOLT
Chemistry

AUTHORS

R. Thomas Myers, Ph.D.
Professor Emeritus of Chemistry
Kent State University
Kent, Ohio

Keith B. Oldham, D.Sc.
Professor Emeritus of Chemistry
Trent University,
Peterborough, Ontario, Canada

Salvatore Tocci
Science Writer
East Hampton, New York

HOLT, RINEHART AND WINSTON

A Harcourt Education Company

Orlando • **Austin** • New York • San Diego • Toronto • London

R. Thomas Myers, Ph.D.
Dr. Myers received his B.S. and Ph.D. in chemistry from West Virginia University in Morgantown, West Virginia. He was an assistant professor of chemistry and department head at Waynesburg College in Waynesburg, Pennsylvania, and an assistant professor at the Colorado School of Mines in Golden, Colorado. He then joined the chemistry faculty at Kent State University in Kent, Ohio, where he is currently a professor emeritus of chemistry.

Keith B. Oldham, D.Sc.
Dr. Oldham received his B.Sc. and Ph.D. in chemistry from the University of Manchester in Manchester, England and performed postdoctoral research at the Noyes Chemical Laboratory at the University of Illinois in Urbana, Illinois. He was awarded a D.Sc. from the University of Manchester for his novel research in the area of electrode processes. He was an assistant lecturer of chemistry at the Imperial College in London and a lecturer in chemistry at the University of Newcastle upon Tyne. Dr. Oldham worked as a scientist for the North American Rockwell Corporation where he performed research for NASA. After 24 years on the faculty, he is now a professor emeritus at Trent University in Peterborough, Canada.

Salvatore Tocci
Salvatore Tocci received his B.A from Cornell University in Ithaca, New York and a Master of Philosophy from the City University of New York in New York City. He was a science teacher and science department chairperson at East Hampton High School in East Hampton, New York, and an adjunct instructor at Syracuse University in Syracuse, New York. He was also an adjunct lecturer at the State University of New York at Stony Brook and a science teacher at Southold High School in Southold, New York. Mr. Tocci is currently a science writer and educational consultant.

CONTRIBUTING WRITERS

Inclusion Specialists

Joan A. Solorio
Special Education Director
Austin Independent School District
Austin, Texas

John A. Solorio
Multiple Technologies Lab Facilitator
Austin Independent School District
Austin, Texas

Lab Safety Consultant

Allen B. Cobb
Science Writer
La Grange, Texas

Lab Tester

Michelle Johnston
Trent University
Peterborough, Ontario, Canada

Teacher Edition Development

Ann Bekebrede
Science Writer
Sherborn, Massachusetts

Elizabeth M. Dabrowski
Science Department Chair
Magnificat High School
Cleveland, Ohio

Frances Jenkins
Science Writer
Sunburg, Ohio

Laura Prescott
Science Writer
Pearland, Texas

Matt Walker
Science Writer
Portland, Oregon

ACADEMIC REVIEWERS

Eric Anslyn, Ph.D.
Professor of Chemistry
Department of Chemistry and Biochemistry
The University of Texas
Austin, Texas

Paul Asimow, Ph.D.
Assistant Professor of Geology and Geochemistry
Division of Geological and Planetary Sciences
California Institute of Technology
Pasadena, California

Nigel Atkinson, Ph.D.
Associate Professor of Neurobiology
Institute for Cellular and Molecular Biology
The University of Texas
Austin, Texas

Scott W. Cowley, Ph.D.
Associate Professor
Department of Chemistry and Geochemistry
Colorado School of Mines
Golden, Colorado

Regina Frey, Ph.D.
Professor of Chemistry
Department of Chemistry
Washington University
St. Louis, Missouri

William B. Guggino, Ph.D.
Professor of Physiology
The Johns Hopkins University
Baltimore, Maryland

Joan Hudson, Ph.D.
Associate Professor of Botany
Sam Houston State University
Huntsville, Texas

Wendy L. Keeney-Kennicutt, Ph.D.
Associate Professor of Chemistry
Department of Chemistry
Texas A&M University
College Station, Texas

Samuel P. Kounaves
Associate Professor of Chemistry
Department of Chemistry
Tufts University
Medford, Massachusetts

Phillip LaRoe
Instructor
Department of Physics and Chemistry
Central Community College
Grande Island, Nebraska

Jeanne L. McHale, Ph.D.
Professor of Chemistry
College of Science
University of Idaho
Moscow, Idaho

Gary Mueller, Ph.D.
Associate Professor of Nuclear Engineering
Department of Engineering
University of Missouri
Rolla, Missouri

Brian Pagenkopf, Ph.D.
Professor of Chemistry
Department of Chemistry and Biochemistry
The University of Texas
Austin, Texas

Charles Scaife, Ph.D.
Chemistry Professor
Department of Chemistry
Union College
Schenectady, New York

Fred Seaman, Ph.D.
Research Scientist and Chemist
Department of Pharmacological Chemistry
The University of Texas
Austin, Texas

Peter Sheridan, Ph.D.
Associate Professor of Chemistry
Department of Chemistry
Colgate University
Hamilton, New York

Spencer Steinberg, Ph.D.
Associate Professor of Environmental Organic Chemistry
Department of Chemistry
University of Nevada
Las Vegas, Nevada

Continued on next page

Aaron Timperman, Ph.D.
Professor of Chemistry
Department of Chemistry
University of West Virginia
Morgantown, West Virginia

Richard S. Treptow, Ph.D.
Professor of Chemistry
Department of Chemistry
and Physics
Chicago State University
Chicago, Illinois

Martin VanDyke, Ph.D.
*Professor Emeritus of
Chemistry*
Front Range Community
College
Westminister, Colorado

Charles Wynn, Ph.D.
Chemistry Assistant Chair
Department of Physical
Sciences
Eastern Connecticut State
University
Willimantic, Connecticut

TEACHER REVIEWERS

David Blinn
Secondary Sciences Teacher
Wrenshall High School
Wrenshall, Minnesota

Robert Chandler
Science Teacher
Soddy-Daisy High School
Soddy-Daisy, Tennessee

Cindy Copolo, Ph.D.
Science Specialist
Summit Solutions
Bahama, North Carolina

Linda Culp
Science Teacher
Thorndale High School
Thorndale, Texas

Chris Diehl
Science Teacher
Belleville High School
Belleville, Michigan

Alonda Droege
Science Teacher
Seattle, Washington

Benjamen Ebersole
Science Teacher
Donnegal High School
Mount Joy, Pennsylvania

Jeffrey L. Engel
Science Teacher
Madison County High School
Athens, Georgia

Stacey Hagberg
Science Teacher
Donnegal High School
Mount Joy, Pennsylvania

Gail Hermann
Science Teacher
Quincy High School
Quincy, Illinois

Donald R. Kanner
*Physics and Chemistry
Instructor*
Lane Technical High School
Chicago, Illinois

Edward Keller
Science Teacher
Morgantown High School
Morgantown, West Virginia

Stewart Lipsky
Science Teacher
Seward Park High School
New York, New York

Mike Lubich
Science Teacher
Maple Town High School
Greensboro, Pennsylvania

Thomas Manerchia
*Environmental Science
Teacher, Retired*
Archmere Academy
Claymont, Delaware

Betsy McGrew
Science Teacher
Star Charter School
Austin, Texas

Jennifer Seelig-Fritz
Science Teacher
North Springs High School
Atlanta, Georgia

Dyanne Semerjibashian
Science Teacher
Star Charter School
Austin, Texas

Linnaea Smith
Science Teacher
Bastrop High School
Bastrop, Texas

Gabriela Waschesky, Ph.D.
*Science and Mathematics
Teacher*
Emery High School
Emeryville, California

(Credits and Acknowledgments continued on p. 908)

Contents

In Brief

v

Contents

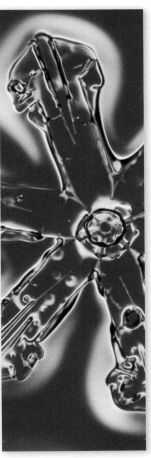

APPENDICES

LABORATORY EXPERIMENTS

 Safety in the Chemistry Laboratory **751**

SAMPLE PROBLEMS

SKILLS Toolkit

FEATURES

Science and Technology

Consumer Focus

Element Spotlight

HOW TO USE YOUR TEXTBOOK

Your Roadmap for Success with *Holt Chemistry*

Read the Objectives

Objectives tell you what you'll need to know.

STUDY TIP Reread the objectives when studying for a test to be sure you know the material.

Study the Key Terms

Key Terms are listed for each section. Learn the definitions of these terms because you will most likely be tested on them. Use the glossary to locate definitions quickly.

STUDY TIP If you don't understand a definition, reread the page where the term is introduced. The surrounding text should help make the definition easier to understand.

Take Notes and Get Organized

Keep a science notebook so that you are ready to take notes when your teacher reviews the material in class. Keep your assignments in this notebook so that you can review them when studying for the chapter test. Appendix B, located in the back of this book, describes a number of Study Skills that can help you succeed in chemistry, including several approaches to note taking.

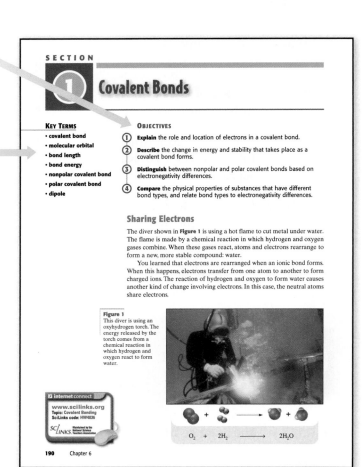

SECTION

1 Covalent Bonds

KEY TERMS
- covalent bond
- molecular orbital
- bond length
- bond energy
- nonpolar covalent bond
- polar covalent bond
- dipole

OBJECTIVES

1. **Explain** the role and location of electrons in a covalent bond.
2. **Describe** the change in energy and stability that takes place as a covalent bond forms.
3. **Distinguish** between nonpolar and polar covalent bonds based on electronegativity differences.
4. **Compare** the physical properties of substances that have different bond types, and relate bond types to electronegativity differences.

Sharing Electrons

The diver shown in **Figure 1** is using a hot flame to cut metal under water. The flame is made by a chemical reaction in which hydrogen and oxygen gases combine. When these gases react, atoms and electrons rearrange to form a new, more stable compound: water.

You learned that electrons are rearranged when an ionic bond forms. When this happens, electrons transfer from one atom to another to form charged ions. The reaction of hydrogen and oxygen to form water causes another kind of change involving electrons. In this case, the neutral atoms share electrons.

Figure 1
This diver is using an oxyhydrogen torch. The energy released by the torch comes from a chemical reaction in which hydrogen and oxygen react to form water.

internet connect
www.scilinks.org
Topic: Covalent Bonding
SciLinks code: HW4036

$$O_2 \ + \ 2H_2 \ \longrightarrow \ 2H_2O$$

190 Chapter 6

Be Resourceful, Use the Web

Internet Connect boxes in your textbook take you to resources that you can use for science projects, reports, and research papers. Go to **scilinks.org,** and type in the SciLinks code to get information on a topic.

Visit go.hrw.com
Find worksheets and other materials that go with your textbook at **go.hrw.com.** Click on the textbook icon and the table of contents to see all of the resources for each chapter.

Next think about the bonds that form between carbon and chlorine and between aluminum and chlorine. The electronegativity difference between C and Cl is 0.6. These two elements form a polar covalent bond. The electronegativity difference between Al and Cl is 1.6. These two elements also form a polar covalent bond. However, the larger difference between Al and Cl means that the bond between these two elements is more polar, with greater partial charges, than the bond between C and Cl is.

Properties of Substances Depend on Bond Type

The type of bond that forms determines the physical and chemical properties of the substance. For example, metals, such as potassium, are very good electric conductors in the solid state. This property is the result of metallic bonding. Metallic bonds are the result of the attraction between the electrons in the outermost energy level of each metal atom and all of the other atoms in the solid metal. The metal atoms are held in the solid because all of the valence electrons are attracted to all of the atoms in the solid. These valence electrons can move easily from one atom to another. They are free to roam around in the solid and can conduct an electric current.

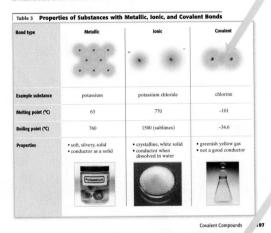

Table 3   Properties of Substances with Metallic, Ionic, and Covalent Bonds

Bond type	Metallic	Ionic	Covalent
Example substance	potassium	potassium chloride	chlorine
Melting point (°C)	63	770	−101
Boiling point (°C)	760	1500 (sublimes)	−34.6
Properties	• soft, silvery, solid • conductor as a solid	• crystalline, white solid • conductor when dissolved in water	• greenish yellow gas • not a good conductor

Covalent Compounds   197

① Section Review

UNDERSTANDING KEY IDEAS

1. Describe the attractive forces and repulsive forces that exist between two atoms as the atoms move closer together.

2. Compare a bond between two atoms to a spring between two students.

3. In what two ways can two atoms share electrons when forming a covalent bond?

4. What happens in terms of energy and stability when a covalent bond forms?

5. How are the partial charges shown in a polar covalent molecule?

6. What information can be obtained by knowing the electronegativity differences between two elements?

7. Why do molecular compounds have low melting points and low boiling points relative to ionic substances?

CRITICAL THINKING

8. Why does the distance between two nuclei in a covalent bond vary?

9. How does a molecular orbital differ from an atomic orbital?

10. How does the strength of a covalent bond relate to bond length?

11. Compare the degree of polarity in HF, HCl, HBr, and HI.

12. Given that it has the highest electronegativity, can a fluorine atom ever form a nonpolar covalent bond? Explain your answer.

13. What does a small electronegativity difference reveal about the strength of a covalent bond?

14. Based on electronegativity values, which bond has the highest degree of ionic character: H—S, Si—Cl, or Cs—Br?

198   Chapter 6

Use the Illustrations and Photos

Art shows complex ideas and processes. Learn to analyze the art so that you better understand the material you read in the text.

Tables and graphs display important information in an organized way to help you see relationships.

A picture is worth a thousand words. Look at the photographs to see relevant examples of science concepts you are reading about.

Answer the Section Reviews

Section Reviews test your knowledge over the main points of the section. Critical Thinking items challenge you to think about the material in greater depth and to find connections that you infer from the text.

STUDY TIP When you can't answer a question, reread the section. The answer is usually there.

Do Your Homework

Your teacher may assign worksheets to help you understand and remember the material in the chapter.

STUDY TIP Don't try to answer the questions without reading the text and reviewing your class notes. A little preparation up front will make your homework assignments a lot easier. Answering the items in the Chapter Review will help prepare you for the chapter test.

Visit Holt Online Learning
If your teacher gives you a special password to log onto the **Holt Online Learning** site, you'll find your complete textbook on the Web. In addition, you'll find some great learning tools and practice quizzes. You'll be able to see how well you know the material from your textbook.

Visit CNN Student News
You'll find up-to-date events in science at **www.cnnstudentnews.com**.

THE SCIENCE OF CHEMISTRY

For one weekend, an ice rink in Tacoma, Washington became a work of art. Thousands of people came to see the amazing collection of ice and lights on display. Huge blocks of ice, each having a mass of about 136 kg, were lit from the inside by lights. The glowing gas in each light made the solid ice shine with color. And as you can see, lights of many different colors were used in the display. In this chapter, you will learn about matter. You will learn about the properties used to describe matter. You will also learn about the changes matter can undergo. Finally, you will learn about classifying matter based on its properties.

START-UP ACTIVITY

SAFETY PRECAUTIONS

Classifying Matter

PROCEDURE

1. Examine the **objects** provided by your teacher.

2. Record in a table observations about each object's individual characteristics.

3. Divide the objects into at least three different categories based on your observations. Be sure that the objects in each category have something in common.

ANALYSIS

1. Describe the basis of your classification for each category you created.

2. Give an example that shows how using these categories makes describing the objects easier.

3. Describe a system of categories that could be used to classify matter. Explain the basis of your categories.

Pre-Reading Questions

① Do you think there are "good chemicals" and "bad chemicals"? If so, how do they differ?

② What are some of the classifications of matter?

③ What is the difference between a chemical change and a physical change?

SECTION

1 What Is Chemistry?

KEY TERMS

- chemical
- chemical reaction
- states of matter
- reactant
- product

OBJECTIVES

1. **Describe** ways in which chemistry is a part of your daily life.

2. **Describe** the characteristics of three common states of matter.

3. **Describe** physical and chemical changes, and give examples of each.

4. **Identify** the reactants and products in a chemical reaction.

5. **List** four observations that suggest a chemical change has occurred.

Working with the Properties and Changes of Matter

Do you think of chemistry as just another subject to be studied in school? Or maybe you feel it is important only to people working in labs? The effects of chemistry reach far beyond schools and labs. It plays a vital role in your daily life and in the complex workings of your world.

Look at **Figure 1.** Everything you see, including the clothes the students are wearing and the food the students are eating, is made of chemicals. The students themselves are made of chemicals! Even things you cannot see, such as air, are made up of chemicals.

Chemistry is concerned with the properties of chemicals and with the changes chemicals can undergo. A **chemical** is any substance that has a definite composition—it's always made of the same stuff no matter where the chemical comes from. Some chemicals, such as water and carbon dioxide, exist naturally. Others, such as polyethylene, are manufactured. Still others, such as aluminum, are taken from natural materials.

chemical

any substance that has a defined composition

Figure 1
Chemicals make up everything you see every day.

You Depend on Chemicals Every Day

Many people think of chemicals in negative terms—as the cause of pollution, explosions, and cancer. Some even believe that chemicals and chemical additives should be banned. But just think what such a ban would mean—after all, everything around you is composed of chemicals. Imagine going to buy fruits and vegetables grown without the use of any chemicals at all. Because water is a chemical, the produce section would be completely empty! In fact, the entire supermarket would be empty because all foods are made of chemicals.

The next time you are getting ready for school, look at the list of ingredients in your shampoo or toothpaste. You'll see an impressive list of chemicals. Without chemicals, you would have nothing to wear. The fibers of your clothing are made of chemicals that are either natural, such as cotton or wool, or synthetic, such as polyester. The air you breathe, the food you eat, and the water you drink are made up of chemicals. The paper, inks, and glue used to make the book you are now reading are chemicals, too. You yourself are an incredibly complex mixture of chemicals.

Chemical Reactions Happen All Around You

You will learn in this course that changes in chemicals—or **chemical reactions**—are taking place around you and inside you. Chemical reactions are necessary for living things to grow and for dead things to decay. When you cook food, you are carrying out a chemical reaction. Taking a photograph, striking a match, switching on a flashlight, and starting a gasoline engine require chemical reactions.

Using reactions to manufacture chemicals is a big industry. **Table 1** lists the top eight chemicals made in the United States. Some of these chemicals may be familiar, and some you may have never heard of. By the end of this course, you will know a lot more about them. Chemicals produced on a small scale are important, too. Life-saving antibiotics, cancer-fighting drugs, and many other substances that affect the quality of your life are also products of the chemical industry.

chemical reaction

the process by which one or more substances change to produce one or more different substances

internet connect
www.scilinks.org
Topic: Chemicals
SciLinks code: HW4030

SC*LINKS* Maintained by the National Science Teachers Association

Table 1	**Top Eight Chemicals Made in the United States (by Weight)**		
Rank	**Name**	**Formula**	**Uses**
1	sulfuric acid	H_2SO_4	production of fertilizer; metal processing; petroleum refining
2	ethene	C_2H_4	production of plastics; ripening of fruits
3	propylene	C_3H_6	production of plastics
4	ammonia	NH_3	production of fertilizer; refrigeration
5	chlorine	Cl_2	bleaching fabrics; purifying water; disinfectant
6	phosphoric acid (anhydrous)	P_2O_5	production of fertilizer; flavoring agent; rustproofing metals
7	sodium hydroxide	$NaOH$	petroleum refining; production of plastics
8	1,2-dichloroethene	$C_2H_2Cl_2$	solvent, particularly for rubber

Source: *Chemical and Engineering News.*

Physical States of Matter

states of matter

the physical forms of matter, which are solid, liquid, gas, and plasma

All matter is made of particles. The type and arrangement of the particles in a sample of matter determine the properties of the matter. Most of the matter you encounter is in one of three **states of matter:** solid, liquid, or gas. **Figure 2** illustrates water in each of these three states at the macroscopic and microscopic levels. *Macroscopic* refers to what you see with the unaided eye. In this text, *microscopic* refers to what you would see if you could see individual atoms.

The microscopic views in this book are models that are designed to show you the differences in the arrangement of particles in different states of matter. They also show you the differences in size, shape, and makeup of particles of chemicals. But don't take these models too literally. Think of them as cartoons. Atoms are not really different colors. And groups of connected atoms, or molecules, do not look lumpy. The microscopic views are also limited in that they often show only a single layer of particles whereas the particles are really arranged in three dimensions. Finally, the models cannot show you that particles are in constant motion.

Figure 2
a Below 0°C, water exists as ice. Particles in a solid are in a rigid structure and vibrate in place.

b Between 0°C and 100°C, water exists as a liquid. Particles in a liquid are close together and slide past one another.

c Above 100°C, water is a gas. Particles in a gas move randomly over large distances.

Water molecule, H_2O

Water molecule, H_2O

Water molecule, H_2O

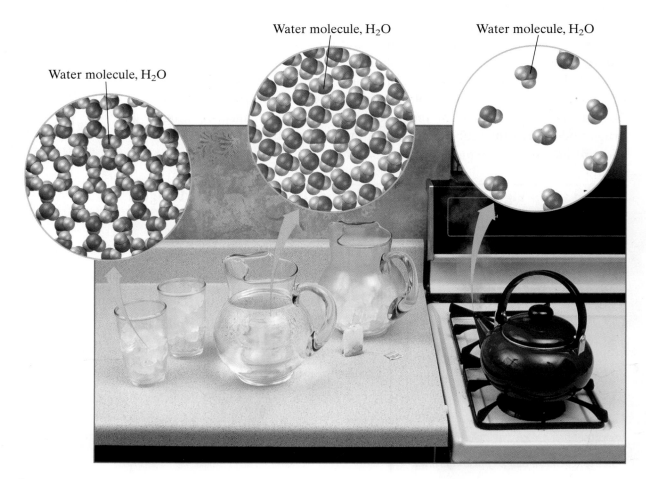

Properties of the Physical States

Solids have fixed volume and shape that result from the way their particles are arranged. Particles that make up matter in the solid state are held tightly in a rigid structure. They vibrate only slightly.

Liquids have fixed volume but not a fixed shape. The particles in a liquid are not held together as strongly as those in a solid. Like grains of sand, the particles of a liquid slip past one another. Thus, a liquid can flow and take the shape of its container.

Gases have neither fixed volume nor fixed shape. Gas particles weakly attract one another and move independently at high speed. Gases will fill any container they occupy as their particles move apart.

There are other states that are beyond the scope of this book. For example, most visible matter in the universe is plasma—a gas whose particles have broken apart and are charged. Bose-Einstein condensates have been described at very low temperatures. A neutron star is also considered by some to be a state of matter.

Changes of Matter

Many changes of matter happen. An ice cube melts. Your bicycle's spokes rust. A red shirt fades. Water fogs a mirror. Milk sours. Scientists who study these and many other events classify them by two broad categories: *physical changes* and *chemical changes*.

Physical Changes

Physical changes are changes in which the identity of a substance doesn't change. However, the arrangement, location, and speed of the particles that make up the substance may change. Changes of state are physical changes. The models in **Figure 2** show that when water changes state, the arrangement of particles changes, but the particles stay water particles. As sugar dissolves in the tea in **Figure 3,** the sugar molecules mix with the tea, but they don't change what they are. The particles are still sugar. Crushing a rock is a physical change because particles separate but do not change identity.

internet connect

www.scilinks.org
Topic: Chemical and
        Physical Changes
SciLinks code: HW4140

SCLINKS. Maintained by the National Science Teachers Association

Figure 3
Dissolving sugar in tea is a physical change.

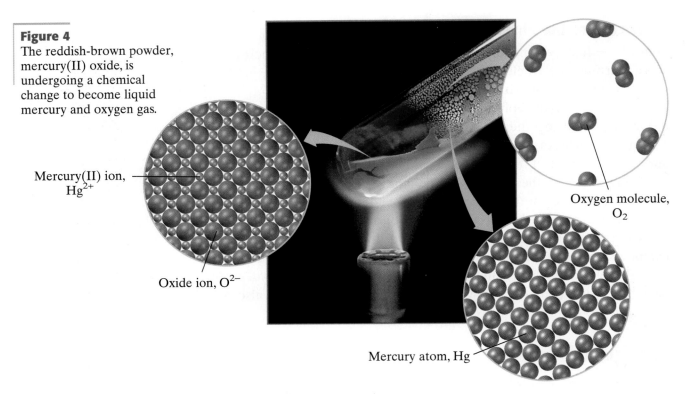

Figure 4
The reddish-brown powder, mercury(II) oxide, is undergoing a chemical change to become liquid mercury and oxygen gas.

Mercury(II) ion, Hg^{2+}

Oxide ion, O^{2-}

Oxygen molecule, O_2

Mercury atom, Hg

Chemical Changes

In a chemical change, the identities of substances change and new substances form. In **Figure 4,** mercury(II) oxide changes into mercury and oxygen as represented by the following word equation:

$$\text{mercury(II) oxide} \longrightarrow \text{mercury} + \text{oxygen}$$

In an equation, the substances on the left-hand side of the arrow are the **reactants.** They are used up in the reaction. Substances on the right-hand side of the arrow are the **products.** They are made by the reaction.

A chemical reaction is a rearrangement of the atoms that make up the reactant or reactants. After rearrangement, those same atoms are present in the product or products. Atoms are not destroyed or created, so mass does not change during a chemical reaction.

reactant

a substance or molecule that participates in a chemical reaction

product

a substance that forms in a chemical reaction

Evidence of Chemical Change

Evidence that a chemical change may be happening generally falls into one of the categories described below and shown in **Figure 5.** The more of these signs you observe, the more likely a chemical change is taking place. But be careful! Some physical changes also have one or more of these signs.

a. **The Evolution of a Gas** The production of a gas is often observed by bubbling, as shown in **Figure 5a,** or by a change in odor.

b. **The Formation of a Precipitate** When two clear solutions are mixed and become cloudy, a precipitate has formed, as shown in **Figure 5b.**

c. **The Release or Absorption of Energy** A change in temperature or the giving off of light energy, as shown in **Figure 5c,** are signs of an energy transfer.

d. **A Color Change in the Reaction System** Look for a different color when two chemicals react, as shown in **Figure 5d.**

Figure 5

a When acetic acid, in vinegar, and sodium hydrogen carbonate, or baking soda, are mixed, the solution bubbles as carbon dioxide forms.

b When solutions of sodium sulfide and cadmium nitrate are mixed, cadmium sulfide, a solid precipitate, forms.

c When aluminum reacts with iron(III) oxide in the clay pot, energy is released as heat and light.

d When phenolphthalein is added to ammonia dissolved in water, a color change from colorless to pink occurs.

 # Section Review

UNDERSTANDING KEY IDEAS

1. Name three natural chemicals and three artificial chemicals that are part of your daily life.

2. Describe how chemistry is a part of your morning routine.

3. Classify the following materials as solid, liquid, or gas at room temperature: milk, helium, granite, oxygen, steel, and gasoline.

4. Describe the motions of particles in the three common states of matter.

5. How does a physical change differ from a chemical change?

6. Give three examples of physical changes.

7. Give three examples of chemical changes.

8. Identify each substance in the following word equation as a reactant or a product.

limestone $\xrightarrow{\text{heat}}$ lime + carbon dioxide

9. Sodium salicylate is made from carbon dioxide and sodium phenoxide. Identify each of these substances as a reactant or a product.

10. List four observations that suggest a chemical change is occurring.

CRITICAL THINKING

11. Explain why neither liquids nor gases have permanent shapes.

12. Steam is sometimes used to melt ice. Is this change physical or chemical?

13. Mass does not change during a chemical change. Is the same true for a physical change? Explain your answer, and give an example.

14. In beaker A, water is heated, bubbles of gas form throughout the water, and the water level in the beaker slowly decreases. In beaker B, electrical energy is added to water, bubbles of gas appear on the ends of the wires in the water, and the water level in the beaker slowly decreases.

    a. What signs of a change are visible in each situation?

    b. What type of change is happening in each beaker? Explain your answer.

Describing Matter

KEY TERMS

- **matter**
- **volume**
- **mass**
- **weight**
- **quantity**
- **unit**
- **conversion factor**
- **physical property**
- **density**
- **chemical property**

matter

anything that has mass and takes up space

volume

a measure of the size of a body or region in three-dimensional space

OBJECTIVES

(1) **Distinguish** between different characteristics of matter, including mass, volume, and weight.

(2) **Identify** and use SI units in measurements and calculations.

(3) **Set up** conversion factors, and use them in calculations.

(4) **Identify** and describe physical properties, including density.

(5) **Identify** chemical properties.

Matter Has Mass and Volume

Matter, the stuff of which everything is made, exists in a dazzling variety of forms. However, matter has a fairly simple definition. **Matter** is anything that has mass and volume. Think about blowing up a balloon. The inflated balloon has more mass and more volume than before. The increase in mass and volume comes from the air that you blew into it. Both the balloon and air are examples of matter.

The Space an Object Occupies Is Its Volume

An object's **volume** is the space the object occupies. For example, this book has volume because it takes up space. Volume can be determined in several different ways. The method used to determine volume depends on the nature of the matter being examined. The book's volume can be found by multiplying the book's length, width, and height. Graduated cylinders are often used in laboratories to measure the volume of liquids, as shown in **Figure 6.** The volume of a gas is the same as that of the container it fills.

Figure 6
To read the liquid level in a graduated cylinder correctly, read the level at the bottom part of the *meniscus,* the curved upper surface of the liquid. The volume shown here is 73.0 mL.

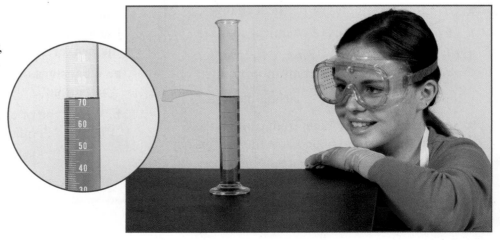

The Quantity of Matter Is the Mass

The **mass** of an object is the quantity of matter contained in that object. Even though a marble is smaller, it has more mass than a ping-pong ball does if the marble contains more matter.

Devices used for measuring mass in a laboratory are called *balances*. Balances can be electronic, as shown in **Figure 7,** or mechanical, such as a triple-beam balance.

Balances also differ based on the precision of the mass reading. The balance in **Figure 7** reports readings to the hundredth place. The balance often found in a school chemistry laboratory is the triple-beam balance. If the smallest scale on the triple-beam balance is marked off in 0.1 g increments, you can be certain of the reading to the tenths place, and you can estimate the reading to the hundredths place. The smaller the markings on the balance, the more decimal places you can have in your measurement.

Mass Is Not Weight

Mass is related to weight, but the two are not identical. Mass measures the quantity of matter in an object. As long as the object is not changed, it will have the same mass, no matter where it is in the universe. On the other hand, the weight of that object is affected by its location in the universe. The weight depends on gravity, while mass does not.

Weight is defined as the force produced by gravity acting on mass. Scientists express forces in *newtons,* but they express mass in *kilograms.* Because gravity can vary from one location to another, the weight of an object can vary. For example, an astronaut weighs about six times more on Earth than he weighs on the moon because the effect of gravity is less on the moon. The astronaut's mass, however, hasn't changed because he is still made up of the same amount of matter.

The force that gravity exerts on an object is proportional to the object's mass. If you keep the object in one place and double its mass, the weight of the object doubles, too. So, measuring weight can tell you about mass. In fact, when you read the word *weigh* in a laboratory procedure, you probably are determining the mass. Check with your teacher to be sure.

mass

a measure of the amount of matter in an object; a fundamental property of an object that is not affected by the forces that act on the object, such as the gravitational force

weight

a measure of the gravitational force exerted on an object; its value can change with the location of the object in the universe

Figure 8 — graduated cylinder photo

Figure 8
This graduated cylinder measures a *quantity,* the volume of a liquid, in a *unit,* the milliliter.

quantity

something that has magnitude, size, or amount

unit

a quantity adopted as a standard of measurement

internet connect

www.scilinks.org
Topic: SI Units
SciLinks code: HW4114

SC/LINKS Maintained by the National Science Teachers Association

Units of Measurement

Terms such as *heavy, light, rough,* and *smooth* describe matter qualitatively. Some properties of matter, such as color and texture, are usually described in this way. But whenever possible, scientists prefer to describe properties in quantitative terms, that is with numbers.

Mass and volume are properties that can be described in terms of numbers. But numbers alone are not enough because their meanings are unclear. For meaningful descriptions, units are needed with the numbers. For example, describing a quantity of sand as 15 kilograms rather than as 15 bucketfuls or just 15 gives clearer information.

When working with numbers, be careful to distinguish between a quantity and its unit. The graduated cylinder shown in **Figure 8,** for example, is used to measure the volume of a liquid in milliliters. *Volume* is the **quantity** being measured. *Milliliters* (abbreviated mL) is the **unit** in which the measured volume is reported.

Scientists Express Measurements in SI Units

Since 1960, scientists worldwide have used a set of units called the *Système Internationale d'Unités* or *SI.* The system is built on the seven base units listed in **Table 2.** The last two find little use in chemistry, but the first five provide the foundation of all chemical measurements.

Base units can be too large or too small for some measurements, so the base units may be modified by attaching prefixes, such as those in **Table 3.** For example, the base unit *meter* is suitable for expressing a person's height. The distance beween cities is more conveniently expressed in kilometers (km), with 1 km being 1000 m. The lengths of many insects are better expressed in millimeters (mm), or one-thousandth of a meter, because of the insects' small size. Additional prefixes can be found in **Appendix A.** Atomic sizes are so small that picometers (pm) are used. A picometer is 0.000 000 000 001 m. The advantage of using prefixes is the ability to use more manageable numbers. So instead of reporting the diameter of a hydrogen atom as 0.000 000 000 120 m, you can report it as 120 pm.

Table 2 **SI Base Units**			
Quantity	**Symbol**	**Unit**	**Abbreviation**
Length	*l*	meter	m
Mass	*m*	kilogram	kg
Time	*t*	second	s
Thermodynamic temperature	*T*	kelvin	K
Amount of substance	*n*	mole	mol
Electric current	*I*	ampere	A
Luminous intensity	I_v	candela	cd

Table 3 SI Prefixes

Prefix	Abbreviation	Exponential multiplier	Meaning	Example using length
Kilo-	k	10^3	1000	1 kilometer (km) = 1000 m
Hecto-	h	10^2	100	1 hectometer (hm) = 100 m
Deka-	da	10^1	10	1 dekameter (dam) = 10 m
		10^0	1	1 meter (m)
Deci-	d	10^{-1}	1/10	1 decimeter (dm) = 0.1 m
Centi-	c	10^{-2}	1/100	1 centimeter (cm) = 0.01 m
Milli-	m	10^{-3}	1/1000	1 millimeter (mm) = 0.001 m

Refer to Appendix A for more SI prefixes.

Converting One Unit to Another

In chemistry, you often need to convert a measurement from one unit to another. One way of doing this is to use a **conversion factor.** A conversion factor is a simple ratio that relates two units that express a measurement of the same quantity. Conversion factors are formed by setting up a fraction that has equivalent amounts on top and bottom. For example, you can construct conversion factors between kilograms and grams as follows:

conversion factor

a ratio that is derived from the equality of two different units and that can be used to convert from one unit to the other

$$1 \text{ kg} = 1000 \text{ g} \text{ can be written as } \frac{1 \text{ kg}}{1000 \text{ g}} \text{ or } \frac{1000 \text{ g}}{1 \text{ kg}}$$

$$0.001 \text{ kg} = 1 \text{ g} \text{ can be written as } \frac{0.001 \text{ kg}}{1 \text{ g}} \text{ or } \frac{1 \text{ g}}{0.001 \text{ kg}}$$

SKILLS Toolkit 1

Using Conversion Factors

1. Identify the quantity and unit given and the unit that you want to convert to.

2. Using the equality that relates the two units, set up the conversion factor that cancels the given unit and leaves the unit that you want to convert to.

3. Multiply the given quantity by the conversion factor. Cancel units to verify that the units left are the ones you want for your answer.

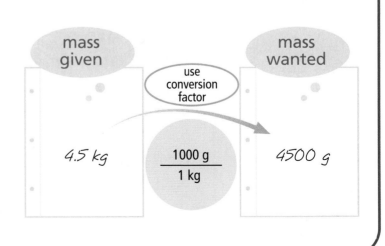

SAMPLE PROBLEM A

Converting Units

Convert 0.851 L to milliliters.

1 Gather information.

- You are given 0.851 L, which you want to convert to milliliters. This problem can be expressed as this equation:

$$? \text{ mL} = 0.851 \text{ L}$$

- The equality that links the two units is 1000 mL = 1 L. (The prefix *milli-* represents 1/1000 of a base unit.)

2 Plan your work.

The conversion factor needed must cancel liters and leave milliliters. Thus, liters must be on the bottom of the fraction and milliliters must be on the top. The correct conversion factor to use is

$$\frac{1000 \text{ mL}}{1 \text{ L}}$$

3 Calculate.

$$? \text{ mL} = 0.851 \cancel{\text{L}} \times \frac{1000 \text{ mL}}{1 \cancel{\text{L}}} = 851 \text{ mL}$$

4 Verify your results.

The unit of liters cancels out. The answer has the unit of milliliters, which is the unit called for in the problem. Because a milliliter is smaller than a liter, the number of milliliters should be greater than the number of liters for the same volume of material. Thus, the answer makes sense because 851 is greater than 0.851.

> **PRACTICE HINT**
>
> Remember that you can cancel only those units that appear in both the top and the bottom of the fractions you multiply together. Be sure to set up your conversion factors so that the unit you want to cancel is in the correct place.

PRACTICE

1. Convert each of the following masses to the units requested.

   a. 0.765 g to kilograms

   b. 1.34 g to milligrams

   c. 34.2 mg to grams

   d. 23 745 kg to milligrams (Hint: Use two conversion factors.)

2. Convert each of the following lengths to the units requested.

   a. 17.3 m to centimeters

   b. 2.56 m to kilometers

   c. 567 dm to meters

   d. 5.13 m to millimeters

3. Which of the following lengths is the shortest, and which is the longest: 1583 cm, 0.0128 km, 17 931 mm, and 14 m?

Derived Units

Many quantities you can measure need units other than the seven basic SI units. These units are derived by multiplying or dividing the base units. For example, *speed* is distance divided by time. The derived unit of speed is meters per second (m/s). A rectangle's *area* is found by multiplying its length (in meters) by its width (also in meters), so its unit is square meters (m^2).

The volume of this book can be found by multiplying its length, width, and height. So the unit of volume is the cubic meter (m^3). But this unit is too large and inconvenient in most labs. Chemists usually use the liter (L), which is one-thousandth of a cubic meter. **Figure 9** shows one liter of liquid and also a cube of one liter volume. Each side of the cube has been divided to show that one liter is exactly 1000 cubic centimeters, which can be expressed in the following equality:

$$1 \text{ L} = 1000 \text{ mL} = 1000 \text{ cm}^3$$

Therefore, a volume of one milliliter (1 mL) is identical to one cubic centimeter (1 cm³).

Properties of Matter

When examining a sample of matter, scientists describe its properties. In fact, when you describe an object, you are most likely describing it in terms of the properties of matter. Matter has many properties. The properties of a substance may be classified as physical or chemical.

Physical Properties

A **physical property** is a property that can be determined without changing the nature of the substance. Consider table sugar, or sucrose. You can see that it is a white solid at room temperature, so color and state are physical properties. It also has a gritty texture. Because changes of state are physical changes, melting point and boiling point are also physical properties. Even the lack of a physical property, such as air being colorless, can be used to describe a substance.

physical property

a characteristic of a substance that does not involve a chemical change, such as density, color, or hardness

Density Is the Ratio of Mass to Volume

The mass and volume of a sample are physical properties that can be determined without changing the substance. But each of these properties changes depending on how much of the substance you have. The **density** of an object is another physical property: the mass of that object divided by its volume. As a result, densities are expressed in derived units such as g/cm^3 or g/mL. Density is calculated as follows:

$$density = \frac{mass}{volume} \quad \text{or} \quad D = \frac{m}{V}$$

The density of a substance is the same no matter what the size of the sample is. For example, the masses and volumes of a set of 10 different aluminum blocks are listed in the table in **Figure 10.** The density of Block 10 is as follows:

$$D = \frac{m}{V} = \frac{36.40 \text{ g}}{13.5 \text{ cm}^3} = 2.70 \text{ g/cm}^3$$

If you divide the mass of any block by the corresponding volume, you will always get an answer close to 2.70 g/cm^3.

The density of aluminum can also be determined by graphing the data, as shown in **Figure 10.** The straight line rising from left to right indicates that mass increases at a constant rate as volume increases. As the volume of aluminum doubles, its mass doubles; as its volume triples, its mass triples, and so on. In other words, the mass of aluminum is directly proportional to its volume.

The slope of the line equals the ratio of mass (from the vertical y-axis) divided by volume (from the horizontal x-axis). You may remember this as "rise over run" from math class. The slope between the two points shown is as follows:

$$slope = \frac{rise}{run} = \frac{29.7 \text{ g} - 10.8 \text{ g}}{11 \text{ cm}^3 - 4 \text{ cm}^3} = \frac{18.9 \text{ g}}{7 \text{ cm}^3} = 2.70 \text{ g/cm}^3$$

As you can see, the value of the slope is the density of aluminum.

Figure 10

The graph of mass versus volume shows a relationship of direct proportionality. Notice that the line has been extended to the origin.

Block number	Mass (g)	Volume (cm³)
1	1.20	0.44
2	3.69	1.39
3	5.72	2.10
4	12.80	4.68
5	15.30	5.71
6	18.80	6.90
7	22.70	8.45
8	26.50	9.64
9	34.00	12.8
10	36.40	13.5

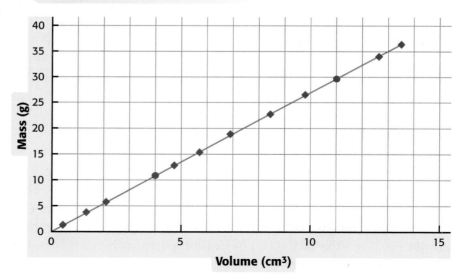

Mass Vs. Volume for Samples of Aluminum

Table 4 Densities of Various Substances

Substance	Density (g/cm³) at 25°C
Hydrogen gas, H_2*	0.000 082 4
Carbon dioxide gas, CO_2*	0.001 80
Ethanol (ethyl alcohol), C_2H_5OH	0.789
Water, H_2O	0.997
Sucrose (table sugar), $C_{12}H_{22}O_{11}$	1.587
Sodium chloride, NaCl	2.164
Aluminum, Al	2.699
Iron, Fe	7.86
Copper, Cu	8.94
Silver, Ag	10.5
Gold, Au	19.3
Osmium, Os	22.6

*at 1 atm

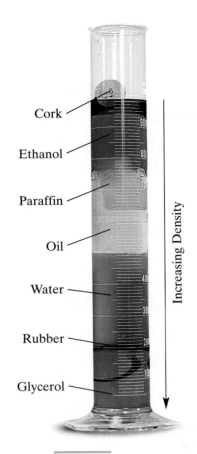

Cork
Ethanol
Paraffin
Oil
Water
Rubber
Glycerol

Increasing Density

Figure 11
Substances float in layers, and the order of the layers is determined by their densities. Dyes have been added to make the liquid layers more visible.

Density Can Be Used to Identify Substances

Because the density of a substance is the same for all samples, you can use this property to help identify substances. For example, suppose you find a chain that appears to be silver on the ground. To find out if it is pure silver, you can take the chain into the lab and use a balance to measure its mass. One way to find the volume is to use the technique of water displacement. Partially fill a graduated cylinder with water, and note the volume. Place the chain in the water, and watch the water level rise. Note the new volume. The difference in water levels is the volume of the chain. If the mass is 199.0 g, and the volume is 20.5 cm³, you can calculate the chain's density as follows:

$$D = \frac{m}{V} = \frac{199.0 \text{ g}}{20.5 \text{ cm}^3} = 9.71 \text{ g/cm}^3$$

Comparing this density with the density of silver in **Table 4,** you can see that your find is not pure silver.

Table 4 lists the densities of a variety of substances. Osmium, a bluish white metal, is the densest substance known. A piece of osmium the size of a football would be too heavy to lift. Whether a solid will float or sink in a liquid depends on the relative densities of the solid and the liquid. **Figure 11** shows several things arranged according to densities, with the most dense on the bottom.

Chemical Properties

You cannot fully describe matter by physical properties alone. You must also describe what happens when matter has the chance to react with other kinds of matter, or the *chemical properties* of matter.

Whereas physical properties can be determined without changing the identity of the substance, **chemical properties** can only be identified by trying to cause a chemical change. Afterward, the substance may have been changed into a new substance.

For example, many substances share the chemical property of reactivity with oxygen. If you have seen a rusty nail or a rusty car, you have seen the result of iron's property of reactivity with oxygen. But gold has a very different chemical property. It does not react with oxygen. This property prevents gold from tarnishing and keeps gold jewelry shiny. If something doesn't react with oxygen, that lack of reaction is also a chemical property.

Not all chemical reactions result from contact between two or more substances. For example, many silver compounds are sensitive to light and undergo a chemical reaction when exposed to light. Photographers rely on silver compounds on film to create photographs. Some sunglasses have silver compounds in their lenses. As a result of this property of the silver compounds, the lenses darken in response to light. Another reaction that involves a single reactant is the reaction you saw earlier in this chapter. The formation of mercury and oxygen when mercury(II) oxide is heated, happens when a single reactant breaks down. Recall that the reaction in this case is described by the following equation:

$$\text{mercury(II) oxide} \longrightarrow \text{mercury} + \text{oxygen}$$

Despite similarities between the names of the products and the reactant, the two products have completely different properties from the starting material, as shown in **Figure 12.**

chemical property

a property of matter that describes a substance's ability to participate in chemical reactions

🔲 internet connect

www.scilinks.org
Topic: Physical/Chemical
Properties
SciLinks code: HW4097

SCiLINKS. Maintained by the National Science Teachers Association

Quick LAB

Thickness of Aluminum Foil

PROCEDURE

1. Using **scissors** and a **metric ruler,** cut a rectangle of **aluminum foil.** Determine the area of the rectangle.

2. Use a **balance** to determine the mass of the foil.

3. Repeat steps **1** and **2** with each brand of aluminum foil available.

ANALYSIS

1. Use the density of aluminum (2.699 g/cm^3) to calculate the volume and the thickness of each piece of foil. Report the thickness in centimeters (cm), meters (m), and micrometers (μm) for each brand of foil. (Hint: 1 μm = 10^{-6}m)

2. Which brand is the thickest?

3. Which unit is the most appropriate unit to use for expressing the thickness of the foil? Explain your reasoning.

Figure 12
The physical and chemical properties of the components of this reaction system are shown. Decomposition of mercury(II) oxide is a chemical change.

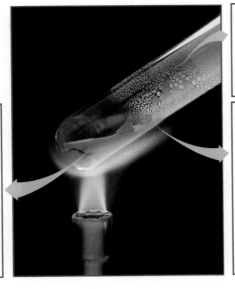

MERCURY(II) OXIDE

Physical properties: Bright red or orange-red, odorless crystalline solid

Chemical properties:
Decomposes when exposed to light or at 500°C to form mercury and oxygen gas; dissolves in dilute nitric acid or hydrochloric acid, but is almost insoluble in water

OXYGEN

Physical properties: Colorless, odorless gas

Chemical properties: Supports combustion; soluble in water

MERCURY

Physical properties: Silver-white, liquid metal; in the solid state, mercury is ductile and malleable and can be cut with a knife

Chemical properties: Combines readily with sulfur at normal temperatures; reacts with nitric acid and hot sulfuric acid; oxidizes to form mercury(II) oxide upon heating in air

② Section Review

UNDERSTANDING KEY IDEAS

1. Name two physical properties that characterize matter.

2. How does mass differ from weight?

3. What derived unit is usually used to express the density of liquids?

4. What SI unit would best be used to express the height of your classroom ceiling?

5. Distinguish between a physical property and a chemical property, and give an example of each.

6. Why is density considered a physical property rather than a chemical property of matter?

7. One inch equals 2.54 centimeters. What conversion factor is useful for converting from centimeters to inches?

PRACTICE PROBLEMS

8. What is the mass, in kilograms, of a 22 000 g bag of fertilizer?

9. Convert each of the following measurements to the units indicated. (Hint: Use two conversion factors if needed.)

   a. 17.3 s to milliseconds

   b. 2.56 mm to kilometers

   c. 567 cg to grams

   d. 5.13 m to kilometers

10. Convert 17.3 cm^3 to liters.

11. Five beans have a mass of 2.1 g. How many beans are in 0.454 kg of beans?

CRITICAL THINKING

12. A block of lead, with dimensions 2.0 dm × 8.0 cm × 35 mm, has a mass of 6.356 kg. Calculate the density of lead in g/cm^3.

13. Demonstrate that kg/L and g/cm^3 are equivalent units of density.

14. In the manufacture of steel, pure oxygen is blown through molten iron to remove some of the carbon impurity. If the combustion of carbon is efficient, carbon dioxide (density = 1.80 g/L) is produced. Incomplete combustion produces the poisonous gas carbon monoxide (density = 1.15 g/L) and should be avoided. If you measure a gas density of 1.77 g/L, what do you conclude?

CONSUMER FOCUS

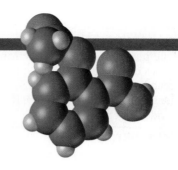

Aspirin

For centuries, plant extracts have been used for treating ailments. The bark of the willow tree was found to relieve pain and reduce fever. Writing in 1760, Edward Stone, an English naturalist and clergyman, reported excellent results when he used "twenty grains of powdered bark dissolved in water and administered every four hours" to treat people suffering from an acute, shiver-provoking illness.

The History of Aspirin

Following up on Stone's research, German chemists isolated a tiny amount of the active ingredient of the willow-bark extract, which they called salicin, from Salix, the botanical name for the willow genus. Researchers in France further purified salicin and converted it to salicylic acid, which proved to be a potent pain-reliever. This product was later marketed as the salt sodium salicylate. Though an effective painkiller, sodium salicylate has the unfortunate side effect of causing nausea and, sometimes, stomach ulcers.

Then back in Germany in the late 1800s, the father of Felix Hoffmann, a skillful organic chemist, developed painful arthritis. Putting aside his research on dyes, the younger Hoffmann looked for a way to

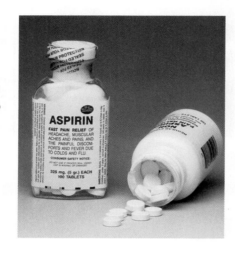

prevent the nauseating effects of salicylic acid. He found that a similar compound, acetylsalicylic acid, was effective in treating pain and fever, while having fewer side effects. Under the name *aspirin*, it has been a mainstay in painkillers for over a century.

The FDA and Product Warning Labels

The Federal Drug Administration requires that all over-the-counter drugs carry a warning label. In fact, when you purchase any product, it is your responsibility as a consumer to check the warning label about the hazards of any chemical it may contain. The label on aspirin bottles warns against giving aspirin to children and teenagers who have chickenpox or severe flu. Some reports suggest that aspirin may play a part in Reye's syndrome, a condition in which the brain swells and the liver malfunctions.

Though side effects and allergic responses are rare, the label warns that aspirin may cause nausea and vomiting and should be avoided late in pregnancy. Because aspirin can interfere with blood clotting, it should not be used by hemophiliacs or following surgery of the mouth.

Questions

1. For an adult, the recommended dosage of 325 mg aspirin tablets is "one or two tablets every four hours, up to 12 tablets per day." In grams, what is the maximum dosage of aspirin an adult should take in one day? Why should you not take 12 tablets at once?

2. Research several over-the-counter painkillers, and write a report of your findings. For each product, compare the active ingredient and the price for a day's treatment.

3. Research Reye's syndrome, and write a report of your findings. Include the causes, symptoms, and risk factors.

How Is Matter Classified?

KEY TERMS

- atom
- pure substance
- element
- molecule
- compound
- mixture
- homogeneous
- heterogeneous

OBJECTIVES

1. **Distinguish** between elements and compounds.
2. **Distinguish** between pure substances and mixtures.
3. **Classify** mixtures as homogeneous or heterogeneous.
4. **Explain** the difference between mixtures and compounds.

Classifying Matter

Everything around you—water, air, plants, and your friends—is made of matter. Despite the many examples of matter, all matter is composed of about 110 different kinds of **atoms.** Even the biggest atoms are so small that it would take more than 3 million of them side by side to span just one millimeter. These atoms can be physically mixed or chemically joined together to make up all kinds of matter.

atom

the smallest unit of an element that maintains the properties of that element

Benefits of Classification

Because matter exists in so many different forms, having a way to classify matter is important for studying it. In a store, such as the nursery in **Figure 13,** classification helps you to find what you want. In chemistry, it helps you to predict what characteristics a sample will have based on what you know about others like it.

Figure 13
Finding the plant you want without the classification scheme adopted by this nursery would be difficult.

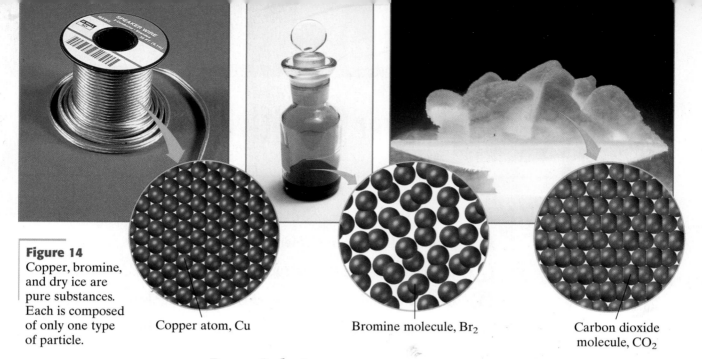

Figure 14
Copper, bromine, and dry ice are pure substances. Each is composed of only one type of particle.

Copper atom, Cu

Bromine molecule, Br_2

Carbon dioxide molecule, CO_2

Pure Substances

Each of the substances shown in **Figure 14** is a **pure substance.** Every pure substance has characteristic properties that can be used to identify it. Characteristic properties can be physical or chemical properties. For example, copper always melts at 1083°C, which is a physical property that is characteristic of copper. There are two types of pure substances: elements and compounds.

pure substance

a sample of matter, either a single element or a single compound, that has definite chemical and physical properties

Elements Are Pure Substances

Elements are pure substances that contain only one kind of atom. Copper and bromine are elements. Each element has its own unique set of physical and chemical properties and is represented by a distinct chemical symbol. **Table 5** shows several elements and their symbols and gives examples of how an element got its symbol.

element

a substance that cannot be separated or broken down into simpler substances by chemical means; all atoms of an element have the same atomic number

Table 5	Element Names, Symbols, and the Symbols' Origins	
Element name	**Chemical symbol**	**Origin of symbol**
Hydrogen	H	first letter of element name
Helium	He	first two letters of element name
Magnesium	Mg	first and third letters of element name
Tin	Sn	from *stannum,* the Latin word for "tin"
Gold	Au	from *aurum,* the Latin word meaning "gold"
Tungsten	W	from *Wolfram,* the German word for "tungsten"
Ununpentium	Uup	first letters of root words that describe the digits of the atomic number; used for elements that have not yet been synthesized or whose official names have not yet been chosen

Refer to Appendix A for an alphabetical listing of element names and symbols.

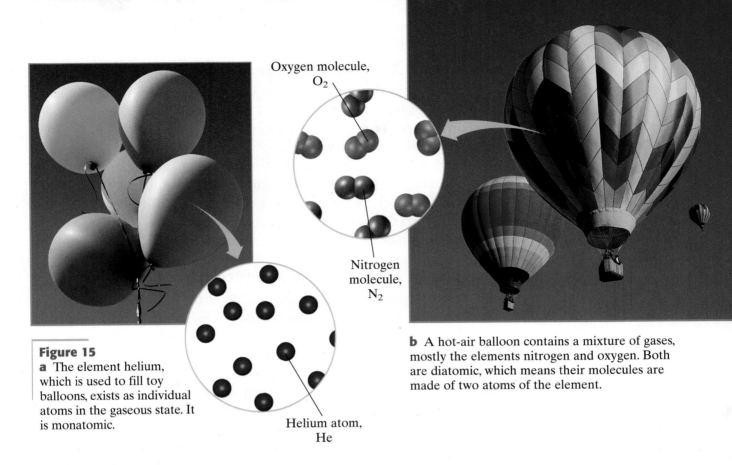

Figure 15
a The element helium, which is used to fill toy balloons, exists as individual atoms in the gaseous state. It is monatomic.

Oxygen molecule, O_2

Nitrogen molecule, N_2

Helium atom, He

b A hot-air balloon contains a mixture of gases, mostly the elements nitrogen and oxygen. Both are diatomic, which means their molecules are made of two atoms of the element.

Elements as Single Atoms or as Molecules

Some elements exist as single atoms. For example, the helium gas in a balloon consists of individual atoms, as shown by the model in **Figure 15a**. Because it exists as individual atoms, helium gas is known as a *monatomic* gas.

Other elements exist as molecules consisting of as few as two or as many as millions of atoms. A **molecule** usually consists of two or more atoms combined in a definite ratio. If an element consists of molecules, those molecules contain just one type of atom. For example, the element nitrogen, found in air, is an example of a molecular element because it exists as two nitrogen atoms joined together, as shown by the model in **Figure 15b**. Oxygen, another gas found in the air, exists as two oxygen atoms joined together. Nitrogen and oxygen are *diatomic* elements. Other diatomic elements are H_2, F_2, Cl_2, Br_2, and I_2.

Some Elements Have More than One Form

Both oxygen gas and ozone gas are made up of oxygen atoms, and are forms of the element oxygen. However, the models in **Figure 16** show that a molecule of oxygen gas, O_2, is made up of two oxygen atoms, and a molecule of ozone, O_3, is made up of three oxygen atoms.

A few elements, including oxygen, phosphorus, sulfur, and carbon, are unusual because they exist as allotropes. An *allotrope* is one of a number of different molecular forms of an element. The properties of allotropes can vary widely. For example, ozone is a toxic, pale blue gas that has a sharp odor. You often smell ozone after a thunderstorm. But oxygen is a colorless, odorless gas essential to most forms of life.

molecule

the smallest unit of a substance that keeps all of the physical and chemical properties of that substance; it can consist of one atom or two or more atoms bonded together

Figure 16
Two forms of the element oxygen are oxygen gas and ozone gas.

O_2

O_3

Compounds Are Pure Substances

Pure substances that are not elements are **compounds.** Compounds are composed of more than one kind of atom. For example, the compound carbon dioxide is composed of molecules that consist of one atom of carbon and two atoms of oxygen.

There may be easier ways of preparing them, but compounds can be made from their elements. On the other hand, compounds can be broken down into their elements, though often with great difficulty. The reaction of mercury(II) oxide described earlier in this chapter is an example of the breaking down of a compound into its elements.

Compounds Are Represented by Formulas

Because every molecule of a compound is made up of the same kinds of atoms arranged the same way, a compound has characteristic properties and composition. For example, every molecule of hydrogen peroxide contains two atoms each of hydrogen and oxygen. To emphasize this ratio, the compound can be represented by an abbreviation or *formula:* H_2O_2. *Subscripts* are placed to the lower right of the element's symbol to show the number of atoms of the element in a molecule. If there is just one atom, no subscript is used. For example, the formula for water is H_2O, not H_2O_1.

Molecular formulas give information only about what makes up a compound. The molecular formula for aspirin is $C_9H_8O_4$. Additional information can be shown by using different models, such as the ones for aspirin shown in **Figure 17.** A *structural formula* shows how the atoms are connected, but the two-dimensional model does not show the molecule's true shape. The distances between atoms and the angles between them are more realistic in a three-dimensional *ball-and-stick model*. However, a *space-filling model* attempts to represent the actual sizes of the atoms and not just their relative positions. A hand-held model can provide even more information than models shown on the flat surface of the page.

Figure 17
These models convey different information about acetylsalicylic acid (aspirin).

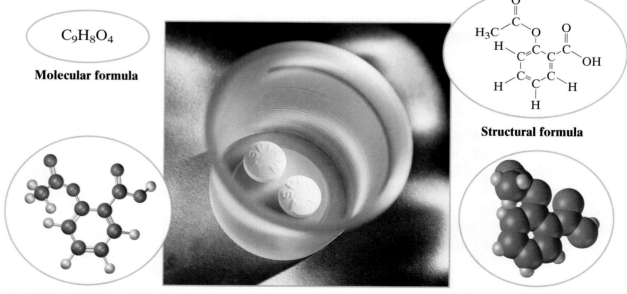

$C_9H_8O_4$

Molecular formula

Structural formula

Ball-and-stick model

Space-filling model

Compounds Are Further Classified

Such a wide variety of compounds exists that scientists classify the compounds to help make sense of them. In later chapters, you will learn that compounds can be classified by their properties, by the type of bond that holds them together, and by whether they are made of certain elements.

Mixtures

A sample of matter that contains two or more pure substances is a **mixture.** Most kinds of food are mixtures, sugar and salt being rare exceptions. Air is a mixture, mostly of nitrogen and oxygen. Water is *not* a mixture of hydrogen and oxygen for two reasons. First, the H and O atoms are chemically bonded together in H_2O molecules, not just physically mixed. Second, the ratio of hydrogen atoms to oxygen atoms is always exactly two to one. In a mixture, such as air, the proportions of the ingredients can vary.

mixture

a combination of two or more substances that are not chemically combined

Mixtures Can Vary in Composition and Properties

A glass of sweetened tea is a mixture. If you have ever had a glass of tea that was too sweet or not sweet enough, you have experienced two important characteristics of mixtures. A mixture does not always have the same balance of ingredients. The proportion of the materials in a mixture can change. Because of this, the properties of the mixture may vary.

For example, pure gold, shown in **Figure 18a,** is often mixed with other metals, usually silver, copper, or nickel, in various proportions to change its density, color, and strength. This solid mixture, or *alloy,* is stronger than pure gold. A lot of jewelry is 18-karat gold, meaning that it contains 18 grams of gold per 24 grams of alloy, or 75% gold by mass. A less expensive, and stronger, alloy is 14-karat gold, shown in **Figure 18b.**

Figure 18

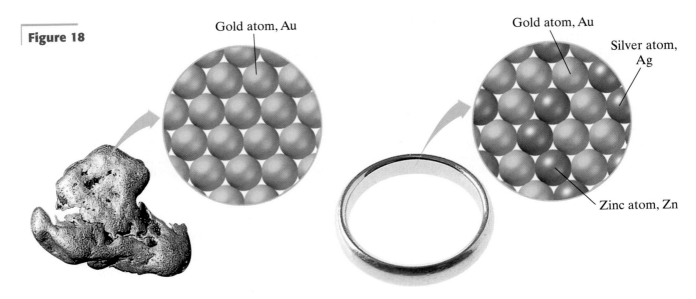

Gold atom, Au

Gold atom, Au

Silver atom, Ag

Zinc atom, Zn

a The gold nugget is a pure substance—gold. Pure gold, also called 24-karat gold, is usually considered too soft for jewelry.

b This ring is 14-karat gold, which is 14/24, or 58.3%, gold. This homogeneous mixture is stronger than pure gold and is often used for jewelry.

Water molecule,
H_2O

Sugar molecule,
$C_{12}H_{22}O_{11}$

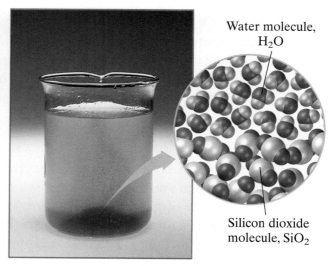

Water molecule,
H_2O

Silicon dioxide
molecule, SiO_2

Figure 19
The mixture of sugar and water on the left is a homogeneous mixture, in which there is a uniform distribution of the two components. Sand and water, on the right, do not mix uniformly, so they form a heterogeneous mixture.

homogeneous

describes something that has a uniform structure or composition throughout

heterogeneous

composed of dissimilar components

Homogeneous Mixtures

Sweetened tea and 14-karat gold are examples of **homogeneous** mixtures. In a homogeneous mixture, the pure substances are distributed uniformly throughout the mixture. Gasoline, syrup, and air are homogeneous mixtures. Their different components cannot be seen—not even using a microscope.

Because of how evenly the ingredients are spread throughout a homogeneous mixture, any two samples taken from the mixture will have the same proportions of ingredients. As a result, the properties of a homogeneous mixture are the same throughout. Look at the homogeneous mixture in **Figure 19.** You cannot see the different materials that make up the mixture because the sugar is mixed evenly throughout the water.

Heterogeneous Mixtures

In **Figure 19** you can clearly see the water and the sand, so the mixture is not homogeneous. It is a **heterogenous** mixture because it contains substances that are not evenly mixed. Different regions of a heterogeneous mixture have different properties. Additional examples of the two types of mixtures are shown in **Table 6.**

Table 6	Examples of Mixtures
Homogeneous	Iced tea—uniform distribution of components; components cannot be filtered out and will not settle out upon standing
	Stainless steel—uniform distribution of components
	Maple syrup—uniform distribution of components; components cannot be filtered out and will not settle out upon standing
Heterogeneous	Orange juice or tomato juice—uneven distribution of components; settles out upon standing
	Chocolate chip pecan cookie—uneven distribution of components
	Granite—uneven distribution of components
	Salad—uneven distribution of components; can be easily separated by physical means

Distinguishing Mixtures from Compounds

A compound is composed of two or more elements chemically joined together. A mixture is composed of two or more substances physically mixed together but not chemically joined. As a result, there are two major differences between mixtures and compounds.

First, the properties of a mixture reflect the properties of the substances it contains, but the properties of a compound often are very different from the properties of the elements that make it up. The oxygen gas that is a component of the mixture air can still support a candle flame. However, the properties of the compound water, including its physical state, do not reflect the properties of hydrogen and oxygen.

Second, a mixture's components can be present in varying proportions, but a compound has a definite composition in terms of the masses of its elements. The composition of milk, for example, will differ from one cow to the next and from day to day. However, the compound sucrose is always exactly 42.107% carbon, 6.478% hydrogen, and 51.415% oxygen no matter what its source is.

Separating Mixtures

One task a chemist often handles is the separation of the components of a mixture based on one or more physical properties. This task is similar to sorting recyclable materials. You can separate glass bottles based on their color and metal cans based on their attraction to a magnet. Techniques used by chemists include filtration, which relies on particle size, and distillation and evaporation, which rely on differences in boiling point.

Separating a Mixture

SAFETY PRECAUTIONS

PROCEDURE

1. Place the **mixture** of iron, sulfur, and salt on a **watch-glass.** Remove the iron from the mixture with the aid of a **magnet.** Transfer the iron to a 50 mL **beaker.**

2. Transfer the sulfur-salt mixture that remains to a second 50 mL **beaker.** Add 25 mL of **water,** and stir with a **glass stirring rod** to dissolve the salt.

3. Place **filter paper** in a **funnel.** Place the end of the funnel into a third 50 mL **beaker.** Filter the mixture and collect the filtrate—the liquid that passes through the filter.

4. Wash the residue in the filter with 15 mL of water, and collect the rinse water with the filtrate.

5. Set up a **ring stand** and a **Bunsen burner.** Evaporate the water from the filtrate.

Stop heating just before the liquid completely disappears.

ANALYSIS

1. What properties did you observe in each of the components of the mixture?

2. How did these properties help you to separate the components of the mixture?

3. Did any of the components share similar properties?

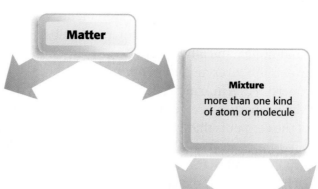

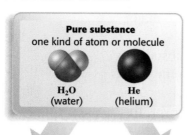

Pure substance
one kind of atom or molecule

H_2O
(water)

He
(helium)

Mixture
more than one kind
of atom or molecule

Element
a single kind of atom

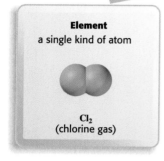

Cl_2
(chlorine gas)

Compound
bonded atoms

CH_4
(methane)

Homogeneous mixture
uniform composition

Sugar Water

Heterogeneous mixture
nonuniform composition

Water

Sand

Section Review

UNDERSTANDING KEY IDEAS

1. What are the two types of pure substances?

2. Define the term *compound*.

3. How does an element differ from a compound?

4. How are atoms and molecules related?

5. What is the smallest number of elements needed to make a compound?

6. What are two differences between compounds and mixtures?

7. Identify each of the following as an element, a compound, a homogeneous mixture, or a heterogeneous mixture.

a. CH_4
b. S_8
c. distilled water
d. salt water
e. CH_2O
f. concrete

8. How is a homogeneous mixture different from a heterogeneous mixture?

CRITICAL THINKING

9. Why is a *monatomic compound* nonsense?

10. Compare the composition of sucrose purified from sugar cane with the composition of sucrose purified from sugar beets. Explain your answer.

11. After a mixture of iron and sulfur are heated and then cooled, a magnet no longer attracts the iron. How would you classify the resulting material? Explain your answer.

12. How could you decide whether a ring was 24-karat gold or 14-karat gold without damaging the ring?

13. To help a child take aspirin, a parent may crush the tablet and add it to applesauce. Is the aspirin-applesauce combination classified as a compound or a mixture? Explain your answer.

14. Four different containers are labeled C + O_2, CO, CO_2, and Co. Based on the labels, classify each as an element, a compound, a homogeneous mixture, or a heterogeneous mixture. Explain your reasoning.

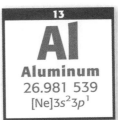

13
Al
Aluminum
26.981 539
$[Ne]3s^2 3p^1$

Element Spotlight

Where Is Al?

Earth's Crust:
8% by mass

Sea Water:
less than 0.1%

Aluminum's Humble Beginnings

In 1881, Charles Martin Hall was a 22-year-old student at Oberlin College, in Ohio. One day, Hall's chemistry professor mentioned in a lecture that anyone who could discover an inexpensive method for making aluminum metal would become rich. Working in a wooden shed and using a cast-iron frying pan, a blacksmith's forge, and homemade batteries, Hall discovered a practical technique for producing aluminum. Hall's process is the basis for the industrial production of aluminum today.

Industrial Uses

- Aluminum is the most abundant metal in Earth's crust. However, it is found in nature only in compounds and never as the pure metal.

- The most important source of aluminum is the mineral bauxite. Bauxite consists mostly of hydrated aluminum oxide.

- Recycling aluminum by melting and reusing it is considerably cheaper than producing new aluminum.

- Aluminum is light, weather-resistant, and easily worked. These properties make aluminum ideal for use in aircraft, cars, cans, window frames, screens, gutters, wire, food packaging, hardware, and tools.

Real-World Connection Recycling just one aluminum can saves enough electricity to run a TV for about four hours.

Aluminum's resistance to corrosion makes it suitable for use outdoors in this statue.

A Brief History

1827: F. Wöhler describes some of the properties of aluminum.

1886: Charles Martin Hall, of the United States, and Paul-Louis Héroult, of France, independently discover the process for extracting aluminum from aluminum oxide.

1800	1900

1824: F. Wöhler, of Germany, isolates aluminum from aluminum chloride.

1854: Henri Saint-Claire Deville, of France, and R. Bunsen, of Germany, independently accomplish the electrolysis of aluminum from sodium aluminum chloride.

Questions

1. Research and identify at least five items that you encounter on a regular basis that are made with aluminum.

2. Research the changes that have occurred in the design and construction of aluminum soft-drink cans and the reasons for the changes. Record a list of items that help illustrate why aluminum is a good choice for this product.

internet connect

www.scilinks.org
Topic: Aluminum
SciLinks code: HW4136

SC*i*LINKS. Maintained by the National Science Teachers Association

1 CHAPTER HIGHLIGHTS

SECTION ONE What Is Chemistry?

- Chemistry is the study of chemicals, their properties, and the reactions in which they are involved.
- Three of the states of matter are solid, liquid, and gas.
- Matter undergoes both physical changes and chemical changes. Evidence can help to identify the type of change.

SECTION TWO Describing Matter

- Matter has both mass and volume; matter thus has density, which is the ratio of mass to volume.
- Mass and weight are not the same thing. Mass is a measure of the amount of matter in an object. Weight is a measure of the gravitational force exerted on an object.
- SI units are used in science to express quantities. Derived units are combinations of the basic SI units.
- Conversion factors are used to change a given quantity from one unit to another unit.
- Properties of matter may be either physical or chemical.

SECTION THREE How Is Matter Classified?

- All matter is made from atoms.
- All atoms of an element are alike.
- Elements may exist as single atoms or as molecules.
- A molecule usually consists of two or more atoms combined in a definite ratio.
- Matter can be classified as a pure substance or a mixture.
- Elements and compounds are pure substances. Mixtures may be homogeneous or heterogeneous.

KEY TERMS

chemical
chemical reaction
states of matter
reactant
product

matter
volume
mass
weight
quantity
unit
conversion factor
physical property
density
chemical property

atom
pure substance
element
molecule
compound
mixture
homogeneous
heterogeneous

KEY SKILLS

Using Conversion Factors
Skills Toolkit 1 p. 13
Sample Problem A p. 14

USING KEY TERMS

1. What is chemistry?

2. What are the common physical states of matter, and how do they differ from one another?

3. Explain the difference between a physical change and a chemical change.

4. Identify the reactants and products in the following word equation:

iron oxide + aluminum ⟶
iron + aluminum oxide

5. During photosynthesis, light energy is captured by plants to make sugar from carbon dioxide and water. In the process, oxygen is also produced. Use this information to explain the terms *chemical reaction, reactant,* and *product.*

6. What units are used to express mass and weight?

7. How does a quantity differ from a unit? Give examples of each in your answer.

8. What is a conversion factor?

9. Explain what derived units are. Give an example of one.

10. Define *density,* and explain why it is considered a physical property rather than a chemical property of matter.

11. Write a brief paragraph that shows that you understand the following terms and the relationships between them: *atom, molecule, compound,* and *element.*

WRITING SKILLS

12. What do the terms *homogeneous* and *heterogeneous* mean?

UNDERSTANDING KEY IDEAS

What Is Chemistry?

13. Your friend mentions that she eats only natural foods because she wants her food to be free of chemicals. What is wrong with this reasoning?

14. Determine whether each of the following substances would be a gas, a liquid, or a solid if found in your classroom.
a. neon
b. mercury
c. sodium bicarbonate (baking soda)
d. carbon dioxide
e. rubbing alcohol

15. Is toasting bread an example of a chemical change? Why or why not?

16. Classify each of the following as a physical change or a chemical change, and describe the evidence that suggests a change is taking place.
a. cracking an egg
b. using bleach to remove a stain from a shirt
c. burning a candle
d. melting butter in the sun

17. Over time, soap scum deposits build up around your bathtub. Unlike soap, the soap scum is a calcium-based compound that cannot be dissolved in water. What evidence do you have that indicates that the formation of soap scum is a chemical change rather than a physical change?

Describing Matter

18. At the top of Mount Everest, the force of gravity is less than at sea level. Describe what would happen to your mass and to your weight if you traveled from sea level to the top of Mount Everest. Explain your answer.

19. Name the five most common SI base units used in chemistry. What quantity is each unit used to express?

20. What derived unit is appropriate for expressing each of the following?
a. rate of water flow
b. speed
c. volume of a room

21. Pick an object you can see right now. List three physical properties of the object that you can observe. Can you also observe a chemical property of the object? Explain your answer.

22. Compare the physical and chemical properties of salt and sugar. What properties do they share? Which properties could you use to distinguish between salt and sugar?

23. What do you need to know to determine the density of a sample of matter?

24. For each pair below, indicate the substance that has the greater density. Explain your answer.
a. penny and cork
b. ice cube and liquid water

25. Substances A and B are colorless, odorless liquids that are nonconductors and flammable. The density of substance A is 0.97 g/mL; the density of substance B is 0.89 g/mL. Are A and B the same substance? Explain your answer.

How Is Matter Classified?

26. Is a compound a pure substance or a mixture? Explain your answer.

27. Iron (Fe) and sulfur (S_8) react to create iron(II) sulfide (FeS). Are iron and sulfur elements, compounds, or mixtures? Is iron(II) sulfide an element, a compound, or a mixture?

28. Which of the following items are homogeneous: a tree, a cheeseburger, flour, paint, and an ice cube?

29. Determine if each material represented below is an element, compound, or mixture, and whether the model illustrates a solid, liquid, or gas.

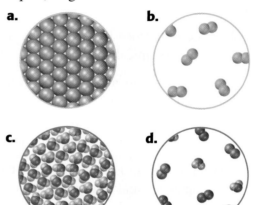

PRACTICE PROBLEMS

Sample Problem A Converting Units

30. Which quantity of each pair is larger?
a. 2400 cm or 2 m
b. 3 L or 3 mL
c. 17 g or 1.7 kg

31. Using **Appendix A,** convert the following measurements to the units specified.
a. 357 mL = ? L　　　**e.** 250 μg = ? g
b. 25 kg = ? mg　　　**f.** 250 μg = ? kg
c. 35 000 cm³ = ? L　**g.** 1.5 s = ? ms
d. 2.46 L = ? cm³　　**h.** 10.5 mol = ? mmol

32. What is the distance in meters between two points that are 150 km apart?

MIXED REVIEW

33. Use particle models to explain why liquids and gases take the shape of their containers.

34. When camping with your family, you boil a pot of water over a campfire. What physical and chemical changes take place during this process? What evidence of a chemical change do you observe?

35. You are given a sample of colorless liquid in a beaker. What type of information could you gather to determine if the liquid is water?

36. Calculate the density of a piece of metal if its mass is 201.0 g and its volume is 18.9 cm^3.

37. The density of CCl$_4$ (carbon tetrachloride) is 1.58 g/mL. What is the mass of 95.7 mL of CCl$_4$?

38 What is the volume of 227 g of olive oil if its density is 0.92 g/mL?

CRITICAL THINKING

39. A white, crystalline material that looks like table salt releases gas when heated under certain conditions. There is no change in the appearance of the solid, but the reactivity of the material changes.
 a. Did a chemical or physical change occur? How do you know?
 b. Was the original material an element or a compound? Explain your answer.

40. A student leaves an uncapped watercolor marker on an open notebook. Later, the student discovers the leaking marker has produced a rainbow of colors on the top page.
 a. Is this an example of a physical change or a chemical change? Explain your answer.
 b. Should the ink be classified as an element, a compound, or a mixture? Explain your answer.

41. A student checks the volume, melting point, and shape of two unlabeled samples of matter and finds that the measurements are identical. From this he concludes that the samples have the same chemical composition. Is this a valid conclusion? What additional information might he collect to test his conclusion?

42. Equal amounts of three liquids—carbon tetrachloride (D = 1.58 g/cm^3), mercury (D = 13.546 g/cm^3), and water (D = 1.00 g/cm^3)—are mixed together. After mixing, they separate to form three distinct layers. Draw and label a sketch showing the order of the layers in a test tube.

ALTERNATIVE ASSESSMENT

43. Your teacher will provide you with a sample of a metallic element. Determine its density. Check references that list the density of metals to identify the sample that you analyzed.

44. Make a poster showing the types of product warning labels that are found on products in your home.

CONCEPT MAPPING

45. Use the following terms to complete the concept map below: *volume, density, matter, physical property,* and *mass.*

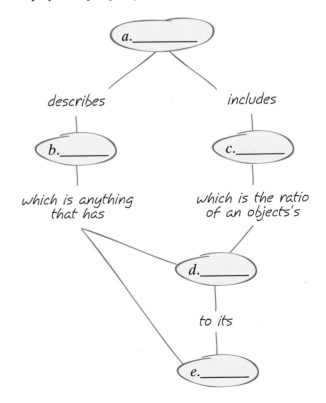

Study the graph below, and answer the questions that follow.
For help in interpreting graphs, see Appendix B, "Study Skills for Chemistry."

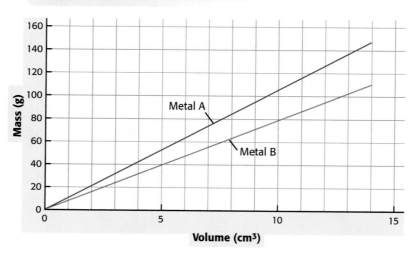

Mass Versus Volume for Two Metals

46. What does the straight line on the graph indicate about the relationship between volume and mass?

47. What does the slope of each line indicate?

48. What is the density of metal A? of metal B?

49. Based on the density values in **Table 4,** what do you think is the identity of metal A? of metal B? Explain your reasoning.

TECHNOLOGY AND LEARNING

50. Graphing Calculator

Graphing Tabular Data

The graphing calculator can run a program that graphs ordered pairs of data, such as temperature versus time. In this problem, you will answer questions based on a graph of temperature versus time that the calculator will create.

Go to Appendix C. If you are using a TI-83 Plus, you can download the program and data sets and run the application as directed. Press the **APPS** key on your calculator, and then choose the application **CHEMAPPS**. Press 1, then highlight **ALL** on the screen, press 1, then highlight **LOAD**, and press 2 to

load the data into your calculator. Quit the application, and then run the program **GRAPH.** A set of data points representing degrees Celsius versus time in minutes will be graphed.

If you are using another calculator, your teacher will provide you with keystrokes and data sets to use.

a. Approximately what would the temperature be at the 16-minute interval?

b. Between which two intervals did the temperature increase the most: between 3 and 5 minutes, between 5 and 8 minutes, or between 8 and 10 minutes?

c. If the graph extended to 20 minutes, what would you expect the temperature to be?

STANDARDIZED TEST PREP

1. Magnesium dissolves in hydrochloric acid to produce magnesium chloride and hydrogen gas. Which of the following represents the reactants in this reaction?
a. magnesium and magnesium chloride
b. hydrochloric acid and hydrogen gas
c. magnesium and hydrochloric acid
d. magnesium chloride and hydrogen gas

2. Matter that has a definite shape and a definite volume is
a. an acid.
b. a base.
c. a solid.
d. a gas.

3. Which of the following pairs of measurements are equal?
a. 1.63 kg and 163 g
b. 0.0704 m and 7.04 mm
c. 0.015 mL and 1.5 L
d. 325 mg and 0.325 g

4. A chemical property of a substance
a. depends on how the substance reacts with other substances.
b. can be described when the substance undergoes a physical change.
c. cannot be quantitatively described.
d. cannot be qualitatively described.

5. The SI base unit that is commonly used in chemistry to describe the amount of a substance is the
a. degree Celsius.
b. gram
c. liter.
d. mole.

6. Matter can be defined as anything that
a. has weight.
b. has mass and volume.
c. can be described in SI units.
d. exhibits both chemical and physical properties.

7. Which of the following is best classified as a homogeneous mixture?
a. pizza
b. blood
c. hot tea
d. copper wire

8. A compound differs from a mixture in that a compound
a. contains only one element.
b. varies in chemical composition depending on the sample size.
c. has a definite composition by mass of the elements it contains.
d. can be classified as either heterogeneous or homogeneous.

9. Which of the following is an element?
a. He
b. $BaCl_2$
c. NaOH
d. CO

10. Determine which of the following measurements is the largest.
a. 0.200 L
b. 0.020 kL
a. 20.0 cL
d. 2000 mL

MATTER AND ENERGY

The photo of the active volcano and the scientists who are investigating it is a dramatic display of matter and energy. Most people who view the photo would consider the volcano and the scientists to be completely different. The scientists seem to be unchanging, while the volcano is explosive and changing rapidly. However, the scientists and the volcano are similar in that they are made of matter and are affected by energy. This chapter will show you the relationship between matter and energy and some of the rules that govern them.

START-UP ACTIVITY

Chemical Changes and Energy

SAFETY PRECAUTIONS

PROCEDURE

1. Place a small **thermometer** completely inside a **jar,** and close the **lid.** Wait 5 min, and record the temperature.

2. While you are waiting to record the temperature, soak one-half of a **steel wool pad** in **vinegar** for 2 min.

3. Squeeze the excess vinegar from the steel wool. Remove the thermometer from the jar, and wrap the steel wool around the bulb of the thermometer. Secure the steel wool to the thermometer with a **rubber band.**

4. Place the thermometer and the steel wool inside the jar, and close the lid. Wait 5 min, and record the temperature.

ANALYSIS

1. How did the temperature change?

2. What do you think caused the temperature to change?

3. Do you think vinegar is a reactant or product? Why?

Pre-Reading Questions

1. When ice melts, what happens to its chemical composition?

2. Name a source of energy for your body.

3. Name some temperature scales.

4. What is a chemical property? What is a physical property?

internet connect

www.scilinks.org
Topic: Matter and Energy
SciLinks code: HW4158

SCiLINKS.
Maintained by the
National Science
Teachers Association

Energy

KEY TERMS

- **energy**
- **physical change**
- **chemical change**
- **evaporation**
- **endothermic**
- **exothermic**
- **law of conservation of energy**
- **heat**
- **kinetic energy**
- **temperature**
- **specific heat**

OBJECTIVES

1 **Explain** that physical and chemical changes in matter involve transfers of energy.

2 **Apply** the law of conservation of energy to analyze changes in matter.

3 **Distinguish** between heat and temperature.

4 **Convert** between the Celsius and Kelvin temperature scales.

Energy and Change

If you ask 10 people what comes to mind when they hear the word *energy,* you will probably get 10 different responses. Some people think of energy in terms of exercising or playing sports. Others may picture energy in terms of a fuel or a certain food.

If you ask 10 scientists what comes to mind when they hear the word *energy,* you may also get 10 different responses. A geologist may think of energy in terms of a volcanic eruption. A biologist may visualize cells using oxygen and sugar in reactions to obtain the energy they need. A chemist may think of a reaction in a lab, such as the one shown in **Figure 1.**

The word *energy* represents a broad concept. One definition of **energy** is the capacity to do some kind of work, such as moving an object, forming a new compound, or generating light. No matter how energy is defined, it is always involved when there is a change in matter.

energy

the capacity to do work

Figure 1
Energy is released in the explosive reaction that occurs between hydrogen and oxygen to form water.

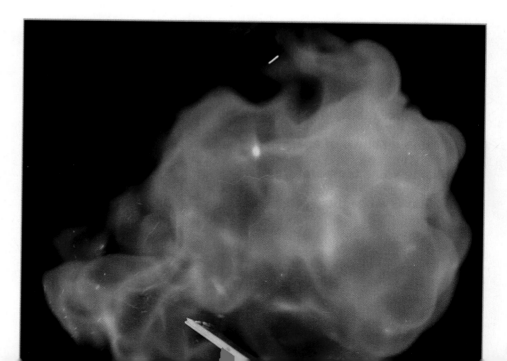

Changes in Matter Can Be Physical or Chemical

Ice melting and water boiling are examples of **physical changes.** A physical change affects only the physical properties of matter. For example, when ice melts and turns into liquid water, you still have the same substance represented by the formula H_2O. When water boils and turns into a vapor, the vapor is still H_2O. Notice that in these examples the chemical nature of the substance does not change; only the physical state of the substance changes to a solid, liquid, or gas.

In contrast, the reaction of hydrogen and oxygen to produce water is an example of a **chemical change.** A chemical change occurs whenever a new substance is made. In other words, a chemical reaction has taken place. You know water is different from hydrogen and oxygen because water has different properties. For example, the boiling points of hydrogen and oxygen at atmospheric pressure are $-252.8°C$ and $-182.962°C$, respectively. The boiling point of water at atmospheric pressure is $100°C$. Hydrogen and oxygen are also much more reactive than water.

Every Change in Matter Involves a Change in Energy

All physical and chemical changes involve a change in energy. Sometimes energy must be supplied for the change in matter to occur. For example, consider a block of ice, such as the one shown in **Figure 2.** As long as the ice remains cold enough, the particles in the solid ice stay in place.

However, if the ice gets warm, the particles will begin to move and vibrate more and more. For the ice to melt, energy must be supplied so that the particles can move past one another. If more energy is supplied and the boiling point of water is reached, the particles of the liquid will leave the liquid's surface through **evaporation** and form a gas. These physical changes require an input of energy. Many chemical changes also require an input of energy.

Sometimes energy is released when a change in matter occurs. For example, energy is released when a vapor turns into a liquid or when a liquid turns into a solid. Some chemical changes also release energy. The explosion that occurs when hydrogen and oxygen react to form water is a release of energy.

physical change

a change of matter from one form to another without a change in chemical properties

chemical change

a change that occurs when one or more substances change into entirely new substances with different properties

evaporation

the change of a substance from a liquid to a gas

Figure 2
Energy is involved when a physical change, such as the melting of ice, happens.

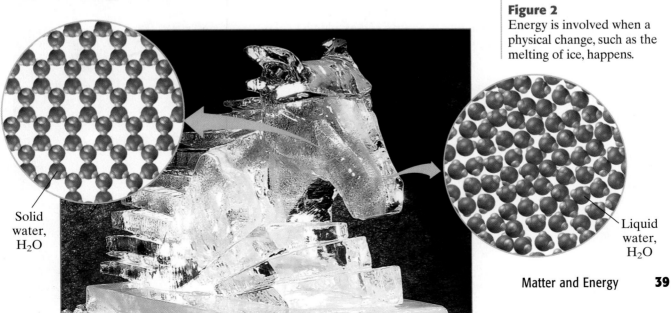

Solid water, H_2O

Liquid water, H_2O

Endothermic and Exothermic Processes

endothermic

describes a process in which heat is absorbed from the environment

Any change in matter in which energy is absorbed is known as an **endothermic** process. The melting of ice and the boiling of water are two examples of physical changes that are endothermic processes.

Some chemical changes are also endothermic processes. **Figure 3** shows a chemical reaction that occurs when barium hydroxide and ammonium nitrate are mixed. Notice in **Figure 3** that these two solids form a liquid, slushlike product. Also, notice the ice crystals that form on the surface of the beaker. As barium hydroxide and ammonium nitrate react, energy is absorbed from the beaker's surroundings. As a result, the beaker feels colder because the reaction absorbs energy as heat from your hand. Water vapor in the air freezes on the surface of the beaker, providing evidence that the reaction is endothermic.

exothermic

describes a process in which a system releases heat into the environment

Any change in matter in which energy is released is an **exothermic** process. The freezing of water and the condensation of water vapor are two examples of physical changes that are exothermic processes.

Recall that when hydrogen and oxygen gases react to form water, an explosive reaction occurs. The vessel in which the reaction takes place becomes warmer after the reaction, giving evidence that energy has been released.

law of conservation of energy

the law that states that energy cannot be created or destroyed but can be changed from one form to another

Endothermic processes, in which energy is absorbed, may make it seem as if energy is being destroyed. Similarly, exothermic processes, in which energy is released, may make it seem as if energy is being created. However, the **law of conservation of energy** states that during any physical or chemical change, the total quantity of energy remains constant. In other words, energy cannot be destroyed or created.

Accounting for all the different types of energy present before and after a physical or chemical change is a difficult process. But measurements of energy during both physical and chemical changes have shown that when energy seems to be destroyed or created, energy is actually being transferred. The difference between exothermic and endothermic processes is whether energy is absorbed or released by the substances involved.

Figure 3
The reaction between barium hydroxide and ammonium nitrate absorbs energy and causes ice crystals to form on the beaker.

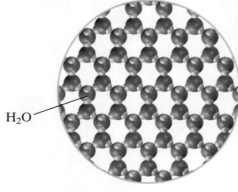

H_2O

Conservation of Energy in a Chemical Reaction

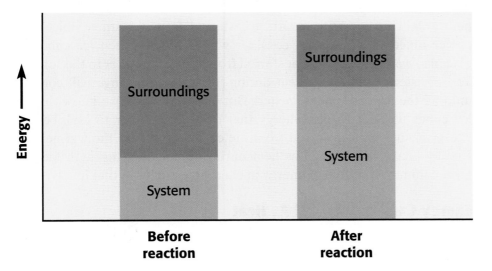

Figure 4
Notice that the energy of the reactants and products increases, while the energy of the surroundings decreases. However, the total energy does not change.

Energy Is Often Transferred

Figure 4 shows the energy changes that take place when barium hydroxide and ammonium nitrate react. To keep track of energy changes, chemists use the terms *system* and *surroundings*. A system consists of all the components that are being studied at any given time. In **Figure 4,** the system consists of the mixture inside the beaker. The surroundings include everything outside the system. In **Figure 4,** the surroundings consist of everything else including the air both inside and outside the beaker and the beaker itself. Keep in mind that the air is made of various gases.

Energy is often transferred back and forth between a system and its surroundings. An exothermic process involves a transfer of energy from a system to its surroundings. An endothermic process involves a transfer of energy from the surroundings to the system. However, in every case, the total energy of the systems and their surroundings remains the same, as shown in **Figure 4.**

Energy Can Be Transferred in Different Forms

Energy exists in different forms, including chemical, mechanical, light, heat, electrical, and sound. The transfer of energy between a system and its surroundings can involve any one of these forms of energy. Consider the process of *photosynthesis*. Light energy is transferred from the sun to green plants. Chlorophyll inside the plant's cells (the system) absorbs energy—the light energy from the sun (the surroundings). This light energy is converted to chemical energy when the plant synthesizes chemical nutrients that serve as the basis for sustaining all life on Earth.

Next, consider what happens when you activate a light stick. Chemicals inside the stick react to release energy in the form of light. This light energy is transferred from the system inside the light stick to the surroundings, generating the light that you see. A variety of animals depend on chemical reactions that generate light, including fish, worms, and fireflies.

Heat

Heat is the energy transferred between objects that are at different temperatures. This energy is always transferred from a warmer object to a cooler object. For example, consider what happens when ice cubes are placed in water. Energy is transferred from the liquid water to the solid ice. The transfer of energy as heat during this physical change will continue until all the ice cubes have melted. But on a warm day, we know that the ice cubes will not release energy that causes the water to boil, because energy cannot be transferred from the cooler objects to the warmer one. Energy is also transferred as heat during chemical changes. In fact, the most common transfers of energy in chemistry are those that involve heat.

Energy Can Be Released As Heat

The worst industrial disaster in U.S. history occurred in April 1947. A cargo ship named the *Grandcamp* had been loaded with fertilizer in Texas City, a Texas port city of 50 000 people. The fertilizer consisted of tons of a compound called *ammonium nitrate*. Soon after the last bags of fertilizer had been loaded, a small fire occurred, and smoke was noticed coming from the ship's cargo hold. About an hour later, the ship exploded.

The explosion was heard 240 km away. An anchor from the ship flew through the air and created a 3 m wide hole in the ground where it landed. Every building in the city was either destroyed or damaged. The catastrophe on the *Grandcamp* was caused by an exothermic chemical reaction that released a tremendous amount of energy as heat.

All of this energy that was released came from the energy that was stored within the ammonium nitrate. Energy can be stored within a chemical substance as chemical energy. When the ammonium nitrate ignited, an exothermic chemical reaction took place and released energy as heat. In addition, the ammonium nitrate explosion generated **kinetic energy,** as shown by the anchor that flew through the air.

Figure 5
Billowing black smoke filled the sky over Texas City in the aftermath of the *Grandcamp* explosion, shown in this aerial photograph.

Energy Can Be Absorbed As Heat

In an endothermic reaction, energy is absorbed by the chemicals that are reacting. If you have ever baked a cake or a loaf of bread, you have seen an example of such a reaction. Recipes for both products require either baking soda or baking powder. Both baking powder and baking soda contain a chemical that causes dough to rise when heated in an oven.

The chemical found in both baking powder and baking soda is sodium bicarbonate. Energy from the oven is absorbed by the sodium bicarbonate. The sodium bicarbonate breaks down into three different chemical substances, sodium carbonate, water vapor, and carbon dioxide gas, in the following endothermic reaction:

$$2NaHCO_3 \rightarrow Na_2CO_3 + H_2O + CO_2$$

The carbon dioxide gas causes the batter to rise while baking, as you can see in **Figure 6.**

Figure 6
Baking a cake or bread is an example of an endothermic reaction, in which energy is absorbed as heat.

Heat Is Different from Temperature

You have learned that energy can be transferred as heat because of a temperature difference. So, the transfer of energy as heat can be measured by calculating changes in temperature. Temperature indicates how hot or cold something is. **Temperature** is actually a measurement of the average kinetic energy of the random motion of particles in a substance.

For example, imagine that you are heating water on a stove to make tea. The water molecules have kinetic energy as they move freely in the liquid. Energy transferred as heat from the stove causes these water molecules to move faster. The more rapidly the water molecules move, the greater their average kinetic energy. As the average kinetic energy of the water molecules increases, the temperature of the water increases. Think of heat as the energy that is transferred from the stove to the water because of a difference in the temperatures of the stove and the water. The temperature change of the water is a measure of the energy transferred as heat.

temperature

a measure of how hot (or cold) something is; specifically, a measure of the average kinetic energy of the particles in an object

Temperature Is Expressed Using Different Scales

Thermometers are usually marked with the Fahrenheit or Celsius temperature scales. However, the Fahrenheit scale is not used in chemistry. Recall that the SI unit for temperature is the *Kelvin*, K. The zero point on the Celsius scale is designated as the freezing point of water. The zero point on the Kelvin scale is designated as *absolute zero*, the temperature at which the minimum average kinetic energies of all particles occur.

In chemistry, you will have to use both the Celsius and Kelvin scales. At times, you will have to convert temperature values between these two scales. Conversion between these two scales simply requires an adjustment to account for their different zero points.

www.scilinks.org
Topic: Temperature Scales
SciLinks code: HW4124

$$t(^\circ C) = T(K) - 273.15 \text{ K} \qquad T(K) = t(^\circ C) + 273.15^\circ C$$

The symbols t and T represent temperatures in degrees Celsius and in kelvins, respectively. Also, notice that a temperature change is the same in kelvins and in Celsius degrees.

Transfer of Heat May Not Affect the Temperature

The transfer of energy as heat does not always result in a change of temperature. For example, consider what happens when energy is transferred to a solid such as ice. Imagine that you have a mixture of ice cubes and water in a sealed, insulated container. A thermometer is inserted into the container to measure temperature changes as energy is added to the ice-water mixture.

As energy is transferred as heat to the ice-water mixture, the ice cubes will start to melt. However, the temperature of the mixture remains at 0°C. Even though energy is continuously being transferred as heat, the temperature of the ice-water mixture does not increase.

Once all the ice has melted, the temperature of the water will start to increase. When the temperature reaches 100°C, the water will begin to boil. As the water turns into a gas, the temperature remains at 100°C, even though energy is still being transferred to the system as heat. Once all the water has vaporized, the temperature will again start to rise.

Notice that the temperature remains constant during the physical changes that occur as ice melts and water vaporizes. What happens to the energy being transferred as heat if the energy does not cause an increase in temperature? The energy that is transferred as heat is actually being used to move molecules past one another or away from one another. This energy causes the molecules in the solid ice to move more freely so that they form a liquid. This energy also causes the water molecules to move farther apart so that they form a gas.

Figure 7 shows the temperature changes that occur as energy is transferred as heat to change a solid into a liquid and then into a gas. Notice that the temperature increases only when the substance is in the solid, liquid, or gaseous states. The temperature does not increase when the solid is changing to a liquid or when the liquid is changing to a gas.

Figure 7
This graph illustrates how temperature is affected as energy is transferred to ice as heat. Notice that much more energy must be transferred as heat to vaporize water than to melt ice.

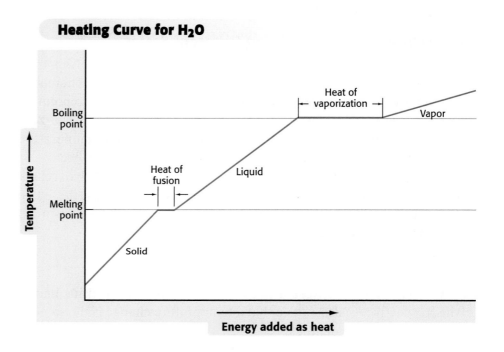

Transfer of Heat Affects Substances Differently

Have you ever wondered why a heavy iron pot gets hot fast but the water in the pot takes a long time to warm up? If you transfer the same quantity of heat to similar masses of different substances, they do not show the same increase in temperature. This relationship between energy transferred as heat to a substance and the substance's temperature change is called the **specific heat.**

The specific heat of a substance is the quantity of energy as heat that must be transferred to raise the temperature of 1 g of a substance 1 K. The SI unit for energy is the *joule* (J). Specific heat is expressed in joules per gram kelvin (J/g·K).

Metals tend to have low specific heats, which indicates that relatively little energy must be transferred as heat to raise their temperatures. In contrast, water has an extremely high specific heat. In fact, it is the highest of most common substances.

During a hot summer day, water can absorb a large quantity of energy from the hot air and the sun and can cool the air without a large increase in the water's temperature. During the night, the water continues to absorb energy from the air. This energy that is removed from the air causes the temperature of the air to drop quickly, while the water's temperature changes very little. This behavior is explained by the fact that air has a low specific heat and water has a high specific heat.

specific heat

the quantity of heat required to raise a unit mass of homogeneous material 1 K or 1°C in a specified way given constant pressure and volume

Section Review

UNDERSTANDING KEY IDEAS

1. What is energy?
2. State the law of conservation of energy.
3. How does heat differ from temperature?
4. What is a system?
5. Explain how an endothermic process differs from an exothermic process.
6. What two temperature scales are used in chemistry?

PRACTICE PROBLEMS

7. Convert the following Celsius temperatures to Kelvin temperatures.
   - **a.** 100°C
   - **c.** 0°C
   - **b.** 785°C
   - **d.** −37°C

8. Convert the following Kelvin temperatures to Celsius temperatures.
   - **a.** 273 K
   - **c.** 0 K
   - **b.** 1200 K
   - **d.** 100 K

CRITICAL THINKING

9. Is breaking an egg an example of a physical or chemical change? Explain your answer.
10. Is cooking an egg an example of a physical or chemical change? Explain your answer.
11. What happens in terms of the transfer of energy as heat when you hold a snowball in your hands?
12. Why is it impossible to have a temperature value below 0 K?
13. If energy is transferred to a substance as heat, will the temperature of the substance always increase? Explain why or why not.

Studying Matter and Energy

KEY TERMS

- **scientific method**
- **hypothesis**
- **theory**
- **law**
- **law of conservation of mass**

scientific method

a series of steps followed to solve problems, including collecting data, formulating a hypothesis, testing the hypothesis, and stating conclusions

Figure 8
Each stage of the scientific method represents a number of different activities. Scientists choose the activities to use depending on the nature of their investigation.

OBJECTIVES

① **Describe** how chemists use the scientific method.

② **Explain** the purpose of controlling the conditions of an experiment.

③ **Explain** the difference between a hypothesis, a theory, and a law.

The Scientific Method

Science is unlike other fields of study in that it includes specific procedures for conducting research. These procedures make up the **scientific method,** which is shown in **Figure 8.** The scientific method is not a series of exact steps, but rather a strategy for drawing sound conclusions.

A scientist chooses the procedures to use depending on the nature of the investigation. For example, a chemist who has an idea for developing a better method to recycle plastics may research scientific articles about plastics, collect information, propose a method to separate the materials, and then test the method. In contrast, another chemist investigating the pollution caused by a trash incinerator would select different procedures. These procedures might include collecting and analyzing samples, interviewing people, predicting the role the incinerator plays in producing the pollution, and conducting field studies to test that prediction.

No matter which approach they use, both chemists are employing the scientific method. Ultimately, the success of the scientific method depends on publishing the results so that others can repeat the procedures and verify the results.

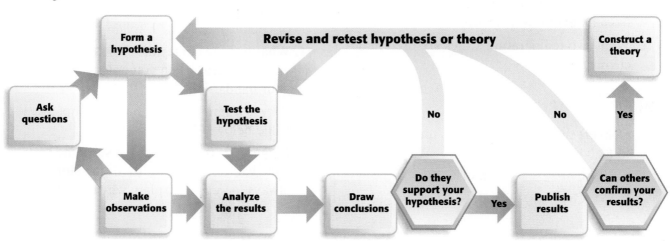

Quick LAB

Using the Scientific Method

PROCEDURE

1. Have someone prepare **five sealed paper bags,** each containing **an item commonly found in a home.**

2. Without opening the bags, try to determine the identity of each item.

3. Test each of your conclusions whenever possible. For example, if you concluded that one of the items is a refrigerator magnet, test it to see if it attracts small metal objects, such as paper clips.

ANALYSIS

1. How many processes that are part of the scientific method shown in **Figure 8** did you use?

2. How many items did you correctly identify?

Experiments Are Part of the Scientific Method

The first scientists depended on rational thought and logic. They rarely felt it was necessary to test their ideas or conclusions, and they did not feel the need to experiment. Gradually, experiments became the crucial test for the acceptance of scientific knowledge. Today, experiments are an important part of the scientific method.

An experiment is the process by which scientific ideas are tested. For example, consider what happens when manganese dioxide is added to a solution of hydrogen peroxide. Tiny bubbles of gas soon rise to the surface of the solution, indicating that a chemical reaction has taken place. Now, consider what happens when a small piece of beef liver is added to a solution of hydrogen peroxide. Tiny gas bubbles are produced. So, you might conclude that the liver contains manganese dioxide. To support your conclusion, you would have to test for the presence of manganese dioxide in the piece of liver.

Experiments May Not Turn Out As Expected

Your tests would reveal that liver does not contain any manganese dioxide. In this case, the results of the experiment did not turn out as you might have expected. Scientists are often confronted by situations in which their results do not turn out as expected. Scientists do not view these results as a failure. Rather, they analyze these results and continue with the scientific method. Unexpected results often give scientists as much information as expected results do. So, unexpected results are as important as expected results.

In this case, the liver might contain a different chemical that acts like manganese dioxide when added to hydrogen peroxide. Additional experiments would reveal that the liver does in fact contain such a chemical. Experimental results can also lead to more experiments. Perhaps the chemical that acts like manganese dioxide can be found in other parts of the body.

internet connect

www.scilinks.org
Topic: Scientific Method
SciLinks code: HW4167

SCiLINKS. Maintained by the National Science Teachers Association

Scientific Discoveries Can Come from Unexpected Observations

Not all discoveries and findings are the results of a carefully worked-out plan based on the scientific method. In fact, some important discoveries and developments have been made simply by accident. An example in chemistry is the discovery of a compound commonly known as Teflon®. You are probably familiar with Teflon as the nonstick coating used on pots and pans, but it has many more applications.

Teflon is used as thermal insulation in clothing, as a component in wall coverings, and as a protective coating on metals, glass, and plastics. Teflon's properties of very low chemical reactivity and very low friction make it valuable in the construction of artificial joints for human limbs. As you can see in **Figure 9,** Teflon is also used as a roofing material.

Teflon was not discovered as a result of a planned series of experiments designed to produce this chemical compound. Rather, it was discovered when a scientist made a simple but puzzling observation.

Teflon Was Discovered by Chance

In 1938, Dr. Roy Plunkett, a chemist employed by DuPont, was trying to produce a new coolant gas to use as a refrigerant. He was hoping to develop a less expensive coolant than the one that was being widely used at that time. His plan was to allow a gas called *tetrafluoroethene* (TFE) to react with hydrochloric acid. To begin his experiment, Plunkett placed a cylinder of liquefied TFE on a balance to record its mass.

He then opened the cylinder to let the TFE gas flow into a container filled with hydrochloric acid. But no TFE came out of the cylinder. Because the cylinder had the same mass as it did when it was filled with TFE, Plunkett knew that none of the TFE had leaked out. He removed the valve and shook the cylinder upside down. Only a few white flakes fell out.

Curious about what had happened, Plunkett decided to analyze the white flakes. He discovered that he had accidentally created the proper conditions for TFE molecules to join together to form a long chain. These long-chained molecules were very slippery. After 10 years of additional research, large-scale manufacturing of these long-chained molecules, known as Teflon or polytetrafluoroethene (PTFE), became practical.

Figure 9
Teflon was used to make the roof of the Hubert H. Humphrey Metrodome in Minneapolis, Minnesota.

Synthetic Dyes Were Also Discovered by Chance

If you have on an article of clothing that is colored, you are wearing something whose history can be traced to another unexpected chemistry discovery. This discovery was made in 1856 by an 18-year-old student named William Perkin, who was in his junior year at London's Royal College of Chemistry.

At that time, England was the world's leading producer of textiles, including those used for making clothing. The dyes used to color the textiles were natural products, extracted from both plants and animals. Only a few colors were available. In addition, the process to get dyes from raw materials was costly. As a result, only the wealthy could afford to wear brightly colored clothes for everyday use.

Mauve, a deep purple, was the color most people wanted for their clothing. In ancient times, only royalty could afford to own clothes dyed a mauve color. In Perkin's time, only the wealthy people could afford mauve.

Making an Unexpected Discovery

At first, Perkin had no interest in brightly colored clothes. Rather, his interest was in finding a way to make quinine, a drug used to treat malaria. At the time, quinine could only be made from the bark of a particular kind of tree. Great Britain needed huge quantities of the drug to treat its soldiers who got malaria in the tropical countries that were part of the British Empire. There was not enough of the drug to keep up with demand.

The only way to get enough quinine was to develop a synthetic version of the drug. During a vacation from college, Perkin was at home experimenting with ways of making synthetic quinine. One of his experiments resulted in a product that was a thick, sticky, black substance. He immediately realized that this attempt to synthesize quinine did not work. Curious about the substance, Perkin washed his reaction vessel with water. But the sticky product would not wash away. Perkin next decided to try cleaning the vessel with an alcohol. What he saw next was an unexpected discovery.

Analyzing an Unexpected Discovery

When Perkin poured an alcohol on the black product, it turned a mauve color. He found a way to extract the purple substance from the black product and determined that his newly discovered substance was perfect for dyeing clothes. He named his accidental discovery "aniline purple," but the fashionable people of Paris soon renamed it mauve.

Perkin became obsessed with his discovery. He left the Royal College of Chemistry and decided to open a factory that could make large amounts of the dye. Within two years, his factory had produced enough dye to ship to the largest maker of silk clothing in London. The color mauve quickly became the most popular color in the fashion industry throughout Europe. Perkin expanded his company and soon started producing other dyes, including magenta and a deep red. As a result of his unexpected discovery, Perkin became a very wealthy man and retired at the age of 36 to devote his time to chemical research. His unexpected discovery also marked the start of the synthetic dye industry.

Figure 10
Through his accidental discovery of aniline purple, William Perkin found an inexpensive way to make the color mauve. His discovery brought on the beginning of the synthetic dye industry.

Scientific Explanations

Questions that scientists seek to answer and problems that they hope to solve often come after they observe something. These observations can be made of the natural world or in a laboratory. A scientist must always make careful observations, not knowing if some totally unexpected result might lead to an interesting finding or important discovery. Consider what would have happened if Plunkett had ignored the white flakes or if Perkin had overlooked the mauve substance.

Once observations have been made, they must be analyzed. Scientists start by looking at all the relevant information or data they have gathered. They look for patterns that might suggest an explanation for the observations. This proposed explanation is called a **hypothesis.** A hypothesis is a reasonable and testable explanation for observations.

hypothesis

a theory or explanation that is based on observations and that can be tested

Chemists Use Experiments to Test a Hypothesis

Once a scientist has developed a hypothesis, the next step is to test the validity of the hypothesis. This testing is often done by carrying out experiments, as shown in **Figure 11.** Even though the results of their experiments were totally unexpected, Plunkett and Perkin developed hypotheses to account for their observations. Both scientists hypothesized that their accidental discoveries might have some practical application. Their next step was to design experiments to test their hypotheses.

To understand what is involved in designing an experiment, consider this example. Imagine that you have observed that your family car has recently been getting better mileage. Perhaps you suggest to your family that their decision to use a new brand of gasoline is the factor responsible for the improved mileage. In effect, you have proposed a hypothesis to explain an observation.

Figure 11
Students conduct experiments to test the validity of their hypotheses.

Scientists Must Identify the Possible Variables

To test the validity of your hypothesis, your next step is to plan your experiments. You must begin by identifying as many factors as possible that could account for your observations. A factor that could affect the results of an experiment is called a *variable*. A scientist changes variables one at a time to see which variable affects the outcome of an experiment.

Several variables might account for the improved mileage you noticed with your family car. The use of a new brand of gasoline is one variable. Driving more on highways, making fewer short trips, having the car's engine serviced, and avoiding quick accelerations are other variables that might have resulted in the improved mileage. To know if your hypothesis is right, the experiment must be designed so that each variable is tested separately. Ideally, the experiments will eliminate all but one variable so that the exact cause of the observed results can be identified.

Each Variable Must Be Tested Individually

Scientists reduce the number of possible variables by keeping all the variables constant except one. When a variable is kept constant from one experiment to the next, the variable is called a *control* and the procedure is called a *controlled experiment*. Consider how a controlled experiment would be designed to identify the variable responsible for the improved mileage.

You would fill the car with the new brand of gasoline and keep an accurate record of how many miles you get per gallon. When the gas tank is almost empty, you would do the same after filling the car with the brand of gasoline your family had been using before. In both trials, you should drive the car under the same conditions. For example, the car should be driven the same number of miles on highways and local streets and at the same speeds in both trials. You then have designed the experiment so that only one variable—the brand of gasoline—is being tested.

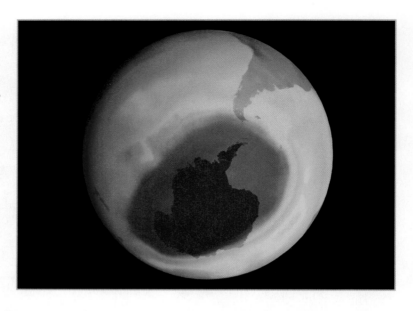

Figure 13
In 1974, scientists proposed a theory to explain the observation of a hole in the ozone layer over Antarctica, which is shown in purple. This hole is about the size of North America.

Data from Experiments Can Lead to a Theory

As early as 1969, scientists observed that the ozone layer was breaking down. Ozone, O_3, is a gas that forms a thin layer high above Earth's surface. This layer shields all living things from most of the sun's damaging ultraviolet light. In 1970, Paul Crutzen, working at the Max Planck Institute for Chemistry, showed the connection between nitrogen oxides and the reduction of ozone in air. In 1974, F. Sherwood Rowland and Mario Molina, two chemists working at the University of California, Irvine, proposed the hypothesis that the release of chlorofluorocarbons (CFCs) into the atmosphere harms the ozone layer. CFCs were being used in refrigerators, air conditioners, aerosol spray containers, and many other consumer products.

Repeated testing has supported the hypothesis proposed by Rowland and Molina. Any hypothesis that withstands repeated testing may become part of a **theory.** In science, a theory is a well-tested explanation of observations. (This is different from common use of the term, which means "a guess.") Because theories are explanations, not facts, they can be disproved but can never be completely proven. In 1995, Crutzen, Rowland, and Molina were awarded the Nobel Prize in chemistry in recognition of their theory of the formation and decomposition of the ozone layer.

Theories and Laws Have Different Purposes

Some facts in science hold true consistently. Such facts are known as laws. A **law** is a statement or mathematical expression that reliably describes a behavior of the natural world. While a theory is an attempt to explain the cause of certain events in the natural world, a scientific law describes the events.

For example, the **law of conservation of mass** states that the products of a chemical reaction have the same mass as the reactants have. This law does not explain why matter in chemical reactions behaves this way; the law simply describes this behavior. In some cases, scientific laws may be reinterpreted as new information is obtained. Keep in mind that a hypothesis *predicts* an event, a theory *explains* it, and a law *describes* it.

theory

an explanation for some phenomenon that is based on observation, experimentation, and reasoning

law

a summary of many experimental results and observations; a law tells how things work

law of conservation of mass

the law that states that mass cannot be created or destroyed in ordinary chemical and physical changes

+ →

Hydrogen molecule   Oxygen atom   Water molecule

Figure 14
Models can be used to show what happens during a reaction between a hydrogen molecule and an oxygen atom.

Models Can Illustrate the Microscopic World of Chemistry

Models play a major role in science. A *model* represents an object, a system, a process, or an idea. A model is also simpler than the actual thing that is modeled. In chemistry, models can be most useful in understanding what is happening at the microscopic level. In this book, you will see numerous illustrations showing models of chemical substances. These models, such as the ones shown in **Figure 14,** are intended to help you understand what happens during physical and chemical changes.

Keep in mind that models are simplified representations. For example, the models of chemical substances that you will examine in this book include various shapes, sizes, and colors. The actual particles of these chemical substances do not have the shapes, sizes, or brilliant colors that are shown in these models. However, these models do show the geometric arrangement of the units, their relative sizes, and how they interact.

One tool that is extremely useful in the construction of models is the computer. Computer-generated models enable scientists to design chemical substances and explore how they interact in virtual reality. A chemical model that looks promising for some practical application, such as treating a disease, might be the basis for the synthesis of the actual chemical.

 # Section Review

CRITICAL THINKING

UNDERSTANDING KEY IDEAS

1. How does a hypothesis differ from a theory?

2. What is the scientific method?

3. Do experiments always turn out as expected? Why or why not?

4. What is a scientific law, and how does it differ from a theory?

5. Why does a scientist include a control in the design of an experiment?

6. Why is there no single set of steps in the scientific method?

7. Describe what is needed for a hypothesis to develop into a theory.

8. Explain the statement "No theory is written in stone."

9. Can a hypothesis that has been rejected be of any value to scientists? Why or why not?

10. How does the phrase "cause and effect" relate to the formation of a good hypothesis?

11. How would a control group be set up to test the effectiveness of a new drug in treating a disease?

12. Suppose you had to test how well two types of soap work. Describe your experiment by using the terms *control* and *variable*.

13. Why is a model made to be simpler than the thing that it represents?

Measurements and Calculations in Chemistry

KEY TERMS

- accuracy
- precision
- significant figure

OBJECTIVES

①   **Distinguish** between accuracy and precision in measurements.

②   **Determine** the number of significant figures in a measurement, and apply rules for significant figures in calculations.

③   **Calculate** changes in energy using the equation for specific heat, and round the results to the correct number of significant figures.

④   **Write** very large and very small numbers in scientific notation.

Accuracy and Precision

When you determine some property of matter, such as density, you are making calculations that are often not the exact values. No value that is obtained from an experiment is exact because all measurements are subject to limits and errors. Human errors, method errors, and the limits of the instrument are a few examples. To reduce the impact of error on their work, scientists always repeat their measurements and calculations a number of times. If their results are not consistent, they will try to identify and eliminate the source of error. What scientists want in their results are *accuracy* and *precision*.

Measurements Must Involve the Right Equipment

Selecting the right piece of equipment to make your measurements is the first step to cutting down on errors in experimental results. For example, the beaker, the buret, and the graduated cylinder shown in **Figure 15** can be used to measure the volume of liquids. If an experimental procedure calls for measuring 8.6 mL of a liquid, which piece of glassware would you use? Obtaining a volume of liquid that is as close to 8.6 mL as possible is best done with the buret. In fact, the buret in **Figure 15** is calibrated to the nearest 0.1 mL.

Even though the buret can measure small intervals, it should not be used for all volume measurements. For example, an experimental procedure may call for using 98 mL of a liquid. In this case, a 100 mL graduated cylinder would be a better choice. An even larger graduated cylinder should be used if the procedure calls for 725 mL of a liquid.

The right equipment must also be selected when making measurements of other values. For example, if the experimental procedure calls for 0.5 g of a substance, using a balance that only measures to the nearest 1 g would introduce significant error.

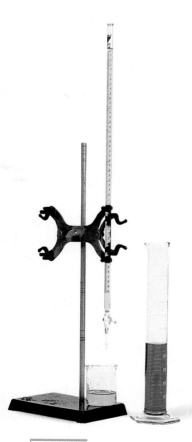

Figure 15
All these pieces of equipment measure volume of liquids, but each is calibrated for different capacities.

Figure 16
a Darts within the bull's-eye mean high accuracy and high precision.

b Darts clustered within a small area but far from the bull's-eye mean low accuracy and high precision.

c Darts scattered around the target and far from the bull's-eye mean low accuracy and low precision.

Accuracy Is How Close a Measurement Is to the True Value

When scientists make and report measurements, one factor they consider is accuracy. The **accuracy** of a measurement is how close the measurement is to the true or actual value. To understand what accuracy is, imagine that you throw four darts separately at a dartboard.

The bull's-eye of the dartboard represents the true value. The closer a dart comes to the bull's-eye, the more accurately it was thrown. **Figure 16a** shows one possible way the darts might land on the dartboard. Notice that all four darts have landed within the bull's-eye. This outcome represents high accuracy.

Accuracy should be considered whenever an experiment is done. Suppose the procedure for a chemical reaction calls for adding 36 mL of a solution. The experiment is done twice. The first time 35.8 mL is added, and the second time 37.2 mL is added. The first measurement was more accurate because 35.8 mL is closer to the true value of 36 mL.

Precision Is How Closely Several Measurements Agree

Another factor that scientists consider when making measurements is precision. **Precision** is the exactness of a measurement. It refers to how closely several measurements of the same quantity made in the same way agree with one another. Again, to understand how precision differs from accuracy, consider how darts might land on a dartboard.

Figure 16b shows another way the four darts might land on the dartboard. Notice that all four darts have hit the target far from the bull's-eye. Because these darts are far from what is considered the true value, this outcome represents low accuracy. However, notice in **Figure 16b** that all four darts have landed very close to one another. The closer the darts land to one another, the more precisely they were thrown. Therefore, **Figure 16b** represents low accuracy but high precision. In **Figure 16c,** the four darts have landed far from the bull's-eye and each in a different spot. This outcome represents low accuracy and low precision.

accuracy

a description of how close a measurement is to the true value of the quantity measured

precision

the exactness of a measurement

Significant Figures

When you make measurements or perform calculations, the way you report a value tells about how you got it. For example, if you report the mass of a sample as 10 g, the mass of the sample may be between 8 g and 12 g or may be between 9.999 g and 10.001 g. However, if you report the mass of a sample as 10.0 g, you are indicating that you used a measuring tool that is precise to the nearest 0.1 g. The mass of the sample can only be between 9.95 g and 10.05 g.

Scientists always report values using significant figures. The **significant figures** of a measurement or a calculation consist of all the digits known with certainty as well as one estimated, or uncertain, digit. Notice that the term *significant* does not mean "certain." The last digit or significant figure reported after a measurement is uncertain or estimated.

Significant Figures Are Essential to Reporting Results

Reporting all measurements in an experiment to the correct number of significant figures is necessary to be sure the results are true. Consider an experiment involving the transfer of energy as heat. Imagine that you use a thermometer calibrated in one-degree increments. Suppose you report a temperature as 37.5°C. The three digits in your reported value are all significant figures. The first two are known with certainty, but the last digit is estimated. You know the temperature is between 37°C and 38°C, and you estimate the temperature to be 37.5°C.

Now assume that you use the thermometer calibrated in one-tenth degree increments. If you report a reading of 36.54°C, the four digits in your reported value are all significant figures. The first three digits are known with certainty, while the last digit is estimated. Using this thermometer, you know the temperature is certainly between 36.5°C and 36.6°C, and estimate it to be 36.54°C.

Figure 17 shows two different thermometers. Notice that the thermometer on the left is calibrated in one-tenth degree increments, while the one on the right is calibrated in one-degree increments.

Figure 17
If the thermometer on the left is used, a reported value contains three certain figures, whereas the thermometer on the right can measure temperature to two certain figures.

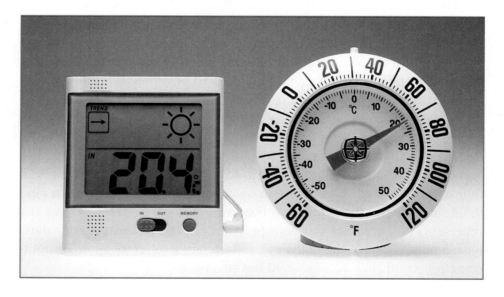

Rules for Determining Significant Figures

1. Nonzero digits are always significant.

- For example, 46.3 m has three significant figures.
- For example, 6.295 g has four significant figures.

2. Zeros between nonzero digits are significant.

- For example, 40.7 L has three significant figures.
- For example, 87 009 km has five significant figures.

3. Zeros in front of nonzero digits are not significant.

- For example, 0.0095 87 m has four significant figures.
- For example, 0.0009 kg has one significant figure.

4. Zeros both at the end of a number and to the right of a decimal point are significant.

- For example, 85.00 g has four significant figures.
- For example, 9.070 000 000 cm has 10 significant figures.

5. Zeros both at the end of a number but to the left of a decimal point may not be significant. If a zero has not been measured or estimated, it is not significant. A decimal point placed after zeros indicates that the zeros are significant.

- For example, 2000 m may contain from one to four significant figures, depending on how many zeros are placeholders. For values given in this book, assume that 2000 m has one significant figure.

Calculators Do Not Identify Significant Figures

When you use a calculator to find a result, you must pay special attention to significant figures to make sure that your result is meaningful. The calculator in **Figure 18** was used to determine the density of isopropyl alcohol, commonly known as rubbing alcohol. The mass of a sample that has a volume of 32.4 mL was measured to be 25.42 g. Remember that the mass and volume of a sample can be used to calulate its density, as shown below.

$$D = \frac{m}{V}$$

The student in **Figure 18** is using a calculator to determine the density of the alcohol by dividing the mass (25.42 g) by the volume (32.4 mL). Notice that the calculator displays the density of the isopropyl alcohol as 0.7845679012 g/mL; the calculator was programmed so that all numbers are significant.

However, the volume was measured to only three significant figures, while the mass was measured to four significant figures. Based on the rules for determining significant figures in calculations described in **Skills Toolkit 1,** the density of the alcohol should be rounded to 0.785 g/mL, or three significant figures.

Figure 18
A calculator does not round the result to the correct number of significant figures.

Rules for Using Significant Figures in Calculations

1. In multiplication and division problems, the answer cannot have more significant figures than there are in the measurement with the smallest number of significant figures. If a sequence of calculations is involved, do not round until the end.

$$12.257 \text{ m}$$
$$\times\ 1.162 \text{ m} \leftarrow \text{four significant figures}$$
$$\overline{14.2426234 \text{ m}^2} \xrightarrow{\text{round off}} 14.24 \text{ m}^2$$

$$8.472 \text{ mL} \overline{)3.05\text{g}} \quad 0.36000944 \text{ g/mL} \xrightarrow{\text{round off}} 0.360 \text{ g/mL}$$
$$\leftarrow \text{three significant figures}$$

2. In addition and subtraction of numbers, the result can be no more certain than the least certain number in the calculation. So, an answer cannot have more dig-

its to the right of the decimal point than there are in the measurement with the smallest number of digits to the right of the decimal. When adding and subtracting you should not be concerned with the total number of significant figures in the values. You should be concerned only with the number of significant figures present to the right of the decimal point.

$$3.95 \text{ g}$$
$$2.879 \text{ g}$$
$$+\ 213.6 \text{ g}$$
$$\overline{220.429 \text{ g}} \xrightarrow{\text{round off}} 220.4 \text{ g}$$

Notice that the answer 220.4 g has four significant figures, whereas one of the values, 3.95 g, has only three significant figures.

3. If a calculation has both addition (or subtraction) and multiplication (or division), round after each operation.

Exact Values Have Unlimited Significant Figures

Some values that you will use in your calculations have no uncertainty. In other words, these values have an unlimited number of significant figures. One example of an exact value is known as a *count value*. As its name implies, a count value is determined by counting, not by measuring. For example, a water molecule contains exactly two hydrogen atoms and exactly one oxygen atom. Therefore, two water molecules contain exactly four hydrogen atoms and two oxygen atoms. There is no uncertainty in these values.

Another value that can have an unlimited number of significant figures is a *conversion factor*. There is no uncertainty in the values that make up this conversion factor, such as 1 m = 1000 mm, because a millimeter is defined as exactly one-thousandth of a meter.

You should ignore both count values and conversion factors when determining the number of significant figures in your calculated results.

Determining the Number of Significant Figures

A student heats 23.62 g of a solid and observes that its temperature increases from 21.6°C to 36.79°C. Calculate the temperature increase per gram of solid.

1 Gather information.

- The mass of the solid is 23.62 g.
- The initial temperature is 21.6°C.
- The final temperature is 36.79°C.

2 Plan your work.

- Calculate the increase in temperature by subtracting the initial temperature (21.6°C) from the final temperature (36.79°C).

  temperature increase = final temperature − initial temperature

- Calculate the temperature increase per gram of solid by dividing the temperature increase by the mass of the solid (23.62 g).

$$\frac{\text{temperature increase}}{\text{gram}} = \frac{\text{temperature increase}}{\text{sample mass}}$$

3 Calculate.

$$36.79°C - 21.6°C = 15.19°C = 15.2°C$$

$$\frac{15.2°C}{23.62\ g} = 0.644\frac{°C}{g} \text{ rounded to three significant figures}$$

4 Verify your results.

- Multiplying the calculated answer by the total number of grams in the solid equals the calculated temperature increase.

$$0.644\frac{°C}{g} \times 23.62\ g = 15.2°C$$
rounded to three significant figures

PRACTICE HINT

Remember that the rules for determining the number of significant figures in multiplication and division problems are different from the rules for determining the number of significant figures in addition and subtraction problems.

PRACTICE

1. Perform the following calculations, and express the answers with the correct number of significant figures.

   a. 0.1273 mL − 0.000008 mL

   b. (12.4 cm × 7.943 cm) + 0.0064 cm^2

   c. (246.83 g/26) − 1.349 g

2. A student measures the mass of a beaker filled with corn oil to be 215.6 g. The mass of the beaker is 110.4 g. Calculate the density of the corn oil if its volume is 114 cm^3.

3. A chemical reaction produces 653 550 kJ of energy as heat in 142.3 min. Calculate the rate of energy transfer in kilojoules per minute.

Specific Heat Depends on Various Factors

Recall that the specific heat is the quantity of energy that must be transferred as heat to raise the temperature of 1 g of a substance by 1 K. The quantity of energy transferred as heat during a temperature change depends on the nature of the material that is changing temperature, the mass of the material, and the size of the temperature change.

For example, consider how the nature of the material changing temperature affects the transfer of energy as heat. One gram of iron that is at 100.0°C is cooled to 50.0°C and transfers 22.5 J of energy to its surroundings. In contrast, 1 g of silver transfers only 11.8 J of energy as heat under the same conditions. Iron has a larger specific heat than silver. Therefore, more energy as heat can be transferred to the iron than to the silver.

Calculating the Specific Heat of a Substance

internet connect

www.scilinks.org
Topic: Specific Heat
SciLinks code: HW4119

SCiLINKS. Maintained by the National Science Teachers Association

Specific heats can be used to compare how different materials absorb energy as heat under the same conditions. For example, the specific heat of iron, which is listed in **Table 1,** is 0.449 J/g·K, while that of silver is 0.235 J/g·K. This difference indicates that a sample of iron absorbs and releases twice as much energy as heat as a comparable mass of silver during the same temperature change does.

Specific heat is usually measured under constant pressure conditions, as indicated by the subscript p in the symbol for specific heat, c_p. The specific heat of a substance at a given pressure is calculated by the following formula:

$$c_p = \frac{q}{m \times \Delta T}$$

In the above equation, c_p is the specific heat at a given pressure, q is the energy transferred as heat, m is the mass of the substance, and ΔT represents the difference between the initial and final temperatures.

Table 1	Some Specific Heats at Room Temperature		
Element	Specific heat (J/g•K)	Element	Specific heat (J/g•K)
Aluminum	0.897	Lead	0.129
Cadmium	0.232	Neon	1.030
Calcium	0.647	Nickel	0.444
Carbon (graphite)	0.709	Platinum	0.133
Chromium	0.449	Silicon	0.705
Copper	0.385	Silver	0.235
Gold	0.129	Water	4.18
Iron	0.449	Zinc	0.388

Calculating Specific Heat

A 4.0 g sample of glass was heated from 274 K to 314 K and was found to absorb 32 J of energy as heat. Calculate the specific heat of this glass.

1 Gather information.

- sample mass (m) = 4.0 g
- initial temperature = 274 K
- final temperature = 314 K
- quantity of energy absorbed (q) = 32 J

2 Plan your work.

- Determine ΔT by calculating the difference between the initial and final temperatures.
- Insert the values into the equation for calculating specific heat.

$$c_p = \frac{32 \text{ J}}{4.0 \text{ g} \times (314 \text{ K} - 274 \text{ K})}$$

3 Calculate.

$$c_p = \frac{32 \text{ J}}{4.0 \text{ g} \times (40 \text{ K})} = 0.20 \text{ J/g·K}$$

4 Verify your results.

The units combine correctly to give the specific heat in J/g·K. The answer is correctly given to two significant figures.

PRACTICE HINT

The equation for specific heat can be rearranged to solve for one of the quantities, if the others are known. For example, to calculate the quantity of energy absorbed or released, rearrange the equation to get $q = c_p \times m \times \Delta T$.

PRACTICE

1. Calculate the specific heat of a substance if a 35 g sample absorbs 48 J as the temperature is raised from 293 K to 313 K.

2. The temperature of a piece of copper with a mass of 95.4 g increases from 298.0 K to 321.1 K when the metal absorbs 849 J of energy as heat. What is the specific heat of copper?

3. If 980 kJ of energy as heat are transferred to 6.2 L of water at 291 K, what will the final temperature of the water be? The specific heat of water is 4.18 J/g·K. Assume that 1.0 mL of water equals 1.0 g of water.

4. How much energy as heat must be transferred to raise the temperature of a 55 g sample of aluminum from 22.4°C to 94.6°C? The specific heat of aluminum is 0.897 J/g·K. Note that a temperature change of 1°C is the same as a temperature change of 1 K because the sizes of the degree divisions on both scales are equal.

PROBLEM SOLVING SKILL

Scientific Notation

Chemists often make measurements and perform calculations using very large or very small numbers. Very large and very small numbers are often written in *scientific notation*. To write a number in scientific notation, first know that every number expressed in scientific notation has two parts. The first part is a number that is between 1 and 10 but that has any number of digits after the decimal point. The second part consists of a power of 10. To write the first part of the number, move the decimal to the right or the left so that only one nonzero digit is to the left of the decimal. Write the second part of the value as an exponent. This part is determined by counting the number of decimal places the decimal point is moved. If the decimal is moved to the right, the exponent is negative. If the decimal is moved to the left, the exponent is positive. For example, 299 800 000 m/s is expressed as 2.998×10^8 m/s in scientific notation. When writing very large and very small numbers in scientific notation, use the correct number of significant figures.

SKILLS Toolkit

3

Using Scientific Notation

1. **In scientific notation, exponents are count values.**

2. **In addition and subtraction problems, all values must have the same exponent before they can be added or subtracted.**

    - $6.2 \times 10^4 + 7.2 \times 10^3 = 62 \times 10^3 + 7.2 \times 10^3 = 69.2 \times 10^3 =$
    $$69 \times 10^3 = 6.9 \times 10^4$$

    - $4.5 \times 10^6 - 2.3 \times 10^5 = 45 \times 10^5 - 2.3 \times 10^5 = 42.7 \times 10^5 =$
    $$43 \times 10^5 = 4.3 \times 10^6$$

3. **In multiplication problems, the first factors of the numbers are multiplied and the exponents of 10 are added.**

    - $(3.1 \times 10^3)(5.01 \times 10^4) = (3.1 \times 5.01) \times 10^{4+3} =$
    $$16 \times 10^7 = 1.6 \times 10^8$$

4. **In division problems, the first factors of the numbers are divided and the exponent of 10 in the denominator is subtracted from the exponent of 10 in the numerator.**

    - $7.63 \times 10^3 / 8.6203 \times 10^4 = 7.63/8.6203 \times 10^{3-4} =$
    $$0.885 \times 10^{-1} = 8.85 \times 10^{-2}$$

Scientific Notation with Significant Figures

1. Use scientific notation to eliminate all placeholding zeros.

- $2400 \longrightarrow 2.4 \times 10^3$ (both zeros are not significant)
- $750\,000. \longrightarrow 7.50000 \times 10^5$ (all zeros are significant)

2. Move the decimal in an answer so that only one digit is to the left, and change the exponent accordingly. The final value must contain the correct number of significant figures.

- $5.44 \times 10^7/8.1 \times 10^4 = 5.44/8.1 \times 10^{7-4} = 0.6716049383 \times 10^3 = 6.7 \times 10^2$ (adjusted to two significant figures)

Section Review

UNDERSTANDING KEY IDEAS

1. How does accuracy differ from precision?
2. Explain the advantage of using scientific notation.
3. When are zeros significant in a value?
4. Why are significant figures important when reporting measurements?
5. Explain how a series of measurements can be precise without being accurate.

PRACTICE PROBLEMS

6. Perform the following calculations, and express the answers using the correct number of significant figures.

   a. 0.8102 m × 3.44 m   **c.** 32.89 g + 14.21 g

   b. $\dfrac{94.20 \text{ g}}{3.167\,22 \text{ mL}}$ **d.** 34.09 L − 1.230 L

7. Calculate the specific heat of a substance when 63 J of energy are transferred as heat to an 8.0 g sample to raise its temperature from 314 K to 340 K.

8. Express the following calculations in the proper number of significant figures. Use scientific notation where appropriate.

   a. 129 g/29.2 mL

   b. (1.551 mm)(3.260 mm)(4.9001 mm)

   c. 35 000 kJ/0.250 s

9. A clock gains 0.020 s/min. How many seconds will the clock gain in exactly six months, assuming 30 days are in each month? Express your answer in scientific notation.

CRITICAL THINKING

10. There are 12 eggs in a carton. How many significant figures does the value 12 have in this case?

11. If you measure the mass of a liquid as 11.50 g and its volume as 9.03 mL, how many significant figures should its density value have? Explain the reason for your answer.

Element Spotlight

Where Is He?

Universe:
about 23% by mass

Earth's crust:
0.000001% by mass

Air:
0.0005% by mass

Deep-sea Diving with Helium

Divers who breathe air while at great undersea depths run the risk of suffering from a condition known as nitrogen narcosis. Nitrogen narcosis can cause a diver to become disoriented and to exercise poor judgment, which leads to dangerous behavior. To avoid nitrogen narcosis, professional divers who work at depths of more than 60 m breathe heliox, a mixture of helium and oxygen, instead of air.

The greatest advantage of heliox is that it does not cause nitrogen narcosis. A disadvantage of heliox is that it removes body heat faster than air does. This effect makes a diver breathing heliox feel chilled sooner than a diver breathing air.

Breathing heliox also affects the voice. Helium is much less dense than nitrogen, so vocal cords vibrate faster in a heliox atmosphere. This raises the pitch of the diver's voice, and makes the diver's voice sound funny. Fortunately, this effect disappears when the diver surfaces and begins breathing air again.

In Florida, divers on the Wakulla Springs project team breathed heliox at depths greater than 90 m.

Industrial Uses

- Helium is used as a lifting gas in balloons and dirigibles.

- Helium is used as an inert atmosphere for welding and for growing high-purity silicon crystals for semiconducting devices.

- Liquid helium is used as a coolant in superconductor research.

Real-World Connection Helium was discovered in the sun before it was found on Earth.

internet connect

www.scilinks.org
Topic: Helium
SciLinks code: HW4171

SCI*LINKS* Maintained by the
National Science
Teachers Association

A Brief History

1888: William Hillebrand discovers that an inert gas is produced when a uranium mineral is dissolved in sulfuric acid.

1908: Ernest Rutherford and Thomas Royds prove that alpha particles emitted during radioactive decay are helium nuclei.

| 1600 | 1700 | 1800 | 1900 |

1868: Pierre Janssen, studies the spectra of a solar eclipse and finds evidence of a new element. Edward Frankland, an English chemist, and Joseph Lockyer, an English astronomer, suggest the name helium.

1894: Sir William Ramsay and Lord Rayleigh discover argon. They suspect that the gas Hillebrand found in 1888 was argon. They repeat his experiment and find that the gas is helium.

Questions

1. Research the industrial, chemical, and commercial uses of helium.

2. Research properties of neon, argon, krypton, and xenon. How are these gases similar to helium? Are they used in a manner similar to helium?

CHAPTER HIGHLIGHTS 2

KEY TERMS

energy
physical change
chemical change
evaporation
endothermic
exothermic
law of conservation
  of energy
heat
kinetic energy
temperature
specific heat

scientific method
hypothesis
theory
law
law of conservation
  of mass

accuracy
precision
significant figure

KEY IDEAS

SECTION ONE Energy

- Energy is the capacity to do work.
- Changes in matter can be chemical or physical. However, only chemical changes produce new substances.
- Every change in matter involves a change in energy.
- Endothermic processes absorb energy. Exothermic processes release energy.
- Energy is always conserved.
- Heat is the energy transferred between objects that are at different temperatures. Temperature is a measure of the average random kinetic energy of the particles in an object.
- Specific heat is the relationship between energy transferred as heat to a substance and a substance's temperature change.

SECTION TWO Studying Matter and Energy

- The scientific method is a strategy for conducting research.
- A hypothesis is an explanation that is based on observations and that can be tested.
- A variable is a factor that can affect an experiment.
- A controlled experiment is an experiment in which variables are kept constant.
- A theory is a well-tested explanation of observations. A law is a statement or mathematical expression that describes the behavior of the world.

SECTION THREE Measurements and Calculations in Chemistry

- Accuracy is the extent to which a measurement approaches the true value of a quantity.
- Precision refers to how closely several measurements that are of the same quantity and that are made in the same way agree with one another.
- Significant figures are digits known with certainty as well as one estimated, or uncertain, digit.
- Numbers should be written in scientific notation.

KEY SKILLS

Rules for Determining Significant Figures
Skills Toolkit 1 p. 57

Rules for Using Significant Figures in Calculations
Skills Toolkit 2 p. 58
Sample Problem A p. 59

Calculating Specific Heat
Sample Problem B p. 61

Scientific Notation in Calculations
Skills Toolkit 3 p. 62

Scientific Notation with Significant Figures
Skills Toolkit 4 p. 63

2 CHAPTER REVIEW

USING KEY TERMS

1. Name two types of energy.

2. State the law of conservation of energy.

3. Define chemical change.

4. What is temperature?

5. What is the difference between heat and temperature?

6. What is the difference between a theory and a law?

7. What is accuracy? What is precision?

8. What are significant figures?

UNDERSTANDING KEY IDEAS

Energy

9. Water evaporates from a puddle on a hot, sunny day faster than on a cold, cloudy day. Explain this phenomenon in terms of interactions between matter and energy.

10. Beaker A contains water at a temperature of 15°C. Beaker B contains water at a temperature of 37°C. Which beaker contains water molecules that have greater average kinetic energy? Explain your answer.

11. What is the difference between a physical change and a chemical change?

Studying Matter and Energy

12. What does a good hypothesis require?

13. Classify the following statements as observation, hypothesis, theory, or law:
    a. A system containing many particles will not go spontaneously from a disordered state to an ordered state.
    b. The substance is silvery white, is fairly hard, and is a good conductor of electricity.
    c. Bases feel slippery in water.
    d. If I pay attention in class, I will succeed in this course.

14. What is a control? What is a variable?

15. Explain the relationship between models and theories.

16. Describe the scientific method.

17. Why is the conservation of energy considered a law, not a theory?

Measurements and Calculations in Chemistry

18. Why it is important to keep track of significant figures?

19. a. If you add several numbers, how many significant figures can the sum have?
    b. If you multiply several numbers, how many significant figures can the product have?

20. How many digits are not known with certainty in a number written with the proper number of significant figures?

21. Perform the following calculations, and express the answers with the correct number of significant figures.
    a. $2.145 + 0.002$
    b. $(9.8 \times 8.934) + 0.0048$
    c. $(172.56/43.8) - 1.825$

22. Which of the following statements contain exact numbers?
    a. There are 12 eggs in a dozen.
    b. Some Major League Baseball pitchers can throw a ball over 140 km/h.
    c. The accident injured 21 people.
    d. The circumference of the Earth at the equator is 40000 km.

e. The tank was filled with 54 L of gas.

f. A nickel has a mass of 5 g.

23. Express 743 000 000 in scientific notation to the following number of significant figures:
    a. one significant figure
    b. two significant figures
    c. four significant figures
    d. seven significant figures

PRACTICE PROBLEMS

Sample Problem A Determining the Number of Significant Figures

24. How many significant figures are there in each of the following measurements?
    a. 0.4004 mL
    c. 1.000 30 km
    b. 6000 g
    d. 400 mm

25. Calculate the sum of 6.078 g and 0.3329 g.

26. Subtract 7.11 cm from 8.2 cm.

27. What is the product of 0.8102 m and 3.44 m?

28. Divide 94.20 g by 3.167 22 mL.

29. How many grams are in 882 μg?

30. The density of gold is 19.3 g/cm^3.
    a. What is the volume, in cubic centimeters, of a sample of gold with mass 0.715 kg?
    b. If this sample of gold is a cube, how long is each edge in centimeters?

31. **a.** Find the number of kilometers made up of 92.25 m.
    b. Convert the answer in kilometers to centimeters.

Sample Problem B Calculating Specific Heat

32. Determine the specific heat of a material if a 35 g sample of the material absorbs 48 J as it is heated from 298 K to 313 K.

33. How much energy is needed to raise the temperature of a 75 g sample of aluminum from 22.4°C to 94.6°C? Refer to **Table 1**.

34. How much energy is needed to raise the temperature of 75 g of gold by 25°C? Refer to **Table 1**.

35. Energy in the amount of 420 J is added to a 35 g sample of water at a temperature of 10.0°C. What is the final temperature of the water? Refer to **Table 1**.

36. How much energy will be transferred as heat to a 15.3 g sample of cadmium when its temperature is raised from 322 K to 363 K?

Skills Toolkit 3 Scientific Notation in Calculations

37. Write the following numbers in scientific notation.
    a. 0.000 673 0
    b. 50 000.0
    c. 0.000 003 010

38. The following numbers are written in scientific notation. Write them in ordinary notation.
    a. 7.050×10^{-3} g
    b. $4.000\ 05 \times 10^{7}$ mg
    c. 2.3500×10^{4} mL

39. Perform the following operation. Express the answer in scientific notation and with the correct number of significant figures.

$$\frac{(6.124\ 33 \times 10^{6} \text{m}^{3})}{(7.15 \times 10^{-3} \text{m})}$$

Skills Toolkit 4 Scientific Notation with Significant Figures

40. Use scientific notation to eliminate all placeholding zeros.
    a. 7500
    b. 92 002 000

41. How many significant figures does 7.324×10^{-3} have?

42. How many significant figures does the answer to $(1.36 \times 10^{-5}) \times (5.02 \times 10^{-2})$ have?

MIXED REVIEW

43. Why can a measured number never be exact?

44. Determine the specific heat of a material if a 78 g sample of the material absorbed 28 J as it was heated from 298 K to 345 K.

45. A piece of copper alloy with a mass of 85.0 g is heated from 30.0°C to 45.0°C. During this process, it absorbs 523 J of energy as heat.
   a. What is the specific heat of this copper alloy?
   b. How much energy will the same sample lose if it is cooled to 25°C?

46. A man finds that he has a mass of 100.6 kg. He goes on a diet, and several months later he finds that he has a mass of 96.4 kg. Express each number in scientific notation, and calculate the number of kilograms the man has lost by dieting.

47. A large office building is 1.07×10^2 m long, 31 m wide, and 4.25×10^2 m high. What is its volume?

48. An object has a mass of 57.6 g. Find the object's density, given that its volume is 40.25 cm^3.

49. A student measures the mass of some sucrose as 0.947 mg. Convert that quantity to grams and to kilograms.

50. Write the following measurements in scientific notation.
   a. 65 900 000 m
   b. 0.0057 km
   c. 22 000 mg
   d. 0.000 003 7 kg

51. Write the following measurements in long form.
   a. 4.5×10^3 g
   b. 6.05×10^{-3} m
   c. 3.115×10^6 km
   d. 1.99×10^{-8} cm

52. Write the following measurements in scientific notation.
   a. 800 000 000 m
   b. 0.000 95 m
   c. 60 200 L
   d. 0.0015 kg

53. Write the following measurements in long form.
   a. 9.8×10^4 mm
   b. 2.38×10^{-7} m
   c. 1.115×10^3 g
   d. 1.5×10^{-10} kg

54. Do the following calculations, and write the answers in scientific notation.
   a. $37\ 000\ 000 \times 7\ 100\ 000$
   b. $0.000\ 312/\ 486$
   c. 4.6×10^4 cm $\times 7.5 \times 10^3$ cm
   d. 8.3×10^6 kg$/\ 2.5 \times 10^9$ cm^3

55. Do the following calculations, and write the answers with the correct number of significant figures.
   a. 15.75 m $\times$ 8.45 m
   b. 5650 L$/$ 27 min
   c. 6271 m$/$ 59.7 s
   d. 0.0058 km $\times$ 0.228 km

56. Explain why the observation that the sun sets in the west could be called a *scientific law.*

57. Calculate the volume of a room with walls that are 3.125 m tall, 4.25 m wide, and 5.75 m long. Write the answer with the correct number of significant figures.

58. You have decided to test the effects of five garden fertilizers by applying some of each to five separate rows of radishes. What is the variable you are testing? What factors should you control? How will you measure the results?

CRITICAL THINKING

59. Suppose a graduated cylinder was not correctly calibrated. How would this affect the results of a measurement? How would it affect the results of a calculation using this measurement? Use the terms *accuracy* and *precision* in your answer.

60. Your friend says that things fall to Earth because of the law of gravitation. Tell what is wrong with your friend's statement, and explain your reasoning.

61. a. The table below contains data from an experiment in which an air sample is subjected to different pressures. Based on this set of observations, propose a hypothesis that could be tested.
  b. What theories can be stated from the data in the table below?
  c. Are the data sufficient for the establishment of a scientific law. Why or why not?

The Results of Compressing an Air Sample

Volume (cm^3)	Pressure (kPa)	Volume × pressure (cm^3 × kPa)
100.0	33.3	3330
50.0	66.7	3340
25.0	133.2	3330
12.5	266.4	3330

62. What components are necessary for an experiment to be valid?

63. Is it possible for a number to be too small or too large to be expressed adequately in the SI system? Explain your answer.

64. Around 1150, King David I of Scotland defined the inch as the width of a man's thumb at the base of the nail. Discuss the practical limitations of this early unit of measurement.

65. How does Einstein's equation $E = mc^2$ seem to contradict both the law of conservation of energy and the law of conservation of mass?

ALTERNATIVE ASSESSMENT

66. Design an experimental procedure for determining the specific heat of a metal.

67. For one week, practice your observation skills by listing chemistry-related events that happen around you. After your list is compiled, choose three events that are especially interesting or curious to you. Label three pocket portfolios, one for each event. As you read the chapters in this textbook, gather information that helps explain these events. Put pertinent notes, questions, figures, and charts in the folders. When you have enough information to explain each phenomenon, write a report and present it in class.

68. Make a poster of scientific laws that you have encountered in previous science courses. What facts were used to support each of these laws?

69. Energy can be transformed from one form to another. For example, light (solar) energy is transformed into chemical energy during photosynthesis. Prepare a list of several different forms of energy. Describe transformations of energy that you encounter on a daily basis. Try to include examples that involve more than one transformation, e.g., light ⟶ chemical ⟶ mechanical. Select one example, and demonstrate the actual transformation to the class.

CONCEPT MAPPING

70. Use the following terms to complete the concept map below: *energy, endothermic, physical change, law of conservation of energy,* and *exothermic.*

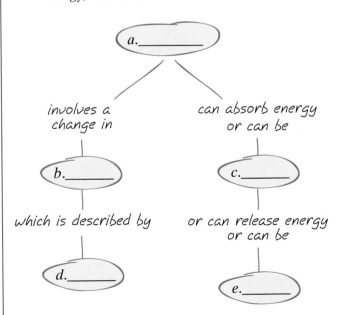

Study the graph below, and answer the questions that follow.
For help in interpreting graphs, see Appendix B, "Study Skills for Chemistry."

71. What is the value for the slope of the curve during the period in which the temperature is equal to the melting point temperature?

72. Is there another period in the graph where the slope equals the value in question 71?

73. Draw the cooling curve for water. Label the axes and the graph.

74. Suppose water could exist in four states of matter at some pressure. Draw what the heating curve for water would look like. Label the axes and the graph.

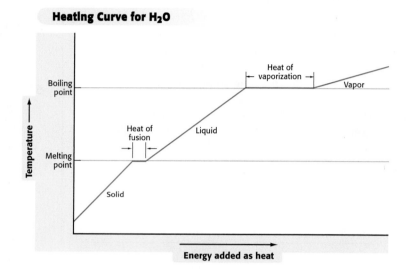

Heating Curve for H$_2$O

TECHNOLOGY AND LEARNING

75. Graphing Calculator

Graphing Celsius and Fahrenheit Temperatures

The graphing calculator can run a program that makes a graph of a given Fahrenheit temperature (on the *x*-axis) and the corresponding Celsius temperature (on the *y*-axis). You can use the **TRACE** button on the calculator to explore this graph and learn more about how the two temperature scales are related.

Go to Appendix C. If you are using a TI-83 Plus, you can download the program **CELSIUS** and run the application as directed. If you are using another calculator, your teacher will provide you with keystrokes and data sets to use. After the graph is displayed, press **TRACE**. An X-shaped cursor on the graph line indicates a specific point. At the

bottom of the screen the values are shown for that point. The one labeled **X=** is the Fahrenheit temperature and the one labeled **Y=** is the Celsius temperature. Use the right and left arrow keys to move the cursor along the graph line to find the answers to these questions.

a. What is the Fahrenheit temperature when the Celsius temperature is zero? (This is where the graph line crosses the horizontal *x*-axis.) What is the significance of this temperature?

b. Human internal body temperature averages 98.6°F. What is the corresponding value on the Celsius scale?

c. Determine the Fahrenheit temperature in your classroom or outside, as given in a weather report. What is the corresponding Celsius temperature?

d. At what temperature are the Celsius and Fahrenheit temperatures the same?

1. The equation $E = mc^2$ shows that
   a. chemical reactions are either exothermic or endothermic.
   b. mass and energy are related.
   c. a hypothesis may develop into a theory.
   d. the kinetic energy of an object relates to its motion.

2. Which of the following measurements contains three significant figures?
   a. 200 mL
   b. 0.02 mL
   c. 20.2 mL
   d. 200.0 mL

3. A control in an experiment
   a. is often not needed.
   b. means that the scientist has everything under control.
   c. is required only if the hypothesis leads to the development of a theory.
   d. allows the scientist to identify the cause of the results in an experiment.

4. A theory differs from a hypothesis in that the former
   a. cannot be disproved.
   b. always leads to the formation of a law.
   c. has been subjected to experimental testing.
   d. represents an educated guess.

5. All measurements in science
   a. must be expressed in scientific notation.
   b. have some degree of uncertainty.
   c. are both accurate and precise.
   d. must include only those digits that are known with certainty.

6. If the temperature outside is 26°C, the temperature would be _____ K.
   a. 26
   b. 273
   c. 299
   d. −247

7. When numbers are multiplied or divided, the answer can have no more
   a. significant figures than there are in the measurement with the smallest number of significant figures.
   b. significant figures than there are in the measurement with the largest number of significant figures.
   c. digits to the right of the decimal point than there are in the measurement with the smallest number of digits to the right of the decimal point.
   d. digits to the right of the decimal point than there are in the measurement with the largest number of digits to the right of the decimal point.

8. Which of the following is not part of the scientific method?
   a. making measurements
   b. introducing bias
   c. making an educated guess
   d. analyzing data

9. The accuracy of a measurement
   a. is how close it is to the true value.
   b. does not depend on the instrument being used to measure the object.
   c. indicates that the measurement is also precise.
   d. is something that scientists rarely achieve.

10. A measurement of 23 465 mg converted to grams equals
    a. 2.3465 g.
    b. 23.465 g.
    c. 234.65 g.
    d. 0.23465 g.

11. A metal sample has a mass of 45.65 g. The volume of the sample is 16.9 cm³. The density of the sample is
    a. 2.7 g/cm³.
    b. 2.70 g/cm³.
    c. 0.370 g/cm³.
    d. 0.37 g/cm³.

CHAPTER

3

ATOMS AND MOLES

Until recently, if you wanted to see an image of atoms, the best you could hope to see was an artists's drawing of atoms. Now, with the help of powerful microscopes, scientists are able to obtain images of atoms. One such microscope is known as the scanning tunneling microscope, which took the image of the nickel atoms shown on the opposite page. As its name implies, this microscope scans a surface, and it can come as close as a billionth of a meter to a surface to get an image. The images that these microscopes provide help scientists understand atoms.

START-UP*ACTIVITY*

SAFETY PRECAUTIONS

Forces of Attraction

PROCEDURE

1. Spread some **salt** and **pepper** on a piece of **paper** that lies on a flat surface. Mix the salt and pepper but make sure that the salt and pepper are not clumped together.

2. Rub a **plastic spoon** with a **wool cloth.**

3. Hold the spoon just above the salt and pepper.

4. Clean off the spoon by using a **towel.** Rub the spoon with the wool cloth and bring the spoon slowly toward the salt and pepper from a distance.

ANALYSIS

1. What happened when you held your spoon right above the salt and pepper? What happened when you brought your spoon slowly toward the salt and pepper?

2. Why did the salt and pepper jump up to the spoon?

3. When the spoon is brought toward the paper from a distance, which is the first substance to jump to the spoon? Why?

CONTENTS 3

Pre-Reading Questions

① **What is an atom?**

② **What particles make up an atom?**

③ **Where are the particles that make up an atom located?**

④ **Name two types of electromagnetic radiation.**

internet connect

www.scilinks.org
Topic: Atoms and Elements
SciLinks code: HW4017

SCILINKS. Maintained by the National Science Teachers Association

Substances Are Made of Atoms

KEY TERMS

- **law of definite proportions**
- **law of conservation of mass**
- **law of multiple proportions**

OBJECTIVES

1. **State** the three laws that support the existence of atoms.

2. **List** the five principles of John Dalton's atomic theory.

Atomic Theory

As early as 400 BCE, a few people believed in an *atomic theory,* which states that atoms are the building blocks of all matter. Yet until recently, even scientists had never seen evidence of atoms. Experimental results supporting the existence of atoms did not appear until more than 2000 years after the first ideas about atoms emerged. The first of these experimental results indicated that all chemical compounds share certain characteristics.

What do you think an atom looks like? Many people think that an atom looks like the diagram in **Figure 1a.** However, after reading this chapter, you will find that the diagram in **Figure 1b** is a better model of an atom.

Recall that a compound is a pure substance composed of atoms of two or more elements that are chemically combined. These observations about compounds and the way that compounds react led to the development of the law of definite proportions, the law of conservation of mass, and the law of multiple proportions. Experimental observations show that these laws also support the current atomic theory.

internet connect

www.scilinks.org
Topic: Development of
Atomic Theory
SciLinks code: HW4148

internet connect

www.scilinks.org
Topic : Current Atomic Theory
SciLinks code: HW4038

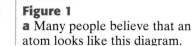

Figure 1
a Many people believe that an atom looks like this diagram.

b This diagram is a better model of the atom.

The Law of Definite Proportions

The **law of definite proportions** states that two samples of a given compound are made of the same elements in exactly the same proportions by mass regardless of the sizes or sources of the samples. Notice the composition of ethylene glycol, as shown in **Figure 2**. Every sample of ethylene glycol is composed of three elements in the following proportions by mass:

51.56% oxygen, 38.70% carbon, and 9.74% hydrogen

The law of definite proportions also states that every molecule of ethylene glycol is made of the same number and types of atoms. A molecule of ethylene glycol has the formula $C_2H_6O_2$, so the law of definite proportions tells you that all other molecules of ethylene glycol have the same formula.

Table salt (sodium chloride) is another example that shows the law of definite proportions. Any sample of table salt consists of two elements in the following proportions by mass:

60.66% chlorine and 39.34% sodium

Every sample of table salt also has the same proportions of ions. As a result, every sample of table salt has the same formula, NaCl.

As chemists of the 18th century began to gather data during their studies of matter, they first began to recognize the law of definite proportions. Their conclusions led to changes in the atomic theory.

Figure 2
a Ethylene glycol is the main component of automotive antifreeze.

b Ethylene glycol is composed of carbon, oxygen, and hydrogen.

c Ethylene glycol is made of exact proportions of these elements regardless of the size of the sample or its source.

Ethylene Glycol Composition by Mass

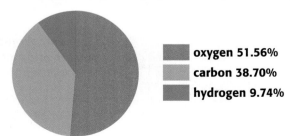

oxygen 51.56%
carbon 38.70%
hydrogen 9.74%

The Law of Conservation of Mass

As early chemists studied more chemical reactions, they noticed another pattern. Careful measurements showed that the mass of a reacting system does not change. The **law of conservation of mass** states that the mass of the reactants in a reaction equals the mass of the products. **Figure 3** shows several reactions that show the law of conservation of mass. For example, notice the combined mass of the sulfur atom and the oxygen molecule equals the mass of the sulfur dioxide molecule.

Also notice that **Figure 3** shows that the sum of the mass of the chlorine molecule and the mass of the phosphorus trichloride molecule is slightly smaller than the mass of the phosphorus pentachloride molecule. This difference is the result of rounding off and of correctly using significant figures.

law of conservation of mass

the law that states that mass cannot be created or destroyed in ordinary chemical and physical changes

Figure 3
The total mass of a system remains the same whether atoms are combined, separated, or rearranged. Here, mass is expressed in kilograms (kg).

Conservation of Mass

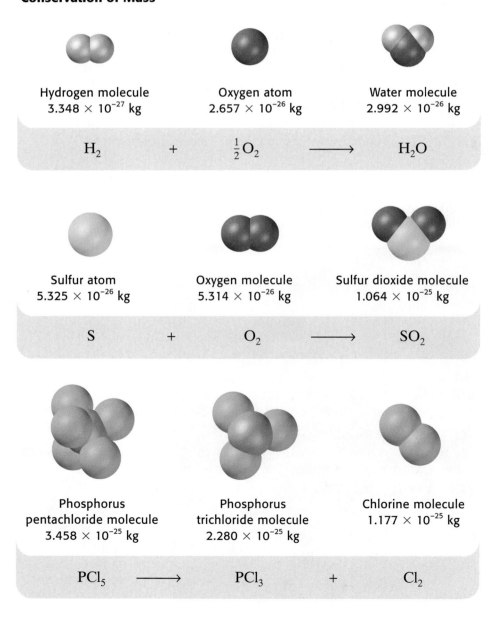

Hydrogen molecule
3.348×10^{-27} kg

Oxygen atom
2.657×10^{-26} kg

Water molecule
2.992×10^{-26} kg

$$H_2 \quad + \quad \tfrac{1}{2}O_2 \quad \longrightarrow \quad H_2O$$

Sulfur atom
5.325×10^{-26} kg

Oxygen molecule
5.314×10^{-26} kg

Sulfur dioxide molecule
1.064×10^{-25} kg

$$S \quad + \quad O_2 \quad \longrightarrow \quad SO_2$$

Phosphorus pentachloride molecule
3.458×10^{-25} kg

Phosphorus trichloride molecule
2.280×10^{-25} kg

Chlorine molecule
1.177×10^{-25} kg

$$PCl_5 \quad \longrightarrow \quad PCl_3 \quad + \quad Cl_2$$

Table 1	Compounds of Nitrogen and Oxygen and the Law of Multiple Proportions						
Name of compound	Description	As shown in figures	Formula	Mass O (g)	Mass N (g)	$\dfrac{\text{Mass O}(g)}{\text{Mass N}(g)}$	
Nitrogen monoxide	colorless gas that reacts readily with oxygen		NO	16.00	14.01	$\dfrac{16.00 \text{ g O}}{14.01 \text{ g N}}$ =	$\dfrac{1.14 \text{ g O}}{1 \text{ g N}}$
Nitrogen dioxide	poisonous brown gas in smog		NO_2	32.00	14.01	$\dfrac{32.00 \text{ g O}}{14.01 \text{ g N}}$ =	$\dfrac{2.28 \text{ g O}}{1 \text{ g N}}$

The Law of Multiple Proportions

Table 1 lists information about the compounds nitrogen monoxide and nitrogen dioxide. For each compound, the table also lists the ratio of the mass of oxygen to the mass of nitrogen. So, 1.14 g of oxygen combine with 1 g of nitrogen when nitrogen monoxide forms. In addition, 2.28 g of oxygen combine with 1 g of nitrogen when nitrogen dioxide forms. The ratio of the masses of oxygen in these two compounds is exactly 1.14 to 2.28 or 1 to 2. This example illustrates **the law of multiple proportions:** If two or more different compounds are composed of the same two elements, the ratio of the masses of the second element (which combines with a given mass of the first element) is always a ratio of small whole numbers.

The law of multiple proportions may seem like an obvious conclusion given the molecules' diagrams and formulas shown. But remember that the early chemists did not know the formulas for compounds. In fact, chemists still have not actually seen these molecules. Scientists think that molecules have these formulas because of these mass data.

law of multiple proportions

the law that states that when two elements combine to form two or more compounds, the mass of one element that combines with a given mass of the other is in the ratio of small whole numbers

Dalton's Atomic Theory

In 1808, John Dalton, an English school teacher, used the Greek concept of the atom and the law of definite proportions, the law of conservation of mass, and the law of multiple proportions to develop an atomic theory. Dalton believed that a few kinds of atoms made up all matter.

According to Dalton, elements are composed of only one kind of atom and compounds are made from two or more kinds of atoms. For example, the element copper consists of only one kind of atom, as shown in **Figure 4.** Notice that the compound iodine monochloride consists of two kinds of atoms joined together. Dalton also reasoned that only whole numbers of atoms could combine to form compounds, such as iodine monochloride. In this way, Dalton revised the early Greek idea of atoms into a scientific theory that could be tested by experiments.

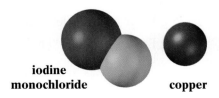

iodine monochloride   **copper**

Figure 4
An element, such as copper, is made of only one kind of atom. In contrast, a compound, such as iodine monochloride, can be made of two or more kinds of atoms.

Dalton's Theory Contains Five Principles

Dalton's atomic theory can be summarized by the following statements:

1. All matter is composed of extremely small particles called *atoms*, which cannot be subdivided, created, or destroyed.

2. Atoms of a given element are identical in their physical and chemical properties.

3. Atoms of different elements differ in their physical and chemical properties.

4. Atoms of different elements combine in simple, whole-number ratios to form compounds.

5. In chemical reactions, atoms are combined, separated, or rearranged but never created, destroyed, or changed.

Dalton's theory explained most of the chemical data that existed during his time. As you will learn later in this chapter, data gathered since Dalton's time shows that the first two principles are not true in all cases.

Today, scientists can divide an atom into even smaller particles. Technology has also enabled scientists to destroy and create atoms. Another feature of atoms that Dalton could not detect is that many atoms will combine with like atoms. Oxygen, for example, is generally found as O_2, a molecule made of two oxygen atoms. Sulfur is found as S_8. Because some parts of Dalton's theory have been shown to be incorrect, his theory has been modified and expanded as scientists learn more about atoms.

① Section Review

UNDERSTANDING KEY IDEAS

1. What is the atomic theory?

2. What is a compound?

3. State the laws of definite proportions, conservation of mass, and multiple proportions.

4. According to Dalton, what is the difference between an element and a compound?

5. What are the five principles of Dalton's atomic theory?

6. Which of Dalton's five principles still apply to the structure of an atom?

CRITICAL THINKING

7. What law is described by the fact that carbon dioxide consists of 27.3% carbon and 72.7% oxygen by mass?

8. What law is described by the fact that the ratio of the mass of oxygen in carbon dioxide to the mass of oxygen in carbon monoxide is 2:1?

9. Three compounds contain the elements sulfur, S, and fluorine, F. How do the following data support the law of multiple proportions?

compound A: 1.188 g F for every 1.000 g S

compound B: 2.375 g F for every 1.000 g S

compound C: 3.563 g F for every 1.000 g S

Structure of Atoms

KEY TERMS

- electron
- nucleus
- proton
- neutron
- atomic number
- mass number
- isotope

OBJECTIVES

① **Describe** the evidence for the existence of electrons, protons, and neutrons, and describe the properties of these subatomic particles.

② **Discuss** atoms of different elements in terms of their numbers of electrons, protons, and neutrons, and define the terms *atomic number* and *mass number*.

③ **Define** *isotope*, and determine the number of particles in the nucleus of an isotope.

Subatomic Particles

Experiments by several scientists in the mid-1800s led to the first change to Dalton's atomic theory. Scientists discovered that atoms can be broken into pieces after all. These smaller parts that make up atoms are called *subatomic particles.* Many types of subatomic particles have since been discovered. The three particles that are most important for chemistry are the electron, the proton, and the neutron.

🔲 internet connect

www.scilinks.org
Topic : Subatomic Particles
SciLinks code: HW4121

SC*i*LINKS. Maintained by the National Science Teachers Association

Electrons Were Discovered by Using Cathode Rays

The first evidence that atoms had smaller parts was found by researchers who were studying electricity, not atomic structure. One of these scientists was the English physicist J. J. Thomson. To study current, Thomson pumped most of the air out of a glass tube. He then applied a voltage to two metal plates, called *electrodes,* which were placed at either end of the tube. One electrode, called the *anode,* was attached to the positive terminal of the voltage source, so it had a positive charge. The other electrode, called a *cathode,* had a negative charge because it was attached to the negative terminal of the voltage source.

Thomson observed a glowing beam that came out of the cathode and struck the anode and the nearby glass walls of the tube. So, he called these rays *cathode rays.* The glass tube Thomson used is known as a *cathode-ray tube* (CRT). CRTs have become an important part of everyday life. They are used in television sets, computer monitors, and radar displays.

Figure 5
The image on a television screen or a computer monitor is produced when cathode rays strike the special coating on the inside of the screen.

An Electron Has a Negative Charge

Thomson knew the rays must have come from the atoms of the cathode because most of the atoms in the air had been pumped out of the tube. Because the cathode ray came from the negatively charged cathode, Thomson reasoned that the ray was negatively charged.

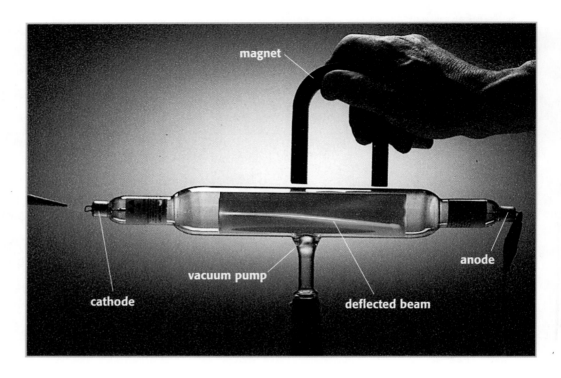

Figure 6
A magnet near the cathode-ray tube causes the beam to be deflected. The deflection indicates that the particles in the beam have a negative charge.

electron

a subatomic particle that has a negative electric charge

He confirmed this prediction by seeing how electric and magnetic fields affected the cathode ray. **Figure 6** shows what Thomson saw when he placed a magnet near the tube. Notice that the beam is deflected by the magnet. Other researchers had shown that moving negative charges are deflected this way.

Thomson also observed that when a small paddle wheel was placed in the path of the rays, the wheel would turn. This observation suggested that the cathode rays consisted of tiny particles that were hitting the paddles of the wheel.

Thomson's experiments showed that a cathode ray consists of particles that have mass and a negative charge. These particles are called **electrons.** **Table 2** lists the properties of an electron. Later experiments, which used different metals for cathodes, confirmed that electrons are a part of atoms of all elements.

Electrons are negatively charged, but atoms have no charge. Therefore, atoms must contain some positive charges that balance the negative charges of the electrons. Scientists realized that positive charges must exist in atoms and began to look for more subatomic particles. Scientists also recognized that atoms must have other particles because an electron was found to have much less mass than an atom does.

Table 2	**Properties of an Electron**				
Name	**Symbol**	**As shown in figures**	**Charge**	**Common charge notation**	**Mass (kg)**
Electron	e, e^-, or $_{-1}^{0}e$		-1.602×10^{-19} C	-1	9.109×10^{-31} kg

Rutherford Discovered the Nucleus

Thomson proposed that the electrons of an atom were embedded in a positively charged ball of matter. His picture of an atom, which is shown in **Figure 7,** was named the *plum-pudding model* because it resembled plum pudding, a dessert consisting of a ball of cake with pieces of fruit in it. Ernest Rutherford, one of Thomson's former students, performed experiments in 1909 that disproved the plum-pudding model of the atom.

Rutherford's team of researchers carried out the experiment shown in **Figure 8.** A beam of small, positively charged particles, called *alpha particles*, was directed at a thin gold foil. The team measured the angles at which the particles were deflected from their former straight-line paths as they came out of the foil.

Rutherford found that most of the alpha particles shot at the foil passed straight through the foil. But a very small number of particles were deflected, in some cases backward, as shown in **Figure 8.** This result greatly surprised the researchers—it was very different from what Thomson's model predicted. As Rutherford said, "It was almost as if you fired a 15-inch shell into a piece of tissue paper and it came back and hit you." After thinking about the startling result for two years, Rutherford finally came up with an explanation. He went on to reason that only a very concentrated positive charge in a tiny space within the gold atom could possibly repel the fast-moving, positively charged alpha particles enough to reverse the alpha particles' direction of travel.

Rutherford also hypothesized that the mass of this positive-charge containing region, called the *nucleus,* must be larger than the mass of the alpha particle. If not, the incoming particle would have knocked the positive charge out of the way. The reason that most of the alpha particles were undeflected, Rutherford argued, was that most parts of the atoms in the gold foil were empty space.

This part of the model of the atom is still considered true today. The **nucleus** is the dense, central portion of the atom. The nucleus has all of the positive charge, nearly all of the mass, but only a very small fraction of the volume of the atom.

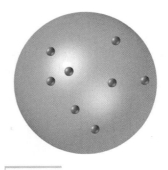

Figure 7
Thomson's model of an atom had negatively charged electrons embedded in a ball of positive charge.

nucleus

an atom's central region, which is made up of protons and neutrons

Figure 8

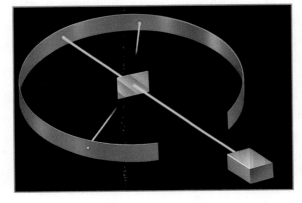

a In the gold foil experiment, small positively charged particles were directed at a thin foil of gold atoms.

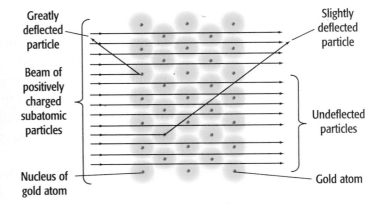

b The pattern of deflected alpha particles supported Rutherford's hypothesis that gold atoms were mostly empty space.

Greatly deflected particle

Slightly deflected particle

Beam of positively charged subatomic particles

Undeflected particles

Nucleus of gold atom

Gold atom

Figure 9
If the nucleus of an atom were the size of a marble, then the whole atom would be about the size of a football stadium.

Protons and Neutrons Compose the Nucleus

By measuring the numbers of alpha particles that were deflected and the angles of deflection, scientists calculated the radius of the nucleus to be less than $\frac{1}{10\,000}$ of the radius of the whole atom. **Figure 9** gives you a better idea of these sizes. Even though the radius of an entire atom is more than 10 000 times larger than the radius of its nucleus, an atom is still extremely small. The unit used to express atomic radius is the picometer (pm). One picometer equals 10^{-12} m.

The positively charged particles that repelled the alpha particles in the gold foil experiments and that compose the nucleus of an atom are called **protons.** The charge of a proton was calculated to be exactly equal in magnitude but opposite in sign to the charge of an electron. Later experiments showed that the proton's mass is almost 2000 times the mass of an electron.

Because protons and electrons have equal but opposite charges, a neutral atom must contain equal numbers of protons and electrons. But solving this mystery led to another: the mass of an atom (except hydrogen atoms) is known to be greater than the combined masses of the atom's protons and electrons. What could account for the rest of the mass? Hoping to find an answer, scientists began to search for a third subatomic particle.

About 30 years after the discovery of the electron, Irene Joliot-Curie (the daughter of the famous scientists Marie and Pierre Curie) discovered that when alpha particles hit a sample of beryllium, a beam that could go through almost anything was produced.

The British scientist James Chadwick found that this beam was not deflected by electric or magnetic fields. He concluded that the particles carried no electric charge. Further investigation showed that these neutral particles, which were named **neutrons,** are part of all atomic nuclei (except the nuclei of most hydrogen atoms).

proton

a subatomic particle that has a positive charge and that is found in the nucleus of an atom; the number of protons of the nucleus is the atomic number, which determines the identity of an element

neutron

a subatomic particle that has no charge and that is found in the nucleus of an atom

Table 3   Properties of a Proton and a Neutron

Name	Symbol	As shown in figures	Charge	Common charge notation	Mass (kg)
Proton	p, p^+, or $^1_{+1}p$		$+1.602 \times 10^{-19}$ C	$+1$	1.673×10^{-27} kg
Neutron	n or ^1_0n		0 C	0	1.675×10^{-27} kg

Protons and Neutrons Can Form a Stable Nucleus

Table 3 lists the properties of a neutron and a proton. Notice that the charge of a neutron is commonly assigned the value 0 while that of a proton is +1. How do protons that are positively charged come together to form a nucleus? In fact, the formation of a nucleus with protons seems impossible if you just consider *Coulomb's law*. Coulomb's law states that the closer two charges are, the greater the force between them. In fact, the force increases by a factor of 4 as the distance is halved. In addition, the larger the two charges are the greater the force between them. If the charges are opposite, they attract one another. If both charges have the same sign, they repel one another.

If you keep Coulomb's law in mind, it is easy to understand why—with the exception of some hydrogen atoms—no atoms have nuclei that are composed of only protons. All protons have a +1 charge. So, the repulsive force between two protons is large when two protons are close together, such as within a nucleus.

Protons, however, do form stable nuclei despite the repulsive force between them. A strong attractive force between these protons overcomes the repulsive force at small distances. Because neutrons also add attractive forces without being subject to repulsive charge-based forces, some neutrons can help stabilize a nucleus. Thus, all atoms that have more than one proton also have neutrons.

☑ internet connect

www.scilinks.org
Topic: Atomic Nucleus
SciLinks Code: HW4014

*SC*LINKS. Maintained by the National Science Teachers Association

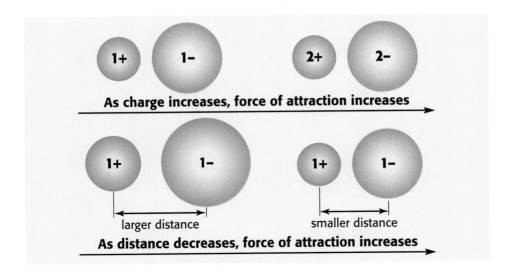

As charge increases, force of attraction increases

larger distance   smaller distance

As distance decreases, force of attraction increases

Figure 10
This figure shows that the larger two charges are, the greater the force between the charges. In addition, the figure shows the smaller the distance between two charges, the greater the force between the charges.

Atomic Number and Mass Number

All atoms consist of protons and electrons. Most atoms also have neutrons. Protons and neutrons make up the small, dense nuclei of atoms. The electrons occupy the space surrounding the nucleus. For example, an oxygen atom has protons and neutrons surrounded by electrons. But that description fits all other atoms, such as atoms of carbon, nitrogen, silver, and gold. How, then, do the atoms of one element differ from those of another element? Elements differ from each other in the number of protons their atoms contain.

Atomic Number Is the Number of Protons of the Nucleus

atomic number

the number of protons in the nucleus of an atom; the atomic number is the same for all atoms of an element

The number of protons that an atom has is known as the atom's **atomic number.** For example, the atomic number of hydrogen is 1 because the nucleus of each hydrogen atom has one proton. The atomic number of oxygen is 8 because all oxygen atoms have eight protons. Because each element has a unique number of protons in its atoms, no two elements have the same atomic number. So an atom whose atomic number is 8 must be an oxygen atom.

To date, scientists have identified 113 elements, whose atomic numbers range from 1 to 114. The element whose atomic number is 113 has yet to be discovered. Note that atomic numbers are always whole numbers. For example, an atom cannot have 2.5 protons.

The atomic number also reveals the number of electrons in an atom of an element. For atoms to be neutral, the number of negatively charged electrons must equal the number of positively charged protons. Therefore, if you know the atomic number of an atom, you immediately know the number of protons and the number of electrons found in that atom. **Figure 11** shows a model of an oxygen atom, whose atomic number is 8 and which has 8 electrons surrounding a nucleus that has 8 protons. The atomic number of gold is 79, so an atom of gold must have 79 electrons surrounding a nucleus of 79 protons. The next step in describing an atom's structure is to find out how many neutrons the atom has.

Figure 11
The atomic number for oxygen, as shown on the periodic table, tells you that the oxygen atom has 8 protons and 8 electrons.

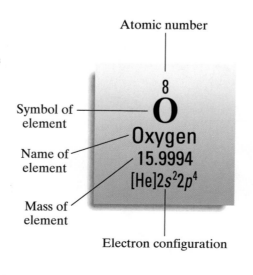

Atomic number

Symbol of element

Name of element

Mass of element

Electron configuration

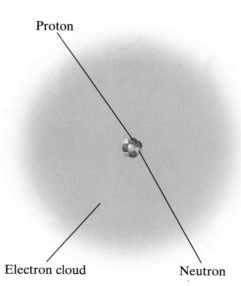

Proton

Electron cloud

Neutron

Mass Number Is the Number of Particles of the Nucleus

Every atomic nucleus can be described not only by its atomic number but also by its mass number. The **mass number** is equal to the total number of particles of the nucleus—that is, the total number of protons and neutrons. For example, a particular atom of neon has a mass number of 20, as shown in **Figure 12.** Therefore, the nucleus of this atom has a total of 20 protons and neutrons. Because the atomic number for an atom of neon is 10, neon has 10 protons. You can calculate the number of neutrons in a neon atom by subtracting neon's atomic number (the number of protons) from neon's mass number (the number of protons and neutrons).

mass number

the sum of the numbers of protons and neutrons of the nucleus of an atom

mass number – atomic number = number of neutrons

In this example, the neon atom has 10 neutrons.

$$\begin{array}{rl} \textit{number of protons and neutrons (mass number)} = & 20 \\ - \textit{ number of protons (atomic number)} = & 10 \\ \hline \textit{number of neutrons} = & 10 \end{array}$$

Unlike the atomic number, which is the same for all atoms of an element, mass number can vary among atoms of a single element. In other words, all atoms of an element have the same number of protons, but they can have different numbers of neutrons. The atomic number of every hydrogen atom is 1, but hydrogen atoms can have mass numbers of 1, 2, or 3. These atoms differ from one another in having 0, 1, and 2 neutrons, respectively. Another example is oxygen. The atomic number of every oxygen atom is 8, but oxygen atoms can have mass numbers of 16, 17, or 18. These atoms differ from one another in having 8, 9, and 10 neutrons, respectively.

Figure 12

The neon atom has 10 protons, 10 neutrons, and 10 electrons. This atom's mass number is 20, or the sum of the numbers of protons and neutrons in the atom.

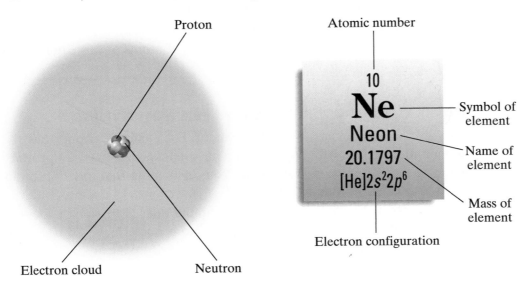

Determining the Number of Particles in an Atom

How many protons, electrons, and neutrons are present in an atom of copper whose atomic number is 29 and whose mass number is 64?

1 Gather information.

- The atomic number of copper is 29.
- The mass number of copper is 64.

2 Plan your work.

- The atomic number indicates the number of protons in the nucleus of a copper atom.
- A copper atom must be electrically neutral, so the number of electrons equals the number of protons.
- The mass number indicates the total number of protons and neutrons in the nucleus of a copper atom.

PRACTICE HINT

Check that the atomic number and the number of protons are the same. Also check that adding the numbers of protons and neutrons equals the mass number.

3 Calculate.

- *atomic number (29) = number of protons = 29*
- *number of protons = number of electrons = 29*
- *mass number (64) − atomic number (29) = number of neutrons = 35*

4 Verify your results.

- *number of protons (29) + number of neutrons (35) = mass number (64)*

PRACTICE

1. How many protons and electrons are in an atom of sodium whose atomic number is 11?

2. An atom has 13 protons and 14 neutrons. What is its mass number?

3. Calculate the mass number for an atom that has 45 neutrons and 35 electrons.

4. An atom of an element has 54 protons. Some of the element's atoms have 77 neutrons, while other atoms have 79 neutrons. What are the atomic numbers and mass numbers of the two types of atoms of this element?

Different Elements Can Have the Same Mass Number

The atomic number identifies an element. For example, copper has the atomic number 29. All copper atoms have nuclei that have 29 protons. Each of these atoms also has 29 electrons. Any atom that has 29 protons must be a copper atom.

In contrast, knowing just the mass number does not help you identify the element. For example, some copper atom nuclei have 36 neutrons. These copper atoms have a mass number of 65. But zinc atoms that have 30 protons and 35 neutrons also have mass numbers of 65.

Atomic Structures Can Be Represented by Symbols

Each element has a name, and the same name is given to all atoms of an element. For example, sulfur is composed of sulfur atoms. Recall that each element also has a symbol, and the same symbol is used to represent one of the element's atoms. Thus, S represents a single atom of sulfur, 2S represents two sulfur atoms, and 8S represents eight sulfur atoms. However, chemists write S_8 to indicate that the eight sulfur atoms are joined together and form a molecule of sulfur, as shown in the model in **Figure 13.**

Atomic number and mass number are sometimes written with an element's symbol. The atomic number always appears on the lower left side of the symbol. For example, the symbols for the first five elements are written with atomic numbers as follows:

$$_1H \quad _2He \quad _3Li \quad _4Be \quad _5B$$

Note that these subscript numbers give no new information. They simply indicate the atomic number of a particular element. On the other hand, mass numbers provide information that specifies particular atoms of an element. Mass numbers are written on the upper left side of the symbol. The following are the symbols of stable atoms of the first five elements written with mass numbers:

$$^1H \quad ^2H \quad ^3He \quad ^4He \quad ^6Li \quad ^7Li \quad ^9Be \quad ^{10}B \quad ^{11}B$$

Both numbers may be written with the symbol. For example, the most abundant kind of each of the first five elements can be represented by the following symbols:

$$^1_1H \quad ^4_2He \quad ^7_3Li \quad ^9_4Be \quad ^{11}_5B$$

An element may be represented by more than one notation. For example, the following notations represent the different atoms of hydrogen:

$$^1_1H \quad ^2_1H \quad ^3_1H$$

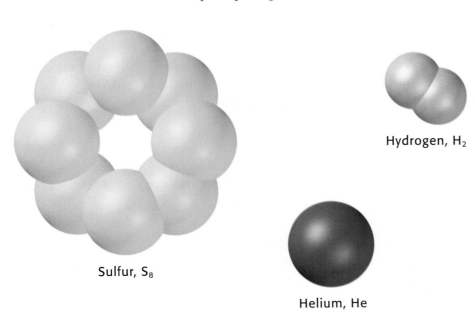

Sulfur, S_8

Hydrogen, H_2

Helium, He

Figure 13
In nature, elemental sulfur exists as eight sulfur atoms joined in a ring, elemental hydrogen exists as a molecule of two hydrogen atoms, and elemental helium exists as single helium atoms.

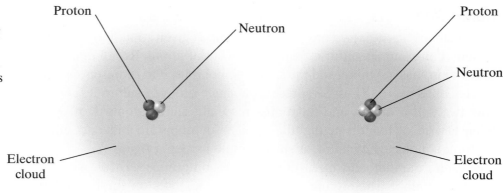

Figure 14
The two stable isotopes of helium are helium-3 and helium-4. The nucleus of a helium-4 atom is known as an *alpha particle*.

Proton
Neutron
Proton
Neutron
Electron cloud
Electron cloud

Helium-3   **Helium-4**

Isotopes of an Element Have the Same Atomic Number

All atoms of an element have the same atomic number and the same number of protons. However, atoms do not necessarily have the same number of neutrons. Atoms of the same element that have different numbers of neutrons are called **isotopes.** The two atoms modeled in **Figure 14** are stable isotopes of helium.

There are two standard methods of identifying isotopes. One method is to write the mass number with a hyphen after the name of an element. For example, the helium isotope shown on the left in **Figure 14** is written helium-3, while the isotope shown on the right is written as helium-4. The second method shows the composition of a nucleus as the isotope's nuclear symbol. Using this method, the notations for the two helium isotopes shown in **Figure 14** are written below.

$$^{3}_{2}\text{He} \quad ^{4}_{2}\text{He}$$

isotope

an atom that has the same number of protons (atomic number) as other atoms of the same element but has a different number of neutrons (atomic mass)

www.scilinks.org
Topic : Atoms and Elements
SciLinks code: HW4017

Notice that all isotopes of an element have the same atomic number. However, their atomic masses are not the same because the number of neutrons of the atomic nucleus of each isotope varies. In the case of helium, both isotopes have two protons in their nuclei. However, helium-3 has one neutron, while helium-4 has two neutrons.

Table 4 lists the four stable isotopes of lead. The least abundant of these isotopes is lead-204, while the most common is lead-208. Why do all lead atoms have 82 protons and 82 electrons?

Table 4   The Stable Isotopes of Lead

Name of atom	Symbol	Number of neutrons	Mass number	Mass (kg)	Abundance (%)
Lead-204	$^{204}_{82}\text{Pb}$	122	204	203.973	1.4
Lead-206	$^{206}_{82}\text{Pb}$	124	206	205.974	24.1
Lead-207	$^{207}_{82}\text{Pb}$	125	207	206.976	22.1
Lead-208	$^{208}_{82}\text{Pb}$	126	208	207.977	52.4

Determining the Number of Particles in Isotopes

Calculate the numbers of protons, electrons, and neutrons in oxygen-17 and in oxygen-18.

1 Gather information.

- The mass numbers for the two isotopes are 17 and 18.

2 Plan your work.

- An oxygen atom must be electrically neutral.

3 Calculate.

- *atomic number = number of protons = number of electrons = 8*
- *mass number – atomic number = number of neutrons*
- For oxygen-17, 17 – 8 = 9 neutrons
- For oxygen-18, 18 – 8 = 10 neutrons

4 Verify your results.

- The two isotopes have the same numbers of protons and electrons and differ only in their numbers of neutrons.

PRACTICE HINT

The only difference between the isotopes of an element is the number of neutrons in the atoms of each isotope.

PRACTICE

1 Chlorine has two stable isotopes, chlorine-35 and chlorine-37. The atomic number of chlorine is 17. Calculate the numbers of protons, electrons, and neutrons each isotope has.

2 Calculate the numbers of protons, electrons, and neutrons for each of the following isotopes of calcium: $^{42}_{20}Ca$ and $^{44}_{20}Ca$.

PROBLEM SOLVING SKILL

② Section Review

UNDERSTANDING KEY IDEAS

1. Describe the differences between electrons, protons, and neutrons.

2. How are isotopes of the same element alike?

3. What subatomic particle was discovered with the use of a cathode-ray tube?

PRACTICE PROBLEMS

4. Write the symbol for element X, which has 22 electrons and 22 neutrons.

5. Determine the numbers of electrons, protons, and neutrons for each of the following:

   a. $^{80}_{35}Br$ **b.** $^{106}_{46}Pd$ **c.** $^{133}_{55}Cs$

6. Calculate the atomic number and mass number of an isotope that has 56 electrons and 82 neutrons.

CRITICAL THINKING

7. Why must there be an attractive force to explain the existence of stable nuclei?

8. Are hydrogen-3 and helium-3 isotopes of the same element? Explain your answer.

Electron Configuration

OBJECTIVES

(1) **Compare** the Rutherford, Bohr, and quantum models of an atom.

(2) **Explain** how the wavelengths of light emitted by an atom provide information about electron energy levels.

(3) **List** the four quantum numbers, and describe their significance.

(4) **Write** the electron configuration of an atom by using the Pauli exclusion principle and the aufbau principle.

Atomic Models

Soon after the atomic theory was widely accepted by scientists, they began constructing models of atoms. Scientists used the information that they had about atoms to build these models. They knew, for example, that an atom has a densely packed nucleus that is positively charged. This conclusion was the only way to explain the data from Rutherford's gold foil experiments.

Building a model helps scientists imagine what may be happening at the microscopic level. For this very same reason, the illustrations in this book provide pictures that are models of chemical compounds to help you understand the relationship between the macroscopic and microscopic worlds. Scientists knew that any model they make may have limitations. A model may even have to be modified or discarded as new information is found. This is exactly what happened to scientists' models of the atom.

Rutherford's Model Proposed Electron Orbits

The experiments of Rutherford's team led to the replacement of the plum-pudding model of the atom with a nuclear model of the atom. Rutherford suggested that electrons, like planets orbiting the sun, revolve around the nucleus in circular or elliptical orbits. **Figure 15** shows Rutherford's model. Because opposite charges attract, the negatively charged electrons should be pulled into the positively charged nucleus. Because Rutherford's model could not explain why electrons did not crash into the nucleus, his model had to be modified.

The Rutherford model of the atom, in turn, was replaced only two years later by a model developed by Niels Bohr, a Danish physicist. The Bohr model, which is shown in **Figure 16,** describes electrons in terms of their energy levels.

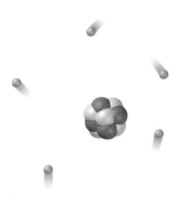

Figure 15
According to Rutherford's model of the atom, electrons orbit the nucleus just as planets orbit the sun.

Bohr's Model Confines Electrons to Energy Levels

According to Bohr's model, electrons can be only certain distances from the nucleus. Each distance corresponds to a certain quantity of energy that an electron can have. An electron that is as close to the nucleus as it can be is in its lowest energy level. The farther an electron is from the nucleus, the higher the energy level that the electron occupies. The difference in energy between two energy levels is known as a *quantum* of energy.

The energy levels in Bohr's model can be compared to the rungs of a ladder. A person can go up and down the ladder only by stepping on the rungs. When standing on the first rung, the person has the lowest potential energy. By climbing to the second rung, the person increases his or her potential energy by a fixed, definite quantity. Because the person cannot stand between the rungs on the ladder, the person's potential energy cannot have a continuous range of values. Instead, the values can be only certain, definite ones. In the same way, Bohr's model states that an electron can be in only one energy level or another, not between energy levels. Bohr also concluded that an electron did not give off energy while in a given energy level.

Figure 16
According to Bohr's model of the atom, electrons travel around the nucleus in specific energy levels.

Electrons Act Like Both Particles and Waves

Thomson's experiments demonstrated that electrons act like particles that have mass. Although the mass of an electron is extremely small, electrons in a cathode ray still have enough mass to turn a paddle wheel.

In 1924, Louis de Broglie pointed out that the behavior of electrons according to Bohr's model was similar to the behavior of waves. For example, scientists knew that any wave confined in space can have only certain frequencies. The frequency of a wave is the number of waves that pass through a given point in one second. De Broglie suggested that electrons could be considered waves confined to the space around a nucleus. As waves, electrons could have only certain frequencies. These frequencies could correspond to the specific energy levels in which electrons are found.

Other experiments also supported the wave nature of electrons. Like light waves, electrons can change direction through diffraction. Diffraction refers to the bending of a wave as the wave passes by the edge of an object, such as a crystal. Experiments also showed that electron beams, like waves, can interfere with each other.

Figure 17 shows the present-day model of the atom, which takes into account both the particle and wave properties of electrons. According to this model, electrons are located in **orbitals,** regions around a nucleus that correspond to specific energy levels. Orbitals are regions where electrons are likely to be found. Orbitals are sometimes called *electron clouds* because they do not have sharp boundaries. When an orbital is drawn, it shows where electrons are most likely to be. Because electrons can be in other places, the orbital has a fuzzy boundary like a cloud.

As an analogy to an electron cloud, imagine the spinning blades of a fan. You know that each blade can be found within the spinning image that you see. However, you cannot tell exactly where any one blade is at a particular moment.

orbital
a region in an atom where there is a high probability of finding electrons

Figure 17
According to the current model of the atom, electrons are found in orbitals.

electromagnetic spectrum

all of the frequencies
or wavelengths of
electromagnetic radiation

Electrons and Light

By 1900, scientists knew that light could be thought of as moving waves that have given frequencies, speeds, and wavelengths.

In empty space, light waves travel at 2.998×10^8 m/s. At this speed, light waves take only 500 s to travel the 150 million kilometers between the sun and Earth. The *wavelength* is the distance between two consecutive peaks or troughs of a wave. The distance of a wavelength is usually measured in meters. The wavelength of light can vary from 10^5 m to less than 10^{-13} m. This broad range of wavelengths makes up the **electromagnetic spectrum,** which is shown in **Figure 18.** Notice in **Figure 18** that our eyes are sensitive to only a small portion of the electromagnetic spectrum. This sensitivity ranges from 700 nm, which is about the value of wavelengths of red light, to 400 nm, which is about the value of wavelengths of violet light.

In 1905, Albert Einstein proposed that light also has some properties of particles. His theory would explain a phenomenon known as the *photoelectric effect.* This effect happens when light strikes a metal and electrons are released. What confused scientists was the observation that for a given metal, no electrons were emitted if the light's frequency was below a certain value, no matter how long the light was on. Yet if light were just a wave, then any frequency eventually should supply enough energy to remove an electron from the metal.

Einstein proposed that light has the properties of both waves and particles. According to Einstein, light can be described as a stream of particles, the energy of which is determined by the light's frequency. To remove an electron, a particle of light has to have at least a minimum energy and therefore a minimum frequency.

Figure 18
The electromagnetic spectrum is composed of light that has a broad range of wavelengths. Our eyes can detect only the visible spectrum.

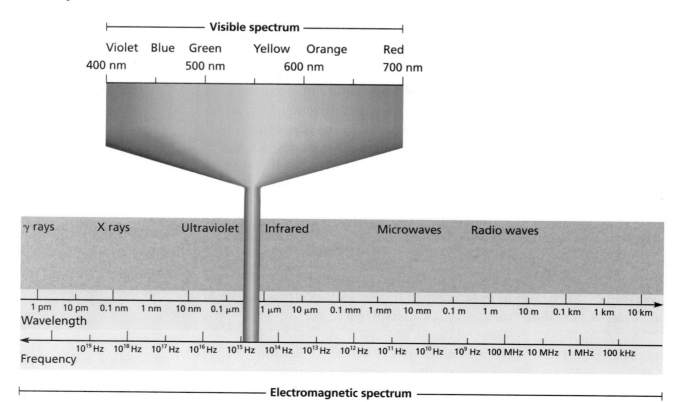

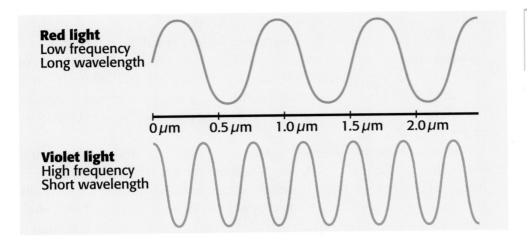

Red light
Low frequency
Long wavelength

0 μm   0.5 μm   1.0 μm   1.5 μm   2.0 μm

Violet light
High frequency
Short wavelength

Figure 19
The frequency and
wavelength of a wave
are inversely related.
As frequency increases,
wavelength decreases.

Light Is an Electromagnetic Wave

When passed through a glass prism, sunlight produces the visible spectrum—all of the colors of light that the human eye can see. You can see from **Figure 18** on the previous page that the visible spectrum is only a tiny portion of the electromagnetic spectrum. The electromagnetic spectrum also includes X rays, ultraviolet and infrared light, microwaves, and radio waves. Each of these electromagnetic waves is referred to as *light,* although we cannot see these wavelengths.

Figure 19 shows the frequency and wavelength of two regions of the spectrum that we see: red and violet lights. If you compare red and violet lights, you will notice that red light has a low frequency and a long wavelength. But violet light has a high frequency and a short wavelength. The frequency and wavelength of a wave are inversely related.

Light Emission

When a high-voltage current is passed through a tube of hydrogen gas at low pressure, lavender-colored light is seen. When this light passes through a prism, you can see that the light is made of only a few colors. This spectrum of a few colors is called a *line-emission spectrum.* Experiments with other gaseous elements show that each element has a line-emission spectrum that is made of a different pattern of colors.

In 1913, Bohr showed that hydrogen's line-emission spectrum could be explained by assuming that the hydrogen atom's electron can be in any one of a number of distinct energy levels. The electron can move from a low energy level to a high energy level by absorbing energy. Electrons at a higher energy level are unstable and can move to a lower energy level by releasing energy. This energy is released as light that has a specific wavelength. Each different move from a particular energy level to a lower energy level will release light of a different wavelength.

Bohr developed an equation to calculate all of the possible energies of the electron in a hydrogen atom. His values agreed with those calculated from the wavelengths observed in hydrogen's line-emission spectrum. In fact, his values matched with the experimental values so well that his atomic model that is described earlier was quickly accepted.

Light Provides Information About Electrons

Normally, if an electron is in a state of lowest possible energy, it is in a **ground state.** If an electron gains energy, it moves to an **excited state.** An electron in an excited state will release a specific quantity of energy as it quickly "falls" back to its ground state. This energy is emitted as certain wavelengths of light, which give each element a unique line-emission spectrum.

Figure 20 shows the wavelengths of light in a line-emission spectrum for hydrogen, through which a high-voltage current was passed. The high-voltage current may supply enough energy to move an electron from its ground state, which is represented by $n = 1$ in **Figure 20,** to a higher excited state for an electron in a hydrogen atom, represented by $n > 1$. Eventually, the electron will lose energy and return to a lower energy level. For example, the electron may fall from the $n = 7$ energy level to the $n = 3$ energy level. Notice in **Figure 20** that when this drop happens, the electron emits a wavelength of infrared light. An electron in the $n = 6$ energy level can also fall to the $n = 2$ energy level. In this case, the electron emits a violet light, which has a shorter wavelength than infrared light does.

Figure 20
An electron in a hydrogen atom can move between only certain energy states, shown as $n = 1$ to $n = 7$. In dropping from a higher energy state to a lower energy state, an electron emits a characteristic wavelength of light.

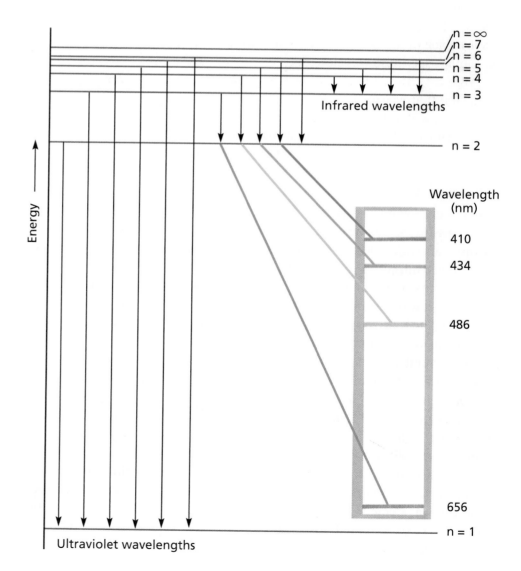

Table 5 Quantum Numbers of the First 30 Atomic Orbitals

n	l	m	Orbital name	Number of orbitals
1	0	0	1s	1
2	0	0	2s	1
2	1	−1, 0, 1	2p	3
3	0	0	3s	1
3	1	−1, 0, 1	3p	3
3	2	−2, −1, 0, 1, 2	3d	5
4	0	0	4s	1
4	1	−1, 0, 1	4p	3
4	2	−2, −1, 0, 1, 2	4d	5
4	3	−3, −2, −1, 0, 1, 2, 3	4f	7

Quantum Numbers

The present-day model of the atom, in which electrons are located in orbitals, is also known as the *quantum model.* According to this model, electrons within an energy level are located in orbitals, regions of high probability for finding a particular electron. However, the model does not explain how the electrons move about the nucleus to create these regions.

To define the region in which electrons can be found, scientists have assigned four **quantum numbers** to each electron. **Table 5** lists the quantum numbers for the first 30 atomic orbitals. The *principal quantum number,* symbolized by n, indicates the main energy level occupied by the electron. Values of n are positive integers, such as 1, 2, 3, and 4. As n increases, the electron's distance from the nucleus and the electron's energy increases.

The main energy levels can be divided into sublevels. These sublevels are represented by the *angular momentum quantum number, l.* This quantum number indicates the shape or type of orbital that corresponds to a particular sublevel. Chemists use a letter code for this quantum number. A quantum number $l = 0$ corresponds to an *s* orbital, $l = 1$ to a *p* orbital, $l = 2$ to a *d* orbital, and $l = 3$ to an *f* orbital. For example, an orbital with $n = 3$ and $l = 1$ is called a 3p orbital, and an electron occupying that orbital is called a 3p electron.

The *magnetic quantum number,* symbolized by m, is a subset of the l quantum number. It also indicates the numbers and orientations of orbitals around the nucleus. The value of m takes whole-number values, depending on the value of l. The number of orbitals includes one *s* orbital, three *p* orbitals, five *d* orbitals, and seven *f* orbitals.

The *spin quantum number,* symbolized by $+\frac{1}{2}$ or $-\frac{1}{2}$ (↑ or ↓), indicates the orientation of an electron's magnetic field relative to an outside magnetic field. A single orbital can hold a maximum of two electrons, which must have opposite spins.

quantum number

a number that specifies the properties of electrons

Electron Configurations

Figure 21 shows the shapes and orientations of the *s*, *p*, and *d* orbitals. Each orbital that is shown can hold a maximum of two electrons. The discovery that two, but no more than two, electrons can occupy a single orbital was made in 1925 by the German chemist Wolfgang Pauli. This rule is known as the **Pauli exclusion principle.**

Another way of stating the Pauli exclusion principle is that no two electrons in the same atom can have the same four quantum numbers. The two electrons can have the same value of *n* by being in the same main energy level. These two electrons can also have the same value of *l* by being in orbitals that have the same shape. And, these two electrons may have the same value of *m* by being in the same orbital. But these two electrons cannot have the same spin quantum number. If one electron has the value of $+\frac{1}{2}$, then the other electron must have the value of $-\frac{1}{2}$.

The arrangement of electrons in an atom is usually shown by writing an **electron configuration.** Like all systems in nature, electrons in atoms tend to assume arrangements that have the lowest possible energies. An electron configuration of an atom shows the lowest-energy arrangement of the electrons for the element.

Pauli exclusion principle

the principle that states that two particles of a certain class cannot be in the exact same energy state

electron configuration

the arrangement of electrons in an atom

s orbital

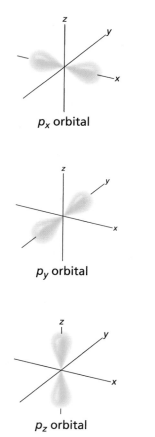

p_x orbital

p_y orbital

p_z orbital

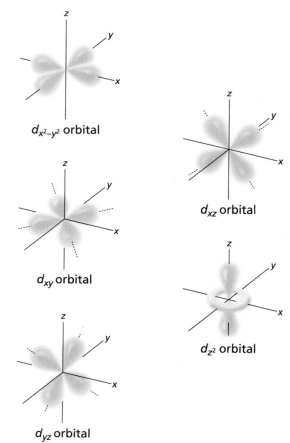

$d_{x^2-y^2}$ orbital

d_{xy} orbital

d_{yz} orbital

d_{xz} orbital

d_{z^2} orbital

Figure 21
a The *s* orbital is spherically shaped. There is one *s* orbital for each value n = 1, 2, 3...of the principal number.

b For each of the values n = 2, 3, 4..., there are three *p* orbitals. All are dumbbell shaped, but they differ in orientation.

c For each of the values n = 3, 4, 5..., there are five *d* orbitals. Four of the five have similar shapes, but differ in orientation.

An Electron Occupies the Lowest Energy Level Available

The Pauli exclusion principle is one rule to help you write an electron configuration for an atom. Another rule is the aufbau principle. *Aufbau* is the German word for "building up." The **aufbau principle** states that electrons fill orbitals that have the lowest energy first.

Recall that the smaller the principal quantum number, the lower the energy. But within an energy level, the smaller the *l* quantum number, the lower the energy. Recall that chemists use letters to represent the *l* quantum number. So, the order in which the orbitals are filled matches the order of energies, which starts out as follows:

$$1s < 2s < 2p < 3s < 3p$$

After this point, the order is less obvious. **Figure 22** shows that the energy of the 3*d* orbitals is slightly higher than the energy of the 4*s* orbitals. As a result, the order in which the orbitals are filled is as follows:

$$1s < 2s < 2p < 3s < 3p < 4s < 3d$$

Additional irregularities occur at higher energy levels.

Can you determine which orbitals electrons of a carbon atom occupy? Two electrons occupy the 1*s* orbital, two electrons occupy the 2*s* orbital, and two electrons occupy the 2*p* orbitals. Now try the same exercise for titanium. Two electrons occupy the 1*s* orbital, two electrons occupy the 2*s* orbital, six electrons occupy the 2*p* orbitals, two electrons occupy the 3*s* orbital, six electrons occupy the 3*p* orbitals, two electrons occupy the 3*d* orbitals, and two electrons occupy the 4*s* orbital.

aufbau principle

the principle that states that the structure of each successive element is obtained by adding one proton to the nucleus of the atom and one electron to the lowest-energy orbital that is available

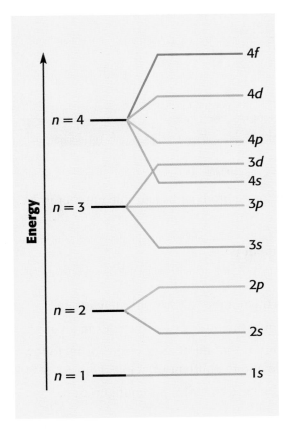

Figure 22
This diagram illustrates how the energy of orbitals can overlap such that 4*s* fills before 3*d*.

An Electron Configuration Is a Shorthand Notation

Based on the quantum model of the atom, the arrangement of the electrons around the nucleus can be shown by the nucleus's electron configuration. For example, sulfur has sixteen electrons. Its electron configuration is written as $1s^2 2s^2 2p^6 3s^2 3p^4$. This line of symbols tells us about these sixteen electrons. Two electrons are in the $1s$ orbital, two electrons are in the $2s$ orbital, six electrons are in the $2p$ orbitals, two electrons are in the $3s$ orbital, and four electrons are in the $3p$ orbitals.

Each element's configuration builds on the previous elements' configurations. To save space, one can write this configuration by using a configuration of a noble gas. The noble gas electron configurations that are often used are the configurations of neon, argon, krypton, and xenon. The neon atom's configuration is $1s^2 2s^2 2p^6$, so the electron configuration of sulfur is written as shown below.

$$[Ne]3s^2 3p^4$$

Does an electron enter the first $3p$ orbital to pair with a single electron that is already there? Or does the electron fill another $3p$ orbital? According to Hund's rule, the second answer is correct. **Hund's rule** states that orbitals of the same n and l quantum numbers are each occupied by one electron before any pairing occurs. For example, sulfur's configuration is shown by the *orbital diagram* below. Electrons are represented by arrows. Note that an electron fills each $3p$ orbital before an electron enters an orbital that is already occupied.

Hund's rule

the rule that states that for an atom in the ground state, the number of unpaired electrons is the maximum possible and these unpaired electrons have the same spin

---

SAMPLE PROBLEM C

Writing Electron Configurations

> **PRACTICE HINT**
>
> Remember that an s orbital holds 2 electrons, three p orbitals hold 6 electrons, and five d orbitals hold 10 electrons.

Write the electron configuration for an atom whose atomic number is 20.

1 Gather information.

- The atomic number of the element is 20.

2 Plan your work.

- The atomic number represents the number of protons in an atom.
- The number of protons must equal the number of electrons in an atom.
- Write the electron configuration for that number of electrons by following the Pauli exclusion principle and the aufbau principle.
- A noble gas configuration can be used to write this configuration.

3 Calculate.

- *atomic number = number of protons = number of electrons = 20*
- According to the aufbau principle, the order of orbital filling is 1s, 2s, 2p, 3s, 3p, 4s, 3d, 4p, and so on.
- The electron configuration for an atom of this element is written as follows:

$$1s^2 2s^2 2p^6 3s^2 3p^6 4s^2$$

- This electron configuration can be abbreviated as follows:

$$[Ar]4s^2$$

4 Verify your results.

- The sum of the superscripts is $(2 + 2 + 6 + 2 + 6 + 2) = 20$. Therefore, all 20 electrons are included in the electron configuration.

PRACTICE

1 Write the electron configuration for an atom of an element whose atomic number is 8.

2 Write the electron configuration for an atom that has 17 electrons.

PROBLEM SOLVING SKILL

---

③ Section Review

UNDERSTANDING KEY IDEAS

1. How does Bohr's model of the atom differ from Rutherford's?

2. What happens when an electron returns to its ground state from its excited state?

3. What does *n* represent in the quantum model of electrons in atoms?

PRACTICE PROBLEMS

4. What is the atomic number of an element whose atom has the following electron configuration: $1s^2 2s^2 2p^6 3s^2 3p^6 3d^2 4s^2$?

5. Write the electron configuration for an atom that has 13 electrons.

6. Write the electron configuration for an atom that has 33 electrons.

7. How many orbitals are completely filled in an atom whose electron configuration is $1s^2 2s^2 2p^6 3s^1$?

CRITICAL THINKING

8. Use the Pauli exclusion principle or the aufbau principle to explain why the following electron configurations are incorrect:

a. $1s^2 2s^3 2p^6 3s^1$

b. $1s^2 2s^2 2p^5 3s^1$

9. Why is a shorter wavelength of light emitted when an electron "falls" from $n = 4$ to $n = 1$ than when an electron "falls" from $n = 2$ to $n = 1$?

10. Calculate the maximum number of electrons that can occupy the third principal energy level.

11. Why do electrons fill the 4s orbital before they start to occupy the 3d orbital?

Counting Atoms

KEY TERMS

• atomic mass

• mole

• molar mass

• Avogadro's number

OBJECTIVES

1. **Compare** the quantities and units for atomic mass with those for molar mass.

2. **Define** *mole,* and explain why this unit is used to count atoms.

3. **Calculate** either mass with molar mass or number with Avogadro's number given an amount in moles.

Atomic Mass

You would not expect something as small as an atom to have much mass. For example, copper atoms have an average mass of only 1.0552×10^{-25} kg.

Each penny in **Figure 23** has an average mass of 3.13×10^{-3} kg and contains copper. How many copper atoms are there in one penny? Assuming that a penny is pure copper, you can find the number of copper atoms by dividing the mass of the penny by the average mass of a single copper atom or by using the following conversion factor:

$$1 \text{ atom Cu}/1.0552 \times 10^{-25} \text{ kg}$$

$$3.13 \times 10^{-3} \text{ kg} \times \frac{1 \text{ atom Cu}}{1.0552 \times 10^{-25} \text{ kg}} = 2.97 \times 10^{22} \text{ Cu atoms}$$

atomic mass

the mass of an atom expressed in atomic mass units

Masses of Atoms Are Expressed in Atomic Mass Units

Obviously, atoms are so small that the gram is not a very convenient unit for expressing their masses. Even the picogram (10^{-12} g) is not very useful. A special mass unit is used to express **atomic mass.** This unit has two names—the atomic mass unit (amu) and the Dalton (Da). In this book, *atomic mass unit* will be used.

But how can you tell what the atomic mass of a specific atom is? When the atomic mass unit was first set up, an atom's mass number was supposed to be the same as the atom's mass in atomic mass units. Mass number and atomic mass units would be the same because a proton and a neutron each have a mass of about 1.0 amu.

For example, a copper-63 atom has an atomic mass of 62.940. A copper-65 atom has an atomic mass of 64.928. (The slight differences from exact values will be discussed in later chapters.)

Another way to determine atomic mass is to check a periodic table, such as the one on the inside cover of this book. The mass shown is an average of the atomic masses of the naturally occurring isotopes. For this reason, copper is listed as 63.546 instead of 62.940 or 64.928.

Figure 23
These pennies are made mostly of copper atoms. Each copper atom has an average mass of 1.0552×10^{-25} kg.

Introduction to the Mole

Most samples of elements have great numbers of atoms. To make working with these numbers easier, chemists created a new unit called the *mole* (mol). A **mole** is defined as the number of atoms in exactly 12 grams of carbon-12. The mole is the SI unit for the amount of a substance.

Chemists use the mole as a counting unit, just as you use the dozen as a counting unit. Instead of asking for 12 eggs, you ask for 1 dozen eggs. Similarly, chemists refer to 1 mol of carbon or 2 mol of iron.

To convert between moles and grams, chemists use the molar mass of a substance. The **molar mass** of an element is the mass in grams of one mole of the element. Molar mass has the unit grams per mol (g/mol). The mass in grams of 1 mol of an element is numerically equal to the element's atomic mass from the periodic table in atomic mass units. For example, the atomic mass of copper to two decimal places is 63.55 amu. Therefore, the molar mass of copper is 63.55 g/mol. **Skills Toolkit 1** shows how to convert between moles and mass in grams using molar mass.

Scientists have also determined the number of particles present in 1 mol of a substance, called **Avogadro's number.** One mole of pure substance contains $6.022\,1367 \times 10^{23}$ particles. To get some idea of how large Avogadro's number is, imagine that every living person on Earth (about 6 billion people) started counting the number of atoms of 1 mol C. If each person counted nonstop at a rate of one atom per second, it would take over 3 million years to count every atom.

Avogadro's number may be used to count any kind of particle, including atoms and molecules. For calculations in this book, Avogadro's number will be rounded to 6.022×10^{23} particles per mole. **Skills Toolkit 2** shows how to use Avogadro's number to convert between amount in moles and the number of particles.

mole

the SI base unit used to measure the amount of a substance whose number of particles is the same as the number of atoms in 12 g of carbon-12

molar mass

the mass in grams of 1 mol of a substance

Avogadro's number

6.022×10^{23}, the number of atoms or molecules in 1 mol

SKILLS Toolkit 1

Determining the Mass from the Amount in Moles

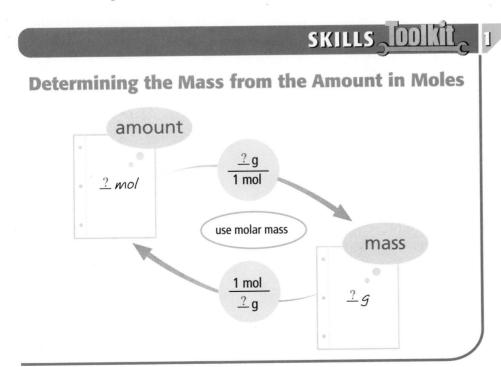

Converting from Amount in Moles to Mass

Determine the mass in grams of 3.50 mol of copper.

1 Gather information.

- amount of Cu = 3.50 mol
- mass of Cu = ? g Cu
- molar mass of Cu = 63.55 g

2 Plan your work.

- First, make a set-up that shows what is given and what is desired.

$$3.50 \text{ mol Cu} \times ? = ? \text{ g Cu}$$

- Use a conversion factor that has g Cu in the numerator and mol Cu in the denominator.

$$3.50 \text{ mol Cu} \times \frac{? \text{ g Cu}}{1 \text{ mol Cu}} = ? \text{ g Cu}$$

3 Calculate.

- The correct conversion factor is the molar mass of Cu, 63.55 g/mol. Place the molar mass in the equation, and calculate the answer. Use the periodic table in this book to find mass numbers of elements.

$$3.50 \text{ mol Cu} \times \frac{63.55 \text{ g Cu}}{1 \text{ mol Cu}} = 222 \text{ g Cu}$$

4 Verify your results.

- To verify that the answer of 222 g is correct, find the number of moles of 222 g of copper.

$$222 \text{ g of Cu} \times \frac{1 \text{ mol Cu}}{63.55 \text{ g Cu}} = 3.49 \text{ mol Cu}$$

The amount of 3.49 mol is close to the 3.50 mol, so the answer of 222 g is reasonable.

PRACTICE

1 What is the mass in grams of 1.00 mol of uranium?

2 What is the mass in grams of 0.0050 mol of uranium?

3 Calculate the number of moles of 0.850 g of hydrogen atoms. What is the mass in grams of 0.850 mol of hydrogen atoms?

4 Calculate the mass in grams of 2.3456 mol of lead. Calculate the number of moles of 2.3456 g of lead.

Determining the Number of Atoms from the Amount in Moles

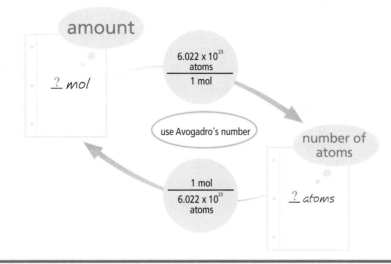

amount

$\underline{?}\ mol$

$\dfrac{6.022 \times 10^{23}\ \text{atoms}}{1\ \text{mol}}$

use Avogadro's number

number of atoms

$\underline{?}\ atoms$

$\dfrac{1\ \text{mol}}{6.022 \times 10^{23}\ \text{atoms}}$

SAMPLE PROBLEM E

Converting from Amount in Moles to Number of Atoms

Determine the number of atoms in 0.30 mol of fluorine atoms.

1 Gather information.

• amount of F = 0.30 mol
• number of atoms of F = ?

2 Plan your work.

• To determine the number of atoms, select the conversion factor that will take you from the amount in moles to the number of atoms.

amount (mol) $\times\ 6.022 \times 10^{23}$ atoms/mol = number of atoms

3 Calculate.

$$0.30\ \cancel{\text{mol F}} \times \frac{6.022 \times 10^{23}\ \text{F atoms}}{1\ \cancel{\text{mol F}}} = 1.8 \times 10^{23}\ \text{F atoms}$$

4 Verify your results.

• The answer has units that are requested in the problem. The answer is also less than 6.022×10^{23} atoms, which makes sense because you started with less than 1 mol.

PRACTICE HINT

Make sure to select the correct conversion factor so that units cancel to give the unit required in the answer.

PRACTICE

1 How many atoms are in 0.70 mol of iron?

2 How many moles of silver are represented by 2.888×10^{23} atoms?

3 How many moles of osmium are represented by 3.5×10^{23} atoms?

PROBLEM SOLVING SKILL

Figure 24
Carbon, which composes diamond, is the basis for the atomic mass scale that is used today.

Chemists and Physicists Agree on a Standard

The atomic mass unit has been defined in a number of different ways over the years. Originally, atomic masses expressed the ratio of the mass of an atom to the mass of a hydrogen atom. Using hydrogen as the standard turned out to be inconvenient because hydrogen does not react with many elements. Early chemists determined atomic masses by comparing how much of one element reacted with another element.

Because oxygen combines with almost all other elements, oxygen became the standard of comparison. The atomic mass of oxygen was defined as exactly 16, and the atomic masses of the other elements were based on this standard. But this choice also led to difficulties. Oxygen exists as three isotopes. Physicists based their atomic masses on the assignment of 16.0000 as the mass of the most common oxygen isotope. Chemists, on the other hand, decided that 16.0000 should be the average mass of all oxygen isotopes, weighted according to the abundance of each isotope. So, to a physicist, the atomic mass of fluorine was 19.0044, but to a chemist, it was 18.9991.

Finally, in 1960, a conference of chemists and physicists agreed on a scale based on an isotope of carbon. Carbon is shown in **Figure 24.** Used by all scientists today, this scale defines the atomic mass unit as exactly one-twelfth of the mass of one carbon-12 atom. As a result, one atomic mass unit is equal to $1.600\ 5402 \times 10^{-27}$ kg. The mass of an atom is indeed quite small.

 Section Review

UNDERSTANDING KEY IDEAS

1. What is atomic mass?
2. What is the SI unit for the amount of a substance that contains as many particles as there are atoms in exactly 12 grams of carbon-12?
3. Which atom is used today as the standard for the atomic mass scale?
4. What unit is used for molar mass?
5. How many particles are present in 1 mol of a pure substance?

PRACTICE PROBLEMS

6. Convert 3.01×10^{23} atoms of silicon to moles of silicon.

7. How many atoms are present in 4.0 mol of sodium?
8. How many moles are represented by 118 g of cobalt? Cobalt has an atomic mass of 58.93 amu.
9. How many moles are represented by 250 g of platinum?
10. Convert 0.20 mol of boron into grams of boron. How many atoms are present?

CRITICAL THINKING

11. What is the molar mass of an element?
12. How is the mass in grams of an element converted to amount in moles?
13. How is the mass in grams of an element converted to number of atoms?

Where Is Be?

Earth's crust:
0.005% by mass

Element Spotlight

Beryllium: An Uncommon Element

Although it is an uncommon element, beryllium has a number of properties that make it very useful. Beryllium has a relatively high melting point (1278°C) and is an excellent conductor of energy as heat and electrical energy. Beryllium transmits X rays extremely well and is therefore used to make "windows" for X-ray devices. All compounds of beryllium are toxic to humans. People who experience prolonged exposure to beryllium dust may contract berylliosis, a disease that can lead to severe lung damage and even death.

Industrial Uses

- The addition of 2% beryllium to copper forms an alloy that is six times stronger than copper is. This alloy is used for nonsparking tools, critical moving parts in jet engines, and components in precision equipment.

- Beryllium is used in nuclear reactors as a neutron reflector and as an alloy with the fuel elements.

Crystals of pure beryllium look very different from the combined form of beryllium in an emerald.

Real-World Connection Emerald and aquamarine are precious forms of the mineral beryl, Be$_3$Al$_2$(SiO$_3$)$_6$.

A Brief History

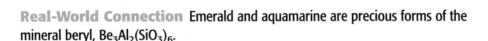

1828: F. Wöhler of Germany gives beryllium its name after he and W. Bussy of France simultaneously isolate the pure metal.

1926: M. G. Corson of the United States discovers that beryllium can be used to age-harden copper-nickel alloys.

| 1800 | 1900 |

1798: R. J. Haüy, a French mineralogist, observes that emeralds and beryl have the same optical properties and therefore the same chemical composition.

1898: P. Lebeau discovers a method of extracting high-purity beryllium by using an electrolytic process.

1942: A Ra-Be source provides the neutrons for Fermi's studies. These studies lead to the construction of a nuclear reactor.

Questions

1. Research how the beryllium and copper alloy is made and what types of equipment are made of this alloy.

2. Research how beryllium is used in nuclear reactors.

3. Research berylliosis and use the information to make a medical information brochure. Be sure to include symptoms, causes, and risk factors in your report.

internet connect

www.scilinks.org
Topic : Beryllium
SciLinks code: HW4021

SCiLINKS® Maintained by the
National Science
Teachers Association

3 CHAPTER HIGHLIGHTS

KEY IDEAS

SECTION ONE Substances Are Made of Atoms

- Three laws support the existence of atoms: the law of definite proportions, the law of conservation of mass, and the law of multiple proportions.
- Dalton's atomic theory contains five basic principles, some of which have been modified.

SECTION TWO Structure of Atoms

- Protons, particles that have a positive charge, and neutrons, particles that have a neutral charge, make up the nuclei of most atoms.
- Electrons, particles that have a negative charge and very little mass, occupy the region around the nucleus.
- The atomic number of an atom is the number of protons the atom has. The mass number of an atom is the number of protons plus the number of neutrons.
- Isotopes are atoms that have the same number of protons but different numbers of neutrons.

SECTION THREE Electron Configuration

- The quantum model describes the probability of locating an electron at any place.
- Each electron is assigned four quantum numbers that describe it. No two electrons of an atom can have the same four quantum numbers.
- The electron configuration of an atom reveals the number of electrons an atom has.

SECTION FOUR Counting Atoms

- The masses of atoms are expressed in atomic mass units (amu). The mass of an atom of the carbon-12 isotope is defined as exactly 12 atomic mass units.
- The mole is the SI unit for the amount of a substance that contains as many particles as there are atoms in exactly 12 grams of carbon-12.
- Avogadro's number, 6.022×10^{23} particles per mole, is the number of particles in a mole.

KEY TERMS

law of definite proportions
law of conservation of mass
law of multiple proportions

electron
nucleus
proton
neutron
atomic number
mass number
isotope

orbital
electromagnetic spectrum
ground state
excited state
quantum number
Pauli exclusion principle
electron configuration
aufbau principle
Hund's rule

atomic mass
mole
molar mass
Avogadro's number

KEY SKILLS

Determining the Number of Particles in an Atom
Sample Problem A p. 86

Determining the Number of Particles in Isotopes
Sample Problem B p. 89

Writing Electron Configurations
Sample Problem C p. 98

Converting Amount in Moles to Mass
Skills Toolkit 1 p. 101
Sample Problem D p. 102

Converting Amount in Moles to Number of Atoms
Skills Toolkit 2 p. 103
Sample Problem E p. 103

CHAPTER REVIEW 3

USING KEY TERMS

1. Define *isotope*.

2. What are neutrons?

3. State the Pauli exclusion principle.

4. What is a cathode?

5. Define *mass number*.

6. What is a line-emission spectrum?

7. Define *ground state*.

8. Define *mole*.

9. State the law of definite proportions.

10. What is an orbital?

11. What is an electron configuration?

12. Which has less mass: an electron or a proton?

UNDERSTANDING KEY IDEAS

Substances Are Made of Atoms

13. What law is illustrated by the fact that ice, water, and steam consist of 88.8% oxygen and 11.2% hydrogen by mass?

14. What law is shown by the fact that the mass of carbon dioxide, which forms as a product of a reaction between oxygen and carbon, equals the combined masses of the carbon and oxygen that reacted?

15. Of the five parts of Dalton's atomic theory, which one(s) have been modified?

16. What is the atomic theory?

Structure of Atoms

17. Which of Dalton's principles was contradicted by the work of J. J. Thomson?

18. What information about an atom is provided by the atom's atomic number? the atom's mass number?

19. How were atomic models developed given that no one had seen an atom?

20. Why are atomic numbers always whole numbers?

21. If all protons have positive charges, how can any atomic nucleus be stable?

22. What observation did Thomson make to suggest that an electron has a negative electric charge?

Electron Configuration

23. Why does the Pauli exclusion principle include the word *exclusion*?

24. How do you use the aufbau principle when you create an electron configuration?

25. Explain what is required to move an electron from the ground state to an excited state.

26. Why can a *p* sublevel hold six electrons while the *s* sublevel can hold no more than two electrons?

27. What did Einstein's explanation about the photoelectric effect state about the nature of light?

28. What do electrons and light have in common?

29. Of a set of four quantum numbers, how many of these numbers can be shared by more than one electron of the same orbital?

30. When does an electron in an atom have the lowest energy?

31. How are the frequency and wavelength of light related?

32. Why does an electron occupy the $4s$ orbital before the $3d$ orbital?

33. The element sulfur has an electron configuration of $1s^2 2s^2 2p^6 3s^2 3p^4$.
a. What does the superscript 6 refer to?
b. What does the letter s refer to?
c. What does the coefficient 3 refer to?

Counting Atoms

34. What is a mole? How is a mole related to Avogadro's number?

35. What significance does carbon-12 have in terms of atomic mass?

36. If the mass of a gold atom is 196.97 amu, what is the atom's molar mass?

37. How many water molecules are present in 1.00 mol of pure water?

38. What advantage is gained by using the mole as a unit when working with atoms?

PRACTICE PROBLEMS

Sample Problem A Determining the Number of Particles in an Atom

39. Calculate the number of neutrons of the atom whose atomic number is 42 and whose mass number is 96.

40. How many electrons are present in an atom of mercury whose atomic number is 80 and whose mass number is 201?

41. Calculate the number of protons of the atom whose mass number is 19 and whose number of neutrons is 10.

42. Calculate the number of electrons of the atom whose mass number is 75 and whose number of neutrons is 42.

Sample Problem B Determining the Number of Particles in Isotopes

43. Write nuclear symbols for isotopes of uranium that have the following numbers of neutrons. The atomic number of uranium is 92.
a. 142 neutrons
b. 143 neutrons
c. 146 neutrons

44. Copy and complete the following table concerning the three isotopes of silicon, whose atomic number is 14.

Isotope	Number of protons	Number of electrons	Number of neutrons
Si-28			
Si-29			
Si-30			

45. Write the symbol for two isotopes of carbon. Both isotopes have six protons. One isotope has six neutrons, while the other has seven neutrons.

46. All barium atoms have 56 protons. One isotope of barium has 74 neutrons, and another isotope has 81 neutrons. Write the symbols for these two isotopes of barium.

Sample Problem C Writing Electron Configurations

47. Write the electron configuration for nickel, whose atomic number is 28. Remember that the $4s$ orbital has lower energy than the $3d$ orbital does and that the d sublevel can hold a maximum of 10 electrons.

48. Write the electron configuration of germanium whose atomic number is 32.

49. How many orbitals are completely filled in an atom that has 12 electrons? The electron configuration is $1s^2 2s^2 2p^6 3s^2$.

50. How many orbitals are completely filled in an atom of an element whose atomic number is 18?

Sample Problem D Converting Amount in Moles to Mass

51. How many moles are represented by each of the following.
  a. 11.5 g Na which has an atomic mass of 22.99 amu
  b. 150 g S which has an atomic mass of 32.07 amu
  c. 5.87 g Ni which has an atomic mass of 58.69 amu

52. Determine the mass in grams represented by 2.50 mol tellurium.

53. What is the mass in grams of 0.0050 mol of hydrogen atoms?

Sample Problem E Converting Amount in Moles to Number of Atoms

54. Calculate the number of atoms in 2.0 g of hydrogen atoms. The atomic mass of hydrogen is 1.01 amu.

55. Calculate the number of atoms present in each of the following:
  a. 2 mol Fe
  b. 40.1 g Ca, which has an atomic mass of 40.08 amu
  c. 4.5 mol of boron-11

56. How many mol of potassium are represented by 7.85×10^{23} potassium atoms?

MIXED REVIEW

57. In the diagram below, indicate which sub-atomic particles would be found in areas *a* and *b*.

a

b

58. What is the approximate atomic mass of an atom if the mass of the atom is
  a. 12 times that of carbon-12?
  b. one-fourth that of carbon-12?

59. What mass of silver, Ag, which has an atomic mass of 107.87 amu, contains the same number of atoms contained in 10.0 g of boron, B, which has an atomic mass of 10.81 amu?

60. Hydrogen's only electron occupies the 1*s* orbital but can be excited to a 4*p* orbital. List all of the orbitals that this electron can occupy as it "falls."

61. What is the electron configuration of zinc?

62. Identify the scientists who proposed each of the models illustrated below.

  a.    **c.**

  b   **d.**

63. Determine the number of grams represented by each of the following:
  a. 3.5 mol of the element carbon, which has an atomic mass of 12.01 amu
  b. 5.0×10^9 atoms of neon, which has an atomic mass of 20.18 amu
  c. 2.25×10^{22} atoms of carbon, which has an atomic mass of 12.01 amu

64. How many atoms are in 0.75 moles of neptunium?

65. How did the results of the gold foil experiment lead Rutherford to recognize the existence of atomic nuclei?

66. Explain why atoms are neutral.

67. Explain Coulomb's law.

68. What is the mass in grams of each of the following?
   a. 0.50 mol of carbon, C
   b. 1.80 mol of calcium

69. Determine the mass in kilograms of 5.50 mol of iron, Fe.

70. What is Avogadro's number?

71. How many moles are present in 11 g of silicon? how many atoms?

72. How many moles are present in 620 grams of lithium? how many atoms?

73. Suppose an atom has a mass of 11 amu and has five electrons. What is this atom's atomic number?

74. Explain why different atoms of the same element always have the same atomic number but can have different mass numbers.

75. What does an element's molar mass tell you about the element?

76. A pure gold bar is made of 19.55 mol of gold. What is the mass of the bar in grams?

77. How does halving the amount of a sample of an element affect the sample's mass?

78. James is holding a balloon that contains 0.54 g of helium gas. How many helium atoms is this?

79. Write the electron configuration of phosphorus.

80. Write the electron configuration of bromine.

81. What are the charges of an electron, a proton, and a neutron?

82. An advertising sign gives off red and green light.
   a. Which light has higher energy?
   b. One of the colors has a wavelength of 680 nm and the other has a wavelength of 500. Which color has which wavelength?

83. Can a stable atom have an orbital which has three electrons? Explain your answer.

CRITICAL THINKING

84. Predict what Rutherford might have observed if he had bombarded copper metal instead of gold metal with alpha particles. The atomic numbers of copper and gold are 29 and 79, respectively.

85. Identify the law that explains why a water molecule in a raindrop falling on Phoenix, Arizona, and a water molecule in the Nile River in Egypt are both made of two hydrogen atoms for every oxygen atom.

86. How would you rewrite Dalton's fourth principle to account for elements such as O_2, P_4, and S_8?

87. Which of Dalton's principles is contradicted by a doctor using radioisotopes to trace chemicals in the body?

88. What would happen to poisonous chlorine gas if the following alterations could be made?
   a. A proton is added to each atom.
   b. A neutron is added to each atom.

89. For hundreds of years, alchemists searched for ways to turn various metals into gold. How would the structure of an atom of $^{202}_{80}Hg$ (mercury) have to be changed for the atom to become an atom of $^{197}_{79}Au$ (gold)?

90. How are quantum numbers like an address? How are they different from an address?

91. The following electron configurations for electrons in the ground state are incorrect. Explain why each configuration is incorrect by using the Pauli exclusion principle or the aufbau principle, and then write the corrected electron configuration.
   a. $1s^2 2s^2 2p^5 3s^2 3p^3$
   b. $1s^2 2s^2 2p^8 3s^1$

92. Which has more atoms: 3.0 g of iron, Fe, or 2.0 grams of sulfur, S?

93. Predict which isotope of nitrogen is more commonly found, nitrogen-14 or nitrogen-15.

94. Suppose you have only 1.9 g of sulfur for an experiment and you must do three trials using 0.030 mol of S each time. Do you have enough sulfur?

95. How many orbitals in an atom can have the following designation?
a. $4p$
b. $7s$
c. $5d$

96. Explain why that if $n = 2$, l cannot be 2.

97. Write the electron configuration of rubidium.

98. Write the electron configuration of tin.

99. The magnetic properties of an element depend on the number of unpaired electrons it has. Explain why iron, Fe, is highly magnetic but neon, Ne, is not.

100. Many elements exist as polyatomic molecules. Use atomic masses to calculate the molecular masses of the following:
a. O_2
b. P_4
c. S_8

101. What do the electron configurations of neon, argon, krypton, xenon, and radon have in common?

102. Answer the following regarding electron configurations of atoms in the fourth period of the periodic table.
a. Which orbitals are filled by transition metals?
b. Which orbitals are filled by nonmetals?

ALTERNATIVE ASSESSMENT

103. So-called neon signs actually contain a variety of gases. Research the different substances used for these signs. Design your own sign on paper, and identify which gases you would use to achieve the desired color scheme.

104. Select one of the essential elements. Check your school library or the Internet for details about the role of each element in the human body and for any guidelines and recommendations about the element.

105. Research several elements whose symbols are inconsistent with their English names. Some examples include silver, Ag; gold, Au; and mercury, Hg. Compare the origin of these names with the origin of the symbols.

106. Research the development of the scanning tunneling microscope, which can be used to make images of atoms. Find out what information about the structure of atoms these microscopes have provided.

CONCEPT MAPPING

107. Use the following terms to complete the concept map below: *proton, atomic number, atomic theory, orbital,* and *electron.*

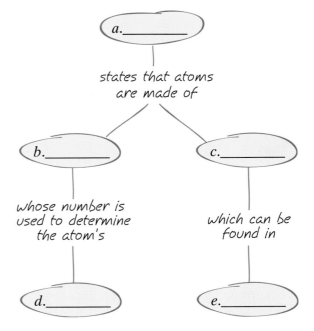

Study the graph below, and answer the questions that follow.
For help in interpreting graphs, see Appendix B, "Study Skills for Chemistry."

108. What represents the ground state in this diagram?

109. Which energy-level changes can be detected by the unaided eye?

110. Does infrared light have more energy than ultraviolet light? Why or why not?

111. Which energy levels represent a hydrogen electron in an excited state?

112. What does the energy level labeled "$n = \infty$" represent?

113. If an electron is beyond the $n = \infty$ level, is the electron a part of the hydrogen atom?

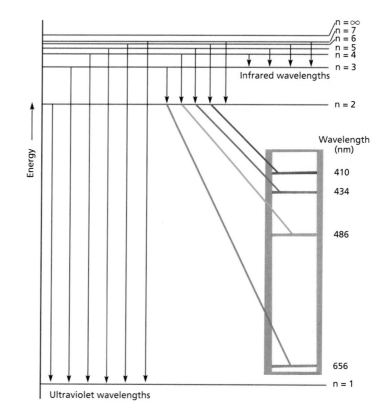

TECHNOLOGY AND LEARNING

114. Graphing Calculator

Calculate Numbers of Protons, Electrons, and Neutrons.

A graphing calculator can run a program that calculates the numbers of protons, electrons, and neutrons given the atomic mass and numbers for an atom. For example, given a calcium-40 atom, you will calculate the numbers of protons, electrons, and neutrons in the atom.

Go to Appendix C. If you are using a TI-83 Plus, you can download the program

NUMBER and data and can run the application as directed. If you are using another calculator, your teacher will provide you with keystrokes and data sets to use. After you have run the program, answer the questions below.

a. Which element has the most protons?

b. How many neutrons does mercury-201 have?

c. Carbon-12 and carbon-14 have the same atomic number. Do they have the same number of neutrons? Why or why not?

1. Which of the following represents an electron configuration for a calcium atom, whose atomic number is 20?
   a. $1s^2 2s^2 2p^6 3s^1 3p^6 4s^2 3d^1$
   b. $1s^2 2s^2 2p^6 3s^2 3p^6 4s^2$
   c. $1s^2 2s^2 2p^6 3s^1 3p^6 4s^3$
   d. $1s^2 2s^2 2p^6 3s^2 3p^6 4s^2 3d^1$

2. Isotopes of an element differ in their
   a. atomic numbers.
   b. electron configurations.
   c. numbers of protons.
   d. numbers of neutrons.

3. The number of protons of an atom equals
   a. the number of neutrons.
   b. the mass number.
   c. the number of electrons.
   d. the number of isotopes.

4. The law of multiple proportions states that
   a. when two elements combine to form more than one compound, the masses of one element that combine with a given mass of the other element are in a ratio of small whole numbers.
   b. elements combine in the same mass ratio in a compound regardless of the quantity of the sample or the source.
   c. elements have different physical and chemical properties.
   d. atoms of elements cannot be created, destroyed, or subdivided when they participate in a chemical reaction.

5. How many neutrons are present in an atom of tin that has an atomic number of 50 and a mass number of 119?
   a. 50   c. 119
   b. 69   d. 169

6. As an electron in an excited state returns to its ground state,
   a. light energy is emitted.
   b. energy is absorbed by the atom.
   c. the atom is likely to undergo spontaneous decay.
   d. the electron configuration of the atom changes.

7. According to quantum theory, an electron
   a. remains in a fixed position.
   b. can replace a proton in the nucleus.
   c. occupies the space around the nucleus only in certain, well-defined orbitals.
   d. has neither mass nor charge.

8. The unit for molar mass is
   a. atoms per mole.
   b. grams per atoms.
   c. g/mol.
   d. mol/g.

9. Which of the following could represent a pair of isotopes?
   a. $^{20}_{8}X$ and $^{20}_{9}X$ c. $^{32}_{15}X$ and $^{34}_{16}X$
   b. $^{44}_{23}X$ and $^{44}_{23}X$ d. $^{40}_{19}X$ and $^{42}_{19}X$

10. Among the phenomena that enabled scientists to infer the existence of atoms was the
    a. Pauli exclusion principle.
    b. law of conservation of mass.
    c. relationship between mass and energy shown by the equation $E = mc^2$.
    d. observation that each element differed in the number of protons in one of its atoms.

11. An important result of Rutherford's work was to establish that
    a. atoms have mass.
    b. electrons have a negative charge.
    c. gold is an element.
    d. the atom is mostly empty space.

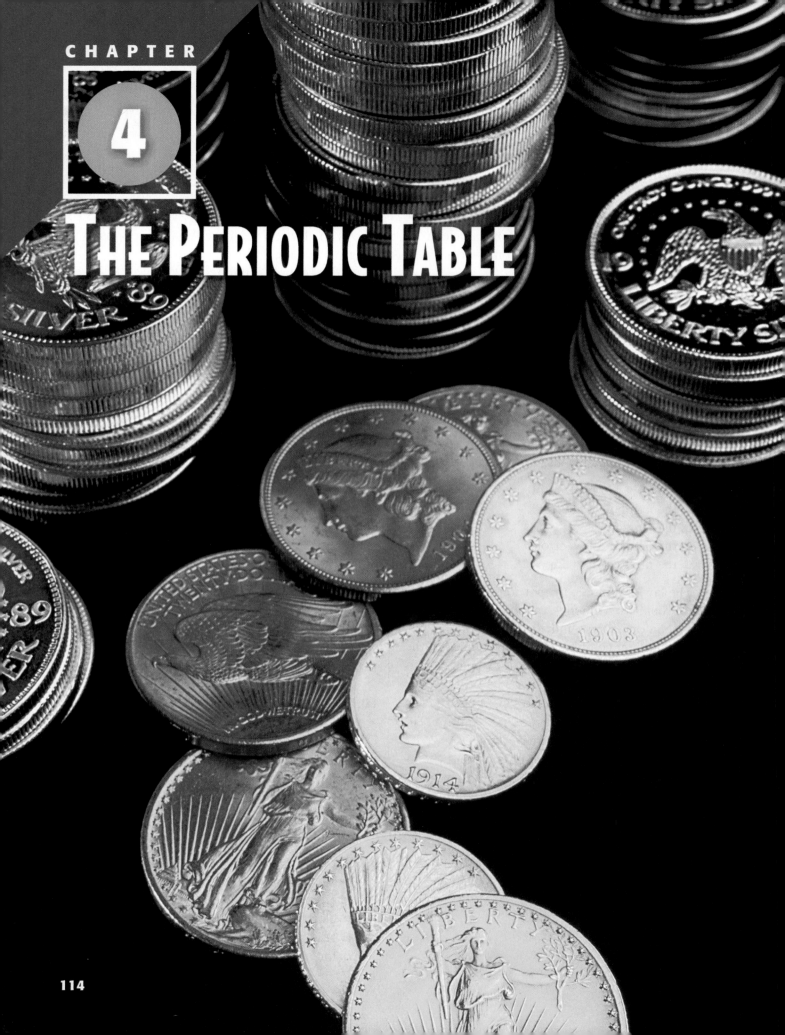

THE PERIODIC TABLE

The United States established its first mint to make silver and gold coins in Philadelphia in 1792. Some of these old gold and silver coins have become quite valuable as collector's items. An 1804 silver dollar recently sold for more than $4 million. A silver dollar is actually 90% silver and 10% copper. Because the pure elements gold and silver are too soft to be used alone in coins, other metals are mixed with them to add strength and durability. These metals include platinum, copper, zinc, and nickel. Metals make up the majority of the elements in the periodic table.

START-UP ACTIVITY

SAFETY PRECAUTIONS

What Is a Periodic Table?

PROCEDURE

1. Sit in your assigned desk according to the seating chart your teacher provides.

2. On the blank chart your teacher gives you, jot down information about yourself—such as name, date of birth, hair color, and height—in the space that represents where you are seated.

3. Find out the same information from as many people sitting around you as possible, and write that information in the corresponding spaces on the seating chart.

ANALYSIS

1. Looking at the information you gathered, try to identify patterns that could explain the order of people in the seating chart. If you cannot yet identify a pattern, collect more information and look again for a pattern.

2. Test your pattern by gathering information from a person you did not talk to before.

3. If the new information does not fit in with your pattern, reevaluate your data to come up with a new hypothesis that explains the patterns in the seating chart.

Pre-Reading Questions

1. Define *element*.

2. What is the relationship between the number of protons and the number of electrons in a neutral atom?

3. As electrons fill orbitals, what patterns do you notice?

internet connect

www.scilinks.org
Topic: The Periodic Table
SciLinks code: HW4094

SC/LINKS. Maintained by the National Science Teachers Association

115

How Are Elements Organized?

KEY TERMS

- **periodic law**
- **valence electron**
- **group**
- **period**

OBJECTIVES

1 **Describe** the historical development of the periodic table.

2 **Describe** the organization of the modern periodic table according to the periodic law.

Topic Link

Refer to the chapter "The Science of Chemistry" for a definition and discussion of elements.

Patterns in Element Properties

Pure elements at room temperature and atmospheric pressure can be solids, liquids, or gases. Some elements are colorless. Others, like the ones shown in **Figure 1,** are colored. Despite the differences between elements, groups of elements share certain properties. For example, the elements lithium, sodium, potassium, rubidium, and cesium can combine with chlorine in a 1:1 ratio to form LiCl, NaCl, KCl, RbCl, and CsCl. All of these compounds are white solids that dissolve in water to form solutions that conduct electricity.

Similarly, the elements fluorine, chlorine, bromine, and iodine can combine with sodium in a 1:1 ratio to form NaF, NaCl, NaBr, and NaI. These compounds are also white solids that can dissolve in water to form solutions that conduct electricity. These examples show that even though each element is different, groups of them have much in common.

Figure 1
The elements chlorine, bromine, and iodine, pictured from left to right, look very different from each other. But each forms a similar-looking white solid when it reacts with sodium.

John Newlands Noticed a Periodic Pattern

Elements vary widely in their properties, but in an orderly way. In 1865, the English chemist John Newlands arranged the known elements according to their properties and in order of increasing atomic mass. He placed the elements in a table.

As he studied his arrangement, Newlands noticed that all of the elements in a given row had similar chemical and physical properties. Because these properties seemed to repeat every eight elements, Newlands called this pattern the *law of octaves*.

This proposed law met with some skepticism when it was first presented, partly because chemists at the time did not know enough about atoms to be able to suggest a physical basis for any such law.

Dmitri Mendeleev Invented the First Periodic Table

In 1869, the Russian chemist Dmitri Mendeleev used Newlands's observation and other information to produce the first orderly arrangement, or periodic table, of all 63 elements known at the time. Mendeleev wrote the symbol for each element, along with the physical and chemical properties and the relative atomic mass of the element, on a card. Like Newlands, Mendeleev arranged the elements in order of increasing atomic mass. Mendeleev started a new row each time he noticed that the chemical properties of the elements repeated. He placed elements in the new row directly below elements of similar chemical properties in the preceding row. He arrived at the pattern shown in **Figure 2.**

Two interesting observations can be made about Mendeleev's table. First, Mendeleev's table contains gaps that elements with particular properties should fill. He predicted the properties of the missing elements.

ПЕРИОДИЧЕСКАЯ СИСТЕМА ЭЛЕМЕНТОВ

ГРУППЫ ЭЛЕМЕНТОВ

РЯДЫ	I	II	III	IV	V	VI	VII	VIII		
I	H 1									
II	Li 7	Be 9,4	B 11	C 12	N 14	O 16	F 19			
III	Na 23	Mg 24	Al 27,4	Si 28	P 31	S 32	Cl 35,5			
IV	K 39	Ca 40	? 45	Ti 50	V 51	Cr 52	Mn 55	Fe 56	Co 59	Ni 59
V	Cu 63,4	Zn 65,2	? 68	? 70	As 75	Se 79,4	Br 80			
VI	Rb 85,4	Sr 87,6	Yt ? 88	Zr 90	Nb 94	Mo 96	? 100	Ru 104,4	Rh 104,4	Pd 106,6
VII	Ag 108	Cd 112	In 113	Sn 118	Sb 122	Te 128?	J 127			
VIII	Cs 133	Ba 137	Di ? 138	Ce ? 140						
IX										
X			Er ? 178	La ? 180	Ta 182	W 186		Pt 197,4	Ir 198	Os 199
XI	Au 197?	Hg 200	Tl 204	Pb 207	Bi 210					
XII				Th 231		U 240				

Figure 2

Mendeleev's table grouped elements with similar properties into vertical columns. For example, he placed the elements highlighted in red in the table—fluorine, chlorine, bromine, and iodine—into the column that he labeled "VII."

Table 1 Predicted Versus Actual Properties for Three Elements

Properties	Ekaaluminum (gallium, discovered 1875)		Ekaboron (scandium, discovered 1877)		Ekasilicon (germanium, discovered 1886)	
	Predicted	Observed	Predicted	Observed	Predicted	Observed
Density	6.0 g/cm^3	5.96 g/cm^3	3.5 g/cm^3	3.5 g/cm^3	5.5 g/cm^3	5.47 g/cm^3
Melting point	low	30°C	*	*	high	900°C
Formula of oxide	Ea_2O_3	Ga_2O_3	Eb_2O_3	Sc_2O_3	EsO_2	GeO_2
Solubility of oxide	*	*	dissolves in acid	dissolves in acid	*	*
Density of oxide	*	*	*	*	4.7 g/cm^3	4.70 g/cm^3
Formula of chloride	*	*	*	*	$EsCl_4$	$GeCl_4$
Color of metal	*	*	*	*	dark gray	grayish white

He also gave these elements provisional names, such as "Ekaaluminum" (the prefix *eka-* means "one beyond") for the element that would come below aluminum. These elements were eventually discovered. As **Table 1** illustrates, their properties were close to Mendeleev's predictions. Although other chemists, such as Newlands, had created tables of the elements, Mendeleev was the first to use the table to predict the existence of undiscovered elements. Because Mendeleev's predictions proved true, most chemists accepted his periodic table of the elements.

Second, the elements do not always fit neatly in order of atomic mass. For example, Mendeleev had to switch the order of tellurium, Te, and iodine, I, to keep similar elements in the same column. At first, he thought that their atomic masses were wrong. However, careful research by others showed that they were correct. Mendeleev could not explain why his order was not always the same.

The Physical Basis of the Periodic Table

About 40 years after Mendeleev published his periodic table, an English chemist named Henry Moseley found a different physical basis for the arrangement of elements. When Moseley studied the lines in the X-ray spectra of 38 different elements, he found that the wavelengths of the lines in the spectra decreased in a regular manner as atomic mass increased. With further work, Moseley realized that the spectral lines correlated to atomic number, not to atomic mass.

When the elements were arranged by increasing atomic number, the discrepancies in Mendeleev's table disappeared. Moseley's work led to both the modern definition of atomic number, and showed that atomic number, not atomic mass, is the basis for the organization of the periodic table.

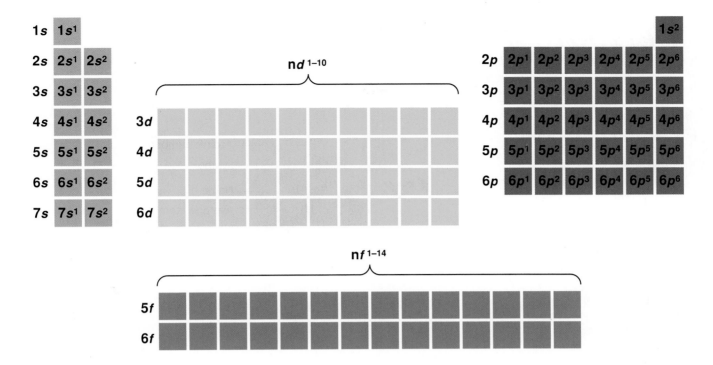

Figure 3
The shape of the periodic table is determined by how electrons fill orbitals. Only the *s* and *p* electrons are shown individually because unlike the *d* and *f* electrons, they fill orbitals sequentially.

The Periodic Law

According to Moseley, tellurium, whose atomic number is 52, belongs before iodine, whose atomic number is 53. Mendeleev had placed these elements in the same order based on their properties. Today, Mendeleev's principle of chemical periodicity is known as the **periodic law,** which states that when the elements are arranged according to their atomic numbers, elements with similar properties appear at regular intervals.

Organization of the Periodic Table

To understand why elements with similar properties appear at regular intervals in the periodic table, you need to examine the electron configurations of the elements. As shown in **Figure 3,** elements in each column of the table have the same number of electrons in their outer energy level. These electrons are called **valence electrons.** It is the valence electrons of an atom that participate in chemical reactions with other atoms, so elements with the same number of valence electrons tend to react in similar ways. Because *s* and *p* electrons fill sequentially, the number of valence electrons in *s*- and *p*-block elements are predictable. For example, atoms of elements in the column on the far left have one valence electron. Atoms of elements in the column on the far right have eight valence electrons (except for helium, which has two). A vertical column on the periodic table is called a **group.** A complete version of the modern periodic table is shown in **Figure 4** on the next two pages.

periodic law

the law that states that the repeating physical and chemical properties of elements change periodically with their atomic number

valence electron

an electron that is found in the outermost shell of an atom and that determines the atom's chemical properties

group

a vertical column of elements in the periodic table; elements in a group share chemical properties

 Topic Link

Refer to the chapter "Atoms and Moles" for a discussion of electron configuration.

Periodic Table of the Elements

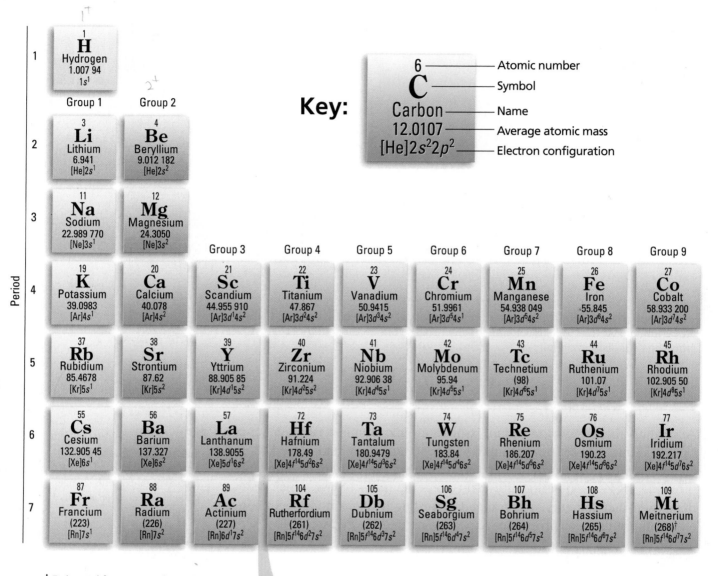

Key:

6	Atomic number
C	Symbol
Carbon	Name
12.0107	Average atomic mass
[He]$2s^2 2p^2$	Electron configuration

Period

Group 1

1	
H	
Hydrogen	
1.007 94	
$1s^1$	

Group 2

Period 2

3	4
Li	**Be**
Lithium	Beryllium
6.941	9.012 182
[He]$2s^1$	[He]$2s^2$

Period 3

11	12
Na	**Mg**
Sodium	Magnesium
22.989 770	24.3050
[Ne]$3s^1$	[Ne]$3s^2$

Group 3 — Group 4 — Group 5 — Group 6 — Group 7 — Group 8 — Group 9

Period 4

19	20	21	22	23	24	25	26	27
K	**Ca**	**Sc**	**Ti**	**V**	**Cr**	**Mn**	**Fe**	**Co**
Potassium	Calcium	Scandium	Titanium	Vanadium	Chromium	Manganese	Iron	Cobalt
39.0983	40.078	44.955 910	47.867	50.9415	51.9961	54.938 049	55.845	58.933 200
[Ar]$4s^1$	[Ar]$4s^2$	[Ar]$3d^1 4s^2$	[Ar]$3d^2 4s^2$	[Ar]$3d^3 4s^2$	[Ar]$3d^5 4s^1$	[Ar]$3d^5 4s^2$	[Ar]$3d^6 4s^2$	[Ar]$3d^7 4s^2$

Period 5

37	38	39	40	41	42	43	44	45
Rb	**Sr**	**Y**	**Zr**	**Nb**	**Mo**	**Tc**	**Ru**	**Rh**
Rubidium	Strontium	Yttrium	Zirconium	Niobium	Molybdenum	Technetium	Ruthenium	Rhodium
85.4678	87.62	88.905 85	91.224	92.906 38	95.94	(98)	101.07	102.905 50
[Kr]$5s^1$	[Kr]$5s^2$	[Kr]$4d^1 5s^2$	[Kr]$4d^2 5s^2$	[Kr]$4d^4 5s^1$	[Kr]$4d^5 5s^1$	[Kr]$4d^6 5s^1$	[Kr]$4d^7 5s^1$	[Kr]$4d^8 5s^1$

Period 6

55	56	57	72	73	74	75	76	77
Cs	**Ba**	**La**	**Hf**	**Ta**	**W**	**Re**	**Os**	**Ir**
Cesium	Barium	Lanthanum	Hafnium	Tantalum	Tungsten	Rhenium	Osmium	Iridium
132.905 45	137.327	138.9055	178.49	180.9479	183.84	186.207	190.23	192.217
[Xe]$6s^1$	[Xe]$6s^2$	[Xe]$5d^1 6s^2$	[Xe]$4f^{14} 5d^2 6s^2$	[Xe]$4f^{14} 5d^3 6s^2$	[Xe]$4f^{14} 5d^4 6s^2$	[Xe]$4f^{14} 5d^5 6s^2$	[Xe]$4f^{14} 5d^6 6s^2$	[Xe]$4f^{14} 5d^7 6s^2$

Period 7

87	88	89	104	105	106	107	108	109
Fr	**Ra**	**Ac**	**Rf**	**Db**	**Sg**	**Bh**	**Hs**	**Mt**
Francium	Radium	Actinium	Rutherfordium	Dubnium	Seaborgium	Bohrium	Hassium	Meitnerium
(223)	(226)	(227)	(261)	(262)	(263)	(264)	(265)	(268)†
[Rn]$7s^1$	[Rn]$7s^2$	[Rn]$6d^1 7s^2$	[Rn]$5f^{14} 6d^2 7s^2$	[Rn]$5f^{14} 6d^3 7s^2$	[Rn]$5f^{14} 6d^4 7s^2$	[Rn]$5f^{14} 6d^5 7s^2$	[Rn]$5f^{14} 6d^6 7s^2$	[Rn]$5f^{14} 6d^7 7s^2$

† Estimated from currently available IUPAC data.

* The systematic names and symbols for elements greater than 109 will be used until the approval of trivial names by IUPAC.

58	59	60	61	62
Ce	**Pr**	**Nd**	**Pm**	**Sm**
Cerium	Praseodymium	Neodymium	Promethium	Samarium
140.116	140.907 65	144.24	(145)	150.36
[Xe]$4f^1 5d^1 6s^2$	[Xe]$4f^3 6s^2$	[Xe]$4f^4 6s^2$	[Xe]$4f^5 6s^2$	[Xe]$4f^6 6s^2$

90	91	92	93	94
Th	**Pa**	**U**	**Np**	**Pu**
Thorium	Protactinium	Uranium	Neptunium	Plutonium
232.0381	231.035 88	238.0289	(237)	(244)
[Rn]$6d^2 7s^2$	[Rn]$5f^2 6d^1 7s^2$	[Rn]$5f^3 6d^1 7s^2$	[Rn]$5f^4 6d^1 7s^2$	[Rn]$5f^6 7s^2$

internet connect

Topic: Periodic Table
Go To: go.hrw.com
Keyword: HN0 PERIODIC

Visit the HRW Web site for updates on the periodic table.

Figure 4

Metals
- Alkali metals
- Alkaline-earth metals
- Transition metals
- Other metals

Nonmetals
- Hydrogen
- Semiconductors (also known as *metalloids*)
- Other nonmetals
- Halogens
- Noble gases

Group 18

| 2 | He | Helium | 4.002 602 | $1s^2$ |

Group 13	Group 14	Group 15	Group 16	Group 17
5 **B** Boron 10.811 $[He]2s^22p^1$	6 **C** Carbon 12.0107 $[He]2s^22p^2$	7 **N** Nitrogen 14.006 74 $[He]2s^22p^3$	8 **O** Oxygen 15.9994 $[He]2s^22p^4$	9 **F** Fluorine 18.998 4032 $[He]2s^22p^5$	10 **Ne** Neon 20.1797 $[He]2s^22p^6$
13 **Al** Aluminum 26.981 538 $[Ne]3s^23p^1$	14 **Si** Silicon 28.0855 $[Ne]3s^23p^2$	15 **P** Phosphorus 30.973 761 $[Ne]3s^23p^3$	16 **S** Sulfur 32.066 $[Ne]3s^23p^4$	17 **Cl** Chlorine 35.4527 $[Ne]3s^23p^5$	18 **Ar** Argon 39.948 $[Ne]3s^23p^6$

Group 10	Group 11	Group 12					
28 **Ni** Nickel 58.6934 $[Ar]3d^84s^2$	29 **Cu** Copper 63.546 $[Ar]3d^{10}4s^1$	30 **Zn** Zinc 65.39 $[Ar]3d^{10}4s^2$	31 **Ga** Gallium 69.723 $[Ar]3d^{10}4s^24p^1$	32 **Ge** Germanium 72.61 $[Ar]3d^{10}4s^24p^2$	33 **As** Arsenic 74.921 60 $[Ar]3d^{10}4s^24p^3$	34 **Se** Selenium 78.96 $[Ar]3d^{10}4s^24p^4$	35 **Br** Bromine 79.904 $[Ar]3d^{10}4s^24p^5$	36 **Kr** Krypton 83.80 $[Ar]3d^{10}4s^24p^6$
46 **Pd** Palladium 106.42 $[Kr]4d^{10}5s^0$	47 **Ag** Silver 107.8682 $[Kr]4d^{10}5s^1$	48 **Cd** Cadmium 112.411 $[Kr]4d^{10}5s^2$	49 **In** Indium 114.818 $[Kr]4d^{10}5s^25p^1$	50 **Sn** Tin 118.710 $[Kr]4d^{10}5s^25p^2$	51 **Sb** Antimony 121.760 $[Kr]4d^{10}5s^25p^3$	52 **Te** Tellurium 127.60 $[Kr]4d^{10}5s^25p^4$	53 **I** Iodine 126.904 47 $[Kr]4d^{10}5s^25p^5$	54 **Xe** Xenon 131.29 $[Kr]4d^{10}5s^25p^6$
78 **Pt** Platinum 195.078 $[Xe]4f^{14}5d^96s^1$	79 **Au** Gold 196.966 55 $[Xe]4f^{14}5d^{10}6s^1$	80 **Hg** Mercury 200.59 $[Xe]4f^{14}5d^{10}6s^2$	81 **Tl** Thallium 204.3833 $[Xe]4f^{14}5d^{10}6s^26p^1$	82 **Pb** Lead 207.2 $[Xe]4f^{14}5d^{10}6s^26p^2$	83 **Bi** Bismuth 208.980 38 $[Xe]4f^{14}5d^{10}6s^26p^3$	84 **Po** Polonium (209) $[Xe]4f^{14}5d^{10}6s^26p^4$	85 **At** Astatine (210) $[Xe]4f^{14}5d^{10}6s^26p^5$	86 **Rn** Radon (222) $[Xe]4f^{14}5d^{10}6s^26p^6$
110 **Uun*** Ununnilium (269)† $[Rn]5f^{14}6d^97s^1$	111 **Uuu*** Unununium (272)† $[Rn]5f^{14}6d^{10}7s^1$	112 **Uub*** Ununbium (277)† $[Rn]5f^{14}6d^{10}7s^2$		114 **Uuq*** Ununquadium (285)† $[Rn]5f^{14}6d^{10}7s^27p^2$				

A team at Lawrence Berkeley National Laboratories reported the discovery of elements 116 and 118 in June 1999. The same team retracted the discovery in July 2001. The discovery of element 114 has been reported but not confirmed.

63 **Eu** Europium 151.964 $[Xe]4f^76s^2$	64 **Gd** Gadolinium 157.25 $[Xe]4f^75d^16s^2$	65 **Tb** Terbium 158.925 34 $[Xe]4f^96s^2$	66 **Dy** Dysprosium 162.50 $[Xe]4f^{10}6s^2$	67 **Ho** Holmium 164.930 32 $[Xe]4f^{11}6s^2$	68 **Er** Erbium 167.26 $[Xe]4f^{12}6s^2$	69 **Tm** Thulium 168.934 21 $[Xe]4f^{13}6s^2$	70 **Yb** Ytterbium 173.04 $[Xe]4f^{14}6s^2$	71 **Lu** Lutetium 174.967 $[Xe]4f^{14}5d^16s^2$
95 **Am** Americium (243) $[Rn]5f^77s^2$	96 **Cm** Curium (247) $[Rn]5f^76d^17s^2$	97 **Bk** Berkelium (247) $[Rn]5f^97s^2$	98 **Cf** Californium (251) $[Rn]5f^{10}7s^2$	99 **Es** Einsteinium (252) $[Rn]5f^{11}7s^2$	100 **Fm** Fermium (257) $[Rn]5f^{12}7s^2$	101 **Md** Mendelevium (258) $[Rn]5f^{13}7s^2$	102 **No** Nobelium (259) $[Rn]5f^{14}7s^2$	103 **Lr** Lawrencium (262) $[Rn]5f^{14}6d^17s^2$

The atomic masses listed in this table reflect the precision of current measurements. (Values listed in parentheses are those of the element's most stable or most common isotope.) In calculations throughout the text, however, atomic masses have been rounded to two places to the right of the decimal.

period

a horizontal row of elements in the periodic table

A horizontal row on the periodic table is called a **period.** Elements in the same period have the same number of occupied energy levels. For example, all elements in Period 2 have atoms whose electrons occupy two principal energy levels, including the 2*s* and 2*p* orbitals. Elements in Period 5 have outer electrons that fill the 5*s*, 5*d*, and 5*p* orbitals.

This correlation between period number and the number of occupied energy levels holds for all seven periods. So a periodic table is not needed to tell to which period an element belongs. All you need to know is the element's electron configuration. For example, germanium has the electron configuration $[Ar]3d^{10}4s^24p^2$. The largest principal quantum number it has is 4, which means germanium has four occupied energy levels. This places it in Period 4.

The periodic table provides information about each element, as shown in the key for **Figure 4.** This periodic table lists the atomic number, symbol, name, average atomic mass, and electron configuration in shorthand form for each element.

In addition, some of the categories of elements are designated through a color code. You may notice that many of the color-coded categories shown in **Figure 4** are associated with a certain group or groups. This shows how categories of elements are grouped by common properties which result from their common number of valence electrons. The next section discusses the different kinds of elements on the periodic table and explains how their electron configurations give them their characteristic properties.

Section Review

UNDERSTANDING KEY IDEAS

1. How can one show that elements that have different appearances have similar chemical properties?

2. Why was the pattern that Newlands developed called the *law of octaves*?

3. What led Mendeleev to predict that some elements had not yet been discovered?

4. What contribution did Moseley make to the development of the modern periodic table?

5. State the periodic law.

6. What do elements in the same period have in common?

7. What do elements in the same group have in common?

CRITICAL THINKING

8. Why can Period 1 contain a maximum of two elements?

9. In which period and group is the element whose electron configuration is $[Kr]5s^1$?

10. Write the outer electron configuration for the Group 2 element in Period 6.

11. What determines the number of elements found in each period in the periodic table?

12. Are elements with similar chemical properties more likely to be found in the same period or in the same group? Explain your answer.

13. How many valence electrons does phosphorus have?

14. What would you expect the electron configuration of element 113 to be?

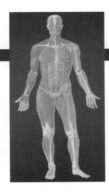

CONSUMER FOCUS

Essential Elements

Four elements—hydrogen, oxygen, carbon, and nitrogen—account for more than 99% of all atoms in the human body.

Good Health Is Elementary

Hydrogen, oxygen, carbon, and nitrogen are the major components of the many different molecules that our bodies need. Likewise, these elements are the major elements in the molecules of the food that we eat.

Another seven elements, listed in **Table 2,** are used by our bodies in substantial quantities, more than 0.1 g per day. These elements are known as *macronutrients* or, more commonly, as *minerals*.

Some elements, known as *trace elements* or *micronutrients,* are necessary for healthy human life, but only in very small amounts. In many cases, humans need less than 15 nanograms, or 15×10^{-9} g, of a particular trace element per day to maintain good health. This means that you need less than 0.0004 g of such trace elements during your entire lifetime!

Table 2   Macronutrients

Element	Symbol	Role in human body chemistry
Calcium	Ca	bones, teeth; essential for blood clotting and muscle contraction
Phosphorus	P	bones, teeth; component of nucleic acids, including DNA
Potassium	K	present as K^+ in all body fluids; essential for nerve action
Sulfur	S	component of many proteins; essential for blood clotting
Chlorine	Cl	present as Cl^- in all body fluids; important to maintaining salt balance
Sodium	Na	present as Na^+ in all body fluids; essential for nerve and muscle action
Magnesium	Mg	in bones and teeth; essential for muscle action

Questions

1. What do the two macronutrients involved in nerve action have in common?
2. You may recognize elements such as arsenic and lead as toxic. Explain how these elements can be nutrients even though they are toxic.

Elements in organic matter

Macronutrients

Trace elements

Group 18

1 H

Group 1 Group 2

2 He

Group 13 Group 14 Group 15 Group 16 Group 17

3 Li	4 Be

5 B	6 C	7 N	8 O	9 F	10 Ne

11 Na	12 Mg

Group 3 Group 4 Group 5 Group 6 Group 7 Group 8 Group 9 Group 10 Group 11 Group 12

13 Al	14 Si	15 P	16 S	17 Cl	18 Ar

19 K	20 Ca	21 Sc	22 Ti	23 V	24 Cr	25 Mn	26 Fe	27 Co	28 Ni	29 Cu	30 Zn	31 Ga	32 Ge	33 As	34 Se	35 Br	36 Kr

37 Rb	38 Sr	39 Y	40 Zr	41 Nb	42 Mo	43 Tc	44 Ru	45 Rh	46 Pd	47 Ag	48 Cd	49 In	50 Sn	51 Sb	52 Te	53 I	54 Xe

55 Cs	56 Ba	57 La	72 Hf	73 Ta	74 W	75 Re	76 Os	77 Ir	78 Pt	79 Au	80 Hg	81 Tl	82 Pb	83 Bi	84 Po	85 At	86 Rn

87 Fr	88 Ra	89 Ac

Tour of the Periodic Table

KEY TERMS

- **main-group element**
- **alkali metal**
- **alkaline-earth metal**
- **halogen**
- **noble gas**
- **transition metal**
- **lanthanide**
- **actinide**
- **alloy**

main-group elements

an element in the *s*-block or *p*-block of the periodic table

OBJECTIVES

① **Locate** the different families of main-group elements on the periodic table, describe their characteristic properties, and relate their properties to their electron configurations.

② **Locate** metals on the periodic table, describe their characteristic properties, and relate their properties to their electron configurations.

The Main-Group Elements

Elements in groups 1, 2, and 13–18 are known as the **main-group elements.** As shown in **Figure 5,** main-group elements are in the *s*- and *p*-blocks of the periodic table. The electron configurations of the elements in each main group are regular and consistent: the elements in each group have the same number of valence electrons. For example, Group 2 elements have two valence electrons. The configuration of their valence electrons can be written as ns^2, where n is the period number. Group 16 elements have a total of six valence electrons in their outermost s and p orbitals. Their valence electron configuration can be written as ns^2np^4.

The main-group elements are sometimes called the *representative elements* because they have a wide range of properties. At room temperature and atmospheric pressure, many are solids, while others are liquids or gases. About half of the main-group elements are metals. Many are extremely reactive, while several are nonreactive. The main-group elements silicon and oxygen account for four of every five atoms found on or near Earth's surface.

Four groups within the main-group elements have special names. These groups are the *alkali metals* (Group 1), the *alkaline-earth metals* (Group 2), the *halogens* (Group 17), and the *noble gases* (Group 18).

Figure 5
Main-group elements have diverse properties and uses. They are highlighted in the groups on the left and right sides of the periodic table.

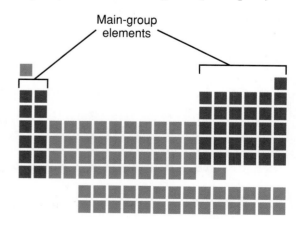

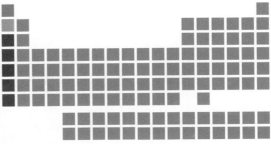

Figure 6
The alkali metals make up the first group of the periodic table. Lithium, pictured here, is an example of an alkali metal.

The Alkali Metals Make Up Group 1

Elements in Group 1, which is highlighted in **Figure 6,** are called **alkali metals.** Alkali metals are so named because they are metals that react with water to make alkaline solutions. For example, potassium reacts vigorously with cold water to form hydrogen gas and the compound potassium hydroxide, KOH. Because the alkali metals have a single valence electron, they are very reactive. In losing its one valence electron, potassium achieves a stable electron configuration.

Alkali metals are usually stored in oil to keep them from reacting with the oxygen and water in the air. Because of their high reactivity, alkali metals are never found in nature as pure elements but are found combined with other elements as compounds. For instance, the salt sodium chloride, NaCl, is abundant in sea water.

Some of the physical properties of the alkali metals are listed in **Table 3.** All these elements are so soft that they can be easily cut with a knife. The freshly cut surface of an alkali metal is shiny, but it dulls quickly as the metal reacts with oxygen and water in the air. Like other metals, the alkali metals are good conductors of electricity.

alkali metal

one of the elements of Group 1 of the periodic table (lithium, sodium, potassium, rubidium, cesium, and francium)

internet connect

www.scilinks.org
Topic: Alkali Metals
SciLinks code: HW4007

SC*LINKS* Maintained by the National Science Teachers Association

Table 3 Physical Properties of Alkali Metals

Element	Flame test	Hardness (Mohs' scale)	Melting Point (°C)	Boiling Point (°C)	Density (g/cm^3)	Atomic radius (pm)
Lithium	red	0.6	180.5	1342	0.53	134
Sodium	yellow	0.4	97.7	883	0.97	154
Potassium	violet	0.5	63.3	759	0.86	196
Rubidium	yellowish violet	0.3	39.3	688	1.53	(216)
Cesium	reddish violet	0.2	28.4	671	1.87	(233)

Refer to Appendix A for more information about the properties of elements, including alkali metals.

Figure 7
The alkaline-earth metals make up the second group of the periodic table. Magnesium, pictured here, is an example of an alkaline-earth metal.

alkaline-earth metal

one of the elements of Group 2 of the periodic table (beryllium, magnesium, calcium, strontium, barium, and radium)

internet connect

www.scilinks.org
Topic: Alkaline-Earth Metals
SciLinks code: HW4008

SCiLINKS. Maintained by the National Science Teachers Association

The Alkaline-Earth Metals Make Up Group 2

Group 2 elements, which are highlighted in **Figure 7,** are called **alkaline-earth metals.** Like the alkali metals, the alkaline-earth metals are highly reactive, so they are usually found as compounds rather than as pure elements. For example, if the surface of an object made from magnesium is exposed to the air, the magnesium will react with the oxygen in the air to form the compound magnesium oxide, MgO, which eventually coats the surface of the magnesium metal.

The alkaline-earth metals are slightly less reactive than the alkali metals. The alkaline-earth metals have two valence electrons and must lose both their valence electrons to get to a stable electron configuration. It takes more energy to lose two electrons than it takes to lose just the one electron that the alkali metals must give up to become stable. Although the alkaline-earth metals are not as reactive, they are harder and have higher melting points than the alkali metals.

Beryllium is found in emeralds, which are a variety of the mineral beryl. Perhaps the best-known alkaline-earth metal is calcium, an important mineral nutrient found in the human body. Calcium is essential for muscle contraction. Bones are made up of calcium phosphate. Calcium compounds, such as limestone and marble, are common in the Earth's crust. Marble is made almost entirely of pure calcium carbonate. Because marble is hard and durable, it is used in sculptures.

halogen

one of the elements of Group 17 of the periodic table (fluorine, chlorine, bromine, iodine, and astatine); halogens combine with most metals to form salts

internet connect

www.scilinks.org
Topic: Halogens
SciLinks code: HW4065

SCiLINKS. Maintained by the National Science Teachers Association

The Halogens, Group 17, Are Highly Reactive

Elements in Group 17 of the periodic table, which are highlighted in **Figure 8** on the next page, are called the **halogens.** The halogens are the most reactive group of nonmetal elements because of their electron configuration. Halogens have seven valence electrons—just one short of a stable configuration. When halogens react, they often gain the one electron needed to have eight valence electrons, a filled outer energy level. Because the alkali metals have one valence electron, they are ideally suited to react with the halogens. For example, the alkali metal sodium easily loses its one valence electron to the halogen chlorine to form the compound sodium chloride, $NaCl$, which is table salt. The halogens react with most metals to produce salts. In fact, the word *halogen* comes from Greek and means "salt maker."

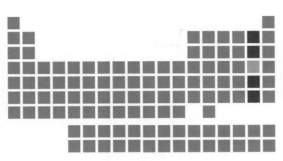

Figure 8
The halogens make up Group 17 of the periodic table. Bromine, one of only two elements that are liquids at room temperature, is an example of a halogen.

The halogens have a wide range of physical properties. Fluorine and chlorine are gases at room temperature, but bromine, depicted in **Figure 8,** is a liquid, and iodine and astatine are solids. The halogens are found in sea water and in compounds found in the rocks of Earth's crust. Astatine is one of the rarest of the naturally occurring elements.

The Noble Gases, Group 18, Are Unreactive

Group 18 elements, which are highlighted in **Figure 9,** are called the **noble gases.** The noble gas atoms have a full set of electrons in their outermost energy level. Except for helium $(1s^2)$, noble gases have an outer-shell configuration of ns^2np^6. From the low chemical reactivity of these elements, chemists infer that this full shell of electrons makes these elements very stable. The low reactivity of noble gases leads to some special uses. Helium, a noble gas, is used to fill blimps because it has a low density and is not flammable.

The noble gases were once called *inert gases* because they were thought to be completely unreactive. But in 1962, chemists were able to get xenon to react, making the compound $XePtF_6$. In 1979, chemists were able to form the first xenon-carbon bonds.

noble gas

an unreactive element of Group 18 of the periodic table (helium, neon, argon, krypton, xenon, or radon) that has eight electrons in its outer level (except for helium, which has two electrons)

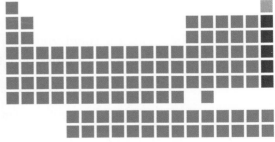

Figure 9
The noble gases make up Group 18 of the periodic table. Helium, whose low density makes it ideal for use in blimps, is an example of a noble gas.

The Periodic Table **127**

Hydrogen Is in a Class by Itself

Hydrogen is the most common element in the universe. It is estimated that about three out of every four atoms in the universe are hydrogen. Because it consists of just one proton and one electron, hydrogen behaves unlike any other element. As shown in **Figure 10,** hydrogen is in a class by itself in the periodic table.

With its one electron, hydrogen can react with many other elements, including oxygen. Hydrogen gas and oxygen gas react explosively to form water. Hydrogen is a component of the organic molecules found in all living things. The main industrial use of hydrogen is in the production of ammonia, NH_3. Large quantities of ammonia are used to make fertilizers.

Most Elements Are Metals

Figure 11 shows that the majority of elements, including many main-group ones, are metals. But what exactly is a metal? You can often recognize a metal by its shiny appearance, but some nonmetal elements, plastics, and minerals are also shiny. For example, a diamond usually has a brilliant luster. However, diamond is a mineral made entirely of the nonmetal element carbon.

Conversely, some metals appear black and dull. An example is iron, which is a very strong and durable metal. Iron is a member of Group 8 and is therefore not a main-group element. Iron belongs to a class of elements called *transition metals.* However, wherever metals are found on the periodic table, they tend to share certain properties.

Figure 11
The regions highlighted in blue indicate the elements that are metals.

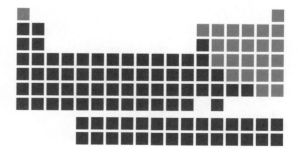

Metals Share Many Properties

All metals are excellent conductors of electricity. Electrical conductivity is the one property that distinguishes metals from the nonmetal elements. Even the least conductive metal conducts electricity 100 000 times better than the best nonmetallic conductor does.

Metals also exhibit other properties, some of which can also be found in certain nonmetal elements. For example, metals are excellent conductors of heat. Some metals, such as manganese and bismuth, are very brittle. Other metals, such as gold and copper, are ductile and malleable. *Ductile* means that the metal can be squeezed out into a wire. *Malleable* means that the metal can be hammered or rolled into sheets. Gold, for example, can be hammered into very thin sheets, called "gold leaf," and applied to objects for decoration.

Transition Metals Occupy the Center of the Periodic Table

The **transition metals** constitute Groups 3 through 12 and are sometimes called the *d*-block elements because of their position in the periodic table, shown in **Figure 12.** Unlike the main-group elements, the transition metals in each group do not have identical outer electron configurations. For example, nickel, Ni, palladium, Pd, and platinum, Pt, are Group 10 metals. However, Ni has the electron configuration $[Ar]3d^84s^2$, Pd has the configuration $[Kr]4d^{10}$, and Pt has the configuration $[Xe]4f^{14}5d^96s^1$. Notice, however, that in each case the sum of the outer *d* and *s* electrons is equal to the group number, 10.

A transition metal may lose different numbers of valence electrons depending on the element with which it reacts. Generally, the transition metals are less reactive than the alkali metals and the alkaline-earth metals are. In fact, some transition metals are so unreactive that they seldom form compounds with other elements. Palladium, platinum, and gold are among the least reactive of all the elements other than the noble gases. These three transition metals can be found in nature as pure elements.

Transition metals, like other metals, are good conductors of heat and electricity. They are also ductile and malleable, as shown in **Figure 12.**

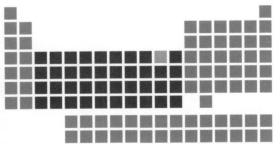

Figure 12
Copper, a transition metal, is used in wiring because it conducts electricity well. Because of its ductility and malleability, it can be formed into wires that bend easily.

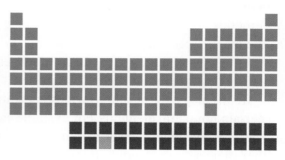

Figure 13
The lanthanides and actinides are placed at the bottom of the periodic table. Uranium, an actinide, is used in nuclear reactors. The collection of uranium-238 kernels is shown here.

Lanthanides and Actinides Fill *f*-orbitals

Part of the last two periods of transition metals are placed toward the bottom of the periodic table to keep the table conveniently narrow, as shown in **Figure 13.** The elements in the first of these rows are called the **lanthanides** because their atomic numbers follow the element lanthanum. Likewise, elements in the row below the lanthanides are called **actinides** because they follow actinium. As one moves left to right along these rows, electrons are added to the 4*f* orbitals in the lanthanides and to the 5*f* orbitals in the actinides. For this reason, the lanthanides and actinides are sometimes called the *f*-block of the periodic table.

The lanthanides are shiny metals similar in reactivity to the alkaline-earth metals. Some lanthanides have practical uses. Compounds of some lanthanide metals are used to produce color television screens.

The actinides are unique in that their nuclear structures are more important than their electron configurations. Because the nuclei of actinides are unstable and spontaneously break apart, all actinides are radioactive. The best-known actinide is uranium.

Other Properties of Metals

The melting points of metals vary widely. Tungsten has the highest melting point, 4322°C, of any element. In contrast, mercury melts at –39°C, so it is a liquid at room temperature. This low melting point, along with its high density, makes mercury useful for barometers.

Metals can be mixed with one or more other elements, usually other metals, to make an **alloy.** The mixture of elements in an alloy gives the alloy properties that are different from the properties of the individual elements. Often these properties eliminate some disadvantages of the pure metal. A common alloy is brass, a mixture of copper and zinc, which is harder than copper and more resistant to corrosion. Brass has a wide range of uses, from inexpensive jewelry to plumbing hardware. Another alloy made from copper is sterling silver. A small amount of copper is mixed with silver to produce sterling silver, which is used for both jewelry and flatware.

lanthanide

a member of the rare-earth series of elements, whose atomic numbers range from 58 (cerium) to 71 (lutetium)

actinide

any of the elements of the actinide series, which have atomic numbers from 89 (actinium, Ac) through 103 (lawrencium, Lr)

alloy

a solid or liquid mixture of two or more metals

Figure 14
Steel is an alloy made of
iron and carbon. When
heated, steel can be worked
into many useful shapes.

Many iron alloys, such as the steel shown in **Figure 14,** are harder,
stronger, and more resistant to corrosion than pure iron. Steel contains
between 0.2% and 1.5% carbon atoms and often has tiny amounts of
other elements such as manganese and nickel. Stainless steel also incor-
porates chromium. Because of its hardness and resistance to corrosion,
stainless steel is an ideal alloy for making knives and other tools.

② Section Review

UNDERSTANDING KEY IDEAS

1. Which group of elements is the most unreactive? Why?

2. Why do groups among the main-group
elements display similar chemical behavior?

3. What properties do the halogens have in
common?

4. Why is hydrogen set apart by itself?

5. How do the valence electron configurations
of the alkali metals compare with each
other?

6. Why are the alkaline-earth metals less
reactive than the alkali metals?

7. In which groups of the periodic table do
the transition metals belong?

8. Why are the nuclear structures of the
actinides more important than the electron
configurations of the actinides?

9. What is an alloy?

CRITICAL THINKING

10. Noble gases used to be called *inert gases.*
What discovery changed that term, and
why?

11. If you find an element in nature in its pure
elemental state, what can you infer about
the element's chemical reactivity?

12. Explain why the transition metals are
sometimes referred to as the *d*-block
elements.

13. Can an element that conducts heat, is
malleable, and has a high melting point be
classified as a metal? Explain your reasoning.

Trends in the Periodic Table

KEY TERMS

- **ionization energy**
- **electron shielding**
- **bond radius**
- **electronegativity**

OBJECTIVES

1. **Describe** periodic trends in ionization energy, and relate them to the atomic structures of the elements.

2. **Describe** periodic trends in atomic radius, and relate them to the atomic structures of the elements.

3. **Describe** periodic trends in electronegativity, and relate them to the atomic structures of the elements.

4. **Describe** periodic trends in ionic size, electron affinity, and melting and boiling points, and relate them to the atomic structures of the elements.

Periodic Trends

The arrangement of the periodic table reveals trends in the properties of the elements. A *trend* is a predictable change in a particular direction. For example, there is a trend in the reactivity of the alkali metals as you move down Group 1. As **Figure 15** illustrates, each of the alkali metals reacts with water. However, the reactivity of the alkali metals varies. At the top of Group 1, lithium is the least reactive, sodium is more reactive, and potassium is still more reactive. In other words, there is a trend toward greater reactivity as you move down the alkali metals in Group 1.

Understanding a trend among the elements enables you to make predictions about the chemical behavior of the elements. These trends in properties of the elements in a group or period can be explained in terms of electron configurations.

Figure 15
Chemical reactivity with water increases from top to bottom for Group 1 elements. Reactions of lithium, sodium, and potassium with water are shown.

Lithium

Sodium

Potassium

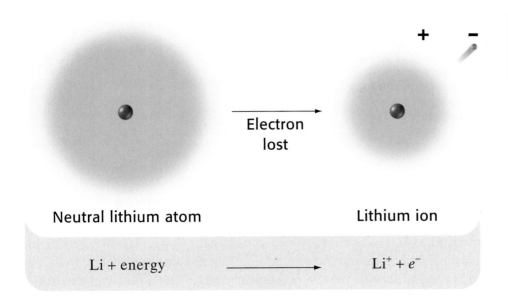

Neutral lithium atom   Electron lost →   Lithium ion

Li + energy $\longrightarrow$ Li$^+$ + e^-

Figure 16
When enough energy is supplied, a lithium atom loses an electron and becomes a positive ion. The ion is positive because its number of protons now exceeds its number of electrons by one.

Ionization Energy

When atoms have equal numbers of protons and electrons, they are electrically neutral. But when enough energy is added, the attractive force between the protons and electrons can be overcome. When this happens, an electron is removed from an atom. The neutral atom then becomes a positively charged ion.

Figure 16 illustrates the removal of an electron from an atom. The energy that is supplied to remove an electron is the **ionization energy** of the atom. This process can be described as shown below.

$$\text{A} + \text{ionization energy} \rightarrow \text{A}^+ + e^-$$

neutral atom   ion   electron

ionization energy

the energy required to remove an electron from an atom or ion

Ionization Energy Decreases as You Move Down a Group

Ionization energy tends to decrease down a group, as **Figure 17** on the next page shows. Each element has more occupied energy levels than the one above it has. Therefore, the outermost electrons are farthest from the nucleus in elements near the bottom of a group.

Similarly, as you move down a group, each successive element contains more electrons in the energy levels between the nucleus and the outermost electrons. These inner electrons shield the outermost electrons from the full attractive force of the nucleus. This **electron shielding** causes the outermost electrons to be held less tightly to the nucleus.

Notice in **Figure 18** on the next page that the ionization energy of potassium is less than that of lithium. The outermost electrons of a potassium atom are farther from its nucleus than the outermost electrons of a lithium atom are from their nucleus. So, the outermost electrons of a lithium atom are held more tightly to its nucleus. As a result, removing an electron from a potassium atom takes less energy than removing one from a lithium atom.

electron shielding

the reduction of the attractive force between a positively charged nucleus and its outermost electrons due to the cancellation of some of the positive charge by the negative charges of the inner electrons

Figure 17
Ionization energy generally decreases down a group and increases across a period, as shown in this diagram. Darker shading indicates higher ionization energy.

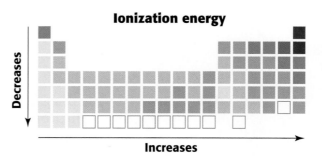

Ionization energy

Decreases

Increases

Ionization Energy Increases as You Move Across a Period

Ionization energy tends to increase as you move from left to right across a period, as **Figure 17** shows. From one element to the next in a period, the number of protons and the number of electrons increase by one each. The additional proton increases the nuclear charge. The additional electron is added to the same outer energy level in each of the elements in the period. A higher nuclear charge more strongly attracts the outer electrons in the same energy level, but the electron-shielding effect from inner-level electrons remains the same. Thus, more energy is required to remove an electron because the attractive force on them is higher.

Figure 18 shows that the ionization energy of neon is almost four times greater than that of lithium. A neon atom has 10 protons in its nucleus and 10 electrons filling two energy levels. In contrast, a lithium atom has 3 protons in its nucleus and 3 electrons distributed in the same two energy levels as those of neon. The attractive force between neon's 10 protons and 10 electrons is much greater than that between lithium's 3 protons and 3 electrons. As a result, the ionization energy of neon is much higher than that of lithium.

Figure 18
Ionization energies for hydrogen and for the main-group elements of the first four periods are plotted on this graph.

Ionization Energies of Main-Block Elements

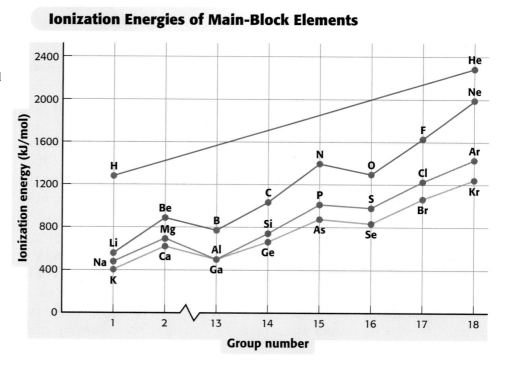

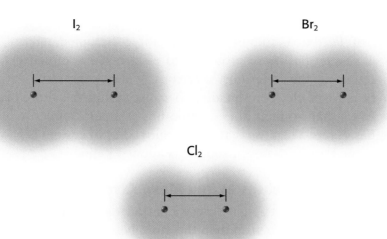

I₂   Br₂

Cl₂

Figure 19
In each molecule, half the distance of the line represents the bond radius of the atom.

Atomic Radius

The exact size of an atom is hard to determine. An atom's size depends on the volume occupied by the electrons around the nucleus, and the electrons do not move in well-defined paths. Rather, the volume the electrons occupy is thought of as an electron cloud, with no clear-cut edge. In addition, the physical and chemical state of an atom can change the size of an electron cloud.

Figure 19 shows one way to measure the size of an atom. This method involves calculating the **bond radius,** the length that is half the distance between the nuclei of two bonded atoms. The bond radius can change slightly depending on what atoms are involved.

bond radius

half the distance from center to center of two like atoms that are bonded together

Atomic Radius Increases as You Move Down a Group

Atomic radius increases as you move down a group, as **Figure 20** shows. As you proceed from one element down to the next in a group, another principal energy level is filled. The addition of another level of electrons increases the size, or atomic radius, of an atom.

Electron shielding also plays a role in determining atomic radius. Because of electron shielding, the effective nuclear charge acting on the outer electrons is almost constant as you move down a group, regardless of the energy level in which the outer electrons are located. As a result, the outermost electrons are not pulled closer to the nucleus. For example, the effective nuclear charge acting on the outermost electron in a cesium atom is about the same as it is in a sodium atom.

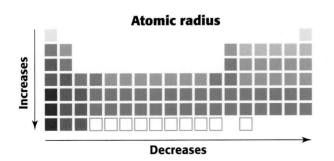

Atomic radius

Increases

Decreases

Figure 20
Atomic radius generally increases down a group and decreases across a period, as shown in this diagram. Darker shading indicates higher atomic radius.

As a member of Period 6, cesium has six occupied energy levels. As a member of Period 3, sodium has only three occupied energy levels. Although cesium has more protons and electrons, the effective nuclear charge acting on the outermost electrons is about the same as it is in sodium because of electron shielding. Because cesium has more occupied energy levels than sodium does, cesium has a larger atomic radius than sodium has. **Figure 21** shows that the atomic radius of cesium is about 230 pm, while the atomic radius of sodium is about 150 pm.

Atomic Radius Decreases as You Move Across a Period

Topic Link

Refer to Appendix A for a chart of relative atomic radii of the elements.

As you move from left to right across a period, each atom has one more proton and one more electron than the atom before it has. All additional electrons go into the same principal energy level—no electrons are being added to the inner levels. As a result, electron shielding does not play a role as you move across a period. Therefore, as the nuclear charge increases across a period, the effective nuclear charge acting on the outer electrons also increases. This increasing nuclear charge pulls the outermost electrons closer and closer to the nucleus and thus reduces the size of the atom.

Figure 21 shows how atomic radii decrease as you move across a period. Notice that the decrease in size is significant as you proceed across groups going from Group 1 to Group 14. The decrease in size then tends to level off from Group 14 to Group 18. As the outermost electrons are pulled closer to the nucleus, they also get closer to one another.

Figure 21
Atomic radii for hydrogen and the main-group elements in Periods 1 through 6 are plotted on this graph.

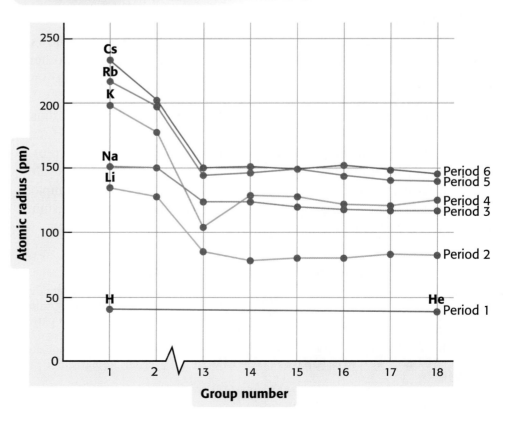

Atomic Radii of Main-Block Elements

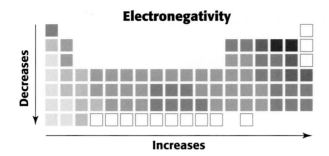

Electronegativity

Decreases ↓ Increases →

Figure 22
Electronegativity tends to decrease down a group and increase across a period, as shown in this diagram. Darker shading indicates higher electronegativity.

Repulsions between these electrons get stronger. Finally, a point is reached where the electrons will not come closer to the nucleus because the electrons would have to be too close to each other. Therefore, the sizes of the atomic radii level off as you approach the end of each period.

Electronegativity

Atoms often bond to one another to form a compound. These bonds can involve the sharing of valence electrons. Not all atoms in a compound share electrons equally. Knowing how strongly each atom attracts bonding electrons can help explain the physical and chemical properties of the compound.

Linus Pauling, one of America's most famous chemists, made a scale of numerical values that reflect how much an atom in a molecule attracts electrons, called **electronegativity** values. Chemical bonding that comes from a sharing of electrons can be thought of as a tug of war. The atom with the higher electronegativity will pull on the electrons more strongly than the other atom will.

Fluorine is the element whose atoms most strongly attract shared electrons in a compound. Pauling arbitrarily gave fluorine an electronegativity value of 4.0. Values for the other elements were calculated in relation to this value.

electronegativity

a measure of the ability of an atom in a chemical compound to attract electrons

Electronegativity Decreases as You Move Down a Group

Electronegativity values generally decrease as you move down a group, as **Figure 22** shows. Recall that from one element to the next one in a group, the principal quantum number increases by one, so another principal energy level is occupied. The more protons an atom has, the more strongly it should attract an electron. Therefore, you might expect that electronegativity increases as you move down a group.

However, electron shielding plays a role again. Even though cesium has many more protons than lithium does, the effective nuclear charge acting on the outermost electron is almost the same in both atoms. But the distance between cesium's sixth principal energy level and its nucleus is greater than the distance between lithium's third principal energy level and its nucleus. This greater distance means that the nucleus of a cesium atom cannot attract a valence electron as easily as a lithium nucleus can. Because cesium does not attract an outer electron as strongly as lithium, it has a smaller electronegativity value.

Electronegativity Versus Atomic Number

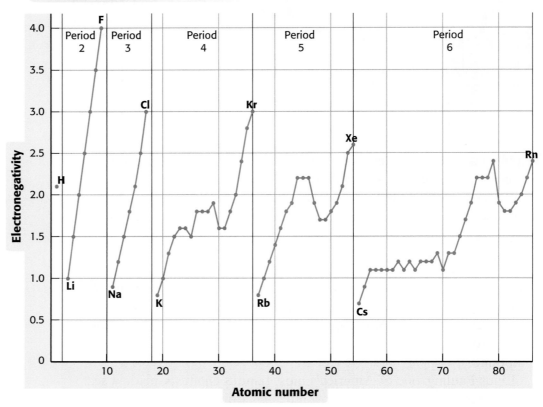

Figure 23
This graph shows electronegativity compared to atomic number for Periods 1 through 6. Electronegativity tends to increase across a period because the effective nuclear charge becomes greater as protons are added.

Electronegativity Increases as You Move Across a Period

As **Figure 23** shows, electronegativity usually increases as you move left to right across a period. As you proceed across a period, each atom has one more proton and one more electron—in the same principal energy level—than the atom before it has. Recall that electron shielding does not change as you move across a period because no electrons are being added to the inner levels. Therefore, the effective nuclear charge increases across a period. As this increases, electrons are attracted much more strongly, resulting in an increase in electronegativity.

Notice in **Figure 23** that the increase in electronegativity across a period is much more dramatic than the decrease in electronegativity down a group. For example, if you go across Period 3, the electronegativity more than triples, increasing from 0.9 for sodium, Na, to 3.2 for chlorine, Cl. In contrast, if you go down Group 1 the electronegativity decreases only slightly, dropping from 0.9 for sodium to 0.8 for cesium, Cs.

This difference can be explained if you look at the changes in atomic structure as you move across a period and down a group. Without the addition of any electrons to inner energy levels, elements from left to right in a period experience a significant increase in effective nuclear charge. As you move down a group, the addition of electrons to inner energy levels causes the effective nuclear charge to remain about the same. The electronegativity drops slightly because of the increasing distance between the nucleus and the outermost energy level.

Other Periodic Trends

You may have noticed that effective nuclear charge and electron shielding are often used in explaining the reasons for periodic trends. Effective nuclear charge and electron shielding also account for two other periodic trends that are related to the ones already discussed: ionic size and electron affinity. Still other trends are seen by examining how melting point and boiling point change as you move across a period or down a group. The trends in melting and boiling points are determined by how electrons form pairs as *d* orbitals fill.

Periodic Trends in Ionic Size and Electron Affinity

Recall that atoms form ions by either losing or gaining electrons. Like atomic size, ionic size has periodic trends. As you proceed down a group, the outermost electrons in ions are in higher energy levels. Therefore, just as atomic radius increases as you move down a group, usually the ionic radius increases as well, as shown in **Figure 24a.** These trends hold for both positive and negative ions.

Metals tend to lose one or more electrons and form a positive ion. As you move across a period, the ionic radii of metal cations tend to decrease because of the increasing nuclear charge. As you come to the nonmetal elements in a period, their atoms tend to gain electrons and form negative ions. **Figure 24a** shows that as you proceed through the anions on the right of a period, ionic radii still tend to decrease because of the anions' increasing nuclear charge.

Neutral atoms can also gain electrons. The energy change that occurs when a neutral atom gains an electron is called the atom's *electron affinity.* This property of an atom is different from electronegativity, which is a measure of an atom's attraction for an electron when the atom is bonded to another atom. **Figure 24b** shows that electron affinity tends to decrease as you move down a group. This trend is due to the increasing effect of electron shielding. In contrast, electron affinity tends to increase as you move across a period because of the increasing nuclear charge.

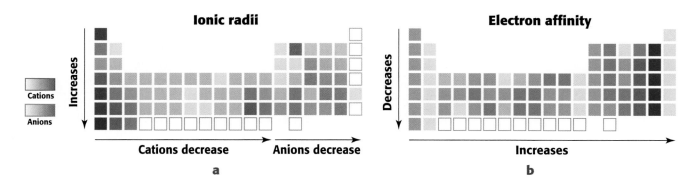

Figure 24
Ionic size tends to increase down groups and decrease across periods. Electron affinity generally decreases down groups and increases across periods.

Periodic Trends in Melting and Boiling Points

The melting and boiling points for the elements in Period 6 are shown in **Figure 25.** Notice that instead of a generally increasing or decreasing trend, melting and boiling points reach two different peaks as *d* and *p* orbitals fill.

Cesium, Cs, has low melting and boiling points because it has only one valence electron to use for bonding. From left to right across the period, the melting and boiling points at first increase. As the number of electrons in each element increases, stronger bonds between atoms can form. As a result, more energy is needed for melting and boiling to occur.

Near the middle of the *d*-block, the melting and boiling points reach a peak. This first peak corresponds to the elements whose *d* orbitals are almost half filled. The atoms of these elements can form the strongest bonds, so these elements have the highest melting and boiling points in this period. For Period 6, the elements with the highest melting and boiling points are tungsten, W, and rhenium, Re.

Figure 25
As you move across Period 6, the periodic trend for melting and boiling points goes through two cycles of first increasing, reaching a peak, and then decreasing.

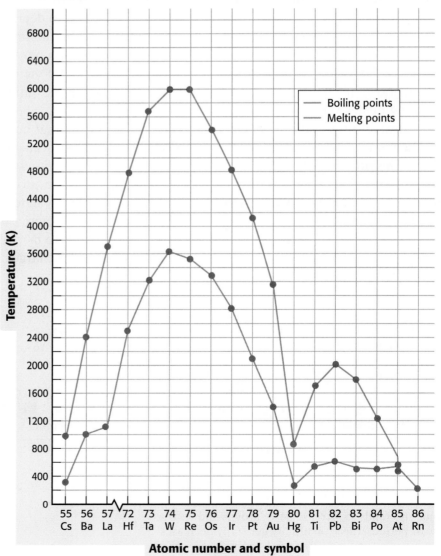

Melting Points and Boiling Points of Period 6 Elements

As more electrons are added, they begin to form pairs within the *d* orbitals. Because of the decrease in unpaired electrons, the bonds that the atoms can form with each other become weaker. As a result, these elements have lower melting and boiling points. The lowest melting and boiling points are reached at mercury, whose *d* orbitals are completely filled. Mercury, Hg, has the second-lowest melting and boiling points in this period. The noble gas radon, Rn, is the only element in Period 6 with a lower boiling point than that of mercury.

As you proceed past mercury, the melting and boiling points again begin to rise as electrons are now added to the *p* orbital. The melting and boiling points continue to rise until they peak at the elements whose *p* orbitals are almost half filled. Another decrease is seen as electrons pair up to fill *p* orbitals. When the noble gas radon, Rn, is reached, the *p* orbitals are completely filled. The noble gases are monatomic and have no bonding forces between atoms. Therefore, their melting and boiling points are unusually low.

③ Section Review

UNDERSTANDING KEY IDEAS

1. What is ionization energy?

2. Why is measuring the size of an atom difficult?

3. What can you tell about an atom that has high electronegativity?

4. How does electron shielding affect atomic size as you move down a group?

5. What periodic trends exist for ionization energy?

6. Describe one way in which *atomic radius* is defined.

7. Explain how the trends in melting and boiling points differ from the other periodic trends.

8. Why do both atomic size and ionic size increase as you move down a group?

9. How is electron affinity different from electronegativity?

10. What periodic trends exist for electronegativity?

11. Why is electron shielding not a factor when you examine a trend across a period?

CRITICAL THINKING

12. Explain why the noble gases have high ionization energies.

13. What do you think happens to the size of an atom when the atom loses an electron? Explain.

14. With the exception of the noble gases, why is an element with a high ionization energy likely to have high electron affinity?

15. Explain why atomic radius remains almost unchanged as you move through Period 2 from Group 14 to Group 18.

16. Helium and hydrogen have almost the same atomic size, yet the ionization energy of helium is almost twice that of hydrogen. Explain why hydrogen has a much higher ionization energy than any element in Group 1 does.

17. Why does mercury, Hg, have such a low melting point? How would you expect mercury's melting point to be different if the *d*-block contained more groups than it does?

18. What exceptions are there in the increase of ionization energies across a period?

Where Did the Elements Come From?

KEY TERMS

- nuclear reaction
- superheavy element

OBJECTIVES

① **Describe** how the naturally occurring elements form.

② **Explain** how a transmutation changes one element into another.

③ **Describe** how particle accelerators are used to create synthetic elements.

Natural Elements

Of all the elements listed in the periodic table, 93 are found in nature. Three of these elements, technetium, Tc, promethium, Pm, and neptunium, Np, are not found on Earth but have been detected in the spectra of stars. The nebula shown in **Figure 26** is one of the regions in the galaxy where new stars are formed and where elements are made.

Most of the atoms in living things come from just six elements. These elements are carbon, hydrogen, oxygen, nitrogen, phosphorus, and sulfur. Scientists theorize that these elements, along with all 93 natural elements, were created in the centers of stars billions of years ago, shortly after the universe formed in a violent explosion.

Figure 26
Three natural elements—technetium, promethium, and neptunium—have been detected only in the spectra of stars.

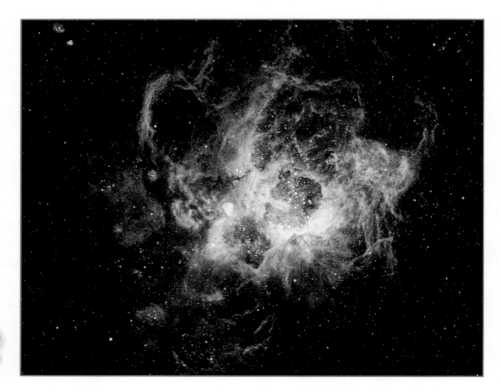

internet connect

www.scilinks.org
Topic: Origin of Elements
SciLinks code: HW4093

SCiLINKS. Maintained by the
National Science
Teachers Association

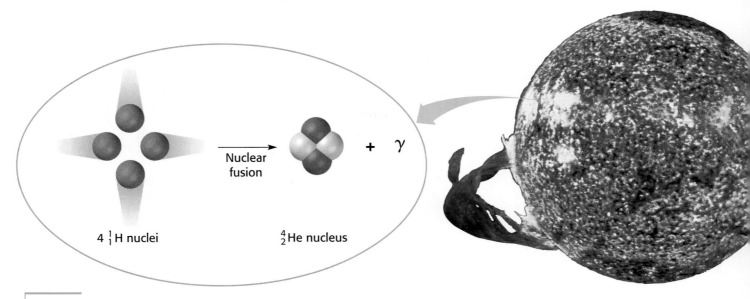

Figure 27
Nuclear reactions like those in the sun can fuse four hydrogen nuclei into one helium nucleus, releasing gamma radiation, γ.

Hydrogen and Helium Formed After the Big Bang

Much of the evidence about the universe's origin points toward a single event: an explosion of unbelievable violence, before which all matter in the universe could fit on a pinhead. This event is known as the *big bang*. Most scientists currently accept this model about the universe's beginnings. Right after the big bang, temperatures were so high that matter could not exist; only energy could. As the universe expanded, it cooled and some of the energy was converted into matter in the form of electrons, protons, and neutrons. As the universe continued to cool, these particles started to join and formed hydrogen and helium atoms.

Over time, huge clouds of hydrogen accumulated. Gravity pulled these clouds of hydrogen closer and closer. As the clouds grew more dense, pressures and temperatures at the centers of the hydrogen clouds increased, and stars were born. In the centers of stars, **nuclear reactions** took place. The simplest nuclear reaction, as shown in **Figure 27,** involves fusing hydrogen nuclei to form helium. Even now, these same nuclear reactions are the source of the energy that we see as the stars' light and feel as the sun's warmth.

nuclear reaction

a reaction that affects the nucleus of an atom

Other Elements Form by Nuclear Reactions in Stars

The mass of a helium nucleus is less than the total mass of the four hydrogen nuclei that fuse to form it. The mass is not really "lost" in this nuclear reaction. Rather, the missing mass is converted into energy. Einstein's equation $E = mc^2$ describes this mass-energy relationship quantitatively. The mass that is converted to energy is represented by m in this equation. The constant c is the speed of light. Einstein's equation shows that fusion reactions release very large amounts of energy. The energy released by a fusion reaction is so great it keeps the centers of the stars at very high temperatures.

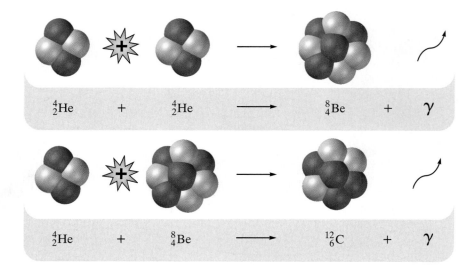

Figure 28
Nuclear reactions can form a beryllium nucleus by fusing helium nuclei. The beryllium nucleus can then fuse with another helium nucleus to form a carbon nucleus.

The temperatures in stars get high enough to fuse helium nuclei with one another. As helium nuclei fuse, elements of still higher atomic numbers form. **Figure 28** illustrates such a process: two helium nuclei fuse to form a beryllium nucleus, and gamma radiation is released. The beryllium nucleus can then fuse with another helium nucleus to form a carbon nucleus. Such repeated fusion reactions can form atoms as massive as iron.

Very massive stars (stars whose masses are more than 100 times the mass of our sun) are the source of heavier elements. When such a star has converted almost all of its core hydrogen and helium into the heavier elements up to iron, the star collapses and then blows apart in an explosion called a *supernova*. All of the elements heavier than iron on the periodic table are formed in this explosion. The star's contents shoot out into space, where they can become part of newly forming star systems.

Transmutations

In the Middle Ages, many early chemists tried to change, or transmute, ordinary metals into gold. Although they made many discoveries that contributed to the development of modern chemistry, their attempts to transmute metals were doomed from the start. These early chemists did not realize that a *transmutation*, whereby one element changes into another, is a nuclear reaction. It changes the nucleus of an atom and therefore cannot be achieved by ordinary chemical means.

Transmutations Are a Type of Nuclear Reaction

Although nuclei do not change into different elements in ordinary chemical reactions, transmutations can happen. Early chemists such as John Dalton had insisted that atoms never change into other elements, so when scientists first encountered transmutations in the 1910s, their results were not always believed.

While studying the passage of high-speed alpha particles (helium nuclei) through water vapor in a cloud chamber, Ernest Rutherford observed some long, thin particle tracks. These tracks matched the ones caused by protons in experiments performed earlier by other scientists.

internet connect

www.scilinks.org
Topic: Alchemy
SciLinks code: HW4006

SCiLINKS Maintained by the National Science Teachers Association

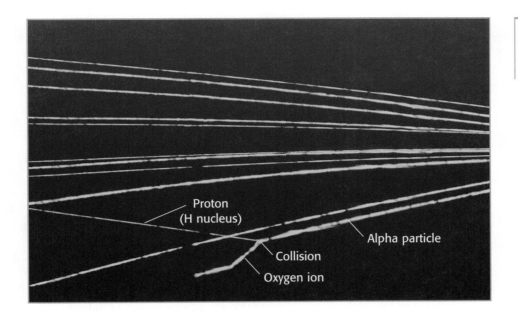

Figure 29
Observe the spot in this cloud-chamber photo where an alpha particle collided with the nucleus of a nitrogen atom. The left track was made by an oxygen atom; the right track, by a proton.

Rutherford reasoned correctly that the atomic nuclei in air were disintegrating upon being struck by alpha particles. He believed that the nuclei in air had disintegrated into the nuclei of hydrogen (protons) plus the nuclei of some other atom.

Two chemists, an American named W. D. Harkins and an Englishman named P.M.S. Blackett, studied this strange phenomenon further. Blackett took photos of 400 000 alpha particle tracks that formed in cloud chambers. He found that 8 of these tracks forked to form a **Y**, as shown in **Figure 29.** Harkins and Blackett concluded that the **Y** formed when an alpha particle collided with a nitrogen atom in air to produce an oxygen atom and a proton, and that a transmutation had thereby occurred.

Synthetic Elements

The discovery that a transmutation had happened started a flood of research. Soon after Harkins and Blackett had observed a nitrogen atom forming oxygen, other transmutation reactions were discovered by bombarding various elements with alpha particles. As a result, chemists have synthesized, or created, more elements than the 93 that occur naturally. These are *synthetic elements*. All of the transuranium elements, or those with more than 92 protons in their nuclei, are synthetic elements. To make them, one must use special equipment, called *particle accelerators,* described below.

The Cyclotron Accelerates Charged Particles

Many of the first synthetic elements were made with the help of a cyclotron, a particle accelerator invented in 1930 by the American scientist E.O. Lawrence. In a cyclotron, charged particles are given one pulse of energy after another, speeding them to very high energies. The particles then collide and fuse with atomic nuclei to produce synthetic elements that have much higher atomic numbers than naturally occurring elements do. However, there is a limit to the energies that can be reached with a cyclotron and therefore a limit to the synthetic elements that it can make.

The Synchrotron Is Used to Create Superheavy Elements

As a particle reaches a speed of about one-tenth the speed of light, it gains enough energy such that the relation between energy and mass becomes an obstacle to any further acceleration. According to the equation $E = mc^2$, the increase in the particle's energy also means an increase in its mass. This makes the particle accelerate more slowly so that it arrives too late for the next pulse of energy from the cyclotron, which is needed to make the particle go faster.

The solution was found with the synchrotron, a particle accelerator that times the pulses to match the acceleration of the particles. A synchrotron can accelerate only a few types of particles, but those particles it can accelerate reach enormous energies. Synchrotrons are now used in many areas of basic research, including explorations into the foundations of matter itself. The Fermi National Accelerator Laboratory in Batavia, IL has a circular accelerator which has a circumference of 4 mi! Subatomic particles are accelerated through this ring to 99.9999% of the speed of light.

Synthetic Element Trivia

Rutherfordium
Discovered by Russian scientists at the Joint Institute for Nuclear Research at Dubna and by scientists at the University of California at Berkeley

Meitnerium
Discovered August 29, 1982, by scientists at the Heavy Ion Research Laboratory in Darmstadt, West Germany; named in honor of Lise Meitner, the Austrian physicist

Mendelevium
Synthesized in 1955 by G. T. Seaborg, A. Ghiorso, B. Harvey, G. R. Choppin, and S. G. Thompson at the University of California, Berkeley; named in honor of the inventor of the periodic system

104	105	106	107	108	109	110	111
Rf	**Db**	**Sg**	**Bh**	**Hs**	**Mt**	**Uun**	**Uuu**
Rutherfordium	Dubnium	Seaborgium	Bohrium	Hassium	Meitnerium	(unnamed)	(unnamed)

93	94	95	96	97	98	99	100	101	102	103
Np	**Pu**	**Am**	**Cm**	**Bk**	**Cf**	**Es**	**Fm**	**Md**	**No**	**Lr**
Neptunium	Plutonium	Americium	Curium	Berkelium	Californium	Einsteinium	Fermium	Mendelevium	Nobelium	Lawrencium

Curium
Synthesized in 1944 by G. T. Seaborg, R.A. James, and A. Ghiorso at the University of California at Berkeley; named in honor of Marie and Pierre Curie

Californium
Synthesized in 1950 by G. T. Seaborg, S. G. Thompson, A. Ghiorso, and K. Street, Jr., at the University of California at Berkeley; named in honor of the state of California

Nobelium
Synthesized in 1958 by A. Ghiorso, G. T. Seaborg, T. Sikkeland, and J. R. Walton; named in honor of Alfred Nobel, discoverer of dynamite and founder of the Nobel Prize

Figure 30
All of the highlighted elements are synthetic. Those shown in orange were created by making moving particles collide with stationary targets. The elements shown in blue were created by making nuclei collide.

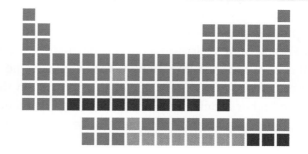

Once the particles have been accelerated, they are made to collide with one another. **Figure 30** shows some of the **superheavy elements** created with such collisions. When a synchrotron is used to create an element, only a very small number of nuclei actually collide. As a result, only a few nuclei may be created in these collisions. For example, only three atoms of meitnerium were detected in the first attempt, and these atoms lasted for only 0.0034 s. Obviously, identifying elements that last for such a short time is a difficult task. Scientists in only a few nations have the resources to carry out such experiments. The United States, Germany, Russia, and Sweden are the locations of the largest such research teams.

One of the recent superheavy elements that scientists report is element 114. To create element 114, Russian scientists took plutonium-244, supplied by American scientists, and bombarded it with accelerated calcium-40 atoms for 40 days. In the end, only a single nucleus was detected. It lasted for 30 seconds before decaying into element 112.

Most superheavy elements exist for only a tiny fraction of a second. Thirty seconds is a very long life span for a superheavy element. This long life span of element 114 points to what scientists have long suspected: that an "island of stability" would be found beginning with element 114. Based on how long element 114 lasted, their predictions may have been correct. However, scientists still must try to confirm that element 114 was in fact created. The results of a single experiment are never considered valid unless the experiments are repeated and produce the same results.

superheavy element

an element whose atomic number is greater than 106

Section Review

UNDERSTANDING KEY IDEAS

1. How and where did the natural elements form?

2. What element is the building block for all other natural elements?

3. What is a synthetic element?

4. What is a transmutation?

5. Why is transmutation classified as a nuclear reaction?

6. How did Ernest Rutherford deduce that he had observed a transmutation in his cloud chamber?

7. How are cyclotrons used to create synthetic elements?

8. How are superheavy elements created?

CRITICAL THINKING

9. Why is the following statement *not* an example of a transmutation? Zinc reacts with copper sulfate to produce copper and zinc sulfate.

10. Elements whose atomic numbers are greater than 92 are sometimes referred to as the transuranium elements. Why?

11. Why must an extremely high energy level be reached before a fusion reaction can take place?

12. If the synchrotron had not been developed, how would the periodic table look?

13. What happens to the mass of a particle as the particle approaches the speed of light?

14. How many different kinds of nuclear reactions must protons go through to produce a carbon atom?

SCIENCE AND TECHNOLOGY

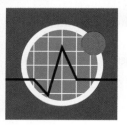

Materials Scientist

A materials scientist is interested in discovering materials that can last through harsh conditions, have unusual properties, or perform unique functions. These materials might include the following: a lightweight plastic that conducts electricity; extremely light but strong materials to construct a space platform; a plastic that can replace iron and aluminum in building automobile engines; a new building material that expands and contracts very little, even in extreme temperatures; or a strong, flexible, but extremely tough material that can replace bone or connective tissue in surgery. Materials engineers develop such materials and discover ways to mold or shape these materials into usable forms. Many materials scientists work in the aerospace industry and develop new materials that can lower the mass of aircraft, rockets, and space vehicles.

internet connect

www.scilinks.org
Topic: Superconductors
SciLinks code: HW4170

SCI**LINKS** Maintained by the
National Science
Teachers Association

Superconductors

Superconductivity Discovered

It has long been known that a metal becomes a better conductor as its temperature is lowered. In 1911, Heike Kamerlingh Onnes, a Dutch physicist, was studying this effect on mercury. When he used liquid helium to cool the metal to about −269°C, an unexpected thing happened—the mercury lost all resistance and became a superconductor. Scientists were excited about this new discovery, but the use of superconductors was severely limited by the huge expense of cooling them to near absolute zero. Scientists began research to find a material that would superconduct at temperatures above −196°C, the boiling point of cheap-to-produce liquid nitrogen.

The strong magnetic field produced by these super-conducting electromagnets can suspend this 8 cm disk.

"High-Temperature" Superconductors

Finally, in 1987 scientists discovered materials that became superconductors when cooled to only −183°C. These "high-temperature" superconductors were not metals but ceramics; usually copper oxides combined with elements such as yttrium or barium.

High-temperature superconductors are used in building very powerful electromagnets that are not limited by resistance or heat. These magnets can be used to build powerful particle accelerators and high-efficiency electric motors and generators. Engineers are working to build a system that will use superconducting electromagnets to levitate a passenger train above its guide rail so that the train can move with little friction and thus save fuel.

Questions

1. How does temperature normally affect electrical conductivity in metals?
2. What happened unexpectedly when mercury was cooled to near absolute zero?
3. How might consumers benefit from the use of superconducting materials?

CHAPTER HIGHLIGHTS 4

KEY IDEAS

SECTION ONE How Are Elements Organized?

- John Newlands, Dmitri Mendeleev, and Henry Moseley contributed to the development of the periodic table.
- The periodic law states that the properties of elements are periodic functions of the elements' atomic numbers.
- In the periodic table, elements are ordered by increasing atomic number. Rows are called *periods*. Columns are called *groups*.
- Elements in the same period have the same number of occupied energy levels. Elements in the same group have the same number of valence electrons.

SECTION TWO Tour of the Periodic Table

main-group element
alkali metal
alkaline-earth metal
halogen
noble gas
transition metal
lanthanide
actinide
alloy

- The main-group elements are Group 1 (alkali metals), Group 2 (alkaline-earth metals), Groups 13–16, Group 17 (halogens), and Group 18 (noble gases).
- Hydrogen is in a class by itself.
- Most elements are metals, which conduct electricity. Metals are also ductile and malleable.
- Transition metals, including the lanthanides and actinides, occupy the center of the periodic table.

SECTION THREE Trends in the Periodic Table

- Periodic trends are related to the atomic structure of the elements.
- Ionization energy, electronegativity, and electron affinity generally increase as you move across a period and decrease as you move down a group.
- Atomic radius and ionic size generally decrease as you move across a period and increase as you move down a group.
- Melting points and boiling points pass through two cycles of increasing, peaking, and then decreasing as you move across a period.

SECTION FOUR Where Did the Elements Come From?

- The 93 natural elements were formed in the interiors of stars. Synthetic elements (elements whose atomic numbers are greater than 93) are made using particle accelerators.
- A transmutation is a nuclear reaction in which one nucleus is changed into another nucleus.

CHAPTER REVIEW

USING KEY TERMS

1. What group of elements do Ca, Be, and Mg belong to?

2. What group of elements easily gains one valence electron?

3. What category do most of the elements of the periodic table fall under?

4. What is the term for the energy released when an atom gains an electron?

5. What are elements 90–103 called?

6. Give an example of a nuclear reaction. Describe the process by which it takes place.

7. What are elements in the first group of the periodic table called?

8. What atomic property affects periodic trends down a group in the periodic table?

9. What two atomic properties have an increasing trend as you move across a period?

10. Write a paragraph describing in your own words how synthetic elements are created. Discuss what modification has to be made to the equipment in order to synthesize super-heavy elements.

   WRITING SKILLS

11. Which group of elements has very high ionization energies and very low electron affinities?

12. How many valence electrons does a fluorine atom have?

13. Give an example of an alloy.

UNDERSTANDING KEY IDEAS

How Are Elements Organized?

14. How was Moseley's arrangement of the elements in the periodic table different from Mendeleev's?

15. What did the gaps on Mendeleev's periodic table represent?

16. Why was Mendeleev's periodic table accepted by most chemists?

17. What determines the horizontal arrangement of the periodic table?

18. Why is barium, Ba, placed in Group 2 and in Period 6?

Tour of the Periodic Table

19. Why is hydrogen in a class by itself?

20. All halogens are highly reactive. What causes these elements to have similar chemical behavior?

21. What property do the noble gases share? How do the electron configurations of the noble gases give them this shared property?

22. How do the electron configurations of the transition metals differ from those of the metals in Groups 1 and 2?

23. Why is carbon, a nonmetal element, added to iron to make nails?

24. If an element breaks when it is struck with a hammer, could it be a metal? Explain.

25. Why are the lanthanides and actinides placed at the bottom of the periodic table?

26. Explain why the main-group elements are also known as representative elements.

Trends in the Periodic Table

27. What periodic trends exist for ionization energy? How does this trend relate to different energy levels?

28. Why don't chemists define atomic radius as the radius of the electron cloud that surrounds a nucleus?

29. How does the periodic trend of atomic radius relate to the addition of electrons?

30. What happens to electron affinity as you move across a period beginning with Group 1? Why do these values change as they do?

31. Identify which trend diagram below describes atomic radius.

a.

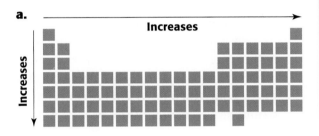

b.

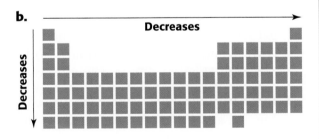

c.

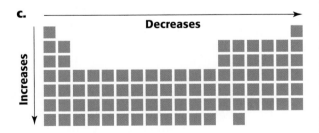

32. What periodic trends exist for electronegativity? Explain the factors involved.

33. Why are the melting and boiling points of mercury almost the lowest of the elements in its period?

Where Did the Elements Come From?

34. How does nuclear fusion generate energy?

35. What happens in the nucleus of an atom when a transmutation takes place?

36. Why are technetium, promethium, and neptunium considered natural elements even though they are not found on Earth?

37. Why must a synchrotron be used to create a superheavy element?

38. What role did supernovae play in creating the natural elements?

39. What is the heaviest element that can be made by normal processes inside a star?

40. What two elements make up most of the matter in a star?

MIXED REVIEW

41. Without looking at the periodic table, identify the period and group in which each of the following elements is located.
   a. $[Rn]7s^1$
   b. $[Ar]4s^2$
   c. $[Ne]3s^2 3p^6$

42. Which of the following ions has the electron configuration of a noble gas: Ca^+ or Cl^-? (Hint: Write the electron configuration for each ion.)

43. When 578 kJ/mol of energy is supplied, Al loses one valence electron. Write the electron configuration of the ion that forms.

44. The electron affinity value of oxygen is −146.1 kJ/mol. This value is the quantity of energy released when an oxygen atom gains an electron. Write the electron configuration of the ion that forms.

45. Without looking at the periodic table, write the complete electron configuration of the element in Group 2, Period 5. (Hint: The 5s orbital has a lower energy than the 4d orbital does.)

46. In the middle ages, many alchemists tried to transform lead into gold. Could gold be made from lead by using the nuclear processes described in this chapter?

47. Name three periodic trends you encounter in your life.

48. Describe some differences between the *s*-block metals and the *d*-block metals.

49. Irène Joliot-Curie created the first artificial radioactive isotope, phosphorus-30, in 1934. She created phosphorus-30 by bombarding aluminum-27, a shiny metal with conductive properties, with helium nuclei. The resulting product was a nonmetal with completely different properties from aluminum. What caused the change in properties?

50. How do the electron configurations of the lanthanide and actinide elements differ from the electron configurations of the other transition metals?

51. Use the periodic table to describe the chemical properties of the following elements:
   a. iodine, I
   b. krypton, Kr
   c. rubidium, Rb

52. The electron configuration of argon differs from those of chlorine and potassium by one electron each. Compare the reactivity of these three elements, and relate them to their electron configurations.

53. What trends were first used to classify the elements? What trends were discovered after the elements were classified in the periodic table?

54. Among the main-group elements, what is the relationship between group number and the number of valence electrons among group members?

55. Why is iodine placed just after tellurium on the periodic table even though its atomic mass is less than the atomic mass of tellurium?

CRITICAL THINKING

56. Consider two main-group elements, A and B. Element A has an ionization energy of 419 kJ/mol. Element B has an ionization energy of 1000 kJ/mol. Which element is more likely to form a cation?

57. Argon differs from both chlorine and potassium by one proton each. Compare the electron configurations of these three elements to explain the reactivity of these elements.

58. Why is it highly unlikely for calcium to form a Ca^+ cation and highly unlikely for sodium to form a Na^{2+} cation?

59. How would you prove that an alloy is a mixture and not a compound?

60. While at an amusement park, you inhale helium from a balloon to make your voice higher pitched. A friend says that helium reacts with and tightens the vocal cords to make your voice have the higher pitch. Could he be correct? Why or why not?

61. A scientist may measure the radius of a bonded atom five different times and may get five different results. The method used is correct, and the instrumentation is working correctly. How are different results possible?

62. In his periodic table, Mendeleev placed Be, Mg, Zn, and Cd in one group and Ca, Sr, Ba, and Pb in another group. Examine the electron configurations of these elements, and explain why Mendeleev grouped the elements this way.

63. The atomic number of yttrium, which follows strontium in the periodic table, exceeds the atomic number of strontium by one. Barium is 18 atomic numbers after strontium but it falls directly beneath strontium in the periodic table. Does strontium share more properties with yttrium or barium? Explain your answer.

64. Examine the following diagram.

Explain why the structure shown on the right was drawn to have a smaller radius than the structure on the left.

65. As of this writing, element 117 has yet to be created. Predict some of the properties that this element might have if it is synthesized.

ALTERNATIVE ASSESSMENT

66. Design an experiment to show that the electrical conductivity of a metal increases as the temperature decreases and decreases as the temperature increases. Test various metals to compare their conductivities at different temperatures. If your teacher approves your design, report your findings to the class after completing your experiments.

67. Select an alloy. You can choose one mentioned in this book or find another one by checking the library or the Internet. Obtain information on how the alloy is made. Obtain information on how the alloy is used for practical purposes.

68. The announcement of a new chemical element is sometimes claimed by two different groups of scientists. Research the history of one such element. What methods were used to create the element? How was the dispute settled so that only one group was recognized as the first to have synthesized the element?

69. Construct a model of a synchrotron. Check the library and Internet for information about synchrotrons. You may want to contact a synchrotron facility directly to find out what is currently being done in the field of synthetic elements.

70. In many labeled foods, the mineral content is stated in terms of the mass of the element, in a stated quantity of food. Examine the product labels of the foods you eat. Determine which elements are represented in your food and what function each element serves in the body. Make a poster of foods that are good sources of minerals that you need.

CONCEPT MAPPING

71. Use the following terms to complete the concept map below: *atomic number, atoms, electrons, periodic table,* and *protons.*

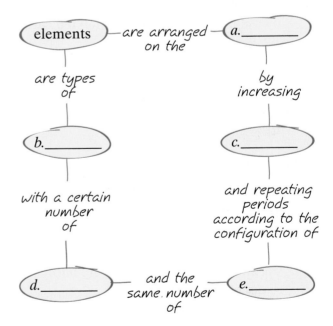

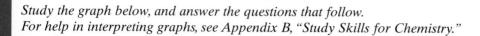
Study the graph below, and answer the questions that follow.
For help in interpreting graphs, see Appendix B, "Study Skills for Chemistry."

72. What relationship is represented in the graph shown?

73. What do the numbers on the *y*-axis represent?

74. In every Period, which Group contains the element with the greatest atomic radius?

75. Why is the axis representing group number drawn the way it is in going from Group 2 to Group 13?

76. Which period shows the greatest change in atomic radius?

77. Notice that the points plotted for the elements in Periods 5 and 6 of Group 2 overlap. What does this overlap indicate?

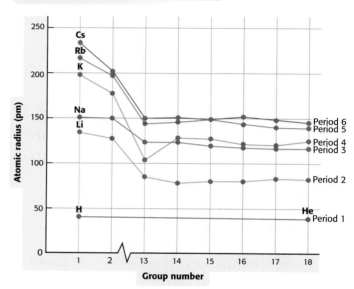

Atomic Radii of Main-Block Elements

TECHNOLOGY AND LEARNING

78. Graphing Calculator

Graphing Atomic Radius Vs. Atomic Number

The graphing calculator can run a program that graphs data such as atomic radius versus atomic number. Graphing the data within the different periods will allow you to discover trends.

Go to Appendix C. If you are using a TI-83 Plus, you can download the program and data sets and run the application as directed. Press the **APPS** key on your calculator, then choose the application **CHEMAPPS**. Press **8**, then highlight **ALL** on the screen, press **1**, then highlight **LOAD** and press **2** to load the data into your calculator. Quit the application, and then run the program **RADIUS**. For

L_1, press **2nd** and **LIST**, and choose **ATNUM**. For L_2, press **2nd** and **LIST** and choose **ATRAD**.

If you are using another calculator, your teacher will provide you with keystrokes and data sets to use.

a. Would you expect any atomic number to have an atomic radius of 20 pm? Explain.

b. A relationship is considered a function if it can pass a vertical line test. That is, if a vertical line can be drawn anywhere on the graph and only pass through one point, the relationship is a function. Does this set of data represent a function? Explain.

c. How would you describe the graphical relationship between the atomic numbers and atomic radii?

STANDARDIZED TEST PREP  4

1. In the modern periodic table, elements are arranged according to
   a. decreasing atomic mass.
   b. Mendeleev's original model.
   c. increasing atomic number.
   d. when they were discovered.

2. Which of the following elements is formed in stars?
   a. curium   **c.** einsteinium
   b. gold   **d.** americium

3. Group 17 elements, the halogens, are the most reactive of the nonmetal elements because they
   a. require only one electron to fill their outer energy level.
   b. have the highest ionization energies.
   c. have the largest atomic radii.
   d. are the farthest to the right in the periodic table.

4. The periodic law states that
   a. the chemical properties of elements can be grouped according to periodicity.
   b. the properties of the elements are functions of their atomic mass.
   c. all elements in the same group have the same number of valence electrons.
   d. all elements with the same number of occupied energy levels must be in the same group.

5. The energy it takes to remove an electron from an atom _____ as you move left to right across a period, from magnesium to chlorine.
   a. generally increases
   b. generally decreases
   c. does not change
   d. varies unpredictably

6. Which of the following is *not* a main-group element?
   a. neon   **c.** oxygen
   b. bromine   **d.** hydrogen

7. _____ is the element with the highest electronegativity.
   a. Oxygen   **c.** Fluorine
   b. Hydrogen   **d.** Carbon

8. To be classified as a metal, an element must
   a. be shiny.
   b. have a full outer energy level of electrons.
   c. conduct electricity.
   d. be malleable and ductile.

9. Because the noble gases have a stable electron configuration, they have
   a. high ionization energies.
   b. high electron affinities.
   c. large atomic radii.
   d. a tendency to form both cations and anions.

10. All elements in Group 2
    a. are also members of Period 2 in the periodic table.
    b. are known as the alkaline-earth metals.
    c. have the same atomic mass.
    d. have a stable electron configuration.

11. As electrons are added to the outer energy levels of atoms within a period, the atoms generally have
    a. increasing radii and increasing ionization energies.
    b. increasing radii and decreasing ionization energies.
    c. decreasing radii and decreasing ionization energies.
    d. decreasing radii and increasing ionization energies.

Ions and Ionic Compounds

The photograph provides a striking view of an ordinary substance—sodium chloride, more commonly known as table salt. Sodium chloride, like thousands of other compounds, is usually found in the form of crystals. These crystals are made of simple patterns of ions that are repeated over and over, and the result is often a beautifully symmetrical shape.

Ionic compounds share many interesting characteristics in addition to the tendency to form crystals. In this chapter you will learn about ions, the compounds they form, and the characteristics that these compounds share.

START-UP ACTIVITY

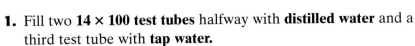

SAFETY PRECAUTIONS

Hard Water

PROCEDURE

1. Fill two **14 × 100 test tubes** halfway with **distilled water** and a third test tube with **tap water.**

2. Add about 1 tsp **Epsom salts** to one of the test tubes containing distilled water to make "hard water." Label the appropriate test tubes "Distilled water," "Tap water," and "Hard water."

3. Add a squirt of **liquid soap** to each test tube. Take one test tube, stopper it with a **cork,** and shake vigorously for 15 s. Repeat with the other two test tubes.

4. Observe the suds produced in each test tube.

ANALYSIS

1. Which water sample produces the most suds? Which produces the least suds?

2. What is meant by the term "hard water"? Is the water from your tap "hard water"?

Pre-Reading Questions

(1) **What is the difference between an atom and an ion?**

(2) **How can an atom become an ion?**

(3) **Why do chemists call table salt sodium chloride?**

(4) **Why do chemists write the formula for sodium chloride as NaCl?**

internet connect

www.scilinks.org
Topic: Crystalline Solids
SciLinks code: HW4037

SCiLINKS. Maintained by the National Science Teachers Association

Simple Ions

KEY TERMS

- octet rule
- ion
- cation
- anion

OBJECTIVES

(1) **Relate** the electron configuration of an atom to its chemical reactivity.

(2) **Determine** an atom's number of valence electrons, and use the octet rule to predict what stable ions the atom is likely to form.

(3) **Explain** why the properties of ions differ from those of their parent atoms.

Chemical Reactivity

Some elements are highly reactive, while others are not. For example, **Figure 1** compares the difference in reactivity between oxygen and neon. Notice that oxygen reacts readily with magnesium, but neon does not. Why is oxygen so reactive while neon is not? How much an element reacts depends on the electron configuration of its atoms. Examine the electron configuration for oxygen.

$$[O] = 1s^2 2s^2 2p^4$$

Notice that the 2p orbitals, which can hold six electrons, have only four. The electron configuration of a neon atom is shown below.

$$[Ne] = 1s^2 2s^2 2p^6$$

Notice that the 2p orbitals in a neon atom are full with six electrons.

Figure 1
Because of its electron configuration, oxygen reacts readily with magnesium (**a**). In contrast, neon's electron configuration makes it unreactive (**b**).

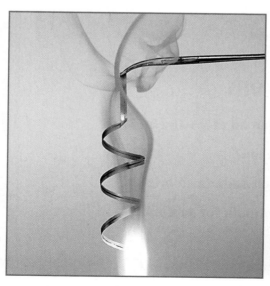

a   **magnesium in oxygen**

b   **magnesium in neon**

Noble Gases Are the Least Reactive Elements

Neon is a member of the noble gases, which are found in Group 18 of the periodic table. The noble gases show almost no chemical reactivity. Because of this, noble gases have a number of uses. For example, helium is used to fill balloons that float in air, which range in size from party balloons to blimps. Like neon, helium will not react with the oxygen in the air. The electron configuration for helium is $1s^2$. The two electrons fill the first energy level, making helium stable.

The other noble gases also have filled outer energy levels. This electron configuration can be written as ns^2np^6 where n represents the outer energy level. Notice that this level has eight electrons. These eight electrons fill the s and p orbitals, making these noble gases stable. In most chemical reactions, atoms tend to match the s and p electron configurations of the noble gases. This tendency is called the **octet rule.**

Alkali Metals and Halogens Are the Most Reactive Elements

Based on the octet rule, an atom whose outer s and p orbitals do not match the electron configurations of a noble gas will react to lose or gain electrons so the outer orbitals will be full. This prediction holds true for the alkali metals, which are some of the most reactive elements. **Figure 2** shows what happens when potassium, an alkali metal, is dropped into water. An explosive reaction occurs immediately, releasing heat and light.

As members of Group 1, alkali metals have only one electron in their outer energy level. When added to water, a potassium atom gives up this electron in its outer energy level. Then, potassium will have the s and p configuration of a noble gas.

$$1s^2 2s^2 2p^6 3s^2 3p^6 4s^1 \longrightarrow 1s^2 2s^2 2p^6 3s^2 3p^6$$

The halogens are also very reactive. As members of Group 17, they have seven electrons in their outer energy level. By gaining just one electron, a halogen will have the s and p configuration of a noble gas. For example, by gaining one electron, chlorine's electron configuration becomes $1s^2 2s^2 2p^6 3s^2 3p^6$.

Topic Link

Refer to the "Periodic Table" chapter for a discussion of the stability of the noble gases.

internet connect

www.scilinks.org
Topic: Inert Gases
SciLinks code: HW4070

SC*LINKS* Maintained by the National Science Teachers Association

octet rule

a concept of chemical bonding theory that is based on the assumption that atoms tend to have either empty valence shells or full valence shells of eight electrons

Figure 2
Alkali metals, such as potassium, react readily with a number of substances, including water.

Valence Electrons

Topic Link

Refer to the "Periodic Table" chapter for more about valence electrons.

You may have noticed that the electron configuration of potassium after it loses one electron is the same as that of chlorine after it gains one. Also, both configurations are the same as that of the noble gas argon.

$$[Ar] = 1s^2 2s^2 2p^6 3s^2 3p^6$$

After reacting, both potassium and chlorine have become stable. The atoms of many elements become stable by achieving the electron configuration of a noble gas. These electrons in the outer energy level are known as valence electrons.

Periodic Table Reveals an Atom's Number of Valence Electrons

It is easy to find out how many valence electrons an atom has. All you have to do is check the periodic table.

For example, **Figure 3** highlights the element magnesium, Mg. The periodic table lists its electron configuration.

$$[Mg] = [Ne]3s^2$$

This configuration shows that a magnesium atom has two valence electrons in the 3s orbital.

Now check the electron configuration of phosphorus, which is also highlighted in **Figure 3.**

$$[P] = [Ne]3s^2 3p^3$$

This configuration shows that a phosphorus atom has five valence electrons. Two valence electrons are in the 3s orbital, and three others are in the 3p orbitals.

Figure 3
The periodic table shows the electron configuration of each element. The number of electrons in the outermost energy level is the number of valence electrons.

Group 1							Group 18
Hydrogen **H**							Helium **He**
	Group 2	Group 13	Group 14	Group 15	Group 16	Group 17	
Lithium **Li**	Beryllium **Be**	Boron **B**	Carbon **C**	Nitrogen **N**	Oxygen **O**	Fluorine **F**	Neon **Ne**
Sodium **Na**	Magnesium **Mg**	Aluminum **Al**	Silicon **Si**	Phosphorus **P**	Sulfur **S**	Chlorine **Cl**	Argon **Ar**

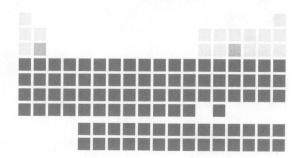

Atoms Gain Or Lose Electrons to Form Stable Ions

Recall that potassium loses its one valence electron so it will have the electron configuration of a noble gas. But why doesn't a potassium atom gain seven more electrons to become stable instead? The reason is the energy that is involved. Removing one electron requires far less energy than adding seven more.

When it gives up one electron to be more stable, a potassium atom also changes in another way. Recall that all atoms are uncharged because they have equal numbers of protons and electrons. For example, a potassium atom has 19 protons and 19 electrons. After giving up one electron, potassium still has 19 protons but only 18 electrons. Because the numbers are not the same, there is a net electrical charge. So the potassium atom becomes an **ion** with a 1+ charge, as shown in **Figure 4.** The following equation shows how a potassium atom forms an ion.

$$K \longrightarrow K^+ + e^-$$

An ion with a positive charge is called a **cation.** A potassium cation has an electron configuration just like the noble gas argon.

$$[K^+] = 1s^2 2s^2 2p^6 3s^2 3p^6 \qquad [Ar] = 1s^2 2s^2 2p^6 3s^2 3p^6$$

In the case of chlorine, far less energy is required for an atom to gain one electron rather than give up its seven valence electrons. By gaining an electron to be more stable, a chlorine atom becomes an ion with a 1– charge, as illustrated in **Figure 4.** The following equation shows the formation of a chlorine ion from a chlorine atom.

$$Cl + e^- \longrightarrow Cl^-$$

An ion with a negative charge is called an **anion.** A chlorine anion has an electron configuration just like the noble gas argon.

$$[Cl^-] = 1s^2 2s^2 2p^6 3s^2 3p^6 \qquad [Ar] = 1s^2 2s^2 2p^6 3s^2 3p^6$$

ion

an atom, radical, or molecule that has gained or lost one or more electrons and has a negative or positive charge

cation

an ion that has a positive charge

anion

an ion that has a negative charge

Figure 4
A potassium atom can lose an electron to become a potassium cation (**a**) with a 1+ charge. After gaining an electron, a chlorine atom becomes a chlorine anion (**b**) with a 1– charge.

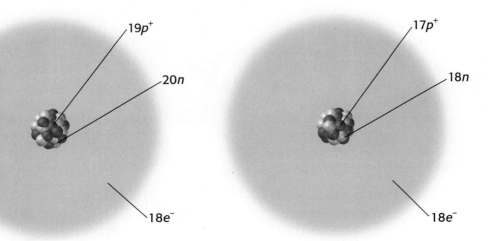

19p^+

20n

18e^-

17p^+

18n

18e^-

a potassium cation, K$^+$

b chloride anion, Cl$^-$

Characteristics of Stable Ions

How does an atom compare to the ion that it forms after it loses or gains an electron? Use of the same name for the atom and the ion that it forms indicates that the nucleus is the same as it was before. Both the atom and the ion have the same number of protons and neutrons. When an atom becomes an ion, it only involves loss or gain of electrons.

Recall that the chemical properties of an atom depend on the number and configuration of its electrons. Therefore, an atom and its ion have different chemical properties. For example, a potassium cation has a different number of electrons from a neutral potassium atom, but the same number of electrons as an argon atom. A chlorine anion also has the same number of electrons as an argon atom. However, it is important to realize that an ion is still quite different from a noble gas. An ion has an electrical charge, so therefore it forms compounds, and also conducts electricity when dissolved in water. Noble gases are very unreactive and have none of these properties.

Figure 5
These are examples of some stable ions that have an electron configuration like that of a noble gas.

Some Ions with Noble-Gas Configurations

Group 1	Group 2	Group 3	Group 13	Group 15	Group 16	Group 17	Group 18 Noble Gases
							Helium **He** $1s^2$
Li$^+$ $1s^2$	**Be^{2+}** $1s^2$			**N^{3-}** $[He]2s^22p^6$	**O^{2-}** $[He]2s^22p^6$	**F$^-$** $[He]2s^22p^6$	**Neon** **Ne** $[He]2s^22p^6$
Na$^+$ $[He]2s^22p^6$	**Mg^{2+}** $[He]2s^22p^6$		**Al^{3+}** $[He]2s^22p^6$	**P^{3-}** $[Ne]3s^23p^6$	**S^{2-}** $[Ne]3s^23p^6$	**Cl$^-$** $[Ne]3s^23p^6$	**Argon** **Ar** $[Ne]3s^23p^6$
K$^+$ $[Ne]3s^23p^6$	**Ca^{2+}** $[Ne]3s^23p^6$	**Sc^{3+}** $[Ne]3s^23p^6$		**As^{3-}** $[Ar]3d^{10}4s^24p^6$	**Se^{2-}** $[Ar]3d^{10}4s^24p^6$	**Br$^-$** $[Ar]3d^{10}4s^24p^6$	**Krypton** **Kr** $[Ar]3d^{10}4s^24p^6$
Rb$^+$ $[Ar]3d^{10}4s^24p^6$	**Sr^{2+}** $[Ar]3d^{10}4s^24p^6$	**Y^{3+}** $[Ar]3d^{10}4s^24p^6$			**Te^{2-}** $[Kr]4d^{10}5s^25p^6$	**I$^-$** $[Kr]4d^{10}5s^25p^6$	**Xenon** **Xe** $[Kr]4d^{10}5s^25p^6$
Cs$^+$ $[Kr]4d^{10}5s^25p^6$	**Ba^{2+}** $[Kr]4d^{10}5s^25p^6$	**La^{3+}** $[Kr]4d^{10}5s^25p^6$					

Each color denotes ions and a noble gas that have the same electron configurations. The small table at right shows the periodic table positions of the ions listed above.

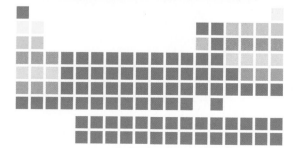

Many Stable Ions Have Noble-Gas Configurations

Potassium and chlorine are not the only atoms that form stable ions with a complete octet of valence electrons. **Figure 5** lists examples of other atoms that form ions with a full octet. For example, examine how calcium, Ca, forms a stable ion. The electron configuration of a calcium atom is written as follows.

$$[Ca] = 1s^2 2s^2 2p^6 3s^2 3p^6 4s^2$$

By giving up its two valence electrons in the $4s$ orbital, calcium forms a stable cation with a 2+ charge that has an electron configuration like that of argon.

$$[Ca^{2+}] = 1s^2 2s^2 2p^6 3s^2 3p^6$$

Some Stable Ions Do Not Have Noble-Gas Configurations

Not all atoms form stable ions with an electron configuration like those of noble gases. As illustrated in **Figure 6,** transition metals often form ions without complete octets. Notice that, with the lone exception of rhenium, Re, these stable ions are all cations. Also notice in **Figure 6** that some elements, mostly transition metals, form stable ions with more than one charge. For example, copper, Cu, can give up one electron, forming a Cu^+ cation. It can also give up two electrons, forming a Cu^{2+} cation. Both the Cu^+ and Cu^{2+} cations are stable even though they do not have noble-gas configurations.

Topic Link

Refer to the "Atoms" chapter for more about electron configuration.

Figure 6
Some stable ions do not have electron configurations like those of the noble gases.

Stable Ions Formed by the Transition Elements and Some Other Metals

Group 4	Group 5	Group 6	Group 7	Group 8	Group 9	Group 10	Group 11	Group 12	Group 13	Group 14
Ti^{2+} Ti^{3+}	V^{2+} V^{3+}	Cr^{2+} Cr^{3+}	Mn^{2+} Mn^{3+}	Fe^{2+} Fe^{3+}	Co^{2+} Co^{3+}	Ni^{2+}	Cu^+ Cu^{2+}	Zn^{2+}	Ga^{2+} Ga^{3+}	Ge^{2+}
		Mo^{3+}	Tc^{2+}			Pd^{2+}	Ag^+ Ag^{2+}	Cd^{2+}	In^+ In^{2+} In^{3+}	Sn^{2+}
Hf^{4+}			Re^{4+} Re^{5+}			Pt^{2+} Pt^{4+}	Au^+ Au^{3+}	Hg_2^{2+} Hg^{2+}	Tl^+ Tl^{3+}	Pb^{2+}

The small table at left shows the periodic table positions of the ions listed above.

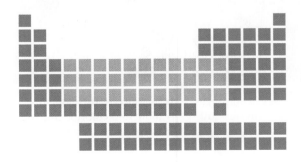

Atoms and Ions

Many atoms form stable ions that have noble-gas configurations. It is important to remember that these elements do not actually *become* noble gases. Having identical electron configurations does not mean that a sodium cation is a neon atom. The sodium cation still has 11 protons and 12 neutrons, like a sodium atom that has not reacted to form an ion. But like a noble-gas atom, a sodium ion is very unlikely to gain or lose any more electrons.

Ions and Their Parent Atoms Have Different Properties

Like potassium and all other alkali metals of Group 1, sodium is extremely reactive. When it is placed in water, a violent reaction occurs, producing heat and light. Like all halogens of Group 17, chlorine is extremely reactive. In fact, atoms of chlorine react with each other to form molecules of chlorine, Cl_2, a poisonous, yellowish green gas. In nature, however, chlorine is almost always found in its anion form as part of ionic compounds.

Because both sodium and chlorine are very reactive, you might expect a violent reaction when these two are brought together. This is exactly what happens. If a small piece of sodium is lowered into a flask filled with chlorine gas, there is a violent reaction that releases both heat and light. After the reaction is complete, all that remains is a white solid. Even though it is formed from two dangerous elements, it is something you probably eat every day—table salt.

Chemists call this salt *sodium chloride*. Sodium chloride is made from sodium cations and chloride anions. As illustrated in **Figure 7,** these ions have very different properties than those of their parent atoms. That is why salt is not as dangerous to have around your house as the elements that make it up. It does not react with water like sodium metal does because salt contains stable sodium ions, not reactive sodium atoms.

Figure 7

Water molecule, H_2O

Chloride ion, Cl^-

Sodium ion, Na^+

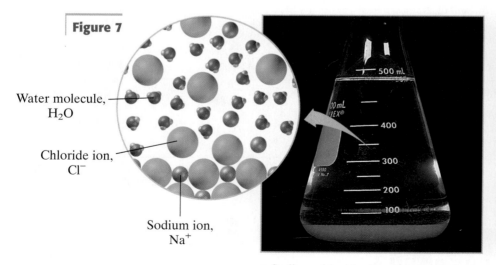

a Sodium chloride dissolves in water to produce unreactive sodium cations and chlorine anions.

b In contrast, the elements sodium and chlorine are very reactive when they are brought together.

Atoms of Metals and Nonmetal Elements Form Ions Differently

Nearly all metals form cations, as can be seen by examining their electron configuration. For example, consider the configuration for the Group 2 metal magnesium, Mg.

$$[\text{Mg}] = 1s^2 2s^2 2p^6 3s^2$$

To have a noble-gas configuration, the atom must either gain six electrons or lose two. Losing two electrons requires less energy than gaining six. Similarly, for all metals, the energy required to remove electrons from atoms to form ions with a noble-gas configuration is always less than the energy required to add more electrons. As a result, the atoms of metals form cations.

In contrast, the atoms of nonmetal elements form anions. Consider the example of oxygen, whose electron configuration is written as follows.

$$[\text{O}] = 1s^2 2s^2 2p^4$$

To have a noble-gas configuration, an oxygen atom must either gain two electrons or lose six. Acquiring two electrons requires less energy than losing six. For other nonmetals, the energy required to add electrons to atoms of nonmetals so that their ions have a noble-gas configuration is always less than the energy required to remove enough electrons. As a result, the atoms of nonmetal elements form anions.

Section Review

UNDERSTANDING KEY IDEAS

1. Explain why the noble gases tend not to react.

2. Where are the valence electrons located in an atom?

3. How does a cation differ from an anion?

4. State the octet rule.

5. Why do the properties of an ion differ from those of its parent atom?

6. Explain why alkali metals are extremely reactive.

7. How can you determine the number of valence electrons an atom has?

8. Explain why almost all metals tend to form cations.

9. Explain why, as a pure element, oxygen is usually found in nature as O_2.

CRITICAL THINKING

10. How could each of the following atoms react to achieve a noble-gas configuration?
    a. iodine
    b. strontium
    c. nitrogen
    d. krypton

11. Write the electron configuration for each of the following ions.
    a. Al^{3+}
    b. Se^{2-}
    c. Sc^{3+}
    d. As^{3-}

12. In what way is an ion the same as its parent atom?

13. To achieve a noble-gas configuration, a phosphorus atom will form a P^{3-} anion rather than forming a P^{5+} cation. Why?

Ionic Bonding and Salts

KEY TERMS

- salt
- lattice energy
- crystal lattice
- unit cell

OBJECTIVES

1. **Describe** the process of forming an ionic bond.

2. **Explain** how the properties of ionic compounds depend on the nature of ionic bonds.

3. **Describe** the structure of salt crystals.

Ionic Bonding

You may think that the material shown in **Figure 8** is very valuable. If you look closely, you will see what appear to be chunks of gold. The object shown in **Figure 8** is actually a mineral called *pyrite,* which does not contain any gold. However, the shiny yellow flakes make many people believe that they have discovered gold. All they have really discovered is a mineral that is made of iron cations and sulfur anions.

Because opposite charges attract, cations and anions should attract one another. This is exactly what happens when an ionic bond is formed. In the case of pyrite, the iron cations and sulfur anions attract one another to form an ionic compound.

Figure 8
The mineral pyrite is commonly called *fool's gold*. Unlike real gold, pyrite is actually quite common in Earth's crust.

Ionic Bonds Form Between Ions of Opposite Charge

To understand how an ionic bond forms, take another look at what happens when sodium and chlorine react to form sodium chloride. Recall that sodium gives up its only valence electron to form a stable Na^+ cation. Chlorine, with seven valence electrons, acquires that electron. As a result, a chlorine atom becomes a stable Cl^- anion.

The force of attraction between the 1+ charge on the sodium cation and the 1– charge on the chloride anion creates the ionic bond in sodium chloride. Recall that sodium chloride is the scientific name for table salt. Chemists call table salt by its scientific name because the word **salt** can actually be used to describe any one of thousands of different ionic compounds. Other salts that are commonly found in a laboratory include potassium chloride and calcium iodide.

All these salts are ionic compounds that are electrically neutral. They are made up of cations and anions that are held together by ionic bonds in a simple, whole-number ratio. For example, sodium chloride consists of sodium cations and chloride anions bonded in a 1:1 ratio. To show this 1:1 ratio, chemists write the formula for sodium chloride as NaCl.

However, the attractions between the ions in a salt do not stop with a single cation and a single anion. These forces are so far reaching that one cation attracts several different anions. At the same time, each anion attracts several different cations. In this way, many ions are pulled together into a tightly packed structure. The tight packing of the ions causes any salt, such as sodium chloride, to have a distinctive crystal structure. The smallest crystal of table salt that you could see would still have more than a billion billion sodium and chloride ions.

salt

an ionic compound that forms when a metal atom or a positive radical replaces the hydrogen of an acid

Transferring Electrons Involves Energy Changes

Recall that ionization energy is the energy that it takes to remove the outermost electron from an atom. In other words, moving a negatively charged electron away from an atom that will become a positively charged ion requires an input of energy before it will take place. In the case of sodium, this process can be written as follows.

$$Na + energy \longrightarrow Na^+ + e^-$$

Recall that electron affinity is the energy needed to add an electron onto a neutral atom. However, some elements, such as chlorine, easily accept extra electrons. For elements like this, energy is released when an electron is added. This process can be written as follows.

$$Cl + e^- \longrightarrow Cl^- + energy$$

But this energy released is less than the energy required to remove an electron from a sodium atom. Then why does an ionic bond form if these steps do not provide enough energy? Adding and removing electrons is only part of forming an ionic bond. The rest of the process of forming a salt supplies more than enough energy to make up the difference so that the overall process releases energy.

internet connect

www.scilinks.org
Topic: Ionic Bonds
SciLinks code: HW4071

SCiLINKS. Maintained by the National Science Teachers Association

lattice energy

the energy associated with constructing a crystal lattice relative to the energy of all constituent atoms separated by infinite distances

Salt Formation Involves Endothermic Steps

The process of forming the salt sodium chloride can be broken down into five steps as shown in **Figure 9** on the following page. Keep in mind that these steps do not really take place in this order. However, these steps, do model what must happen for an ionic bond to form between sodium cations and chloride anions.

The starting materials are sodium metal and chlorine gas. Energy must be supplied to make the solid sodium metal into a gas. This process takes energy and can be written as follows.

$$\text{Na(solid)} + \text{energy} \longrightarrow \text{Na(gas)}$$

Recall that energy is also required to remove an electron from a gaseous sodium atom.

$$\text{Na(gas)} + \text{energy} \longrightarrow \text{Na}^+\text{(gas)} + e^-$$

No energy is required to convert chlorine into the gaseous state because it is already a gas. However, chlorine gas consists of two chlorine atoms that are bonded to one another. Therefore, energy must be supplied to separate these chlorine atoms so that they can react with sodium. This third process can be written as follows.

$$\text{Cl–Cl(gas)} + \text{energy} \longrightarrow \text{Cl(gas)} + \text{Cl(gas)}$$

To this point, the first three steps have all been endothermic. These steps have produced sodium cations and chlorine atoms.

Salt Formation Also Involves Exothermic Steps

As **Figure 9** illustrates, the next step adds an electron to a chlorine atom to form an anion. This is the first step that releases energy. Recall that this step cannot supply enough energy to remove an electron from a sodium atom. Obviously, this step cannot produce nearly enough energy to drive the first three steps.

The chief driving force for the formation of the salt is the last step, in which the separated ions come together to form a crystal held together by ionic bonds. When a cation and anion form an ionic bond, it is an exothermic process. Energy is released.

$$\text{Na}^+\text{(gas)} + \text{Cl}^-\text{(gas)} \longrightarrow \text{NaCl(solid)} + \text{energy}$$

The energy released when ionic bonds are formed is called the **lattice energy.** This energy is released when the crystal structure of a salt is formed as the separated ions bond. In the case of sodium chloride, the lattice energy is greater than the energy needed for the first three steps. Without this energy, there would not be enough energy to make the overall process spontaneous. Lattice energy is the key to salt formation.

The value of the lattice energy is different if other cations and anions form the salt. For example, Na^+ ions can form salts with anions of any of the halogens. The lattice energy values for each of these salts are about the same. However, when magnesium cations, Mg^{2+}, form salts, these values are much higher than the values for salts of sodium. This large difference in lattice energy is due to the fact that ions with greater charge are more strongly attracted to the oppositely charged ions in the crystal. The lattice energy value for magnesium oxide is almost five times greater than that for sodium chloride.

If energy is released when ionic bonds are formed, then energy must be supplied to break these bonds and separate the ions. In the case of sodium chloride, the needed energy can come from water. As a result, a sample of sodium chloride dissolves when it is added to a glass of water. As the salt dissolves, the Na^+ and Cl^- ions separate as the ionic bonds between them are broken. Because of its much higher lattice energy, magnesium oxide does not dissolve well in water. In this case, the energy that is available in a glass of water is significantly less than the lattice energy of the magnesium oxide. There is not enough energy to separate the Mg^{2+} and O^{2-} ions from one another.

Figure 9
The reaction between $Na(s)$ and $Cl_2(g)$ to form sodium chloride can be broken down into steps. More energy is released overall than is absorbed.

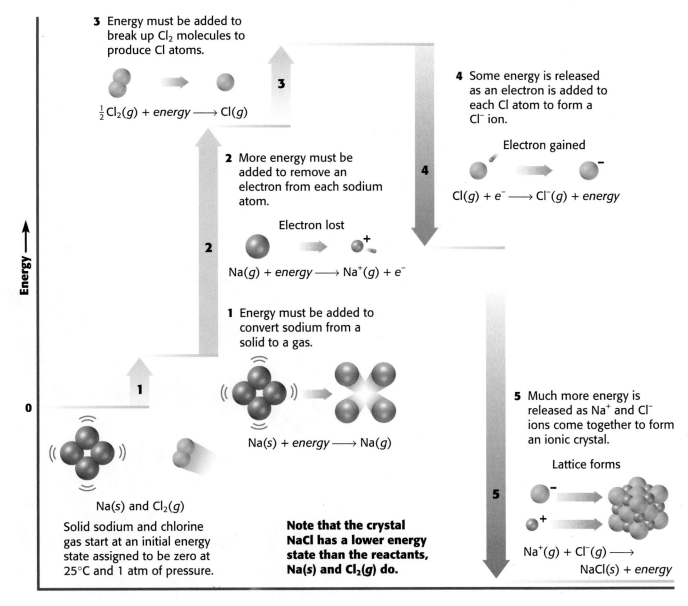

Energy

3 Energy must be added to break up Cl_2 molecules to produce Cl atoms.

$\frac{1}{2}Cl_2(g) + energy \longrightarrow Cl(g)$

4 Some energy is released as an electron is added to each Cl atom to form a Cl^- ion.

Electron gained

$Cl(g) + e^- \longrightarrow Cl^-(g) + energy$

2 More energy must be added to remove an electron from each sodium atom.

Electron lost

$Na(g) + energy \longrightarrow Na^+(g) + e^-$

1 Energy must be added to convert sodium from a solid to a gas.

$Na(s) + energy \longrightarrow Na(g)$

5 Much more energy is released as Na^+ and Cl^- ions come together to form an ionic crystal.

Lattice forms

$Na^+(g) + Cl^-(g) \longrightarrow NaCl(s) + energy$

Na(s) and Cl₂(g)

Solid sodium and chlorine gas start at an initial energy state assigned to be zero at 25°C and 1 atm of pressure.

Note that the crystal NaCl has a lower energy state than the reactants, Na(s) and Cl₂(g) do.

Ions and Ionic Compounds **169**

Ionic Compounds

Recall that salts are ionic compounds made of cations and anions. Many of the rocks and minerals in Earth's crust are made of cations and anions held together by ionic bonds. The ratio of cations to anions is always such that an ionic compound has no overall charge. For example, in sodium chloride, for every Na^+ cation, there is a Cl^- anion to balance the charge. In magnesium oxide, for every Mg^{2+} cation, there is an O^{2-} anion. Ionic compounds also share certain other chemical and physical properties.

Ionic Compounds Do Not Consist of Molecules

Figure 10 shows sodium chloride, an ionic compound, being added to water, a molecular compound. If you could look closely enough into the water, you would find individual water molecules, each made of two hydrogen atoms and one oxygen atom. The pot would be filled with many billions of these individual H_2O molecules.

Recall that the smallest crystal of table salt that you could see contains many billions of sodium and chloride ions all held together by ionic bonds. However, if you could look closely enough into the salt, all you would see are many Na^+ and Cl^- ions all bonded together to form a crystal. There are no NaCl molecules.

Elements in Groups 1 and 2 reacting with elements in Groups 16 and 17 will almost always form ionic compounds and not molecular compounds. Therefore, the formula CaO likely indicates an ionic compound because Ca is a Group 2 metal and O is a Group 16 nonmetal. In contrast, the formula ICl likely indicates a molecular compound because both I and Cl are members of Group 17.

However, you cannot be absolutely sure that something is made of ions or molecules just by looking at its formula. That determination must be made in the laboratory.

Figure 10
Salt is often added to water for flavor when pasta is being cooked.

Ionic Bonds Are Strong

Both repulsive and attractive forces exist within a salt crystal. The repulsive forces include those between like-charged ions. Within the crystal, each Na^+ ion repels the other Na^+ ions. The same is true for the Cl^- ions. Another repulsive force exists between the electrons of ions that are close together, even if the ions have opposite charges.

The attractive forces include those between the positively charged nuclei of one ion and the electrons of other nearby ions. In addition, attractive forces exist between oppositely charged ions. These forces involve more than a single Na^+ ion and a single Cl^- ion. Within the crystal, each sodium cation is surrounded by six chloride anions. At the same time, each chloride anion is surrounded by six sodium cations. As a result, the attractive force between oppositely charged ions is significantly greater in a crystal than it would be if the sodium cations and chloride anions existed only in pairs.

Overall, the attractive forces are significantly stronger than the repulsive forces, so ionic bonds are very strong.

Ionic Compounds Have Distinctive Properties

All ionic compounds share certain properties because of the strong attraction between their ions. Compare the boiling point of sodium chloride (1413°C) with that of water, a molecular compound (100°C). Similarly, most other ionic compounds have high melting and boiling points, as you can see in **Table 1.**

To melt, ions cannot be in fixed locations. Because each ion in these compounds forms strong bonds to neighboring ions, considerable energy is required to free them. Still more energy is needed to move ions out of the liquid state and cause boiling.

As a result of their high boiling points, ionic compounds are rarely gaseous at room temperature, while many molecular compounds are. Ice, for example, will eventually melt and then vaporize. In contrast, salt will remain a solid no matter how long it remains at room temperature.

internet connect

www.scilinks.org
Topic: Ionic Compounds
SciLinks code: HW4072

SC*LINKS* Maintained by the National Science Teachers Association

Table 1 Melting and Boiling Points of Compounds

Compound name	Formula	Type of compound	Melting point °C	K	Boiling point °C	K
Magnesium fluoride	MgF_2	ionic	1261	1534	2239	2512
Sodium chloride	NaCl	ionic	801	1074	1413	1686
Calcium iodide	CaI_2	ionic	784	1057	1100	1373
Iodine monochloride	ICl	covalent	27	300	97	370
Carbon tetrachloride	CCl_4	covalent	−23	250	77	350
Hydrogen fluoride	HF	covalent	−83	190	20	293
Hydrogen sulfide	H_2S	covalent	−86	187	−61	212
Methane	CH_4	covalent	−182	91	−164	109

Liquid and Dissolved Salts Conduct Electric Current

To conduct an electric current, a substance must satisfy two conditions. First, the substance must contain charged particles. Second, those particles must be free to move. Because ionic compounds are composed of charged particles, you might expect that they could be good conductors. While particles in a solid have some vibrational motion, they remain in fixed locations, as shown by the model in **Figure 11a.** Therefore, ionic solids, such as salts, generally are not conductors of electric current because the ions cannot move.

However, when the ions can move about, salts are excellent electrical conductors. This is possible when a salt melts or dissolves. When a salt melts, the ions that make up the crystal can freely move past each other, as **Figure 11b** illustrates. Molten salts are good conductors of electric current, although they do not conduct as well as metals. Similarly, if a salt dissolves in water, its ions are no longer held tightly in a crystal. Because the ions are free to move, as shown by the model in **Figure 11c,** the solution can conduct electric current.

As often happens in chemistry, there are exceptions to this rule. There is a small class of ionic compounds that can allow charges to move through their crystals. The lattices of these compounds have an unusually open structure, so certain ions can move past others, jumping from one site to another. One of these salts, zirconium oxide, is used in a device that controls emissions from the exhaust of automobiles.

Figure 11

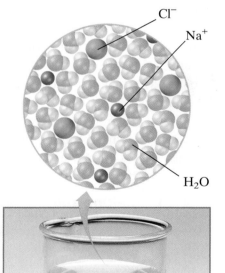

Cl^- Na^+ Cl^- Na^+ Cl^- Na^+ H_2O

a As a solid, an ionic compound has charged particles that are held in fixed positions and cannot conduct electric current.

b When melted, an ionic compound conducts electric current because its charged particles move about more freely.

c When dissolved, an ionic compound conducts electric current because its charged particles move freely.

Salts Are Hard and Brittle

Like most other ionic compounds, table salt is fairly hard and brittle. *Hard* means that the crystal is able to resist a large force applied to it. *Brittle* means that when the applied force becomes too strong to resist, the crystal develops a widespread fracture rather than a small dent. Both of these properties can be attributed to the patterns in which the cations and anions are arranged in all salt crystals.

The ions in a crystal are arranged in a repeating pattern, forming layers. Each layer is positioned so that a cation is next to an anion in the next layer. As long as the layers stay in a fixed position relative to one another, the attractive forces between oppositely charged ions will resist motion. As a result, the ionic compound will be hard, and it will take a lot of energy to break all the bonds between layers of ions.

However, if a force causes one layer to move slightly, ions with the same charge will be positioned next to each other. The cations in one layer are now lined up with other cations in a nearby layer. In the same way, anions from one layer are lined up with other anions in a nearby layer. Because the anions are next to each other, the like charges will repel each other and the layers will split apart. This is why all salts shatter along a line extending through the crystal known as a cleavage plane.

internet connect

www.scilinks.org
Topic: Salt Properties
SciLinks code: HW4113

SCI
LINKS Maintained by the
National Science
Teachers Association

SKILLS Toolkit 1

How to Identify a Compound as Ionic

You can carry out the following procedures in a laboratory to determine if a substance is an ionic compound.

- Examine the substance. All ionic compounds are solid at room temperature. If the substance is a liquid or gas, then it is not an ionic compound. However, if it is a solid, then it *may* or *may not* be an ionic compound.

- Tap the substance gently. Ionic compounds are hard and brittle. If it is an ionic compound, then it should not break apart easily. If it does break apart, the substance should fracture into tinier crystals and not crumble into a powder.

- Heat a sample of the substance. Ionic compounds generally have high melting and boiling points.

- If the substance melts, use a conductivity apparatus to determine if the melted substance conducts electric current. Ionic compounds are good conductors of electric current in the liquid state.

- Dissolve a sample of the substance in water. Use a conductivity apparatus to see if it conducts electric current. Ionic compounds conduct electric current when dissolved in water.

Salt Crystals

The ions in a salt crystal form repeating patterns, with each ion held in place because the attractive forces are stronger than the repulsive ones. The way the ions are arranged is the same in a number of different salts. Not all salts, however, have the same crystal structure as sodium chloride. Despite their differences, the crystals of all salts are made of simple repeating units. These repeating units are arranged in a salt to form a **crystal lattice.** These arrangements of repeating units within a salt are the reason for the crystal shape that can be seen in most salts.

crystal lattice

the regular pattern in which a crystal is arranged

Crystal Structure Depends on the Sizes and Ratios of Ions

As the formula for sodium chloride, NaCl, indicates, there is a 1:1 ratio of sodium cations and chlorine anions. Recall that the attractions in sodium chloride involve more than a single cation and a single anion. **Figure 12a** illustrates the crystal lattice structure of sodium chloride. Within the crystal, each Na^+ ion is surrounded by six Cl^- ions, and, in turn, each Cl^- ion is surrounded by six Na^+ ions. Because this arrangement does not hold for the edges of the crystal, the edges are locations of weak points.

The arrangement of cations and anions to form a crystal lattice depends on the size of the ions. Another factor that affects how the crystal forms is the ratio of cations to anions. Not all salts have a 1:1 ratio of cations to anions as found in sodium chloride. For example, the salt calcium fluoride has one Ca^{2+} ion for every two F^- ions. A Ca^{2+} ion is larger than an Na^+ ion, and an F^- ion is smaller than a Cl^- ion. Because of the size differences of its ions and their ratio in the salt, the crystal lattice structure of calcium fluoride is different from that of sodium chloride. As illustrated in **Figure 12b,** each calcium ion is surrounded by eight fluoride ions. At the same time, each fluoride ion is surrounded by four calcium ions. This is very different from the arrangement of six oppositely charged ions around any given positive or negative ion in a crystal of NaCl.

Figure 12
The crystal structure of sodium chloride (**a**) is not the same as that of calcium fluoride (**b**) because of the differences in the sizes of their ions and the cation-anion ratio making up each salt.

a   sodium chloride   b   calcium fluoride

Salts Have Ordered Packing Arrangements

Salts vary in the types of ions from which they are made. Salts also vary in the ratio of the ions that make up the crystal lattice. Despite these differences, all salts are made of simple repeating units. The smallest repeating unit in a crystal lattice is called a **unit cell.**

The ways in which a salt's unit cells are arranged are determined by a technique called X-ray diffraction crystallography. First, a salt is bombarded with X rays. Then, the X rays that strike ions in the salt are deflected, while X rays that do not strike ions pass straight through the crystal lattice without stopping. The X rays form a pattern on exposed film. By analyzing this pattern, scientists can calculate the positions that the ions in the salt must have in order to cause the X rays to make such a pattern. After this work, scientists can then make models to show how the ions are arranged in the unit cells of the salt.

Analysis of many different salts show that the salts all have ordered packing arrangements, such as those described earlier for NaCl and CaF$_2$. Another example is the salt cesium chloride, where the ratio of cations to anions is 1:1 just as it is in sodium chloride. However, the size of a cesium cation is larger than that of a sodium cation. As a result, the structure of the crystal lattice is different. In sodium chloride, a sodium cation is surrounded by six chloride anions. In cesium chloride, a cesium cation is surrounded by eight chloride anions. The bigger cation has more room around it, so more anions can cluster around it.

unit cell

the smallest portion of a crystal lattice that shows the three-dimensional pattern of the entire lattice

② Section Review

UNDERSTANDING KEY IDEAS

1. What force holds together the ions in a salt?

2. Describe how an ionic bond forms.

3. Why are ionic solids hard?

4. Why are ionic solids brittle?

5. Explain why lattice energy is the key to the formation of a salt.

6. Why do ionic crystals conduct electric current in the liquid phase or when dissolved in water but do not conduct electric current in the solid phase?

CRITICAL THINKING

7. Crystals of the ionic compound calcium fluoride have a different structure from that of the ionic compound calcium chloride. Suggest a reason for this difference.

8. Explain why each of the following pairs is not likely to form an ionic bond.

   a. chlorine and bromine

   b. potassium and helium

   c. sodium and lithium

9. The lattice energy for sodium iodide is 700 kJ/mol, while that for calcium sulfide is 2775 kJ/mol. Which of these salts do you predict has the higher melting point? Explain.

10. The electron affinity for chlorine has a negative value, indicating that the atom readily accepts another electron. Why does a chlorine atom readily accept another electron?

11. Use **Figure 9** on page 169 to describe how the formation of calcium chloride would differ from that of sodium chloride. (Hint: Compare the electron configurations of each atom.)

Names and Formulas of Ionic Compounds

OBJECTIVES

1 **Name** cations, anions, and ionic compounds.

2 **Write** chemical formulas for ionic compounds such that an overall neutral charge is maintained.

3 **Explain** how polyatomic ions and their salts are named and how their formulas relate to their names.

Naming Ionic Compounds

You may recall that chemists call table salt *sodium chloride*. In fact, they have a name for every salt. With thousands of different salts, you might think that it would be hard to remember the names of all of them. But naming salts is very easy, especially for those that are made of a simple cation and a simple anion. These kinds of salts are known as binary ionic compounds. The adjective *binary* indicates that the compound is made up of just two elements.

Rules for Naming Simple Ions

Simple cations borrow their names from the names of the elements. For example, K^+ is known as the potassium ion, and Zn^{2+} is known as the zinc ion. When an element forms two or more ions, the ion names include roman numerals to indicate charge. In the case of copper, Cu, the names of the two ions are written as follows.

$$Cu^+ \text{ copper(I) ion} \qquad Cu^{2+} \text{ copper(II) ion}$$

When we read the names of these ions out loud, we say "copper one ion" or "copper two ion."

The name of a simple anion is also formed from the name of the element, but it ends in *-ide*. Thus, Cl^- is the chloride ion, O^{2-} is the oxide ion, and P^{3-} is the phosphide ion.

The Names of Ions Are Used to Name an Ionic Compound

Naming binary ionic compounds is simple. The name is made up of just two words: the name of the cation followed by the name of the anion.

NaCl sodium chloride	$CuCl_2$ copper(II) chloride
ZnS zinc sulfide	Mg_3N_2 magnesium nitride
K_2O potassium oxide	Al_2S_3 aluminum sulfide

Writing Ionic Formulas

Ionic compounds never have an excess of positive or negative charges. To maintain this balance the total positive and negative charges must be the same. Because both ions in sodium chloride carry a single charge, this compound is made up of equal numbers of the ions Na^+ and Cl^-. As you have seen, the formula for sodium chloride is written as NaCl to show this one-to-one ratio. The cation in zinc sulfide has a 2+ charge and the anion has a 2– charge. Again there is a one-to-one ratio in the salt. Zinc sulfide has the formula ZnS.

Compounds Must Have No Overall Charge

You must take care when writing the formula for an ionic compound where the charges of the cation and anion differ. Consider the example of magnesium nitride. The magnesium ion, Mg^{2+}, has two positive charges, and the nitride ion, N^{3-}, has three negative charges. The cations and anions must be combined in such a way that there are the same number of negative charges and positive charges. Three Mg^{2+} cations are needed for every two N^{3-} anions for electroneutrality. That way, there are six positive charges and six negative charges. Subscripts are used to denote the three magnesium ions and two nitride ions. Therefore, the formula for magnesium nitride is Mg_3N_2.

SKILLS Toolkit 2

Writing the Formula of an Ionic Compound

Follow the following steps when writing the formula of a binary ionic compound, such as iron(III) oxide.

- Write the symbol and charges for the cation and anion. Refer to **Figures 5** and **6** earlier in the chapter for the charges on the ions. The roman numeral indicates which cation iron forms.

    symbol for iron(III): Fe^{3+} symbol for oxide: O^{2-}

- Write the symbols for the ions side by side, beginning with the cation.

    $$Fe^{3+}O^{2-}$$

- To determine how to get a neutral compound, look for the lowest common multiple of the charges on the ions. The lowest common multiple of 3 and 2 is 6. Therefore, the formula should indicate six positive charges and six negative charges.

For six positive charges, you need two Fe^{3+} ions because $2 \times 3+ = 6+$.

For six negative charges, you need three O^{2-} ions because $3 \times 2- = 6-$.

Therefore the ratio of Fe^{3+} to O^{2-} is 2Fe:3O. The formula is written as follows.

$$Fe_2O_3$$

Polyatomic Ions

Fertilizers have potassium, nitrogen, and phosphorus in a form that dissolves easily in water so that plants can absorb them. The potassium in fertilizer is in an ionic compound called *potassium carbonate*. Two ionic compounds in the fertilizer contain the nitrogen—ammonium nitrate and ammonium sulfate. The phosphorus supplied is in another ionic compound, calcium dihydrogen phosphate.

These compounds in fertilizer are made of cations and anions in a ratio so there is no overall charge, like all other ionic compounds. But instead of ions made of a single atom, these compounds contain groups of atoms that are ions.

Many Atoms Can Form One Ion

The adjective *simple* describes an ion formed from a single atom. A simple ion could also be called *monatomic*, which means "one-atom." Just as the prefix *mon-* means "one," the prefix *poly-* means "many." The term **polyatomic ion** means a charged group of two or more bonded atoms that can be considered a single ion. A polyatomic ion as a whole forms ionic bonds in the same way that simple ions do.

Unlike simple ions, most polyatomic ions are made of atoms of several elements. However, polyatomic ions are like simple ones in that their charge is either positive or negative. Consider the polyatomic ion ammonium, NH_4^+, found in many fertilizers. Ammonium is made of one nitrogen and four hydrogen atoms. These atoms have a combined total of 11 protons but only 10 electrons. So the ammonium ion has a 1+ charge overall. This charge is not found on any single atom. Instead, it is spread across this group of atoms, which are bonded together.

The Names of Polyatomic Ions Can Be Complicated

Naming polyatomic ions is not as easy as naming simple cations and anions. Even so, there are rules you can follow to help you remember how to name some of them.

Many polyatomic ions contain oxygen. The endings *-ite* and *-ate* indicate the presence of oxygen. Examples include sulfite, nitrate, and acetate. Often there are several polyatomic ions that differ only in the number of oxygen atoms present. For example, the formulas for two polyatomic ions made from sulfur and oxygen are SO_3^{2-} and SO_4^{2-}. In such cases, the one with less oxygen takes the *-ite* ending, so SO_3^{2-} is named *sulfite*. The ion with more oxygen takes the *-ate* ending, so SO_4^{2-} is named *sulfate*. For the same reason, NO_2^- is named *nitrite*, and NO_3^- is named *nitrate*.

The presence of hydrogen is often indicated by an ion's name starting with *hydrogen*. The prefixes *mono-* and *di-* are also used. Thus, HPO_4^{2-} and $H_2PO_4^-$ are monohydrogen phosphate and dihydrogen phosphate ions, respectively. The prefix *thio-* means "replace an oxygen with a sulfur" in the formula, as in potassium thiosulfate, $K_2S_2O_3$, compared with potassium sulfate, K_2SO_4. **Table 2** lists the names and formulas for some common polyatomic ions. Notice that some are made of more than one atom of the same element, such as peroxide, O_2^{2-}.

polyatomic ion

an ion made of two or more atoms

Table 2
Some Polyatomic Ions

Ion name	Formula
Acetate	CH_3COO^-
Ammonium	NH_4^+
Carbonate	CO_3^{2-}
Chromate	CrO_4^{2-}
Cyanide	CN^-
Dichromate	$Cr_2O_7^{2-}$
Hydroxide	OH^-
Nitrate	NO_3^-
Nitrite	NO_2^-
Permanganate	MnO_4^-
Peroxide	O_2^{2-}
Phosphate	PO_4^{3-}
Sulfate	SO_4^{2-}
Sulfite	SO_3^{2-}
Thiosulfate	$S_2O_3^{2-}$

Naming Compounds with Polyatomic Ions

Follow these steps when naming an ionic compound that contains one or more polyatomic ions, such as K_2CO_3.

- Name the cation. Recall that a cation is simply the name of the element. In this formula, K is potassium that forms a singly charged cation, K^+, of the same name.

- Name the anion. Recall that salts are electrically neutral. Because there are two K^+ cations present in this salt, these two positive charges must be balanced by two negative charges. Therefore, the polyatomic anion in this salt must be CO_3^{2-}. You may find it helpful to think of the formula as follows, although it is not written this way.

$$(K^+)_2(CO_3^{2-})$$

If you check **Table 2,** you will see that the CO_3^{2-} polyatomic ion is called carbonate.

- Name the salt. Recall that the name of a salt is just the names of the cation and anion. The salt K_2CO_3 is potassium carbonate.

SAMPLE PROBLEM A

Formula of a Compound with a Polyatomic Ion

What is the formula for iron(III) chromate?

1 Gather information.

- Use **Figure 6,** found earlier in this chapter, to determine the formula and charge for the iron(III) cation.

$$Fe^{3+}$$

- Use **Table 2,** found earlier in this chapter, to determine the formula and charge for the chromate polyatomic ion.

$$CrO_4^{2-}$$

2 Plan your work.

- Because all ionic compounds are electrically neutral, the total charges of the cations and anions must be equal. To balance the charges, find the least common multiple of the ions' charges. The least common multiple of 2 and 3 is 6. To get six positive charges, you need two Fe^{3+} ions.

$$2 \times 3 = 6+$$

To get six negative charges, you need three CrO_4^{2-} ions.

$$3 \times 2 = 6-$$

> **PRACTICE HINT**
>
> Sometimes parentheses must be placed around the polyatomic cation, as in the formula $(NH_4)_2CrO_4$.

continued on next page

3 Calculate.

- The formula must indicate that two Fe^{3+} ions and three CrO_4^{2-} ions are present. Parentheses are used whenever a polyatomic ion is present more than once. The formula for iron(III) chromate is written as follows.

$$Fe_2(CrO_4)_3$$

Notice that the parentheses show that everything inside the parentheses is tripled by the subscript *3* outside.

4 Verify your result.

- The formula includes the correct symbols for the cation and polyatomic anion.
- The formula reflects that the salt is electrically neutral.

PRACTICE

1 Write the formulas for the following ionic compounds.

a. calcium cyanide

b. rubidium thiosulfate

c. calcium acetate

d. ammonium sulfate

---

3 Section Review

UNDERSTANDING KEY IDEAS

1. In what ways are polyatomic ions like simple ions? In what ways are they different?

2. Why must roman numerals be used when naming certain ionic compounds?

3. What do the endings *-ite* and *-ate* indicate about a polyatomic ion?

4. Explain how calcium, Ca^{2+}, and phosphate, PO_4^{3-}, can make a compound with electroneutrality.

CRITICAL THINKING

5. Name the compounds represented by the following formulas.

a. $Ca(NO_2)_2$

b. $Fe(OH)_3$

c. $(NH_4)_2Cr_2O_7$

d. $CuCH_3COO$

6. Write the formulas for the following ionic compounds made of simple ions.

a. sodium oxide

b. magnesium phosphide

c. silver(I) sulfide

d. niobium(V) chloride

7. Name the following binary ionic compounds. If the metal forms more than one cation, be sure to denote the charge.

a. Rb_2O **b.** FeF_2 **c.** K_3N

PRACTICE PROBLEMS

8. Write formulas for the following compounds.

a. mercury(II) sulfate

b. lithium thiosulfate

c. ammonium phosphate

d. potassium permanganate

Na
Sodium
22.989 77
[Ne]3s^1

Element Spotlight

Where Is Na?

Earth's Crust
2.36% by mass
Seventh most abundant element
Fifth most abundant metal

Sea Water
30.61% of all dissolved materials
1.03% by mass, taking the water
into account

A Major Nutritional Mineral

Sodium is important in the regulation of fluid balance within the body. Most sodium in the diet comes from the use of table salt, NaCl, to season and preserve foods. Sodium is also supplied by compounds such as sodium carbonate and sodium hydrogen carbonate in baked goods. Sodium benzoate is a preservative in carbonated beverages. Sodium citrate and sodium glutamate are used in packaged foods as flavor additives.

In ancient Rome, salt was so scarce and highly prized that it was used as a form of payment. Today, however, salt is plentiful in the diet. Many people must limit their intake of sodium as a precaution against high blood pressure, heart attacks, and strokes.

Food labels list the amount of sodium contained in each serving.

Industrial Uses

- Common table salt is the most important commercial sodium compound. It is used in ceramic glazes, metallurgy, soap manufacture, home water softeners, highway de-icing, herbicides, fire extinguishers, and resins.

- The United States produces about 42.1 million metric tons of sodium chloride per year.

- Elemental sodium is used in sodium vapor lamps for lighting highways, stadiums, and other buildings.

- Liquid sodium is used to cool liquid-metal fast-breeder nuclear reactors.

Real-World Connection For most people, the daily intake of sodium should not exceed 2400 mg.

A Brief History

1807: Sir Humphry Davy isolates sodium by the electrolysis of caustic soda (NaOH) and names the metal.

1990: The Nutrition Labeling and Education Act defines a Daily Reference Value for sodium to be listed in the Nutrition Facts portion of a food label.

| 1200 | 1800 | 1900 | 2000 |

1251: The Wieliczka Salt Mine, located in Krakow, Poland, is started. The mine is still in use today.

1930: Sodium vapor lamps are first used for street lighting.

1940: The Food and Nutrition Board of the National Research Council develops the first Recommended Dietary Allowances.

internet connect

www.scilinks.org
Topic: Sodium
SciLinks code: HW4173

SCiLINKS. Maintained by the National Science Teachers Association

Questions

1. What is one possible consequence of too much sodium in the diet?

2. What is the chemical formula of the most important commercial sodium compound?

CHAPTER HIGHLIGHTS

SECTION ONE   Simple Ions

- Atoms may gain or lose electrons to achieve an electron configuration identical to that of a noble gas.
- Alkali metals and halogens are very reactive when donating and accepting electrons from one another.
- Electrons in the outermost energy level are known as valence electrons.
- Ions are electrically charged particles that have different chemical properties than their parent atoms.

octet rule
ion
cation
anion

SECTION TWO   Ionic Bonding and Salts

- The opposite charges of cations and anions attract to form a tightly packed substance of bonded ions called a *crystal lattice*.
- Salts have high melting and boiling points and do not conduct electric current in the solid state, but they do conduct electric current when melted or when dissolved in water.
- Salts are made of unit cells that have an ordered packing arrangement.

salt
lattice energy
crystal lattice
unit cell

SECTION THREE   Names and Formulas of Ionic Compounds

- Ionic compounds are named by joining the cation and anion names.
- Formulas for ionic compounds are written to show their balance of overall charge.
- A polyatomic ion is a group of two or more atoms bonded together that functions as a single unit.
- Parentheses are used to group polyatomic ions in a chemical formula with a subscript.

polyatomic ion

How to Identify a Compound as Ionic	Writing the Formula of an Ionic Compound	Naming Compounds with Polyatomic Ions	Formula of a Compound with a Polyatomic Ion
Skills Toolkit 1 p. 173	Skills Toolkit 2 p. 177	Skills Toolkit 3 p. 179	Sample Problem A p. 179

CHAPTER REVIEW 5

USING KEY TERMS

1. How is an ion different from its parent atom?

2. What does a metal atom need to do in order to form a cation?

3. What does a nonmetal element need to do to form an anion?

4. Explain how the octet rule describes how atoms form stable ions.

5. Explain how you can tell from an element's number of valence electrons whether the element is more likely to form a cation or an anion.

6. Why is lattice energy the key to forming an ionic bond?

7. Explain why it is appropriate to group a polyatomic ion in parentheses in a chemical formula, if more than one of that ion is present in the formula.

UNDERSTANDING KEY IDEAS

Simple Ions

8. The electron configuration for arsenic, As, is $[Ar]3d^{10}4s^2 4p^3$. How many valence electrons does an As atom have? Write the symbol for the ion it forms to achieve a noble-gas configuration.

9. Explain why magnesium forms Mg^{2+} cations and not Mg^{6-} anions.

10. Explain why the properties of an ion differ from its parent atom.

11. How does the octet rule help predict the chemical reactivity of an element?

12. Why are the halogens so reactive?

13. If helium does not obey the octet rule, then why do its atoms not react?

14. How do the stable ions of the transition metals differ from those formed by the metals in Groups 1 and 2?

15. Explain why metals tend to form cations, while nonmetals tend to form anions.

16. Which of the following diagrams illustrates the electron diagram for a potassium ion found in the nerve cells of your body? (Hint: potassium's atomic number is 19.)

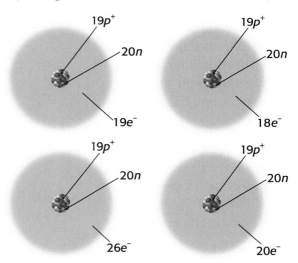

Ionic Bonding and Salts

17. Why do most ionic compounds have such high melting and boiling points?

18. Explain the importance of lattice energy in the formation of a salt.

19. Explain why a salt usually dissolves slowly.

20. What is the relationship between lattice energy and the strength of an ionic bond?

21. Why can't an ionic bond form between potassium and magnesium?

Names and Formulas of Ionic Compounds

22. What is the difference between the chlorite ion and the chlorate ion?

23. How are simple cations and anions named?

24. Why is peroxide, O_2^{2-}, considered a polyatomic ion and not a simple ion?

25. Identify and name the cations and anions that make up the following ionic compounds and indicate the charge on each ion.
  a. $NaNO_3$ **c.** $(NH_4)_2CrO_4$
  b. K_2SO_3 **d.** $Al_2(SO_4)_3$

26. Name the compounds represented by the following formulas.
  a. $Cu_3(PO_4)_2$ **c.** Cu_2O
  b. $Fe(NO_3)_3$ **d.** CuO

PRACTICE PROBLEMS

PROBLEM SOLVING SKILL

27. Write formulas for the following ionic compounds.
  a. lithium sulfate
  b. strontium nitrate
  c. ammonium acetate
  d. titanium(III) sulfate

28. Write formulas for the following ionic compounds.
  a. silver nitrate
  b. calcium thiosulfate
  c. barium hydroxide
  d. cobalt(III) phosphate

29. Complete the table below, and then use it to answer the questions that follow.

Element	Ion	Name of ion
Barium	Ba^{2+}	
Chlorine		chloride
Chromium	Cr^{3+}	
Fluorine	F^-	
Manganese		manganese(II)
Oxygen		oxide

Write the formula for the following substances:
  a. manganese chloride
  b. chromium(III) fluoride
  c. barium oxide

MIXED REVIEW

30. Write formulas for the following ionic compounds.
  a. barium nitrate
  b. calcium fluoride
  c. potassium iodide
  d. ammonium hydroxide

31. Name the following polyatomic ions.
  a. O_2^{2-} **c.** NH_4^+
  b. CrO_4^{2-} **d.** CO_3^{2-}

32. Complete the table below.

Atom	Ion	Noble-gas configuration of ion
S		
Be		
I		
Rb		
O		
Sr		
F		

33. Write formulas for the following polyatomic ions.
  a. cyanide   **c.** nitrite
  b. sulfate   **d.** permanganate

34. Name the following ionic compounds.
  a. $AlPO_4$ **c.** KCN
  b. $SrSO_4$ **d.** CuS

35. Determine the number of valence electrons in the following atoms.
  a. Al   **c.** Si
  b. Rb   **d.** F

CRITICAL THINKING

36. Why are most metals found in nature as ores and not as pure metals?

37. Why can't sodium gain a positive charge by acquiring a proton in its nucleus?

38. The electron configuration for a lithium atom is $1s^2 2s^1$. The configuration for an iodine atom is $1s^2 2s^2 2p^6 3s^2 3p^6 4s^2 3d^{10} 4p^6 5s^2 4d^{10} 5p^5$.

Write the electron configurations for the ions that form lithium iodide, a salt used in photography.

39. Why are there no rules for naming Group 18 ions?

40. Determine the ratios of cations to anions that are most likely in the formulas for ionic substances of the following elements:
a. an alkali metal and a halogen
b. an alkaline-earth metal and a halogen
c. an alkali metal and a member of Group 16
d. an alkaline-earth metal and a member of Group 16

41. Sodium chloride is prepared by reacting sodium with chlorine gas. Another way to prepare it is to react sodium hydroxide with hydrochloric acid and allow the water to evaporate. Will the method of preparation affect the type of unit cell and crystalline structure of sodium chloride? Explain.

42. Compound B has lower melting and boiling points than compound A does. At the same temperature, compound B vaporizes faster and to a greater extent than compound A. If only one of these compounds is ionic, which one would you expect it to be? Why?

ALTERNATIVE ASSESSMENT

43. Ions play an important physiological role in your body, such as nerve impulse conduction and muscle contraction. Select one such ion, and prepare a report detailing its function. Be sure to include recent medical information, including any diseases associated with a deficiency of the ion.

44. A number of homes have "hard water," which, as you learned in the Start-Up Activity, does not produce as many soap suds as water that contains fewer ions. Such homes often have water conditioners that remove the ions from the water, making it "softer" and more likely to produce soapsuds. Research how such water softeners operate by checking the Internet or by contacting a company that sells such devices. Design an experiment to test the effectiveness of the softener in removing ions from water.

45. Devise a set of criteria that you can use to classify the following substances as ionic or nonionic compounds: $CaCO_3$, Cu, H_2O, $NaBr$, and C (graphite). Show your criteria to your teacher. If your plan is approved, carry out the analysis and report your results to the class.

CONCEPT MAPPING

46. Build a concept map about ionic bonding. Be sure to include the following terms in your concept map: *atoms*, *valence electrons*, *ions*, *cations*, *anions*, and *ionic compounds*.

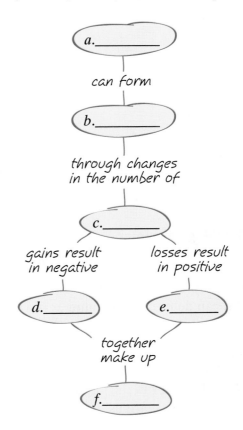

Study the graph below, and answer the questions that follow.
For help in interpreting graphs, see Appendix B, "Study Skills for Chemistry."

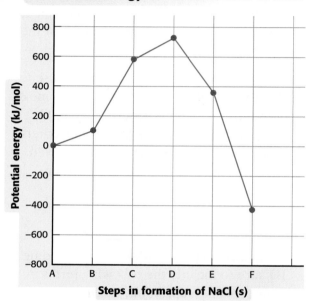

Potential Energy in the Formation of NaCl

Potential energy (kJ/mol)

Steps in formation of NaCl (s)

The graph shows the changes in potential energy that occur when an ionic bond forms between Na(s) and $Cl_2(g)$. The reactants, solid sodium and chlorine gas, start at an initial energy state that is assigned a value of zero at 25°C and 1 atm of pressure.

47. In terms of energy, what do the steps from point A to point D have in common?

48. What do the steps from point D to point F have in common?

49. What is occurring between points D and E?

50. Write the word equation to show what happens between points B and C when electrons are removed from 1 mol of sodium atoms.

51. Which portion of this graph represents the lattice energy involved in the formation of an ionic bond between sodium and chlorine?

52. Calculate the quantity of energy released when 2.5 mol of NaCl form.

TECHNOLOGY AND LEARNING

53. Graphing Calculator

Calculating the Number of Valence Electrons

The graphing calculator can run a program that can determine the number of valence electrons in an atom, given its atomic number.

Go to Appendix C. If you are using a TI-83 Plus, you can download the program **VALENCE** and run the application as directed. If you are using another calculator,

your teacher will provide you with keystrokes to use. After you have run the program, answer these questions.

How many valence electrons are there in the following atoms?

a. Rutherfordium, Rf, atomic number 104

b. Gold, Au, atomic number 79

c. Molybdenum, Mo, atomic number 42

d. Indium, In, atomic number 49

STANDARDIZED TEST PREP

1. The correct formula for a copper atom that has lost two electrons is
   a. Co^+.
   b. Co^{2+}.
   c. Cu^{2-}.
   d. Cu^{2+}.

2. Which of the following atoms can achieve the same electron configuration as a noble gas when the atom forms an ion?
   a. argon
   b. potassium
   c. nickel
   d. iron

3. In forming NaCl, energy is required to
   a. change chlorine to a gas.
   b. add an electron to a chlorine atom.
   c. remove an electron from a sodium atom.
   d. bring together the sodium ions and chloride ions.

4. Which of the following is not a characteristic of a salt?
   a. hardness
   b. a high melting point
   c. an ability to conduct electric current in the molten state
   d. an ability to conduct electric current in the solid state

5. Which elements on the periodic table are most likely to form two or more ions with different charges?
   a. halogens
   b. alkali metals
   c. transition metals
   d. noble gases

6. Which of the following compounds does not contain a polyatomic ion?
   a. sodium carbonate
   b. sodium sulfate
   c. sodium sulfite
   d. sodium sulfide

7. The correct formula for ammonium phosphate is
   a. $(NH_4)_3PO_4$.
   b. $(NH_4)_2PO_4$.
   c. NH_4PO_4.
   d. $NH_4(PO_4)_2$.

8. When writing the formula for a compound that contains a polyatomic ion,
   a. write the anion's formula first.
   b. use superscripts to show the number of polyatomic ions present.
   c. use parentheses if the number of polyatomic ions is greater than 1.
   d. always place the polyatomic ion in parentheses.

9. Compounds that are electrically neutral must have equal numbers of
   a. anions and cations.
   b. positive and negative charges.
   c. molecules.
   d. ionic bonds.

10. The correct name for NH_4CH_3COO is
    a. ammonium carbonate.
    b. ammonium hydroxide.
    c. ammonium acetate.
    d. ammonium nitrate.

11. Which of the following is the correct formula for iron(III) sulfate?
    a. Fe_3SO_4
    b. $Fe_3(SO_4)_2$
    c. $Fe_2(SO_4)_3$
    d. $3FeSO_4$

12. How many valence electrons are indicated by the following electron configuration? $1s^2 2s^2 2p^6 3s^2 3p^5$
    a. two
    b. five
    c. three
    d. seven

13. Metals tend to lose one or more electrons to form
    a. cations.
    b. polyatomic ions
    c. anions.
    d. unit cells.

COVALENT COMPOUNDS

Natural rubber comes from tropical trees. It is soft and sticky, so it has little practical use. However, while experimenting with rubber in 1839, Charles Goodyear dropped a mixture of sulfur and natural rubber on a hot stove by mistake. The heated rubber became tough and elastic because of the formation of covalent bonds. The resulting compound was *vulcanized rubber*, which is strong enough to make up a basketball that can take a lot of hard bounces.

START-UP ACTIVITY

SAFETY PRECAUTIONS

Ionic Versus Covalent

PROCEDURE

1. Clean and dry **three test tubes.** Place a small amount of **paraffin wax** into the first test tube. Place an equal amount of **table salt** into the second test tube. Place an equal amount of **sugar** into the third test tube.

2. Fill a **plastic-foam cup** halfway with **hot water.** Place the test tubes into the water. After 3 min, remove the test tubes from the water. Observe the contents of the test tubes, and record your observations.

3. Place a small amount of each substance on a **watch glass.** Crush each substance with a **spatula.** Record your observations.

4. Add **10 mL deionized water** to each test tube. Use a **stirring rod** to stir each test tube. Using a **conductivity device** (watch your teacher perform the conductivity tests), record the conductivity of each mixture.

ANALYSIS

1. Summarize the properties you observed for each compound.

2. Ionic bonding is present in many compounds that are brittle, have a high melting point, and conduct electric current when dissolved in water. Covalent bonding is present in many compounds that are not brittle, have a low melting point, and do not conduct electric current when mixed with water. Identify the type of bonding present in paraffin wax, table salt, and sugar.

Pre-Reading Questions

① **What determines whether two atoms will form a bond?**

② **How can a hydrogen atom, which has one valence electron, bond with a chlorine atom, which has seven valence electrons?**

③ **What happens in terms of energy after a hydrogen atom bonds with a chlorine atom?**

internet connect

www.scilinks.org
Topic: Rubber
SciLinks code: HW4128

SCILINKS Maintained by the National Science Teachers Association

Covalent Bonds

KEY TERMS

- **covalent bond**
- **molecular orbital**
- **bond length**
- **bond energy**
- **nonpolar covalent bond**
- **polar covalent bond**
- **dipole**

OBJECTIVES

(1) **Explain** the role and location of electrons in a covalent bond.

(2) **Describe** the change in energy and stability that takes place as a covalent bond forms.

(3) **Distinguish** between nonpolar and polar covalent bonds based on electronegativity differences.

(4) **Compare** the physical properties of substances that have different bond types, and relate bond types to electronegativity differences.

Sharing Electrons

The diver shown in **Figure 1** is using a hot flame to cut metal under water. The flame is made by a chemical reaction in which hydrogen and oxygen gases combine. When these gases react, atoms and electrons rearrange to form a new, more stable compound: water.

You learned that electrons are rearranged when an ionic bond forms. When this happens, electrons transfer from one atom to another to form charged ions. The reaction of hydrogen and oxygen to form water causes another kind of change involving electrons. In this case, the neutral atoms share electrons.

Figure 1
This diver is using an oxyhydrogen torch. The energy released by the torch comes from a chemical reaction in which hydrogen and oxygen react to form water.

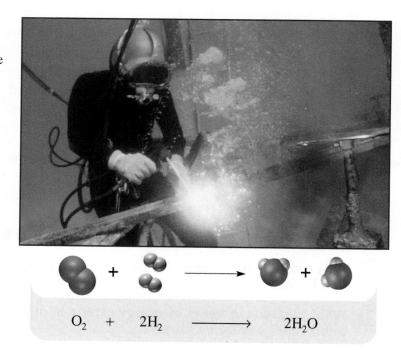

$$O_2 \quad + \quad 2H_2 \quad \longrightarrow \quad 2H_2O$$

internet connect

www.scilinks.org
Topic: Covalent Bonding
SciLinks code: HW4036

SCiLINKS. Maintained by the National Science Teachers Association

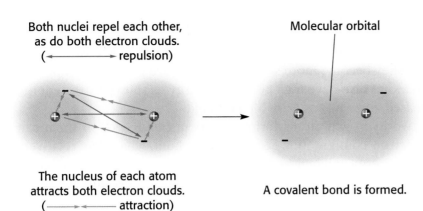

Both nuclei repel each other, as do both electron clouds.
($\longleftrightarrow$ repulsion)

The nucleus of each atom attracts both electron clouds.
($\longrightarrow\leftarrow$ attraction)

Molecular orbital

A covalent bond is formed.

Figure 2
The positive nucleus of one hydrogen atom attracts the electron of the other atom. At the same time, the two atoms' positive nuclei repel each other. The two electron clouds also repel each other.

Forming Molecular Orbitals

The simplest example of sharing electrons is found in diatomic molecules, such as hydrogen, H_2. **Figure 2** shows the attractive and repulsive forces that exist when two hydrogen atoms are near one another. When these forces are balanced, the two hydrogen atoms form a bond. Because both atoms are of the same element, the attractive force of each atom is the same. Thus, neither atom will remove the electron from the other atom. Instead of transferring electrons to each other, the two hydrogen atoms share the electrons.

The result is a H_2 molecule that is more stable than either hydrogen atom is by itself. The H_2 molecule is stable because each H atom has a shared pair of electrons. This shared pair gives both atoms a stability similar to that of a helium configuration. Helium is stable because its atoms have filled orbitals.

The sharing of a pair of electrons is the bond that holds the two hydrogen atoms together. When two atoms share electrons, they form a **covalent bond.** The shared electrons move in the space surrounding the nuclei of the two hydrogen atoms. The space that these shared electrons move within is called a **molecular orbital.** As shown in **Figure 2,** a molecular orbital is made when two atomic orbitals overlap. Sugar and water, shown in **Figure 3,** have molecules with covalent bonds.

covalent bond

a bond formed when atoms share one or more pairs of electrons

molecular orbital

the region of high probability that is occupied by an individual electron as it travels with a wavelike motion in the three-dimensional space around one of two or more associated nuclei

Figure 3
The sugar, $C_{12}H_{22}O_{11}$, and water, H_2O, in the tea are examples of covalent, or molecular, compounds.

Covalent Compounds **191**

Energy and Stability

Most individual atoms have relatively low stability. (Noble gases are the exception.) They become more stable when they are part of a compound. Unbonded atoms also have high potential energy, as shown by the energy that is released when atoms form a compound.

After two hydrogen atoms form a covalent bond, each of them can have an electron configuration like that of helium, which has relatively low potential energy and high stability. Thus, bonding causes a decrease in energy for the atoms. This energy is released to the atoms' surroundings.

Energy Is Released When Atoms Form a Covalent Bond

Figure 4 shows the potential energy changes that take place as two hydrogen atoms come near one another. In part (a) of the figure, the distance between the two atoms is large enough that there are no forces between their nuclei. At this distance, the potential energy of the atoms is arbitrarily set at zero.

In part (b) of the figure, the potential energy decreases as the attractive electric force pulls the two atoms closer together. As the potential energy goes down, the system gives off energy. In other words, energy is released as the attractive force pulls the atoms closer. Eventually, the atoms get close enough that the attractive forces between the electrons of one atom and the nucleus of the other atom are balanced by the repulsive force caused by the two positively charged nuclei as they are forced closer together. The two hydrogen atoms are now covalently bonded. In part (c) of the figure, the two atoms have bonded, and they are at their lowest potential energy. If they get any closer, repulsive forces will take over between the nuclei.

Potential Energy Determines Bond Length

bond length

the distance between two bonded atoms at their minimum potential energy; the average distance between the nuclei of two bonded atoms

Part (c) of **Figure 4** shows that when the two bonded hydrogen atoms are at their lowest potential energy, the distance between them is 75 pm. This distance is considered the length of the covalent bond between two hydrogen atoms. The distance between two bonded atoms at their minimum potential energy is known as the **bond length.**

Figure 4
Two atoms form a covalent bond at a distance where attractive and repulsive forces balance. At this point, the potential energy is at a minimum.

Potential Energy Curve for H₂

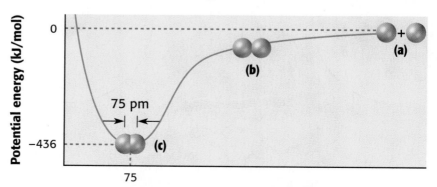

Distance between hydrogen nuclei (pm)

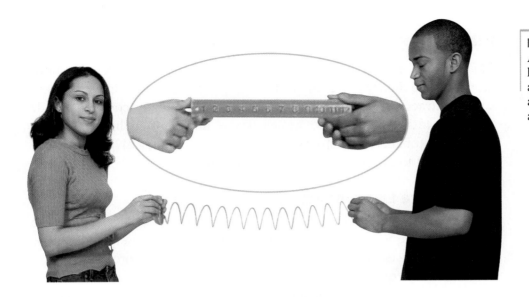

Figure 5
A covalent bond is more like a flexible spring than a rigid ruler, because the atoms can vibrate back and forth.

Bonded Atoms Vibrate, and Bonds Vary in Strength

Models often incorrectly show covalent bonds as rigid "sticks." If these bonds were in fact rigid, then the nuclei of the bonded atoms would be at a fixed distance from one another. Because the ruler held by the students in the top part of **Figure 5** is rigid, the students are at a fixed distance from one another.

However, a covalent bond is more flexible, like two students holding a spring. The two nuclei vibrate back and forth. As they do, the distance between them constantly changes. The bond length is in fact the average distance between the two nuclei.

At a bond length of 75 pm, the potential energy of H_2 is –436 kJ/mol. This means that 436 kJ of energy is released when 1 mol of bonds form. It also means that 436 kJ of energy must be supplied to break the bonds and separate the hydrogen atoms in 1 mol of H_2 molecules. The energy required to break a bond between two atoms is the **bond energy. Table 1** lists the energies and lengths of some common bonds in order of decreasing bond energy. Note that the bonds that have the highest bond energies (the "strongest" bonds) usually involve the elements H or F. Also note that stronger bonds generally have shorter bond lengths.

bond energy

the energy required to break the bonds in 1 mol of a chemical compound

Table 1	**Bond Energies and Bond Lengths for Single Bonds**				
	Bond energy (kJ/mol)	**Bond length (pm)**		**Bond energy (kJ/mol)**	**Bond length (pm)**
H—F	570	92	**H—I**	299	161
C—F	552	138	**C—Br**	280	194
O—O	498	121	**Cl—Cl**	243	199
H—H	436	75	**C—I**	209	214
H—Cl	432	127	**Br—Br**	193	229
C—Cl	397	177	**F—F**	159	142
H—Br	366	141	**I—I**	151	266

Electronegativity and Covalent Bonding

The example in which two hydrogen atoms bond is simple because both atoms are the same. Also, each one has a single proton and a single electron, so the attractions are easy to identify. However, many covalent bonds form between two different atoms. These atoms often have different attractions for shared electrons. In such cases, electronegativity values are a useful tool to predict what kind of bond will form.

Atoms Share Electrons Equally or Unequally

Figure 6 lists the electronegativity values for several elements. In a molecule such as H_2, the values of the two atoms in the bond are equal. Because each one attracts the bonding electrons with the same force, they share the electrons equally. A **nonpolar covalent bond** is a covalent bond in which the bonding electrons in the molecular orbital are shared equally.

What happens when the electronegativity values are not the same? If the values differ significantly, the two atoms form a different type of covalent bond. Think about a carbon atom bonding with an oxygen atom. The O atom has a higher electronegativity and attracts the bonding electrons more than the C atom does. As a result, the two atoms share the bonding electrons, but unequally. This type of bond is a **polar covalent bond.** In a polar covalent bond, the shared electrons, which are in a molecular orbital, are more likely to be found nearer to the atom whose electronegativity is higher.

If the difference in electronegativity values of the two atoms is great enough, the atom with the higher value may remove an electron from the other atom. An ionic bond will form. For example, the electronegativity difference between magnesium and oxygen is great enough for an O atom to remove two electrons from a Mg atom. **Figure 7** shows a model of how to classify bonds based on electronegativity differences. Keep in mind that the boundaries between bond types are arbitrary. This model is just one way that you can classify bonds. You can also classify bonds by looking at the characteristics of the substance.

Topic Link

Refer to the "Periodic Table" chapter for more about electronegativity.

nonpolar covalent bond

a covalent bond in which the bonding electrons are equally attracted to both bonded atoms

polar covalent bond

a covalent bond in which a shared pair of electrons is held more closely by one of the atoms

Figure 6
Fluorine has an electronegativity of 4.0, the highest value of any element.

Electronegativities

H 2.2																	
Li 1.0	Be 1.6											B 2.0	C 2.6	N 3.0	O 3.4	F 4.0
Na 0.9	Mg 1.3											Al 1.6	Si 1.9	P 2.2	S 2.6	Cl 3.2
K 0.8	Ca 1.0	Sc 1.4	Ti 1.5	V 1.6	Cr 1.7	Mn 1.6	Fe 1.9	Co 1.9	Ni 1.9	Cu 2.0	Zn 1.7	Ga 1.8	Ge 2.0	As 2.2	Se 2.5	Br 3.0
Rb 0.8	Sr 1.0	Y 1.2	Zr 1.3	Nb 1.6	Mo 2.2	Tc 1.9	Ru 2.2	Rh 2.3	Pd 2.2	Ag 1.9	Cd 1.7	In 1.8	Sn 1.9	Sb 2.0	Te 2.1	I 2.7
Cs 0.8	Ba 0.9	La 1.1	Hf 1.3	Ta 1.5	W 2.4	Re 1.9	Os 2.2	Ir 2.2	Pt 2.3	Au 2.5	Hg 2.0	Tl 1.8	Pb 2.1	Bi 2.0	Po 2.0	At 2.2
Fr 0.7	Ra 0.9	Ac 1.1														

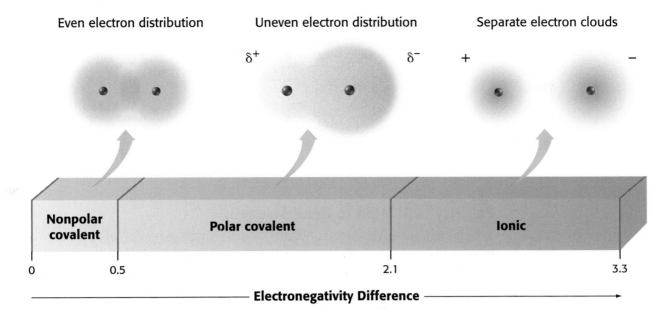

Even electron distribution   Uneven electron distribution   Separate electron clouds

δ^+ δ^- +   −

| Nonpolar covalent | Polar covalent | Ionic |

0   0.5   2.1   3.3

⟶ **Electronegativity Difference** ⟶

Polar Molecules Have Positive and Negative Ends

Hydrogen fluoride, HF, in solution is used to etch glass, such as the vase shown in **Figure 8.** The difference between the electronegativity values of hydrogen and fluorine shows that H and F atoms form a polar covalent bond. The word *polar* suggests that this bond has ends that are in some way opposite one another, like the two poles of a planet, a magnet, or a battery. In fact, the ends of the HF molecule have opposite partial charges.

The electronegativity of fluorine (4.0) is much higher than that of hydrogen (2.2). Therefore, the shared electrons are more likely to be found nearer to the fluorine atom. For this reason, the fluorine atom in the HF molecule has a partial negative charge. In contrast, the shared electrons are less likely to be found nearer to the hydrogen atom. As a result, the hydrogen atom in the HF molecule has a partial positive charge. A molecule in which one end has a partial positive charge and the other end has a partial negative charge is called a **dipole.** The HF molecule is a dipole.

To emphasize the dipole nature of the HF molecule, the formula can be written as $H^{\delta+}F^{\delta-}$. The symbol δ is a lowercase Greek *delta,* which is used in science and math to mean *partial.* With polar molecules, such as HF, the symbol δ^+ is used to show a partial positive charge on one end of the molecule. Likewise, the symbol δ^- is used to show a partial negative charge on the other end.

Although δ^+ means a positive charge, and δ^- means a negative charge, these symbols do not mean that the bond between hydrogen and fluorine is ionic. An electron is not transferred completely from hydrogen to fluorine, as in an ionic bond. Instead, the atoms share a pair of electrons, which makes the bond covalent. However, the shared pair of electrons is more likely to be found nearer to the fluorine atom. This unequal distribution of charge makes the bond polar covalent.

Figure 7
Electronegativity differences can be used to predict the properties of a bond. Note that there are no distinct boundaries between the bond types—the distinction is arbitrary.

dipole

a molecule or part of a molecule that contains both positively and negatively charged regions

Figure 8
Hydrogen fluoride, HF, is an acid that is used to etch beautiful patterns in glass.

Covalent Compounds   **195**

Table 2	Electronegativity Difference for Hydrogen Halides	
Molecule	**Electronegativity difference**	**Bond energy**
H—F	1.8	570 kJ/mol
H—Cl	1.0	432 kJ/mol
H—Br	0.8	366 kJ/mol
H—I	0.5	298 kJ/mol

Polarity Is Related to Bond Strength

When examining the electronegativity differences between elements, you may notice a connection between electronegativity difference, the polarity of a bond, and the strength of that bond. The greater the difference between the electronegativity values of two elements joined by a bond, the greater the polarity of the bond. In addition, greater electronegativity differences tend to be associated with stronger bonds. Of the compounds listed in **Table 2,** H—F has the greatest electronegativity difference and thus the greatest polarity. Notice that H—F also requires the largest input of energy to break the bond and therefore has the strongest bond.

Electronegativity and Bond Types

You have learned that when sodium and chlorine react, an electron is removed from Na and transferred to Cl to form Na$^+$ and Cl$^-$ ions. These ions form an ionic bond. However, when hydrogen and oxygen gas react, their atoms form a polar covalent bond by sharing electrons. How do you know which type of bond the atoms will form? Differences in electronegativity values provide one model that can tell you.

Bonds Can Be Classified by Bond Character

Figure 7 shows the relationship between electronegativity differences and the type of bond that forms between two elements. Notice the general rule that can be used to predict the type of bond that forms. If the difference in electronegativity is between 0 and 0.5, the bond is probably nonpolar covalent. If the difference in electronegativity is between 0.5 and 2.1, the bond is considered polar covalent. If the difference is larger than 2.1, then the bond is usually ionic. Remember that this method of classifying bonds is just one model. Another general rule states that covalent bonds tend to form between nonmetals, while a nonmetal and a metal will form an ionic bond.

You can see how electronegativity differences provide information about bond character. Think about the bonds that form between the ions sodium and fluoride and between the ions calcium and oxide. The electronegativity difference between Na and F is 3.1. Therefore, they form an ionic bond. The electronegativity difference between Ca and O is 2.4. They also form an ionic bond. However, the larger electronegativity difference between Na and F means that the bond between them has a higher percentage of ionic character.

Next think about the bonds that form between carbon and chlorine and between aluminum and chlorine. The electronegativity difference between C and Cl is 0.6. These two elements form a polar covalent bond. The electronegativity difference between Al and Cl is 1.6. These two elements also form a polar covalent bond. However, the larger difference between Al and Cl means that the bond between these two elements is more polar, with greater partial charges, than the bond between C and Cl is.

Properties of Substances Depend on Bond Type

The type of bond that forms determines the physical and chemical properties of the substance. For example, metals, such as potassium, are very good electric conductors in the solid state. This property is the result of metallic bonding. Metallic bonds are the result of the attraction between the electrons in the outermost energy level of each metal atom and all of the other atoms in the solid metal. The metal atoms are held in the solid because all of the valence electrons are attracted to all of the atoms in the solid. These valence electrons can move easily from one atom to another. They are free to roam around in the solid and can conduct an electric current.

Table 3 **Properties of Substances with Metallic, Ionic, and Covalent Bonds**

Bond type	Metallic	Ionic	Covalent
Example substance	potassium	potassium chloride	chlorine
Melting point (°C)	63	770	−101
Boiling point (°C)	760	1500 (sublimes)	−34.6
Properties	• soft, silvery, solid • conductor as a solid	• crystalline, white solid • conductor when dissolved in water	• greenish yellow gas • not a good conductor

In ionic substances, the overall attraction between all the cations and anions is very strong. Ionic compounds, such as potassium chloride, KCl, are made up of many K^+ and Cl^- ions. Each ion is held into place by many oppositely charged neighbors, so the forces—the ionic bonds—that hold them together are very strong and hard to break.

In molecular substances, such as Cl_2, the molecules are held together by sharing electrons. The shared electrons are attracted to the two bonding atoms, and they have little attraction for the atoms of other nearby molecules. Therefore, the attractive forces between separate Cl_2 molecules are very small compared to the attractive forces between the ions in KCl.

The difference in the strength of attraction between the basic units of ionic and molecular substances gives rise to different properties in the two types of substances. For example, the stronger the force between the ions or molecules of a substance in a liquid state, the more energy is required for the substance to change into a gas. **Table 3** shows that the strong forces in ionic substances, such as KCl, account for the high melting and boiling points they have compared to molecular substances, such as Cl_2. You will learn more about this relationship in a later chapter. The table also compares the conductivity of each substance.

Section Review

UNDERSTANDING KEY IDEAS

1. Describe the attractive forces and repulsive forces that exist between two atoms as the atoms move closer together.

2. Compare a bond between two atoms to a spring between two students.

3. In what two ways can two atoms share electrons when forming a covalent bond?

4. What happens in terms of energy and stability when a covalent bond forms?

5. How can the partial charges be shown in a polar covalent molecule?

6. What information can be obtained by knowing the electronegativity differences between two elements?

7. Why do molecular compounds have low melting points and low boiling points relative to ionic substances?

CRITICAL THINKING

8. Why does the distance between two nuclei in a covalent bond vary?

9. How does a molecular orbital differ from an atomic orbital?

10. How does the strength of a covalent bond relate to bond length?

11. Compare the degree of polarity in HF, HCl, HBr, and HI.

12. Given that it has the highest electronegativity, can a fluorine atom ever form a nonpolar covalent bond? Explain your answer.

13. What does a small electronegativity difference reveal about the strength of a covalent bond?

14. Based on electronegativity values, which bond has the highest degree of ionic character: H—S, Si—Cl, or Cs—Br?

② Drawing and Naming Molecules

KEY TERMS

- **valence electron**
- **Lewis structure**
- **unshared pair**
- **single bond**
- **double bond**
- **triple bond**
- **resonance structure**

OBJECTIVES

① **Draw** Lewis structures to show the arrangement of valence electrons among atoms in molecules and polyatomic ions.

② **Explain** the differences between single, double, and triple covalent bonds.

③ **Draw** resonance structures for simple molecules and polyatomic ions, and recognize when they are required.

④ **Name** binary inorganic covalent compounds by using prefixes, roots, and suffixes.

Lewis Electron-Dot Structures

Both ionic and covalent bonds involve **valence electrons,** the electrons in the outermost energy level of an atom. In 1920, G. N. Lewis, the American chemist shown in **Figure 9,** came up with a system to represent the valence electrons of an atom. This system—known as electron-dot diagrams or **Lewis structures** —uses dots to represent valence electrons. Lewis's system is a valuable model for covalent bonding. However, these diagrams do not show the actual locations of the valence electrons. They are models that help you to keep track of valence electrons.

Lewis Structures Model Covalently Bonded Molecules

A Lewis structure shows only the valence electrons in an atom or molecule. The nuclei and the electrons of the inner energy levels (if any) of an atom are represented by the symbol of the element. With only one valence electron, a hydrogen atom has the electron configuration $1s^1$. When drawing hydrogen's Lewis structure, you represent the nucleus by the element's symbol, H. The lone valence electron is represented by a dot.

H·

When two hydrogen atoms form a nonpolar covalent bond, they share two electrons. These two electrons are represented by a pair of dots between the symbols.

H:H

This Lewis structure represents a stable hydrogen molecule in which both atoms share the same pair of electrons.

valence electron

> an electron that is found in the outermost shell of an atom and that determines the atom's chemical properties

Lewis structure

> a structural formula in which electrons are represented by dots; dot pairs or dashes between two atomic symbols represent pairs in covalent bonds

Figure 9
G. N. Lewis (1875–1946) not only came up with important theories of bonding but also gave a new definition to acids and bases.

Figure 10
An electron configuration shows all of the electrons of an atom, while the Lewis structure, above, shows only the valence electrons.

Table 4 Lewis Structures of the Second-Period Elements

Element	Electron configuration	Number of valence electrons	Lewis structure (for bonding)
Li	$1s^2 2s^1$	1	Li·
Be	$1s^2 2s^2$	2	Be·
B	$1s^2 2s^2 2p^1$	3	·B·
C	$1s^2 2s^2 2p^2$	4	·C·
N	$1s^2 2s^2 2p^3$	5	:N·
O	$1s^2 2s^2 2p^4$	6	:O·
F	$1s^2 2s^2 2p^5$	7	:F·
Ne	$1s^2 2s^2 2p^6$	8	:Ne:

Lewis Structures Show Valence Electrons

The Lewis structure of a chlorine atom shows only the atom's seven valence electrons. Its Lewis structure is written with three pairs of electrons and one unpaired electron around the element's symbol, as shown below and in **Figure 10**.

:Cl·

Table 4 shows the Lewis structures of the elements in the second period of the periodic table as they would appear in a bond. Notice that as you go from element to element across the period, you add a dot to each side of the element's symbol. You do not begin to pair dots until all four sides of the element's symbol have a dot.

An element with an octet of valence electrons, such as that found in the noble gas Ne, has a stable configuration. When two chlorine atoms form a covalent bond, each atom contributes one electron to a shared pair. With this shared pair, both atoms can have a stable octet. This tendency of bonded atoms to have octets of valence electrons is called the *octet rule*.

:Cl:Cl:

unshared pair

a nonbonding pair of electrons in the valence shell of an atom; also called *lone pair*

single bond

a covalent bond in which two atoms share one pair of electrons

Each chlorine atom in Cl_2 has three pairs of electrons that are not part of the bond. These pairs are called **unshared pairs** or *lone pairs*. The pair of dots that represents the shared pair of electrons can also be shown by a long dash. Both notations represent a **single bond.**

:Cl:Cl: or :Cl—Cl:

For Lewis structures of bonded atoms, you may want to keep in mind that when the dots for the valence electrons are placed around the symbol, each side must contain an unpaired electron before any side can contain a pair of electrons. For example, see the Lewis structure for a carbon atom below.

$$\cdot \overset{\displaystyle \cdot}{C} \cdot$$

The electrons can pair in any order. However, any unpaired electrons are usually filled in to show how they will form a covalent bond. For example, think about the bonding between hydrogen and chlorine atoms.

$$\text{H} \cdot \; + \; \cdot \ddot{\underset{\cdot\cdot}{Cl}} \colon \; \longrightarrow \; \text{H} \!-\! \ddot{\underset{\cdot\cdot}{Cl}} \colon$$

SKILLS Toolkit 1

Drawing Lewis Structures with Many Atoms

1. Gather information.
- Draw a Lewis structure for each atom in the compound. When placing valence electrons around an atom, place one electron on each side before pairing any electrons.
- Determine the total number of valence electrons in the compound.

2. Arrange the atoms.
- Arrange the Lewis structure to show how the atoms bond in the molecule.
- Halogen and hydrogen atoms *often* bind to only one other atom and are *usually* at an end of the molecule.
- Carbon is often placed in the center of the molecule.
- You will find that, with the exception of carbon, the atom with the lowest electronegativity is often the central atom.

3. Distribute the dots.
- Distribute the electron dots so that each atom, except for hydrogen, beryllium, and boron, satisfies the octet rule.

4. Draw the bonds.
- Change each pair of dots that represents a shared pair of electrons to a long dash.

5. Verify the structure.
- Count the number of electrons surrounding each atom. Except for hydrogen, beryllium, and boron, all atoms must satisfy the octet rule. Check that the number of valence electrons is still the same number you determined in step 1.

Drawing Lewis Structures with Single Bonds

Draw a Lewis structure for CH_3I.

1 Gather information.

Draw each atom's Lewis structure, and count the total number of valence electrons.

$$\cdot \overset{\cdot}{\underset{\cdot}{C}} \cdot \quad H \cdot \quad H \cdot \quad H \cdot \quad \cdot \overset{\cdot\cdot}{\underset{\cdot\cdot}{I}} \colon \qquad\qquad \text{number of dots: 14}$$

2 Arrange the atoms.

Arrange the Lewis structure so that carbon is the central atom.

$$\begin{array}{c} H \\ H\ C\ I \\ H \end{array}$$

PRACTICE HINT

You may have to try several Lewis structures until you get one in which all of the atoms, except hydrogen, beryllium, and boron, obey the octet rule.

3 Distribute the dots.

Distribute one bonding pair of electrons between each of the bonded atoms. Then, distribute the remaining electrons, in pairs, around the remaining atoms to form an octet for each atom.

$$\begin{array}{c} H \\ H \colon \overset{\cdot\cdot}{\underset{\cdot\cdot}{C}} \colon \overset{\cdot\cdot}{\underset{\cdot\cdot}{I}} \colon \\ H \end{array}$$

4 Draw the bonds.

Change each pair of dots that represents a shared pair of electrons to a long dash.

$$\begin{array}{c} H \\ | \\ H - C - \overset{\cdot\cdot}{\underset{\cdot\cdot}{I}} \colon \\ | \\ H \end{array}$$

5 Verify the structure.

Carbon and iodine have 8 electrons, and hydrogen has 2 electrons. The total number of valence electrons is still 14.

PRACTICE

1 Draw the Lewis structures for H_2S, CH_2Cl_2, NH_3, and C_2H_6.

2 Draw the Lewis structure for methanol, CH_3OH. First draw the CH_3 part, and then add O and H.

Lewis Structures for Polyatomic Ions

Lewis structures are also helpful in describing polyatomic ions, such as the ammonium ion, NH_4^+. An ammonium ion, shown in **Figure 11,** forms when ammonia, NH_3, is combined with a substance that easily gives up a hydrogen ion, H^+. To draw the Lewis structure of NH_4^+, first draw the structure of NH_3. With five valence electrons, a nitrogen atom can make a stable octet by forming three covalent bonds, one with each hydrogen atom. Then add H^+, which is simply the nucleus of a hydrogen atom, or a proton, and has no electrons to share. The H^+ can form a covalent bond with NH_3 by bonding with the unshared pair on the nitrogen atom.

$$
\begin{array}{c}
H \\
H:\overset{\displaystyle \cdot\cdot}{N}:H
\end{array}
+ H^+ \longrightarrow
\left[\begin{array}{c}
H \\
H:\overset{\displaystyle }{N}:H \\
H
\end{array}\right]^+
$$

The Lewis structure is enclosed in brackets to show that the positive charge is distributed over the entire ammonium ion.

$$
\left[\begin{array}{c}
H \\
H:\overset{\displaystyle }{N}:H \\
H
\end{array}\right]^+
$$

Ammonium ion

Figure 11
Smelling salts often have an unstable ionic compound made of two polyatomic ions: ammonium and carbonate.

SAMPLE PROBLEM B

Drawing Lewis Structures for Polyatomic Ions

Draw a Lewis structure for the sulfate ion, SO_4^{2-}.

1 Gather information.

When counting the total number of valence electrons, add two additional electrons to account for the 2– charge on the ion.

$:\overset{\displaystyle \cdot}{\underset{\displaystyle }{S}}\cdot$ $:\overset{\displaystyle \cdot\cdot}{\underset{\displaystyle }{O}}\cdot$ $:\overset{\displaystyle \cdot\cdot}{\underset{\displaystyle }{O}}\cdot$ $:\overset{\displaystyle \cdot\cdot}{\underset{\displaystyle }{O}}\cdot$ $:\overset{\displaystyle \cdot\cdot}{\underset{\displaystyle }{O}}\cdot$ number of dots: $30 + 2 = 32$

2 Arrange the atoms. Distribute the dots.

Sulfur has the lowest electronegativity, so it is the central atom. Distribute the 32 dots so that there are 8 dots around each atom.

$$
\begin{array}{c}
:\overset{\displaystyle \cdot\cdot}{O}: \\
:\overset{\displaystyle }{O}:\overset{\displaystyle }{S}:\overset{\displaystyle }{O}: \\
:\underset{\displaystyle \cdot\cdot}{O}:
\end{array}
$$

3 Draw the bonds. Verify the structure.

- Change each bonding pair to a long dash. Place brackets around the ion and a 2– charge outside the bracket to show that the charge is spread out over the entire ion.
- There are 32 valence electrons, and each O and S has an octet.

$$
\left[\begin{array}{c}
:\overset{\displaystyle \cdot\cdot}{O}: \\
| \\
:\overset{\displaystyle }{O}-S-\overset{\displaystyle }{O}: \\
| \\
:\underset{\displaystyle \cdot\cdot}{O}:
\end{array}\right]^{2-}
$$

PRACTICE HINT

- If the polyatomic ion has a negative charge, add the appropriate number of valence electrons. (For example, the net charge of 2– on SO_4^{2-} means that there are two more electrons than in the neutral atoms.)
- If the polyatomic ion has a positive charge, subtract the appropriate number of valence electrons. (For example, the net charge of 1+ on H_3O^+ means that there is one fewer electron than in the neutral atoms.)

PRACTICE

1 Draw the Lewis structure for ClO_3^-.

2 Draw the Lewis structure for the hydronium ion, H_3O^+.

Multiple Bonds

Atoms can share more than one pair of electrons in a covalent bond. Think about a nonpolar covalent bond formed between two oxygen atoms in an O_2 molecule. Each O has six valence electrons, as shown below.

$$:\ddot{O}\cdot \quad \cdot\ddot{O}:$$

If these oxygen atoms together shared only one pair of electrons, each atom would have only seven electrons. The octet rule would not be met.

Bonds with More than One Pair of Electrons

To make an octet, each oxygen atom needs two more electrons to be added to its original six. To add two electrons, each oxygen atom must share two electrons with the other atom so that the two atoms share four electrons. The covalent bond formed by the sharing of two pairs of electrons is a **double bond,** shown in the Lewis structures below.

$$:\ddot{O}::\ddot{O}: \quad \text{or} \quad :\ddot{O}=\ddot{O}:$$

Atoms will form a single or a multiple bond depending on what is needed to make an octet. While two O atoms form a double bond in O_2, an O atom forms a single bond with each of two H atoms in a water molecule.

$$\overset{H}{\underset{\cdot\cdot}{\ddot{O}:}}H \quad \text{or} \quad :\overset{H}{\underset{\cdot\cdot}{\ddot{O}}}-H$$

Another example of a molecule that has a double bond is ethene, C_2H_4, shown in **Figure 12.** Each H atom forms a single bond with a C atom. Each C atom below has two electrons that are not yet part of a bond.

$$\overset{H}{H:\ddot{C}\cdot} \quad \overset{H}{\cdot\ddot{C}:H}$$

With only six electrons, each C atom needs two more electrons to have an octet. The only way to complete the octets is to form a **double bond.**

$$\overset{H\ H}{H:\ddot{C}::\ddot{C}:H} \quad \text{or} \quad \overset{H\ H}{H-C=C-H}$$

Figure 12
Most plants have a hormone called *ethene,* C_2H_4. Tomatoes release ethene, also called *ethylene,* as they ripen.

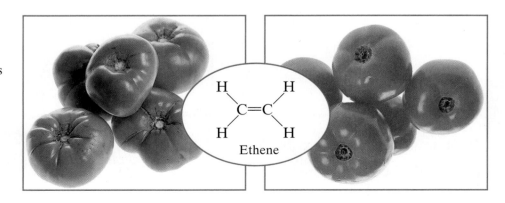

Carbon, oxygen, and nitrogen atoms often form double bonds by sharing two pairs of electrons. Carbon and nitrogen atoms may even share three pairs of electrons to form a **triple bond.** Think about the molecule N_2. With five valence electrons, each N atom needs three more electrons for a stable octet. Each N atom contributes three electrons to form three bonding pairs. The two N atoms form a triple bond by sharing these three pairs of electrons, or a total of six electrons. Because the two N atoms share the electrons equally, the triple bond is a nonpolar covalent bond.

triple bond

a covalent bond in which two atoms share three pairs of electrons

:N::N: or :N≡N:

SAMPLE PROBLEM C

Drawing Lewis Structures with Multiple Bonds

Draw a Lewis structure for formaldehyde, CH_2O.

1 Gather information.

Draw each atom's Lewis structure, and count the total dots.

·Ċ·   H·   H·   ·Ö: total dots: 12

2 Arrange the atoms. Distribute the dots.

- Arrange the atoms so that carbon is the central atom.
- Distribute one pair of dots between each of the atoms. Then, starting with the outside atoms, distribute the rest of the dots, in pairs, around the atoms. You will run out of electrons before all of the atoms have an octet (left structure). C does not have an octet, so there must be a multiple bond. To obtain an octet for C, move one of the unshared pairs from the O atom to between the O and the C (right structure).

incorrect: :Ö:
H:Ċ:H   correct: :O:
H:Ċ:H

3 Draw the bonds. Verify the structure.

- Change each pair of dots that represents a shared pair of electrons to a long dash. Two pairs of dots represent a double bond.
- C and O atoms both have eight electrons, and each H atom has two electrons. The total number of valence electrons is still 12.

:O:
‖
H—C—H

PRACTICE HINT

- Begin with a single pair of dots between each pair of bonded atoms. If no arrangement of single bonds provides a Lewis structure whose atoms satisfy the octet rule, the molecule might have multiple bonds.
- N and C can form single bonds or combinations of single and double or triple bonds.

PRACTICE

1 Draw the Lewis structures for carbon dioxide, CO_2, and carbon monoxide, CO.

2 Draw the Lewis structures for ethyne, C_2H_2, and hydrogen cyanide, HCN.

PROBLEM SOLVING SKILL

Resonance Structures

Some molecules, such as ozone, O_3, cannot be represented by a single Lewis structure. Ozone has two Lewis structures, as shown below.

$$\ddot{O}=\ddot{O}-\ddot{O}: \leftrightarrow :\ddot{O}-\ddot{O}=\ddot{O}$$

Each O atom follows the octet rule, but the two structures use different arrangements of the single and double bonds. So which structure is correct? Neither structure is correct by itself. When a molecule has two or more possible Lewis structures, the two structures are called **resonance structures.** You place a double-headed arrow between the structures to show that the actual molecule is an average of the two possible states.

Another molecule that has resonance structures is sulfur dioxide, SO_2, shown in **Figure 13.** Sulfur dioxide released into the atmosphere is partly responsible for acid precipitation. The actual structure of SO_2 is an average, or a *resonance hybrid,* of the two structures. Although you draw the structures as if the bonds change places again and again, the bonds do not in fact move back and forth. The actual bonding is a mixture of the two extremes represented by each of the Lewis structures.

Naming Covalent Compounds

Covalent compounds made of two elements are named by using a method similar to the one used to name ionic compounds. Think about how the covalent compound SO_2 is named. The first element named is usually the first one written in the formula, in this case *sulfur*. Sulfur is the less-electronegative element. The second element named has the ending *-ide,* in this case *oxide.*

However, unlike the names for ionic compounds, the names for covalent compounds must often distinguish between two different molecules made of the same elements. For example, SO_2 and SO_3 cannot both be called *sulfur oxide.* These two compounds are given different names based on the number of each type of atom in the compound.

Prefixes Indicate How Many Atoms Are in a Molecule

The system of prefixes shown in **Table 5** is used to show the number of atoms of each element in the molecule. SO_2 and SO_3 are distinguished from one another by the use of prefixes in their names. With only two oxygen atoms, SO_2 is named *sulfur dioxide.* With three oxygen atoms, SO_3 is named *sulfur trioxide.* The following example shows how to use the system of prefixes to name P_2S_5.

Figure 13
You can draw resonance structures for sulfur dioxide, SO_2, a chemical that can add to air pollution.

resonance structure

in chemistry, any one of two or more possible configurations of the same compound that have identical geometry but different arrangements of electrons

Topic Link

Refer to the "Ions and Ionic Compounds" chapter for more about naming ionic compounds.

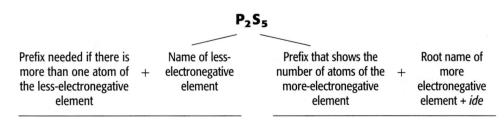

P_2S_5

Prefix needed if there is more than one atom of the less-electronegative element	+	Name of less-electronegative element		Prefix that shows the number of atoms of the more-electronegative element	+	Root name of more electronegative element + *ide*
diphosphorus				**pentasulfide**		

Table 5   Prefixes for Naming Covalent Compounds

Prefix	Number of atoms	Example	Name
mono-	1	CO	carbon monoxide
di-	2	SiO_2	silicon dioxide
tri-	3	SO_3	sulfur trioxide
tetra-	4	SCl_4	sulfur tetrachloride
penta-	5	$SbCl_5$	antimony pentachloride

Refer to Appendix A for a more complete list of prefixes.

■ internet connect ≡

www.scilinks.org
Topic: Naming Compounds
SciLinks code: HW4081

SCiLINKS. Maintained by the National Science Teachers Association

Prefixes are added to the first element in the name only if the molecule contains more than one atom of that element. So, N_2O is named *dinitrogen oxide*, S_2F_{10} is named *disulfur decafluoride*, and P_4O_6 is named *tetraphosphorus hexoxide*. If the molecule contains only one atom of the first element given in the formula, the prefix *mono-* is left off. Both SO_2 and SO_3 have only one S atom each. Therefore, the names of both start with the word *sulfur*. Note that the vowels *a* and *o* are dropped from a prefix that is added to a word begining with a vowel. For example, CO is carbon monoxide, not carbon monooxide. Similarly, N_2O_4 is named *dinitrogen tetroxide,* not *dinitrogen tetraoxide*.

② Section Review

UNDERSTANDING KEY IDEAS

1. Which electrons do a Lewis structure show?

2. In a polyatomic ion, where is the charge located?

3. How many electrons are shared by two atoms that form a triple bond?

4. What do resonance structures represent?

5. How do the names for SO_2 and SO_3 differ?

PRACTICE PROBLEMS

6. Draw a Lewis structure for an atom that has the electron configuration $1s^2 2s^2 2p^6 3s^2 3p^3$.

7. Draw Lewis structures for each compound:
   a. BrF
   b. $N(CH_3)_3$
   c. Cl_2O
   d. ClO_2^-

8. Draw three resonance structures for SO_3.

9. Name the following compounds.
   a. SnI_4
   b. N_2O_3
   c. PCl_3
   d. CSe_2

10. Write the formula for each compound:
   a. phosphorus pentabromide
   b. diphosphorus trioxide
   c. arsenic tribromide
   d. carbon tetrachloride

CRITICAL THINKING

11. Compare and contrast the Lewis structures for krypton and radon.

12. Do you always follow the octet rule when drawing a Lewis structure? Explain.

13. What is incorrect about the name *monosulfur dioxide* for the compound SO_3?

Molecular Shapes

OBJECTIVES

① **Predict** the shape of a molecule using VSEPR theory.

② **Associate** the polarity of molecules with the shapes of molecules, and relate the polarity and shape of molecules to the properties of a substance.

Determining Molecular Shapes

Lewis structures are two-dimensional and do not show the three-dimensional shape of a molecule. However, the three-dimensional shape of a molecule is important in determining the molecule's physical and chemical properties. Sugar, or sucrose, is an example. Sucrose has a shape that fits certain nerve receptors on the tongue. Once stimulated, the nerves send signals to the brain, and the brain interprets these signals as sweetness. Inside body cells, sucrose is processed for energy.

People who want to avoid sucrose in their diet often use a sugar substitute, such as sucralose, shown in **Figure 14.** These substitutes have shapes similar to that of sucrose, so they can stimulate the nerve receptors in the same way that sucrose does. However, sucralose has a different chemical makeup than sucrose does and cannot be processed by the body.

Figure 14
Sucralose is chemically very similar to sucrose. Both have the same three-dimensional shape. However, three Cl atoms have been substituted in sucralose, so the body cannot process it.

CO is linear.

a Molecules made up of only two atoms, such as CO, have a linear shape.

SO$_2$ is bent.   CO$_2$ is linear

b Although SO$_2$ and CO$_2$ have the same numbers of atoms, they have different shapes because the numbers of electron groups surrounding the central atoms differ.

Figure 15
Molecules with three or fewer atoms have shapes that are in a flat plane.

A Lewis Structure Can Help Predict Molecular Shape

The shape of a molecule made of only two atoms, such as H$_2$ or CO, is easy to determine. As shown in **Figure 15,** only a linear shape is possible when there are two atoms. Determining the shapes of molecules made of more than two atoms is more complicated. Compare carbon dioxide, CO$_2$, and sulfur dioxide, SO$_2$. Both molecules are made of three atoms. Although the molecules have similar formulas, their shapes are different. Notice that CO$_2$ is linear, while SO$_2$ is bent.

Obviously, the formulas CO$_2$ and SO$_2$ do not provide any information about the shapes of these molecules. However, there is a model that can be used to predict the shape of a molecule. This model is based on the **valence shell electron pair repulsion (VSEPR) theory.** Using this model, you can predict the shape of a molecule by examining the Lewis structure of the molecule.

Electron Pairs Can Determine Molecular Shape

According to the VSEPR theory, the shape of a molecule is determined by the valence electrons surrounding the central atom. For example, examine the Lewis structure for CO$_2$.

$$\ddot{O}=C=\ddot{O}$$

Notice the two double bonds around the central carbon atom. Because of their negative charge, electrons repel each other. Therefore, the two shared pairs that form each double bond repel each other and remain as far apart as possible. These two sets of two shared pairs are farthest apart when they are on opposite sides of the carbon atom. Thus, the shape of a CO$_2$ molecule is linear. You'll read about SO$_2$'s bent shape later.

Now think about what happens when the central atom is surrounded by three shared pairs. Look at the Lewis structure for BF$_3$, which has boron, an example of an atom that does not always obey the octet rule.

$$:\ddot{F}-B-\ddot{F}:$$
$$|$$
$$:\ddot{F}:$$

Notice the three single bonds around the central boron atom. Like three spokes of a wheel, these shared pairs of electrons extend from the central boron atom. The three F atoms, each of which has three unshared pairs, will repel each other and will be at a maximum distance apart. This molecular shape is known as *trigonal planar,* as shown in **Figure 16.**

valence shell electron pair repulsion (VSEPR) theory

a theory that predicts some molecular shapes based on the idea that pairs of valence electrons surrounding an atom repel each other

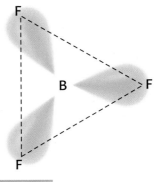

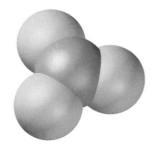

Figure 16
Trigonal planar molecules, such as BF$_3$, are flat structures in which three atoms are evenly spaced around the central atom.

Next, think about what happens when the central atom is surrounded by four shared pairs of electrons. Examine the Lewis structure for methane, CH_4, shown below.

$$
\begin{array}{c}
H \\
| \\
H-C-H \\
| \\
H
\end{array}
$$

Notice that four single bonds surround the central carbon atom. On a flat plane the bonds are not as far apart as they can be. Instead, the four shared pairs are farthest apart when each pair of electrons is positioned at the corners of a tetrahedron, as shown in **Figure 17.** Only the electron clouds around the central atom are shown.

In CO_2, BF_3, and CH_4, all of the valence electrons of the central atom form shared pairs. What happens to the shape of a molecule if the central atom has an unshared pair? Tin(II) chloride, $SnCl_2$, gives an example. Examine the Lewis structure for $SnCl_2$, shown below.

$$
\begin{array}{c}
:Sn-\ddot{C}l: \\
| \\
:\ddot{C}l:
\end{array}
$$

Notice that the central tin atom has two shared pairs and one unshared pair of electrons. In VSEPR theory, unshared pairs occupy space around a central atom, just as shared pairs do. The two shared pairs and one unshared pair of the tin atom cause the shape of the $SnCl_2$ molecule to be bent, as shown in **Figure 17.**

The unshared pairs of electrons *influence* the shape of a molecule but are not visible in the space-filling model. For example, the shared and unshared pairs of electrons in $SnCl_2$ form a trigonal planar geometry, but the molecule has a bent shape. The bent shape of SO_2, shown in **Figure 15** on the previous page, is also due to unshared pairs. However, in the case of SO_2, there are two unshared pairs.

Figure 17
The electron clouds around the central atom help determine the shape of a molecule.

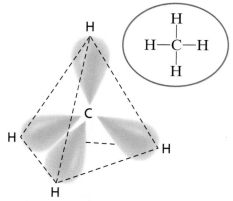

a A molecule whose central atom is surrounded by four shared pairs of electrons, such as CH_4, has a tetrahedral shape.

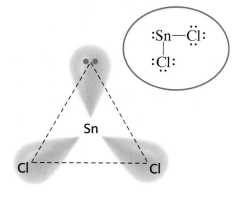

b A molecule whose central atom is surrounded by two shared pairs and one unshared pair, such as $SnCl_2$, has a bent shape.

Predicting Molecular Shapes

Determine the shapes of NH_3 and H_2O.

1 Gather information.

Draw the Lewis structures for NH_3 and H_2O.

$$\begin{array}{cc} H & H \\ | & | \\ H-\overset{\cdot\cdot}{N}-H & :\overset{\cdot\cdot}{O}-H \end{array}$$

2 Count the shared and unshared pairs.

Count the number of shared and unshared pairs of electrons around each central atom.

NH₃ has three shared pairs and one unshared pair.

H₂O has two shared pairs and two unshared pairs.

3 Apply VSEPR theory.

- Use VSEPR theory to find the shape that allows the shared and unshared pairs of electrons to be spaced as far apart as possible.
- The ammonia molecule will have the shape of a pyramid. This geometry is called *trigonal pyramidal*.

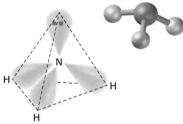

- The water molecule will have a bent shape.

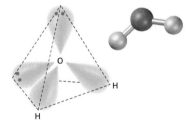

4 Verify the Structure.

For both molecules, be sure that all atoms, except hydrogen, obey the octet rule.

PRACTICE HINT

- Keep in mind that the geometry is difficult to show on the printed page because the atoms are arranged in three dimensions.
- If the sum of the shared and unshared pairs of electrons in each molecule is four, the electron pairs have *tetrahedral* geometry. However, the shape of the molecule is based on the number of shared pairs of electrons present. That is, the shape is based only on the position of the atoms and not on the position of the unshared pairs of electrons.

PRACTICE

Predict the shapes of the following molecules and polyatomic ions.

1 **a.** NH_2Cl **c.** NO_3^-

b. $NOCl$ **d.** NH_4^+

PROBLEM SOLVING SKILL

Figure 18

Molecules of both water and carbon dioxide have polar bonds. The symbol ←→ shows a dipole.

a Because CO_2 is linear, the molecule is nonpolar.

Carbon dioxide, CO_2
(no molecular dipole)

b Because H_2O has a bent shape, the molecule is polar.

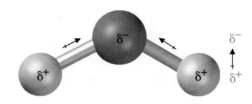

Water, H_2O
(overall molecular dipole)

Molecular Shape Affects a Substance's Properties

A molecule's shape affects both the physical and chemical properties of the substance. Recall that both sucrose and sucralose have a shape that allows each molecule to fit into certain nerve endings on the tongue and stimulate a sweet taste. If bending sucrose or sucralose molecules into a different shape were possible, the substances might not taste sweet. Shape determines many other properties. One property that shape determines is the polarity of a molecule.

Shape Affects Polarity

The polarity of a molecule that has more than two atoms depends on the polarity of each bond and the way the bonds are arranged in space. For example, compare CO_2 and H_2O. Oxygen has a higher electronegativity than carbon does, so each oxygen atom in CO_2 attracts electrons more strongly. Therefore, the shared pairs of electrons are more likely to be found near each oxygen atom than near the carbon atom. Thus, the double bonds between carbon and oxygen are polar. As shown in **Figure 18,** each oxygen atom has a partial negative charge, while the carbon atom has a partial positive charge.

Notice also that CO_2 has a linear shape. This shape determines the overall polarity of the molecule. The polarities of the double bonds extend from the carbon atom in opposite directions. As a result, they cancel each other and CO_2 is nonpolar overall even though the individual covalent bonds are polar.

Now think about H_2O. Oxygen has a higher electronegativity than hydrogen does, so oxygen attracts the shared pairs more strongly than either hydrogen does. As a result, each covalent bond between hydrogen and oxygen is polar. The O atom has a partial negative charge, while each H atom has a partial positive charge. Notice also that H_2O has a bent shape. Because the bonds are at an angle to each other, their polarities do not cancel each other. As a result, H_2O is polar.

You can think of a molecule's overall polarity in the same way that you think about forces on a cart. If you and a friend pull on a wheeled cart in equal and opposite directions—you pull the cart westward and your friend pulls the cart eastward—the cart does not move. The pull forces cancel each other in the same way that the polarities on the CO_2 molecule cancel each other. What happens if the two of you pull with equal force but in nonopposite directions? If you pull the cart northward and your friend pulls it westward, the cart moves toward the northwest. Because the cart has a net force applied to it, it moves. The water molecule has a net partial positive charge on the H side and a net negative charge on the O side. As a result, the molecule has an overall charge and is therefore polar.

Polarity Affects Properties

Because CO_2 molecules are nonpolar, the attractive force between them is very weak. In contrast, the attractive force between polar H_2O molecules is much stronger. The H atoms (with partial positive charges) attract the O atoms (with partial negative charges) on other water molecules. The attractive force between polar water molecules contributes to the greater amount of energy required to separate these polar molecules. The polarity of water molecules also adds to their attraction to positively and negatively charged objects. Other properties realted to polarity and molecular shape will be discussed in a later chapter.

③ Section Review

UNDERSTANDING KEY IDEAS

1. In VSEPR theory, what information about a central atom do you need in order to predict the shape of a molecule?

2. What is the only shape that a molecule made up of two atoms can have?

3. Explain how Lewis structures help predict the shape of a molecule.

4. Explain how a molecule that has polar bonds can be nonpolar.

5. Give one reason why water molecules are attracted to each other.

PRACTICE PROBLEMS

6. Determine the shapes of Br_2 and HBr. Which molecule is more polar and why?

7. Use VSEPR theory to determine the shapes of each of the following.
   a. SCl_2
   b. PF_3
   c. NCl_3
   d. NH_4^+

8. Predict the shape of the CCl_4 molecule. Is the molecule polar or nonpolar? Explain your answer.

CRITICAL THINKING

9. Can a molecule made up of three atoms have a linear shape? Explain your answer.

10. Why is knowing something about the shape of a molecule important?

11. The electron pairs in a molecule of NH_3 form a tetrahedron. Why does the NH_3 molecule have a trigonal pyramidal shape rather than a tetrahedral shape?

14
Si
Silicon
28.0855
$[Ne]3s^23p^2$

Element Spotlight

Many integrated circuit chips can be made on the same silicon wafer. The wafer will be cut up into individual chips.

internet connect

www.scilinks.org
Topic: Silicon
SciLinks code: HW4116

SCiLINKS Maintained by the National Science Teachers Association

Silicon and Semiconductors

Silicon's most familiar use is in the production of microprocessor chips. Computer microprocessor chips are made from thin slices, or wafers, of a pure silicon crystal. The wafers are doped with elements such as boron, phosphorus, and arsenic to confer semiconducting properties on the silicon. A photographic process places patterns for several chips onto one wafer. Gaseous compounds of metals are allowed to diffuse into the open spots in the pattern, and then the pattern is removed. This process is repeated several times to build up complex microdevices on the surface of the wafer. When the wafer is finished and tested, it is cut into individual chips.

Industrial Uses

- Silicon and its compounds are used to add strength to alloys of aluminum, magnesium, copper, and other metals.

- When doped with elements of Group 13 or Group 15, silicon is a semiconductor. This property is important in the manufacture of computer chips and photovoltaic cells.

- Quartz (silicon dioxide) crystals are used for piezoelectric crystals for radio-frequency control oscillators and digital watches and clocks.

Real-World Connection Organic compounds containing silicon, carbon, chlorine, and hydrogen are used to make silicone polymers, which are used in water repellents, electrical insulation, hydraulic fluids, lubricants, and caulks.

A Brief History

1854: Henri S. C. Deville prepares crystalline silicon.

1943: Commercial production of silicone rubber, oils, and greases begins in the United States.

1800	1900	2000

1811: Joseph Louis Gay-Lussac and Louis Thenard prepare impure amorphous silicon from silicon tetrafluoride.

1824: Jöns Jacob Berzelius prepares pure amorphous silicon and is credited with the discovery of the element.

1904: F. S. Kipping produces the first silicone compound.

1958: Jack Kilby and Robert Noyce produce the first integrated circuit on a silicon chip.

Questions

1. Research and identify five items that you encounter on a regular basis and that are constructed by using silicon.

2. Research piezoelectric materials, and identify how piezoelectric materials that contain silicon are used in science and industry.

KEY IDEAS

SECTION ONE Covalent Bonds

• Covalent bonds form when atoms share pairs of electrons.
• Atoms have less potential energy and more stability after they form a covalent bond.
• The greater the electronegativity difference, the greater the polarity of the bond.
• The physical and chemical properties of a compound are related to the compound's bond type.

SECTION TWO Drawing and Naming Molecules

• In a Lewis structure, the element's symbol represents the atom's nucleus and inner-shell electrons, and dots represent the atom's valence electrons.
• Two atoms form single, double, and triple bonds depending on the number of electron pairs that the atoms share.
• Some molecules have more than one valid Lewis structure. These structures are called *resonance structures.*
• Molecular compounds are named using the elements' names, a system of prefixes, and *-ide* as the ending for the second element in the compound.

SECTION THREE Molecular Shapes

• VSEPR theory states that electron pairs in the valence shell stay as far apart as possible.
• VSEPR theory can be used to predict the shape of a molecule.
• Molecular shapes predicted by VSEPR theory include linear, bent, trigonal planar, tetrahedral, and trigonal pyramidal.
• The shape of a molecule affects the molecule's physical and chemical properties.

KEY SKILLS

6 CHAPTER REVIEW

USING KEY TERMS

1. How are bond length and potential energy related?

2. Describe the difference between a shared pair and an unshared pair of electrons.

3. How are the inner-shell electrons represented in a Lewis structure?

4. What term is used to describe the situation when two or more correct Lewis structures represent a molecule?

5. How is VSEPR theory useful?

6. Describe a molecular dipole.

7. Why is the electronegativity of an element important?

8. What type of bond results if two atoms share six electrons?

9. Contrast a polar covalent bond and a nonpolar covalent bond.

10. Describe how bond energy is related to the breaking of covalent bonds.

UNDERSTANDING KEY IDEAS

Covalent Bonds

11. How does a covalent bond differ from an ionic bond?

12. How are bond energy and bond strength related?

13. Why is a spring a better model than a stick for a covalent bond?

14. Describe the energy changes that take place when two atoms form a covalent bond.

15. Predict whether the bonds between the following pairs of elements are ionic, polar covalent, or nonpolar covalent.
    a. Na—F
    b. H—I
    c. N—O
    d. Al—O
    e. S—O
    f. H—H
    g. Cl—Br
    h. Sr—O

16. Where are the bonding electrons between two atoms?

17. How do the attractive and repulsive forces between two atoms compare when the atoms form a covalent bond?

18. Arrange the following diatomic molecules in order of increasing bond polarity.
    a. I—Cl
    b. H—F
    c. H—Br

19. What determines the electron distribution between two atoms in a bond?

20. Explain why the melting and boiling points of covalent compounds are usually lower than those of ionic compounds.

Drawing and Naming Molecules

21. Draw the Lewis structures for boron, nitrogen, and phosphorus.

22. Describe a weakness of using Lewis structures to model covalent compounds.

23. What do the dots in a Lewis structure represent?

24. How does a Lewis structure show a bond between two atoms that share four electrons?

25. Why are resonance structures used to model certain molecules?

26. Name the following covalent compounds.
  a. SF_4
  b. XeF_4
  c. PBr_5
  d. N_2O_5
  e. Si_3N_4
  f. PBr_3
  g. Np_3O_8

Molecular Shapes

27. Why do electron pairs around a central atom stay as far apart as possible?

28. Name the following molecular shapes.

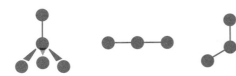

29. Two molecules have different shapes but the same composition. Can you conclude that they have the same physical and chemical properties? Explain you answer.

30. Why is VSEPR theory *not* needed to predict the shape of HCl?

31. a. What causes H_2O to have a bent shape rather than a linear shape?
  b. How does this bent shape relate to the polarity of the water molecule?

PRACTICE PROBLEMS

Sample Problem A Drawing Lewis Structures with Single Bonds

32. Draw Lewis structures for the following molecules. Remember that hydrogen can form only a single bond.
  a. NF_3 **d.** CCl_2F_2
  b. CH_3OH **e.** $HOCl$
  c. ClF

Sample Problem B Drawing Lewis Structures for Polyatomic Ions

33. Draw Lewis structures for the following polyatomic ions.
  a. OH^-
  b. O_2^{2-}
  c. NO^{2-}
  d. NO^{2+}
  e. AsO_4^{3-}

Sample Problem C Drawing Lewis Structures with Multiple Bonds

34. Draw Lewis structures for the following molecules.
  a. O_2
  b. CS_2
  c. N_2O

Sample Problem D Predicting Molecular Shapes

35. Determine the shapes of the following compounds.
  a. CF_4
  b. Cl_2O

36. Draw the shapes of the following polyatomic ions.
  a. NH_4^+
  b. OCl^-
  c. CO_3^{2-}

MIXED REVIEW

37. Draw Lewis structures for the following molecules.
  a. a refrigerant that has one C atom, one H atom, one Br atom, and two F atoms
  b. a natural-gas ingredient that has two C atoms and six H atoms

38. a. Determine the shapes of SCl_2, PF_3, and NCl_3.
  b. Which of these molecules has the greatest polarity?

39. Draw three resonance structures for NO_3^-.

40. Draw Lewis structures for the following polyatomic ions.
  a. CO_3^{2-}
  b. O_2^{2-}
  c. PO_4^{3-}

41. Name the following compounds, draw their Lewis structures, and determine their shapes.
  a. $SiCl_4$
  b. BCl_3
  c. NBr_3

42. a. How do ionic and covalent bonding differ?
  b. How does an ionic compound differ from a molecular compound?

43. How does the bond that forms between Ba and Br demonstrate that the boundaries between bond types as determined by electronegativity differences are not clear cut?

44. Explain why a halogen is unlikely to form a double bond with another element.

45. a. Draw the Lewis structure for $BeCl_2$.
  b. What is unusual about the Be atom in this compound?
  c. Label your Lewis structure to show the partial charge distribution.
  d. What type of bond do Be and Cl form?
  e. What is the shape of the $BeCl_2$ molecule based on VSEPR theory?
  f. Is $BeCl_2$ a polar or nonpolar molecule? Explain your choice.

46. According to VSEPR theory, what molecular shapes are associated with the following types of molecules?
  a. AB
  b. AB_2
  c. AB_3
  d. AB_4

47. What types of atoms tend to form the following types of bonding?
  a. ionic
  b. covalent
  c. metallic

CRITICAL THINKING

48. What is the difference between a dipole and electronegativity difference?

49. Why does F generally form covalent bonds with great polarity?

50. Explain what is wrong with the following Lewis structures, and then correct each one.

  a. H—H—S̈:

  b. 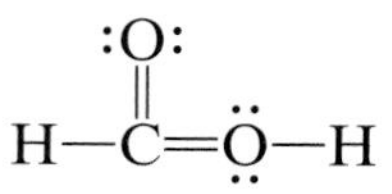
  :O:
  ‖
  H—C=Ö—H

  c.
  :C̈l:
  ‖
  N
  :C̈l   C̈l:

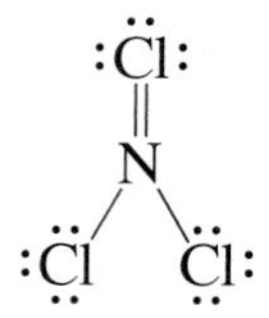

51. Unlike other elements, noble gases are relatively inert. When noble gases do react, they do not follow the octet rule. Examine the following Lewis structure for the molecule XeO_2F_2.

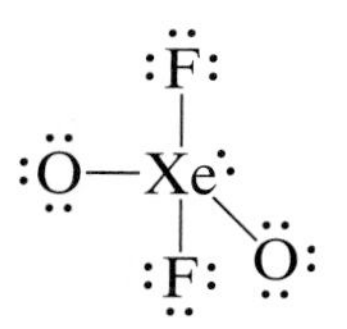

  :F̈:
  |
  :Ö—Xe·
  | ·· :Ö:
  :F̈:

  a. Explain why the valence electrons of Xe do not follow the octet rule.
  b. How many unshared pairs of electrons are in this molecule?
  c. How many electrons make up all of the shared pairs in this molecule?

52. Ionic compounds tend to have higher boiling points than covalent substances do. Both ammonia, NH_3, and methane, CH_4, are covalent compunds, yet the boiling point of ammonia is 130°C higher than that of methane. What might account for this large difference?

53. Draw the Lewis structure of NH_4^+. Examine this structure to explain why this five-atom group exists only as a cation.

54. The length of a covalent bond varies depending on the type of bond formed. Triple bonds are generally shorter than double bonds, and double bonds are generally shorter than single bonds. Predict how the lengths of the C—C bond in the following molecules compare.

  a. C_2H_6
  b. C_2H_4
  c. C_2H_2

ALTERNATIVE ASSESSMENT

55. Natural rubber consists of long chains of carbon and hydrogen atoms covalently bonded together. When Goodyear accidentally dropped a mixture of sulfur and rubber on a hot stove, the energy joined these chains together to make vulcanized rubber. Vulcan was the Roman god of fire. The carbon-hydrogen chains in vulcanized rubber are held together by two sulfur atoms that form covalent bonds between the chains. These covalent bonds are commonly called *disulfide bridges*. Explore other molecules that have such disulfide bridges. Present your findings to the class.

56. Searching for the perfect artificial sweetener—great taste with no Calories—has been the focus of chemical research for some time. Molecules such as sucralose, aspartamine, and saccharin owe their sweetness to their size and shape. One theory holds that any sweetener must have three sites that fit into the proper taste buds on the tongue. This theory is appropriately known as the *triangle theory*. Research artificial sweeteners to develop a model to show how the triangle theory operates.

57. Computer Skill RasMol™ is a free software program for examining the shapes of molecules. Use a search engine on the Internet to locate the RasMol site. Download a copy of the program, which includes several models of large molecules. Present your findings to the class.

58. Devise a set of criteria that will allow you to classify the following substances as covalent, ionic, or metallic: $CaCO_3$, Cu, H_2O, $NaBr$, and C (graphite). Show your criteria to your teacher.

59. Identify 10 common substances in and around your home, and indicate whether you would expect these substances to contain covalent, ionic, or metallic bonds.

CONCEPT MAPPING

60. Use the following terms to complete the concept map below: *valence electrons, nonpolar, covalent compounds, polar, dipoles,* and *Lewis structures.*

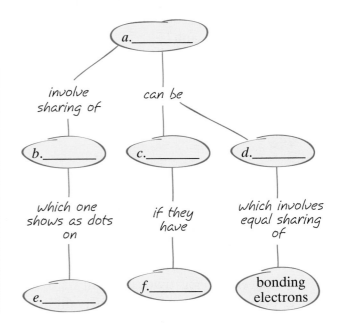

Study the graph below, and answer the questions that follow.
For help in interpreting graphs, see Appendix B, "Study Skills for Chemistry."

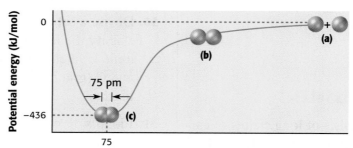

Potential Energy Curve for H$_2$

61. What do the blue spheres represent on this graph?

62. What are the coordinates of the minimum (the lowest point) of the graph?

63. What relationship does the graph describe?

64. What is significant about the distance between the hydrogen nuclei at the lowest point on the graph?

65. When the distance between the hydrogen nuclei is greater than 75 pm is the slope positive or is it negative?

66. Miles measures the energy required to hold two magnets apart at varying distances. He notices that it takes less and less energy to hold the magnets apart as the distance between them increases. Compare the results of Miles's experiment with the data given in the graph.

TECHNOLOGY AND LEARNING

67. Graphing Calculator

Classifing Bonding Type

The graphing calculator can run a program that classifies bonding between atoms according to the difference between the atoms' electronegativities. Use this program to determine the electronegativity difference between bonded atoms and to classify bonding type.

Go to Appendix C. If you are using a TI-83 Plus, you can download the program

BONDTYPE and data sets and run the application as directed. If you are using another calculator, your teacher will provide you with the keystrokes and data sets to use. After you have graphed the data sets, answer the questions below.

a. Which element pair or pairs have a pure covalent bond?

b. What type of bond does the pair H and O have?

c. What type of bond does the pair Ca and O have?

1. A chemical bond results from the mutual attraction of the nuclei for
   a. electrons.
   b. neutrons.
   c. protons.
   d. dipoles.

2. A polar covalent bond is likely to form between two atoms that
   a. differ in electronegativity.
   b. are of similar size.
   c. differ in state of matter.
   d. have the same number of electrons.

3. The Lewis structure of methane, CH_4, has
   a. two double bonds.
   b. one triple bond.
   c. four single bonds.
   d. two single bonds and two double bonds.

4. In a polar covalent bond, the electronegativity of the two atoms is
   a. exactly the same.
   b. so different that one atom takes an electron away from the other atom.
   c. different, but the electrons are still shared by both atoms.
   d. zero.

5. Because ammonia, NH_3, has an unshared pair of electrons, its molecular shape is
   a. tetrahedral.
   b. trigonal pyramidal.
   c. bent.
   d. linear.

6. The energy released in the formation of a covalent bond is the difference between the potential energy of the separated atoms (zero) and the
   a. maximum potential energy.
   b. kinetic energy of the atom.
   c. minimum potential energy.
   d. bond length expressed in picometers.

7. Multiple covalent bonds may be found in atoms that contain carbon, nitrogen, or
   a. helium.
   b. argon.
   c. hydrogen.
   d. oxygen.

8. VSEPR theory states that the repulsion between electron pairs surrounding an atom causes
   a. these pairs to be as far apart as possible.
   b. an electron sea to form.
   c. a bond to break.
   d. a covalent bond to form.

9. In writing a Lewis structure for a negative polyatomic ion, you must add one electron for each unit of
   a. positive charge.
   b. negative charge.
   c. mass.
   d. electronegativity.

10. To draw a Lewis structure, you do not need to know
    a. the number of valence electrons for each atom.
    b. the types of atoms in the molecule.
    c. the number of atoms in the molecule.
    d. bond energies.

11. The polarity of a bond is related to the
    a. Lewis structure of each atom.
    b. electronegativity difference between the two atoms.
    c. number of shared pairs that are present.
    d. number of unshared pairs that are present.

12. In Lewis structures, the atomic symbol represents
    a. valence electrons.
    b. atomic number.
    c. the nucleus and inner-shell electrons.
    d. a stable octet of electrons.

THE MOLE AND CHEMICAL COMPOSITION

Galaxies have hundreds of billions of stars. The universe may have as many as sextillion stars—that's 1000 000 000 000 000 000 000 (or 1×10^{21}) stars. Such a number is called *astronomical* because it is so large that it usually refers only to vast quantities such as those described in astronomy. Can such a large number describe quantities that are a little more down to Earth? It certainly can. In fact, it takes an even larger number to describe the number of water molecules in a glass of water! In this chapter, you will learn about the *mole*, a unit used in chemistry to make working with such large quantities a little easier.

START-UP *ACTIVITY*

Counting Large Numbers

PROCEDURE

1. Count out exactly **200 small beads.** Using a **stopwatch,** record the amount of time it takes you to count them.

2. Your teacher will tell you the approximate number of small beads in 1 g. Knowing that number, calculate the mass of 200 small beads. Record the mass that you have calculated.

3. Use a **balance** to determine the mass of the 200 small beads that you counted in step 1. Compare this mass with the mass you calculated in step 2.

4. Using the mass you calculated in step 2 and a balance, measure out **another 200 small beads.** Record the amount of time it takes you to count small beads when using this counting method.

5. Count the number of **large beads** in 1 g.

ANALYSIS

1. Which method of counting took the most time?

2. Which method of counting do you think is the most accurate?

3. In a given mass, how does the number of large beads compare with the number of small beads? Explain your results.

Pre-Reading Questions

① **What are some things that are sold by weight instead of by number?**

② **Which would need a larger package, a kilogram of pencils or a kilogram of drinking straws?**

③ **If you counted one person per second, how many hours would it take to count the 6 billion people now in the world?**

🖪 **internet** connect ≡

www.scilinks.org
Topic : Galaxies
SciLinks code: HW4062

SC*i*LINKS. Maintained by the National Science Teachers Association

① Avogadro's Number and Molar Conversions

KEY TERMS

- mole
- Avogadro's number
- molar mass

OBJECTIVES

① **Identify** the mole as the unit used to count particles, whether atoms, ions, or molecules.

② **Use** Avogadro's number to convert between amount in moles and number of particles.

③ **Solve** problems converting between mass, amount in moles, and number of particles using Avogadro's number and molar mass.

Avogadro's Number and the Mole

mole

the SI base unit used to measure the amount of a substance whose number of particles is the same as the number of atoms of carbon in exactly 12 grams of carbon-12

Atoms, ions, and molecules are very small, so even tiny samples have a huge number of particles. To make counting such large numbers easier, scientists use the same approach to represent the number of ions or molecules in a sample as they use for atoms. The SI unit for amount is called the **mole** (mol). A mole is the number of atoms in exactly 12 grams of carbon-12.

Avogadro's number

6.022×10^{23}, the number of atoms or molecules in 1.000 mol

The number of particles in a mole is called **Avogadro's number,** or Avogadro's constant. One way to determine this number is to count the number of particles in a small sample and then use mass or particle size to find the amount in a larger sample. This method works only if all of the atoms in the sample are identical. Thus, scientists measure Avogadro's number using a sample that has atoms of only one isotope.

Figure 1
The particles in a mole can be atoms, molecules, or ions. Examples of a variety of molar quantities are given. Notice that the volume and mass of a molar quantity varies from substance to substance.

MOLAR QUANTITIES OF SOME SUBSTANCES

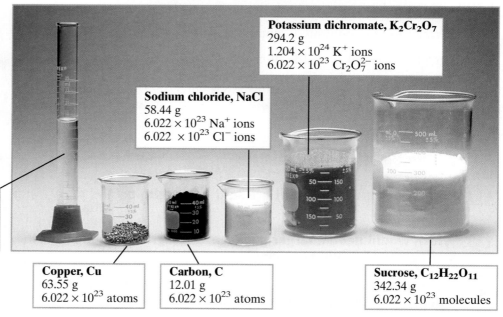

Potassium dichromate, $K_2Cr_2O_7$
294.2 g
1.204×10^{24} K^+ ions
6.022×10^{23} $Cr_2O_7^{2-}$ ions

Sodium chloride, NaCl
58.44 g
6.022×10^{23} Na^+ ions
6.022×10^{23} Cl^- ions

Water, H_2O
18.02 g
6.022×10^{23}
molecules

Copper, Cu
63.55 g
6.022×10^{23} atoms

Carbon, C
12.01 g
6.022×10^{23} atoms

Sucrose, $C_{12}H_{22}O_{11}$
342.34 g
6.022×10^{23} molecules

Table 1	Counting Units
Unit	**Example**
1 dozen	12 objects
1 score	20 objects
1 roll	50 pennies
1 gross	144 objects
1 ream	500 sheets of paper
1 hour	3600 seconds
1 mole	6.022×10^{23} particles

Figure 2
You can use mass to count out a roll of new pennies; 50 pennies are in a roll. One roll has a mass of about 125 g.

The most recent measurement of Avogadro's number shows that it is $6.02214199 \times 10^{23}$ units/mole. In this book, the measurement is rounded to 6.022×10^{23} units/mol. Avogadro's number is used to count any kind of particle, as shown in **Figure 1.**

The Mole Is a Counting Unit

Keep in mind that the mole is used to count out a given number of particles, whether they are atoms, molecules, formula units, ions, or electrons. The mole is used in the same way that other, more familiar counting units, such as those in **Table 1,** are used. For example, there are 12 eggs in one dozen eggs. You might want to know how many eggs are in 15 dozen. You can calculate the number of eggs by using a conversion factor as follows.

$$15 \text{ dozen eggs} \times \frac{12 \text{ eggs}}{1 \text{ dozen eggs}} = 180 \text{ eggs}$$

Figure 2 shows another way that you can count objects: by using mass.

Exploring the Mole

SAFETY PRECAUTIONS

PROCEDURE

1. Use a **periodic table** to find the atomic mass of the following substances: **graphite (carbon), iron filings, sulfur powder, aluminum foil,** and **copper wire.**

2. Use a **balance** to measure out 1 mol of each substance.

3. Use **graduated beakers** to find the approximate volume in 1 mol of each substance.

ANALYSIS

1. Which substance has the greatest atomic mass?

2. Which substance has the greatest mass in 1 mol?

3. Which substance has the greatest volume in 1 mol?

4. Does the mass of a mole of a substance relate to the substance's atomic mass?

5. Does the volume of a mole of a substance relate to the substance's atomic mass?

Converting Between Amount in Moles and Number of Particles

1. Decide which quantity you are given: amount (in moles) or number of particles (in atoms, molecules, formula units, or ions).

2. If you are converting from amount to number of particles (going left to right), use the top conversion factor.

3. If you are converting from number of particles to amount (going right to left), use the bottom conversion factor.

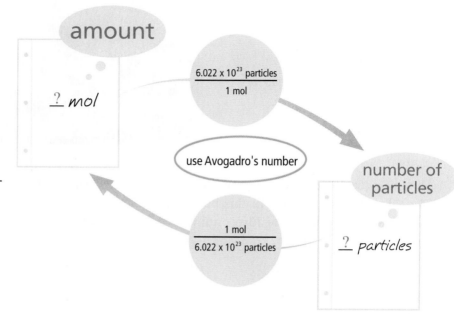

amount

$\underline{?}$ mol

$$\frac{6.022 \times 10^{23} \text{ particles}}{1 \text{ mol}}$$

use Avogadro's number

number of particles

$$\frac{1 \text{ mol}}{6.022 \times 10^{23} \text{ particles}}$$

$\underline{?}$ particles

Amount in Moles Can Be Converted to Number of Particles

A conversion factor begins with a definition of a relationship. The definition of one mole is

$$6.022 \times 10^{23} \text{ particles} = 1 \text{ mol}$$

If two quantities are equal and you divide one by the other, the factor you get is equal to 1. The following equation shows how this relationship is true for the definition of the mole.

$$\frac{6.022 \times 10^{23} \text{ particles}}{1 \text{ mol}} = 1$$

The factor on the left side of the equation is a conversion factor. The reciprocal of a conversion factor is also a conversion factor and is also equal to one, so the following is true.

$$\frac{6.022 \times 10^{23} \text{ particles}}{1 \text{ mol}} = \frac{1 \text{ mol}}{6.022 \times 10^{23} \text{ particles}} = 1$$

Because a conversion factor is equal to 1, it can multiply any quantity without changing the quantity's value. Only the units are changed.

These conversion factors can be used to convert between a number of moles of substance and a corresponding number of molecules. For example, imagine that you want to convert 2.66 mol of a compound into the corresponding number of molecules. How do you know which conversion factor to use? **Skills Toolkit 1** can help.

Choose the Conversion Factor That Cancels the Given Units

Take the amount (in moles) that you are given, shown in **Skills Toolkit 1** on the left, and multiply it by the conversion factor, shown in the top green circle, to get the number of particles, shown on the right. The calculation is as follows:

$$2.66 \; \cancel{mol} \times \frac{6.022 \times 10^{23} \; \text{molecules}}{1 \; \cancel{mol}} = 1.60 \times 10^{24} \; \text{molecules}$$

You can tell which of the two conversion factors to use, because the needed conversion factor should cancel the units of the given quantity to give you the units of the answer or the unknown quantity.

SKILLS Toolkit 2

Working Practice Problems

PROBLEM SOLVING SKILL

1. Gather information.
- Read the problem carefully.
- List the quantities and units given in the problem.
- Determine what value is being asked for (the answer) and the units it will need.

2. Plan your work.
Write the value of the given quantity times a question mark (which stands for a conversion factor) and then the equals sign, followed by another question mark (which stands for the answer) and the units of the answer. For example:

$$4.2 \; \text{mol} \; CO_2 \times ? = ? \; \text{molecules} \; CO_2$$

3. Calculate.
- Determine the conversion factor(s) needed to change the units of the given quantity to the units of the answer. Write the conversion factor(s) in the order you need them to cancel units.
- Cancel units, and check that the units that remain are the same on both sides and are the units desired for the answer.
- Calculate and round off the answer to the correct number of significant figures.
- Report your answer with correct units.

4. Verify your result.
- Verify your answer by estimating. One way to do so is to round off the numbers in the setup and make a quick calculation.
- Make sure your answer is reasonable. For example, if the number of atoms is less than one, the answer cannot possibly be correct.

STUDY TIP

WORKING PROBLEMS
If you have difficulty working practice problems, review the outline of procedures in **Skills Toolkit 2.** You may also refer back to the sample problems.

Converting Amount in Moles to Number of Particles

Find the number of molecules in 2.5 mol of sulfur dioxide.

1 Gather information.
- amount of SO_2 = 2.5 mol
- 1 mol of any substance = 6.022×10^{23} particles
- number of molecules of SO_2 = ? molecules

2 Plan your work.
The setup is: 2.5 mol SO_2 × ? = ? molecules SO_2

3 Calculate.
You are converting from the unit *mol* to the unit *molecules*. The conversion factor must have the units of *molecules/mol*. **Skills Toolkit 1** shows that this means you use 6.022×10^{23} molecules/1 mol.

$$2.5 \text{ mol } SO_2 \times \frac{6.022 \times 10^{23} \text{ molecules } SO_2}{1 \text{ mol } SO_2} = 1.5 \times 10^{24} \text{ molecules } SO_2$$

4 Verify your result.
The units cancel correctly. The answer is greater than Avogadro's number, as expected, and has two significant figures.

PRACTICE

1 How many ions are there in 0.187 mol of Na^+ ions?

2 How many atoms are there in 1.45×10^{-17} mol of arsenic?

3 How many molecules are there in 4.224 mol of acetic acid, $C_2H_4O_2$?

4 How many formula units are there in 5.9 mol of NaOH?

Number of Particles Can Be Converted to Amount in Moles

Notice in **Skills Toolkit 1** that the reverse calculation is similar but that the conversion factor is inverted to get the correct units in the answer. Look at the following problem. How many moles are 2.54×10^{22} iron(III) ions, Fe^{3+}?

$$2.54 \times 10^{22} \text{ ions } Fe^{3+} \times ? = ? \text{ mol } Fe^{3+}$$

Multiply by the conversion factor that cancels the unit of *ions* and leaves the unit of *mol*. (That is, you use the conversion factor that has the units that you want to get on top and the units that you want to get rid of on the bottom.)

$$2.54 \times 10^{22} \text{ ions } Fe^{3+} \times \frac{1 \text{ mol } Fe^{3+}}{6.022 \times 10^{23} \text{ ions } Fe^{3+}} = 0.0422 \text{ mol } Fe^{3+}$$

This answer makes sense, because you started with fewer than Avogadro's number of ions, so you have less than one mole of ions.

Converting Number of Particles to Amount in Moles

A sample contains 3.01×10^{23} molecules of sulfur dioxide, SO_2.
Determine the amount in moles.

1 Gather information.
- number of molecules of $SO_2 = 3.01 \times 10^{23}$ molecules
- 1 mol of any substance $= 6.022 \times 10^{23}$ particles
- amount of $SO_2 = ?$ mol

2 Plan your work.
The setup is similar to the calculation in **Sample Problem A.**

$$3.01 \times 10^{23} \text{ molecules } SO_2 \times ? = ? \text{ mol } SO_2$$

3 Calculate.
The conversion factor is used to remove the unit of *molecules* and
introduce the unit of *mol.*

$$3.01 \times 10^{23} \text{ molecules } SO_2 \times \frac{1 \text{ mol } SO_2}{6.022 \times 10^{23} \text{ molecules } SO_2} = 0.500 \text{ mol } SO_2$$

4 Verify your result.
There are fewer than 6.022×10^{23} (Avogadro's number) of SO_2
molecules, so it makes sense that the result is less than 1 mol. Three
is the correct number of significant figures.

PRACTICE HINT

Always check your
answer for the correct
number of significant
figures.

PRACTICE

1 How many moles of xenon do 5.66×10^{23} atoms equal?

2 How many moles of silver nitrate do 2.888×10^{15} formula units equal?

3 A biologist estimates that there are 2.7×10^{17} termites on Earth.
How many moles of termites is this?

4 How many moles do 5.66×10^{25} lithium ions, Li^+, equal?

5 Determine the number of moles of each specified atom or ion in the
given samples of the following compounds. (Hint: The formula tells you
how many atoms or ions are in each molecule or formula unit.)

a. O atoms in 3.161×10^{21} molecules of CO_2

b. C atoms in 3.161×10^{21} molecules of CO_2

c. O atoms in 2.222×10^{24} molecules of NO

d. K^+ ions in 5.324×10^{16} formula units of KNO_2

e. Cl^- ions in 1.000×10^{14} formula units of $MgCl_2$

f. N atoms in 2.000×10^{14} formula units of $Ca(NO_3)_2$

g. O atoms in 4.999×10^{25} formula units of $Mg_3(PO_4)_2$

Molar Mass Relates Moles to Grams

In chemistry, you often need to know the mass of a given number of moles of a substance or the number of moles in a given mass. Fortunately, the mole is defined in a way that makes figuring out either of these easy.

Amount in Moles Can Be Converted to Mass

molar mass

the mass in grams of one mole of a substance

The mole is the SI unit for amount. The **molar mass**, or mass in grams of one mole of an element or compound, is numerically equal to the atomic mass of monatomic elements and the formula mass of compounds and diatomic elements. To find a monatomic element's molar mass, use the atomic mass, but instead of having units of *amu*, the molar mass will have units of *g/mol*. So, the molar mass of carbon is 12.01 g/mol, and the molar mass of iron is 55.85 g/mol. How to find the molar mass of compounds and diatomic elements is shown in the next section.

You use molar masses as conversion factors in the same way you use Avogadro's number. The right side of **Skills Toolkit 3** shows how the *amount* in moles relates to the *mass* in grams of a substance. Suppose you must find the mass of 3.50 mol of copper. You will use the molar mass of copper. By checking the periodic table, you find the atomic mass of copper, 63.546 amu, which you round to 63.55 amu. So, in calculations with copper, use 63.55 g/mol.

The Mole Plays a Central Part in Chemical Conversions

You know how to convert from number of particles to amount in moles and how to convert from amount in moles to mass. Now you can use the same methods one after another to convert from *number of particles* to *mass*. **Skills Toolkit 3** shows the two-part process for this conversion. One step common to many problems in chemistry is converting to amount in moles. **Sample Problem C** shows how to convert from number of particles to the mass of a substance by first converting to amount in moles.

3 SKILLS Toolkit

Converting Between Mass, Amount, and Number of Particles

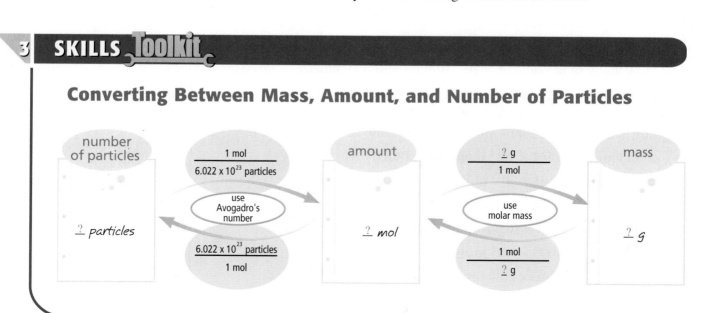

Converting Number of Particles to Mass

Find the mass in grams of 2.44×10^{24} atoms of carbon, whose molar mass is 12.01 g/mol.

1 Gather information.
- number of atoms C = 2.44×10^{24} atoms
- molar mass of carbon = 12.01 g/mol
- amount of C = ? mol
- mass of the sample of carbon = ? g

2 Plan your work.
- **Skills Toolkit 3** shows that to convert from number of atoms to mass in grams, you must first convert to amount in moles.
- To find the amount in moles, select the conversion factor that will take you from number of atoms to amount in moles.

$$2.44 \times 10^{24} \text{ atoms} \times ? = ? \text{ mol}$$

- Multiply the number of atoms by the following conversion factor:

$$\frac{1 \text{ mol}}{6.022 \times 10^{23} \text{ atoms}}$$

- To find the mass in grams, select the conversion factor that will take you from amount in moles to mass in grams.

$$? \text{ mol} \times ? = ? \text{ g}$$

- Multiply the amount in moles by the following conversion factor:

$$\frac{12.01 \text{ g C}}{1 \text{ mol}}$$

3 Calculate.
Solve and cancel identical units in the numerator and denominator.

$$2.44 \times 10^{24} \text{ atoms} \times \frac{1 \text{ mol}}{6.022 \times 10^{23} \text{ atoms}} \times \frac{12.01 \text{ g C}}{1 \text{ mol}} = 48.7 \text{ g C}$$

4 Verify your result.
The answer has the units requested in the problem.

PRACTICE HINT

Make sure to select the correct conversion factors so that units cancel to get the unit required in the answer.

PRACTICE

Given molar mass, find the mass in grams of each of the following substances:

1. 2.11×10^{24} atoms of copper (molar mass of Cu = 63.55 g/mol)
2. 3.01×10^{23} formula units of NaCl (molar mass of NaCl = 58.44 g/mol)
3. 3.990×10^{25} molecules of CH_4 (molar mass of CH_4 = 16.05 g/mol)
4. 4.96 mol titanium (molar mass of Ti = 47.88 g/mol)

Mass Can Be Converted to Amount in Moles

Converting from mass to number of particles is the reverse of the operation in the previous problem. This conversion is also shown in **Skills Toolkit 3,** but this time you are going from right to left and using the bottom conversion factors.

Sample Problem D shows how to convert the mass of a substance to amount (mol) and then convert amount to the number of particles. Notice that the problem is the reverse of **Sample Problem C.**

SAMPLE PROBLEM D

Converting Mass to Number of Particles

Find the number of molecules present in 47.5 g of glycerol, $C_3H_8O_3$. The molar mass of glycerol is 92.11 g/mol.

1 Gather information.
- mass of the sample of $C_3H_8O_3$ = 47.5 g
- molar mass of $C_3H_8O_3$ = 92.11 g/mol
- amount of $C_3H_8O_3$ = ? mol
- number of molecules $C_3H_8O_3$ = ? molecules

2 Plan your work.
- **Skills Toolkit 3** shows that you must first find the amount in moles.
- To determine the amount in moles, select the conversion factor that will take you from mass in grams to amount in moles.

$$47.5 \text{ g} \times ? = ? \text{ mol}$$

- Multiply mass by the conversion factor $\dfrac{1 \text{ mol}}{92.11 \text{ g } C_3H_8O_3}$

- To determine the number of particles, select the conversion factor that will take you from amount in moles to number of particles.

$$? \text{ mol} \times ? = ? \text{ molecules}$$

- Multiply amount by the conversion factor $\dfrac{6.022 \times 10^{23} \text{ molecules}}{1 \text{ mol}}$

3 Calculate.

$$47.5 \text{ g } C_3H_8O_3 \times \frac{1 \text{ mol}}{92.11 \text{ g } C_3H_8O_3} \times \frac{6.022 \times 10^{23} \text{ molecules}}{1 \text{ mol}} = 3.11 \times 10^{23} \text{ molecules}$$

4 Verify your result.
The answer has the units requested in the problem.

PRACTICE HINT

Because no elements have a molar mass less than one, the number of grams in a sample of a substance will always be larger than the number of moles of the substance. Thus, when you convert from grams to moles, you will get a smaller number. And the opposite is true for the reverse calculation.

PRACTICE

1 Find the number of atoms in 237 g Cu (molar mass of Cu = 63.55 g/mol).

2 Find the number of ions in 20.0 g Ca^{2+} (molar mass of Ca^{2+} = 40.08 g/mol).

3 Find the number of atoms in 155 mol of arsenic.

1	11	17
H	**Na**	**Cl**
Hydrogen	Sodium	Chlorine
1.007 94	22.989 770	35.4527
$1s^1$	$[Ne]3s^1$	$[Ne]3s^23p^5$
1.01 g/mol	22.99 g/mol	35.45 g/mol

Figure 3
Round molar masses from the periodic table to two significant figures to the right of the decimal point.

Remember to Round Consistently

Calculators may report many figures. However, an answer must never be given to more figures than is appropriate. If the given amount has only two significant figures, then you must round the calculated number off to two significant figures. Also, keep in mind that many numbers are exact. In the definition of the mole, the chosen amount is *exactly* 12 grams of the carbon-12 isotope. Such numbers are not considered when rounding. **Figure 3** shows how atomic masses are rounded in this text.

internet connect
www.scilinks.org
Topic : Significant Figures
SciLinks code: HW4115

SCiLINKS. Maintained by the National Science Teachers Association

 # Section Review

UNDERSTANDING KEY IDEAS

1. What is the definition of a mole?

2. How many particles are there in one mole?

3. Explain how Avogadro's number can give two conversion factors.

4. Which will have the greater number of ions, 1 mol of nickel(II) or 1 mol of copper(I)?

5. Without making a calculation, is 1.11 mol Pt more or less than 6.022×10^{23} atoms?

PRACTICE PROBLEMS

6. Find the number of molecules or ions.
 a. 2.00 mol Fe^{3+} **c.** 0.25 mol K^+
 b. 4.5 mol BCl_3 **d.** 6.022 mol O_2

7. Find the number of sodium ions, Na^+.
 a. 3.00 mol Na_2CO_3
 b. 3.00 mol $Na_4P_2O_7$
 c. 5.12 mol $NaNO_3$

8. Find the number of moles.
 a. 3.01×10^{23} molecules H_2O
 b. 1.000×10^{23} atoms C
 c. 5.610×10^{22} ions Na^+

9. Find the mass in grams.
 a. 4.30×10^{16} atoms He, 4.00 g/mol
 b. 5.710×10^{23} molecules CH_4, 16.05 g/mol
 c. 3.012×10^{24} ions Ca^{2+}, 40.08 g/mol

10. Find the number of molecules or ions.
 a. 1.000 g I^-, 126.9 g/mol
 b. 3.5 g Cu^{2+}, 63.55 g/mol
 c. 4.22 g SO_2, 64.07 g/mol

11. What is the mass of 6.022×10^{23} molecules of ibuprofen (molar mass of 206.31 g/mol)?

12. Find the mass in grams.
 a. 4.01×10^{23} atoms Ca, 40.08 g/mol
 b. 4.5 mol boron-11, 11.01 g/mol
 c. 1.842×10^{19} ions Na^+, 22.99 g/mol

13. Find the number of molecules.
 a. 2.000 mol H_2, 2.02 g/mol
 b. 4.01 g HF, 20.01 g/mol
 c. 4.5 mol $C_6H_{12}O_6$, 180.18 g/mol

CRITICAL THINKING

14. Why do we use carbon-12 rather than ordinary carbon as the basis for the mole?

15. Use **Skills Toolkit 1** to explain how a number of atoms is converted into amount in moles.

Atomic Mass and ical Formulas

OBJECTIVES

1. **Use** a periodic table or isotopic composition data to determine the average atomic masses of elements.

2. **Infer** information about a compound from its chemical formula.

3. **Determine** the molar mass of a compound from its formula.

Average Atomic Mass and the Periodic Table

You have learned that you can use atomic masses on the periodic table to find the molar mass of elements. Many of these values on the periodic table are close to whole numbers. However, most atomic masses are written to at least three places past the decimal.

Why are the atomic masses of most elements on the periodic table not exact whole numbers? One reason is that the masses reported are *relative* atomic masses. To understand relative masses, think about the setup in **Figure 4.** Eight pennies have the same mass as five nickels do. Thus, you could say that a single penny has a relative mass of 0.625 "nickel masses." Just as you can find the mass of a penny compared with the mass of a nickel, scientists have determined the masses of the elements relative to each other. Remember that atomic mass is given in units of *amu.* This means that it reflects an atom's mass relative to the mass of a carbon-12 atom. So, now you may ask why carbon's atomic mass on the periodic table is not exactly 12.

Most Elements Are Mixtures of Isotopes

You remember that *isotopes* are atoms that have different numbers of neutrons than other atoms of the same element do. So, isotopes have different atomic masses. The periodic table reports **average atomic mass,** a weighted average of the atomic mass of an element's isotopes. A *weighted average* takes into account the relative importance of each number in the average. Thus, if there is more of one isotope in a typical sample, it affects the average atomic mass more than an isotope that is less abundant does.

For example, carbon has two stable isotopes found in nature, carbon-12 and carbon-13. The average atomic mass of carbon takes into account the masses of both isotopes and their relative abundance. So, while the atomic mass of a carbon-12 atom is exactly 12 amu, any carbon sample will include enough carbon-13 atoms that the average mass of a carbon atom is 12.0107 amu.

Topic Link

Refer to the "Atoms and Moles" chapter for a discussion of atomic mass and isotopes.

average atomic mass

the weighted average of the masses of all naturally occurring isotopes of an element

Figure 4
You can determine the mass of a penny relative to the mass of a nickel; eight pennies have the same mass as five nickels.

Like carbon, most elements are a mixture of isotopes. In most cases, the fraction of each isotope is the same no matter where the sample comes from. Most average atomic masses can be determined to several decimal places. However, some elements have different percentages of isotopes depending on the source of the sample. This is true of *native* lead, or lead that occurs naturally on Earth. The average atomic mass of lead is given to only one decimal place because its composition varies so much from one sample to another.

If you know the abundance of each isotope, you can calculate the average atomic mass of an element. For example, the average atomic mass of native copper is a weighted average of the atomic masses of two isotopes, shown in **Figure 5.** The following sample problem shows how this calculation is made from data for the abundance of each of native copper's isotopes.

Figure 5
Native copper is a mixture of two isotopes. Copper-63 contributes 69.17% of the atoms, and copper-65 the remaining 30.83%.

SAMPLE PROBLEM E

Calculating Average Atomic Mass

The mass of a Cu-63 atom is 62.94 amu, and that of a Cu-65 atom is 64.93 amu. Using the data in **Figure 5,** find the average atomic mass of Cu.

1 Gather information.
- atomic mass of a Cu-63 atom = 62.94 amu
- abundance of Cu-63 = 69.17%
- atomic mass of Cu-65 = 64.93 amu
- abundance of Cu-65 = 30.83%
- average atomic mass of Cu = ? g

2 Plan your work.
The average atomic mass of an element is the sum of the contributions of the masses of each isotope to the total mass. This type of average is called a *weighted average.* The contribution of each isotope is equal to its atomic mass multiplied by the fraction of that isotope. (To change a percentage into a fraction, divide it by 100.)

Isotope	Percentage	Decimal fraction	Contribution
Copper-63	69.17%	0.6917	62.94×0.6917
Copper-65	30.83%	0.3083	64.93×0.3083

3 Calculate.
Average atomic mass is the sum of the individual contributions:

$$(62.94 \text{ amu} \times 0.6917) + (64.93 \text{ amu} \times 0.3083) = 63.55 \text{ amu}$$

4 Verify your results.
- The answer lies between 63 and 65, and the result is closer to 63 than it is to 65. This is expected because the isotope 63 makes a larger contribution to the average.
- Compare your answer with the value in the periodic table.

Practice problems on next page

www.scilinks.org
Topic: Isotopes
SciLinks code: HW4073

PRACTICE HINT

In calculating average atomic masses, remember that the resulting value must be greater than the lightest isotope and less than the heaviest isotope.

PRACTICE

1 Calculate the average atomic mass for gallium if 60.00% of its atoms have a mass of 68.926 amu and 40.00% have a mass of 70.925 amu.

2 Calculate the average atomic mass of oxygen. Its composition is 99.76% of atoms with a mass of 15.99 amu, 0.038% with a mass of 17.00 amu, and 0.20% with a mass 18.00 amu.

Chemical Formulas and Moles

Until now, when you needed to perform molar conversions, you were given the molar mass of compounds in a sample. Where does this molar mass of compounds come from? You can determine the molar mass of compounds the same way that you find the molar mass of individual elements—by using the periodic table.

Formulas Express Composition

The first step to finding a compound's molar mass is understanding what a chemical formula tells you. It tells you which elements, as well as how much of each, are present in a compound. The formula KBr shows that the compound is made up of potassium and bromide ions in a 1:1 ratio. The formula H_2O shows that water is made up of hydrogen and oxygen atoms in a 2:1 ratio. These ratios are shown in **Figure 6.**

You have learned that covalent compounds, such as water and hexachloroethane, consist of molecules as units. Formulas for covalent compounds show both the elements and the number of atoms of each element in a molecule. Hexachloroethane has the formula C_2Cl_6. Each molecule has 8 atoms covalently bonded to each other. Ionic compounds aren't found as molecules, so their formulas do not show numbers of atoms. Instead, the formula shows the simplest ratio of cations and anions.

Figure 6
Although any sample of a compound has many atoms and ions, the chemical formula gives a ratio of those atoms or ions.

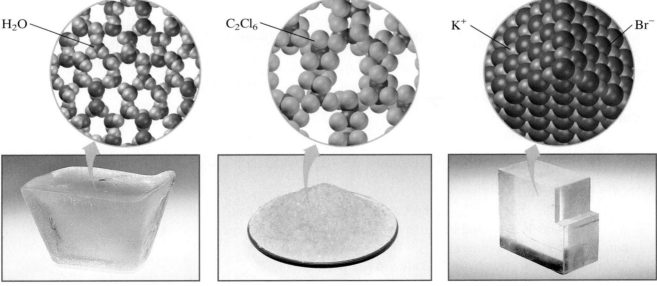

Water, H_2O Hexachloroethane, C_2Cl_6 Potassium bromide, KBr

Figure 7

The formula for a polyatomic ionic compound is the simplest ratio of cations to anions.

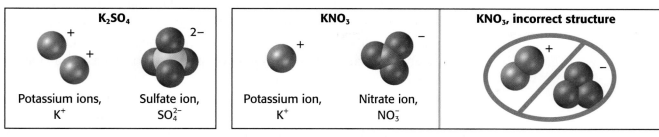

K₂SO₄

Potassium ions, K⁺   Sulfate ion, SO₄²⁻

KNO₃

Potassium ion, K⁺   Nitrate ion, NO₃⁻

KNO₃, incorrect structure

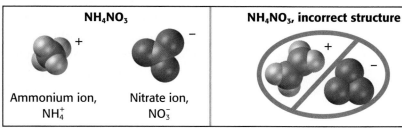

(NH₄)₂SO₄

Ammonium ions, NH₄⁺   Sulfate ion, SO₄²⁻

NH₄NO₃

Ammonium ion, NH₄⁺   Nitrate ion, NO₃⁻

NH₄NO₃, incorrect structure

a Elements in polyatomic ions are bound together in a group and carry a characteristic charge.

b The formula for a compound with polyatomic ions shows how the atoms in each ion are bonded together.

c You cannot move atoms from one polyatomic ion to the next.

Formulas Give Ratios of Polyatomic Ions

The meaning of a formula does not change when polyatomic ions are involved. Potassium nitrate has the formula KNO_3. Just as the formula KBr indicates a 1:1 ratio of K^+ cations to Br^- anions, the formula KNO_3 indicates a ratio of one K^+ cation to one NO_3^- anion.

When a compound has polyatomic ions, such as those in **Figure 7,** look for the cations and anions. Formulas can tell you which elements make up polyatomic ions. For example, in the formula KNO_3, NO_3 is a nitrate ion, NO_3^-. KNO_3 does not have a KN^+ and an O_3^- ion. Similarly, the formula of ammonium nitrate is written NH_4NO_3, because NH_4 in a formula stands for the ammonium ion, NH_4^+, and NO_3 stands for a nitrate ion, NO_3^-. If it were written as $H_4N_2O_3$, the number of atoms would be correct. However, the formula would no longer clearly show which ions were in the substance and how many there were. The formula NH_4NO_3 shows that ammonium nitrate is made up of ammonium and nitrate ions in a 1:1 ratio.

Formulas Are Used to Calculate Molar Masses

A formula tells you what atoms (or ions) are present in an element or compound. So, from a formula you can find the mass of a mole of the substance, or its molar mass. The simplest formula for most elements is simply that element's symbol. For example, the symbol for silver is Ag. The molar mass of elements whose formulas are this simple equals the atomic mass of the element expressed in g/mol. So, the molar mass of silver is 107.87 g/mol. Diatomic elements have twice the number of atoms in each molecule, so their molecules have molar masses that are twice the molar mass of each atom. For example, the molar mass of Br_2 molecules is two times the molar mass of Br atoms (2×79.90 g/mol = 159.80 g/mol).

🔲 internet connect

www.scilinks.org
Topic : Chemical Formulas
SciLinks code: HW4028

SCi
LINKS. Maintained by the National Science Teachers Association

Let's say you want to determine the molar mass of a molecular compound. You must use the periodic table to find the molar mass of more than one element. The molar mass of a molecular compound is the sum of the masses of all the atoms in it expressed in g/mol. For example, one mole of H_2O molecules will have two moles of H and one mole of O. Thus, the compound's molar mass is equal to two times the molar mass of a H atom plus the molar mass of an O atom, or 18.02 g (2×1.01 g + 16.00 g).

Scientists also use the simplest formula to represent one mole of an ionic compound. They often use the term *formula unit* when referring to ionic compounds, because they are not found as single molecules. A formula unit of an ionic compound represents the simplest ratio of cations to anions. A formula unit of KBr is made up of one K^+ ion and one Br^- ion. One mole of an ionic compound has 6.022×10^{23} of these formula units. As with molecular compounds, the molar mass of an ionic compound is the sum of the masses of all the atoms in the formula expressed in g/mol. **Table 2** compares the formula units and molar masses of three ionic compounds. **Sample Problem F** shows how to calculate the molar mass of barium nitrate.

Table 2   Calculating Molar Mass for Ionic Compounds

Formula	Formula unit	Calculation of molar mass
$ZnCl_2$	Cl^- Zn^{2+} Cl^-	1 Zn = 1×65.39 g/mol = 65.39 g/mol + 2 Cl = 2×35.45 g/mol = 70.90 g/mol $ZnCl_2$ = 136.29 g/mol
$ZnSO_4$	Zn^{2+} SO_4^{2-}	1 Zn = 1×65.39 g/mol = 65.39 g/mol 1 S = 1×32.07 g/mol = 32.07 g/mol + 4 O = 4×16.00 g/mol = 64.00 g/mol $ZnSO_4$ = 161.46 g/mol
$(NH_4)_2SO_4$	NH_4^+ NH_4^+ SO_4^{2-}	2 N = 2×14.01 g/mol = 28.02 g/mol 8 H = 8×1.01 g/mol = 8.08 g/mol 1 S = 1×32.07 g/mol = 32.07 g/mol + 4 O = 4×16.00 g/mol = 64.00 g/mol $(NH_4)_2SO_4$ = 132.17 g/mol

Calculating Molar Mass of Compounds

Find the molar mass of barium nitrate.

1 Gather information.
- simplest formula of ionic barium nitrate: $Ba(NO_3)_2$
- molar mass of $Ba(NO_3)_2 = ?$ g/mol

2 Plan your work.
- Find the number of moles of each element in 1 mol $Ba(NO_3)_2$. Each mole has:

    1 mol Ba
    2 mol N
    6 mol O

- Use the periodic table to find the molar mass of each element in the formula.

    molar mass of Ba = 137.33 g/mol
    molar mass of N = 14.01 g/mol
    molar mass of O = 16.00 g/mol

3 Calculate.
- Multiply the molar mass of each element by the number of moles of each element. Add these masses to get the total molar mass of $Ba(NO_3)_2$.

$$
\begin{aligned}
\text{mass of 1 mol Ba} &= 1 \times 137.33 \text{ g/mol} = 137.33 \text{ g/mol}\\
\text{mass of 2 mol N} &= 2 \times 14.01 \text{ g/mol} = 28.02 \text{ g/mol}\\
+ \text{mass of 6 mol O} &= 6 \times 16.00 \text{ g/mol} = 96.00 \text{ g/mol}\\
\hline
\text{molar mass of } Ba(NO_3)_2 &= 261.35 \text{ g/mol}
\end{aligned}
$$

4 Verify your result.
- The answer has the correct units. The sum of the molar masses of elements can be approximated as $140 + 30 + 100 = 270$, which is close to the calculated value.

> **PRACTICE HINT**
>
> Use the same methods for molecular compounds, but use the molecular formula in place of a formula unit.

PRACTICE

1 Find the molar mass for each of the following compounds:

  a. CsI   **c.** $C_{12}H_{22}O_{11}$ **e.** $HC_2H_3O_2$

  b. $CaHPO_4$ **d.** I_2 **f.** $Mg_3(PO_4)_2$

2 Write the formula and then find the molar mass.

  a. sodium hydrogen carbonate   **e.** iron(III) hydroxide

  b. cerium hexaboride   **f.** tin(II) chloride

  c. magnesium perchlorate   **g.** tetraphosphorus decoxide

  d. aluminum sulfate   **h.** iodine monochloride

continued on next page

PRACTICE

3 a. Find the molar mass of toluene, $C_6H_5CH_3$.

 b. Find the number of moles in 7.51 g of toluene.

4 a. Find the molar mass of cisplatin, $PtCl_2(NH_3)_2$, a cancer therapy chemical.

 b. Find the mass of 4.115×10^{21} formula units of cisplatin.

② Section Review

UNDERSTANDING KEY IDEAS

1. What is a weighted average?

2. On the periodic table, the average atomic mass of carbon is 12.01 g. Why is it not exactly 12.00?

3. What is the simplest formula for cesium carbonate?

4. What ions are present in cesium carbonate?

5. What is the ratio of N and H atoms in NH_3?

6. What is the ratio of calcium and chloride ions in $CaCl_2$?

7. Why is the simplest formula used to determine the molar mass for ionic compounds?

PRACTICE PROBLEMS

8. Calculate the average atomic mass of chromium. Its composition is: 83.79% with a mass of 51.94 amu; 9.50% with a mass of 52.94 amu; 4.35% with a mass of 49.95 amu; 2.36% with a mass of 53.94 amu.

9. Element X has two isotopes. One has a mass of 10.0 amu and an abundance of 20.0%. The other has a mass of 11.0 amu and an abundance of 80.0%. Estimate the average atomic mass. What element is it?

10. Find the molar mass.

 a. CsCl **d.** $(NH_4)_2HPO_4$

 b. $KClO_3$ **e.** $C_2H_5NO_2$

 c. $C_6H_{12}O_6$

11. Determine the formula, the molar mass, and the number of moles in 2.11 g of each of the following compounds.

 a. strontium sulfide

 b. phosphorus trifluoride

 c. zinc acetate

 d. mercury(II) bromate

 e. calcium nitrate

12. Find the molar mass and the mass of 5.0000 mol of each of the following compounds.

 a. calcium acetate, $Ca(C_2H_3O_2)_2$

 b. iron(II) phosphate, $Fe_3(PO_4)_2$

 c. saccharin, $C_7H_5NO_3S$, a sweetener

 d. acetylsalicylic acid, $C_9H_8O_4$, or aspirin

CRITICAL THINKING

13. In the periodic table, the atomic mass of fluorine is given to 9 significant figures, whereas oxygen is given to only 6. Why? (Hint: fluorine has only one isotope.)

14. **Figure 6** shows many K^+ and Br^- ions. Why is the formula not written as $K_{20}Br_{20}$?

15. Why don't scientists use HO as the formula for hydrogen peroxide, H_2O_2?

16. a. How many atoms of H are in a formula unit of $(NH_4)_2SO_4$?

 b. How many atoms of H are in 1 mol of $(NH_4)_2SO_4$?

Formulas and Percentage Composition

KEY TERMS

- **percentage composition**
- **empirical formula**
- **molecular formula**

OBJECTIVES

① **Determine** a compound's empirical formula from its percentage composition.

② **Determine** the molecular formula or formula unit of a compound from its empirical formula and its formula mass.

③ **Calculate** percentage composition of a compound from its molecular formula or formula unit.

Using Analytical Data

Scientists synthesize new compounds for many uses. Once they make a new product, they must check its identity. One way is to carry out a chemical analysis that provides a **percentage composition.** For example, in 1962, two chemists made a new compound from xenon and fluorine. Before 1962, scientists thought that xenon did not form compounds. The scientists analyzed their surprising find. They found that it had a percentage composition of 63.3% Xe and 36.7% F, which is the same as that for the formula XeF_4. Percentage composition not only helps verify a substance's identity but also can be used to compare the ratio of masses contributed by the elements in two substances, as in **Figure 8.**

percentage composition

the percentage by mass of each element in a compound

📶 internet connect

www.scilinks.org
Topic : Percentage
        Composition
SciLinks code: HW4131

SC*LINKS*. Maintained by the National Science Teachers Association

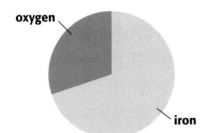

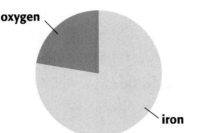

iron(III) oxide, Fe₂O₃

iron	69.9%
oxygen	30.1%

iron(II) oxide, FeO

iron	77.7%
oxygen	22.3%

Figure 8
Iron forms two different compounds with oxygen. The two compounds have different ratios of atoms and therefore have different percentage compositions and different properties.

empirical formula

a chemical formula that shows the composition of a compound in terms of the relative numbers and kinds of atoms in the simplest ratio

Empirical formula NH_2O

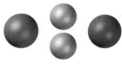

Actual formula NH_4NO_2

Space-filling model

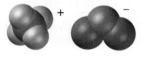

Figure 9
The empirical formula for ammonium nitrite is NH_2O. Its actual formula has 1 ammonium ion, NH_4^+, and 1 nitrite ion, NO_2^-.

Determining Empirical Formulas

Data for percentage composition allow you to calculate the simplest ratio among the atoms found in a compound. The **empirical formula** shows this simplest ratio. For example, ammonium nitrite, shown in **Figure 9,** has the actual formula NH_4NO_2 and is made up of ammonium ions, NH_4^+, and nitrite ions, NO_2^-, in a 1:1 ratio.

But if a chemist does an elemental analysis, she will find the empirical formula to be NH_2O, because it shows the *simplest ratio* of the elements. For some other compounds, the empirical formula and the actual formula are the same.

Let's say that you want to find an empirical formula from the percentage composition. First, convert the mass percentage of each element to grams. Second, convert from grams to moles using the molar mass of each element as a conversion factor. (Keep in mind that a formula for a compound can be read as a number of atoms or as a number of moles.) Third, as shown in **Sample Problem G,** compare these amounts in moles to find the simplest whole-number ratio among the elements in the compound.

To find this ratio, divide each amount by the smallest of all the amounts. This process will give a subscript of 1 for the atoms present in the smallest amount. Finally, you may need to multiply by a number to convert all subscripts to the smallest whole numbers. The final numbers you get are the subscripts in the empirical formula. For example, suppose the subscripts were 1.33, 2, and 1. Multiplication by 3 gives subscripts of 4, 6, and 3.

SAMPLE PROBLEM G

Determining an Empirical Formula from Percentage Composition

Chemical analysis of a liquid shows that it is 60.0% C, 13.4% H, and 26.6% O by mass. Calculate the empirical formula of this substance.

1 Gather information.
- percentage C = 60.0%
- percentage H = 13.4%
- percentage O = 26.6%
- empirical formula = $C_?H_?O_?$

2 Plan your work.
- Assume that you have a 100.0 g sample of the liquid, and convert the percentages to grams.

$$\text{for C: } 60.0\% \times 100.0 \text{ g} = 60.0 \text{ g C}$$

$$\text{for H: } 13.4\% \times 100.0 \text{ g} = 13.4 \text{ g H}$$

$$\text{for O: } 26.6\% \times 100.0 \text{ g} = 26.6 \text{ g O}$$

- To convert the mass of each element into the amount in moles, you must multiply by the proper conversion factor, which is the reciprocal of the molar mass. Find molar mass by using the periodic table.

molar mass of C: 12.01 g/mol

molar mass of H: 1.01 g/mol

molar mass of O: 16.00 g/mol

3 Calculate.

- Calculate the amount in moles of C, H, and O. Round the answers to the correct number of significant figures.

$$60.0 \text{ g C} \times \frac{1 \text{ mol C}}{12.01 \text{ g C}} = 5.00 \text{ mol C}$$

$$13.4 \text{ g H} \times \frac{1 \text{ mol H}}{1.01 \text{ g H}} = 13.3 \text{ mol H}$$

$$26.6 \text{ g O} \times \frac{1 \text{ mol O}}{16.00 \text{ g O}} = 1.66 \text{ mol O}$$

- At this point the formula can be written as $C_5H_{13.3}O_{1.66}$, but you know that subscripts in chemical formulas are usually whole numbers.
- To begin the conversion to whole numbers, divide all subscripts by the smallest subscript, 1.66. This will make at least one of the subscripts a whole number, 1.

$$\frac{5.00 \text{ mol C}}{1.66} = 3.01 \text{ mol C}$$

$$\frac{13.3 \text{ mol H}}{1.66} = 8.01 \text{ mol H}$$

$$\frac{1.66 \text{ mol O}}{1.66} = 1.00 \text{ mol O}$$

- These numbers can be assumed to be the whole numbers 3, 8 and 1. The empirical formula is therefore C_3H_8O.

4 Verify your result.

Verify your answer by calculating the percentage composition of C_3H_8O. If the result agrees with the composition stated in the problem, then the formula is correct.

PRACTICE HINT

When you get fractions for the first calculation of subscripts, think about how you can turn these into whole numbers. For example:
- the subscript 1.33 is roughly $1\frac{1}{3}$, so it will give the whole number 4 when multiplied by 3
- the subscript 0.249 is roughly $\frac{1}{4}$, so it will give the whole number 1 when multiplied by 4
- the subscript 0.74 is roughly $\frac{3}{4}$, so it will give the whole number 3 when multiplied by 4

PRACTICE

Determine the empirical formula for each substance.

1 A dead alkaline battery is found to contain a compound of Mn and O. Its analysis gives 69.6% Mn and 30.4% O.

2 A compound is 38.77% Cl and 61.23% O.

3 Magnetic iron oxide is 72.4% iron and 27.6% oxygen.

4 A liquid compound is 18.0% C, 2.26% H, and 79.7% Cl.

Molecular Formulas Are Multiples of Empirical Formulas

molecular formula

a chemical formula that shows the number and kinds of atoms in a molecule, but not the arrangement of the atoms

The formula for an ionic compound shows the simplest whole-number ratio of the large numbers of ions in a crystal of the compound. The formula $Ca_3(PO_4)_2$ shows that the ratio of Ca^{2+} ions to PO_4^{3-} ions is 3:2.

Molecular compounds, on the other hand, are made of single molecules. Some molecular compounds have the same molecular and empirical formulas. Examples are water, H_2O, and nitric acid, HNO_3. But for many molecular compounds the **molecular formula** is a whole-number multiple of the empirical formula. Both kinds of formulas are just two different ways of representing the composition of the same molecule.

The molar mass of a compound is equal to the molar mass of the empirical formula times a whole number, n. There are several experimental techniques for finding the molar mass of a molecular compound even though the compound's chemical composition and formula are unknown. If you divide the experimental molar mass by the molar mass of the empirical formula, you can figure out the value of n needed to scale the empirical formula up to give the molecular formula.

Think about the three compounds in **Table 3**—formaldehyde, acetic acid, and glucose, which is shown in **Figure 10**. Each has the empirical formula CH_2O. However, acetic acid has a molecular formula that is twice the empirical formula. The molecular formula for glucose is six times the empirical formula. The relationship is shown in the following equation.

$$n(\text{empirical formula}) = \text{molecular formula}$$

In general, the molecular formula is a whole-number multiple of the empirical formula. For formaldehyde, $n = 1$, for acetic acid, $n = 2$, and for glucose, $n = 6$. In some cases, n may be a very large number.

Figure 10
The formula for glucose, which is found in many sports drinks, is $C_6H_{12}O_6$.

Table 3 **Comparing Empirical and Molecular Formulas**				
Compound	Empirical formula	Molecular formula	Molar mass (g)	Space-filling model
Formaldehyde	CH_2O	CH_2O • same as empirical formula • $n = 1$	30.03	
Acetic acid	CH_2O	$C_2H_4O_2$ $(HC_2H_3O_2)$ • 2 × empirical formula • $n = 2$	60.06	
Glucose	CH_2O	$C_6H_{12}O_6$ • 6 × empirical formula • $n = 6$	180.18	

Determining a Molecular Formula from an Empirical Formula

The empirical formula for a compound is P_2O_5. Its experimental molar mass is 284 g/mol. Determine the molecular formula of the compound.

1 Gather information.
- empirical formula = P_2O_5
- molar mass of compound = 284 g/mol
- molecular formula = ?

2 Plan your work.
- Find the molar mass of the empirical formula using the molar masses of the elements from the periodic table.

$$\text{molar mass of P} = 30.97 \text{ g/mol}$$
$$\text{molar mass of O} = 16.00 \text{ g/mol}$$

3 Calculate.
- Find the molar mass of the empirical formula, P_2O_5.

$$
\begin{array}{ll}
2 \times \text{molar mass of P} = & 61.94 \text{ g/mol} \\
+ \, 5 \times \text{molar mass of O} = & 80.00 \text{ g/mol} \\
\hline
\text{molar mass of } P_2O_5 = & 141.94 \text{ g/mol}
\end{array}
$$

- Solve for n, the factor multiplying the empirical formula to get the molecular formula.

$$n = \frac{\text{experimental molar mass of compound}}{\text{molar mass of empirical formula}}$$

- Substitute the molar masses into this equation, and solve for n.

$$n = \frac{284 \text{ g/mol}}{141.94 \text{ g/mol}} = 2.00 = 2$$

- Multiply the empirical formula by this factor to get the answer.

$$n(\text{empirical formula}) = 2(P_2O_5) = P_4O_{10}$$

4 Verify your result.
- The molar mass of P_4O_{10} is 283.88 g/mol. It is equal to the experimental molar mass.

PRACTICE HINT

In some cases, you can figure out the factor n by just looking at the numbers. For example, let's say you noticed that the experimental molar mass was almost exactly twice as much as the molar mass of the empirical formula (as in this problem). That means n must be 2.

PRACTICE

1 A compound has an experimental molar mass of 78 g/mol. Its empirical formula is CH. What is its molecular formula?

2 A compound has the empirical formula CH_2O. Its experimental molar mass is 90.0 g/mol. What is its molecular formula?

3 A brown gas has the empirical formula NO_2. Its experimental molar mass is 46 g/mol. What is its molecular formula?

PROBLEM SOLVING SKILL

Chemical Formulas Can Give Percentage Composition

If you know the chemical formula of any compound, then you can calculate the percentage composition. From the subscripts, you can determine the mass contributed by each element and add these to get the molar mass. Then, divide the mass of each element by the molar mass. Multiply by 100 to find the percentage composition of that element.

Think about the two compounds shown in **Figure 11.** Carbon dioxide, CO_2, is a harmless gas that you exhale, while carbon monoxide, CO, is a poisonous gas present in car exhaust. The percentage composition of carbon dioxide, CO_2, is calculated as follows.

$$
\begin{aligned}
1 \text{ mol} \times 12.01 \text{ g C/mol} &= 12.01 \text{ g C} \\
+ 2 \text{ mol} \times 16.00 \text{ g O/mol} &= 32.00 \text{ g O} \\
\hline
\text{mass of 1 mol } CO_2 &= 44.01 \text{ g}
\end{aligned}
$$

$$
\% \text{ C in } CO_2 = \frac{12.01 \text{ g C}}{44.01 \text{ g } CO_2} \times 100 = 27.29\%
$$

$$
\% \text{ O in } CO_2 = \frac{32.00 \text{ g O}}{44.01 \text{ g } CO_2} \times 100 = 72.71\%
$$

The percentage composition of carbon monoxide, CO, is calculated as follows.

$$
\begin{aligned}
1 \text{ mol} \times 12.01 \text{ g C/mol} &= 12.01 \text{ g C} \\
+ 1 \text{ mol} \times 16.00 \text{ g O/mol} &= 16.00 \text{ g O} \\
\hline
\text{mass of 1 mol } CO &= 28.01 \text{ g}
\end{aligned}
$$

$$
\% \text{ C in } CO = \frac{12.01 \text{ g C}}{28.01 \text{ g } CO} \times 100 = 42.88\%
$$

$$
\% \text{ O in } CO = \frac{16.00 \text{ g O}}{28.01 \text{ g } CO} \times 100 = 57.12\%
$$

Figure 11
Carbon monoxide and carbon dioxide are both made up of the same elements, but they have different percentage compositions.

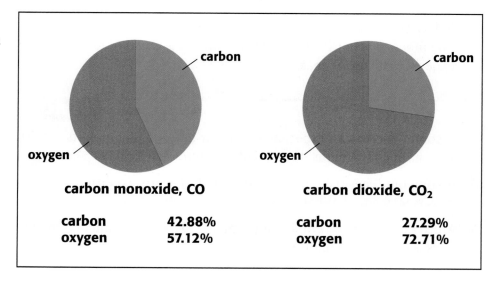

carbon monoxide, CO

| carbon | 42.88% |
| oxygen | 57.12% |

carbon dioxide, CO₂

| carbon | 27.29% |
| oxygen | 72.71% |

Using a Chemical Formula to Determine Percentage Composition

Calculate the percentage composition of copper(I) sulfide, a copper ore called chalcocite.

1 Gather information.
- name and formula of the compound: copper(I) sulfide, Cu_2S
- percentage composition: %Cu = ?, %S = ?

2 Plan your work.
To determine the molar mass of copper(I) sulfide, find the molar mass of the elements copper and sulfur using the periodic table.

$$\text{molar mass of Cu} = 63.55 \text{ g/mol}$$
$$\text{molar mass of S} = 32.07 \text{ g/mol}$$

3 Calculate.
- Find the masses of 2 mol Cu and 1 mol S. Use these masses to find the molar mass of Cu_2S.

$$2 \text{ mol} \times 63.55 \text{ g Cu/mol} = 127.10 \text{ g Cu}$$
$$\underline{+\ 1 \text{ mol} \times 32.07 \text{ g S/mol} =\ \ \ 32.07 \text{ g S}}$$
$$\text{molar mass of } Cu_2S = 159.17 \text{ g/mol}$$

- Calculate the fraction that each element contributes to the total mass. Do this by dividing the total mass contributed by that element by the total mass of the compound. Convert the fraction to a percentage by multiplying by 100.

$$\text{mass \% Cu} = \frac{\text{mass of 2 mol Cu}}{\text{molar mass of } Cu_2S} \times 100$$

$$\text{mass \% S} = \frac{\text{mass of 1 mol S}}{\text{molar mass of } Cu_2S} \times 100$$

- Substitute the masses into the equations above. Round the answers you get on the calculator to the correct number of significant figures.

$$\text{mass \% Cu} = \frac{127.10 \text{ g Cu}}{159.17 \text{ g } Cu_2S} \times 100 = 79.852\% \text{ Cu}$$

$$\text{mass \% S} = \frac{32.07 \text{ g S}}{159.17 \text{ g } Cu_2S} \times 100 = 20.15\% \text{ S}$$

4 Verify your result.
- Add the percentages. The sum should be near 100%.

$$79.852\% + 20.15\% = 100.00\%$$

Practice problems on next page

PRACTICE HINT

Sometimes, rounding gives a sum that differs slightly from 100%. This is expected. (However, if you find a sum that differs significantly, such as 112%, you have made an error.)

1. Calculate the percentage composition of Fe_3C, a compound in cast iron.

2. Calculate the percentage of both elements in sulfur dioxide.

3. Calculate the percentage composition of ammonium nitrate, NH_4NO_3.

4. Calculate the percentage composition of each of the following:

   a. $SrBr_2$ **c.** $Mg(CN)_2$
   b. $CaSO_4$ **d.** $Pb(CH_3COO)_2$

5. **a.** Calculate the percentage of each element in acetic acid, $HC_2H_3O_2$, and glucose, $C_6H_{12}O_6$.

   b. These two substances have the same empirical formula. What would you expect the percentage composition of the empirical formula to be?

③ Section Review

UNDERSTANDING KEY IDEAS

1. a. Suppose you know that a compound is 11.2% H and 88.8% O. What information do you need to determine the empirical formula?

b. What additional information do you need to determine the molecular formula?

2. Isooctane has the molecular formula C_8H_{18}. What is its empirical formula?

3. What information do you need to calculate the percentage composition of CF_4?

PRACTICE PROBLEMS

4. Determine the empirical formula.

a. The analysis of a compound shows that it is 9.2% B and 90.8% Cl.

b. An analysis shows that a compound is 50.1% S and 49.9% O.

c. The analysis of a compound shows that it is 27.0% Na, 16.5% N, and 56.5% O.

5. The experimental molar mass of the compound in item 4b is 64 g/mol. What is the compound's molecular formula?

6. Determine the formula, and then calculate the percentage composition.

a. calcium sulfate

b. silicon dioxide

c. silver nitrate

d. nitrogen monoxide

7. Calculate the percentage composition.

a. silver acetate, $AgC_2H_3O_2$

b. lead(II) chlorate, $Pb(ClO_3)_2$

c. iron(III) sulfate, $Fe_2(SO_4)_3$

d. copper(II) sulfate, $CuSO_4$

CRITICAL THINKING

8. When you determine the empirical formula of a compound from analytical data, you seldom get exact whole numbers for the subscripts. Explain why.

9. An amino acid has the molecular formula $C_2H_5NO_2$. What is the empirical formula?

10. A compound has the empirical formula CH_2O. Its experimental molar mass is 45 g/mol. Is it possible to calculate the molecular formula with the information given?

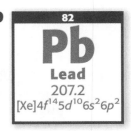

Where Is Pb?

Earth's crust:
< 0.01% by mass

Element Spotlight

Get the Lead Out

Humans have known for many centuries that lead is toxic, but it is still used in many common materials. High levels of lead were used in white paints until the 1940s. Since then, the lead compounds in paints have gradually been replaced with less toxic titanium dioxide. However, many older buildings still have significant amounts of lead paint, and many also have lead solder in their water pipes.

Lead poisoning is caused by the absorption of lead through the digestive tract, lungs, or skin. Children living in older homes are especially susceptible to lead poisoning. Children eat paint chips that contain lead because the paint has a sweet taste.

The hazards of lead poisoning can be greatly reduced by introducing programs that increase public awareness, removing lead-based paint from old buildings, and screening children for lead exposure.

Industrial Uses

- The largest industrial use of lead is in the manufacture of storage batteries.
- Solder used for joining metals is often an alloy of lead and tin.
- Other lead alloys are used to make bearings for gasoline and diesel engines, type metal for printing, corrosion-resistant cable coverings, and ammunition.
- Lead sheets and lead bricks are used to shield workers and sensitive objects from X rays.

Real-World Connection Lead inside the human body interferes with the production of red blood cells and can cause damage to the kidneys, liver, brain, and other organs.

Posters such as this one are part of public-awareness programs to reduce the hazards of lead poisoning.

A Brief History

600 BCE: Lead ore deposits are discovered near Athens; they are mined until the second century CE.

1977: The U.S. government restricts lead content in paint.

| 3000 BCE | 1000 BCE | 1 CE | 1000 CE |

3000 BCE: Egyptians refine and use lead to make art figurines.

60 BCE: Romans begin making lead pipes, lead sheets for waterproofing roofs, and lead crystal.

Questions

1. How do you perform tests for lead in paint, soil, and water? Present a report that explains how the tests work.

2. Research the laws regarding the recycling of storage batteries that contain lead.

☑ internet connect

www.scilinks.org
Topic : Lead
SciLinks code: HW4074

SCiLINKS Maintained by the National Science Teachers Association

7 CHAPTER HIGHLIGHTS

SECTION ONE Avogadro's Number and Molar Conversions

- Avogadro's number, 6.022×10^{23} units/mol, is the number of units (atoms, ions, molecules, formula units, etc.) in 1 mol of any substance.
- Avogadro's number is used to convert from number of moles to number of particles or vice versa.
- Conversions between moles and mass require the use of molar mass.
- The molar mass of a monatomic element is the number of grams numerically equal to the atomic mass on the periodic table.

mole
Avogadro's number
molar mass

SECTION TWO Relative Atomic Mass and Chemical Formulas

- The average atomic mass of an element is the average mass of the element's isotopes, weighted by the percentage of their natural abundance.
- Chemical formulas reveal composition. The subscripts in the formula give the number of atoms of a given element in a molecule or formula unit of a compound or diatomic element.
- Formulas are used to calculate molar masses of compounds.

average atomic mass

SECTION THREE Formulas and Percentage Composition

- Percentage composition gives the relative contribution of each element to the total mass of one molecule or formula unit.
- An empirical formula shows the elements and the smallest whole-number ratio of atoms or ions that are present in a compound. It can be found by using the percentage composition.
- The molecular formula is determined from the empirical formula and the experimentally determined molar mass.
- Chemical formulas can be used to calculate percentage composition.

percentage composition
empirical formula
molecular formula

Working Practice Problems
Skills Toolkit 2 p. 227

Converting Between Amount in Moles and Number of Particles
Skills Toolkit 1 p. 226
Sample Problem A p. 228
Sample Problem B p. 229

Converting Between Mass, Amount, and Number of Particles
Skills Toolkit 3 p. 230
Sample Problem C p. 231
Sample Problem D p. 232

Calculating Average Atomic Mass
Sample Problem E p. 235

Calculating Molar Mass of Compounds
Sample Problem F p. 239

Determining an Empirical Formula from Percentage Composition
Sample Problem G p. 242

Determining a Molecular Formula from an Empirical Formula
Sample Problem H p. 245

Using a Chemical Formula to Determine Percentage Composition
Sample Problem I p. 247

CHAPTER REVIEW 7

USING KEY TERMS

1. Distinguish between Avogadro's number and the mole.

2. What term is used to describe the mass in grams of 1 mol of a substance?

3. Why is the ratio between the empirical formula and the molecular formula a whole number?

4. A scientist isolates a compound with the formula CH_2O. She later discovers that its true formula is actually $C_6H_{12}O_6$. What might account for the differences in the two formulas?

5. What do you need to calculate the percentage composition of a substance?

6. Explain the difference between atomic mass and average atomic mass.

UNDERSTANDING KEY IDEAS

Avogadro's Number and Molar Conversions

7. What particular isotope is the basis for defining the atomic mass unit and the mole?

8. How is Avogadro's number related to moles?

9. How would you determine the number of molecules in 3 mol of oxygen, O_2?

10. What conversion factor do you use in converting number of moles into number of formula units?

11. How is molar mass of an element used to convert from number of moles to mass in grams?

12. What result do you get when you multiply the number of moles of a sample by the following conversion factor?

$$\frac{\text{g of element}}{1 \text{ mol element}}$$

13. You convert 10 mol of a substance to grams. Is the number in the answer larger or smaller than 10 g?

Relative Atomic Mass and Chemical Formulas

14. Which has the greater number of molecules: 10 g of N_2 or 10 g of O_2?

15. How is average atomic mass determined from isotopic masses?

16. For an element, what is the relationship between atomic mass and molar mass?

17. How do you determine the molar mass of a compound?

Formulas and Percentage Composition

18. What information does percentage composition reveal about a compound?

19. Compare the empirical formula of a compound with the molecular formula of the compound.

20. Summarize briefly the process of using empirical formula and the value for experimental molar mass to determine the molecular formula.

21. When you calculate the percentage composition of a compound from both the empirical formula and the molecular formula, why are the two results identical?

PRACTICE PROBLEMS

Sample Problem A Converting Amount in Moles to Number of Particles

22. How many sodium ions in 2.00 mol of NaCl?

23. How many molecules in 2.00 mol of sucrose, $C_{12}H_{22}O_{11}$?

24. How many molecules of carbon dioxide exit your lungs when you exhale 5.00×10^{-2} mol of carbon dioxide, CO_2?

25. How many atoms are in the 1.25×10^{-2} mol of mercury within the bulb of a thermometer?

26. Assume that you have 2.59 mol of aluminum. How many atoms of aluminum do you have?

27. How many ions are there in a solution with 9.656 mol of Ni^{2+}?

28. a. How many formula units are there in a 3.12 mol sample of $MgCl_2$?
   b. How many Cl^- ions are there in the sample?

Sample Problem B Converting Number of Particles to Amount in Moles

29. How many moles of magnesium oxide are there in 2.50×10^{25} formula units of MgO?

30. How many moles of gold are there in 1.00 L of sea water if there are 1.50×10^{17} atoms of gold in the sample?

31. A sample has 7.51×10^{24} molecules of benzene, C_6H_6. How many moles is this?

32. How many moles are in a sample having 9.3541×10^{13} particles?

33. How many moles of sodium ions are there in a sample of salt water that contains 4.11×10^{22} Na^+ ions?

34. How many moles are equal to 3.6×10^{23} molecules of oxygen gas, O_2?

Sample Problem C Converting Number of Particles to Mass

35. How many grams are in the 1.204×10^{20} atoms of phosphorus used to coat your television screen?

36. How many grams are present in 4.336×10^{24} formula units of table salt, NaCl, whose molar mass is 58.44 g/mol?

37. A scientist collects a sample that has 2.00×10^{14} molecules of carbon dioxide gas. How many grams is this, given that the molar mass of CO_2 is 44.01 g/mol?

38. What is the mass in grams of a sample of $Fe_2(SO_4)_3$ that contains 3.59×10^{23} sulfate ions, SO_4^{2-} ? The molar mass of $Fe_2(SO_4)_3$ is 399.91 g/mol.

39. Calculate the mass in grams of 2.55 mol of oxygen gas, O_2 (molar mass of O_2 = 32.00 g/mol).

40. How many grams of Ne are in a neon sign that contains 0.0450 mol of neon gas?

41. How many grams are in 2.7 mol of table salt, NaCl (molar mass of NaCl = 58.44 g/mol)?

42. Calculate the mass of each of the following samples:
   a. 0.500 mol I_2 (molar mass of I_2 = 253.80 g/mol)
   b. 2.82 mol PbS (molar mass of PbS = 239.3 g/mol)
   c. 4.00 mol of C_4H_{10} (molar mass of C_4H_{10} = 58.14 g/mol)
   d. 0.300 mol of $Al_2(SO_4)_3$ (molar mass of $Al_2(SO_4)_3$ = 342.17 g/mol)
   e. 0.222 mol of $CuSO_4$ (molar mass of $CuSO_4$ = 159.62 g/mol)

43. How many grams are in each of the following samples?
   a. 1.000 mol NaCl (molar mass of NaCl = 58.44 g/mol)
   b. 2.000 mol H_2O (molar mass of H_2O = 18.02 g/mol)

c. 3.5 mol $Ca(OH)_2$ (molar mass of $Ca(OH)_2 = 74.10$ g/mol)

d. 0.625 mol $Ba(NO_3)_2$ (molar mass of $Ba(NO_3)_2 = 261.35$ g/mol)

Sample Problem D Converting Mass to Number of Particles

44. How many atoms of gold are there in a pure gold ring with a mass of 10.6 g?

45. How many formula units are there in 302.48 g of zinc chloride, $ZnCl_2$? The molar mass of zinc chloride is 136.29 g/mol.

46. Naphthalene, $C_{10}H_8$, an ingredient in mothballs, has a molar mass of 128.18 g/mol. How many molecules of naphthalene are in a mothball that has 2.000 g of naphthalene.

47. How many atoms of aluminum are there in 125 g of aluminum foil?

48. How many moles of water are there if a pipet delivers 99.7 g of water (molar mass of water = 18.02 g/mol)?

49. How many moles of compound are in each of the following samples:

a. 6.60 g $(NH_4)_2SO_4$ (molar mass of $(NH_4)_2SO_4 = 132.17$ g/mol)

b. 4.5 kg of $Ca(OH)_2$ (molar mass of $Ca(OH)_2 = 74.10$ g/mol)

c. 7.35 g of H_2SO_4 (molar mass of $H_2SO_4 = 98.09$ g/mol)

50. Ibuprofen, $C_{13}H_{18}O_2$, an active ingredient in pain relievers has a molar mass of 206.31 g/mol. How many moles of ibuprofen are in a bottle that contains 33 g of ibuprofen?

51. How many moles of $NaNO_2$ are there in a beaker that contains 0.500 kg of $NaNO_2$ (molar mass of $NaNO_2 = 69.00$ g/mol)?

52. How many moles of calcium are in one cup of milk that contains 290.0 mg of calcium?

53. How many moles of propane are in a pressure container that has 2.55 kg of propane, C_3H_8 (molar mass of $C_3H_8 = 44.11$ g/mol)?

Sample Problem E Calculating Average Atomic Mass

54. Naturally occurring silver is composed of two isotopes: Ag-107 is 51.35% with a mass of 106.905092 amu, and the rest is Ag-109 with a mass of 108.9044757 amu. Calculate the average atomic mass of silver.

55. The element bromine is distributed between two isotopes. The first, amounting to 50.69%, has a mass of 78.918 amu. The second, amounting to 49.31%, has a mass of 80.916 amu. Calculate the average atomic mass of bromine.

56. Antimony has two isotopes. The first, amounting to 57.3%, has a mass of 120.9038 amu. The second, amounting to 42.7%, has a mass of 122.9041 amu. What is the average atomic mass of antimony?

57. Calculate the average atomic mass of iron. Its composition is 5.90% with a mass of 53.94 amu, 91.72% with a mass of 55.93 amu, 2.10% with a mass of 56.94 amu, and 0.280% with a mass of 57.93 amu.

Sample Problem F Calculating Molar Mass of Compounds

58. Find the molar mass of the following compounds:
a. lithium chloride
b. sodium sulfate
c. copper(I) cyanide
d. propylene, C_3H_6
e. potassium dichromate
f. magnesium nitrate
g. magnetite, Fe_3O_4
h. uracil, $C_4H_4N_2O_2$
i. tetrasulfur tetranitride
j. cesium tribromide
k. caffeine, $C_8H_{10}N_4O_2$

59. What is the molar mass of the phosphate ion, PO_4^{3-}?

60. Find the molar mass of isopropyl alcohol, C_3H_7OH, used as rubbing alcohol.

61. What is the molar mass of the amino acid glycine, $C_2H_5NO_2$?

Sample Problem G Determining an Empirical Formula from Percentage Composition

62. A compound of silver has the following analytical composition: 63.50% Ag, 8.25% N, and 28.25% O. Calculate the empirical formula.

63. An oxide of phosphorus is 56.34% phosphorus, and the rest is oxygen. Calculate the empirical formula for this compound.

64. Determine the empirical formula for a compound made up of 52.11% carbon, 13.14% hydrogen, and oxygen.

Sample Problem H Determining a Molecular Formula from an Empirical Formula

65. The empirical formula of the anticancer drug altretamine is $C_3H_6N_2$. The experimental molar mass is 210 g/mol. What is its molecular formula?

66. Benzene has the empirical formula CH and an experimental molar mass of 78 g/mol. What is its molecular formula?

67. Determine the molecular formula for a compound with the empirical formula CoC_4O_4 and a molar mass of 341.94 g/mol.

68. Oleic acid has the empirical formula $C_9H_{17}O$. If the experimental molar mass is 282 g/mol, what is the molecular formula of oleic acid?

69. Pyrazine has the empirical formula C_2H_2N. Its experimental formula mass is 80.1 amu. What is the molecular formula?

Sample Problem I Using a Chemical Formula to Determine Percentage Composition

70. Determine the percentage composition of the following compounds:
   a. ammonium nitrate, NH_4NO_3, a common fertilizer
   b. tin(IV) oxide, SnO_2, an ingredient in fingernail polish
   c. $YBa_2Cu_3O_7$, a superconductor

71. What percentage of ammonium carbonate, $(NH_4)_2CO_3$, an ingredient in smelling salts, is the ammonium ion, NH_4^+?

72. Some antacids use compounds of calcium, a mineral that is often lacking in the diet. What is the percentage composition of calcium carbonate, a common antacid ingredient?

73. Determine the percentage composition of ammonium oxalate, $(NH_4)_2C_2O_4$.

MIXED REVIEW

74. Calculate the number of moles in each of the following samples:
   a. 8.2 g of sodium phosphate
   b. 6.66 g of calcium nitrate
   c. 8.22 g of sulfur dioxide

75. There are exactly 1000 mg in 1 g. A cup of hot chocolate has 35.0 mg of sodium ions, Na^+. One cup of milk has 290 mg of calcium ions, Ca^{2+}.
   a. How many moles of sodium ions are in the cup of hot chocolate?
   b. How many moles of calcium ions are in the milk?

76. How many atoms are there in a platinum ring made of 0.0466 mol of platinum?

77. Cyclopentane has the molecular formula C_5H_{10}. How many moles of hydrogen atoms are there in 4 moles of cyclopentane?

78. A 1.344 g sample of a compound contains 0.365 g Na, 0.221 g N, and 0.758 g O. What is its percentage composition? Calculate its empirical formula.

79. The naturally occurring silicon in sand has three isotopes; 92.23% is made up of atoms with a mass of 27.9769 amu, 4.67% is made up of atoms with a mass of 28.9765 amu, and 3.10% is made up of atoms with a mass of 29.9738 amu. Calculate the average atomic mass of silicon.

80. How many atoms of Fe are in the formula Fe_3C? How many moles of Fe are in one mole of Fe_3C?

81. Shown below are the structures for two sugars, glucose and fructose.

    a. What is the molar mass of glucose?

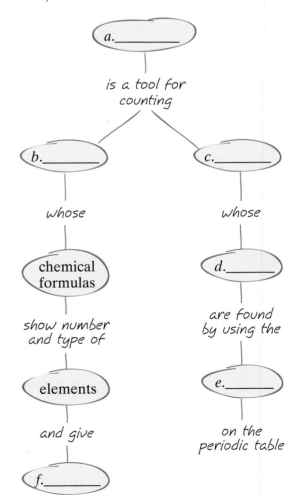

glucose

    b. What is the molar mass of fructose?

fructose

82. Chlorine gas is a diatomic molecule, Cl_2. There are 6.00 mol of chlorine atoms in a sample of chlorine gas. How many moles of chlorine gas molecules is this?

83. Which yields a higher percentage of pure aluminum per gram, aluminum phosphate or aluminum chloride?

84. a. Calculate the molar mass of trinitro-toluene, TNT. The chemical formula for TNT is $C_6H_2CH_3(NO_2)_3$.
    b. Determine the percentage composition of TNT.

CRITICAL THINKING

85. Use **Skills Toolkit 1** and **Skills Toolkit 3** to make your own graphic model for finding the mass of one atom, reported in grams.

86. Your calculation of the percentage composition of a compound gives 66.9% C and 29.6% H. Is the calculation correct? Explain.

87. Imagine you are a farmer, using NH_3 and NH_4NO_3 as sources of nitrogen. NH_3 costs $0.50 per kg, and NH_4NO_3 costs $0.25 per kg. Use percentage composition to decide which is the best buy for your money.

ALTERNATIVE ASSESSMENT

88. Research methods scientists initially used to find Avogadro's number. Then compare these methods with modern methods.

89. The most accurate method for determining the mass of an element involves a *mass spectrometer*. This instrument is also used to determine the isotopic composition of a natural element. Find out more about how a mass spectrometer works. Draw a model of how it works. Present the model to the class.

CONCEPT MAPPING

90. Use the following terms to complete the concept map below: *atoms, average atomic mass, molecules, mole, percentage composition,* and *molar masses.*

a._____

is a tool for counting

b._____   c._____

whose   whose

chemical formulas   d._____

show number and type of   are found by using the

elements   e._____

and give   on the periodic table

f._____

Study the graphs below, and answer the questions that follow.
For help in interpreting graphs, see Appendix B, "Study Skills for Chemistry."

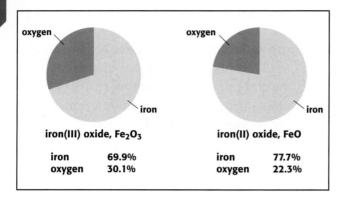

iron(III) oxide, Fe_2O_3

| iron | 69.9% |
| oxygen | 30.1% |

iron(II) oxide, FeO

| iron | 77.7% |
| oxygen | 22.3% |

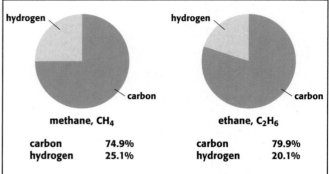

methane, CH_4

| carbon | 74.9% |
| hydrogen | 25.1% |

ethane, C_2H_6

| carbon | 79.9% |
| hydrogen | 20.1% |

91. What do the slices of the pie represent?

92. What do the pie charts show about different compounds that are made up of the same elements?

93. Which has a higher percentage of oxygen, iron(II) oxide or iron(III) oxide?

94. Carlita has 30.0 g of oxygen and 70.0 g of iron. Can she make more FeO or Fe_2O_3 using only the reactants that she has?

95. a. Determine the percentage composition of propane, C_3H_8.
b. Make a pie chart for propane using a protractor to draw the correct sizes of the pie slices. (Hint: A circle has 360°. To draw the correct angle for each slice, multiply each percentage by 360°.)
c. Compare the charts for methane, ethane, and propane. How do the slices for carbon and hydrogen differ for each chart?

TECHNOLOGY AND LEARNING

96. Graphing Calculator

Calculating the Molar Mass of a Compound

The graphing calculator can run a program that calculates the molar mass of a compound given the chemical formula for the compound. This program will prompt for the number of elements in the formula, the number of atoms of each element in the formula, and the atomic mass of each element in the formula. It then can be used to find the molar masses of various compounds.

Go to Appendix C. If you are using a TI-83 Plus, you can download the program **MOLMASS** and data sets and run the application as directed. If you are using another calculator, your teacher will provide you with the keystrokes and data sets to use. After you have graphed the data, answer the questions below.

a. What is the molar mass of $BaTiO_3$?

b. What is the molar mass of $PbCl_2$?

c. What is the molar mass of NH_4NO_3?

1. Element A has two isotopes. One has an atomic mass of 120 amu and constitutes 60%; the other has an atomic mass of 122 amu and constitutes 40%. The average atomic mass is
   a. less than 120.
   c. exactly 121.
   b. less than 121.
   d. more than 121.

2. One mole of NaCl has 6.022×10^{23}
   a. atoms.
   c. formula units.
   b. molecules.
   d. electrons.

3. The molecular formula for acetylene is C_2H_2. The molecular formula for benzene is C_6H_6. The empirical formula for both is
   a. CH.
   c. C_6H_6.
   b. C_2H_2.
   d. $(CH)_2$.

4. If the empirical formula of a compound is known, then
   a. the compound's true formula is also known.
   b. the compound's percentage composition can be calculated.
   c. the arrangement of the compound's atoms is also known.
   d. the molecular mass of the compound can be determined.

5. To calculate the average atomic mass of an element, which of the following must you know?
   a. the atomic number and the atomic mass of each isotope
   b. the atomic mass and the symbol of each isotope
   c. the number of atoms present in each isotope
   d. the atomic mass and relative abundance of each isotope

6. Which of the following compounds has the highest percentage of oxygen?
   a. CH_4O
   c. H_2O
   b. CO_2
   d. Na_2CO_3

7. The units for molar mass are
   a. g/mol.
   c. g/atoms.
   b. atoms/mol.
   d. mol/g.

8. A compound's molar mass is calculated by
   a. adding the number of atoms present in the compound.
   b. adding the atomic masses of all the atoms in the compound and expressing that value in g/mol.
   c. multiplying Avogadro's number times the number of atoms present in the compound.
   d. dividing the number of grams of the sample by Avogadro's number.

9. The empirical formula for a compound that is 1.2% H, 42.0% Cl, and 56.8% O is
   a. HClO.
   c. $HClO_3$.
   b. $HClO_2$.
   d. $HClO_4$.

10. Which of the following shows the percentage composition of H_2SO_4?
    a. 2.5% H, 39.1% S, 58.5% O
    b. 2.1% H, 32.7% S, 65.2% O
    c. 28.6% H, 14.3% S, 57.1% O
    d. 33.3% H, 16.7% S, 50% O

11. You know the empirical formula for a compound. What further data do you need to determine the molecular formula?
    a. Avogadro's number
    b. the formula unit
    c. the molar mass
    d. the number of isotopes

12. To calculate the number of moles of a compound in a sample, which of the following must you know?
    a. atomic mass and symbol of each isotope
    b. mass and molar mass of the compound
    c. atomic mass and Avogadro's number
    d. formula and Avogadro's number

CHEMICAL EQUATIONS AND REACTIONS

With an eruption of flames and hot gases, a space shuttle leaves the ground on its way into orbit. The brightness and warmth of the flame clearly indicates that a change is occurring. From the jets of the shuttle itself, blue flames emerge. These flames are the result of a reaction between hydrogen and oxygen. The sight is awesome and beautiful. In this chapter, you will learn about chemical reactions, such as the ones that send a space shuttle into space.

START-UP ACTIVITY

Observing a Chemical Reaction

SAFETY PRECAUTIONS

PROCEDURE

1. Place about 5 g (1 tsp) of **baking soda** into a **sealable plastic bag.**

2. Place about 5 mL (1 tsp) of **vinegar** into a **plastic film canister.** Secure the lid.

3. Place the canister into the bag. Squeeze the air out of the bag, and tightly seal the bag.

4. Use a **balance** to determine the total mass of the bag and the bag's contents. Make a note of this value.

5. Open the canister without opening the bag, and allow the vinegar and baking soda to mix.

6. When the reaction has stopped, measure and record the total mass of the bag and the bag's contents.

ANALYSIS

1. What evidence shows that a chemical reaction has taken place?

2. Compare the masses of the bag and its contents before and after the reaction. What does this result demonstrate about chemical reactions?

Pre-Reading Questions

1. What are some signs that a chemical change may be taking place?

2. What are the reactants of a reaction? What are the products of a reaction?

3. Describe the law of conservation of mass.

4. Define the terms *synthesis* and *decomposition,* and describe what you would expect to happen in each of these types of reactions.

Describing Chemical Reactions

KEY TERMS

- **chemical reaction**
- **chemical equation**

OBJECTIVES

1. **List** evidence that suggests that a chemical reaction has occurred and evidence that proves that a chemical reaction has occurred.

2. **Describe** a chemical reaction by using a word equation and a formula equation.

3. **Interpret** notations in formula equations, such as those relating to states of matter or reaction conditions.

Chemical Change

You witness chemical changes taking place in iron that rusts, in milk that turns sour, and in a car engine that burns gasoline. The processes of digestion and respiration in your body are the result of chemical changes.

A **chemical reaction** is the process by which one or more substances change into one or more new substances whose chemical and physical properties differ from those of the original substances. In any chemical reaction, the original substances, which can be elements or compounds, are known as *reactants*. The substances created are called *products*. A common example of a chemical reaction is shown in **Figure 1.**

Evidence of a Chemical Reaction

It's not always easy to tell that a chemical change is happening, but there are some signs to look for, which are summarized in **Table 1.** For example, certain signs indicate that wood burning in a campfire is undergoing a chemical change. Smoke rises from the wood, and a hissing sound is made. Energy that lights up the campsite and warms the air around the fire is released. The surface of the wood changes color as the wood burns. Eventually, all that remains of the firewood is a grey, powdery ash.

In **Figure 2,** you can see copper reacting with nitric acid. Again, several clues suggest that a chemical reaction is taking place. The color of the solution changes from colorless to blue. The solution bubbles and fizzes as a gas forms. The copper seems to be used up as the reaction continues.

Sometimes, the evidence for a chemical change is indirect. When you place a new battery in a flashlight, you don't see any changes in the battery. However, when you turn the flashlight on, electrical energy causes the filament in the bulb to heat up and emit light. This release of electrical energy is a clue that a chemical reaction is taking place in the battery. Although these signs suggest a change may be chemical, they do not prove that the change is chemical.

chemical reaction

the process by which one or more substances change to produce one or more different substances

Figure 1
Chemical changes occur as wood burns. Two products formed are carbon dioxide and water.

Table 1 Evidence of Chemical Change

Changes in energy	Formation of new substances
release of energy as heat	formation of a gas
release of energy as light	formation of a precipitate (an insoluble solid)
production of sound	change in color
reduction or increase of temperature	change in odor
absorption or release of electrical energy	

Chemical Reaction Versus Physical Change

For proof of a chemical change, you need a chemical analysis to show that at least one new substance forms. The properties of the new substance—such as density, melting point, or boiling point—must differ from those of the original substances.

Even when evidence suggests a chemical change, you can't be sure immediately. For example, when paints mix, the color of the resulting paint differs from the color of the original paints. But the change is physical—the substances making up the paints have not changed. When you boil water, the water absorbs energy and a gas forms. But the gas still consists of water molecules, so a new substance has not formed. Even though they demonstrate some of the signs of a chemical change, all changes of state, including evaporation, condensation, melting, and freezing, are physical changes.

Figure 2
When copper reacts with nitric acid, several signs of a reaction are seen. A toxic, brown gas is produced, and the color of the solution changes.

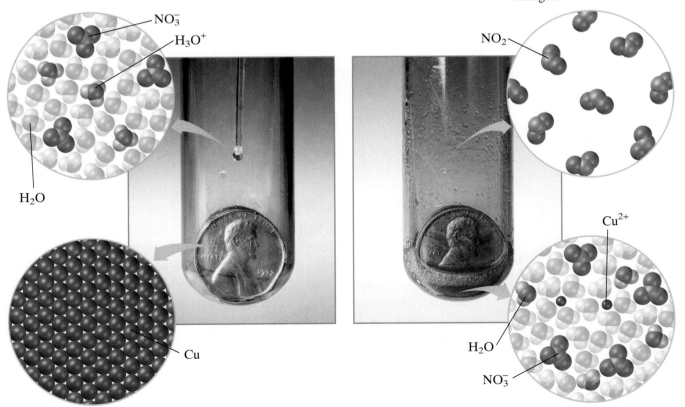

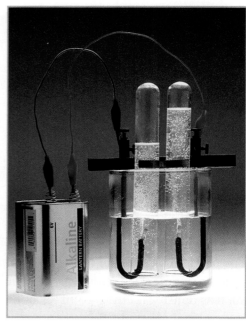

Figure 3
Energy is released as the elements sodium and chlorine react to form the compound sodium chloride. Breaking down water into hydrogen and oxygen requires the input of electrical energy.

Reactions and Energy Changes

Chemical reactions either release energy or absorb energy as they happen, as shown in **Figure 3.** A burning campfire and burning natural gas are examples of reactions that release energy. Natural gas, which is mainly methane, undergoes the following reaction:

methane + oxygen ⟶ carbon dioxide + water + energy

Notice that when energy is released, it can be considered a product of the reaction.

If the energy required is not too great, some other reactions that absorb energy will occur because they take energy from their surroundings. An example is the decomposition of dinitrogen tetroxide, which occurs at room temperature.

dinitrogen tetroxide + energy ⟶ nitrogen dioxide

Notice that when energy is absorbed, it can be considered a reactant of the reaction.

Reactants Must Come Together

You cannot kick a soccer ball unless your shoe contacts the ball. Chemical reactions are similar. Molecules and atoms of the reactants must come into contact with each other for a reaction to take place. Think about what happens when a safety match is lighted, as shown in **Figure 4.** One reactant, potassium chlorate ($KClO_3$) is on the match head. The other reactant, phosphorus, P_4, is on the striking surface of the matchbox. The reaction begins when the two substances come together by rubbing the match head across the striking surface. If the reactants are kept apart, the reaction will not happen. Under most conditions, safety matches do not ignite by themselves.

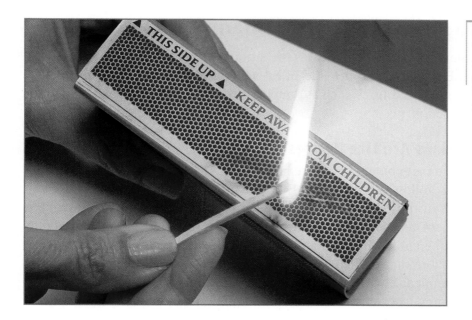

Figure 4
The reactants $KClO_3$ (on the match head) and P_4 (on the striking surface) must be brought together for a safety match to ignite.

Constructing a Chemical Equation

You know that symbols represent elements, and formulas represent compounds. In the same way, equations are used to represent chemical reactions. A correctly written **chemical equation** shows the chemical formulas and relative amounts of all reactants and products. Constructing a chemical equation usually begins with writing a word equation. This word equation contains the names of the reactants and of the products separated by an arrow. The arrow means "forms" or "produces." Then, the chemical formulas are substituted for the names. Finally, the equation is balanced so that it obeys the law of conservation of mass. The numbers of atoms of each element must be the same on both sides of the arrow.

chemical equation

a representation of a chemical reaction that uses symbols to show the relationship between the reactants and the products

Writing a Word Equation or a Formula Equation

The first step in writing a chemical equation is to write a word equation. To write the word equation for a reaction, you must write down the names of the reactants and separate the names with plus signs. An arrow is used to separate the reactants from the products. Then, the names of the products are written to the right of the arrow and are separated by plus signs. The word equation for the reaction of methane with oxygen to form carbon dioxide and water is written as follows:

$$\text{methane} + \text{oxygen} \longrightarrow \text{carbon dioxide} + \text{water}$$

To convert this word equation into a formula equation, use the formulas for the reactants and for the products. The formulas for methane, oxygen, carbon dioxide, and water replace the words in the word equation to make a formula equation. The word *methane* carries no quantitative meaning, but the formula CH_4 means a molecule of methane. This change gives the unbalanced formula equation below. The question marks indicate that we do not yet know the number of molecules of each substance.

$$?CH_4 + ?O_2 \longrightarrow ?CO_2 + ?H_2O$$

Equations and Reaction Information

A chemical equation indicates the amount of each substance in the reaction. But it can also provide other valuable information about the substances or conditions, such as temperature or pressure, that are needed for the reaction.

Equations Are Like Recipes

Imagine that you need to bake brownies for a party. Of course, you would want to follow a recipe closely to be sure that your brownies turn out right. You must know which ingredients to use and how much of each ingredient to use. Special instructions, such as whether the ingredients should be chilled or at room temperature when you mix them, are also provided in the recipe.

Chemical equations have much in common with a recipe. Like a recipe, any instructions shown in an equation can help you or a chemist be sure the reaction turns out the way it should, as shown in **Figure 5**. A balanced equation indicates the relative amounts of reactants and products in the reaction. As discussed below, even more information can be shown by an equation.

Figure 5
The equation for the reaction between baking soda and vinegar provides a lot of information about the reaction.

$$NaHCO_3(s) + HC_2H_3O_2(aq) \longrightarrow NaC_2H_3O_2(aq) + CO_2(g) + H_2O(l)$$

Equations Can Show Physical States and Reaction Conditions

The recipe for brownies will specify whether each ingredient should be used in a solid or liquid form. The recipe also may state that the batter should bake at 400°F for 20 min. Additional instructions tell what to do if you are baking at high elevation. Chemical equations are similar. Equations for chemical reactions often list the physical state of each reactant and the conditions under which the reaction takes place.

Look closely at the equation that represents the reaction of baking soda with vinegar.

$$NaHCO_3(s) + HC_2H_3O_2(aq) \longrightarrow NaC_2H_3O_2(aq) + CO_2(g) + H_2O(l)$$

Baking soda, sodium hydrogen carbonate, is a solid, so the formula is followed by the symbol (s). Vinegar, the other reactant, is acetic acid dissolved in water—an aqueous solution. Sodium acetate, one of the products, remains in aqueous solution. So, the formulas for vinegar and sodium acetate are followed by the symbol (aq). Another product, carbon dioxide, is a gas and is marked with the symbol (g). Finally, water is produced in the liquid state, so its formula is followed by the symbol (l).

When information about the conditions of the reaction is desired, the arrow is a good place to show it. Several symbols are used to show the conditions under which a reaction happens. Consider the preparation of ammonia in a commercial plant.

$$N_2(g) + 3H_2(g) \underset{\text{catalyst}}{\overset{350°C, \ 25\ 000 \ kPa}{\rightleftharpoons}} 2NH_3(g)$$

The double arrow indicates that reactions occur in both the forward and reverse directions and that the final result is a mixture of all three substances. The temperature at which the reaction occurs is 350°C. The pressure at which the reaction occurs, 25 000 kPa, is also shown above the arrow. A catalyst is used to speed the reaction, so the catalyst is mentioned, too. Other symbols used in equations are shown in **Table 2**.

Table 2	State Symbols and Reaction Conditions
Symbol	**Meaning**
$(s), (l), (g)$	substance in the solid, liquid, or gaseous state
(aq)	substance in aqueous solution (dissolved in water)
$\longrightarrow$	"produces" or "yields," indicating result of reaction
$\rightleftharpoons$	reversible reaction in which products can reform into reactants; final result is a mixture of products and reactants
$\overset{\Delta}{\longrightarrow}$ or $\overset{heat}{\longrightarrow}$	reactants are heated; temperature is not specified
$\overset{Pd}{\longrightarrow}$	name or chemical formula of a catalyst, added to speed a reaction

Refer to Appendix A to see more symbols used in equations.

When to Use the Symbols

Although chemical equations can be packed with information, most of the ones you will work with will show only the formulas of reactants and products. However, sometimes you need to know the states of the substances. Recognizing and knowing the symbols used will help you understand these equations better. And learning these symbols now will make learning new information that depends on these symbols easier.

Section Review

UNDERSTANDING KEY IDEAS

1. What is a chemical reaction?

2. What is the only way to prove that a chemical reaction has occurred?

3. When water boils on the stove, does a chemical change or a physical change take place?

4. Give four examples of evidence that suggests that a chemical change probably is occurring.

5. When propane gas, C_3H_8, is burned with oxygen, the products are carbon dioxide and water. Write an unbalanced formula equation for the reaction.

6. Assume that liquid water forms in item 5. Write a formula equation for the reaction that shows the physical states of all compounds.

7. What does "Mn" above the arrow in a formula equation mean?

8. What symbol is used in a chemical equation to indicate the phrase "reacts with"?

9. Solid silicon and solid magnesium chloride form when silicon tetrachloride gas reacts with magnesium metal. Write a word equation and an unbalanced formula equation. Include all of the appropriate notations.

10. Magnesium oxide forms from magnesium metal and oxygen gas. Write a word equation and an unbalanced formula equation. Include all of the appropriate notations.

CRITICAL THINKING

11. Describe evidence that burning gasoline in an engine is a chemical reaction.

12. Describe evidence that chemical reactions take place during a fireworks display.

13. The directions on a package of an epoxy glue say to mix small amounts of liquid from two separate tubes. Either liquid alone does not work as a glue. Should the liquids be considered reactants? Explain your answer.

14. When sulfur is heated until it melts and then is allowed to cool, beautiful yellow crystals form. How can you prove that this change is physical?

15. Besides the reactant, what is needed for the electrolysis experiment that breaks down water?

16. Write the word equation for the electrolysis of water, and indicate the physical states and condition(s) of the reaction.

17. For each of the following equations, write a sentence that describes the reaction, including the physical states and reaction conditions.

a. $Zn(s) + 2HCl(aq) \longrightarrow ZnCl_2(aq) + H_2(g)$

b. $CaCl_2(aq) + Na_2CO_3(aq) \longrightarrow$
$$CaCO_3(s) + 2NaCl(aq)$$

c. $NaOH(aq) + HCl(aq) \longrightarrow$
$$NaCl(aq) + H_2O(l)$$

d. $CaCO_3(s) \xrightarrow{\Delta} CaO(s) + CO_2(g)$

Balancing Chemical Equations

KEY TERMS

- coefficient

OBJECTIVES

1 **Relate** the conservation of mass to the rearrangement of atoms in a chemical reaction.

2 **Write** and interpret a balanced chemical equation for a reaction, and relate conservation of mass to the balanced equation.

Reactions Conserve Mass

A basic law of science is the law of conservation of mass. This law states that in ordinary chemical or physical changes, mass is neither created nor destroyed. If you add baking soda to vinegar, they react to release carbon dioxide gas, which escapes into the air. But if you collect all of the products of the reaction, you find that their total mass is the same as the total mass of the reactants.

Topic Link

Refer to the "The Science of Chemistry" chapter for more information about the law of conservation of mass.

Reactions Rearrange Atoms

This law is based on the fact that the products and the reactants of a reaction are made up of the same number and kinds of atoms. The atoms are just rearranged and connected differently. Look at the formula equation for the reaction of sodium with water.

$$?Na + ?H_2O \longrightarrow ?NaOH + ?H_2$$

The same types of atoms appear in both the reactants and products. However, **Table 3** shows that the number of each type of atom is not the same on both sides of the equation. To show that a reaction satisfies the law of conservation of mass, its equation must be *balanced*.

Table 3 Counting Atoms in an Equation

	Reactants	Products	Balanced?
Unbalanced formula equation	$Na + H_2O$	$NaOH + H_2$	
Sodium atoms	1	1	yes
Hydrogen atoms	2	3	no
Oxygen atoms	1	1	yes

Balancing Equations

To balance an equation, you need to make the number of atoms for each element the same on the reactants' side and on the products' side. But there is a catch. You cannot change the formulas of any of the substances. For example, you could not change CO_2 to CO_3. You can only place numbers called *coefficients* in front of the formulas. A **coefficient** multiplies the number of atoms of each element in the formula that follows. For example, the formula H_2O represents 2 atoms of hydrogen and 1 atom of oxygen. But $2H_2O$ represents 2 molecules of water, for a total of 4 atoms of hydrogen and 2 atoms of oxygen. The formula $3Ca(NO_3)_2$ represents 3 calcium atoms, 6 nitrogen atoms, and 18 oxygen atoms. Look at **Skills Toolkit 1** as you balance equations.

coefficient

a small whole number that appears as a factor in front of a formula in a chemical equation

---

1 SKILLS Toolkit

Balancing Chemical Equations

1. Identify reactants and products.
- If no equation is provided, identify the reactants and products and write an unbalanced equation for the reaction. (You may find it helpful to write a word equation first.)
- If not all chemicals are described in the problem, try to predict the missing chemicals based on the type of reaction.

2. Count atoms.
- Count the number of atoms of each element in the reactants and in the products, and record the results in a table.
- Identify elements that appear in only one reactant and in only one product, and balance the atoms of those elements first. Delay the balancing of atoms (often hydrogen and oxygen) that appear in more than one reactant or product.
- If a polyatomic ion appears on both sides of the equation, treat it as a single unit in your counts.

3. Insert coefficients.
- Balance atoms one element at a time by inserting coefficients.
- Count atoms of each element frequently as you try different coefficients. Watch for elements whose atoms become unbalanced as a result of your work.
- Try the odd-even technique (explained later in this section) if you see an even number of a particular atom on one side of an equation and an odd number of that atom on the other side.

4. Verify your results.
- Double-check to be sure that the numbers of atoms of each element are equal on both sides of the equation.

Balancing an Equation

Balance the equation for the reaction of iron(III) oxide with hydrogen to form iron and water.

1 Identify reactants and products.

Iron(III) oxide and hydrogen are the reactants. Iron and water are the products. The unbalanced formula equation is

$$Fe_2O_3 + H_2 \longrightarrow Fe + H_2O$$

2 Count atoms.

	Reactants $Fe_2O_3 + H_2$	Products $Fe + H_2O$	Balanced?
Unbalanced formula equation	$Fe_2O_3 + H_2$	$Fe + H_2O$	
Iron atoms	2	1	no
Oxygen atoms	3	1	no
Hydrogen atoms	2	2	yes

3 Insert coefficients.

Add a coefficient of 2 in front of Fe to balance the iron atoms.

$$Fe_2O_3 + H_2 \longrightarrow 2Fe + H_2O$$

Add a coefficient of 3 in front of H_2O to balance the oxygen atoms.

$$Fe_2O_3 + H_2 \longrightarrow 2Fe + 3H_2O$$

Now there are two hydrogen atoms in the reactants and six in the products. Add a coefficient of 3 in front of H_2.

$$Fe_2O_3 + 3H_2 \longrightarrow 2Fe + 3H_2O$$

4 Verify your results.

There are two iron atoms, three oxygen atoms, and six hydrogen atoms on both sides of the equation, so it is balanced.

PRACTICE HINT

One way to know what coefficient to use is to find a lowest common multiple. In this example, there were six hydrogen atoms in the products and two in the reactants. The lowest common multiple of 6 and 2 is 6, so a coefficient of 3 in the reactants balances the atoms.

PRACTICE

Write a balanced equation for each of the following.

1 $P_4 + O_2 \longrightarrow P_2O_5$

2 $C_3H_8 + O_2 \longrightarrow CO_2 + H_2O$

3 $Ca_2Si + Cl_2 \longrightarrow CaCl_2 + SiCl_4$

4 Silicon reacts with carbon dioxide to form silicon carbide, SiC, and silicon dioxide.

Balanced Equations Show Mass Conservation

The balanced equation for the reaction of sodium with water is

$$2Na + 2H_2O \longrightarrow 2NaOH + H_2$$

Each side of the equation has two atoms of sodium, four atoms of hydrogen, and two atoms of oxygen. The reactants and the products are made up of the same atoms so they must have equal masses. So a balanced equation shows the conservation of mass.

Never Change Subscripts to Balance an Equation

If you needed to write a balanced equation for the reaction of H_2 with O_2 to form H_2O, you might start with this formula equation:

$$H_2 + O_2 \longrightarrow H_2O$$

To balance this equation, some people may want to change the formula of the product to H_2O_2.

$$H_2 + O_2 \longrightarrow H_2O_2$$

Although the equation is balanced, the product is no longer water, but hydrogen peroxide. Look at the models and equations in **Figure 6** to understand the problem. The first equation was balanced correctly by adding coefficients. As expected, the model shows the correct composition of the water molecules formed by the reaction. The second equation was incorrectly balanced by changing a subscript. The model shows that the change of a subscript changes the composition of the substance. As a result, the second equation no longer shows the formation of water, but that of hydrogen peroxide. When balancing equations, never change subscripts. Keep this in mind as you learn about the odd-even technique for balancing equations.

Figure 6
Use coefficients to balance an equation. Never change subscripts.

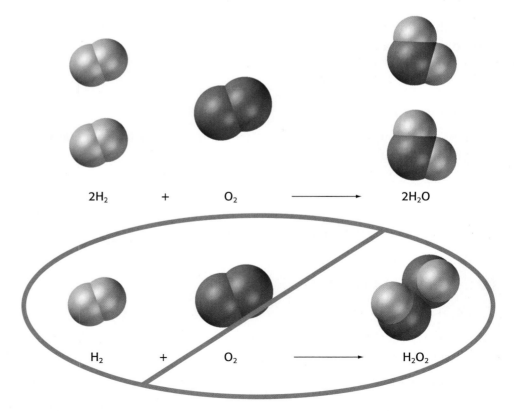

$$2H_2 \quad + \quad O_2 \quad \longrightarrow \quad 2H_2O$$

$$H_2 \quad + \quad O_2 \quad \longrightarrow \quad H_2O_2$$

The Odd-Even Technique

The reaction of ammonia with oxygen produces nitrogen monoxide and water vapor. Write a balanced equation for this reaction.

1 Identify reactants and products.

The unbalanced formula equation is

$$NH_3 + O_2 \longrightarrow NO + H_2O$$

2 Count atoms.

	Reactants	Products	Balanced?
Unbalanced formula equation	$NH_3 + O_2$	$NO + H_2O$	
Nitrogen atoms	1	1	yes
Hydrogen atoms	3	2	no
Oxygen atoms	2	2	yes

The odd-even technique uses the fact that multiplying an odd number by 2 always results in an even number.

3 Insert coefficients.

A 2 in front of NH_3 gives an even number of H atoms. Add coefficients to NO and H_2O to balance the H atoms and N atoms.

$$2NH_3 + O_2 \longrightarrow 2NO + 3H_2O$$

For oxygen, double *all* coefficients to have an even number of O atoms on both sides and keep the other atoms balanced.

$$4NH_3 + 2O_2 \longrightarrow 4NO + 6H_2O$$

Change the coefficient for O_2 to 5 to balance the oxygen atoms.

$$4NH_3 + 5O_2 \longrightarrow 4NO + 6H_2O$$

4 Verify your results.

There are four nitrogen atoms, twelve hydrogen atoms, and ten oxygen atoms on both sides of the equation, so it is balanced.

PRACTICE HINT

Watch for cases in which all atoms in an equation are balanced except one, which has an odd number on one side of the equation and an even number on the other side. Multiplying all coefficients by 2 will result in an even number of atoms for the unbalanced atoms while keeping the rest balanced.

PRACTICE

Write a balanced chemical equation for each of the following.

1 $C_2H_2 + O_2 \longrightarrow CO_2 + H_2O$

2 $Fe(OH)_2 + H_2O_2 \longrightarrow Fe(OH)_3$

3 $FeS_2 + Cl_2 \longrightarrow FeCl_3 + S_2Cl_2$

PROBLEM SOLVING SKILL

Polyatomic Ions Can Be Balanced as a Unit

So far, you've balanced equations by balancing individual atoms one at a time. However, balancing some equations is made easier because groups of atoms can be balanced together. This is especially true in the case of polyatomic ions, such as NO_3^-. Often a polyatomic ion appears in both the reactants and the products without changing. The atoms within such ions are not rearranged during the reaction. The polyatomic ion can be counted as a single unit that appears on both sides of the equation. Of course, when you think that you have finished balancing an equation, checking each atom by itself is still helpful.

Look at **Figure 7.** The sulfate ion appears in both the reactant sulfuric acid and in the product aluminum sulfate. You could look at the sulfate ion as a single unit to make balancing the equation easier. Looking at the balanced equation, you can see that there are three sulfate ions on the reactants' side and three on the products' side.

In balancing the equation for the reaction between sodium phosphate and calcium nitrate, you can consider the nitrate ion and the phosphate ion each to be a unit. The resulting balanced equation is

$$2Na_3PO_4 + 3Ca(NO_3)_2 \longrightarrow 6NaNO_3 + Ca_3(PO_4)_2$$

Count the atoms of each element to make sure that the equation is balanced.

Figure 7
In the reaction of aluminum with sulfuric acid, sulfate ions are part of both the reactants and the products.

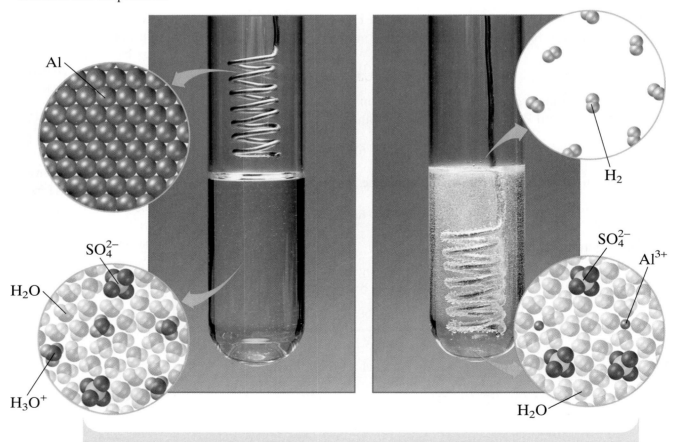

$$2Al(s) + 3H_2SO_4(aq) \longrightarrow Al_2(SO_4)_3(aq) + 3H_2(g)$$

Polyatomic Ions as a Group

Aluminum reacts with arsenic acid, $HAsO_3$, to form H_2 and aluminum arsenate. Write a balanced equation for this reaction.

1 Identify reactants and products.

The unbalanced formula equation is

$$Al + HAsO_3 \longrightarrow H_2 + Al(AsO_3)_3$$

2 Count atoms.

	Reactants	Products	Balanced?
Unbalanced formula equation	$Al + HAsO_3$	$H_2 + Al(AsO_3)_3$	
Aluminum atoms	1	1	yes
Hydrogen atoms	1	2	no
Arsenate ions	1	3	no

Because the arsenate ion appears on both sides of the equation, consider it a single unit while balancing.

3 Insert coefficients.

Change the coefficient of $HAsO_3$ to 3 to balance the arsenate ions.

$$Al + 3HAsO_3 \longrightarrow H_2 + Al(AsO_3)_3$$

Double all coefficients to keep the other atoms balanced and to get an even number of hydrogen atoms on each side.

$$2Al + 6HAsO_3 \longrightarrow 2H_2 + 2Al(AsO_3)_3$$

Change the coefficient of H_2 to 3 to balance the hydrogen atoms.

$$2Al + 6HAsO_3 \longrightarrow 3H_2 + 2Al(AsO_3)_3$$

4 Verify your results.

There are 2 aluminum atoms, 6 hydrogen atoms, 6 arsenic atoms, and 18 oxygen atoms on both sides of the equation, so it is balanced.

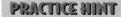

PRACTICE HINT

If you consider polyatomic ions as single units, be sure to count the atoms of each element when you double-check your work.

PRACTICE

Write a balanced equation for each of the following.

1 $HgCl_2 + AgNO_3 \longrightarrow Hg(NO_3)_2 + AgCl$

2 $Al + Hg(CH_3COO)_2 \longrightarrow Al(CH_3COO)_3 + Hg$

3 Calcium phosphate and water are produced when calcium hydroxide reacts with phosphoric acid.

Practice Makes Perfect

You have learned a few techniques that you can use to help you approach balancing equations logically. But don't think that you are done. The more you practice balancing equations, the faster and better you will become. The best way to discover more tips to help you balance equations is to practice a lot! As you learn about the types of reactions in the next section, be aware that these types can provide tips that make balancing equations even easier.

② Section Review

UNDERSTANDING KEY IDEAS

1. What fundamental law is demonstrated in balancing equations?

2. What is meant by a balanced equation?

3. When balancing an equation, should you adjust the subscripts or the coefficients?

PRACTICE PROBLEMS

4. Write each of the following reactions as a word equation, an unbalanced formula equation, and finally as a balanced equation.

   a. When heated, potassium chlorate decomposes into potassium chloride and oxygen.

   b. Silver sulfide forms when silver and sulfur, S_8, react.

   c. Sodium hydrogen carbonate breaks down to form sodium carbonate, carbon dioxide, and water vapor.

5. Balance the following equations.

   a. $ZnS + O_2 \longrightarrow ZnO + SO_2$

   b. $Fe_2O_3 + CO \longrightarrow Fe + CO_2$

   c. $AgNO_3 + AlCl_3 \longrightarrow AgCl + Al(NO_3)_3$

   d. $Ni(ClO_3)_2 \longrightarrow NiCl_2 + O_2$

6. Balance the following equations.

   a. $(NH_4)_2Cr_2O_7 \longrightarrow Cr_2O_3 + N_2 + H_2O$

   b. $NH_3 + CuO \longrightarrow N_2 + Cu + H_2O$

   c. $Na_2SiF_6 + Na \longrightarrow Si + NaF$

   d. $C_4H_{10} + O_2 \longrightarrow CO_2 + H_2O$

CRITICAL THINKING

7. Use diagrams of particles to explain why four atoms of phosphorus can produce only two molecules of diphosphorus trioxide, even when there is an excess of oxygen atoms.

8. Which numbers in the reactants and products in the following equation are coefficients, and which are subscripts?

$$2Al + 3H_2SO_4 \longrightarrow Al_2(SO_4)_3 + 3H_2$$

9. Write a balanced equation for the formation of water from hydrogen and oxygen. Use the atomic mass of each element to determine the mass of each molecule in the equation. Use these masses to show that the equation demonstrates the law of conservation of mass.

10. A student writes the equation below as the balanced equation for the reaction of iron with chlorine. Is this equation correct? Explain.

$$Fe(s) + Cl_3(g) \longrightarrow FeCl_3(s)$$

Classifying Chemical Reactions

KEY TERMS

- **combustion reaction**
- **synthesis reaction**
- **decomposition reaction**
- **activity series**
- **double-displacement reaction**

OBJECTIVES

1. **Identify** combustion reactions, and write chemical equations that predict the products.

2. **Identify** synthesis reactions, and write chemical equations that predict the products.

3. **Identify** decomposition reactions, and write chemical equations that predict the products.

4. **Identify** displacement reactions, and use the activity series to write chemical equations that predict the products.

5. **Identify** double-displacement reactions, and write chemical equations that predict the products.

Reaction Types

So far in this book, you have learned about a lot of chemical reactions. But they are just a few of the many that take place. To make learning about reations simpler, it is helpful to classify them and to start with a few basic types. Consider a grocery store as an example of how classification makes things simpler. A store may have thousands of items. Even if you have never been to a particular store before, you should be able to find everything you need. Because similar items are grouped together, you know what to expect when you start down an aisle.

Look at the reaction shown in **Figure 8.** The balanced equation for this reaction is

$$2Al + Fe_2O_3 \longrightarrow 2Fe + Al_2O_3$$

By classifying chemical reactions into several types, you can more easily predict what products are likely to form. You will also find that reactions in each type follow certain patterns, which should help you balance the equations more easily.

The five reaction types that you will learn about in this section are not the only ones. Additional types are discussed in other chapters, and there are others beyond the scope of this book. In addition, reactions can belong to more than one type. There are even reactions that do not fit into any type. The value in dividing reactions into categories is not to force each reaction to fit into a single type but to help you see patterns and similarities in reactions.

Figure 8
Knowing which type of reaction occurs between aluminum and iron(III) oxide could help you predict that iron is produced.

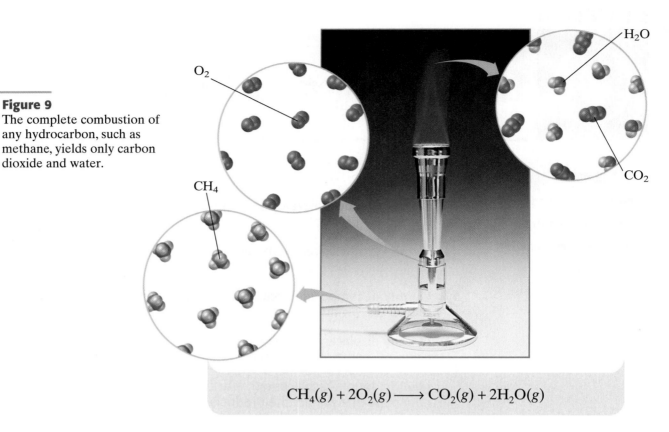

Figure 9
The complete combustion of any hydrocarbon, such as methane, yields only carbon dioxide and water.

$$CH_4(g) + 2O_2(g) \longrightarrow CO_2(g) + 2H_2O(g)$$

Combustion Reactions

Combustion reactions are often used to generate energy. Much of our electrical energy is generated in power plants that work because of the combustion of coal. Combustion of hydrocarbons (as in gasoline) provides energy used in transportation—on the land, in the sea, and in the air. For our purposes, a **combustion reaction** is the reaction of a carbon-based compound with oxygen. The products are carbon dioxide and water vapor. An example of a combustion reaction is shown in **Figure 9.**

Many of the compounds in combustion reactions are called *hydrocarbons* because they are made of only carbon and hydrogen. Propane is a hydrocarbon that is often used as a convenient portable fuel for lanterns and stoves. The balanced equation for the combustion of propane is shown below.

$$C_3H_8 + 5O_2 \longrightarrow 3CO_2 + 4H_2O$$

Some compounds, such as alcohols, are made of carbon, hydrogen, and oxygen. In the combustion of these compounds, carbon dioxide and water are still made. For example, the fuel known as gasohol is a mixture of gasoline and ethanol, an alcohol. The balanced chemical equation for the combustion of ethanol is shown below.

$$CH_3CH_2OH + 3O_2 \longrightarrow 2CO_2 + 3H_2O$$

When enough oxygen is not available, the combustion reaction is incomplete. Carbon monoxide and unburned carbon (soot), as well as carbon dioxide and water vapor are made.

combustion reaction

the oxidation reaction of an organic compound, in which heat is released

☑ internet connect

www.scilinks.org
Topic: Combustion
SciLinks code: HW4033

SCiLINKS. Maintained by the National Science Teachers Association

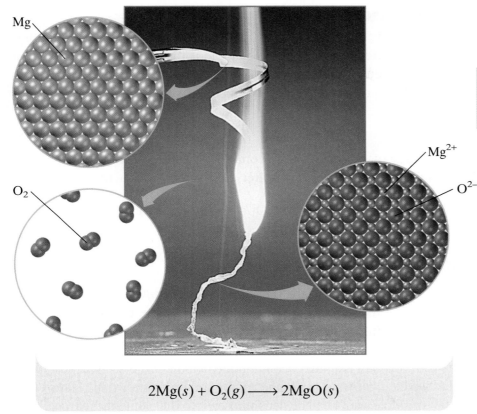

$$2Mg(s) + O_2(g) \longrightarrow 2MgO(s)$$

Figure 10
When the elements magnesium and oxygen react, they combine to form the binary compound magnesium oxide.

Synthesis Reactions

The word *synthesis* comes from a Greek word that means "to put together." In the case of a **synthesis reaction,** a single compound forms from two or more reactants. If you see a chemical equation that has only one product, the reaction is a synthesis reaction. The reactants in many of these reactions are two elements or two small compounds.

synthesis reaction

a reaction in which two or more substances combine to form a new compound

Two Elements Form a Binary Compound

If the reactants in an equation are two elements, the only way in which they can react is to form a binary compound, which is composed of two elements. Often, when a metal reacts with a nonmetal, electrons are transferred and an ionic compound is formed. You can use the charges of the ions to predict the formula of the compound formed. Metals in Groups 1 and 2 lose one electron and two electrons, respectively. Nonmetals in Groups 16 and 17 gain two electrons and one electron, respectively. Using the charges on the ions, you can predict the formula of the product of a synthesis reaction, such as the one in **Figure 10.**

Nonmetals on the far right of the periodic table can react with one another to form binary compounds. Often, more than one compound could form, however, so predicting the product of these reactions is not always easy. For example, carbon and oxygen can combine to form carbon dioxide or carbon monoxide, as shown below.

$$C + O_2 \longrightarrow CO_2 \qquad\qquad 2C + O_2 \longrightarrow 2CO$$

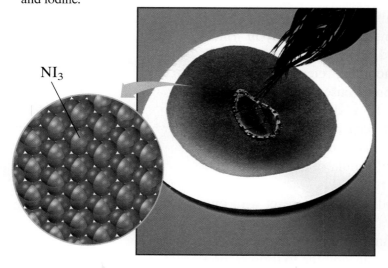

STUDY TIP

WORKING WITH A PARTNER

If you can explain difficult concepts to a study partner, then you know that you understand them yourself.

- Make flashcards that contain examples of chemical reactions. Quiz each other on reaction types by using the flashcards. Explain how you identified each type.

  Refer to Appendix B for other studying strategies.

decomposition reaction

a reaction in which a single compound breaks down to form two or more simpler substances

Two Compounds Form a Ternary Compound

Two compounds can combine to form a ternary compound, a compound composed of three elements. One example is the reaction of water and a Group 1 or Group 2 metal oxide to form a metal hydroxide. An example is the formation of "slaked lime," or calcium hydroxide.

$$CaO(s) + H_2O(l) \longrightarrow Ca(OH)_2(s)$$

Some oxides of nonmetals can combine with water to produce acids. Carbon dioxide combines with water to form carbonic acid.

$$CO_2(g) + H_2O(l) \longrightarrow H_2CO_3(aq)$$

Decomposition Reactions

Decomposition reactions are the opposite of synthesis reactions—they have only one reactant. In a **decomposition reaction,** a single compound breaks down, often with the input of energy, into two or more elements or simpler compounds.

If your reactant is a binary compound, then the products will most likely be the two elements that make the compound up, as shown in **Figure 11.** In another example, water can be decomposed into the elements hydrogen and oxygen through the use of electrical energy.

$$2H_2O(l) \xrightarrow{\text{electricity}} 2H_2(g) + O_2(g)$$

The gases produced are very pure and are used for special purposes, such as in hospitals. But these gases are very expensive because of the energy needed to make them. Experiments are underway to make special solar cells in which sunlight is used to decompose water.

Figure 11
Nitrogen triiodide is a binary compound that decomposes into the elements nitrogen and iodine.

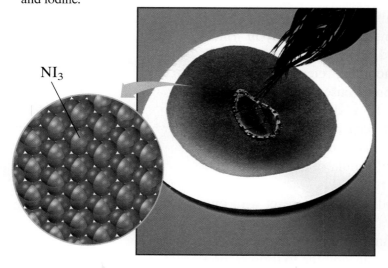
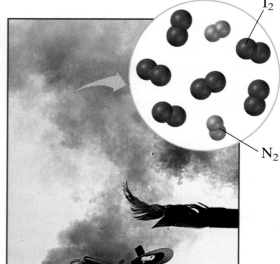

NI_3

I_2

N_2

$$2NI_3(s) \longrightarrow N_2(g) + 3I_2(g)$$

Compounds made up of three or more elements usually do not decompose into those elements. Instead, each compound that consists of a given polyatomic ion will break down in the same way. For example, a metal carbonate, such as $CaCO_3$ in limestone, decomposes to form a metal oxide and carbon dioxide.

$$CaCO_3(s) \xrightarrow{\text{heat}} CaO(s) + CO_2(g)$$

Many of the synthesis reactions that form metal hydroxides and acids can be reversed to become decomposition reactions.

SAMPLE PROBLEM D

Predicting Products

Predict the product(s) and write a balanced equation for the reaction of potassium with chlorine.

1 Gather information.

Because the reactants are two elements, the reaction is most likely a synthesis. The product will be a binary compound.

2 Plan your work.

Potassium, a Group 1 metal, will lose one electron to become a 1+ ion. Chlorine, a Group 17 nonmetal, gains one electron to form a 1– ion. The formula for the product will be KCl. The unbalanced formula equation is

$$K + Cl_2 \longrightarrow KCl$$

3 Calculate.

Place a coefficient of 2 in front of KCl and also K.

$$2K + Cl_2 \longrightarrow 2KCl$$

4 Verify your results.

The final equation has two atoms of each element on each side, so it is balanced.

PRACTICE HINT

Look for hints about the type of reaction. If the reactants are two elements or simple compounds, the reaction is probably a synthesis reaction. The reaction of oxygen with a hydrocarbon is a combustion reaction. If there is only one reactant, it is a decomposition reaction.

PRACTICE

Predict the product(s) and write a balanced equation for each of the following reactions.

1 the reaction of butane, C_4H_{10}, with oxygen

2 the reaction of water with calcium oxide

3 the reaction of lithium with oxygen

4 the decomposition of carbonic acid

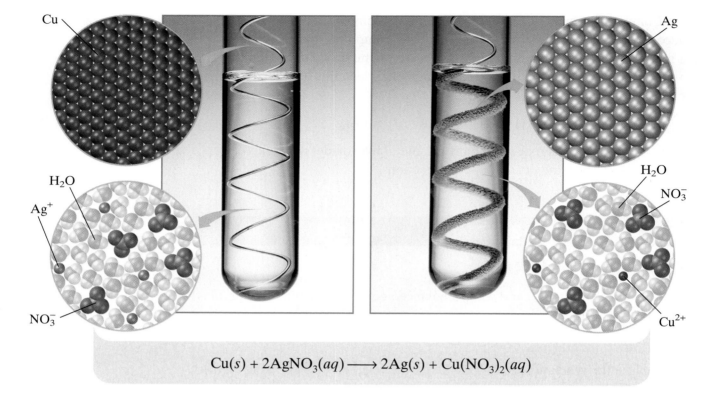

$$Cu(s) + 2AgNO_3(aq) \longrightarrow 2Ag(s) + Cu(NO_3)_2(aq)$$

Figure 12
Copper is the more active metal and displaces silver from the silver nitrate solution. So copper is higher on the activity series than silver is. The Cu^{2+} formed gives the solution a blue color.

Displacement Reactions

When aluminum foil is dipped into a solution of copper(II) chloride, reddish copper metal forms on the aluminum and the solution loses its blue color. It is as if aluminum atoms and copper ions have switched places to form aluminum ions and copper atoms.

$$2Al(s) + 3CuCl_2(aq) \longrightarrow 2AlCl_3(aq) + 3Cu(s)$$

In this *displacement reaction,* a single element reacts with a compound and displaces another element from the compound. The products are a different element and a different compound than the reactants are. In general, a metal may displace another metal (or hydrogen), while a non-metal may displace only another nonmetal.

The Activity Series Ranks Reactivity

activity series

a series of elements that have similar properties and that are arranged in descending order of chemical activity

Results of experiments, such as the one in **Figure 12,** in which displacement reactions take place are summarized in the **activity series,** a portion of which is shown in **Table 4.** In the activity series, elements are arranged in order of activity with the most active one at the top. In general, an element can displace those listed below it from compounds in solution, but not those listed above it. Thus, you can use the activity series to make predictions about displacement reactions. You could also predict that no reaction would happen, such as when silver is put into a copper(II) nitrate solution.

When a metal is placed in water, the reactivity information in the activity series helps you tell if hydrogen is displaced. If the metal is active enough for this to happen, a metal hydroxide and hydrogen gas form.

Table 4   Activity Series

Element	Reactivity
K Ca Na	react with cold water and acids to replace hydrogen; react with oxygen to form oxides
Mg Al Zn Fe	react with steam (but not with cold water) and acids to replace hydrogen; react with oxygen to form oxides
Ni Pb	do not react with water; react with acids to replace hydrogen; react with oxygen to form oxides
H_2 Cu	react with oxygen to form oxides
Ag Au	fairly unreactive; form oxides only indirectly

Refer to Appendix A for a more complete activity series of metals and of halogens.

SKILLS Toolkit 2

Using the Activity Series

1. Identify the reactants.
- Determine whether the single element is a metal or a halogen.
- Determine the element that might be displaced from the compound if a displacement reaction occurs.

2. Check the activity series.
- Determine whether the single element or the element that might be displaced from the compound is more active. The more active element is higher on the activity series.
- For a metal reacting with water, determine whether the metal can replace hydrogen from water in that state.

3. Write the products, and balance the equation.
- If the more active element is already part of the compound, then no reaction will occur.
- Otherwise, the more active element will displace the less active element.

4. Verify your results.
- Double-check to be sure that the equation is balanced.

Determining Products by Using the Activity Series

Magnesium is added to a solution of lead(II) nitrate. Will a reaction happen? If so, write the equation and balance it.

1 Identify the reactants.

Magnesium will attempt to displace lead from lead(II) nitrate.

2 Check the activity series.

Magnesium is more active than lead and displaces it.

3 Write the products, and balance the equation.

A reaction will occur. Lead is displaced by magnesium.

$$Mg + Pb(NO_3)_2 \longrightarrow Pb + Mg(NO_3)_2$$

4 Verify your results.

The equation is balanced.

PRACTICE

For the following situations, write a balanced equation if a reaction happens. Otherwise write "no reaction."

1 Aluminum is dipped into a zinc nitrate solution.

2 Sodium is placed in cold water.

3 Gold is added to a solution of calcium chloride.

Quick LAB

Balancing Equations by Using Models

PROCEDURE

1. Use **toothpicks** and **gumdrops of at least four different colors** (representing atoms of different elements)

to make models of the substances in each equation below.

2. For each reaction below, use your models to determine the

products, if needed, and then balance the equation.

ANALYSIS

Use your models to classify each reaction by type.

a. $H_2 + Cl_2 \longrightarrow HCl$

b. $Mg + O_2 \longrightarrow MgO$

c. $C_2H_6 + O_2 \longrightarrow CO_2 + H_2O$

d. $KI + Br_2 \longrightarrow KBr + I_2$

e. $H_2CO_3 \longrightarrow CO_2 + H_2O$

f. $Ca + H_2O \longrightarrow Ca(OH)_2 + H_2$

g. $KClO_3 \longrightarrow KCl + O_2$

h. $CH_4 + O_2 \longrightarrow CO_2 + H_2O$

i. $Zn + HCl \longrightarrow \underline{\hspace{1cm}}$

j. $H_2O \xrightarrow{electricity} \underline{\hspace{1cm}}$

k. $C_3H_8 + O_2 \longrightarrow \underline{\hspace{1cm}}$

l. $BaO + H_2O \longrightarrow \underline{\hspace{1cm}}$

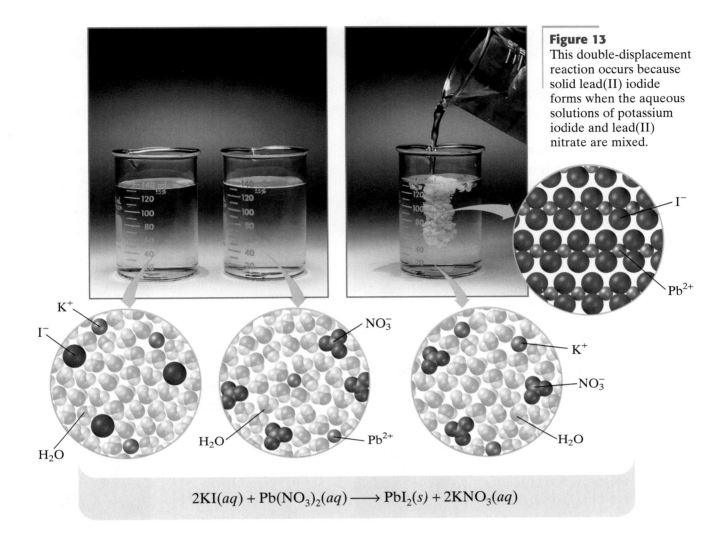

Figure 13
This double-displacement reaction occurs because solid lead(II) iodide forms when the aqueous solutions of potassium iodide and lead(II) nitrate are mixed.

$$2KI(aq) + Pb(NO_3)_2(aq) \longrightarrow PbI_2(s) + 2KNO_3(aq)$$

Double-Displacement Reactions

Figure 13 shows the result of the reaction between KI and $Pb(NO_3)_2$. The products are a yellow precipitate of PbI_2 and a colorless solution of KNO_3. From the equation, it appears as though the parts of the compounds just change places. Early chemists called this a **double-displacement reaction.** It occurs when two compounds in aqueous solution appear to exchange ions and form two new compounds. For this to happen, one of the products must be a solid precipitate, a gas, or a molecular compound, such as water. Water is often written as HOH in these equations.

For example, when dilute hydrochloric acid and sodium hydroxide are mixed, little change appears to happen. However, by looking at the equation for the reaction, you can see that liquid water, a molecular compound, forms.

$$HCl(aq) + NaOH(aq) \longrightarrow HOH(l) + NaCl(aq)$$

Although this type of formula equation is not the best description, the term *double-displacement reaction* is still in use. A better way to represent these reactions is to use a net ionic equation, which will be covered in the next section.

double-displacement reaction

a reaction in which a gas, a solid precipitate, or a molecular compound forms from the apparent exchange of atoms or ions between two compounds

Identifying Reactions and Predicting Products

1. Is there only one reactant?

If the answer is no, go to step 2.
If the answer is yes, you have a decomposition reaction.

- A binary compound generally breaks into its elements.
- A ternary compound breaks according to the guidelines given earlier in this section.

2. Are the reactants two elements or two simple compounds?

If the answer is no, go to step 3.
If the answer is yes, you probably have a synthesis reaction.

- If both reactants are elements, the product is a binary compound. For a metal reacting with a nonmetal, use the expected charges to predict the formula of the compound.
- If the reactants are compounds, the product will be a single ternary compound according to the guidelines given earlier in this section.

3. Are the reactants oxygen and a hydrocarbon?

If the answer is no, go to step 4.
If the answer is yes, you have a combustion reaction.

- The products of a combustion reaction are carbon dioxide and water.

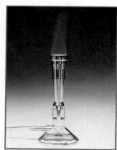

4. Are the reactants an element and a compound other than a hydrocarbon?

If the answer is no, go to step 5.
If the answer is yes, you probably have a displacement reaction.

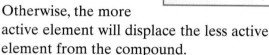

- Use the activity series to determine the activities of the elements.
- If the more active element is already part of the compound, no reaction will occur. Otherwise, the more active element will displace the less active element from the compound.

5. Are the reactants two compounds composed of ions?

If the answer is no, go back to step 1 because you might have missed the proper category.
If the answer is yes, you probably have a double-displacement reaction.

- Write formulas for the possible products by forming two new compounds from the ions available.
- Determine if one of the possible products is a solid precipitate, a gas, or a molecular compound, such as water. If neither product qualifies in the above categories, no reaction occurs. Use the rules below to determine whether a substance will be an insoluble solid.

All compounds of Group 1 and NH_4^+ are soluble.

All nitrates are soluble.

All halides, except those of Ag^+ and Pb^{2+}, are soluble.

All sulfates, except Ca^{2+}, Sr^{2+}, Ba^{2+}, Hg_2^{2+}, and Pb^{2+}, are soluble.

All carbonates, except those of Group 1 and NH_4^+, are insoluble.

More Types to Come

This section has been a short introduction to the classification of chemical reactions. Even so, you now have the tools, summarized in **Skills Toolkit 3,** to predict the products of hundreds of reactions. Keep the reaction types in mind as you continue your study of chemistry. And as you learn about other reaction types, think about how they relate to the five types described here.

Section Review

UNDERSTANDING KEY IDEAS

1. Why is the formation of a ternary compound also a synthesis reaction?

2. When a binary compound is the only reactant, what are the products most likely to be?

3. Explain how synthesis and decomposition reactions can be the reverse of one another.

4. What two compounds form when hydrocarbons burn completely?

5. Explain how to use the activity series to predict chemical behavior.

6. In which part of the periodic table are the elements at the top of the activity series?

7. What must be produced for a double-displacement reaction to occur?

PRACTICE PROBLEMS

8. Balance each of the equations below, and indicate the type of reaction for each equation.

a. $Cl_2(g) + NaBr(aq) \longrightarrow NaCl(aq) + Br_2(l)$

b. $CaO(s) + H_2O(l) \longrightarrow Ca(OH)_2(aq)$

c. $Ca(ClO_3)_2(s) \longrightarrow CaCl_2(s) + O_2(g)$

d. $AgNO_3(aq) + K_2SO_4(aq) \longrightarrow$
$Ag_2SO_4(s) + KNO_3(aq)$

e. $Zn(s) + CuBr_2(aq) \longrightarrow ZnBr_2(aq) + Cu(s)$

f. $C_8H_{18}(l) + O_2(g) \longrightarrow CO_2(g) + H_2O(g)$

9. Predict whether a reaction would occur when the materials indicated are brought together. For each reaction that would occur, complete and balance the equation.

a. $Ag(s) + H_2O(l)$

b. $Mg(s) + Cu(NO_3)_2(aq)$

c. $Al(s) + O_2(g)$

d. $H_2SO_4(aq) + KOH(aq)$

10. Predict the products, write a balanced equation, and identify the type of reaction for each of the following reactions.

a. $HgO \longrightarrow$

b. $C_3H_7OH + O_2 \longrightarrow$

c. $Zn + CuSO_4 \longrightarrow$

d. $BaCl_2 + Na_2SO_4 \longrightarrow$

e. $Zn + F_2 \longrightarrow$

f. $C_5H_{10} + O_2 \longrightarrow$

CRITICAL THINKING

11. When will a displacement reaction *not* occur?

12. Explain why the terms *synthesis* and *decomposition* are appropriate names for their respective reaction types.

13. Platinum is used for jewelry because it does not corrode. Where would you expect to find platinum on the activity series?

14. Will a reaction occur when copper metal is dipped into a solution of silver nitrate? Explain.

Writing Net Ionic Equations

OBJECTIVES

1. **Write** total ionic equations for reactions in aqueous solutions.

2. **Identify** spectator ions and write net ionic equations for reactions in aqueous solutions.

Ionic Equations

When ionic compounds dissolve in water, the ions separate from each other and spread throughout the solution. Thus, the formulas $KI(aq)$ and $Pb(NO_3)_2(aq)$ are actually aqueous ions, as shown below.

$$KI(aq) = K^+(aq) + I^-(aq)$$

$$Pb(NO_3)_2(aq) = Pb^{2+}(aq) + 2NO_3^-(aq)$$

www.scilinks.org
Topic: Precipitation Reactions
SciLinks code: HW4160

Notice that when lead(II) nitrate dissolves, there are two nitrate ions for every lead ion, so a coefficient of 2 is used for NO_3^-. The reaction between KI and $Pb(NO_3)_2$ can be described by the chemical equation below.

$$2KI(aq) + Pb(NO_3)_2(aq) \longrightarrow PbI_2(s) + 2KNO_3(aq)$$

However, it is more correct to describe the reaction by using a *total ionic equation* as shown below. When you write a total ionic equation, make sure that both the mass and the electric charge are conserved.

$$2K^+(aq) + 2I^-(aq) + Pb^{2+}(aq) + 2NO_3^-(aq) \longrightarrow$$
$$PbI_2(s) + 2K^+(aq) + 2NO_3^-(aq)$$

But even this equation is not the best way to view the reaction.

Identifying Spectator Ions

When two solutions are mixed, all of the ions are present in the combined solution. In many cases, some of the ions will react with each other. However, some ions do not react. These **spectator ions** remain unchanged in the solution as aqueous ions. In the equation above, the K^+ and NO_3^- ions appear as aqueous ions both on the reactants' side and on the products' side. Because K^+ and NO_3^- ions are spectator ions in the above reaction, they can be removed from the total ionic equation. What remains are the substances that do change during the reaction.

spectator ions

ions that are present in a solution in which a reaction is taking place but that do not participate in the reaction

$$2\cancel{K^+}(aq) + 2I^-(aq) + Pb^{2+}(aq) + 2\cancel{NO_3^-}(aq) \longrightarrow$$
$$PbI_2(s) + 2\cancel{K^+}(aq) + 2\cancel{NO_3^-}(aq)$$

Chemical equation:

$$K_2SO_4(aq) + Ba(NO_3)_2(aq) \longrightarrow 2KNO_3(aq) + BaSO_4(s)$$

Total ionic equation:

$$2K^+(aq) + SO_4^{2-}(aq) + Ba^{2+}(aq) + 2NO_3^-(aq) \longrightarrow 2K^+(aq) + 2NO_3^-(aq) + BaSO_4(s)$$

Net ionic equation:

$$SO_4^{2-}(aq) + Ba^{2+}(aq) \longrightarrow BaSO_4(s)$$

Writing Net Ionic Equations

The substances that remain once the spectator ions are removed from the chemical equation make an equation that shows only the net change. This is called a *net ionic equation*. The one for the reaction of KI with $Pb(NO_3)_2$ is shown below.

$$2I^-(aq) + Pb^{2+}(aq) \longrightarrow PbI_2(s)$$

Figure 14 shows the process of determining the net ionic equation for another reaction.

The net ionic equation above is the same as the one for the reaction between NaI and $Pb(ClO_3)_2$. Both compounds are soluble, and their aqueous solutions contain iodide and lead(II) ions, which would form lead(II) iodide. So, the net change is the same.

Net ionic equations can also be used to describe displacement reactions. For example, Zn reacts with a solution of $CuSO_4$ and displaces the copper ion, Cu^{2+}, as shown in the total ionic equation

$$Zn(s) + Cu^{2+}(aq) + SO_4^{2-}(aq) \longrightarrow Cu(s) + Zn^{2+}(aq) + SO_4^{2-}(aq)$$

Only the sulfate ion remains unchanged and is a spectator ion. Thus, the net ionic equation is as follows:

$$Zn(s) + Cu^{2+}(aq) \longrightarrow Cu(s) + Zn^{2+}(aq)$$

Figure 14
For the reaction of potassium sulfate with barium nitrate, the net ionic equation shows that aqueous barium and sulfate ions join to form solid, insoluble barium sulfate.

4

SKILLS Toolkit

Writing Net Ionic Equations

1. List what you know.
- Identify each chemical described as a reactant or product.
- Identify the type of reaction taking place.

2. Write a balanced equation.
- Use the type of reaction to predict products, if necessary.
- Write a formula equation, and balance it. Include the physical state for each substance. Use the rules below with double-displacement reactions to determine whether a substance is an insoluble solid.

> All compounds of Group 1 and NH_4^+ are soluble.
>
> All nitrates are soluble.
>
> All halides, except those of Ag^+ and Pb^{2+}, are soluble.
>
> All sulfates, except Ca^{2+}, Sr^{2+}, Ba^{2+}, Hg_2^{2+}, and Pb^{2+}, are soluble.
>
> All carbonates, except those of Group 1 and NH_4^+, are insoluble.

3. Write the total ionic equation.
- Write separated aqueous ions for each aqueous ionic substance in the chemical equation.
- Do not split up any substance that is a solid, liquid, or gas.

4. Find the net ionic equation.
- Cancel out spectator ions, and write whatever remains as the net ionic equation.
- Double-check that the equation is balanced with respect to atoms and electric charge.

Check Atoms and Charge

Balanced net ionic equations are no different than other equations in that the numbers and kinds of atoms must be the same on each side of the equation. However, you also need to check that the sum of the charges for the reactants equals the sum of the charges for the products. As an example, recall the net ionic equation from **Figure 14.**

$$SO_4^{2-}(aq) + Ba^{2+}(aq) \longrightarrow BaSO_4(s)$$

One barium atom is on both sides of the equation, and one sulfate ion is on both sides of the equation. The sum of the charges is zero both in the reactants and in the products. Each side of a net ionic equation can have a net charge that is not zero. For example, the net ionic equation below has a net charge of 2+ on each side and is balanced.

$$Zn(s) + Cu^{2+}(aq) \longrightarrow Zn^{2+}(aq) + Cu(s)$$

Section Review

UNDERSTANDING KEY IDEAS

1. Explain why the term *spectator ions* is used.

2. What chemicals are present in a net ionic equation?

3. Is the following a correct net ionic equation? Explain.

$$Na^+(aq) + Cl^-(aq) \longrightarrow NaCl(aq)$$

4. Identify the spectator ion(s) in the following reaction:

$$MgSO_4(aq) + 2AgNO_3(aq) \longrightarrow$$
$$Ag_2SO_4(s) + Mg(NO_3)_2(aq)$$

5. Use the rules from **Skills Toolkit 4** to explain how to determine the physical states of the products in item 4.

PRACTICE PROBLEMS

6. Write a total ionic equation for each of the following unbalanced formula equations:

   a. $Br_2(l) + NaI(aq) \longrightarrow NaBr(aq) + I_2(s)$

   b. $Ca(OH)_2(aq) + HCl(aq) \longrightarrow$
   $$CaCl_2(aq) + H_2O(l)$$

   c. $Mg(s) + AgNO_3(aq) \longrightarrow$
   $$Ag(s) + Mg(NO_3)_2(aq)$$

   d. $AgNO_3(aq) + KBr(aq) \longrightarrow$
   $$AgBr(s) + KNO_3(aq)$$

   e. $Ni(s) + Pb(NO_3)_2(aq) \longrightarrow$
   $$Ni(NO_3)_2(aq) + Pb(s)$$

   f. $Ca(s) + H_2O(l) \longrightarrow Ca(OH)_2(aq) + H_2(g)$

7. Identify the spectator ions, and write a net ionic equation for each reaction in item 6.

8. Predict the products for each of the following reactions. If no reaction happens, write "no reaction." Write a total ionic equation for each reaction that does happen.

   a. $AuCl_3(aq) + Ag(s) \longrightarrow$

   b. $AgNO_3(aq) + CaCl_2(aq) \longrightarrow$

   c. $Al(s) + NiSO_4(aq) \longrightarrow$

   d. $Na(s) + H_2O(l) \longrightarrow$

   e. $AgNO_3(aq) + NaCl(aq) \longrightarrow$

9. Identify the spectator ions, and write a net ionic equation for each reaction that happens in item 8.

10. Write a total ionic equation for each of the following reactions:

    a. silver nitrate + sodium sulfate

    b. aluminum + nickel(II) iodide

    c. potassium sulfate + calcium chloride

    d. magnesium + copper(II) bromide

    e. lead(II) nitrate + sodium chloride

11. Identify the spectator ions, and write a net ionic equation for each reaction in item 10.

CRITICAL THINKING

12. Why is K^+ always a spectator ion?

13. Do net ionic equations always obey the rule of conservation of charge? Explain.

14. Suppose a drinking-water supply contains Ba^{2+}. Using solubility rules, write a net ionic equation for a double-displacement reaction that indicates how Ba^{2+} might be removed.

15. Explain why no reaction occurs if a double-displacement reaction has four spectator ions.

16. Explain why more than one reaction can have the same net ionic equation. Provide at least two reactions that have the same net ionic equation.

CONSUMER FOCUS

Fire Extinguishers

A fire is a combustion reaction. Three things are needed for a combustion reaction: a fuel, oxygen, and an ignition source. If any one of these three is absent, combustion cannot occur. One goal in fighting a fire is to remove one or more of these parts. Many extinguishers are designed to cool the burning material (to hinder ignition) or to prevent air and oxygen from reaching it.

Types of Fires

Each type of fire requires different firefighting methods. Class A fires involve solid fuels, such as wood. Class B fires involve a liquid or a gas, such as gasoline or natural gas. Class C fires involve the presence of a "live" electric circuit. Class D fires involve burning metals.

The type of extinguisher is keyed to the type of fire. Extinguishers for Class A fires often use water. The water cools the fuel so that it does not react as readily. The steam that is produced helps displace the oxygen-containing air around the fire. Carbon dioxide extinguishers can also be used. Because carbon dioxide is denser than air, it forms a layer underneath the air and cuts off the O_2 supply. Water cannot be used on Class B fires.

Because water is usually denser than the fuel, it sinks below the fuel. Carbon dioxide is preferred for Class B fires.

Dry Chemical Extinguishers

Class C fires involving a "live" electric circuit can also be extinguished by CO_2. Water cannot be used because of the danger of electric shock. Some Class C fire extinguishers contain a dry chemical that smothers the fire by interrupting the chain reaction that is occurring. For example, a competing reaction may take place with the contents of the fire extinguisher and the intermediates of the reaction. Class C fire extinguishers usually contain compounds such as ammonium dihydrogen phosphate, $NH_4H_2PO_4$, or sodium hydrogen carbonate, $NaHCO_3$.

Finally, Class D fires involve burning metals. These fires cannot be extinguished with CO_2 or water because these compounds may react with some hot metals. For these fires, nonreactive dry powders are used to cover the metal and to keep it separate from oxygen. One kind of powder contains finely ground sodium chloride crystals mixed with a special polymer that allows the crystals to adhere to any surface, even a vertical one.

Questions

1. Identify the type of fire extinguisher available in your laboratory. On what classes of fires should it be used? Record the steps needed to use the fire extinguisher.
2. Explain why a person whose clothing has caught fire is likely to make the situation worse by running. Explain why wrapping a person in a fire blanket can help extinguish the flames.

internet connect

www.scilinks.org
Topic: Fire Extinguishers
SciLinks code: HW4059

SCiLINKS   Maintained by the
National Science
Teachers Association

CHAPTER HIGHLIGHTS

8

KEY TERMS

chemical reaction
chemical equation

coefficient

combustion reaction
synthesis reaction
decomposition reaction
activity series
double-displacement
  reaction

spectator ions

KEY IDEAS

SECTION ONE Describing Chemical Reactions

- In a chemical reaction, atoms rearrange to form new substances.
- A chemical analysis is the only way to prove that a reaction has occurred.
- Symbols are used in chemical equations to identify the physical states of substances and the physical conditions during a chemical reaction.

SECTION TWO Balancing Chemical Equations

- A word equation is translated into a formula equation to describe the change of reactants into products.
- The masses, numbers, and types of atoms are the same on both sides of a balanced equation.
- Coefficients in front of the formulas of reactants and products are used to balance an equation. Subscripts cannot be changed.

SECTION THREE Classifying Chemical Reactions

- In a combustion reaction, a carbon-based compound reacts with oxygen to form carbon dioxide and water.
- In a synthesis reaction, two reactants form a single product.
- In a decomposition reaction, a single reactant forms two or more products.
- In a displacement reaction, an element displaces an element from a compound. The activity series is used to determine if a reaction will happen.
- In a double-displacement reaction, the ions of two compounds switch places such that two new compounds form. One of the products must be a solid, a gas, or a molecular compound, such as water, for a reaction to occur.

SECTION FOUR Writing Net Ionic Equations

- A total ionic equation shows all aqueous ions for a reaction.
- Spectator ions do not change during a reaction and can be removed from the total ionic equation.
- Net ionic equations show only the net change of a reaction and are the best way to describe displacement and double-displacement reactions.

KEY SKILLS

Balancing an Equation
Skills Toolkit 1 p. 268
Sample Problem A p. 269

The Odd-Even Technique
Sample Problem B p. 271

Polyatomic Ions as a Group
Sample Problem C p. 273

Predicting Products
Sample Problem D p. 279
Skills Toolkit 3 p. 284

**Determining Products by
Using the Activity Series**
Skills Toolkit 2 p. 281
Sample Problem E p. 282

Writing Net Ionic Equations
Skills Toolkit 4 p. 288

8 CHAPTER REVIEW

USING KEY TERMS

1. Describe the relationship between a synthesis reaction and a decomposition reaction.

2. How does a coefficient in front of a formula affect the number of each type of atom in the formula?

3. Define each of the following terms:
a. *decomposition reaction*
b. *double-displacement reaction*
c. *spectator ions*
d. *activity series*

4. How does the term *spectator* apply to spectator ions?

5. How does a coefficient differ from a subscript?

6. Give an example of a word equation, a formula equation, and a chemical equation.

UNDERSTANDING KEY IDEAS

Describing Chemical Reactions

7. A student writes the following statement in a lab report: "During the reaction, the particles of the reactants are lost. The reaction creates energy and particles of the products."
a. Explain the scientific inaccuracies in the student's statement.
b. How could the student correct the inaccurate statement?

8. Write an unbalanced chemical equation for each of the following.
a. Aluminum reacts with oxygen to produce aluminum oxide.

b. Phosphoric acid, H_3PO_4, is produced through the reaction between tetraphosphorus decoxide and water.
c. Iron(III) oxide reacts with carbon monoxide to produce iron and carbon dioxide.

9. Write the symbol used in a chemical equation to represent each of the following:
a. an aqueous solution
b. heated
c. a reversible reaction
d. a solid
e. at a temperature of 25°C

10. Write an unbalanced formula equation for each of the following. Include symbols for physical states in the equation.
a. solid zinc sulfide + oxygen gas $\longrightarrow$ solid zinc oxide + sulfur dioxide gas
b. aqueous hydrochloric acid + solid magnesium hydroxide $\longrightarrow$ aqueous magnesium chloride + liquid water
c. aqueous nitric acid + aqueous calcium hydroxide $\longrightarrow$ aqueous calcium nitrate + liquid water

11. Calcium oxide, CaO, is an ingredient in cement mixes. When water is added, the mixture warms up and calcium hydroxide, $Ca(OH)_2$, forms.
a. Is there any evidence of a chemical reaction?
b. In the reaction above, how can you prove that a chemical reaction has taken place?

12. Evaporating ocean water leaves a mixture of salts. Is this a chemical change? Explain.

13. Translate the following chemical equation into a sentence:

$$CH_4(g) + 2O_2(g) \longrightarrow CO_2(g) + 2H_2O(g)$$

Balancing Chemical Equations

14. Explain what is meant when one says that an equation is balanced.

15. How does the process of balancing an equation illustrate the law of conservation of mass?

16. In balancing a chemical equation, why can you change coefficients, but not subscripts?

17. Differentiate between formula equations and balanced chemical equations.

18. The white paste that lifeguards rub on their nose to prevent sunburn contains zinc oxide, $ZnO(s)$, as an active ingredient. Zinc oxide is produced by burning zinc sulfide.

$$2ZnS(s) + 3O_2(g) \longrightarrow 2ZnO(s) + 2SO_2(g)$$

a. What is the coefficient for sulfur dioxide?
b. What is the subscript for oxygen gas?
c. How many atoms of oxygen react?
d. How many atoms of oxygen appear in the total number of sulfur dioxide molecules?

Classifying Chemical Reactions

19. What are some of the characteristics of each of these five common chemical reactions?
a. combustion
b. synthesis
c. decomposition
d. displacement
e. double-displacement

20. Write a general equation for each of the five types of reaction described in item 19.

21. Explain the difference between displacement and double-displacement reactions.

22. What is an activity series?

23. When would a displacement reaction cause no reaction?

24. What must form in order for a double-displacement reaction to occur?

25. What are the products of the complete combustion of a hydrocarbon?

Writing Net Ionic Equations

26. How do total and net ionic equations differ?

27. Which ions in a total ionic equation are called *spectator ions*? Why?

28. Explain why a net ionic equation is the best way to represent a double-displacement reaction.

29. The saline solution used to soak contact lenses is primarily NaCl dissolved in water. Which of the following ways to represent the solution is *not* correct?
a. $NaCl(aq)$
b. $NaCl(s)$
c. $Na^+(aq) + Cl^-(aq)$

30. How should each of the following substances be represented in a total ionic equation?
a. $KCl(aq)$
b. $H_2O(l)$
c. $Cu(NO_3)_2(aq)$
d. $AgCl(s)$
e. $NH_4ClO_3(aq)$

PRACTICE PROBLEMS

Sample Problem A Balancing an Equation

31. Balance each of the following:
a. $H_2 + Cl_2 \longrightarrow HCl$
b. $Al + Fe_2O_3 \longrightarrow Al_2O_3 + Fe$
c. $Ba(ClO_3)_2 \longrightarrow BaCl_2 + O_2$
d. $Cu + HNO_3 \longrightarrow Cu(NO_3)_2 + NO + H_2O$

32. Write a balanced equation for each of the following:
a. iron(III) oxide + magnesium $\longrightarrow$ magnesium oxide + iron
b. nitrogen dioxide + water $\longrightarrow$ nitric acid + nitrogen monoxide
c. silicon tetrachloride + water $\longrightarrow$ silicon dioxide + hydrochloric acid
d. ammonium dichromate $\longrightarrow$ nitrogen + chromium(III) oxide + water

Sample Problem B The Odd-Even Technique

33 Balance each of the following:
  a. $Fe + O_2 \longrightarrow Fe_2O_3$
  b. $H_2O_2 \longrightarrow H_2O + O_2$
  c. $C_8H_{18} + O_2 \longrightarrow CO_2 + H_2O$
  d. $Al + F_2 \longrightarrow AlF_3$

34. Write a balanced equation for each of the following:
  a. propanol (C_3H_7OH) + oxygen $\longrightarrow$
          carbon dioxide + water
  b. aluminum + iron(II) nitrate $\longrightarrow$
          aluminum nitrate + iron
  c. iron(III) hydroxide $\longrightarrow$
          iron(III) oxide + water
  d. lead(IV) oxide $\longrightarrow$ lead(II) oxide + oxygen

Sample Problem C Polyatomic Ions as a Group

35. Balance each of the following:
  a. $Zn + Pb(NO_3)_2 \longrightarrow Pb + Zn(NO_3)_2$
  b. $NH_4CH_3COO + AgNO_3 \longrightarrow$
          $NH_4NO_3 + AgCH_3COO$
  c. $H_2C_2O_4 + NaOH \longrightarrow Na_2C_2O_4 + H_2O$
  d. $Al + CuSO_4 \longrightarrow Al_2(SO_4)_3 + Cu$

36. Write a balanced equation for each of the following:
  a. copper(II) sulfate + ammonium sulfide $\longrightarrow$
          copper(II) sulfide + ammonium sulfate
  b. nitric acid + barium hydroxide $\longrightarrow$
          water + barium nitrate
  c. iron(III) nitrate + lithium hydroxide $\longrightarrow$
          lithium nitrate + iron(III) hydroxide
  d. barium chloride + phosphoric acid $\longrightarrow$
          barium phosphate + hydrochloric acid

Sample Problem D Predicting Products

37. Complete and balance the equation for each of the following synthesis reactions.
  a. $Zn + O_2 \longrightarrow$ **c.** $Cl_2 + K \longrightarrow$
  b. $F_2 + Mg \longrightarrow$ **d.** $H_2 + I_2 \longrightarrow$

38. Complete and balance the equation for the decomposition of each of the following.
  a. $HgO \longrightarrow$ **c.** $AgCl \longrightarrow$
  b. $H_2O \longrightarrow$ **d.** $KOH \longrightarrow$

39. Complete and balance the equation for the complete combustion of each of the following.
  a. C_3H_6 **c.** CH_3OH
  b. C_5H_{12} **d.** $C_{12}H_{22}O_{11}$

40. Each of the following reactions is a synthesis, decomposition, or combustion reaction. For each reaction, determine the type of reaction and complete and balance the equation.
  a. $C_3H_8 + O_2 \longrightarrow$
  b. $CuCl_2 \longrightarrow$
  c. $Mg + O_2 \longrightarrow$
  d. $Na_2CO_3 \longrightarrow$
  e. $Ba(OH)_2 \longrightarrow$
  f. $C_2H_5OH + O_2 \longrightarrow$

Sample Problem E Determining Products by Using the Activity Series

41. Using the activity series in **Appendix A,** predict whether each of the possible reactions listed below will occur. For the reactions that will occur, write the products and balance the equation.
  a. $Mg(s) + CuCl_2(aq) \longrightarrow$
  b. $Pb(NO_3)_2(aq) + Zn(s) \longrightarrow$
  c. $KI(aq) + Cl_2(g) \longrightarrow$
  d. $Cu(s) + FeSO_4(aq) \longrightarrow$

42. Using the activity series in **Appendix A,** predict whether each of the possible reactions listed below will occur. For the reactions that will occur, write the products and balance the equation.
  a. $H_2O(l) + Ba(s) \longrightarrow$
  b. $Ca(s) + O_2(g) \longrightarrow$
  c. $O_2(g) + Au(s) \longrightarrow$
  d. $Al(s) + O_2(g) \longrightarrow$

Skills Toolkit 3 Identifying Reactions and Predicting Products

43. Identify the type of reaction for each of the following. Then, predict products for the reaction and balance the equation. If no reaction occurs, write "no reaction."
  a. $C_2H_6 + O_2 \longrightarrow$
  b. $H_2SO_4 + Al \longrightarrow$
  c. $N_2 + Mg \longrightarrow$

d. $Na_2CO_3 \longrightarrow$

e. $Mg(NO_3)_2 + Na_2SO_4 \longrightarrow$

44. Identify the type of reaction for each of the following. Then, predict products for the reaction, and balance the equation. If no reaction occurs, write "no reaction."

  a. water + lithium $\longrightarrow$

  b. calcium + bromine $\longrightarrow$

  c. silver nitrate + hydrochloric acid $\longrightarrow$

  d. hydrogen iodide $\longrightarrow$

45. Identify the type of reaction for each of the following. Then, predict products for the reaction, and balance the equation. If no reaction occurs, write "no reaction."

  a. ethanol (C_2H_5OH) + oxygen $\longrightarrow$

  b. nitric acid + lithium hydroxide $\longrightarrow$

  c. sodium nitrate + calcium chloride $\longrightarrow$

  d. lead(II) nitrate + sodium carbonate $\longrightarrow$

Skills Toolkit 4 Writing Net Ionic Equations

46. Write a total ionic equation and a net ionic equation for each of the following reactions.

  a. $HCl(aq) + NaOH(aq) \longrightarrow$
  $$NaCl(aq) + H_2O(l)$$

  b. $Mg(s) + 2HCl(aq) \longrightarrow MgCl_2(aq) + H_2(g)$

  c. $CdCl_2(aq) + Na_2CO_3(aq) \longrightarrow$
  $$2NaCl(aq) + CdCO_3(s)$$

  d. $Mg(s) + Zn(NO_3)_2(aq) \longrightarrow$
  $$Zn(s) + Mg(NO_3)_2(aq)$$

47. Identify the spectator ions in each reaction in item 46.

48. Predict the products and write a net ionic equation for each of the following reactions. If no reaction occurs, write "no reaction."

  a. $K_2CO_3(aq) + CaCl_2(aq) \longrightarrow$

  b. $Na_2SO_4(aq) + AgNO_3(aq) \longrightarrow$

  c. $NH_4Cl(aq) + AgNO_3(aq) \longrightarrow$

  d. $Pb(s) + ZnCl_2(aq) \longrightarrow$

49. Identify the spectator ions in each reaction in item 48.

MIXED REVIEW

50. For each photograph below, identify evidence of a chemical change, then identify the type of reaction shown. Support your answer.

a.

b.

51. Balance the following equations.

  a. $CaH_2(s) + H_2O(l) \longrightarrow$
  $$Ca(OH)_2(aq) + H_2(g)$$

  b. $CH_3CH_2CCH(g) + Br_2(l) \longrightarrow$
  $$CH_3CH_2CBr_2CHBr_2(l)$$

  c. $Pb^{2+}(aq) + OH^-(aq) \longrightarrow Pb(OH)_2(s)$

  d. $NO_2(g) + H_2O(l) \longrightarrow HNO_3(aq) + NO(g)$

52. Write and balance each of the following equations, and then identify each equation by type.

  a. hydrogen + iodine $\longrightarrow$ hydrogen iodide

  b. lithium + water $\longrightarrow$
  $$\text{lithium hydroxide} + \text{hydrogen}$$

  c. mercury(II) oxide $\longrightarrow$ mercury + oxygen

  d. copper + chlorine $\longrightarrow$ copper(II) chloride

53. Write a balanced equation, including all of the appropriate notations, for each of the following reactions.

  a. Steam reacts with solid carbon to form the gases carbon monoxide and hydrogen.

b. Heating ammonium nitrate in aqueous solution forms dinitrogen monoxide gas and liquid water.

c. Nitrogen dioxide gas forms from the reaction of nitrogen monoxide gas and oxygen gas.

54. Methanol, CH_3OH, is a clean-burning fuel.
a. Write a balanced chemical equation for the synthesis of methanol from carbon monoxide and hydrogen gas.
b. Write a balanced chemical equation for the complete combustion of methanol.

55. Use the activity series to predict whether the following reactions are possible. Explain your answers.
a. $Ni(s) + MgSO_4(aq) \longrightarrow$
$$NiSO_4(aq) + Mg(s)$$
b. $3Mg(s) + Al_2(SO_4)_3(aq) \longrightarrow$
$$3MgSO_4(aq) + 2Al(s)$$
c. $Pb(s) + 2H_2O(l) \longrightarrow$
$$Pb(OH)_2(aq) + H_2(g)$$

56. Iron(III) chloride, $FeCl_3$, is a chemical used in photography. It can be produced by reacting iron and chlorine. Identify the correct chemical equation from the choices below, and then explain what is wrong with the other two choices.
a. $Fe(s) + Cl_3(g) \longrightarrow FeCl_3(s)$
b. $2Fe(s) + 3Cl_2(g) \longrightarrow 2FeCl_3(s)$
c. $Fe(s) + 3Cl_2(g) \longrightarrow Fe2Cl_3(s)$

57. Write the balanced equation for each of the following:
a. the complete combustion of propane gas, C_3H_8
b. the decomposition of magnesium carbonate
c. the synthesis of platinum(IV) fluoride from platinum and fluorine gas
d. the reaction of zinc with lead(II) nitrate

58. Predict the products for each of the following reactions. Write a total ionic equation and a net ionic equation for each reaction. If no reaction occurs, write "no reaction."
a. $Li_2CO_3(aq) + BaBr_2(aq) \longrightarrow$

b. $Na_2SO_4(aq) + Sr(NO_3)_2(aq) \longrightarrow$
c. $Al(s) + NiCl_2(aq) \longrightarrow$
d. $K_2CO_3(aq) + FeCl_3(aq) \longrightarrow$

59. Identify the spectator ions in each reaction in item 58.

CRITICAL THINKING

60. The following equations are incorrect in some way. Identify and correct each error, and then balance each equation.
a. $Li + O_2 \longrightarrow LiO_2$
b. $MgCO_3 \longrightarrow Mg + C + 3O_2$
c. $NaI + Cl_2 \longrightarrow NaCl + I$
d. $AgNO_3 + CaCl_2 \longrightarrow Ca(NO_3) + AgCl_2$
e. $3Mg + 2FeBr_3 \longrightarrow Fe_2Mg_3 + 3Br_2$

61. Although cesium is not listed in the activity series in this chapter, predict where cesium would appear based on its position in the periodic table.

62. Create an activity series for the hypothetical elements A, J, Q, and Z by using the reaction information provided below.

$$A + ZX \longrightarrow AX + Z$$
$$J + ZX \longrightarrow \text{no reaction}$$
$$Q + AX \longrightarrow QX + A$$

63. When wood burns, the ash weighs much less than the original wood did. Explain why the law of conservation of mass is not violated in this situation.

64. Write the total and net ionic equations for the reaction in which the antacid $Al(OH)_3$ neutralizes the stomach acid HCl. Identify the type of reaction.
a. Identify the spectator ions in this reaction.
b. What would be the advantages of using $Al(OH)_3$ as an antacid rather than $NaHCO_3$, which undergoes the following reaction with stomach acid?

$$NaHCO_3(aq) + HCl(aq) \longrightarrow$$
$$NaCl(aq) + H_2O(l) + CO_2(g)$$

65. The images below represent the reactants of a chemical reaction. Study the images, then answer the items that follow.

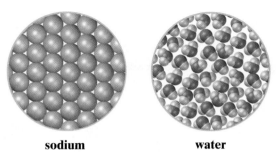

sodium   water

  a. Write a balanced chemical equation for the reaction that shows the states of all substances.

  b. What type of reaction is this?

ALTERNATIVE ASSESSMENT

66. Using the materials listed below, describe a procedure that would enable you to organize the metals in order of reactivity. The materials are pieces of aluminum, chromium, and magnesium and solutions of aluminum chloride, chromium(III) chloride, and magnesium chloride.

67. Design an experiment for judging the value and efficacy of different antacids. Include $NaHCO_3$, $Mg(OH)_2$, $CaCO_3$, and $Al(OH)_3$ in your tests. Discover which one neutralizes the most acid and what byproducts form. Show your experiment to your teacher. If your experiment is approved, obtain the necessary chemicals from your teacher and test your procedure.

68. For one day, record situations that suggest that a chemical change has occurred. Identify the reactants and the products, and state whether there is proof of a chemical reaction. Classify each of the chemical reactions according to the five common reaction types discussed in the chapter.

69. Research safety tips for dealing with fires. Create a poster or brochure about fire safety in which you explain both these tips and their basis in science.

70. Many products are labeled "biodegradable." Choose several biodegradable items on the market, and research the decomposition reactions that occur. Take into account any special conditions that must occur for the substance to biodegrade. Present your information to the class to help inform the students about which products are best for the environment.

71. Develop an analogy to help illustrate each of the five types of reactions described in this chapter. Write the analogy as though you were explaining what happens in the reaction to a classmate who was having trouble understanding. You might consider starting your analogy with the phrase "A _____ reaction is like . . ."

CONCEPT MAPPING

72. Use the following terms to complete the concept map below: *a synthesis reaction, a decomposition reaction, coefficients, a chemical reaction,* and *a chemical equation.*

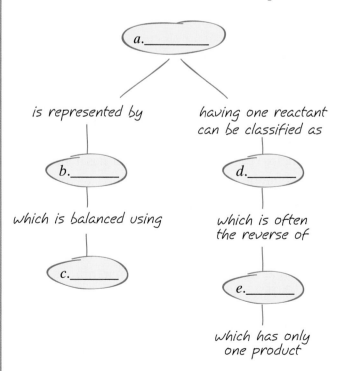

Study the graph below, and answer the questions that follow.
For help in interpreting graphs, see Appendix B, "Study Skills for Chemistry."

73. Which halogen has the shortest single bond with hydrogen?

74. What is the difference in length between an H–Br bond and an H–I bond?

75. Describe the trend in bond length as you move down the elements in Group 17 on the periodic table.

76. Based on this graph, what conclusion can be drawn about the relative sizes of halogen atoms? Could you draw the same conclusion if an atom of an element other than hydrogen was bonded to an atom of each halogen?

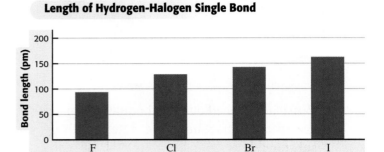

Length of Hydrogen-Halogen Single Bond

TECHNOLOGY AND LEARNING

77. Graphing Calculator

Least Common Multiples When writing chemical formulas or balancing a chemical equation, being able to identify the least common multiple of a set of numbers can often help. Your graphing calculator has a least common multiple function that can compare two numbers. On a TI-83 Plus or similar graphing calculator, press **MATH** ➢ **8**. The screen should read "lcm(." Next, enter one number and then a comma followed by the other number and a closing parenthesis. Press **ENTER,** and the calculator will show the least common multiple of the pair you entered.

Use this function as needed to find the answers to the following questions.

a. Tin(IV) sulfate contains Sn^{4+} and SO_4^{2-} ions. Use the least common multiple of 2 and 4 to determine the empirical formula for this compound.

b. Aluminum ferrocyanide contains Al^{3+} ions and $Fe(CN)_6^{4-}$ ions. Use the least common multiple of 3 and 4 to determine the empirical formula for this compound.

c. Balance the following **unbalanced** equation.

$$\_\_P_4O_{10}(s) + \_\_H_2O(g) \longrightarrow \_\_H_3PO_4(aq)$$

d. Balance the following **unbalanced** equation.

$$\_\_KMnO_4(aq) + \_\_MnCl_2(aq) + 2H_2O(l) \longrightarrow$$
$$\_\_MnO_2(s) + 4HCl(aq) + 2KCl(aq)$$

e. The combustion of octane, C_8H_{18}, and oxygen, O_2, is one of many reactions that occur in a car's engine. The products are CO_2 and H_2O. Balance the equation for the combustion reaction. (Hint: Balance oxygen last, and use the least common multiple of the number of oxygen atoms on the products' side and on the reactants' side to help balance the equation.)

1. According to the law of conservation of mass, the total mass of the reacting substances is
  a. always more than the total mass of the products.
  b. always less than the total mass of the products.
  c. sometimes more and sometimes less than the total mass of the products.
  d. always equal to the total mass of the products.

2. In a chemical equation, the symbol (*aq*) indicates that the substance is
  a. water.
  b. dissolved in water.
  c. an acid.
  d. insoluble.

3. The ratio of chlorine to hydrogen chloride in the reaction $H_2(g) + Cl_2(g) \longrightarrow 2HCl(g)$ is
  a. 1:1.
  c. 2:1.
  b. 1:2.
  d. 2:2.

4. A common interpretation is that the coefficients in a chemical equation
  a. show the number of grams of each substance that would react.
  b. indicate the number of molecules of each substance.
  c. are the molar masses of the substances.
  d. show the valence electrons for each atom.

5. A reaction in which the ions of two compounds exchange places in aqueous solution to form two new compounds is called a
  a. synthesis reaction.
  b. double-displacement reaction.
  c. decomposition reaction.
  d. combustion reaction.

6. To balance a chemical equation, you may adjust the
  a. coefficients.
  b. subscripts.
  c. formulas of the products.
  d. number of products.

7. The use of a double arrow in a chemical equation indicates that the reaction
  a. is reversible.
  b. requires heat.
  c. is written backward.
  d. has not been confirmed in the laboratory.

8. In a word equation, the plus sign, +, represents the word
  a. *and.*
  c. *produce.*
  b. *heat.*
  d. *yield.*

9. The ions that do not participate in a reaction and that do not appear in a net ionic equation are called
  a. reactants.
  c. total ions.
  b. spectator ions.
  d. products.

10. The reaction type that has only one reactant is
  a. decomposition.
  c. synthesis.
  b. displacement.
  d. oxidation.

11. A substance followed by which symbol should be written as separated ions in a total ionic equation?
  a. (*s*)
  c. (*g*)
  b. (*l*)
  d. (*aq*)

12. Elements in the activity series are arranged by
  a. atomic number.
  b. position in the family.
  c. ionization energy.
  d. experimentally determined order of reactivity.

STOICHIOMETRY

To play a standard game of chess, each side needs the proper number of pieces and pawns. Unless you find all of them—a king, a queen, two bishops, two knights, two rooks, and eight pawns—you cannot start the game. In chemical reactions, if you do not have every reactant, you will not be able to start the reaction. In this chapter you will look at amounts of reactants present and calculate the amounts of other reactants or products that are involved in the reaction.

START-UP ACTIVITY

All Used Up

PROCEDURE

1. Use a **balance** to find the mass of **8 nuts** and the mass of **5 bolts**.

2. Attach 1 nut (N) to 1 bolt (B) to assemble a nut-bolt (NB) model. Make as many NB models as you can. Record the number of models formed, and record which material was used up. Take the models apart.

3. Attach 2 nuts to 1 bolt to assemble a nut-nut-bolt (N_2B) model. Make as many N_2B models as you can. Record the number of models formed, and record which material was used up. Take the models apart.

ANALYSIS

1. Using the masses of the starting materials (the nuts and the bolts), could you predict which material would be used up first? Explain.

2. Write a balanced equation for the "reaction" that forms NB. How can this equation help you predict which component runs out?

3. Write a balanced equation for the "reaction" that forms N_2B. How can this equation help you predict which component runs out?

4. If you have 18 bolts and 26 nuts, how many models of NB could you make? of N_2B?

Pre-Reading Questions

① A recipe calls for one cup of milk and three eggs per serving. You quadruple the recipe because you're expecting guests. How much milk and eggs do you need?

② A bicycle mechanic has 10 frames and 16 wheels in the shop. How many complete bicycles can he assemble using these parts?

③ List at least two conversion factors that relate to the mole.

Calculating Quantities in Reactions

KEY TERMS

• stoichiometry

OBJECTIVES

(1) **Use** proportional reasoning to determine mole ratios from a balanced chemical equation.

(2) **Explain** why mole ratios are central to solving stoichiometry problems.

(3) **Solve** stoichiometry problems involving mass by using molar mass.

(4) **Solve** stoichiometry problems involving the volume of a substance by using density.

(5) **Solve** stoichiometry problems involving the number of particles of a substance by using Avogadro's number.

Balanced Equations Show Proportions

If you wanted homemade muffins, like the ones in **Figure 1,** you could make them yourself—if you had the right things. A recipe for muffins shows how much of each ingredient you need to make 12 muffins. It also shows the proportions of those ingredients. If you had just a little flour on hand, you could determine how much of the other things you should use to make great muffins. The proportions also let you adjust the amounts to make enough muffins for all your classmates.

A balanced chemical equation is very similar to a recipe in that the coefficients in the balanced equation show the proportions of the reactants and products involved in the reaction. For example, consider the reaction for the synthesis of water.

$$2H_2 + O_2 \rightarrow 2H_2O$$

On a very small scale, the coefficients in a balanced equation represent the numbers of particles for each substance in the reaction. For the equation above, the coefficients show that two molecules of hydrogen react with one molecule of oxygen and form two molecules of water.

Calculations that involve chemical reactions use the proportions from balanced chemical equations to find the quantity of each reactant and product involved. As you learn how to do these calculations in this section, you will assume that each reaction goes to completion. In other words, all of the given reactant changes into product. For each problem in this section, assume that there is more than enough of all other reactants to completely react with the reactant given. Also assume that every reaction happens perfectly, so that no product is lost during collection. As you will learn in the next section, this usually is not the case.

Figure 1
In using a recipe to make muffins, you are using proportions to determine how much of each ingredient is needed.

Relative Amounts in Equations Can Be Expressed in Moles

Just as you can interpret equations in terms of particles, you can interpret them in terms of moles. The coefficients in a balanced equation also represent the moles of each substance. For example, the equation for the synthesis of water shows that 2 mol H_2 react with 1 mol O_2 to form 2 mol H_2O. Look at the equation below.

$$2C_8H_{18} + 25O_2 \rightarrow 16CO_2 + 18H_2O$$

This equation shows that 2 molecules C_8H_{18} react with 25 molecules O_2 to form 16 molecules CO_2 and 18 molecules H_2O. And because Avogadro's number links molecules to moles, the equation also shows that 2 mol C_8H_{18} react with 25 mol O_2 to form 16 mol CO_2 and 18 mol H_2O.

In this chapter you will learn to determine how much of a reactant is needed to produce a given quantity of product, or how much of a product is formed from a given quantity of reactant. The branch of chemistry that deals with quantities of substances in chemical reactions is known as **stoichiometry.**

■ internet connect
www.scilinks.org
Topic: Chemical Equations
SciLinks code: HW4141

SCi
LINKS. Maintained by the National Science Teachers Association

stoichiometry

the proportional relationship between two or more substances during a chemical reaction

The Mole Ratio Is the Key

If you normally buy a lunch at school each day, how many times would you need to "brown bag" it if you wanted to save enough money to buy a CD player? To determine the answer, you would use the units of dollars to bridge the gap between a CD player and school lunches. In stoichiometry problems involving equations, the unit that bridges the gap between one substance and another is the mole.

The coefficients in a balanced chemical equation show the relative numbers of moles of the substances in the reaction. As a result, you can use the coefficients in conversion factors called *mole ratios*. Mole ratios bridge the gap and can convert from moles of one substance to moles of another, as shown in **Skills Toolkit 1.**

SKILLS Toolkit 1

Converting Between Amounts in Moles

1. Identify the amount in moles that you know from the problem.

2. Using coefficients from the balanced equation, set up the mole ratio with the known substance on bottom and the unknown substance on top.

3. Multiply the original amount by the mole ratio.

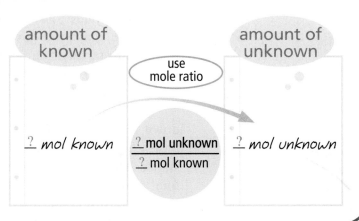

Using Mole Ratios

Consider the reaction for the commercial preparation of ammonia.

$$N_2 + 3H_2 \longrightarrow 2NH_3$$

How many moles of hydrogen are needed to prepare 312 moles of ammonia?

1 Gather information.

- amount of NH_3 = 312 mol
- amount of H_2 = ? mol
- From the equation: 3 mol H_2 = 2 mol NH_3.

2 Plan your work.

The mole ratio must cancel out the units of mol NH_3 given in the problem and leave the units of mol H_2. Therefore, the mole ratio is

$$\frac{3 \text{ mol } H_2}{2 \text{ mol } NH_3}$$

3 Calculate.

$$? \text{ mol } H_2 = 312 \text{ mol } NH_3 \times \frac{3 \text{ mol } H_2}{2 \text{ mol } NH_3} =$$

$$468 \text{ mol } H_2$$

4 Verify your result.

- The answer is larger than the initial number of moles of ammonia. This is expected, because the conversion factor is greater than one.
- The number of significant figures is correct because the coefficients 3 and 2 are considered to be exact numbers.

PRACTICE

1 Calculate the amounts requested if 1.34 mol H_2O_2 completely react according to the following equation.

$$2H_2O_2 \rightarrow 2H_2O + O_2$$

a. moles of oxygen formed
b. moles of water formed

2 Calculate the amounts requested if 3.30 mol Fe_2O_3 completely react according to the following equation.

$$Fe_2O_3 + 2Al \rightarrow 2Fe + Al_2O_3$$

a. moles of aluminum needed
b. moles of iron formed
c. moles of aluminum oxide formed

PRACTICE HINT

The mole ratio must always have the unknown substance on top and the substance given in the problem on bottom for units to cancel correctly.

Getting into Moles and Getting out of Moles

Substances are usually measured by mass or volume. As a result, before using the mole ratio you will often need to convert between the units for mass and volume and the unit *mol*. Yet each stoichiometry problem has the step in which moles of one substance are converted into moles of a second substance using the mole ratio from the balanced chemical equation. Follow the steps in **Skills Toolkit 2** to understand the process of solving stoichiometry problems.

The thought process in solving stoichiometry problems can be broken down into three basic steps. First, change the units you are given into moles. Second, use the mole ratio to determine moles of the desired substance. Third, change out of moles to whatever unit you need for your final answer. And if you are given moles in the problem or need moles as an answer, just skip the first step or the last step! As you continue reading, you will be reminded of the conversion factors that involve moles.

Solving Stoichiometry Problems

You can solve all types of stoichiometry problems by following the steps outlined below.

1. Gather information.

- If an equation is given, make sure the equation is balanced. If no equation is given, write a balanced equation for the reaction described.
- Write the information provided for the given substance. If you are not given an amount in moles, determine the information you need to change the given units into moles and write it down.
- Write the units you are asked to find for the unknown substance. If you are not asked to find an amount in moles, determine the information you need to change moles into the desired units, and write it down.
- Write an equality using substances and their coefficients that shows the relative amounts of the substances from the balanced equation.

2. Plan your work.

- Think through the three basic steps used to solve stoichiometry problems: change to moles, use the mole ratio, and change out of moles. Know which conversion factors you will use in each step.
- Write the mole ratio you will use in the form:

$$\frac{\text{moles of unknown substance}}{\text{moles of given substance}}$$

3. Calculate.

- Write a question mark with the units of the answer followed by an equals sign and the quantity of the given substance.
- Write the conversion factors—including the mole ratio—in order so that you change the units of the given substance to the units needed for the answer.
- Cancel units and check that the remaining units are the required units of the unknown substance.
- When you have finished your calculations, round off the answer to the correct number of significant figures. In the examples in this book, only the final answer is rounded off.
- Report your answer with correct units and with the name or formula of the substance.

4. Verify your result.

- Verify your answer by estimating. You could round off the numbers in the setup in step 3 and make a quick calculation. Or you could compare conversion factors in the setup and decide whether the answer should be bigger or smaller than the initial value.
- Make sure your answer is reasonable. For example, imagine that you calculate that 725 g of a reactant is needed to form 5.3 mg (0.0053 g) of a product. The large difference in these quantities should alert you that there may be an error and that you should double-check your work.

Figure 2
These tanks store ammonia for use as fertilizer. Stoichiometry is used to determine the amount of ammonia that can be made from given amounts of H_2 and N_2.

Problems Involving Mass, Volume, or Particles

Figure 2 shows a few of the tanks used to store the millons of metric tons of ammonia made each year in the United States. Stoichiometric calculations are used to determine how much of the reactants are needed and how much product is expected. However, the calculations do not start and end with moles. Instead, other units, such as liters or grams, are used. Mass, volume, or number of particles can all be used as the starting and ending quantities of stoichiometry problems. Of course, the key to each of these problems is the mole ratio.

For Mass Calculations, Use Molar Mass

The conversion factor for converting between mass and amount in moles is the molar mass of the substance. The molar mass is the sum of atomic masses from the periodic table for the atoms in a substance. **Skills Toolkit 3** shows how to use the molar mass of each substance involved in a stoichiometry problem. Notice that the problem is a three-step process. The mass in grams of the given substance is converted into moles. Next, the mole ratio is used to convert into moles of the desired substance. Finally, this amount in moles is converted into grams.

3 SKILLS Toolkit

Solving Mass-Mass Problems

mass of known		amount of known		amount of unknown		mass of unknown
	use molar mass		use mole ratio		use molar mass	
? g known	$\dfrac{1 \text{ mol}}{? \text{ g}}$	? mol known	$\dfrac{? \text{ mol unknown}}{? \text{ mol known}}$	? mol unknown	$\dfrac{? \text{ g}}{1 \text{ mol}}$	? g unknown

Problems Involving Mass

What mass of NH_3 can be made from 1221 g H_2 and excess N_2?

$$N_2 + 3H_2 \longrightarrow 2NH_3$$

1 Gather information.

- mass of H_2 = 1221 g H_2
- molar mass of H_2 = 2.02 g/mol
- mass of NH_3 = ? g NH_3
- molar mass of NH_3 = 17.04 g/mol
- From the balanced equation: 3 mol H_2 = 2 mol NH_3.

2 Plan your work.

- To change grams of H_2 to moles, use the molar mass of H_2.
- The mole ratio must cancel out the units of mol H_2 given in the problem and leave the units of mol NH_3. Therefore, the mole ratio is

$$\frac{2 \text{ mol } NH_3}{3 \text{ mol } H_2}$$

- To change moles of NH_3 to grams, use the molar mass of NH_3.

3 Calculate.

$$? \text{ g } NH_3 = 1221 \text{ g } H_2 \times \frac{1 \text{ mol } H_2}{2.02 \text{ g } H_2} \times \frac{2 \text{ mol } NH_3}{3 \text{ mol } H_2} \times \frac{17.04 \text{ g } NH_3}{1 \text{ mol } NH_3} =$$

$$6867 \text{ g } NH_3$$

4 Verify your result.

The units cancel to give the correct units for the answer. Estimating shows the answer should be about 6 times the original mass.

PRACTICE HINT

Remember to check both the units and the substance when canceling. For example, 1221 g H_2 cannot be converted to moles by multiplying by 1 mol NH_3/17.04 g NH_3. The units of grams in each one cannot cancel because they involve different substances.

PRACTICE

Use the equation below to answer the questions that follow.

$$Fe_2O_3 + 2Al \longrightarrow 2Fe + Al_2O_3$$

1 How many grams of Al are needed to completely react with 135 g Fe_2O_3?

2 How many grams of Al_2O_3 can form when 23.6 g Al react with excess Fe_2O_3?

3 How many grams of Fe_2O_3 react with excess Al to make 475 g Fe?

4 How many grams of Fe will form when 97.6 g Al_2O_3 form?

PROBLEM SOLVING SKILL

Solving Volume-Volume Problems

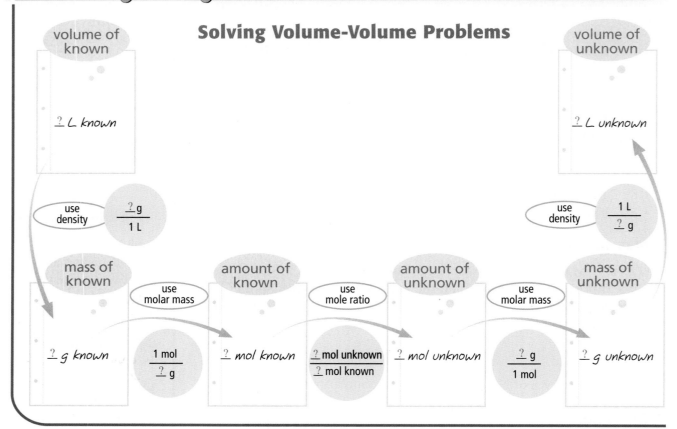

For Volume, You Might Use Density and Molar Mass

When reactants are liquids, they are almost always measured by volume. So, to do calculations involving liquids, you add two more steps to the sequence of mass-mass problems—the conversions of volume to mass and of mass to volume. Five conversion factors—two densities, two molar masses, and a mole ratio—are needed for this type of calculation, as shown in **Skills Toolkit 4.**

To convert from volume to mass or from mass to volume of a substance, use the density of the substance as the conversion factor. Keep in mind that the units you want to cancel should be on the bottom of your conversion factor.

There are ways other than density to include volume in stoichiometry problems. For example, if a substance in the problem is a gas at standard temperature and pressure (STP), use the molar volume of a gas to change directly between volume of the gas and moles. The molar volume of a gas is 22.41 L/mol for any gas at STP. Also, if a substance in the problem is in aqueous solution, then use the concentration of the solution to convert the volume of the solution to the moles of the substance dissolved. This procedure is especially useful when you perform calculations involving the reaction between an acid and a base. Of course, even in these problems, the basic process remains the same: change to moles, use the mole ratio, and change to the desired units.

Problems Involving Volume

What volume of H_3PO_4 forms when 56 mL $POCl_3$ completely react? (density of $POCl_3$ = 1.67 g/mL; density of H_3PO_4 = 1.83 g/mL)

$$POCl_3(l) + 3H_2O(l) \longrightarrow H_3PO_4(l) + 3HCl(g)$$

1 **Gather information.**

- volume $POCl_3$ = 56 mL $POCl_3$
- density $POCl_3$ = 1.67 g/mL
- volume H_3PO_4 = ?
- density H_3PO_4 = 1.83 g/mL
- molar mass $POCl_3$ = 153.32 g/mol
- molar mass H_3PO_4 = 98.00 g/mol
- From the equation: 1 mol $POCl_3$ = 1 mol H_3PO_4.

2 **Plan your work.**

- To change milliliters of $POCl_3$ to moles, use the density of $POCl_3$ followed by its molar mass.
- The mole ratio must cancel out the units of mol $POCl_3$ given in the problem and leave the units of mol H_3PO_4. Therefore, the mole ratio is

$$\frac{1 \text{ mol } H_3PO_4}{1 \text{ mol } POCl_3}$$

- To change out of moles of H_3PO_4 into milliliters, use the molar mass of H_3PO_4 followed by its density.

3 **Calculate.**

$$? \text{ mL } H_3PO_4 = 56 \text{ mL } POCl_3 \times \frac{1.67 \text{ g } POCl_3}{1 \text{ mL } POCl_3} \times \frac{1 \text{ mol } POCl_3}{153.32 \text{ g } POCl_3} \times$$

$$\frac{1 \text{ mol } H_3PO_4}{1 \text{ mol } POCl_3} \times \frac{98.00 \text{ g } H_3PO_4}{1 \text{ mol } H_3PO_4} \times \frac{1 \text{ mL } H_3PO_4}{1.83 \text{ g } H_3PO_4} = 33 \text{ mL } H_3PO_4$$

4 **Verify your result.**

The units of the answer are correct. Estimating shows the answer should be about two-thirds of the original volume.

PRACTICE HINT

Do not try to memorize the exact steps of every type of problem. For long problems like these, you might find it easier to break the problem into three steps rather than solving all at once. Remember that whatever you are given, you need to change to moles, then use the mole ratio, then change out of moles to the desired units.

PRACTICE

Use the densities and balanced equation provided to answer the questions that follow. (density of C_5H_{12} = 0.620 g/mL; density of C_5H_8 = 0.681 g/mL; density of H_2 = 0.0899 g/L)

$$C_5H_{12}(l) \longrightarrow C_5H_8(l) + 2H_2(g)$$

1 How many milliliters of C_5H_8 can be made from 366 mL C_5H_{12}?

2 How many liters of H_2 can form when 4.53×10^3 mL C_5H_8 form?

3 How many milliliters of C_5H_{12} are needed to make 97.3 mL C_5H_8?

4 How many milliliters of H_2 can be made from 1.98×10^3 mL C_5H_{12}?

5

SKILLS Toolkit

Solving Particle Problems

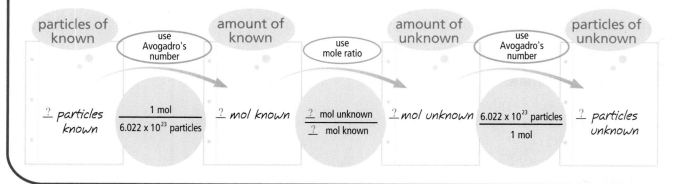

Topic Link

Refer to the chapter "The Mole and Chemical Composition" for more information about Avogadro's number and molar mass.

For Number of Particles, Use Avogadro's Number

Skills Toolkit 5 shows how to use Avogadro's number, 6.022×10^{23} particles/mol, in stoichiometry problems. If you are given particles and asked to find particles, Avogadro's number cancels out! For this calculation you use only the coefficients from the balanced equation. In effect, you are interpreting the equation in terms of the number of particles again.

SAMPLE PROBLEM D

Problems Involving Particles

How many grams of C_5H_8 form from 1.89×10^{24} molecules C_5H_{12}?

$$C_5H_{12}(l) \longrightarrow C_5H_8(l) + 2H_2(g)$$

PRACTICE HINT

Expect more problems like this one that do not exactly follow any single **Skills Toolkit** in this chapter. These problems will combine steps from one or more problems, but all will still use the mole ratio as the key step.

1 **Gather information.**
- quantity of $C_5H_{12} = 1.89 \times 10^{24}$ molecules
- Avogadro's number = 6.022×10^{23} molecules/mol
- mass of C_5H_8 = ? g C_5H_8
- molar mass of C_5H_8 = 68.13 g/mol
- From the balanced equation: 1 mol C_5H_{12} = 1 mol C_5H_8.

2 **Plan your work.**
Set up the problem using Avogadro's number to change to moles, then use the mole ratio, and finally use the molar mass of C_5H_8 to change to grams.

3 **Calculate.**

$$? \text{ g } C_5H_8 = 1.89 \times 10^{24} \text{ molecules } C_5H_{12} \times \frac{1 \text{ mol } C_5H_{12}}{6.022 \times 10^{23} \text{ molecules } C_5H_{12}} \times$$

$$\frac{1 \text{ mol } C_5H_8}{1 \text{ mol } C_5H_{12}} \times \frac{68.13 \text{ g } C_5H_8}{1 \text{ mol } C_5H_8} = 214 \text{ g } C_5H_8$$

4 **Verify your result.**
The units cancel correctly, and estimating gives 210.

Use the equation provided to answer the questions that follow.

$$Br_2(l) + 5F_2(g) \longrightarrow 2BrF_5(l)$$

1 How many molecules of BrF_5 form when 384 g Br_2 react with excess F_2?

2 How many molecules of Br_2 react with 1.11×10^{20} molecules F_2?

Many Problems, Just One Solution

Although you could be given many different problems, the solution boils down to just three steps. Take whatever you are given, and find a way to change it into moles. Then, use a mole ratio from the balanced equation to get moles of the second substance. Finally, find a way to convert the moles into the units that you need for your final answer.

 # Section Review

UNDERSTANDING KEY IDEAS

1. What conversion factor is present in almost all stoichiometry calculations?

2. For a given substance, what information links mass to moles? number of particles to moles?

3. What conversion factor will change moles CO_2 to grams CO_2? moles H_2O to molecules H_2O?

PRACTICE PROBLEMS

4. Use the equation below to answer the questions that follow.

$$Br_2 + Cl_2 \longrightarrow 2BrCl$$

a. How many moles of BrCl form when 2.74 mol Cl_2 react with excess Br_2?

b. How many grams of BrCl form when 239.7 g Cl_2 react with excess Br_2?

c. How many grams of Br_2 are needed to react with 4.53×10^{25} molecules Cl_2?

5. The equation for burning C_2H_2 is

$$2C_2H_2(g) + 5O_2(g) \longrightarrow 4CO_2(g) + 2H_2O(g)$$

a. If 15.9 L C_2H_2 react at STP, how many moles of CO_2 are produced? (Hint: At STP, 1 mol = 22.41 L for any gas.)

b. How many milliliters of CO_2 (density = 1.977 g/L) can be made when 59.3 mL O_2 (density = 1.429 g/L) react?

CRITICAL THINKING

6. Why do you need to use amount in moles to solve stoichiometry problems? Why can't you just convert from mass to mass?

7. LiOH and NaOH can each react with CO_2 to form the metal carbonate and H_2O. These reactions can be used to remove CO_2 from the air in a spacecraft.

a. Write a balanced equation for each reaction.

b. Calculate the grams of NaOH and of LiOH that remove 288 g CO_2 from the air.

c. NaOH is less expensive per mole than LiOH. Based on your calculations, explain why LiOH is used during shuttle missions rather than NaOH.

Limiting Reactants and Percentage Yield

OBJECTIVES

1. **Identify** the limiting reactant for a reaction and use it to calculate theoretical yield.

2. **Perform** calculations involving percentage yield.

Limiting Reactants and Theoretical Yield

To drive a car, you need gasoline in the tank and oxygen from the air. When the gasoline runs out, you can't go any farther even though there is still plenty of oxygen. In other words, the gasoline limits the distance you can travel because it runs out and the reaction in the engine stops.

In the previous section, you assumed that 100% of the reactants changed into products. And that is what should happen theoretically. But in the real world, other factors, such as the amounts of all reactants, the completeness of the reaction, and product lost in the process, can limit the yield of a reaction. The analogy of assembling homecoming mums for a fund raiser, as shown in **Figure 3,** will help you understand that whatever is in short supply will limit the quantity of product made.

Figure 3
The number of mums these students can assemble will be limited by the component that runs out first.

The Limiting Reactant Forms the Least Product

The students assembling mums use one helmet, one flower, eight blue ribbons, six white ribbons, and two bells to make each mum. As a result, the students cannot make any more mums once any one of these items is used up. Likewise, the reactants of a reaction are seldom present in ratios equal to the mole ratio in the balanced equation. So one of the reactants is used up first. For example, one way to to make H_2 is

$$Zn + 2HCl \longrightarrow ZnCl_2 + H_2$$

If you combine 0.23 mol Zn and 0.60 mol HCl, would they react completely? Using the coefficients from the balanced equation, you can predict that 0.23 mol Zn can form 0.23 mol H_2, and 0.60 mol HCl can form 0.30 mol H_2. Zinc is called the **limiting reactant** because the zinc limits the amount of product that can form. The zinc is used up first by the reaction. The HCl is the **excess reactant** because there is more than enough HCl present to react with all of the Zn. There will be some HCl left over after the reaction stops.

Again, think of the mums, and look at **Figure 4.** The supplies at left are the available reactants. The products formed are the finished mums. The limiting reactant is the flowers because they are completely used up first. The ribbons, helmets, and bells are excess reactants because there are some of each of these items left over, at right. You can determine the limiting reactant by calculating the amount of product that each reactant could form. Whichever reactant would produce the least amount of product is the limiting reactant.

limiting reactant

the substance that controls the quantity of product that can form in a chemical reaction

excess reactant

the substance that is not used up completely in a reaction

Figure 4
The flowers are in short supply. They are the limiting reactant for assembling these homecoming mums.

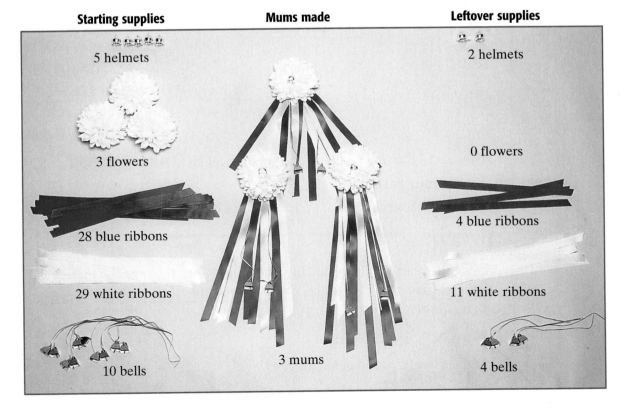

Starting supplies

5 helmets

3 flowers

28 blue ribbons

29 white ribbons

10 bells

Mums made

3 mums

Leftover supplies

2 helmets

0 flowers

4 blue ribbons

11 white ribbons

4 bells

Determine Theoretical Yield from the Limiting Reactant

So far you have done only calculations that assume reactions happen perfectly. The maximum quantity of product that a reaction could theoretically make if everything about the reaction works perfectly is called the *theoretical yield*. The theoretical yield of a reaction should always be calculated based on the limiting reactant.

In the reaction of Zn with HCl, the theoretical yield is 0.23 mol H_2 even though the HCl could make 0.30 mol H_2.

SAMPLE PROBLEM E

Limiting Reactants and Theoretical Yield

Identify the limiting reactant and the theoretical yield of phosphorous acid, H_3PO_3, if 225 g of PCl_3 is mixed with 123 g of H_2O.

$$PCl_3 + 3H_2O \longrightarrow H_3PO_3 + 3HCl$$

1 Gather information.

- mass PCl_3 = 225 g PCl_3
- mass H_2O = 123 g H_2O
- mass H_3PO_3 = ? g H_3PO_3
- molar mass PCl_3 = 137.32 g/mol
- molar mass H_2O = 18.02 g/mol
- molar mass H_3PO_3 = 82.00 g/mol
- From the balanced equation:
  1 mol PCl_3 = 1 mol H_3PO_3 and 3 mol H_2O = 1 mol H_3PO_3.

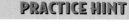

PRACTICE HINT

Whenever a problem gives you quantities of two or more reactants, you must determine the limiting reactant and use it to determine the theoretical yield.

2 Plan your work.

Set up problems that will calculate the mass of H_3PO_3 you would expect to form from each reactant.

3 Calculate.

$$? \text{ g } H_3PO_3 = 225 \text{ g } PCl_3 \times \frac{1 \text{ mol } PCl_3}{137.32 \text{ g } PCl_3} \times \frac{1 \text{ mol } H_3PO_3}{1 \text{ mol } PCl_3} \times \frac{82.00 \text{ g } H_3PO_3}{1 \text{ mol } H_3PO_3} =$$
$$134 \text{ g } H_3PO_3$$

$$? \text{ g } H_3PO_3 = 123 \text{ g } H_2O \times \frac{1 \text{ mol } H_2O}{18.02 \text{ g } H_2O} \times \frac{1 \text{ mol } H_3PO_3}{3 \text{ mol } H_2O} \times \frac{82.00 \text{ g } H_3PO_3}{1 \text{ mol } H_3PO_3} =$$
$$187 \text{ g } H_3PO_3$$

PCl_3 is the limiting reactant. The theoretical yield is 134 g H_3PO_3.

4 Verify your result.

The units of the answer are correct, and estimating gives 128.

PRACTICE

Using the reaction above, identify the limiting reactant and the theoretical yield (in grams) of HCl for each pair of reactants.

1 3.00 mol PCl_3 and 3.00 mol H_2O

2 75.0 g PCl_3 and 75.0 g H_2O

3 1.00 mol of PCl_3 and 50.0 g of H_2O

Limiting Reactants and the Food You Eat

In industry, the cheapest reactant is often used as the excess reactant. In this way, the expensive reactant is more completely used up. In addition to being cost-effective, this practice can be used to control which reactions happen. In the production of cider vinegar from apple juice, the apple juice is first kept where there is no oxygen so that the microorganisms in the juice break down the sugar, glucose, into ethanol and carbon dioxide. The resulting solution is hard cider.

Having excess oxygen in the next step allows the organisms to change ethanol into acetic acid, resulting in cider vinegar. Because the oxygen in the air is free and is easy to get, the makers of cider vinegar constantly pump air through hard cider as they make it into vinegar. Ethanol, which is not free, is the limiting reactant and is used up in the reaction.

Cost is also used to choose the excess reactant when making banana flavoring, isopentyl acetate. Acetic acid is the excess reactant because it costs much less than isopentyl alcohol.

$$CH_3COOH + C_5H_{11}OH \longrightarrow CH_3COOC_5H_{11} + H_2O$$

acetic acid + isopentyl alcohol $\longrightarrow$ isopentyl acetate + water

As shown in **Figure 5,** when compared mole for mole, isopentyl alcohol is more than twice as expensive as acetic acid. When a large excess of acetic acid is present, almost all of the isopentyl alcohol reacts.

Choosing the excess and limiting reactants based on cost is also helpful in areas outside of chemistry. In making the homecoming mums, the flower itself is more expensive than any of the other materials, so it makes sense to have an excess of ribbons and charms. The expensive flowers are the limiting reactant.

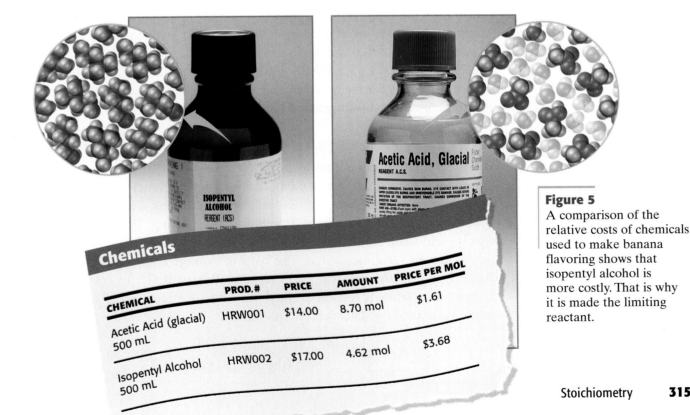

Chemicals

CHEMICAL	PROD.#	PRICE	AMOUNT	PRICE PER MOL
Acetic Acid (glacial) 500 mL	HRW001	$14.00	8.70 mol	$1.61
Isopentyl Alcohol 500 mL	HRW002	$17.00	4.62 mol	$3.68

Figure 5
A comparison of the relative costs of chemicals used to make banana flavoring shows that isopentyl alcohol is more costly. That is why it is made the limiting reactant.

Table 1 Predictions and Results for Isopentyl Acetate Synthesis

Reactants	Formula	Mass present	Amount present	Amount left over
Isopentyl alcohol	$C_5H_{11}OH$	500.0 g	5.67 mol (limiting reactant)	0.0 mol
Acetic acid	CH_3COOH	1.25×10^3 g	20.8 mol	15.1 mol

Products	Formula	Amount expected	Theoretical yield (mass expected)	Actual yield (mass produced)
Isopentyl acetate	$CH_3COOC_5H_{11}$	5.67 mol	738 g	591 g
Water	H_2O	5.67 mol	102 g	81.6 g

Actual Yield and Percentage Yield

Although equations tell you what *should* happen in a reaction, they cannot always tell you what *will* happen. For example, sometimes reactions do not make all of the product predicted by stoichiometric calculations, or the theoretical yield. In most cases, the **actual yield,** the mass of product actually formed, is less than expected. Imagine that a worker at the flavoring factory mixes 500.0 g isopentyl alcohol with 1.25×10^3 g acetic acid. The actual and theoretical yields are summarized in **Table 1.** Notice that the actual yield is less than the mass that was expected.

There are several reasons why the actual yield is usually less than the theoretical yield in chemical reactions. Many reactions do not completely use up the limiting reactant. Instead, some of the products turn back into reactants so that the final result is a mixture of reactants and products. In many cases the main product must go through additional steps to purify or separate it from other chemicals. For example, banana flavoring must be *distilled,* or isolated based on its boiling point. Solid compounds, such as sugar, must be recrystallized. Some of the product may be lost in the process. There also may be other reactions, called *side reactions*, that can use up reactants without making the desired product.

Determining Percentage Yield

The ratio relating the actual yield of a reaction to its theoretical yield is called the *percentage yield* and describes the efficiency of a reaction. Calculating a percentage yield is similar to calculating a batting average. A batter might get a hit every time he or she is at bat. This is the "theoretical yield." But no player has gotten a hit every time. Suppose a batter gets 41 hits in 126 times at bat. The batting average is 41 (the actual hits) divided by 126 (the possible hits theoretically), or 0.325. In the example described in **Table 1,** the theoretical yield for the reaction is 738 g. The actual yield is 591 g. The percentage yield is

$$\text{percentage yield} = \frac{591 \text{ g (actual yield)}}{738 \text{ g (theoretical yield)}} \times 100 = 80.1\%$$

actual yield

the measured amount of a product of a reaction

Calculating Percentage Yield

Determine the limiting reactant, the theoretical yield, and the percentage yield if 14.0 g N_2 are mixed with 9.0 g H_2, and 16.1 g NH_3 form.

$$N_2 + 3H_2 \longrightarrow 2NH_3$$

1 **Gather information.**

- mass N_2 = 14.0 g N_2
- mass H_2 = 9.0 g H_2
- theoretical yield of NH_3 = ? g NH_3
- actual yield of NH_3 = 16.1 g NH_3
- From the balanced equation:
  1 mol N_2 = 2 mol NH_3 and 3 mol H_2 = 2 mol NH_3.

- molar mass N_2 = 28.02 g/mol
- molar mass H_2 = 2.02 g/mol
- molar mass NH_3 = 17.04 g/mol

2 **Plan your work.**

Set up problems that will calculate the mass of NH_3 you would expect to form from each reactant.

3 **Calculate.**

$$? \text{ g } NH_3 = 14.0 \text{ g } N_2 \times \frac{1 \text{ mol } N_2}{28.02 \text{ g } N_2} \times \frac{2 \text{ mol } NH_3}{1 \text{ mol } N_2} \times \frac{17.04 \text{ g } NH_3}{1 \text{ mol } NH_3} = 17.0 \text{ g } NH_3$$

$$? \text{ g } NH_3 = 9.0 \text{ g } H_2 \times \frac{1 \text{ mol } H_2}{2.02 \text{ g } H_2} \times \frac{2 \text{ mol } NH_3}{3 \text{ mol } H_2} \times \frac{17.04 \text{ g } NH_3}{1 \text{ mol } NH_3} = 51 \text{ g } NH_3$$

- The smaller quantity made, 17.0 g NH_3, is the theoretical yield so the limiting reactant is N_2.
- The percentage yield is calculated:

$$\text{percentage yield} = \frac{16.1 \text{ g (actual yield)}}{17.0 \text{ g (theoretical yield)}} \times 100 = 94.7\%$$

4 **Verify your result.**

The units of the answer are correct. The percentage yield is less than 100%, so the final calculation is probably set up correctly.

PRACTICE HINT

If an amount of product actually formed is given in a problem, this is the reaction's actual yield.

PRACTICE

Determine the limiting reactant and the percentage yield for each of the following.

1 14.0 g N_2 react with 3.15 g H_2 to give an actual yield of 14.5 g NH_3.

2 In a reaction to make ethyl acetate, 25.5 g CH_3COOH react with 11.5 g C_2H_5OH to give a yield of 17.6 g $CH_3COOC_2H_5$.

$$CH_3COOH + C_2H_5OH \longrightarrow CH_3COOC_2H_5 + H_2O$$

3 16.1 g of bromine are mixed with 8.42 g of chlorine to give an actual yield of 21.1 g of bromine monochloride.

Determining Actual Yield

Although the actual yield can only be determined experimentally, a close estimate can be calculated if the percentage yield for a reaction is known. The percentage yield in a particular reaction is usually fairly consistent. For example, suppose an industrial chemist determined the percentage yield for six tries at making banana flavoring and found the results were 80.0%, 82.1%, 79.5%, 78.8%, 80.5%, and 81.9%. In the future, the chemist can expect a yield of around 80.5%, or the average of these results.

If the chemist has enough isopentyl alcohol to make 594 g of the banana flavoring theoretically, then an actual yield of around 80.5% of that, or 478 g, can be expected.

SAMPLE PROBLEM G

Calculating Actual Yield

How many grams of $CH_3COOC_5H_{11}$ should form if 4808 g are theoretically possible and the percentage yield for the reaction is 80.5%?

1 Gather information.
- theoretical yield of $CH_3COOC_5H_{11}$ = 4808 g $CH_3COOC_5H_{11}$
- actual yield of $CH_3COOC_5H_{11}$ = ? g $CH_3COOC_5H_{11}$
- percentage yield = 80.5%

2 Plan your work.

Use the percentage yield and the theoretical yield to calculate the actual yield expected.

PRACTICE HINT

The actual yield should always be less than the theoretical yield. A wrong answer that is greater than the theoretical yield can result if you accidentally reverse the actual and theoretical yields.

3 Calculate.

$$80.5\% = \frac{\text{actual yield}}{4808 \text{ g}} \times 100$$

actual yield = 4808 g × 0.805 = 3.87×10^3 g $CH_3COOC_5H_{11}$

4 Verify your result.

The units of the answer are correct. The actual yield is less than the theoretical yield, as it should be.

PRACTICE

1 The percentage yield of NH_3 from the following reaction is 85.0%. What actual yield is expected from the reaction of 1.00 kg N_2 with 225 g H_2?

$$N_2 + 3H_2 \longrightarrow 2NH_3$$

2 If the percentage yield is 92.0%, how many grams of CH_3OH can be made by the reaction of 5.6×10^3 g CO with 1.0×10^3 g H_2?

$$CO + 2H_2 \longrightarrow CH_3OH$$

3 Suppose that the percentage yield of BrCl is 90.0%. How much BrCl can be made by reacting 338 g Br_2 with 177 g Cl_2?

Section Review

UNDERSTANDING KEY IDEAS

1. Distinguish between limiting reactant and excess reactant in a chemical reaction.

2. How do manufacturers decide which reactant to use in excess in a chemical reaction?

3. How do you calculate the percentage yield of a chemical reaction?

4. Give two reasons why a 100% yield is not obtained in actual chemical manufacturing processes.

5. How do the values of the theoretical and actual yields generally compare?

PRACTICE PROBLEMS

6. A chemist reacts 8.85 g of iron with an excess of hydrogen chloride to form hydrogen gas and iron(II) chloride. Calculate the theoretical yield and the percentage yield of hydrogen if 0.27 g H_2 are collected.

7. Use the chemical reaction below to answer the questions that follow.

$$P_4O_{10} + H_2O \longrightarrow H_3PO_4$$

a. Balance the equation.

b. Calculate the theoretical yield if 100.0 g P_4O_{10} react with 200.0 g H_2O.

c. If the actual mass recovered is 126.2 g H_3PO_4, what is the percentage yield?

8. Titanium dioxide is used as a white pigment in paints. If 3.5 mol $TiCl_4$ reacts with 4.5 mol O_2, which is the limiting reactant? How many moles of each product are produced? How many moles of the excess reactant remain?

$$TiCl_4 + O_2 \longrightarrow TiO_2 + 2Cl_2$$

9. If 1.85 g Al reacts with an excess of copper(II) sulfate and the percentage yield of Cu is 56.6%, what mass of Cu is produced?

10. Quicklime, CaO, can be prepared by roasting limestone, $CaCO_3$, according to the chemical equation below. When 2.00×10^3 g of $CaCO_3$ are heated, the actual yield of CaO is 1.05×10^3 g. What is the percentage yield?

$$CaCO_3(s) \longrightarrow CaO(s) + CO_2(g)$$

11. Magnesium powder reacts with steam to form magnesium hydroxide and hydrogen gas.

a. Write a balanced equation for this reaction.

b. What is the percentage yield if 10.1 g Mg reacts with an excess of water and 21.0 g $Mg(OH)_2$ is recovered?

c. If 24 g Mg is used and the percentage yield is 95%, how many grams of magnesium hydroxide should be recovered?

12. Use the chemical reaction below to answer the questions that follow.

$$CuO(s) + H_2(g) \longrightarrow Cu(s) + H_2O(g)$$

a. What is the limiting reactant when 19.9 g CuO react with 2.02 g H_2?

b. The actual yield of copper was 15.0 g. What is the percentage yield?

c. How many grams of Cu can be collected if 20.6 g CuO react with an excess of hydrogen with a yield of 91.0%?

CRITICAL THINKING

13. A chemist reacts 20 mol H_2 with 20 mol O_2 to produce water. Assuming all of the limiting reactant is converted to water in the reaction, calculate the amount of each substance present after the reaction.

14. A pair of students performs an experiment in which they collect 27 g CaO from the decomposition of 41 g $CaCO_3$. Are these results reasonable? Explain your answer using percentage yield.

 Stoichiometry and Cars

OBJECTIVES

(1) **Relate** volume calculations in stoichiometry to the inflation of automobile safety air bags.

(2) **Use** the concept of limiting reactants to explain why fuel-air ratios affect engine performance.

(3) **Compare** the efficiency of pollution-control mechanisms in cars using percentage yield.

Stoichiometry and Safety Air Bags

So far you have examined stoichiometry in a number of chemical reactions, including making banana flavoring and ammonia. Now it is time to look at stoichiometry in terms of something a little more familiar— a car. Stoichiometry is important in many aspects of automobile operation and safety. First, let's look at how stoichiometry can help keep you safe should you ever be in an accident. Air bags have saved the lives of many people involved in accidents. And the design of air bags requires an understanding of stoichiometry.

An Air Bag Could Save Your Life

Air bags are designed to protect people in a car from being hurt during a high-speed collision. When inflated, air bags slow the motion of a person so that he or she does not strike the steering wheel, windshield, or dashboard with as much force.

Stoichiometry is used by air-bag designers to ensure that air bags do not underinflate or overinflate. Bags that underinflate do not provide enough protection, and bags that overinflate can cause injury by bouncing the person back with too much force. Therefore, the chemicals must be present in just the right proportions. To protect riders, air bags must inflate within one-tenth of a second after impact. The basic components of most systems that make an air bag work are shown in **Figure 6.** A front-end collision transfers energy to a crash sensor that causes an igniter to fire. The igniter provides the energy needed to start a very fast reaction that produces gas in a mixture called the *gas generant.* The igniter also raises the temperature and pressure within the inflator (a metal vessel) so that the reaction happens fast enough to fill the bag before the rider strikes it. A high-efficiency filter keeps the hot reactants and the solid products away from the rider, and additional chemicals are used to make the products safer.

Air-Bag Design Depends on Stoichiometric Precision

The materials used in air bags are constantly being improved to make air bags safer and more effective. Many different materials are used. One of the first gas generants used in air bags is still in use in some systems. It is a solid mixture of sodium azide, NaN_3, and an oxidizer. The gas that inflates the bag is almost pure nitrogen gas, N_2, which is produced in the following decomposition reaction.

$$2NaN_3(s) \longrightarrow 2Na(s) + 3N_2(g)$$

However, this reaction does not inflate the bag enough, and the sodium metal is dangerously reactive. Oxidizers such as ferric oxide, Fe_2O_3, are included, which react rapidly with the sodium. Energy is released, which heats the gas and causes the gas to expand and fill the bag.

$$6Na(s) + Fe_2O_3(s) \longrightarrow 3Na_2O(s) + 2Fe(s) + \text{energy}$$

One product, sodium oxide, Na_2O, is extremely corrosive. Water vapor and CO_2 from the air react with it to form less harmful $NaHCO_3$.

$$Na_2O(s) + 2CO_2(g) + H_2O(g) \longrightarrow 2NaHCO_3(s)$$

The mass of gas needed to fill an air bag depends on the density of the gas. Gas density depends on temperature. To find the amount of gas generant to put into each system, designers must know the stoichiometry of the reactions and account for changes in temperature and thus the density of the gas.

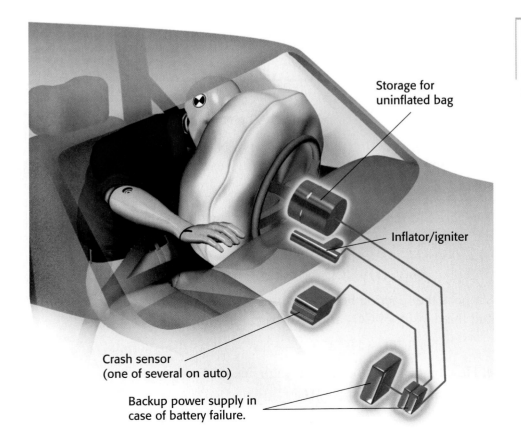

Storage for
uninflated bag

Inflator/igniter

Crash sensor
(one of several on auto)

Backup power supply in
case of battery failure.

Figure 6
Inflating an air bag requires a rapid series of events, eventually producing nitrogen gas to inflate the air bag.

SAMPLE PROBLEM H

Air-Bag Stoichiometry

Assume that 65.1 L N_2 inflates an air bag to the proper size. What mass of NaN_3 must be used? (density of N_2 = 0.92 g/L)

1 Gather information.

- Write a balanced chemical equation

$$2NaN_3(s) \longrightarrow 2Na(s) + 3N_2(g)$$

- volume of N_2 = 65.1 L N_2
- density of N_2 = 0.92 g/L
- molar mass of N_2 = 28.02 g/mol
- mass of reactant = ? g NaN_3
- molar mass of NaN_3 = 65.02 g/mol
- From the balanced equation: 2 mol NaN_3 = 3 mol N_2.

2 Plan your work.

Start with the volume of N_2, and change it to moles using density and molar mass. Then use the mole ratio followed by the molar mass of NaN_3.

3 Calculate.

$$? \text{ g NaN}_3 = 65.1 \text{ L } N_2 \times \frac{0.92 \text{ g } N_2}{1 \text{ L } N_2} \times \frac{1 \text{ mol } N_2}{28.02 \text{ g } N_2} \times$$

$$\frac{2 \text{ mol NaN}_3}{3 \text{ mol } N_2} \times \frac{65.02 \text{ g NaN}_3}{1 \text{ mol NaN}_3} = 93 \text{ g NaN}_3$$

4 Verify your result.

The number of significant figures is correct. Estimating gives 90.

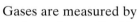

PRACTICE HINT

Gases are measured by volume, just as liquids are. In problems with volume, you can use the density to convert to mass and the molar mass to convert to moles. Then use the mole ratio, just as in any other stoichiometry problem.

PRACTICE

1 How many grams of Na form when 93 g NaN_3 react?

2 The Na formed during the breakdown of NaN_3 reacts with Fe_2O_3. How many grams of Fe_2O_3 are needed to react with 35.3 g Na?

$$6Na(s) + Fe_2O_3(s) \longrightarrow 3Na_2O(s) + 2Fe(s)$$

3 The Na_2O formed in the above reaction is made less harmful by the reaction below. How many grams of $NaHCO_3$ are made from 44.7 g Na_2O?

$$Na_2O(s) + 2CO_2(g) + H_2O(g) \longrightarrow 2NaHCO_3(s)$$

4 Suppose the reaction below was used to fill a 65.1 L air bag with CO_2 and the density of CO_2 at the air bag temperature is 1.35 g/L.

$$NaHCO_3 + HC_2H_3O_2 \longrightarrow NaC_2H_3O_2 + CO_2 + H_2O$$

a. How many grams of $NaHCO_3$ are needed?

b. How many grams of $HC_2H_3O_2$ are needed?

Stoichiometry and Engine Efficiency

The efficiency of a car's engine depends on having the correct stoichiometric ratio of gasoline and oxygen. Although gasoline used in automobiles is a mixture, it can be treated as if it were pure isooctane, one of the many compounds whose formula is C_8H_{18}. (This compound has a molar mass that is about the same as the weighted average of the compounds in actual gasoline.) The other reactant in gasoline combustion is oxygen, which is about 21% of air by volume. The reaction for gasoline combustion can be written as follows.

$$2C_8H_{18}(g) + 25O_2(g) \longrightarrow 16CO_2(g) + 18H_2O(g)$$

Engine Efficiency Depends on Reactant Proportions

For efficient combustion, the above two reactants must be mixed in a mole ratio that is close to the one shown in the balanced chemical equation, that is 2:25, or 1:12.5. If there is not enough of either reactant, the engine might stall. For example, if you pump the gas pedal too many times before starting, the mixture of reactants in the engine will contain an excess of gasoline, and the lack of oxygen may prevent the mixture from igniting. This is referred to as "flooding the engine." On the other hand, if there is too much oxygen and not enough gasoline, the engine will stall just as if the car were out of gas.

Although the best stoichiometric mixture of fuel and oxygen is 1:12.5 in terms of moles, this is not the best mixture to use all the time. **Figure 7** shows a model of a carburetor controlling the fuel-oxygen ratio in an engine that is starting, idling, and running at normal speeds. Carburetors are often used in smaller engines, such as those in lawn mowers. Computer-controlled fuel injectors have taken the place of carburetors in car engines.

Figure 7
The fuel-oxygen ratio changes depending on what the engine is doing.

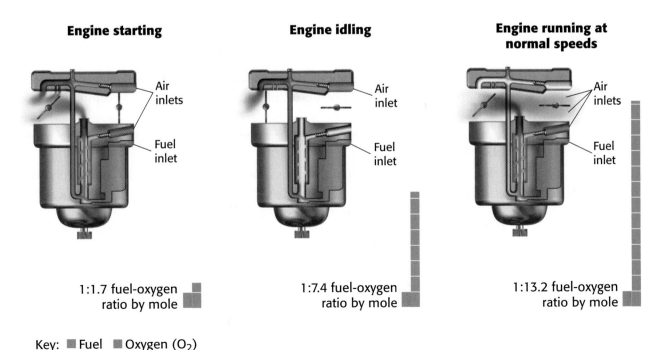

Engine starting	Engine idling	Engine running at normal speeds
Air inlets	Air inlet	Air inlets
Fuel inlet	Fuel inlet	Fuel inlet
1:1.7 fuel-oxygen ratio by mole	1:7.4 fuel-oxygen ratio by mole	1:13.2 fuel-oxygen ratio by mole

Key: ■ Fuel ■ Oxygen (O_2)

Air-Fuel Ratio

A cylinder in a car's engine draws in 0.500 L of air. How many milliliters of liquid isooctane should be injected into the cylinder to completely react with the oxygen present? The density of isooctane is 0.692 g/mL, and the density of oxygen is 1.33 g/L. Air is 21% oxygen by volume.

1 **Gather information.**

- Write a balanced equation for the chemical reaction.

$$2C_8H_{18} + 25O_2 \longrightarrow 16CO_2 + 18H_2O$$

- volume of air = 0.500 L air
- percentage of oxygen in air: 21% by volume
- Organize the data in a table.

Reactant	Formula	Molar mass	Density	Volume
Oxygen	O_2	32.00 g/mol	1.33 g/L	? L
Isooctane	C_8H_{18}	114.26 g/mol	0.692 g/mL	? mL

- From the balanced equation: 2 mol C_8H_{18} = 25 mol O_2.

2 **Plan your work.**

Use the percentage by volume of O_2 in air to find the volume of O_2. Then set up a volume-volume problem.

3 **Calculate.**

$$? \text{ mL } C_8H_{18} = 0.500 \text{ L air} \times \frac{21 \text{ L } O_2}{100 \text{ L air}} \times \frac{1.33 \text{ g } O_2}{1 \text{ L } O_2} \times \frac{1 \text{ mol } O_2}{32.00 \text{ g } O_2} \times$$

$$\frac{2 \text{ mol } C_8H_{18}}{25 \text{ mol } O_2} \times \frac{114.26 \text{ g } C_8H_{18}}{1 \text{ mol } C_8H_{18}} \times \frac{1 \text{ mL } C_8H_{18}}{0.692 \text{ g } C_8H_{18}} = 5.76 \times 10^{-2} \text{ mL } C_8H_{18}$$

4 **Verify your result.**

The denominator is about 10 times larger than the numerator, so the answer in mL should be about one-tenth of the original volume in L.

PRACTICE

1 A V-8 engine has eight cylinders each having a 5.00×10^2 cm³ capacity. How many cycles are needed to completely burn 1.00 mL of isooctane? (One cycle is the firing of all eight cylinders.)

2 How many milliliters of isooctane are burned during 25.0 cycles of a V-6 engine having six cylinders each having a 5.00×10^2 cm³ capacity?

3 Methyl alcohol, CH_3OH, with a density of 0.79 g/mL, can be used as fuel in race cars. Calculate the volume of air needed for the complete combustion of 51.0 mL CH_3OH.

Table 2 Clean Air Act Standards for 1996 Air Pollution

Pollutant	Cars	Light trucks	Motorcycles
Hydrocarbons	0.25 g/km	0.50 g/km	5.0 g/km
Carbon monoxide	2.1 g/km	2.1–3.1 g/km, depending on truck size	12 g/km
Oxides of nitrogen (NO, NO_2)	0.25 g/km	0.25–0.68 g/km, depending on truck size	not regulated

Stoichioimetry and Pollution Control

Automobiles are the primary source of air pollution in many parts of the world. The Clean Air Act was enacted in 1968 to address the issue of smog and other forms of pollution caused by automobile exhaust. This act has been amended to set new, more restrictive emission-control standards for automobiles driven in the United States. **Table 2** lists the standards for pollutants in exhaust set in 1996 by the U.S. Environmental Protection Agency.

The Fuel-Air Ratio Influences the Pollutants Formed

The equation for the combustion of "isooctane" shows most of what happens when gasoline burns, but it does not tell the whole story. For example, if the fuel-air mixture does not have enough oxygen, some carbon monoxide will be produced instead of carbon dioxide. When a car is started, there is less air, so fairly large amounts of carbon monoxide are formed, and some unburned fuel (hydrocarbons) also comes out in the exhaust. In cold weather, an engine needs more fuel to start, so larger amounts of unburned hydrocarbons and carbon monoxide come out as exhaust. These hydrocarbons are involved in forming smog. So the fuel-air ratio is a key factor in determining how much pollution forms.

Another factor in auto pollution is the reaction of nitrogen and oxygen at the high temperatures inside the engine to form small amounts of highly reactive nitrogen oxides, including NO and NO_2.

$$N_2(g) + O_2(g) \longrightarrow 2NO(g)$$

$$2NO(g) + O_2(g) \longrightarrow 2NO_2(g)$$

One of the Clean Air Act standards limits the amount of nitrogen oxides that a car can emit. These compounds react with oxygen to form another harmful chemical, ozone, O_3.

$$NO_2(g) + O_2(g) \longrightarrow 2NO(g) + O_3(g)$$

Because these reactions are started by energy from the sun's ultraviolet light, they form what is referred to as *photochemical smog*. The harmful effects of photochemical smog are caused by very small concentrations of pollutants, including unburned hydrocarbon fuel.

🗗 internet connect

www.scilinks.org
Topic: Air Pollution
SciLinks code: HW4133

SCi LINKS
Maintained by the
National Science
Teachers Association

Meeting the Legal Limits Using Stoichiometry

Automobile manufacturers use stoichiometry to predict when adjustments will be necessary to keep exhaust emissions within legal limits. Because the units in **Table 2** are *grams per kilometer*, auto manufacturers must consider how much fuel the vehicle will burn to move a certain distance. Automobiles with better gas mileage will use less fuel per kilometer, resulting in lower emissions per kilometer.

Catalytic Converters Can Help

All cars that are currently manufactured in the United States are built with catalytic converters, like the one shown in **Figure 8,** to treat the exhaust gases before they are released into the air. Platinum, palladium, or rhodium in these converters act as catalysts and increase the rate of the decomposition of NO and of NO_2 into N_2 and O_2, harmless gases already found in the atmosphere. Catalytic converters also speed the change of CO into CO_2 and the change of unburned hydrocarbons into CO_2 and H_2O. These hydrocarbons are involved in the formation of ozone and smog, so it is important that unburned fuel does not come out in the exhaust.

Catalytic converters perform at their best when the exhaust gases are hot and when the ratio of fuel to air in the engine is very close to the proper stoichiometric ratio. Newer cars include on-board computers and oxygen sensors to make sure the proper fuel-air ratio is automatically maintained, so that the engine and the catalytic converter work at top efficiency.

Figure 8
Catalytic converters are used to decrease nitrogen oxides, carbon monoxide, and hydrocarbons in exhaust. Leaded gasoline and extreme temperatures decrease their effectiveness.

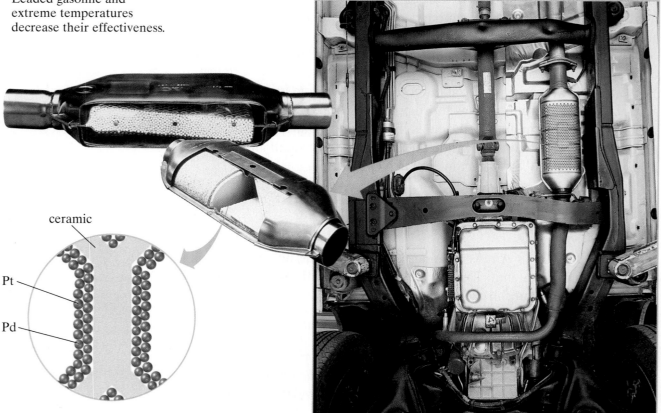

ceramic

Pt

Pd

Calculating Yields: Pollution

What mass of ozone, O_3, can be produced from 3.50 g of NO_2 contained in a car's exhaust? The equation is as follows.

$$NO_2(g) + O_2(g) \longrightarrow NO(g) + O_3(g)$$

1 Gather information.

- mass of NO_2 = 3.50 g NO_2
- mass of O_3 = ? g O_3
- From the balanced equation: 1 mol NO_2 = 1 mol O_3.
- molar mass of NO_2 = 46.01 g/mol
- molar mass of O_3 = 48.00 g/mol

2 Plan your work.

This is a mass-mass problem.

3 Calculate.

$$? \text{ g } O_3 = 3.50 \text{ g } NO_2 \times \frac{1 \text{ mol } NO_2}{46.01 \text{ g } NO_2} \times \frac{1 \text{ mol } O_3}{1 \text{ mol } NO_2} \times \frac{48.00 \text{ g } O_3}{1 \text{ mol } O_3} = 3.65 \text{ g } O_3$$

4 Verify your result.

The denominator and numerator are almost equal, so the mass of product is almost the same as the mass of reactant.

PRACTICE

1 A catalytic converter combines 2.55 g CO with excess O_2. What mass of CO_2 forms?

③ Section Review

UNDERSTANDING KEY IDEAS

1. What is the main gas in an air bag that is inflated using the NaN_3 reaction?

2. How do you know that the correct mole ratio of isooctane to oxygen is 1:12.5?

3. What do the catalysts in the catalytic converters accomplish?

4. Give at least two results of too little air being in a running engine.

PRACTICE PROBLEMS

5. Assume that 22.4 g of NaN_3 react completely in an air bag. What mass of Na_2O is produced after complete reaction of the Na with Fe_2O_3?

6. Na_2O eventually reacts with CO_2 and H_2O to form $NaHCO_3$. What mass of $NaHCO_3$ is formed when 44.4 g Na_2O completely react?

CRITICAL THINKING

7. Why are nitrogen oxides in car exhaust, even though there is no nitrogen in the fuel?

8. Why not use the following reaction to produce N_2 in an air bag?

$$NH_3(g) + O_2(g) \longrightarrow N_2(g) + H_2O(g)$$

9. Just after an automobile is started, you see water dripping off the end of the tail pipe. Is this normal? Why or why not?

CHAPTER HIGHLIGHTS

SECTION ONE Calculating Quantities in Reactions

- Reaction stoichiometry compares the amounts of substances in a chemical reaction.

- Stoichiometry problems involving reactions can always be solved using mole ratios.

- Stoichiometry problems can be solved using three basic steps. First, change what you are given into moles. Second, use a mole ratio based on a balanced chemical equation. Third, change to the units needed for the answer.

SECTION TWO Limiting Reactants and Percentage Yield

- The limiting reactant is a reactant that is consumed completely in a reaction.

- The theoretical yield is the amount of product that can be formed from a given amount of limiting reactant.

- The actual yield is the amount of product collected from a real reaction.

- Percentage yield is the actual yield divided by the theoretical yield multiplied by 100. It is a measure of the efficiency of a reaction.

SECTION THREE Stoichiometry and Cars

- Stoichiometry is used in designing air bags for passenger safety.

- Stoichiometry is used to maximize a car's fuel efficiency.

- Stoichiometry is used to minimize the pollution coming from the exhaust of an auto.

KEY TERMS

stoichiometry

**limiting reactant
excess reactant
actual yield**

KEY SKILLS

Using Mole Ratios
Skills Toolkit 1 p. 303
Sample Problem A p. 304

Solving Stoichiometry Problems
Skills Toolkit 2 p. 305

Problems Involving Mass
Skills Toolkit 3 p. 306
Sample Problem B p. 307

Problems Involving Volume
Skills Toolkit 4 p. 308
Sample Problem C p. 309

Problems Involving Particles
Skills Toolkit 5 p. 310
Sample Problem D p. 310

Limiting Reactants and Theoretical Yield
Sample Problem E p. 314

Calculating Percentage Yield
Sample Problem F p. 317

Calculating Actual Yield
Sample Problem G p. 318

Air-Bag Stoichiometry
Sample Problem H p. 322

Air-Fuel Ratio
Sample Problem I p. 324

Calculating Yields: Pollution
Sample Problem J p. 327

USING KEY TERMS

1. Define *stoichiometry*.

2. Compare the limiting reactant and the excess reactant for a reaction.

3. Compare the actual yield and the theoretical yield from a reaction.

4. How is percentage yield calculated?

5. Why is the term *limiting* used to describe the limiting reactant?

UNDERSTANDING KEY IDEAS

Calculating Quantities in Reactions

6. Why is it necessary to use mole ratios in solving stoichiometry problems?

7. What is the key conversion factor needed to solve all stoichiometry problems?

8. When you multiply by a mole ratio, the unit of mol does not change. What is the purpose of using the mole ratio?

9. Why is a balanced chemical equation needed to solve stoichiometry problems?

10. Use the balanced equation below to write mole ratios for the situations that follow.

$$2H_2(g) + O_2(g) \longrightarrow 2H_2O(g)$$

a. calculating mol H_2O given mol H_2
b. calculating mol O_2 given mol H_2O
c. calculating mol H_2 given mol O_2

11. What are two conversions in stoichiometry problems for which you use the molar mass of a substance?

12. Why can density be used in solving stoichiometry problems involving volume?

13. Write the conversion factor needed to convert from g O_2 to L O_2 if the density of O_2 is 1.429 g/L.

14. What conversion factor is used to convert from volume of a gas directly to moles at STP?

15. How is Avogadro's number used as a conversion factor in solving problems involving number of particles?

16. A student uses the equation as written below to calculate the amount of O_2 that reacts with a given amount of CO. Is it likely that the student would get the correct answer? Explain.

$$CO(g) + O_2(g) \longrightarrow CO_2(g)$$

17. Describe a general plan for solving all stoichiometry problems in three steps.

Limiting Reactants and Percentage Yield

18. Explain why cost is often a major factor in choosing a limiting reactant.

19. Give two reasons why the actual yield from chemical reactions is less than 100%.

20. Describe the relationship between the limiting reactant and the theoretical yield.

21. Identify the limiting reactant and the excess reactant in the following situations:
a. firewood burning in a campfire
b. stomach acid reacting with a tablet of $Mg(OH)_2$
c. iron in a nail rusting in the presence of oxygen and water
d. NO_2 gas reacting with oxygen and water vapor in air to produce acid precipitation

Stoichiometry and Cars

22. What are three areas of a car's operation or design that depend on stoichiometry?

23. Describe what might happen if too much or too little gas generant is used in an air bag.

24. Why is the ratio of fuel to air in a car's engine important in controlling pollution?

25. Under what conditions will exhaust from a car's engine contain high levels of carbon monoxide?

26. What is the function of the catalytic converter in the exhaust system?

PRACTICE PROBLEMS

Sample Problem A Using Mole Ratios

27. The chemical equation for the formation of water is

$$2H_2 + O_2 \longrightarrow 2H_2O$$

a. If 3.3 mol O_2 are used, how many moles of H_2 are needed?

b. How many moles O_2 must react with excess H_2 to form 6.72 mol H_2O?

c. If you wanted to make 8.12 mol H_2O, how many moles of H_2 would you need?

28. The reaction between hydrazine, N_2H_4, and dinitrogen tetroxide is sometimes used in rocket propulsion. Balance the equation below, then use it to answer the following questions.

$$N_2H_4(l) + N_2O_4(l) \longrightarrow N_2(g) + H_2O(g)$$

a. How many moles H_2O are produced as 1.22×10^3 mol N_2 are formed?

b. How many moles N_2H_4 must react with 1.45×10^3 mol N_2O_4?

c. If 2.13×10^3 mol N_2O_4 completely react, how many moles of N_2 form?

29. Aluminum reacts with oxygen to form aluminum oxide.

a. How many moles of O_2 are needed to react with 1.44 mol of aluminum?

b. How many moles of aluminum oxide can be made if 5.23 mol Al completely react?

c. If 2.98 mol O_2 react completely, how many moles of Al_2O_3 can be made?

Sample Problem B Problems Involving Mass

30. Calcium carbide, CaC_2, reacts with water to form acetylene.

$$CaC_2(s) + 2H_2O(l) \longrightarrow C_2H_2(g) + Ca(OH)_2(s)$$

a. How many grams of water are needed to react with 485 g of calcium carbide?

b. How many grams of CaC_2 could make 23.6 g C_2H_2?

c. If 55.3 g $Ca(OH)_2$ are formed, how many grams of water reacted?

31. Oxygen can be prepared by heating potassium chlorate.

$$2KClO_3(s) \longrightarrow 2KCl(s) + 3O_2(g)$$

a. What mass of O_2 can be made from heating 125 g of $KClO_3$?

b. How many grams of $KClO_3$ are needed to make 293 g O_2?

c. How many grams of KCl could form from 20.8 g $KClO_3$?

32. How many grams of aluminum oxide can be formed by the reaction of 38.8 g of aluminum with oxygen?

33. Ozone, O_3, changes into oxygen. Write the balanced equation. How many grams of oxygen can be obtained from 2.22 mol O_3?

Sample Problem C Problems Involving Volume

34. Use the equation provided to answer the questions that follow. The density of oxygen gas is 1.428 g/L.

$$2KClO_3(s) \longrightarrow 2KCl(s) + 3O_2(g)$$

a. What volume of oxygen can be made from 5.00×10^{-2} mol of $KClO_3$?

b. How many grams $KClO_3$ must react to form 42.0 mL O_2?

c. How many milliliters of O_2 will form at STP from 55.2 g $KClO_3$?

35. Hydrogen peroxide, H_2O_2, decomposes to form water and oxygen.

   a. How many liters of O_2 can be made from 342 g H_2O_2 if the density of O_2 is 1.428 g/L?

   b. The density of H_2O_2 is 1.407 g/mL, and the density of O_2 is 1.428 g/L. How many liters of O_2 can be made from 55 mL H_2O_2?

   c. How many liters of O_2 will form at STP if 22.5 g H_2O_2 react?

Sample Problem D Problems Involving Particles

36. Use the equation provided to answer the questions that follow.

$$2NO + O_2 \longrightarrow 2NO_2$$

   a. How many molecules of NO_2 can form from 1.11 mol O_2 and excess NO?

   b. How many molecules of NO will react with 25.7 g O_2?

   c. How many molecules of O_2 are needed to make 3.76×10^{22} molecules NO_2?

37. Use the equation provided to answer the questions that follow.

$$2Na + 2H_2O \longrightarrow 2NaOH + H_2$$

   a. How many molecules of H_2 could be made from 27.6 g H_2O?

   b. How many atoms of Na will completely react with 12.9 g H_2O?

   c. How many molecules of H_2 could form when 6.59×10^{20} atoms Na react?

Sample Problem E Limiting Reactants and Theoretical Yield

38. In the reaction shown below, 4.0 mol of NO is reacted with 4.0 mol O_2.

$$2NO + O_2 \longrightarrow 2NO_2$$

   a. Which is the excess reactant, and which is the limiting reactant?

   b. What is the theoretical yield, in units of mol, of NO_2?

39. In the reaction shown below, 64 g CaC_2 is reacted with 64 g H_2O.

$$CaC_2(s) + 2H_2O(l) \longrightarrow C_2H_2(g) + Ca(OH)_2(s)$$

   a. Which is the excess reactant, and which is the limiting reactant?

   b. What is the theoretical yield of C_2H_2?

   c. What is the theoretical yield of $Ca(OH)_2$?

40. In the reaction shown below, 28 g of nitrogen are reacted with 28 g of hydrogen.

$$N_2(g) + 3H_2(g) \longrightarrow 2NH_3(g)$$

   a. Which is the excess reactant, and which is the limiting reactant?

   b. What is the theoretical yield of ammonia?

   c. How many grams of the excess reactant remain?

Sample Problem F Calculating Percentage Yield

41. Reacting 991 mol of SiO_2 with excess carbon yields 30.0 kg of SiC. What is the percentage yield?

$$SiO_2 + 3C \longrightarrow SiC + 2CO$$

42. If 156 g of sodium nitrate react, and 112 g of sodium nitrite are recovered, what is the percentage yield?

$$2NaNO_3(s) \longrightarrow 2NaNO_2(s) + O_2(g)$$

43. If 185 g of magnesium are recovered from the decomposition of 1000.0 g of magnesium chloride, what is the percentage yield?

Sample Problem G Calculating Actual Yield

44. How many grams of $NaNO_2$ form when 256 g $NaNO_3$ react? The yield is 91%.

$$2NaNO_3(s) \longrightarrow 2NaNO_2(s) + O_2(g)$$

45. How many grams of Al form from 9.73 g of aluminum oxide if the yield is 91%?

$$Al_2O_3 + 3C \longrightarrow 2Al + 3CO$$

46. Iron and CO are made by heating 4.56 kg of iron ore, Fe_2O_3, and carbon. The yield of iron is 88%. How many kilograms of iron are made?

Sample Problem H Air-Bag Stoichiometry

47. Assume that 44.3 g Na_2O are formed during the inflation of an air bag. How many liters of CO_2 (density = 1.35 g/L) are needed to completely react with the Na_2O?

$$Na_2O(s) + 2CO_2(g) + H_2O(g) \longrightarrow 2NaHCO_3(s)$$

48. Assume that 59.5 L N_2 with a density of 0.92 g/L are needed to fill an air bag.

$$2NaN_3(s) \longrightarrow 2Na(s) + 3N_2(g)$$

  a. What mass of NaN_3 is needed to form this volume of nitrogen?
  b. How many liters of N_2 are actually made from 65.7 g NaN_3 if the yield is 94%?
  c. What mass of NaN_3 is actually needed to form 59.5 L N_2?

Sample Problem I Air-Fuel Ratio

49. Write a balanced equation for the combustion of octane, C_8H_{18}, with oxygen to obtain carbon dioxide and water. What is the mole ratio of oxygen to octane?

50. What mass of oxygen is required to burn 688 g of octane, C_8H_{18}, completely?

51. How many liters of O_2, density 1.43 g/L, are needed for the complete combustion of 1.00 L C_8H_{18}, density 0.700 g/mL?

Sample Problem J Calculating Yields: Pollution

52. Nitrogen dioxide from exhaust reacts with oxygen to form ozone. What mass of ozone could be formed from 4.55 g NO_2? If only 4.58 g O_3 formed, what is the percentage yield?

$$NO_2(g) + O_2(g) \longrightarrow NO(g) + O_3(g)$$

53. How many grams CO_2 form from the complete combustion of 1.00 L C_8H_{18}, density 0.700 g/mL? If only 1.90×10^3 g CO_2 form, what is the percentage yield?

MIXED REVIEW

54. The following reaction can be used to remove CO_2 breathed out by astronauts in a spacecraft.

$$2LiOH(s) + CO_2(g) \longrightarrow Li_2CO_3(s) + H_2O(l)$$

  a. How many grams of carbon dioxide can be removed by 5.5 mol LiOH?
  b. How many milliliters H_2O (density = 0.997 g/mL) could form from 25.7 g LiOH?
  c. How many molecules H_2O could be made when 3.28 g CO_2 react?

55. Use the equation below to answer the questions that follow.

$$2NaN_3(s) \longrightarrow 2Na(s) + 3N_2(g)$$

  a. Write the six mole ratios that can be obtained from this equation.
  b. How many moles N_2 can be formed from 2.7 mol NaN_3?
  c. How many grams Na can be made from 31.1 g NaN_3?
  d. How many molecules N_2 are made when 3.24×10^{23} atoms Na are formed?

56. When 55 g NO is mixed with 35 g O_2, 31 g NO_2 are recovered. The density of NO is 1.3388 g/L, and that of NO_2 is 2.053 g/L.

$$2NO(g) + O_2(g) \longrightarrow 2NO_2(g)$$

  a. Which is the limiting reactant?
  b. What is the theoretical yield from this reaction in grams NO_2?
  c. What is the percentage yield?

57. How many liters N_2, density 0.92 g/L, can be made by the decomposition of 2.05 g NaN_3?

$$2NaN_3(s) \longrightarrow 2Na(s) + 3N_2(g)$$

58. The percentage yield of nitric acid is 95%. If 9.88 kg of nitrogen dioxide react, what mass of nitric acid is isolated?

$$3NO_2(g) + H_2O(g) \longrightarrow 2HNO_3(aq) + NO(g)$$

59. If you get 25.3 mi/gal, what mass of carbon dioxide is produced by the complete combustion of C_8H_{18} if you drive 5.40 mi? (Hint: 1 gal = 3.79 L; density of octane = 0.700 g/mL)

CRITICAL THINKING

60. Nitrogen monoxide, NO, reacts with oxygen to form nitrogen dioxide. Then the nitrogen dioxide reacts with oxygen to form nitrogen monoxide and ozone. Write the balanced equations. What is the theoretical yield in grams of ozone from 4.55 g of nitrogen monoxide with excess O_2? (Hint: First calculate the theoretical yield for NO_2, then use that value to calculate the yield for ozone.)

61. Explain the stoichiometry involved in blowing air on the base of a dwindling campfire to keep the coals burning.

62. Why would it be unreasonable for an amendment to the Clean Air Act to call for 0% pollution emissions from cars with combustion engines?

63. Use stoichiometry to explain the following problems that a lawn mower may have.
   a. A lawn mower fails to start because the engine floods.
   b. A lawn mower stalls after starting cold and idling.

64. Air is inexpensive and easily available. Give some reasons why air would not be good to use to fill an air bag.

ALTERNATIVE ASSESSMENT

65. Design an experiment to measure the percentage yields for the reactions listed below. If your teacher approves your design, get the necessary materials, and carry out your plan.
   a. $Zn(s) + 2HCl(aq) \longrightarrow ZnCl_2(aq) + H_2(g)$
   b. $2NaHCO_3(s) \longrightarrow$
   $Na_2CO_3(s) + H_2O(g) + CO_2(g)$
   c. $CaCl_2(aq) + Na_2CO_3(aq) \longrightarrow$
   $CaCO_3(s) + 2NaCl(aq)$
   d. $NaOH(aq) + HCl(aq) \longrightarrow$
   $NaCl(aq) + H_2O(l)$
   (Note: use only dilute NaOH and HCl, less concentrated than 1.0 mol/L.)

66. Your teacher will give you an index card specifying a volume of a gas. Reactants to make the gas will also be listed. Describe exactly how you would make the gas from the reactants. Include a method of collecting the gas without allowing it to mix with the air. Then specify how much of each reactant you need.

67. Calculate the theoretical yield (in kg) of carbon dioxide emitted by a car in one year, assuming 1.20×10^4 mi/y, 25 mi/gal, and octane, C_8H_{18}, as the fuel, 0.700 g/mL. (1 gal = 3.79 L)

68. Ozone is helpful in the upper atmosphere, but we worry about ozone formed from auto exhaust. Research the role of auto exhaust in the formation of ozone near the surface of Earth. Determine the hazards posed by ozone at Earth's surface.

CONCEPT MAPPING

69. Use the following terms to complete the concept map below: *stoichiometry*, *excess reactant*, *theoretical yield*, and *mole ratio*.

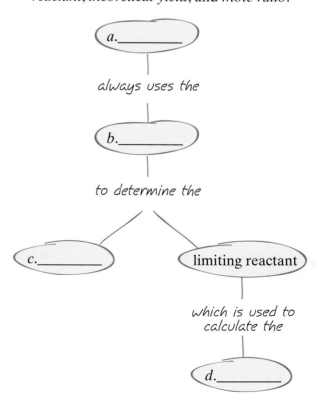

Study the graph below, and answer the questions that follow.
For help in interpreting graphs, see Appendix B, "Study Skills for Chemistry."

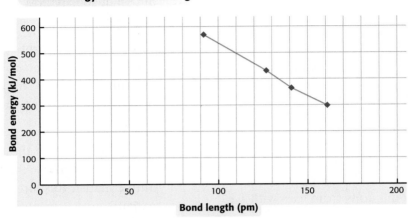

Bond Energy Versus Bond Length

70. Describe the relationship between bond length and bond energy.

71. Estimate the bond energy of a bond of length 100 pm.

72. If the trend of the graph continues, what bond length will have an energy of 200 kJ/mol?

73. The title of the graph does not provide much information about the contents of the graph. What additional information would be useful to better understand and use this graph?

TECHNOLOGY AND LEARNING

74. Graphing Calculator

Calculating Percentage Yield of a Chemical Reaction

The graphing calculator can run a program that calculates the percentage yield of a chemical reaction when you enter the actual yield and the theoretical yield. Using an example in which the actual yield is 38.8 g and the theoretical yield is 53.2 g, you will calculate the percentage yield. First, the program will carry out the calculation. Then you can use it to make other calculations.

Go to Appendix C. If you are using a TI-83 Plus, you can download the program **YIELD** and data and run the application as directed.

If you are using another calculator, your teacher will provide you with keystrokes and data sets to use. After you have run the program, answer the questions.

Note: all answers are written with three significant figures.

a. What is the percentage yield when the actual yield is 27.3 g and the theoretical yield is 44.6 g?

b. What is the percentage yield when the actual yield is 5.40 g and the theoretical yield is 9.20 g?

c. What actual yield/theoretical yield pair produced the largest percentage yield?

1. Stoichiometry problems may require the use of a
  a. table of bond energies.
  b. Lewis structure.
  c. chart of electron configurations.
  d. mole ratio.

2. In the chemical equation $A + B \longrightarrow C + D$, if you know the mass of A, you can determine
  a. the mass of any of the other reactants and products.
  b. only the mass of C and D combined.
  c. only the mass of B.
  d. only the mass of A and B combined.

3. To solve a mass-mass stoichiometry problem, you must know the
  a. coefficients of the balanced equation.
  b. phases of the reactants and products.
  c. rate at which the reaction occurs.
  d. chemical names of the reactants and products.

4. For the reaction below, how many moles of N_2 are required to produce 18 mol NH_3?

$$N_2 + 3H_2 \longrightarrow 2NH_3$$

  a. 9   **c.** 18
  b. 27   **d.** 36

5. If a chemical reaction involving substances A and B stops when B is completely used up, then B is referred to as the
  a. excess reactant.
  b. primary reactant.
  c. limiting reactant.
  d. primary product.

6. How much product is collected during a chemical reaction is called the
  a. mole ratio.   **c.** percentage yield.
  b. theoretical yield.   **d.** actual yield.

7. If a chemist calculates the maximum amount of product that might be obtained in a chemical reaction, he or she is calculating the
  a. theoretical yield.
  b. mole ratio.
  c. percentage yield.
  d. actual yield.

8. Knowing the mole ratio of a reactant and product in a chemical reaction allows one to determine
  a. the energy released in the reaction.
  b. the mass of the product produced from a known mass of reactant.
  c. the speed of the reaction.
  d. whether the reaction was reversible.

9. In stoichiometry, chemists are mainly concerned with
  a. the types of bonds found in compounds.
  b. energy changes occurring in chemical reactions.
  c. mass relationships in chemical reactions.
  d. speed with which chemical reactions occur.

10. What is the mole ratio of CO_2 to $C_6H_{12}O_6$ in the reaction below?

$$6CO_2 + 6H_2O \longrightarrow C_6H_{12}O_6 + 6O_2$$

  a. 1:2
  b. 1:1
  c. 6:1
  d. 1:4

11. In a chemical reaction, the reactant remaining after all of the limiting reactant is completely used up is referred to as the
  a. product.
  b. controlling reactant.
  c. excess reactant.
  d. catalyst.

CAUSES OF CHANGE

Achemical reaction can release or absorb energy and can increase or decrease disorder. The forest fire is a chemical reaction in which cellulose and oxygen form carbon dioxide, water, and other chemicals. This reaction also releases energy and increases disorder because the reaction generates energy as heat and breaks down the long molecules found in living trees into smaller and simpler molecules, such as carbon dioxide, CO_2, and water, H_2O.

START-UP*ACTIVITY*

Heat Exchange

SAFETY PRECAUTIONS

PROCEDURE

1. Fill a **film canister** three-fourths full of **hot water.** Insert the **thermometer apparatus** prepared by your teacher in the hot water.

2. Fill a **250 ml beaker** one-third full of **cool water.** Insert another **thermometer apparatus** in the cool water, and record the water's temperature.

3. Record the temperature of the water in the film canister. Place the film canister in the cool water. Record the temperature measured by each thermometer every 30 s.

4. When the two temperatures are nearly the same, stop and graph your data. Plot temperature versus time on the graph. Remember to write "Time" on the *x*-axis and "Temperature" on the *y*-axis.

ANALYSIS

1. How can you tell that energy is transferred? Is energy transferred to or from the hot water?

2. Predict what the final temperatures would become after a long time.

Pre-Reading Questions

① **Can a chemical reaction generate energy as heat?**

② **Name two types of energy.**

③ **What is specific heat?**

④ **Does a thermometer measure temperature or heat?**

internet connect

www.scilinks.org
Topic: U.S. National Parks
SciLinks code: HW4126

SCiLINKS. Maintained by the National Science Teachers Association

Energy Transfer

OBJECTIVES

1. **Define** *enthalpy.*

2. **Distinguish** between heat and temperature.

3. **Perform** calculations using molar heat capacity.

Energy as Heat

A sample can transfer energy to another sample. Some examples of energy transfer are the electric current in a wire, a beam of light, a moving piston, and a flame used by a welder as shown in **Figure 1.** One of the simplest ways energy is transferred is as **heat.**

Though energy has many different forms, all energy is measured in units called *joules* (J). So, the amount of energy that one sample transfers to another sample as heat is measured in joules. Energy is never created or destroyed. The amount of energy transferred from one sample must be equal to the amount of energy received by a second sample. Therefore, the total energy of the two samples remains exactly the same.

heat

the energy transferred between objects that are at different temperatures

Figure 1

A welder uses an exothermic combustion reaction to create a high-temperature flame. The iron piece then absorbs energy from the flame.

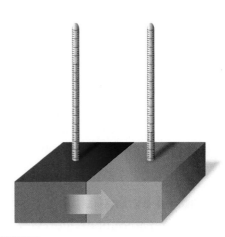

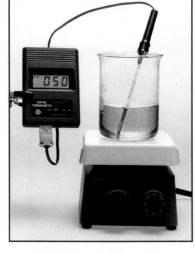

Figure 2
a Energy is always transferred from a warmer sample to a cooler sample, as the thermometers show.

b Even though both beakers receive the same amount of energy, the beakers do not have the same amount of liquid. So, the beaker on the left has a temperature of 30°C, and the beaker on the right has a temperature of 50°C.

Temperature

When samples of different temperatures are in contact, energy is transferred from the sample that has the higher **temperature** to the sample that has the lower temperature. **Figure 1** shows a welder at work; he is placing a high-temperature flame very close to a low-temperature piece of metal. The flame transfers energy as heat to the metal. The welder wants to increase the temperature of the metal so that it will begin to melt. Then, he can fuse this piece of metal with another piece of metal.

If no other process occurs, the temperature of a sample increases as the sample absorbs energy, as shown in **Figure 2a.** The temperature of a sample depends on the average kinetic energy of the sample's particles. The higher the temperature of a sample is, the faster the sample's particles move.

The temperature increase of a sample also depends on the mass of the sample. For example, the liquids in both the beakers in **Figure 2b** were initially 10.0°C, and equal quantities of energy were transferred to each beaker. The temperature increase in the beaker on the left is only about one-half of the temperature increase in the beaker on the right, because the beaker on the left has twice as much liquid in it.

Heat and Temperature are Different

You know that heat and temperature are different because you know that when two samples at different temperatures are in contact, energy can be transferred as heat. Heat and temperature differ in other ways. Temperature is an *intensive property,* which means that the temperature of a sample does not depend on the amount of the sample. However, heat is an *extensive property* which means that the amount of energy transferred as heat by a sample depends on the amount of the sample. So, water in a glass and water in a pitcher can have the same temperature. But the water in the pitcher can transfer more energy as heat to another sample because the water in the pitcher has more particles than the water in the glass.

temperature

a measure of how hot (or cold) something is; specifically, a measure of the average kinetic energy of the particles in an object

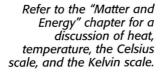
Topic Link

Refer to the "Matter and Energy" chapter for a discussion of heat, temperature, the Celsius scale, and the Kelvin scale.

Figure 3
The boiling in a kettle on a stove shows several physical and chemical processes: a combustion reaction, conduction, and a change of state.

enthalpy

the sum of the internal energy of a system plus the product of the system's volume multiplied by the pressure that the system exerts on its surroundings

Figure 3 shows a good example of the relationship between heat and temperature. The controlled combustion in the burner of a gas stove transfers energy as heat to the metal walls of the kettle. The temperature of the kettle walls increases. As a result, the hot walls of the kettle transfer energy to the cool water in the kettle. This energy transferred as heat raises the water's temperature to 100°C. The water boils, and steam exits from the kettle's spout. If the burner on the stove was turned off, the burner would no longer transfer energy to the kettle. Eventually, the kettle and the water would have equal temperatures, and the kettle would not transfer energy as heat to the water.

A Substance's Energy Can Be Measured by Enthalpy

All matter contains energy. Measuring the total amount of energy present in a sample of matter is impossible, but changes in energy content can be determined. These changes are determined by measuring the energy that enters or leaves the sample of matter. If 73 J of energy enter a piece of silver and no change in pressure occurs, we know that the enthalpy of the silver has increased by 73 J. **Enthalpy,** which is represented by the symbol H, is the total energy content of a sample. If pressure remains constant, the enthalpy increase of a sample of matter equals the energy as heat that is received. This relationship remains true even when a chemical reaction or a change of state occurs.

A Sample's Enthalpy Includes the Kinetic Energy of Its Particles

The particles in a sample are in constant motion. In other words, these particles have kinetic energy. You know that the enthalpy of a sample is the energy that a sample has. So, the enthalpy of a sample also includes the total kinetic energy of its particles.

Imagine a gold ring being cooled. As the ring transfers energy as heat to its surroundings, there is a decrease in the motions of the atoms that make up the gold ring. The kinetic energies of the atoms decrease. As the total kinetic energy decreases, the enthalpy of the ring decreases. This decrease in the kinetic energy is observed as a decrease in temperature.

You may think that all the atoms in the ring have the same kinetic energy. However, some of the atoms of the gold ring move faster than other atoms in the ring. Therefore, both the total and average kinetic energies of a substance's particles are important to chemistry, because these quantities account for every particle's kinetic energy.

What happens to the motions of the gold atoms if the ring is cooled to absolute zero ($T = 0.00$ K)? The atoms still move! However, the average and total kinetic energies of the atoms at 0.00 K are the *minimum* average and total kinetic energies these atoms can have. This idea is true of any substance and its particles. The minimum average and total kinetic energies of particles that make up a substance occur at 0.00 K.

How can the enthalpy change of a sample be calculated? Enthalpy changes can be calculated by using several different methods. The next section discusses molar heat capacity, which will be used to determine the enthalpy change of a sample.

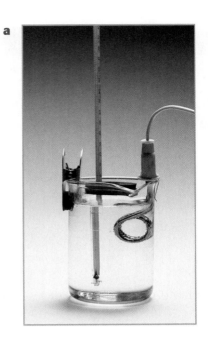

a

b

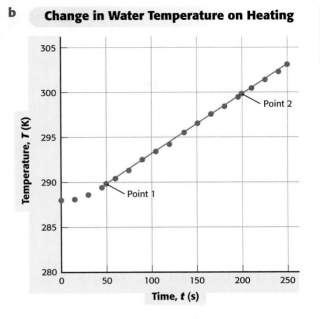

Change in Water Temperature on Heating

Temperature, *T* (K)

Point 2

Point 1

Time, *t* (s)

Figure 4
a This figure shows apparatus used for determining the molar heat capacity of water by supplying energy at a known constant rate and recording the temperature rise.
b The graph shows the data points from the experiment. The red points are not data points; they were used in the calculation of the line's slope.

Molar Heat Capacity

The *molar heat capacity* of a pure substance is the energy as heat needed to increase the temperature of 1 mol of the substance by 1 K. Molar heat capacity has the symbol C and the unit J/K•mol. Molar heat capacity is accurately measured only if no other process, such as a chemical reaction, occurs.

The following equation shows the relationship between heat and molar heat capacity, where q is the heat needed to increase the temperature of n moles of a substance by ΔT.

$$q = nC\Delta T$$

heat = (amount in moles)(molar heat capacity)(change in temperature)

Experiments and analyses that are similar to **Figure 4** determine molar heat capacity. **Figure 4a** shows 20.0 mol of water, a thermometer, and a 100 W heater in a beaker. The temperature of the water is recorded every 15 s for 250 s. The data are graphed in **Figure 4b.**

The slope of the straight line that is drawn to closely match the data points can be used to determine water's molar heat capacity. During 150 s, the interval between $t = 50$ s and $t = 200$ s, the temperature of the water increased by 9.9 K. The value of the slope is calculated below.

$$\text{slope} = \frac{y_2 - y_1}{x_2 - x_1} = \frac{\Delta T}{\Delta t} = \frac{9.9 \text{ K}}{150 \text{ s}} = 0.066 \text{ K/s}$$

To calculate the molar heat capacity of water, you need to know the heater's power rating multiplied by the amount of time the heater warmed the water. This is because watts are equal to joules per second. So, C for H_2O can be determined by using the following equation. Also notice that Δt divided by ΔT is the inverse of the slope calculated above.

$$C = \frac{q}{n\Delta T} = \frac{1.00 \times 10^2 \text{ J/s}}{n(\text{slope})} = \frac{1.00 \times 10^2 \text{ J/s}}{(20.0 \text{ mol})(0.066 \text{ K/s})} = 76 \text{ J/K•mol}$$

internet connect

www.scilinks.org
Topic : Heat Transfer
SciLinks code: HW4068

*SCi*LINKS. Maintained by the
National Science
Teachers Association

SAMPLE PROBLEM A

Calculating the Molar Heat Capacity of a Sample

Determine the energy as heat needed to increase the temperature of 10.0 mol of mercury by 7.5 K. The value of C for mercury is 27.8 J/K•mol.

PRACTICE HINT

Always convert temperatures to the Kelvin scale before carrying out calculations in this chapter. Notice that in molar heat capacity problems, you will never multiply heat by molar heat capacity. If you did multiply, the joules would not cancel.

1 Gather information.

The amount of mercury is 10.0 mol.
C for Hg = 27.8 J/K•mol
$\Delta T = 7.5$ K

2 Plan your work.

Use the values that are given in the problem and the equation $q = nC\Delta T$ to determine q.

3 Calculate.

$q = nC\Delta T$
$q = (10.0\ \cancel{mol})(27.8\ J/\cancel{K}\bullet\cancel{mol})(7.5\ \cancel{K})$
$q = 2085$ J
The answer should only have two significant figures, so it is reported as 2100 J or 2.1×10^3 J.

4 Verify your results.

The calculation yields an energy as heat with the correct unit, *joules*. This result supports the idea that the answer is the energy as heat needed to raise 10.0 mol Hg 7.5 K.

PRACTICE

PROBLEM SOLVING SKILL

1. The molar heat capacity of tungsten is 24.2 J/K•mol. Calculate the energy as heat needed to increase the temperature of 0.40 mol of tungsten by 10.0 K.

2. Suppose a sample of NaCl increased in temperature by 2.5 K when the sample absorbed 1.7×10^2 J of energy as heat. Calculate the number of moles of NaCl if the molar heat capacity is 50.5 J/K•mol.

3. Calculate the energy as heat needed to increase the temperature of 0.80 mol of nitrogen, N_2, by 9.5 K. The molar heat capacity of nitrogen is 29.1 J/K•mol.

4. A 0.07 mol sample of octane, C_8H_{18}, absorbed 3.5×10^3 J of energy. Calculate the temperature increase of octane if the molar heat capacity of octane is 254.0 J/K•mol.

Table 1 Molar Heat Capacities of Elements and Compounds

Element	C (J/K•mol)	Compound	C (J/K•mol)
Aluminum, Al(s)	24.2	Aluminum chloride, $AlCl_3(s)$	92.0
Argon, Ar(g)	20.8	Barium chloride, $BaCl_2(s)$	75.1
Helium, He(g)	20.8	Cesium iodide, CsI(s)	51.8
Iron, Fe(s)	25.1	Octane, $C_8H_{18}(l)$	254.0
Mercury, Hg(l)	27.8	Sodium chloride, NaCl(s)	50.5
Nitrogen, $N_2(g)$	29.1	Water, $H_2O(g)$	36.8
Silver, Ag(s)	25.3	Water, $H_2O(l)$	75.3
Tungsten W(s)	24.2	Water, $H_2O(s)$	37.4

Molar Heat Capacity Depends on the Number of Atoms

The molar heat capacities of a variety of substances are listed in **Table 1.** One mole of tungsten has a mass of 184 g, while one mole of aluminum has a mass of only about 27 g. So, you might expect that much more heat is needed to change the temperature of 1 mol W than is needed to change the temperature of 1 mol Al. This is not true, however. Notice that the molar heat capacities of all of the metals are nearly the same. The temperature of 1 mol of *any* solid metal is raised 1 K when the metal absorbs about 25 J of heat. The reason the temperature is raised is that the energy is absorbed by increasing the kinetic energy of the atoms in the metal, and every metal has exactly the same number of atoms in one mole.

Notice in **Table 1** that the same "about 25 joule" rule also applies to the molar heat capacities of solid ionic compounds. One mole barium chloride has three times as many ions as atoms in 1 mol of metal. So, you expect the molar heat capacity for $BaCl_2$ to be $C = 3 \times 25$ J/K•mol. The value in **Table 1,** 75.1 J/K•mol, is similar to this prediction.

Molar Heat Capacity Is Related to Specific Heat

The specific heat of a substance is represented by c_p and is the energy as heat needed to raise the temperature of one gram of substance by one kelvin. Remember that molar heat capacity of a substance, C, has a similar definition except that molar heat capacity is related to moles of a substance not to the mass of a substance. Because the molar mass is the mass of 1 mol of a substance, the following equation is true.

$$M \text{ (g/mol)} \times c_p \text{ (J/K•g)} = C \text{ (J/K•mol)}$$

(molar mass)(specific heat) = (molar heat capacity)

Topic Link

Refer to the "Matter and Energy" chapter for a discussion of specific heat.

Heat Results in Disorderly Particle Motion

When a substance receives energy in the form of heat, its enthalpy increases and the kinetic energy of the particles that make up the substance increases. The direction in which any particle moves is not related to the direction in which its neighboring particles move. The motions of these particles are *random*.

Suppose the substance was a rubber ball and you kicked the ball across a field. The energy that you gave the ball produces a different result than heat because the energy caused the particles in the ball to move together and in the same direction. The kinetic energy that you gave the particles in the ball is not random but is *concerted*.

Do you notice any relationships between energy and motion? Heat often produces disorderly particle motion. Other types of energy can produce orderly motion or orderly positioning of particles.

Section Review

UNDERSTANDING KEY IDEAS

1. What is heat?
2. What is temperature?
3. How does temperature differ from heat?
4. What is the enthalpy of a substance?
5. Define *molar heat capacity*.
6. How does molar heat capacity differ from specific heat?
7. How is the Kelvin temperature scale different from the Celsius and Fahrenheit scales?

PRACTICE PROBLEMS

8. Calculate the molar heat capacity of diamond, given that 63 J were needed to heat a 1.2 g of diamond by 1.0×10^2 K.

9. Use the molar heat capacity for aluminum from **Table 1** to calculate the amount of energy needed to raise the temperature of 260.5 g of aluminum from 0°C to 125°C.

10. Use the molar heat capacity for iron from **Table 1** to calculate the amount of energy needed to raise the temperature of 260.5 g of iron from 0°C to 125°C.

11. A sample of aluminum chloride increased in temperature by 3.5 K when the sample absorbed 1.67×10^2 J of energy. Calculate the number of moles of aluminum chloride in this sample. Use **Table 1.**

12. Use **Table 1** to determine the final temperature when 2.5×10^2 J of energy as heat is transferred to 0.20 mol of helium at 298 K.

13. Predict the final temperature when 1.2 kJ of energy as heat is transferred from 1.0×10^2 mL of water at 298 K.

14. Use **Table 1** to determine the specific heat of silver.

15. Use **Table 1** to determine the specific heat of sodium chloride.

CRITICAL THINKING

16. Why is a temperature difference the same in Celsius and Kelvin?

17. Predict the molar heat capacities of PbS(s) and Ag_2S(s).

18. Use **Table 1** to predict the molar heat capacity of $FeCl_3$(s).

19. Use your answer from item 18 to predict the specific heat of $FeCl_3$(s).

Using Enthalpy

KEY TERMS

• thermodynamics

OBJECTIVES

① **Define** *thermodynamics*.

② **Calculate** the enthalpy change for a given amount of substance for a given change in temperature.

Molar Enthalpy Change

Because enthalpy is the total energy of a system, it is an important quantity. However, the only way to measure energy is through a change. In fact, there's no way to determine the true value of *H*. But ΔH can be measured as a change occurs. The enthalpy change for one mole of a pure substance is called *molar enthalpy change*. The blacksmith in **Figure 5** is causing a molar enthalpy change by heating the iron horseshoe. Though describing a physical change by a chemical equation is unusual, the blacksmith's work could be described as follows.

$$Fe(s,\ 300\ K) \longrightarrow Fe(s,\ 1100\ K) \quad \Delta H = 20.1\ kJ/mol$$

This equation indicates that when 1 mol of solid iron is heated from 27°C to 827°C, its molar enthalpy increases by 20 100 joules.

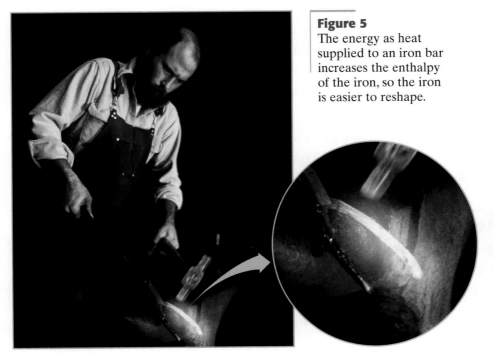

Figure 5
The energy as heat supplied to an iron bar increases the enthalpy of the iron, so the iron is easier to reshape.

Molar Heat Capacity Governs the Changes

The iron that the blacksmith uses does not change state and is not involved in a chemical reaction. So, the change in enthalpy of the iron horseshoe represents only a change in the kinetic energy of the iron atoms. When a pure substance is only heated or cooled, the amount of heat involved is the same as the enthalpy change. In other words, $\Delta H = q$ for the heating or cooling of substances. So the molar enthalpy change is related to the molar heat capacity by the following equation.

$$\text{molar enthalpy change} = C\Delta T$$
$$\text{molar enthalpy change} = (\text{molar heat capacity})(\text{temperature change})$$

Note that this equation does not apply to chemical reactions or changes of state.

SAMPLE PROBLEM B

Calculating Molar Enthalpy Change for Heating

How much does the molar enthalpy change when ice warms from $-5.4\,°C$ to $-0.2\,°C$?

1 Gather information.

$T_{\text{initial}} = -5.4\,°C = 267.8\ \text{K}$ and $T_{\text{final}} = -0.2\,°C = 273.0\ \text{K}$
For $H_2O(s)$, $C = 37.4\ \text{J/K·mol}$.

2 Plan your work.

The change in temperature is $\Delta T = T_{\text{final}} - T_{\text{initial}} = 5.2\ \text{K}$.
Because there is no reaction and the ice does not melt, you can use the equation below to determine the molar enthalpy change.

$$\Delta H = C\Delta T$$

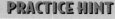

PRACTICE HINT

Remember that molar enthalpy change has units of kJ/mol.

3 Calculate.

$$\Delta H = C(\Delta T) = \left(37.4\ \frac{\text{J}}{\text{K·mol}}\right)(5.2\ \text{K}) = 1.9 \times 10^2\ \frac{\text{J}}{\text{mol}}$$

The molar enthalpy change is 0.19 kJ/mol.

4 Verify your results.

The C of ice is about 40 J/K·mol and its temperature change is about 5 K, so you should expect a molar enthalpy increase of about 200 J/mol, which is close to the calculated answer.

PRACTICE

1 Calculate the molar enthalpy change of $H_2O(l)$ when liquid water is heated from $41.7\,°C$ to $76.2\,°C$.

2 Calculate the ΔH of NaCl when it is heated from $0.0\,°C$ to $100.0\,°C$.

3 Calculate the molar enthalpy change when tungsten is heated by 15 K.

Calculating the Molar Enthalpy Change for Cooling

Calculate the molar enthalpy change when an aluminum can that has a temperature of 19.2°C is cooled to a temperature of 4.00°C.

1 Gather information.

For Al, $C = 24.2$ J/K•mol.

$T_{initial} = 19.2°C = 292$ K

$T_{final} = 4.00°C = 277$ K

2 Plan your work.

The change in temperature is calculated by using the following equation.

$$\Delta T = T_{final} - T_{initial} =$$
$$277 \text{ K} - 292 \text{ K} = -15 \text{ K}$$

To determine the molar enthalpy change, use the equation $\Delta H = C\Delta T$.

3 Calculate.

$$\Delta H = C\Delta T$$
$$\Delta H = (24.2 \text{ J/K•mol})(-15 \text{ K}) = -360 \text{ J/mol}$$

4 Verify your results.

The calculation shows the molar enthalpy change has units of joules per mole. The enthalpy value is negative, which indicates a cooling process.

PRACTICE HINT

Remember that the Δ notation always represents initial value subtracted from the final value, even if the initial value is larger than the final value.

PRACTICE

1 The molar heat capacity of Al(s) is 24.2 J/K•mol. Calculate the molar enthalpy change when Al(s) is cooled from 128.5°C to 22.6°C.

2 Lead has a molar heat capacity of 26.4 J/K•mol. What molar enthalpy change occurs when lead is cooled from 302°C to 275°C?

3 Calculate the molar enthalpy change when mercury is cooled 10 K. The molar heat capacity of mercury is 27.8 J/K•mol.

PROBLEM SOLVING SKILL

Enthalpy Changes of Endothermic or Exothermic Processes

Notice the molar enthalpy change for **Sample Problem B.** This enthalpy change is positive, which means that the heating of a sample requires energy. So, the heating of a sample is an endothermic process. In contrast, the cooling of a sample releases energy or has a negative enthalpy change and is an exothermic process, such as the process in **Sample Problem C.** In fact, you can use enthalpy changes to determine if a process is endothermic or exothermic. Processes that have positive enthalpy changes are endothermic and processes that have negative enthalpy changes are exothermic.

Enthalpy of a System of Several Substances

You have read about how a substance's enthalpy changes when the substance receives energy as heat. Enthalpy changes can be found for a system of substances, such as the reaction shown in **Figure 6.** In this figure, hydrogen gas reacts with bromine liquid to form the gas hydrogen bromide, HBr, and to generate energy as heat. Energy transfers out of this system in the form of heat because the enthalpy of the product 2HBr is less than the enthalpy of the reactants H_2 and Br_2. Or, the enthalpy of 2HBr is less than the enthalpy of H_2 and Br_2, so the enthalpy change is negative for this reaction. This negative enthalpy change reveals that the reaction is exothermic.

Enthalpy is the first of three thermodynamic properties that you will encounter in this chapter. **Thermodynamics** is a science that examines various processes and the energy changes that accompany the processes. By studying and measuring thermodynamic properties, chemists have learned to predict whether a chemical reaction can occur and what kind of energy change it will have.

thermodynamics

the branch of science concerned with the energy changes that accompany chemical and physical changes

Figure 6
When hydrogen gas and bromine liquid react, hydrogen bromide gas is formed and energy is released.

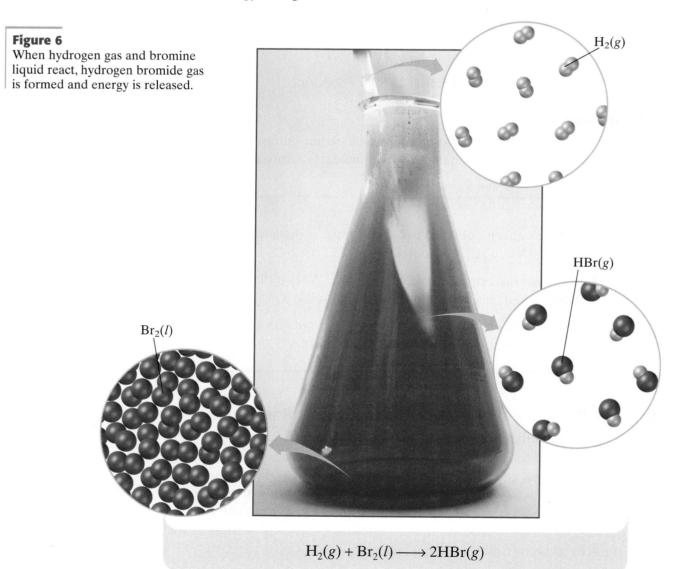

$H_2(g)$

$HBr(g)$

$Br_2(l)$

$$H_2(g) + Br_2(l) \longrightarrow 2HBr(g)$$

Writing Equations for Enthalpy Changes

Do you remember the equation that represents the molar enthalpy change when the iron horseshoe is heated?

$$\text{Fe}(s, 300 \text{ K}) \longrightarrow \text{Fe}(s, 1100 \text{ K}) \quad \Delta H = 20.1 \text{ kJ/mol}$$

Just as an equation can be written for the enthalpy change in the blacksmith's iron, an equation can be written for the enthalpy change that occurs during a change of state or a chemical reaction. The thermodynamics of changes of state are discussed in the chapter entitled "States and Intermolecular Forces." An example of an equation for a chemical reaction is the following equation for the hydrogen and bromine reaction.

$$\text{H}_2(g, 298 \text{ K}) + \text{Br}_2(l, 298 \text{ K}) \longrightarrow 2\text{HBr}(g, 298 \text{ K}) \quad \Delta H = -72.8 \text{ kJ}$$

Notice that the enthalpy change for this reaction and other chemical reactions are written using the symbol ΔH. Also, notice that the negative enthalpy change indicates the reaction is exothermic. Enthalpy changes that are involved in chemical reactions are the subject of section three of this chapter.

② Section Review

UNDERSTANDING KEY IDEAS

1. Name and define the quantity represented by H.
2. During a heating or cooling process, how are changes in enthalpy and temperature related?
3. What is thermodynamics?

PRACTICE PROBLEMS

4. A block of ice is cooled from −0.5°C to −10.1°C. Calculate the temperature change, ΔT, in degrees Celsius and in kelvins.
5. Calculate the molar enthalpy change when a block of ice is heated from −8.4°C to −5.2°C.
6. Calculate the molar enthalpy change when $\text{H}_2\text{O}(l)$ is cooled from 48.3°C to 25.2°C.

7. The molar heat capacity of benzene, $\text{C}_6\text{H}_6(l)$, is 136 J/K•mol. Calculate the molar enthalpy change when the temperature of $\text{C}_6\text{H}_6(l)$ changes from 19.7°C to 46.8°C.
8. The molar heat capacity of diethyl ether, $(\text{C}_2\text{H}_5)_2\text{O}(l)$, is 172 J/K•mol. What is the temperature change if the molar enthalpy change equals −186.9 J/mol?
9. If the enthalpy of 1 mol of a compound decreases by 428 J when the temperature decreases by 10.0 K, what is the compound's molar heat capacity?

CRITICAL THINKING

10. Under what circumstances could the enthalpy of a system be increased without the temperature rising?
11. What approximate enthalpy increase would you expect if you heated one mole of a solid metal by 40 K?

Changes in Enthalpy During Chemical Reactions

3

KEY TERMS

- calorimetry
- calorimeter
- Hess's law

OBJECTIVES

1. **Explain** the principles of calorimetry.

2. **Use** Hess's law and standard enthalpies of formation to calculate ΔH.

Topic Link

Refer to the "Science of Chemistry" chapter for a discussion of endothermic and exothermic reactions.

Changes in Enthalpy Accompany Reactions

Changes in enthalpy occur during reactions. A change in enthalpy during a reaction depends on many variables, but temperature is one of the most important variables. To standardize the enthalpies of reactions, data are often presented for reactions in which both reactants and products have the *standard thermodynamic temperature* of 25.00°C or 298.15 K.

Chemists usually present a thermodynamic value for a chemical reaction by using the chemical equation, as in the example below.

$$\tfrac{1}{2}H_2(g) + \tfrac{1}{2}Br_2(l) \longrightarrow HBr(g) \quad \Delta H = -36.4 \text{ kJ}$$

This equation shows that when 0.5 mol of H_2 reacts with 0.5 mol of Br_2 to produce 1 mol HBr and all have a temperature of 298.15 K, the enthalpy decreases by 36.4 kJ.

Remember that reactions that have negative enthalpy changes are exothermic, and reactions that have positive enthalpy changes are endothermic.

Figure 7
The combustion of wood generates energy as heat and cooks the food in the pan.

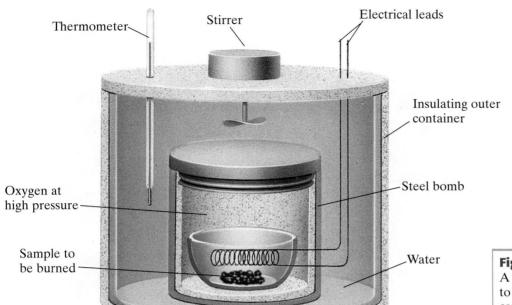

Thermometer Stirrer Electrical leads

Insulating outer container

Steel bomb

Oxygen at high pressure

Sample to be burned

Water

Figure 8
A bomb calorimeter is used to measure enthalpy changes caused by combustion reactions.

Chemical Calorimetry

For the H_2 and Br_2 reaction, in which ΔH is negative, the total energy of the reaction decreases. Energy cannot disappear, so what happens to the energy? The energy is released as heat by the system. If the reaction was endothermic, energy in the form of heat would be absorbed by the system and the enthalpy would increase.

The experimental measurement of an enthalpy change for a reaction is called **calorimetry**. Combustion reactions, such as the reaction in **Figure 7**, are always exothermic. The enthalpy changes of combustion reactions are determined using a bomb **calorimeter**, such as the one shown in **Figure 8**. This instrument is a sturdy, steel vessel in which the sample is ignited electrically in the presence of high-pressure oxygen. The energy from the combustion is absorbed by a surrounding water bath and by other parts of the calorimeter. The water and the other parts of the calorimeter have known specific heats. So, a measured temperature increase can be used to calculate the energy released in the combustion reaction and then the enthalpy change. In **Figure 7**, the combustion of 1.00 mol of carbon yields 393.5 kJ of energy.

$$C(s) + O_2(g) \longrightarrow CO_2(g) \quad \Delta H = -393.5 \text{ kJ}$$

calorimetry

the measurement of heat-related constants, such as specific heat or latent heat

calorimeter

a device used to measure the heat absorbed or released in a chemical or physical change

Nutritionists Use Bomb Calorimetry

Inside the pressurized oxygen atmosphere of a bomb calorimeter, most organic matter, including food, fabrics, and plastics, will ignite easily and burn rapidly. Some samples of matter may even explode, but the strong walls of the calorimeter contain the explosions. Sample sizes are chosen so that there is excess oxygen during the combustion reactions. Under these conditions, the reactions go to completion and produce carbon dioxide, water, and possibly other compounds.

Nutritionists, such as the nutritionist shown in **Figure 9,** use bomb calorimetry to measure the energy content of foods. To measure the energy, nutritionists assume that all the combustion energy is available to the body as we digest food. For example, consider table sugar, $C_{12}H_{22}O_{11}$, also known as sucrose. Its molar mass is 342.3 g/mol. When 3.423 grams of sugar are burned in a bomb calorimeter, the 1.505 kg of the calorimeter's water bath increased in temperature by 3.524°C. The enthalpy change can be calculated and is shown below.

$$C_{12}H_{22}O_{11}(s) + 12O_2(g) \longrightarrow 12CO_2(g) + 11H_2O(l) \quad \Delta H = -2226 \text{ kJ}$$

When enthalpy changes are reported in this way, a coefficient in the chemical equation indicates the number of moles of a substance. So, the equation above describes the enthalpy change when 1 mol of sucrose reacts with 12 mol of oxygen to produce 12 mol of carbon dioxide and 11 mol of liquid water, at 298.15 K.

Calorimetric measurements can be made with very high precision. In fact, most thermodynamic quantities are known to many significant figures.

Adiabatic Calorimetry Is Another Strategy

Instead of using a water bath to absorb the energy generated by a chemical reaction, *adiabatic calorimetry* uses an insulating vessel. The word *adiabatic* means "not allowing energy to pass through." So, no energy can enter or escape this type of vessel. As a result, the reaction mixture increases in temperature if the reaction is exothermic or decreases in temperature if the reaction is endothermic. If the system's specific heat is known, the reaction enthalpy can be calculated. Adiabatic calorimetry is used for reactions that are not ignited, such as for reactions in aqueous solution.

Hess's Law

Any two processes that both start with the same reactants in the same state and finish with the same products in the same state will have the same enthalpy change. This statement is the basis for **Hess's law,** which states that the overall enthalpy change in a reaction is equal to the sum of the enthalpy changes for the individual steps in the process.

Consider the following reaction, the synthesis of 4 mol of phosphorus pentachloride, PCl_5, when phosphorus is burned in excess chlorine.

$$P_4(s) + 10Cl_2(g) \longrightarrow 4PCl_5(g) \quad \Delta H = -1596 \text{ kJ}$$

Phosphorus pentachloride may also be prepared in a two-step process.

$$\text{Step 1: } P_4(s) + 6Cl_2(g) \longrightarrow 4PCl_3(g) \quad \Delta H = -1224 \text{ kJ}$$
$$\text{Step 2: } PCl_3(g) + Cl_2(g) \longrightarrow PCl_5(g) \quad \Delta H = -93 \text{ kJ}$$

However, the second reaction must take place four times for each occurrence of the first reaction in the two-step process. This two-step process is more accurately described by the following equations.

$$P_4(s) + 6Cl_2(g) \longrightarrow 4PCl_3(g) \quad \Delta H = -1224 \text{ kJ}$$
$$4PCl_3(g) + 4Cl_2(g) \longrightarrow 4PCl_5(g) \quad \Delta H = 4(-93 \text{ kJ}) = -372 \text{ kJ}$$

So, the total change in enthalpy by the two-step process is as follows:

$$(-1224 \text{ kJ}) + (-372 \text{ kJ}) = -1596 \text{ kJ}$$

This enthalpy change, ΔH, for the two-step process is the same as the enthalpy change for the direct route of the formation of PCl_5. This example is in agreement with Hess's law.

Hess's law

the law that states that the amount of heat released or absorbed in a chemical reaction does not depend on the number of steps in the reaction

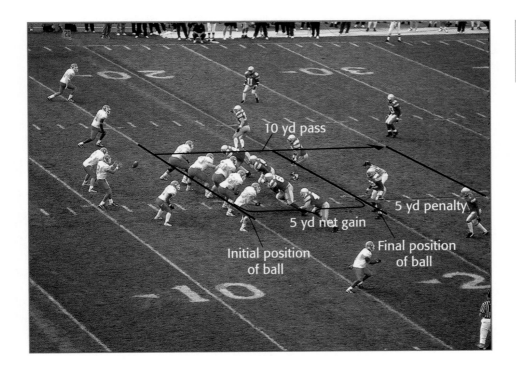

Figure 10
In football, as in Hess's law, only the initial and final conditions matter. A team that gains 10 yards on a pass play but has a five-yard penalty, has the same net gain as the team that gained only 5 yards.

Using Hess's Law and Algebra

Chemical equations can be manipulated using rules of algebra to get a desired equation. When equations are added or subtracted, enthalpy changes must be added or subtracted. And when equations are multiplied by a constant, the enthalpy changes must also be multiplied by that constant. For example, the enthalpy of the formation of CO, when CO_2 and solid carbon are reactants, is found using the equations below.

$$2C(s) + O_2(g) \longrightarrow 2CO(g) \quad \Delta H = -221 \text{ kJ}$$
$$C(s) + O_2(g) \longrightarrow CO_2(g) \quad \Delta H = -393 \text{ kJ}$$

You cannot simply add these equations because CO_2 would not be a reactant. But if you subtract or reverse the second equation, carbon dioxide will be on the correct side of the equation. This process is shown below.

$$-C(s) - O_2(g) \longrightarrow -CO_2(g) \quad \Delta H = -(-393 \text{ kJ})$$
$$CO_2(g) \longrightarrow C(s) + O_2(g) \quad \Delta H = 393 \text{ kJ}$$

So, reversing an equation causes the enthalpy of the new reaction to be the negative of the enthalpy of the original reaction. Now add the two equations to get the equation for the formation of CO by using CO_2 and C.

$$2C(s) + O_2(g) \longrightarrow 2CO(g) \quad \Delta H = -221 \text{ kJ}$$
$$CO_2(g) \longrightarrow C(s) + O_2(g) \quad \Delta H = 393 \text{ kJ}$$
$$\overline{2C(s) + O_2(g) + CO_2(g) \longrightarrow 2CO(g) + C(s) + O_2(g) \quad \Delta H = 172 \text{ kJ}}$$

Oxygen and carbon that appear on both sides of the equation can be canceled. So, the final equation is as shown below.

$$C(s) + CO_2(g) \longrightarrow 2CO(g) \quad \Delta H = 172 \text{ kJ}$$

Standard Enthalpies of Formation

The enthalpy change in forming 1 mol of a substance from elements in their standard states is called the *standard enthalpy of formation* of the substance, ΔH_f^0. Many values of ΔH_f^0 are listed in **Table 2.** Note that the values of the standard enthalpies of formation for elements are 0. From a list of standard enthalpies of formation, the enthalpy change of any reaction for which data is available can be calculated. For example, the following reaction can be considered to take place in four steps.

$$SO_2(g) + NO_2(g) \longrightarrow SO_3(g) + NO(g) \quad \Delta H = ?$$

Two of these steps convert the reactants into their elements. Notice that the reverse reactions for the formations of SO_2 and NO_2 are used. So, the standard enthalpies of formation for these reverse reactions are the negative of the standard enthalpies of formation for SO_2 and NO_2.

$$SO_2(g) \longrightarrow \tfrac{1}{8}S_8(s) + O_2(g) \quad \Delta H = -\Delta H_f^0 = -(-296.8 \text{ kJ/mol})$$
$$NO_2(g) \longrightarrow \tfrac{1}{2}N_2(g) + O_2(g) \quad \Delta H = -\Delta H_f^0 = -(33.1 \text{ kJ/mol})$$

Table 2   Standard Enthalpies of Formation

Substance	ΔH^0_f(kJ/mol)	Substance	ΔH^0_f(kJ/mol)
$Al_2O_3(s)$	−1676.0	$H_2O(g)$	−241.8
$CaCO_3(s)$	−1206.9	$H_2O(l)$	−285.8
$CaO(s)$	−634.9	$Na^+(g)$	609.4
$Ca(OH)_2(s)$	−985.2	$NaBr(s)$	−361.1
$C_2H_6(g)$	−83.8	$Na_2CO_3(s)$	−1130.7
$CH_4(g)$	−74.9	$NO(g)$	90.3
$CO(g)$	−110.5	$NO_2(g)$	33.1
$CO_2(g)$	−393.5	$Pb(s)$	0
$Fe_2O_3(s)$	−825.5	$SO_2(g)$	−296.8
$H_2(g)$	0	$SO_3(g)$	−395.8
$Hg(l)$	0	$ZnO(s)$	−348.3

Refer to Appendix A for more standard enthalpies of formation.

The two other steps, which are listed below reform those elements into the products.

$$\tfrac{1}{8}S_8(s) + \tfrac{3}{2}O_2(g) \longrightarrow SO_3(g) \quad \Delta H^0_f = -395.8 \text{ kJ/mol}$$
$$\tfrac{1}{2}N_2(g) + \tfrac{1}{2}O_2(g) \longrightarrow NO(g) \quad \Delta H^0_f = 90.3 \text{ kJ/mol}$$

In fact, the enthalpy change of any reaction can be determined in the same way—the reactants can be converted to their elements, and the elements can be recombined into the products. Why? Hess's law states that the overall enthalpy change of a reaction is the same, whether for a single-step process or a multiple step one. If you apply this rule, the exothermic reaction that forms sulfur trioxide and nitrogen oxide has the enthalpy change listed below.

$$SO_2(g) + NO_2(g) \longrightarrow SO_3(g) + NO(g)$$
$$\Delta H = (\Delta H^0_{f,\, NO} + \Delta H^0_{f,\, SO_3}) + (-\Delta H^0_{f,\, NO_2} - \Delta H^0_{f,\, SO_2})$$
$$\Delta H = (90.3 \text{ kJ/mol} - 395.8 \text{ kJ/mol}) + (-33.1 \text{ kJ/mol} + 296.8 \text{ kJ/mol}) =$$
$$-41.8 \text{ kJ/mol}$$

When using standard enthalpies of formation to determine the enthalpy change of a chemical reaction, remember the following equation.

$$\Delta H_{\text{reaction}} = \Delta H_{\text{products}} - \Delta H_{\text{reactants}}$$

SAMPLE PROBLEM D

Calculating a Standard Enthalpy of Formation

Calculate the standard enthalpy of formation of pentane, C_5H_{12}, using the given information.

(1) $C(s) + O_2(g) \longrightarrow CO_2(g)$ $\Delta H_f^0 = -393.5$ kJ/mol

(2) $H_2(g) + \frac{1}{2}O_2(g) \longrightarrow H_2O(l)$ $\Delta H_f^0 = -285.8$ kJ/mol

(3) $C_5H_{12}(g) + 8O_2(g) \longrightarrow 5CO_2(g) + 6H_2O(l)$ $\Delta H = -3535.6$ kJ/mol

1 **Gather information.**

The equation for the standard enthalpy of formation is

$$5C(s) + 6H_2(g) \longrightarrow C_5H_{12}(g) \quad \Delta H_f^0 = ?$$

2 **Plan your work.**

C_5H_{12} is a product, so reverse the equation (3) and the sign of ΔH. Multiply equation (1) by 5 to give 5C as a reactant. Multiply equation (2) by 6 to give $6H_2$ as a reactant.

3 **Calculate.**

(1) $5C(s) + 5O_2(g) \longrightarrow 5CO_2(g)$ $\Delta H = 5(-393.5$ kJ/mol)

(2) $6H_2(g) + 3O_2(g) \longrightarrow 6H_2O(l)$ $\Delta H = 6(-285.8$ kJ/mol)

(3) $5CO_2(g) + 6H_2O(l) \longrightarrow C_5H_{12}(g) + 8O_2(g)$ $\Delta H = 3536.6$ kJ/mol

$$\overline{5C(s) + 6H_2(g) \longrightarrow C_5H_{12}(g) \quad \Delta H_f^0 = -145.7 \text{ kJ/mol}}$$

4 **Verify your results.**

The unnecessary reactants and products cancel to give the correct equation.

PRACTICE HINT

A positive ΔH means that the reaction has absorbed energy or that the reaction is endothermic. A negative ΔH means that the reaction has released energy or that the reaction is exothermic.

PRACTICE

1 Calculate the enthalpy change for the following reaction.

$$NO(g) + \frac{1}{2}O_2(g) \longrightarrow NO_2(g)$$

2 Calculate the enthalpy change for the combustion of methane gas, CH_4, to form $CO_2(g)$ and $H_2O(l)$.

SAMPLE PROBLEM E

Calculating a Reaction's Change in Enthalpy

Calculate the change in enthalpy for the reaction below by using data from **Table 2**.

$$2H_2(g) + 2CO_2(g) \longrightarrow 2H_2O(g) + 2CO(g)$$

Then, state whether the reaction is exothermic or endothermic.

1 Gather information.

Standard enthalpies of formation for the products are as follows:
For $H_2O(g)$, $\Delta H_f^0 = -241.8$ kJ/mol. For $CO(g)$, $\Delta H_f^0 = -110.5$ kJ/mol.
Standard enthalpies of formation for the reactants are as follows:
For $H_2(g)$, $\Delta H_f^0 = 0$ kJ/mol. For $CO_2(g)$, $\Delta H_f^0 = -393.5$ kJ/mol.

2 Plan your work.

The general rule is $\Delta H = \Delta H(\text{products}) - \Delta H(\text{reactants})$. So,
$\Delta H = (\text{mol } H_2O(g))\, \Delta H_f^0 \text{ (for } H_2O(g)) + (\text{mol } CO(g))\, \Delta H_f^0 \text{ (for } CO(g)) - (\text{mol } H_2(g))\, \Delta H_f^0 \text{ (for } H_2(g)) - (\text{mol } CO_2(g))\, \Delta H_f^0 \text{ (for } CO_2(g))$.

3 Calculate.

$\Delta H = (2 \text{ mol})(-241.8 \text{ kJ/mol}) + (2 \text{ mol})(-110.5 \text{ kJ/mol}) - (2 \text{ mol})(0 \text{ kJ/mol}) - (2 \text{ mol})(-393.5 \text{ kJ/mol}) = 82.4$ kJ
Because the enthalpy change is positive, the reaction is endothermic.

4 Verify your results.

The enthalpy of the reactants, -787 kJ, is more negative than that of the products, -704.6 kJ, and shows that the total energy of the reaction increases by 82.4 kJ.

> **PRACTICE HINT**
>
> Always be sure to check the states of matter when you use standard enthalpy of formation data. $H_2O(g)$ and $H_2O(l)$ have different values.

PRACTICE

1. Use data from **Table 2** to calculate ΔH for the following reaction.

$$C_2H_6(g) + \frac{7}{2}O_2(g) \longrightarrow 2CO_2(g) + 3H_2O(g)$$

2. The exothermic reaction known as lime slaking is $CaO(s) + H_2O(l) \longrightarrow Ca(OH)_2(s)$. Calculate ΔH from the data in **Table 2.**

③ Section Review

UNDERSTANDING KEY IDEAS

1. What is the standard thermodynamic temperature?
2. Why does $\Delta H_f^0 = 0$ for elements, as listed in **Table 2**?
3. How do bomb calorimetry and adiabatic calorimetry differ?

PRACTICE PROBLEMS

4. Use **Table 2** to calculate ΔH for the decomposition of calcium carbonate into calcium oxide and carbon dioxide.
5. What enthalpy change accompanies the reaction $2Al(s) + 3H_2O(l) \longrightarrow Al_2O_3(s) + 3H_2(g)$?

CRITICAL THINKING

6. **Table 2** includes two entries for water. What does the difference between the two values represent?
7. What general conclusion can you draw from observing that most standard enthalpies of formation are negative?

4 Order and Spontaneity

KEY TERMS

• entropy

• Gibbs energy

OBJECTIVES

(1) **Define** *entropy*, and discuss the factors that influence the sign and magnitude of ΔS for a chemical reaction.

(2) **Describe** *Gibbs energy*, and discuss the factors that influence the sign and magnitude of ΔG.

(3) **Indicate** whether ΔG values describe spontaneous or nonspontaneous reactions.

Entropy

entropy

a measure of the randomness or disorder of a system

Some reactions happen easily, but others do not. For example, sodium and chlorine react when they are brought together. However, nitrogen and oxygen coexist in the air you breathe without forming poisonous nitrogen monoxide, NO. One factor you can use to predict whether reactions will occur is enthalpy. A reaction is more likely to occur if it is accompanied by a *decrease in enthalpy* or if ΔH is negative.

But a few processes that are endothermic can occur easily. Why? Another factor known as entropy can determine if a process will occur. **Entropy,** *S,* is a measure of the disorder in a system and is a thermodynamic property. Entropy is not a form of energy and has the units joules per kelvin, J/K. A process is more likely to occur if it is accompanied by an *increase in entropy*; that is, ΔS is positive.

Figure 11

a Crystals of potassium permanganate, $KMnO_4$, are dropped into a beaker of water and dissolve to produce the $K^+(aq)$ and $MnO_4^-(aq)$ ions.

b Diffusion causes entropy to increase and leads to a uniform solution.

Table 3 Standard Entropy Changes for Some Reactions

Reaction	Entropy change, ΔS (J/K)
$CaCO_3(s) + 2H_3O^+(aq) \longrightarrow Ca^{2+}(aq) + CO_2(g) + 3H_2O(l)$	138
$NaCl(s) \longrightarrow Na^+(aq) + Cl^-(aq)$	43
$N_2(g) + O_2(g) \longrightarrow 2NO(g)$	25
$CH_4(g) + 2O_2(g) \longrightarrow CO_2(g) + 2H_2O(g)$	−5
$2Na(s) + Cl_2(g) \longrightarrow 2NaCl(s)$	−181
$2NO_2(g) \longrightarrow N_2O_4(g)$	−176

Factors That Affect Entropy

If you scatter a handful of seeds, you have dispersed them. You have created a more disordered arrangement. In the same way, as molecules or ions become dispersed, their disorder increases and their entropy increases. In **Figure 11,** the intensely violet permanganate ions, $MnO_4^-(aq)$ are initially found only in a small volume of solution. But they gradually spread until they occupy the whole beaker. You can't see the potassium $K^+(aq)$ ions because these ions are colorless, but they too have dispersed. This process of dispersion is called *diffusion* and causes the increase in entropy.

Entropy also increases as solutions become more dilute or when the pressure of a gas is reduced. In both cases, the molecules fill larger spaces and so become more disordered. Entropies also increase with temperature, but this effect is not great unless a phase change occurs.

The entropy can change during a reaction. The entropy of a system can increase when the total number of moles of product is greater than the total number of moles of reactant. Entropy can increase in a system when the total number of particles in the system increases. Entropy also increases when a reaction produces more gas particles, because gases are more disordered than liquids or solids.

Table 3 lists the entropy changes of some familiar chemical reactions. Notice that entropy decreases as sodium chloride forms: 2 mol of sodium combine with 1 mol of chlorine to form 2 mol of sodium chloride.

$$2Na(s) + Cl_2(g) \longrightarrow 2NaCl(s) \quad \Delta S = -181 \text{ J/K}$$

This decrease in entropy is because of the order present in crystalline sodium chloride.

Also notice that the entropy increases when 1 mol of sodium chloride dissolves in water to form 1 mol of aqueous sodium ions and 1 mol of aqueous chlorine ions.

$$NaCl(s) \longrightarrow Na^+(aq) + Cl^-(aq) \quad \Delta S = 43 \text{ J/K}$$

This increase in entropy is because of the order lost when a crystalline solid dissociates to form ions.

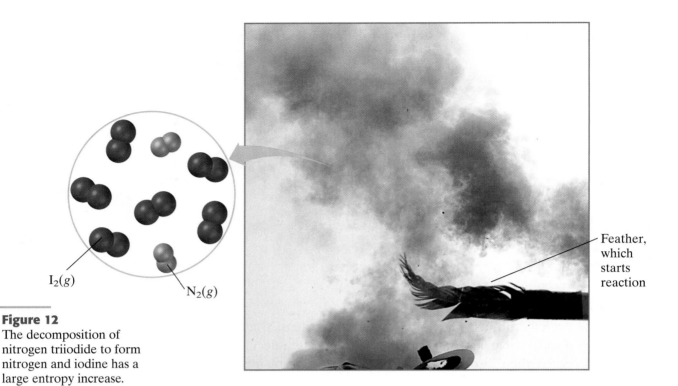

$I_2(g)$

$N_2(g)$

Figure 12
The decomposition of nitrogen triiodide to form nitrogen and iodine has a large entropy increase.

Table 4 Standard Entropies of Some Substances

Substance	S^0 (J/K•mol)
C(s) (graphite)	5.7
CO(g)	197.6
$CO_2(g)$	213.8
$H_2(g)$	130.7
$H_2O(g)$	188.7
$H_2O(l)$	70.0
$Na_2CO_3(s)$	135.0
$O_2(g)$	205.1

Refer to Appendix A for more standard entropies.

Hess's Law Also Applies to Entropy

The decomposition of nitrogen triiodide to form nitrogen and iodine in **Figure 12** creates 4 mol of gas from 2 mol of a solid.

$$2NI_3(s) \longrightarrow N_2(g) + 3I_2(g)$$

This reaction has such a large entropy increase that the reaction proceeds once the reaction is initiated by a mechanical shock.

Molar entropy has the same unit, J/K•mol, as molar heat capacity. In fact, molar entropies can be calculated from molar heat capacity data.

Entropies can also be calculated by using Hess's law and entropy data for other reactions. This statement means that you can manipulate chemical equations using rules of algebra to get a desired equation. But remember that when equations are added or subtracted, entropy changes must be added or subtracted. And when equations are multiplied by a constant, the entropy changes must also be multiplied by that constant. Finally, atoms and molecules that appear on both sides of the equation can be canceled.

The *standard entropy* is represented by the symbol S^0 and some standard entropies are listed in **Table 4.** The standard entropy of the substance is the entropy of 1 mol of a substance at a standard temperature, 298.15 K. Unlike having standard enthalpies of formation equal to 0, elements can have standard entropies that have values other than zero. You should also know that most standard entropies are positive; this is not true of standard enthalpies of formation.

The entropy change of a reaction can be calculated by using the following equation.

$$\Delta S_{reaction} = S_{products} - S_{reactants}$$

Hess's Law and Entropy

Use **Table 4** to calculate the entropy change that accompanies the following reaction.

$$\tfrac{1}{2}H_2(g) + \tfrac{1}{2}CO_2(g) \longrightarrow \tfrac{1}{2}H_2O(g) + \tfrac{1}{2}CO(g)$$

1 Gather information.

Products: $H_2O(g) + CO(g)$
Reactants: $H_2(g) + CO_2(g)$

2 Plan your work.

The general rule is $\Delta S = \Delta S(\text{products}) - \Delta S(\text{reactants})$.
So, $\Delta S = (\text{mol } H_2O(g))\, S^0 \text{ (for } H_2O(g)) + (\text{mol } CO(g))\, S^0$
(for $CO(g)$) $- (\text{mol } H_2(g))\, S^0$ (for $H_2(g)$) $- (\text{mol } CO_2(g))\, S^0$ (for $CO_2(g)$).

The standard entropies from **Table 4** are as follows:
For H_2O, $S^0 = 188.7$ J/K•mol. For CO, $S^0 = 197.6$ J/K•mol.
For H_2, $S^0 = 130.7$ J/K•mol. For CO_2, $S^0 = 213.8$ J/K•mol.

3 Calculate.

Substitute the values into the equation for ΔS.

$$\Delta S = \left[\left(\tfrac{1}{2}\text{ mol}\right)(188.7\text{ J/K•mol}) + \left(\tfrac{1}{2}\text{ mol}\right)(197.6\text{ J/K•mol}) - \right.$$
$$\left. \left(\tfrac{1}{2}\text{ mol}\right)(130.7\text{ J/K•mol}) - \left(\tfrac{1}{2}\text{ mol}\right)(213.8\text{ J/K•mol})\right] = 94.35\text{ J/K} +$$
$$98.8\text{ J/K} - 65.35\text{ J/K} - 106.9\text{ J/K} = 193.1\text{ J/K} - 172.2\text{ J/K} = 20.9\text{ J/K}$$

4 Verify your results.

The sum of the standard entropies of gaseous water and carbon monoxide is larger than the sum of the standard entropies of gaseous hydrogen and carbon dioxide. So, the ΔS for this reaction should be positive.

> **PRACTICE HINT**
>
> Always check the signs of entropy values. Standard entropies are almost always positive, while standard entropies of formation are positive and negative.

PRACTICE

1. Find the change in entropy for the reaction below by using **Table 4** and that S^0 for $CH_3OH(l)$ is 126.8 J/K•mol.

$$CO(g) + 2H_2(g) \longrightarrow CH_3OH(l)$$

2. What is the entropy change for

$$\tfrac{1}{2}CO(g) + H_2(g) \longrightarrow \tfrac{1}{2}CH_3OH(l)?$$

3. Use data from **Table 3** to calculate the entropy change for the following reaction:

$$2Na(s) + Cl_2(g) \longrightarrow 2Na^+(aq) + 2Cl^-(aq)$$

Gibbs Energy

You have learned that the tendency for a reaction to occur depends on both ΔH and ΔS. If ΔH is negative and ΔS is positive for a reaction, the reaction will likely occur. If ΔH is positive and ΔS is negative for a reaction, the reaction will *not* occur. How can you predict what will happen if ΔH and ΔS are both positive or both negative?

Josiah Willard Gibbs, a professor at Yale University, answered that question by proposing another thermodynamic quantity, which now bears his name. **Gibbs energy** is represented by the symbol G and is defined by the following equation.

$$G = H - TS$$

Another name for Gibbs energy is *free energy*.

Gibbs energy

the energy in a system that is available for work

Gibbs Energy Determines Spontaneity

When the term *spontaneous* is used to describe reactions, it has a different meaning than the meaning that we use to describe other events. A spontaneous reaction is one that does occur or is likely to occur without continuous outside assistance, such as input of energy. A nonspontaneous reaction will never occur without assistance. The avalanche shown in **Figure 13** is a good example of a spontaneous process. On mountains during the winter, an avalanche may or may not occur, but it always *can* occur. The return of the snow from the bottom of the mountain to the mountaintop is a nonspontaneous event, because this event will not happen without aid.

A reaction is spontaneous if the Gibbs energy change is negative. If a reaction has a ΔG greater than 0, the reaction is nonspontaneous. If a reaction has a ΔG of exactly zero, the reaction is at equilibrium.

Figure 13
An avalanche is a spontaneous process driven by an increase in disorder and a decrease in energy.

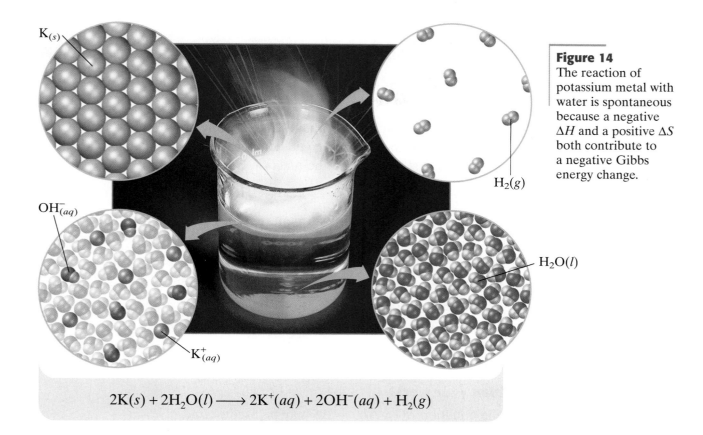

$$2K(s) + 2H_2O(l) \longrightarrow 2K^+(aq) + 2OH^-(aq) + H_2(g)$$

Entropy and Enthalpy Determine Gibbs Energy

Reactions that have large negative ΔG values often release energy and increase disorder. The vigorous reaction of potassium metal and water shown in **Figure 14** is an example of this type of reaction. The reaction is described by the following equation.

$$2K(s) + 2H_2O(l) \longrightarrow 2K^+(aq) + 2OH^-(aq) + H_2(g)$$
$$\Delta H = -392 \text{ kJ} \quad \Delta S = 0.047 \text{ kJ/K}$$

The change in Gibbs energy for the reaction above is calculated below.

$$\Delta G = \Delta H - T\Delta S =$$
$$-392 \text{ kJ} - (298.15 \text{ K})(0.047 \text{ kJ/K}) = -406 \text{ kJ}$$

Notice that the reaction of potassium and water releases energy and increases disorder. This example and **Sample Problem G** show how to determine ΔG values at 25°C by using ΔH and ΔS data. However, you can calculate ΔG in another way because lists of standard Gibbs energies of formation exist, such as **Table 5**.

The standard Gibbs energy of formation, ΔG_f^0, of a substance is the change in energy that accompanies the formation of 1 mol of the substance from its elements at 298.15 K. These standard Gibbs energies of formation can be used to find the ΔG for any reaction in exactly the same way that ΔH_f^0 data were used to calculate the enthalpy change for any reaction. Hess's law also applies when calculating ΔG.

$$\Delta G_{reaction} = \Delta G_{products} - \Delta G_{reactants}$$

Table 5 Standard Gibbs Energies of Formation

Substance	ΔG_f^0 (kJ/mol)
$Ca(s)$	0
$CaCO_3(s)$	−1128.8
$CaO(s)$	−604.0
$CaCl_2(s)$	−748.1
$CH_4(g)$	−50.7
$CO_2(g)$	−394.4
$CO(g)$	−137.2
$H_2(g)$	0
$H_2O(g)$	−228.6
$H_2O(l)$	−237.2

Refer to Appendix A for more standard Gibbs energies of formation.

SAMPLE PROBLEM G

Calculating a Change in Gibbs Energy from ΔH and ΔS

Given that the changes in enthalpy and entropy are –139 kJ and 277 J/K respectively for the reaction given below, calculate the change in Gibbs energy. Then, state whether the reaction is spontaneous at 25°C.

$$C_6H_{12}O_6(aq) \longrightarrow 2C_2H_5OH(aq) + 2CO_2(g)$$

This reaction represents the fermentation of glucose into ethanol and carbon dioxide, which occurs in the presence of enzymes provided by yeast cells. This reaction is used in baking.

1 Gather information.

$\Delta H = -139$ kJ
$\Delta S = 277$ J/K
$T = 25°C = (25 + 273.15)$ K = 298 K
$\Delta G = ?$

2 Plan your work.

The equation $\Delta G = \Delta H - T\Delta S$ may be used to find ΔG.
If ΔG is positive, the reaction is nonspontaneous. If ΔG is negative, the reaction is spontaneous.

3 Calculate.

$\Delta G = \Delta H - T\Delta S = (-139 \text{ kJ}) - (298 \text{ K})(277 \text{ J/K})$

$= (-139 \text{ kJ}) - (298 \text{ K})(0.277 \text{ kJ/K})$

$= (-139 \text{ kJ}) - (83 \text{ kJ}) = -222 \text{ kJ}$

The negative sign of ΔG shows that the reaction is spontaneous.

4 Verify your results.

The calculation was not necessary to prove the reaction is spontaneous, because each requirement for spontaneity—a negative ΔH and a positive ΔS—was met. In addition, the reaction occurs in nature without a source of energy, so the reaction must be spontaneous.

PRACTICE HINT

Enthalpies and Gibbs energies are generally expressed in kilojoules, but entropies are usually stated in joules (not kilojoules) per kelvin. Remember to divide all entropy values expressed in joules by 1000.

PRACTICE

1 A reaction has a ΔH of –76 kJ and a ΔS of –117 J/K. Is the reaction spontaneous at 298.15 K?

2 A reaction has a ΔH of 11 kJ and a ΔS of 49 J/K. Calculate ΔG at 298.15 K. Is the reaction spontaneous?

3 The gas-phase reaction of H_2 with CO_2 to produce H_2O and CO has a $\Delta H = 11$ kJ and a $\Delta S = 41$ J/K. Is the reaction spontaneous at 298.15 K? What is ΔG?

Calculating a Gibbs Energy Change Using ΔG_f^0 Values

Use **Table 5** to calculate ΔG for the following water-gas reaction.

$$C(s) + H_2O(g) \longrightarrow CO(g) + H_2(g)$$

Is this reaction spontaneous?

1 Gather information.

For $H_2O(g)$, $\Delta G_f^0 = -228.6$ kJ/mol.
For $CO(g)$, $\Delta G_f^0 = -137.2$ kJ/mol.
For $H_2(g)$, $\Delta G_f^0 = 0$ kJ/mol.
For $C(s)$ (graphite), $\Delta G_f^0 = 0$ kJ/mol.

2 Plan your work.

The following simple relation may be used to find the total change in Gibbs energy.

$$\Delta G = \Delta G(\text{products}) - \Delta G(\text{reactants})$$

If ΔG is positive, the reaction is nonspontaneous. If ΔG is negative, the reaction is spontaneous.

3 Calculate.

$\Delta G = \Delta G(\text{products}) - \Delta G(\text{reactants}) =$
$[(\text{mol } CO(g))(\Delta G_f^0 \text{ for } CO(g)) + (\text{mol } H_2(g))(\Delta G_f^0 \text{ for } H_2(g))] -$
$[(\text{mol } C(s))(\Delta G_f^0 \text{ for } C(s)) + (\text{mol } H_2O(g))(\Delta G_f^0 \text{ for } H_2O(g))] =$
$[(1 \text{ mol})(-137.2 \text{ kJ/mol}) + (1 \text{ mol})(0 \text{ kJ/mol})] - [(1 \text{ mol})(0 \text{ kJ/mol}) -$
$(1 \text{ mol})(-228.6 \text{ kJ/mol})] = (-137.2 + 228.6) \text{ kJ} = 91.4 \text{ kJ}$

The reaction is nonspontaneous under standard conditions.

4 Verify your results.

The ΔG_f^0 values in this problem show that water has a Gibbs energy that is 91.4 kJ lower than the Gibbs energy of carbon monoxide. Therefore, the reaction would increase the Gibbs energy by 91.4 kJ. Processes that lead to an increase in Gibbs energy never occur spontaneously.

PRACTICE HINT

Because many thermodynamic data are negative, it is easy to use the incorrect signs while performing calculations. Pay attention to signs, and check them frequently.

The ΔG_f^0 values for elements are always zero.

PRACTICE

1 Use **Table 5** to calculate the Gibbs energy change that accompanies the following reaction.

$$C(s) + O_2(g) \longrightarrow CO_2(g)$$

Is the reaction spontaneous?

2 Use **Table 5** to calculate the Gibbs energy change that accompanies the following reaction.

$$CaCO_3(s) \longrightarrow CaO(s) + CO_2(g)$$

Is the reaction spontaneous?

PROBLEM SOLVING SKILL

Table 6   Relating Enthalpy and Entropy Changes to Spontaneity

ΔH	ΔS	ΔG	Is the reaction spontaneous?
Negative	positive	negative	yes, at all temperatures
Negative	negative	either positive or negative	only if $T < \Delta H/\Delta S$
Positive	positive	either positive or negative	only if $T > \Delta H/\Delta S$
Positive	negative	positive	never

Predicting Spontaneity

Does temperature affect spontaneity? Consider the equation for ΔG.

$$\Delta G = \Delta H - T\Delta S$$

The terms ΔH and ΔS change very little as temperature changes, but the presence of T in the equation for ΔG indicates that temperature may greatly affect ΔG. **Table 6** summarizes the four possible combinations of enthalpy and entropy changes for any chemical reaction. Suppose a reaction has both a positive ΔH value and a positive ΔS value. If the reaction occurs at a low temperature, the value for $T\Delta S$ will be small and will have little impact on the value of ΔG. The value of ΔG will be similar to the value of ΔH and will have a positive value. But when the same reaction proceeds at a high enough temperature, $T\Delta S$ will be larger than ΔH and ΔG will be negative. So, increasing the temperature of a reaction can make a nonspontaneous reaction spontaneous.

Figure 15
Photosynthesis, the non-spontaneous conversion of carbon dioxide and water into carbohydrate and oxygen, is made possible by light energy.

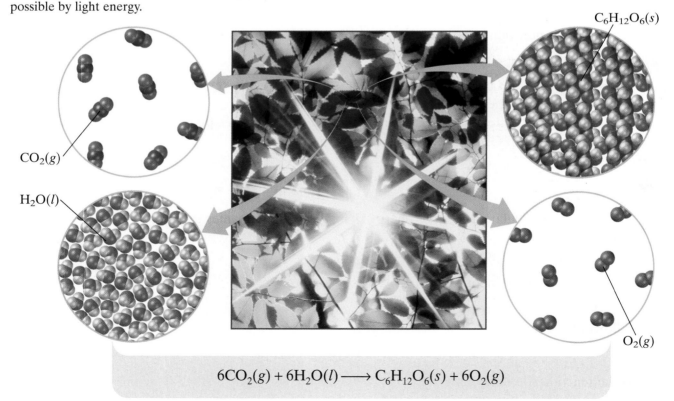

$C_6H_{12}O_6(s)$

$CO_2(g)$

$H_2O(l)$

$O_2(g)$

$$6CO_2(g) + 6H_2O(l) \longrightarrow C_6H_{12}O_6(s) + 6O_2(g)$$

Can a nonspontaneous reaction ever occur? A nonspontaneous reaction cannot occur unless some form of energy is added to the system. **Figure 15** shows that the nonspontaneous reaction of photosynthesis occurs with outside assistance. During photosynthesis, light energy from the sun is used to drive the nonspontaneous process. This reaction is described by the equation below.

$$6CO_2(g) + 6H_2O(l) \longrightarrow C_6H_{12}O_6(s) + 6O_2(g) \quad \Delta H = 2870 \text{ kJ/mol}$$

 # Section Review

UNDERSTANDING KEY IDEAS

1. What aspect of a substance contributes to a high or a low entropy?

2. What is diffusion? Give an example.

3. Name three thermodynamic properties, and give the relationship between them.

4. What signs of ΔH, ΔS, and ΔG favor spontaneity?

5. What signs of ΔH, ΔS, and ΔG favor nonspontaneity?

6. How can the Gibbs energy change of a reaction can be calculated?

PRACTICE PROBLEMS

7. The standard entropies for the following substances are 210.8 J/K•mol for NO(g), 240.1 J/K•mol for NO$_2$(g), and 205.1 J/K•mol for O$_2$(g). Determine the entropy for the reaction below.

$$2NO(g) + O_2(g) \longrightarrow 2NO_2(g)$$

8. Suppose $X(s) + 2Y_2(g) \longrightarrow XY_4(g)$ has a $\Delta H = -74.8$ kJ and a $\Delta S = -80.8$ J/K. Calculate ΔG for this reaction at 298.15 K.

9. Use **Table 5** to determine whether the reaction below is spontaneous.

$$CaCl_2(s) + H_2O(g) \longrightarrow CaO(s) + 2HCl(g)$$

The standard Gibbs energy of formation for HCl(g) is −95.3 kJ/mol.

10. Calculate the Gibbs energy change for the reaction $2CO(g) \longrightarrow C(s) + CO_2(g)$. Is the reaction spontaneous?

11. Calculate the Gibbs energy change for the reaction $CO(g) \rightarrow \frac{1}{2}C(s) + \frac{1}{2}CO_2(g)$? How does this result differ from the result in item 10?

CRITICAL THINKING

12. A reaction is endothermic and has a $\Delta H = 8$ kJ. This reaction occurs spontaneously at 25°C. What must be true about the entropy change?

13. You are looking for a method of making chloroform, CHCl$_3$(l). The standard Gibbs energy of formation for HCl(g) is −95.3 kJ/mol and the standard Gibbs energy of formation for CHCl$_3$(l) is −73.66 kJ/mol. Use **Table 5** to decide which of the following reactions should be investigated.

$$2C(s) + H_2(g) + 3Cl_2(g) \longrightarrow 2CHCl_3(l)$$

$$C(s) + HCl(g) + Cl_2(g) \longrightarrow CHCl_3(l)$$

$$CH_4(g) + 3Cl_2(g) \longrightarrow CHCl_3(l) + 3HCl(g)$$

$$CO(g) + 3HCl(g) \longrightarrow CHCl_3(l) + H_2O(l)$$

14. If the reaction $X \longrightarrow Y$ is spontaneous, what can be said about the reaction $Y \longrightarrow X$?

15. At equilibrium, what is the relationship between ΔH and ΔS?

16. If both ΔH and ΔS are negative, how does temperature affect spontaneity?

SCIENCE AND TECHNOLOGY

Hydrogen-Powered Cars

Hydrogen As Fuel

When you think of fuel, you probably think of gasoline or nuclear fuel. But did you know that scientists have been studying ways to use the energy generated by the following reaction?

$$H_2(g) + \tfrac{1}{2}O_2(g) \longrightarrow H_2O(l)\ \Delta H_f^0 = -285.8 \text{ kJ/mol}$$

Engineers use fuel cells that drive an electrochemical reaction, which converts hydrogen or hydrogen-containing materials and oxygen into water, electrical energy, and energy as heat. Fuel cells have already been used by NASA to provide space crews with electrical energy and drinking water. In the future, electrical energy for buildings, ships, submarines, and vehicles may be obtained using the reaction of hydrogen and oxygen to form water.

Fuel cells that use hydrogen are used to power cars, such as the car in the photo.

Cars That Are Powered by Hydrogen Fuel Cells

Many car manufacturers are researching ways to mass produce vehicles that are powered by hydrogen fuel cells. Some hydrogen-powered cars that manufacturers have already developed can reach speeds of over 150 km/h (90 mi/h). These types of cars will also travel 400 to 640 km (250 to 400 mi) before refueling.

These cars have many benefits. Fuel cells have an efficiency of 50 to 60%, which is about twice as efficient as internal combustion engines. These cells are also safe for the environment because they can produce only water as a by product. Unfortunately, fuel cells are expensive because they contain expensive materials, such as platinum.

Questions

1. Research electrical energy and its sources. Which source is the most environmentally safe? Which source is the cheapest? Which source is the most efficient?
2. Research your favorite type of car. How does this car run? How far can this car travel before refueling? What pollutants does this car produce?

KEY TERMS

heat
enthalpy
temperature

thermodynamics

calorimetry
calorimeter
Hess's law

entropy
Gibbs energy

KEY IDEAS

SECTION ONE Energy Transfer

- Heat is energy transferred from a region at one temperature to a region at a lower temperature.
- Temperature depends on the average kinetic energy of the atoms.
- The molar heat capacity of an element or compound is the energy as heat needed to increase the temperature of 1 mol by 1 K.

SECTION TWO Using Enthalpy

- The enthalpy of a system can be its total energy.
- When *only* temperature changes, the change in molar enthalpy is represented by $\Delta H = C\Delta T$.

SECTION THREE Changes in Enthalpy During Reactions

- Calorimetry measures the enthalpy change, which is represented by ΔH, during a chemical reaction.
- Reactions that have positive ΔH are endothermic; reactions that have negative ΔH are exothermic.
- Hess's law indicates that the thermodynamic changes for any particular process are the same, whether the changes are treated as a single reaction or a series of steps.

$$\Delta H_{reaction} = \Delta H_{products} - \Delta H_{reactants}$$

SECTION FOUR Order and Spontaneity

- The entropy of a system reflects the system's disorder.

$$\Delta S_{reaction} = S_{products} - S_{reactants}$$

- Gibbs energy is defined by $G = H - TS$.
- The sign of ΔG determines spontaneity.

$$\Delta G_{reaction} = \Delta G_{products} - \Delta G_{reactants}$$

KEY SKILLS

Calculating the Molar Heat Capacity of a Sample
Sample Problem A p. 342

Calculating Molar Enthalpy Change for Heating
Sample Problem B p. 346

Calculating the Molar Enthalpy Change for Cooling
Sample Problem C p. 347

Calculating a Standard Enthalpy of Formation
Sample Problem D p. 356

Calculating a Reaction's Change in Enthalpy
Sample Problem E p. 356

Hess's Law and Entropy
Sample Problem F p. 361

Calculating Changes in Gibbs Energy
Sample Problem G p. 364
Sample Problem H p. 365

USING KEY TERMS

1. What is dependent on the average kinetic energy of the atoms in a substance?

2. Define *heat*.

3. Name a device used for measuring enthalpy changes.

4. Do most exothermic reactions have a negative or positive change in Gibbs energy?

5. What quantities do the Gibbs energy of a reaction depend on?

6. What is a spontaneous reaction?

7. What is entropy?

8. Define *Gibbs energy*, and explain its usefulness.

UNDERSTANDING KEY IDEAS

Energy Transfer

9. Distinguish between heat and temperature.

10. What advantage does the Kelvin temperature scale have?

11. How can you tell which one of two samples will release energy in the form of heat when the two samples are in contact?

12. What is molar heat capacity, and how can it be measured?

Using Enthalpy

13. What is molar enthalpy change?

14. What influences the changes in molar enthalpy?

15. Name two processes for which you could determine an enthalpy change.

Changes in Enthalpy During Reactions

16. Define *standard molar enthalpy of formation*.

17. Explain the meanings of H, ΔH, and ΔH_f^0.

18. State Hess's law. How is it used?

19. Describe a bomb calorimeter.

20. Which thermodynamic property of a food is of interest to nutritionists? Why?

21. What is adiabatic calorimetry?

22. What is the value for the standard molar enthalpy of formation for an element?

Order and Spontaneity

23. Why is entropy described as an extensive property?

24. Explain why Josiah Willard Gibbs invented a new thermodynamic quantity.

25. Explain how a comprehensive table of standard Gibbs energies of formation can be used to determine the spontaneity of any chemical reaction.

26. What information is needed to be certain that a chemical reaction is nonspontaneous?

PRACTICE PROBLEMS

Sample Problem A Calculating the Molar Heat Capacity of a Substance

27. You need 70.2 J to raise the temperature of 34.0 g of ammonia, $NH_3(g)$, from 23.0°C to 24.0°C. Calculate the molar heat capacity of ammonia.

28. Calculate C for indium metal given that 1.0 mol In absorbs 53 J during the following process.

$$In(s, 297.5 \text{ K}) \longrightarrow In(s, 299.5 \text{ K})$$

Sample Problem B Calculating the Molar Enthalpy Change for Heating

29. Calculate ΔH when 1.0 mol of nitrogen is heated from 233 K to 475 K.

30. What is the change in enthalpy when 11.0 g of liquid mercury is heated by 15°C?

Sample Problem C Calculating the Molar Enthalpy Change for Cooling

31. Calculate ΔH when 1.0 mol of argon is cooled from 475 K to 233 K.

32. What enthalpy change occurs when 112.0 g of barium chloride experiences a change of temperature from 15°C to −30°C.

Sample Problem D Calculating a Standard Enthalpy of Formation

33. The diagram below represents an interpretation of Hess's law for the following reaction.

$$Sn(s) + 2Cl_2(g) \longrightarrow SnCl_4(l)$$

Use the diagram to determine ΔH for each step and the net reaction.

$$Sn(s) + Cl_2(g) \longrightarrow SnCl_2(l) \qquad \Delta H = ?$$

$$SnCl_2(s) + Cl_2(g) \longrightarrow SnCl_4(l) \qquad \Delta H = ?$$

$$Sn(s) + 2Cl_2(g) \longrightarrow SnCl_4(l) \qquad \Delta H = ?$$

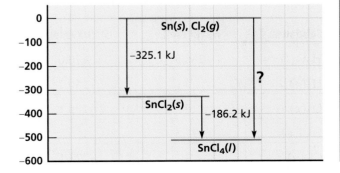

Sample Problem E Calculating a Reaction's Change in Enthalpy

34. Use tabulated values of standard enthalpies of formation to calculate the enthalpy change accompanying the reaction $4Al(s) + 6H_2O(l) \longrightarrow 2Al_2O_3(s) + 6H_2(g)$. Is the reaction exothermic?

35. The reaction $2Fe_2O_3(s) + 3C(s) \longrightarrow 4Fe(s) + 3CO_2(g)$ is involved in the smelting of iron. Use ΔH_f^0 values to calculate the enthalpy change during the production of 1 mol of iron.

36. For glucose, $\Delta H_f^0 = -1263$ kJ/mol. Calculate the enthalpy change when 1 mol of $C_6H_{12}O_6(s)$ combusts to form $CO_2(g)$ and $H_2O(l)$.

Sample Problem F Hess's Law and Entropy

37. Given the entropy change for the first two reactions below, calculate the entropy change for the third reaction below.

$$S_8(s) + 8O_2(g) \longrightarrow 8SO_2(g) \ \Delta S = 89 \text{ J/K}$$

$$2SO_2(s) + O_2(g) \longrightarrow 2SO_3(g) \ \Delta S = -188 \text{ J/K}$$

$$S_8(s) + 12O_2(g) \longrightarrow 8SO_3(g) \ \Delta S = ?$$

38. The standard entropies for the following substances are 26.9 J/K•mol for $MgO(s)$, 213.8 J/K•mol for $CO_2(g)$, and 65.7 J/K•mol for $MgCO_3(s)$. Determine the entropy for the reaction below.

$$MgCO_3(s) \longrightarrow MgO(s) + CO_2(g)$$

Sample Problem G, Sample Problem H Calculating Changes in Gibbs Energy

39. A reaction has $\Delta H = -356$ kJ and $\Delta S = -36$ J/K. Calculate ΔG at 25°C to confirm that the reaction is spontaneous.

40. A reaction has $\Delta H = 98$ kJ and $\Delta S = 292$ J/K. Investigate the spontaneity of the reaction at room temperature. Would increasing the temperature have any effect on the spontaneity of the reaction?

41. Use tabulated values of standard Gibbs energies to calculate ΔG for the following reaction.

$$4CO(g) + 2H_2O(g) \longrightarrow CH_4(g) + 3CO_2(g)$$

Is the reaction spontaneous under standard conditions?

42. The sugars glucose, $C_6H_{12}O_6(aq)$, and sucrose, $C_{12}H_{22}O_{11}(aq)$, have ΔG_f^0 values of −915 kJ and −1551 kJ respectively. Is the hydrolysis reaction, $C_{12}H_{22}O_{11}(aq) + H_2O(l) \longrightarrow 2C_6H_{12}O_6(aq)$, likely to occur?

MIXED REVIEW

43. How are the coefficients in a chemical equation used to determine the change in a thermodynamic property during a chemical reaction?

44. Is the following reaction exothermic? The standard enthalpy of formation for $CH_2O(g)$ is approximately −109 kJ/mol.

$$CH_2O(g) + CO_2(g) \longrightarrow H_2O(g) + 2CO(g)$$

45. Nitrogen dioxide, $NO_2(g)$, has a standard enthalpy of formation of 33.1 kJ/mol. Dinitrogen tetroxide, $N_2O_4(g)$, has a standard enthalpy of formation of 9.1 kJ/mol. What is the enthalpy change for synthesizing 1.0 mol of N_2O_4 from NO_2?

46. If the molar heat capacity and the temperature change of a substance are known, why is the mass of a substance not needed to calculate its molar enthalpy change?

47. Predict whether ΔS is positive or negative for the following reaction.

$$Ag^+(aq) + Cl^-(aq) \longrightarrow AgCl(s)$$

48. Explain why $AlCl_3$ has a molar heat capacity that is approximately four times the molar heat capacity of a metallic crystal.

49. At high temperatures, does enthalpy or entropy have a greater effect on a reaction's Gibbs energy?

50. Calculate the enthalpy of formation for sulfur dioxide, SO_2, from its elements, sulfur and oxygen. Use the balanced chemical equation and the following information.

$$S(s) + \tfrac{3}{2}O_2(g) \longrightarrow SO_3(g) \quad \Delta H = -395.8 \text{ kJ/mol}$$
$$2SO_2(g) + O_2(g) \longrightarrow 2SO_3(g) \quad \Delta H = -198.2 \text{ kJ/mol}$$

51. For 1 mol of $HCl(g)$, the $\Delta G_f^0 = -92.307$ kJ/mol. Use this information to calculate ΔG for the following reaction.

$$H_2(g) + Cl_2(g) \longrightarrow 2HCl(g)$$

Is the reaction spontaneous?

52. The photosynthesis reaction is shown below.

$$6CO_2(g) + 6H_2O(l) \longrightarrow C_6H_{12}O_6(s) + 6O_2(g)$$

Calculate ΔG using the following information for the reaction.

$$\Delta H = 2870 \text{ kJ} \quad \Delta S = 259 \text{ J/K} \quad T = 297 \text{ K}$$

53. Using the following values, compute the ΔG value for each reaction and predict whether they will occur spontaneously.

Reaction	ΔH (kJ)	Temperature	ΔS (J/K)
1	+125	293 K	+35
2	−85.2	127 K	+125
3	−275	500°C	+45

54. Hydrogen gas can be prepared for use in cars in several ways, such as by the decomposition of water or hydrogen chloride.

$$2H_2O(l) \longrightarrow 2H_2(g) + O_2(g)$$
$$2HCl(g) \longrightarrow H_2(g) + Cl_2(g)$$

Use the following data to determine whether these reactions can occur spontaneously at 25°C. Assume that ΔH and ΔS are constant.

Substance	H_f^0 (kJ/mol)	S^0 (J/K•mol)
$H_2O(l)$	−285.8	+70.0
$H_2(g)$	0	+130.7
$O_2(g)$	0	+205.1
$HCl(g)$	−95.3	+186.9
$Cl_2(g)$	0	+223.1

55. Find the change in entropy for the reaction below.

$$NH_4NO_3(s) \longrightarrow NH_4NO_3(aq, 1m)$$

The standard entropy for $NH_4NO_3(s)$ is 151.1 J/K•mol and the standard entropy for $NH_4NO_3(aq, 1m)$ is 259.8 J/K•mol.

CRITICAL THINKING

56. Why are the specific heats of $F_2(g)$ and $Br_2(g)$ very different, whereas their molar heat capacities are very similar?

57. Look at the two pictures below this question. Which picture appears to have more order? Why? Are there any similarities between the order of marbles and the entropy of particles?

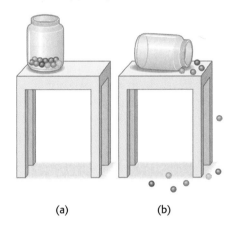

(a)   (b)

58. In what way is specific heat a more useful quantity than molar heat capacity? In what way is molar heat capacity a more useful quantity than specific heat?

59. Why must nutritionists make corrections to bomb calorimetric data if a food contains cellulose or other indigestible fibers?

60. Why are most entries in **Table 2** negative?

61. Why are most entries in **Table 4** negative?

62. Give examples of situations in which **(a)** the entropy is low; **(b)** the entropy is high.

63. Under what circumstances might a nonspontaneous reaction occur?

64. What does ΔT represent?

ALTERNATIVE ASSESSMENT

65. Design an experiment to measure the molar heat capacities of zinc and copper. If your teacher approves the design, obtain the materials needed and conduct the experiment. When you are finished, compare your experimental values with those from a chemical handbook or other reference source.

66. Develop a procedure to measure the ΔH of the following reaction. If your teacher approves your procedure, test your procedure by measuring the ΔH value of the reaction. Determine the accuracy of your method by comparing your ΔH with the accepted ΔH value.

$$CH_3COONa(s) \longrightarrow Na^+(aq) + CH_3COO^-(aq)$$

CONCEPT MAPPING

67. Use the following terms to complete the concept map below: *calorimeter, enthalpy, entropy, Gibbs energy*, and *Hess's Law*

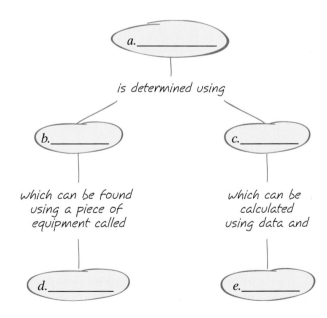

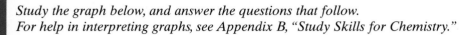

Study the graph below, and answer the questions that follow.
For help in interpreting graphs, see Appendix B, "Study Skills for Chemistry."

68. How would the slope differ if you were to cool the water at the same rate that graph shows the water was heated?

69. What would a slope of zero indicate about the temperature of water during heating?

70. Calculate the slope given the following data.

$y_2 = 3.3$ K $\qquad x_2 = 50$ s

$y_1 = 5.6$ K $\qquad x_1 = 30$ s

71. Calculate the slope given the following data.

$y_2 = 63.7$ mL $\qquad x_2 = 5$ s

$y_1 = 43.5$ mL $\qquad x_1 = 2$ s

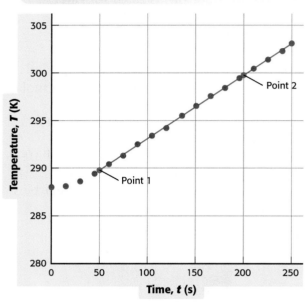

Change in Water Temperature on Heating

TECHNOLOGY AND LEARNING

72. Graphing Calculator

Calculating the Gibbs-Energy Change

The graphing calculator can run a program that calculates the Gibbs-energy change, given the temperature, T, change in enthalpy, ΔH, and change in entropy, ΔS. Given that the temperature is 298 K, the change in enthalpy is 131.3 kJ/mol, and the change in entropy is 0.134 kJ/(mol•K), you can calculate Gibbs-energy change in kilojoules per mole. Then use the program to make calculations.

Go to Appendix C. If you are using a TI-83 Plus, you can download the program **ENERGY** data and run the application as directed. If you are using another

calculator, your teacher will provide you with keystrokes and data sets to use. After you have run the program, answer the following questions.

a. What is the Gibbs-energy change given a temperature of 300 K, a change in enthalpy of 132 kJ/mol, and a change in entropy of 0.086 kJ/(mol•K)?

b. What is the Gibbs-energy change given a temperature of 288 K, a change in enthalpy of 115 kJ/mol, and a change in entropy of 0.113 kJ/(mol•K)?

c. What is the Gibbs-energy change given a temperature of 298 K, a change in enthalpy of 181 kJ/mol, and a change in entropy of 0.135 kJ/(mol•K)?

1. An adiabatic calorimeter can be used to determine the _____ of a reaction.
   a. ΔG
   b. ΔS
   c. ΔH
   d. molar Gibbs energy

2. The temperature of a substance is a measurement of the
   a. total kinetic energy of its atoms.
   b. entropy of the substance.
   c. average kinetic energy of its atoms.
   d. molar heat capacity of the substance.

3. A reaction has a Gibbs energy of –92 kJ. The reaction
   a. is nonspontaneous.
   b. does not have entropy.
   c. does not have free energy.
   d. is spontaneous.

4. Which of the following is not a form of energy?
   a. heat
   b. enthalpy
   c. entropy
   d. kinetic energy

5. Which of the following terms best describes the process of heating a substance?
   a. enthalpy
   b. endothermic
   c. exothermic
   d. standard entropy

6. When carrying out thermodynamic calculations, always
   a. convert joules to calories.
   b. convert temperatures that are in Celsius to temperatures that are in kelvins.
   c. determine the ΔH value first.
   d. convert moles to grams.

7. The following equation
   $$2H_2(g) + O_2(g) \longrightarrow 2H_2O(s) \quad \Delta H = -560 \text{ kJ}$$
   reveals that the standard molar enthalpy of formation of solid water is
   a. –560 kJ
   b. 560 kJ
   c. –280 kJ
   d. 560 J/K

8. Which of the following two conditions will favor a spontaneous reaction?
   a. an increase in entropy and a decrease in enthalpy
   b. an increase in entropy and an increase in enthalpy
   c. a decrease in entropy and a decrease in enthalpy
   d. a decrease in entropy and an increase in enthalpy

9. Photosynthesis is a nonspontaneous reaction, as reflected by
   a. $\Delta H > 0$.
   b. $\Delta S < 0$.
   c. $\Delta G < 0$.
   d. $\Delta G > 0$.

10. The gasification of coal is a method of producing methane by the following reaction.
    $$C(s) + 2H_2(g) \longrightarrow CH_4(g) \quad \Delta H = ?$$
    Find ΔH by using the enthalpy changes in the following combustion reactions
    $$C(s) + O_2(g) \longrightarrow CO_2(g) \quad \Delta H = -393 \text{ kJ}$$
    $$2H_2(g) + O_2(g) \longrightarrow 2H_2O(l) \quad \Delta H = -572 \text{ kJ}$$
    $$CH_4(g) + 2O_2(g) \longrightarrow$$
    $$CO_2(g) + 2H_2O(l) \quad \Delta H = -891 \text{ kJ}$$
    a. 74 kJ
    b. –74 kJ
    c. 1856 kJ
    d. –1856 kJ

11. Which of the following is negative in an exothermic reaction?
    a. ΔT
    b. ΔH
    c. ΔS
    d. ΔG

12. Decreasing the temperature of a nonspontaneous reaction may turn the nonspontaneous reaction into a spontaneous reaction if
    a. $\Delta H > 0$ and $\Delta S > 0$.
    b. $\Delta H > 0$ and $\Delta S < 0$.
    c. $\Delta H < 0$ and $\Delta S > 0$.
    d. $\Delta H < 0$ and $\Delta S < 0$.

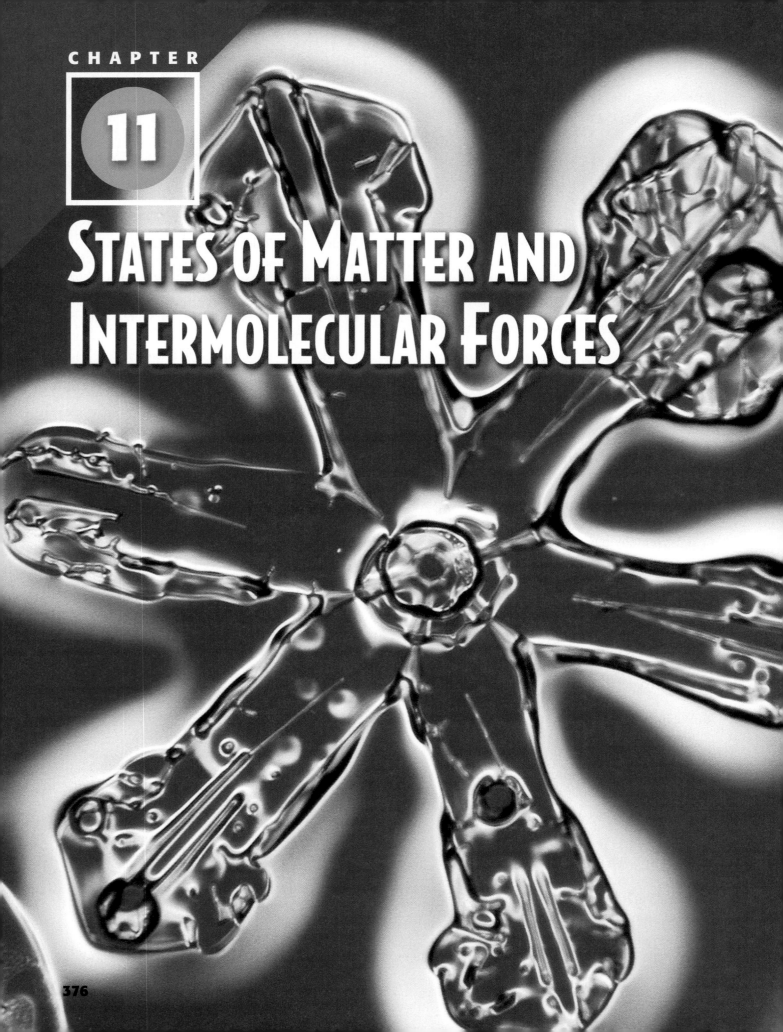

STATES OF MATTER AND INTERMOLECULAR FORCES

Where does a snowflake's elegant structure come from? Snowflakes may not look exactly alike, but all are made up of ice crystals that have a hexagonal arrangement of molecules. This arrangement is due to the shape of water molecules and the attractive forces between them. Water molecules are very polar and form a special kind of attraction, called a *hydrogen bond*, with other water molecules. Hydrogen bonding is just one of the *intermolecular forces* that you will learn about in this chapter.

START-UP*ACTIVITY*

Heating Curve for Water

SAFETY PRECAUTIONS

PROCEDURE

1. Place several **ice cubes** in a **250 mL beaker.** Fill the beaker halfway with **water.** Place the beaker on a **hot plate.** Using a **ring stand,** clamp a **thermometer** so that it is immersed in the ice water but not touching the bottom or sides of the beaker. Record the temperature of the ice water after the temperature has stopped changing.

2. Turn on the hot plate and heat the ice water. Using a **stirring rod,** carefully stir the water as the ice melts.

3. Observe the water as it is heated. Continue stirring. Record the temperature of the water every 30 s. Note the time at which the ice is completely melted. Also note when the water begins to boil.

4. Allow the water to boil for several minutes, and continue to record the temperature every 30 s. Turn off the hot plate, and allow the beaker of water to cool.

5. Is your graph a straight line? If not, where does the slope change?

ANALYSIS

1. Make a graph of temperature as a function of time.

2. What happened to the temperature of the ice water as you heated the beaker?

3. What happened to the temperature of the water after the water started boiling?

CONTENTS

Pre-Reading Questions

(1) **Name two examples each of solids, liquids, and gases.**

(2) **What happens when you heat an ice cube?**

(3) **What force is there between oppositely charged objects?**

internet connect

www.scilinks.org
Topic: Snowflakes
SciLinks code: HW4129

SC*i*LINKS. Maintained by the National Science Teachers Association

States and State Changes

KEY TERMS

- **surface tension**
- **evaporation**
- **boiling point**
- **condensation**
- **melting**
- **melting point**
- **freezing**
- **freezing point**
- **sublimation**

Topic Link

Refer to the chapter "The Science of Chemistry" for a discussion of states of matter.

OBJECTIVES

① **Relate** the properties of a state to the energy content and particle arrangement of that state of matter.

② **Explain** forces and energy changes involved in changes of state.

States of Matter

Have you ever had candy apples like those shown in **Figure 1**? Or have you had strawberries dipped in chocolate? When you make these treats, you can see a substance in two states. The fruit is dipped into the liquid candy or chocolate to coat it. But the liquid becomes solid when cooled. However, the substance has the same identity—and delicious taste—in both states. Most substances, such as the mercury shown in **Figure 2,** can be in three states: solid, liquid, and gas. The physical properties of each state come from the arrangement of particles.

Solid Particles Have Fixed Positions

The particles in a solid are very close together and have an orderly, fixed arrangement. They are held in place by the attractive forces that are between all particles. Because solid particles can vibrate only in place and do not break away from their fixed positions, solids have fixed volumes and shapes. That is, no matter what container you put a solid in, the solid takes up the same amount of space. Solids usually exist in crystalline form. Solid crystals can be very hard and brittle, like salt, or they can be very soft, like lead. Another example of a solid is ice, the solid state of water.

Figure 1
When you make candy apples, you see a substance in two states. The warm liquid candy becomes a solid when cooled.

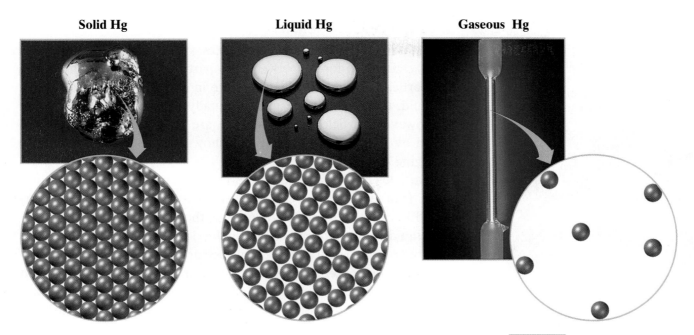

Solid Hg **Liquid Hg** **Gaseous Hg**

Liquid Particles Can Move Easily Past One Another

If you add energy as heat to ice, the ice will melt and become liquid water. In other words, the highly ordered crystals of ice will break apart to form the random arrangement of liquid particles. Liquid particles are also held close together by attractive forces. Thus, the density of a liquid substance is similar to that of the solid substance. However, liquid particles have enough energy to be able to move past each other readily, which allows liquids to flow. That is, liquids are fluids. Some liquids can flow very readily, such as water or gasoline. Other liquids, such as molasses, are thicker and very *viscous* and flow very slowly. Like solids, liquids have fixed volumes. However, while solids keep the same shape no matter the container, liquids flow to take the shape of the lower part of a container. Because liquid particles can move past each other, they are noticeably affected by forces between particles, which gives them special properties.

Liquid Forces Lead to Surface Wetting and Capillary Action

Why does water bead up on a freshly waxed car? Liquid particles can have *cohesion,* attraction for each other. They can also have *adhesion,* attraction for particles of solid surfaces. The balance of these forces determines whether a liquid will wet a solid surface. For example, water molecules have a high cohesion for each other and a low adhesion to particles in car wax. Thus, water drops tend to stick together rather than stick to the car wax.

Water has a greater adhesion to glass than to car wax. The forces of adhesion and cohesion will pull water up a narrow glass tube, called a *capillary tube,* shown in **Figure 3.** The adhesion of the water molecules to the molecules that make up the glass tube pulls water molecules up the sides of the tube. The molecules that are pulled up the glass pull other water molecules with them because of cohesion. The water rises up the tube until the weight of the water above the surface level balances the upward force caused by adhesion and cohesion.

Figure 2
Mercury is the only metal that is a liquid at room temperature, but when cooled below −40°C, it freezes to a solid. At 357°C, it boils and becomes a gas.

Topic Link

Refer to the "Ions and Ionic Compounds" chapter for a discussion of crystal structure.

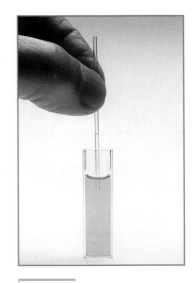

Figure 3
Capillary action, which moves water up through a narrow glass tube, also allows water to move up the roots and stems of plants.

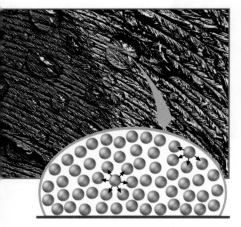

Figure 4
a Water does not wet the feather, because the water's particles are not attracted to the oily film on the feather's surface.

b A drop of water on a surface has particles that are attracted to each other.

surface tension

the force that acts on the surface of a liquid and that tends to minimize the area of the surface

Liquids Have Surface Tension

Why are water drops rounded? Substances are liquids instead of gases because the cohesive forces between the particles are strong enough to pull the particles together so that they are in contact. Below the surface of the liquid, the particles are pulled equally in all directions by these forces. However, particles at the surface are pulled only sideways and downward by neighboring particles, as shown in the model of a water drop in **Figure 4.**

The particles on the surface have a net force pulling them down into the liquid. It takes energy to oppose this net force and increase the surface area. Energy must be added to increase the number of particles at the surface. Liquids tend to decrease energy by decreasing surface area. The tendency of liquids to decrease their surface area to the smallest size possible is called **surface tension.** Surface tension accounts for many liquid properties. Liquids tend to form spherical shapes, because a sphere has the smallest surface area for a given volume. For example, rain and fog droplets are spherical.

Gas Particles Are Essentially Independent

Gas particles are much farther apart than the particles in solids and liquids. They must go far before colliding with each other or with the walls of a container. Because gas particles are so far apart, the attractive forces between them do not have a great effect. They move almost independently of one another. So, unlike solids and liquids, gases fill whatever container they are in. Thus, the shape, volume, and density of an amount of gas depend on the size and shape of the container.

Because gas particles can move around freely, gases are fluids and can flow easily. When you breathe, you can feel how easily the gases that make up air can flow to fill your lungs. Examples of gases include carbon dioxide, a gas that you exhale, and helium, a gas that is used to fill balloons. You will learn more about gases in the "Gases" chapter.

Quick LAB

SAFETY PRECAUTIONS

Wetting a Surface

PROCEDURE
1. Wash **plastic, steel,** and **glass plates** well by using **dilute detergent,** and rinse them completely. Do not touch the clean surfaces.
2. Using a **toothpick,** put a small drop of **water** on each

plate. Observe the shape of the drops from the side.

ANALYSIS
1. On which surface does the water spread the most?
2. On which surface does the water spread the least?

3. What can you conclude about the adhesion of water for plastic, steel, and glass?
4. Explain your observations in terms of wetting.

Changing States

The hardening of melted candy on an apple is just one example of how matter changes states. *Freezing* is the change of state in which a liquid becomes a solid. You can observe freezing when you make ice cubes in the freezer. *Melting* is the change of state in which a solid becomes a liquid. For example, a solid wax candle melts when it is lit.

Evaporation—the change of state in which a liquid becomes a gas—takes place when water boils in a pot or evaporates from damp clothing. Gases can become liquids. *Condensation* is the change of state in which a gas becomes a liquid. For example, water vapor in the air can condense onto a cold glass or onto grass as dew in the morning.

But solids can evaporate, too. A thin film of ice on the edges of a windshield can become a gas by *sublimation* as the car moves through the air. Gases become solids by a process sometimes called *deposition*. For example, frost can form on a cold, clear night from water vapor in the air. **Figure 5** shows these six state changes. All state changes are physical changes, because the identity of the substance does not change, while the physical form of the substance does change.

Temperature, Energy, and State

All matter has energy related to the energy of the rapid, random motion of atom-sized particles. This energy of random motion increases as temperature increases. The higher the temperature is, the greater the average kinetic energy of the particles is. As temperature increases, the particles in solids vibrate more rapidly in their fixed positions. Like solid particles, liquid particles vibrate more rapidly as temperature increases, but they can also move past each other more quickly. Increasing the temperature of a gas causes the free-moving particles to move more rapidly and to collide more often with one another.

Generally, adding energy to a substance will increase the substance's temperature. But after a certain point, adding more energy will cause a substance to experience a change of state instead of a temperature increase.

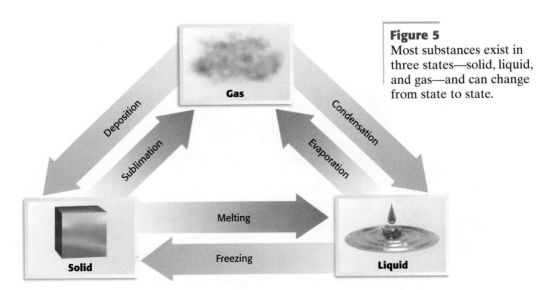

Figure 5
Most substances exist in three states—solid, liquid, and gas—and can change from state to state.

Figure 6
A runner sweats when the body heats as a result of exertion. As sweat evaporates, the body is cooled.

evaporation

the change of a substance from a liquid to a gas

boiling point

the temperature and pressure at which a liquid and a gas are in equilibrium

condensation

the change of state from a gas to a liquid

Figure 7
On a cool night, when humidity is high, water vapor condenses to the liquid state.

Liquid Evaporates to Gas

If you leave an uncovered pan of water standing for a day or two, some of the water disappears. Some of the molecules have left the liquid and gone into the gaseous state. Because even neutral particles are attracted to each other, energy is required to separate them. If the liquid particles gain enough energy of movement, they can escape from the liquid. But where does the energy come from? The liquid particles gain energy when they collide with each other. Sometimes, a particle is struck by several particles at once and gains a large amount of energy. This particle can then leave the liquid's surface through **evaporation.** Because energy must be added to the water, evaporation is an endothermic process. This is why people sweat when they are hot and when they exercise, as shown in **Figure 6.** The evaporation of sweat cools the body.

You may have noticed that a puddle of water on the sidewalk evaporates more quickly on a hot day than on a cooler day. The reason is that the hotter liquid has more high-energy molecules. These high-energy molecules are more likely to gain the extra energy needed to become gas particles more rapidly.

Think about what happens when you place a pan of water on a hot stove. As the liquid is heated, its temperature rises and it evaporates more rapidly. Eventually, it reaches a temperature at which bubbles of vapor rise to the surface, and the temperature of the liquid remains constant. This temperature is the **boiling point.** Why doesn't all of the liquid evaporate at once at the boiling point? The answer is that it takes a large amount of energy to move a molecule from the liquid state to the gaseous state.

Gas Condenses to Liquid

Now, think about what happens if you place a glass lid over a pan of boiling water. You will see liquid form on the underside of the lid. Instead of escaping from the closed pan, the water vapor formed from boiling hits the cooler lid and forms liquid drops through **condensation.** Energy is transferred as heat from the gas particles to the lid. The gas particles no longer have enough energy to overcome the attractive forces between them, so they go into the liquid state. Condensation takes place on a cool night and forms dew on plants, as shown in **Figure 7.** Because energy is released from the water, condensation is an exothermic process.

Figure 8
When the temperature drops below freezing, farmers spray water on the orange trees. Energy is released by water as the water freezes, which warms the oranges and keeps the crop from freezing.

Solid Melts to Liquid

As a solid is heated, the particles vibrate faster and faster in their fixed positions. Their energy of random motion increases. Eventually, a temperature is reached such that some of the molecules have enough energy to break out of their fixed positions and move around. At this point, the solid is **melting.** That is, the solid is becoming a liquid. As long as both the newly formed liquid and the remaining solid are in contact, the temperature will not change. This temperature is the **melting point** of the solid. The energy of random motion is the same for both states. Energy must be absorbed for melting to happen, so melting is endothermic.

melting

the change of state in which a solid becomes a liquid by adding heat or changing pressure

melting point

the temperature and pressure at which a solid becomes a liquid

Liquid Freezes to Solid

The opposite process takes place when **freezing,** shown in **Figure 8,** takes place. As a liquid is cooled, the movement of particles becomes slower and slower. The particles' energy of random motion decreases. Eventually, a temperature is reached such that the particles are attracted to each other and pulled together into the fixed positions of the solid state, and the liquid crystallizes. This exothermic process releases energy—an amount equal to what is added in melting. As long as both states are present, the temperature will not change. This temperature is the **freezing point** of the liquid. Note that the melting point and freezing point are the same for pure substances.

freezing

the change of state in which a liquid becomes a solid as heat is removed

freezing point

the temperature at which a liquid substance freezes

Solid Sublimes to Gas

The particles in a solid are constantly vibrating. Some particles have higher energy than others. Particles with high enough energy can escape from the solid. This endothermic process is called **sublimation.** Sublimation is similar to evaporation. One difference is that it takes more energy to move a particle from a solid into a gaseous state than to move a particle from a liquid into a gaseous state. A common example of sublimation takes place when mothballs are placed in a chest, as shown on the next page in **Figure 9.** The solid naphthalene crystals in mothballs sublime to form naphthalene gas, which surrounds the clothing and keeps moths away.

sublimation

the process in which a solid changes directly into a gas (the term is sometimes also used for the reverse process)

Figure 9
Molecules of naphthalene sublime from the surface of the crystals in the mothball.

Gas Deposits to Solid

The reverse of sublimation is often called *deposition*. Molecules in the gaseous state become part of the surface of a crystal. Energy is released in the exothermic process. The energy released in deposition is equal to the energy required for sublimation. A common example of deposition is the formation of frost on exposed surfaces during a cold night when the temperature is below freezing. In a laboratory, you may see iodine gas deposit as solid crystals onto the surface of a sealed container.

Section Review

UNDERSTANDING KEY IDEAS

1. Describe what happens to the shape and volume of a solid, a liquid, and a gas when you place each into separate, closed containers.

2. What is surface tension?

3. You heat a piece of iron from 200 to 400 K. What happens to the atoms' energy of random motion?

4. When water boils, bubbles form at the base of the container. What gas has formed?

5. What two terms are used to describe the temperature at which solids and liquids of the same substance exist at the same time?

6. How are sublimation and evaporation similar?

7. Describe an example of deposition.

CRITICAL THINKING

8. The densities of the liquid and solid states of a substance are often similar. Explain.

9. How could you demonstrate evaporation?

10. How could you demonstrate boiling point?

11. You are boiling potatoes on a gas stove, and your friend suggests turning up the heat to cook them faster. Will this idea work?

12. A dehumidifier takes water vapor from the air by passing the moist air over a set of cold coils to perform a state change. How does a dehumidifier work?

13. Water at 50°C is cooled to −10°C. Describe what will happen.

14. How could you demonstrate melting point?

15. Explain why changes of state are considered physical transitions and not chemical processes.

KEY TERMS

- **intermolecular forces**
- **dipole-dipole forces**
- **hydrogen bond**
- **London dispersion force**

OBJECTIVES

① **Contrast** ionic and molecular substances in terms of their physical characteristics and the types of forces that govern their behavior.

② **Describe** dipole-dipole forces.

③ **Explain** how a hydrogen bond is different from other dipole-dipole forces and how it is responsible for many of water's properties.

④ **Describe** London dispersion forces, and relate their strength to other types of attractions.

Comparing Ionic and Covalent Compounds

Particles attract each other, so it takes energy to overcome the forces holding them together. If it takes high energy to separate the particles of a substance, then it takes high energy to cause that substance to go from the liquid to the gaseous state. The boiling point of a substance is a good measure of the strength of the forces that hold the particles together. Melting point also relates to attractive forces between particles. Most covalent compounds melt at lower temperatures than ionic compounds do. As shown in **Table 1,** ionic substances with small ions tend to be solids that have high melting points, and covalent substances tend to be gases and liquids or solids that have low melting points.

Table 1 Comparing Ionic and Molecular Substances

Type of substance	Common use	State at room temperature	Melting point (°C)	Boiling point (°C)
Ionic substances				
Potassium chloride, KCl	salt substitute	solid	770	sublimes at 1500
Sodium chloride, NaCl	table salt	solid	801	1413
Calcium fluoride, CaF_2	water fluoridation	solid	1423	2500
Covalent substances				
Methane, CH_4	natural gas	gas	−182	−164
Ethyl acetate, $CH_3COOCH_2CH_3$	fingernail polish	liquid	−84	77
Water, H_2O	(many)	liquid	0	100
Heptadecane, $C_{17}H_{36}$	wax candles	solid	22	302

Oppositely Charged Ions Attract Each Other

Ionic substances generally have much higher forces of attraction than covalent substances. Recall that ionic substances are made up of separate ions. Each ion is attracted to all ions of opposite charge. For small ions, these attractions hold the ions tightly in a crystal lattice that can be disrupted only by heating the crystal to very high temperatures.

The strength of ionic forces depends on the size of the ions and the amount of charge. Ionic compounds with small ions have high melting points. If the ions are larger, then the distances between them are larger and the forces are weaker. This effect helps explain why potassium chloride, KCl, melts at a lower temperature than sodium chloride, NaCl, does. Now compare ions that differ by the amount of charge they have. If the ions have larger charges, then the ionic force is larger than the ionic forces of ions with smaller charges. This effect explains why calcium fluoride, CaF_2 melts at a higher temperature than NaCl does.

Intermolecular Forces Attract Molecules to Each Other

For covalent substances, forces that act between molecules are called **intermolecular forces.** They can be dipole-dipole forces or London dispersion forces. Both forces are short-range and decrease rapidly as molecules get farther apart. Because the forces are effective only when molecules are near each other, they do not have much of an impact on gases. A substance with weak attractive forces will be a gas because there is not enough attractive force to hold molecules together as a liquid or a solid.

The forces that hold the molecules together act only between neighboring molecules. The forces may be weak; some molecular substances boil near absolute zero. For example, hydrogen gas, H_2, boils at −252.8°C. The forces may be strong; some molecular substances have very high boiling points. For example, coronene, $C_{24}H_{12}$, boils at 525°C.

Dipole-Dipole Forces

In **dipole-dipole forces,** the positive end of one molecule attracts the negative end of a neighboring molecule. Bonds are polar because atoms of differing electronegativity are bonded together. The greater the difference in electronegativity in a diatomic molecule, the greater the polarity is.

Dipole-Dipole Forces Affect Melting and Boiling Points

When polar molecules get close and attract each other, the force is significant if the degree of polarity is fairly high. When molecules are very polar, the dipole-dipole forces are very significant. Remember that the boiling point of a substance tells you something about the forces between the molecules. For example, **Table 2** shows that the polar compound 1-propanol, C_3H_7OH, boils at 97.4°C. The less polar compound of similar size, 1-propanethiol, C_3H_7SH, boils at 67.8°C. However, the nonpolar compound butane, C_4H_{10}, also of similar size, boils at −0.5°C. The more polar the molecules are, the stronger the dipole-dipole forces between them, and thus, the higher the boiling point.

Topic Link

Refer to the "Ions and Ionic Compounds" chapter for a discussion of crystal lattices.

intermolecular forces

the forces of attraction between molecules

dipole-dipole forces

interactions between polar molecules

Topic Link

Refer to the "Covalent Compounds" chapter for a discussion of dipoles.

Table 2 Comparing Dipole-Dipole Forces

Substance	Boiling point (°C)	Polarity	State at room temperature	Structure
1-propanol, C_3H_7OH	97.4	polar	liquid	H H H H—C—C—C—O—H H H H
1-propanethiol, C_3H_7SH	67.8	less polar	liquid	H H H H—C—C—C—S—H H H H
Butane, C_4H_{10}	−0.5	nonpolar	gas	H H H H H—C—C—C—C—H H H H H
Water, H_2O	100.0	polar	liquid	O H H
Hydrogen sulfide, H_2S	−60.7	less polar	gas	S H H
Ammonia, NH_3	−33.35	polar	gas	H H—N—H
Phosphine, PH_3	−87.7	less polar	gas	H H—P—H

Hydrogen Bonds

Compare the boiling points of H_2O and H_2S, shown in **Table 2.** These molecules have similar sizes and shapes. However, the boiling point of H_2O is much higher than that of H_2S. A similar comparison of NH_3 with PH_3 can be made. The greater the polarity of a molecule, the higher the boiling point is. However, when hydrogen atoms are bonded to very electronegative atoms, the effect is even more noticeable.

Compare the boiling points and electronegativity differences of the hydrogen halides, shown in **Table 3.** As the electronegativity difference increases, the boiling point increases. The boiling points increase somewhat from HCl to HBr to HI but increase a lot more for HF. What accounts for this jump? The answer has to do with a special form of dipole-dipole forces, called a **hydrogen bond.**

hydrogen bond

the intermolecular force occurring when a hydrogen atom that is bonded to a highly electronegative atom of one molecule is attracted to two unshared electrons of another molecule

☑ internet connect

www.scilinks.org
Topic: Hydrogen Bonding
SciLinks code: HW4069

SCiLINKS. Maintained by the National Science Teachers Association

Table 3 Boiling Points of the Hydrogen Halides

Substance	HF	HCl	HBr	HI
Boiling point (°C)	20	−85	−67	−35
Electronegativity difference	1.8	1.0	0.8	0.5

Hydrogen Bonds Form with Electronegative Atoms

Strong hydrogen bonds can form with a hydrogen atom that is covalently bonded to very electronegative atoms in the upper-right part of the periodic table: nitrogen, oxygen, and fluorine. When a hydrogen atom bonds to an atom of N, O, or F, the hydrogen atom has a large, partially positive charge. The partially positive hydrogen atom of polar molecules can be attracted to the unshared pairs of electrons of neighboring molecules. For example, the hydrogen bonds shown in **Figure 10** result from the attraction of the hydrogen atoms in the H–N and H–O bonds of one DNA strand to the unshared pairs of electrons in the complementary DNA strand. These hydrogen bonds hold together the complementary strands of DNA, which contain the body's genetic information.

Hydrogen Bonds Are Strong Dipole-Dipole Forces

It is not just electronegativity difference that accounts for the strength of hydrogen bonds. One reason that hydrogen bonds are such strong dipole-dipole forces is because the hydrogen atom is small and has only one electron. When that electron is pulled away by a highly electronegative atom, there are no more electrons under it. Thus, the single proton of the hydrogen nucleus is partially exposed. As a result, hydrogen's proton is strongly attracted to the unbonded pair of electrons of other molecules. The combination of the large electronegativity difference (high polarity) and hydrogen's small size accounts for the strength of the hydrogen bond.

Figure 10
Hydrogen bonding between base pairs on adjacent molecules of DNA holds the two strands together. Yet the force is not so strong that the strands cannot be separated.

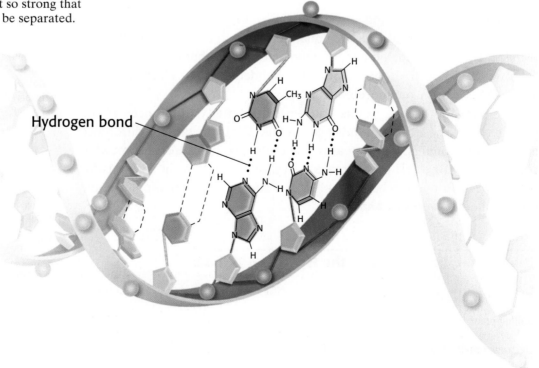

Hydrogen bond

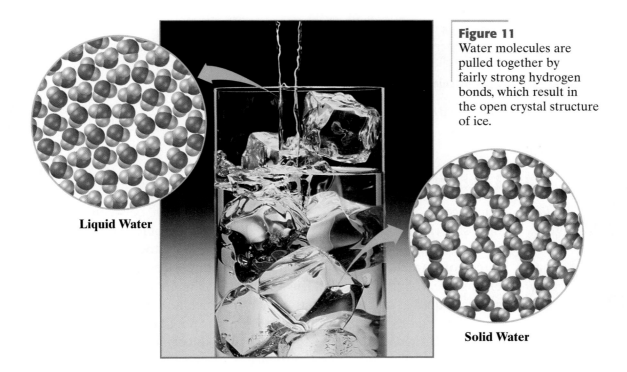

Figure 11
Water molecules are pulled together by fairly strong hydrogen bonds, which result in the open crystal structure of ice.

Liquid Water

Solid Water

Hydrogen Bonding Explains Water's Unique Properties

The energy of hydrogen bonds is lower than that of normal chemical bonds but can be stronger than that of other intermolecular forces. Hydrogen bonding can account for many properties. **Figure 11** shows an example of hydrogen bonding that involves oxygen. Water has unique properties. These unique properties are the result of hydrogen bonding.

Water is different from most other covalent compounds because of how much it can participate in strong hydrogen bonding. In water, two hydrogen atoms are bonded to oxygen by polar covalent bonds. Each hydrogen atom can form hydrogen bonds with neighboring molecules. Because of the water molecule's ability to form multiple hydrogen bonds at once, the intermolecular forces in water are strong.

Another different characteristic of water results from hydrogen bonding and the shape of a water molecule. Unlike most solids, which are denser than their liquids, solid water is less dense than liquid water and floats in liquid water. The angle between the two H atoms in water is 104.5°. This angle is very close to the tetrahedral angle of 109.5°. When water forms solid ice, the angle in the molecules causes the special geometry of molecules in the crystal shown in **Figure 11**. Ice crystals have large amounts of open space, which causes ice to have a low density.

The unusual density difference between liquid and solid water explains many important phenomena in the natural world. For example, because ice floats on water, ponds freeze from the top down and not from the bottom up. Thus, fish can survive the winter in water under an insulating layer of ice. Because water expands when it freezes, water seeping into the cracks of rock or concrete can cause considerable damage due to fracturing. You should never freeze water-containing foods in glass containers, which may break when the water freezes and expands.

Table 4 Boiling Points of the Noble Gases

Substance	He	Ne	Ar	Kr	Xe	Rn
Boiling point (°C)	−269	−246	−186	−152	−107	−62
Number of electrons	2	10	18	36	54	86

Figure 12
The nonpolar molecules in gasoline are held together by London dispersion forces, so it is not a gas at room temperature.

London dispersion force

the intermolecular attraction resulting from the uneven distribution of electrons and the creation of temporary dipoles

Figure 13
Temporary dipoles in molecules cause forces of attraction between the molecules.

London Dispersion Forces

Some compounds are ionic, and forces of attraction between ions of opposite charge cause the ions to stick together. Some molecules are polar, and dipole-dipole forces hold polar molecules together. But what forces of attraction hold together nonpolar molecules and atoms? For example, gasoline, shown in **Figure 12,** contains nonpolar octane, C_8H_{18}, and is a liquid at room temperature. Why isn't octane a gas? Clearly, some sort of intermolecular force allows gasoline to be a liquid.

In 1930, the German chemist Fritz W. London came up with an explanation. Nonpolar molecules experience a special form of dipole-dipole force called **London dispersion force.** In dipole-dipole forces, the negative part of one molecule attracts the positive region of a neighboring molecule. However, in London dispersion forces, there is no special part of the molecule that is always positive or negative.

London Dispersion Forces Exist Between Nonpolar Molecules

In general, the strength of London dispersion forces between nonpolar particles increases as the molar mass of the particles increases. This is because generally, as molar mass increases, so does the number of electrons in a molecule. Consider the boiling point of the noble gases, as shown in **Table 4.** Generally, as boiling point increases, so does the number of electrons in the atoms. For groups of similar atoms and molecules, such as the noble gases or hydrogen halides, London dispersion forces are roughly proportional to the number of electrons present.

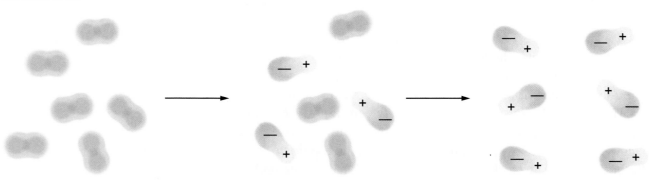

a Nonpolar molecules can become momentarily polar.

b The instantaneous dipoles that form cause adjacent molecules to polarize.

c These London dispersion forces cause the molecules to attract each other.

London Dispersion Forces Result from Temporary Dipoles

How do electrons play a role in London dispersion forces? The answer lies in the way that electrons move and do not stay still. The electrons in atoms and molecules can move. They not only move about in orbitals but also can move from one side of an atom to the other. When the electrons move toward one side of an atom or molecule, that side becomes momentarily negative and the other side becomes momentarily positive. If the positive side of a momentarily charged molecule moves near another molecule, the positive side can attract the electrons in the other molecule. Or the negative side of the momentarily charged molecule can push the electrons of the other molecule away. The temporary dipoles that form attract each other, as shown in **Figure 13,** and make temporary dipoles form in other molecules. When molecules are near each other, they always exert an attractive force because electrons can move.

Properties Depend on Types of Intermolecular Force

Compare the properties of an ionic substance, NaCl, with those of a nonpolar substance, I_2, as shown in **Figure 14.** The differences in the properties of the substances are related to the differences in the types of forces that act within each substance. Because ionic, polar covalent, and nonpolar covalent substances are different in electron distribution, they are different in the types of attractive forces that they experience.

While nonpolar molecules can experience only London dispersion forces, polar molecules experience both dipole-dipole forces and London dispersion forces. Determining how much each force adds to the overall force of attraction between polar molecules is not easy. London dispersion forces also exist between ions in ionic compounds, but they are quite small relative to ionic forces and can almost always be overlooked.

Figure 14
Forces between ions are generally much stronger than the forces between molecules, so the melting points of ionic substances tend to be higher.

a In the sodium chloride crystal, each ion is strongly attracted to six oppositely charged ions. NaCl has a melting point of 801°C.

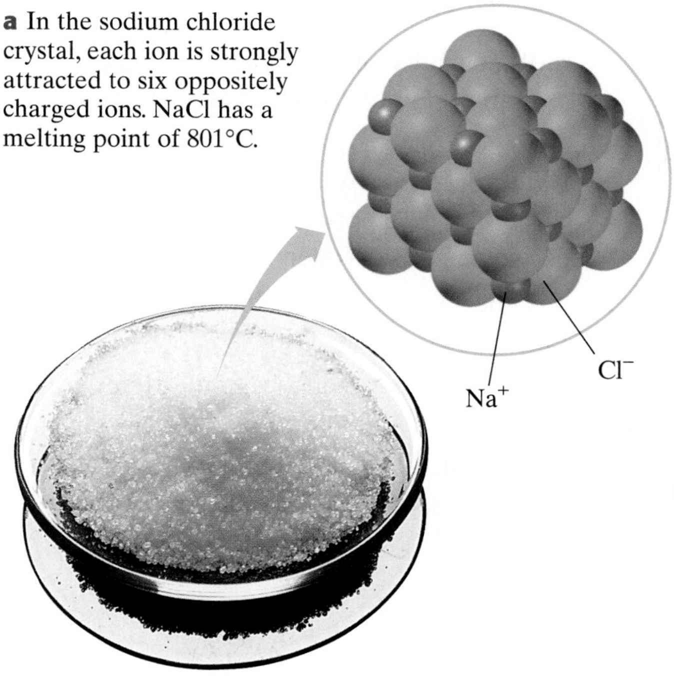

Na$^+$

Cl$^-$

b In the iodine crystals, the particles are neutral molecules that are not as strongly attracted to each other. I_2 has a melting point of 114°C.

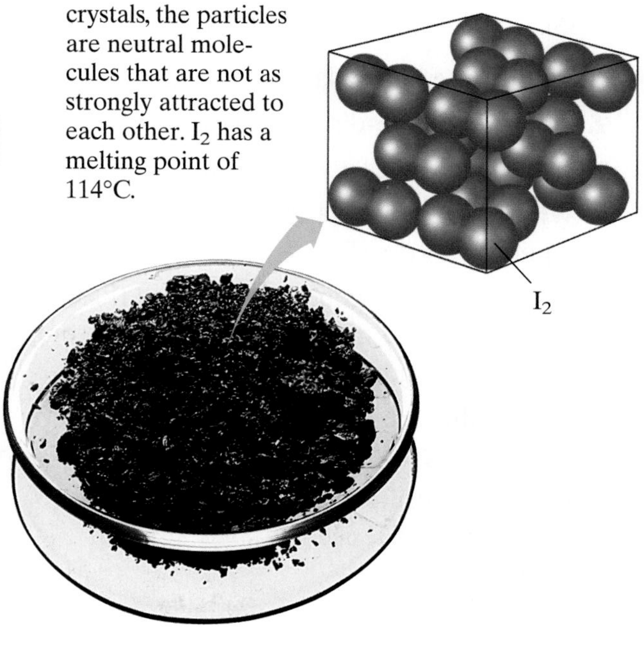

I_2

Figure 15
a The polyatomic ionic compound 1-butylpyridinium nitrate is a liquid solvent at room temperature. The large size of the cations keeps the ionic forces from having a great effect.

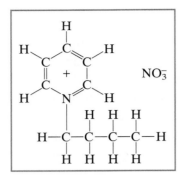

b A molecule of coronene, $C_{24}H_{12}$, is very large, yet its flat shape allows it to have relatively strong London dispersion forces.

Particle Size and Shape Also Play a Role

Dipole-dipole forces are generally stronger than London dispersion forces. However, both of these forces between molecules are usually much weaker than ionic forces in crystals. There are exceptions. One major factor is the size of the atoms, ions, or molecules. The larger the particles are, the farther apart they are and the smaller the effects of the attraction are. If an ionic substance has very large ions—especially if the ions are not symmetrical—the ionic substance's melting point can be very low. A few ionic compounds are even liquid at room temperature, such as 1-butylpyridinium nitrate, shown in **Figure 15.**

The shape of the particles can also play a role in determining the strength of attraction. For example, coronene molecules, $C_{24}H_{12}$, are very large. However, they are flat, so they can come close together and the attractive forces have a greater effect. Thus, the boiling point of nonpolar coronene is almost as high as that of some ionic compounds.

② Section Review

UNDERSTANDING KEY IDEAS

1. What force holds NaCl units together?
2. Describe dipole-dipole forces.
3. What force gives water unique properties?
4. Why does ice have a lower density than liquid water does?
5. Explain why oxygen, nitrogen, and fluorine are elements in molecules that form strong hydrogen bonds.
6. How is the strength of London dispersion forces related to the number of electrons?
7. How do intermolecular forces affect whether a substance is a solid at room temperature?

CRITICAL THINKING

8. **a.** Which is nonpolar: CF_4 or CH_2F_2?
   b. Which substance likely has a higher boiling point? Explain your answer.
9. Are the London dispersion forces between water molecules weaker or stronger than the London dispersion forces between molecules of hydrogen sulfide, H_2S?
10. NH_3 has a much higher boiling point than PH_3 does. Explain.
11. Why does argon boil at a higher temperature than neon does?
12. Which will have the higher melting point, KF or KNO_3? Explain your answer.

SECTION

3 Energy of State Changes

OBJECTIVES

1. **Define** the molar enthalpy of fusion and the molar enthalpy of vaporization, and **identify** them for a substance by using a heating curve.

2. **Describe** how enthalpy and entropy of a substance relate to state.

3. **Predict** whether a state change will take place by using Gibbs energy.

4. **Calculate** melting and boiling points by using enthalpy and entropy.

5. **Explain** how pressure affects the entropy of a gas and affects changes between the liquid and vapor states.

Enthalpy, Entropy, and Changes of State

Adding enough energy to boil a pan of water takes a certain amount of time. Removing enough energy to freeze a tray of ice cubes also takes a certain amount of time. At that rate, you could imagine that freezing the water that makes up the iceberg in **Figure 16** would take a *very* long time.

Enthalpy is the total energy of a system. *Entropy* measures a system's disorder. The energy added during melting or removed during freezing is called the *enthalpy of fusion*. (*Fusion* means *melting*.) Particle motion is more random in the liquid state, so as a solid melts, the entropy of its particles increases. This increase is the *entropy of fusion*. As a liquid evaporates, a lot of energy is needed to separate the particles. This energy is the *enthalpy of vaporization*. (*Vaporization* means *evaporation*.) Particle motion is much more random in a gas than in a liquid. A substance's *entropy of vaporization* is much larger than its entropy of fusion.

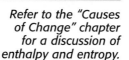

Refer to the "Causes of Change" chapter for a discussion of enthalpy and entropy.

Figure 16
Melting an iceberg would take a great amount of enthalpy of fusion.

393

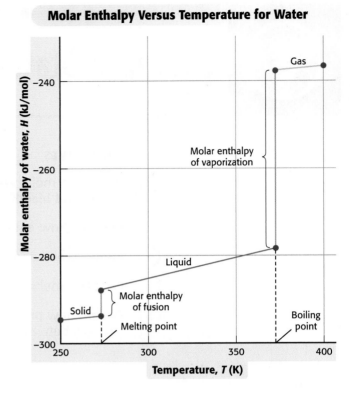

Figure 17
Energy is added to 1 mol of ice. At the melting point and boiling point, the temperatures remain constant and large changes in molar enthalpy take place.

Molar Enthalpy Versus Temperature for Water

Enthalpy and Entropy Changes for Melting and Evaporation

Enthalpy and entropy change as energy in the form of heat is added to a substance, as shown with water in **Figure 17.** The graph starts with 1 mol of solid ice at 250 K (−23°C). The ice warms to 273.15 K. The enthalpy, H, increases slightly during this process. At 273.15 K, the ice begins to melt. As long as both ice and liquid water are present, the temperature remains at 273.15 K. The energy added is the *molar enthalpy of fusion* (ΔH_{fus}), which is 6.009 kJ/mol for ice. ΔH_{fus} is the difference in enthalpy between solid and liquid water at 273.15 K as shown in the following equation:

$$\Delta H_{fus} = H_{\text{(liquid at melting point)}} - H_{\text{(solid at melting point)}}$$

After the ice melts, the temperature of the liquid water increases as energy is added until the temperature reaches 373.15 K.

At 373.15 K, the water boils. If the pressure remains constant, so does the temperature as long as the two states (liquid and gas) are present. The energy added is the *molar enthalpy of vaporization* (ΔH_{vap}), 40.67 kJ/mol. ΔH_{vap} is the difference in enthalpy between liquid and gaseous water at 373.15 K and is defined in the following equation:

$$\Delta H_{vap} = H_{\text{(vapor at boiling point)}} - H_{\text{(liquid at boiling point)}}$$

After all of the liquid water has evaporated, the energy added increases the temperature of the water vapor.

Like water, almost all substances can be in the three common states of matter. **Table 5** lists the molar enthalpies and entropies of fusion and vaporization for some elements and compounds. Because intermolecular forces are not significant in the gaseous state, most substances have similar values for *molar entropy of vaporization,* ΔS_{vap}.

Table 5 — Molar Enthalpies and Entropies of Fusion and Vaporization

Substance	T_{mp} (K)	ΔH_{fus} (kJ/mol)	ΔS_{fus} (J/mol•K)	T_{bp} (K)	ΔH_{vap} (kJ/mol)	ΔS_{vap} (J/mol•K)
Nitrogen, N_2	63	0.71	11.3	77	5.6	72.2
Hydrogen sulfide, H_2S	188	23.8	126.6	214	18.7	87.4
Bromine, Br_2	266	10.57	39.8	332	30.0	90.4
Water, H_2O	273	6.01	22.0	373	40.7	108.8
Benzene, C_6H_6	279	9.95	35.7	353	30.7	87.0
Lead, Pb	601	4.77	7.9	2022	179.5	88.8

Gibbs Energy and State Changes

As you have learned, the relative values of H and S determine whether any process, including a state change, will take place. The following equation describes a change in Gibbs energy.

$$\Delta G = \Delta H - T\Delta S$$

You may recall that a process is spontaneous if ΔG is negative. That is, the process can take place with a decrease in Gibbs energy. If ΔG is positive, then a process will not take place unless an outside source of energy drives the process. If ΔG is zero, then the system is said to be in a state of *equilibrium*. At equilibrium, the forward and reverse processes are happening at the same rate. For example, when solid ice and liquid water are at equilibrium, ice melts at the same rate that water freezes. You will learn more about equilibrium in the next section.

Enthalpy and Entropy Determine State

At normal atmospheric pressure, water freezes at 273.15 K (0.00°C). At this pressure, pure water will not freeze at any temperature above 273.15 K. Likewise, pure ice will not melt at any temperature below 273.15 K. This point may be proven by looking at ΔG just above and just below the normal freezing point of water. At the normal freezing point, the enthalpy of fusion of ice is 6.009 kJ/mol, or 6009 J/mol. For changes in state that take place at constant temperature, the entropy change, ΔS, is $\Delta H/T$. Thus, ΔS is (6009 J/mol)/(273.15 K) = 22.00 J/mol•K for the melting of ice.

Now let us calculate the Gibbs energy change for the melting of ice at 273.00 K. For this change, ΔH is positive (energy is absorbed), and ΔS is also positive (greater degree of disorder).

$$\Delta G = \Delta H - T\Delta S = +6009 \text{ J/mol} - (273.00 \text{ K} \times +22.00 \text{ J/mol•K})$$
$$= +6009 \text{ J/mol} - 6006 \text{ J/mol} = +3 \text{ J/mol}$$

Because ΔG is positive, the change will not take place on its own. The ordered state of ice is preferred at this temperature, which is below the normal freezing point.

Similarly, think about the possibility of water freezing at 273.30 K. The ΔH is now negative (energy is released). The ΔS is also negative (greater degree of order in the crystal).

$$\Delta G = \Delta H - T\Delta S = -6009 \text{ J/mol} - (273.30 \text{ K} \times -22.00 \text{ J/mol} \cdot \text{K})$$
$$= -6009 \text{ J/mol} - 6013 \text{ J/mol} = +4 \text{ J/mol}$$

ΔG is positive, so the water will not freeze. The disordered state of liquid water is preferred at 273.30 K, which is above the melting point.

Determining Melting and Boiling Points

For a system at the melting point, a solid and a liquid are in equilibrium, so ΔG is zero. Thus, $\Delta H = T\Delta S$. Rearranging the equation, you get the following relationship, in which *mp* means melting point and *fus* means fusion.

$$T_{mp} = \frac{\Delta H_{fus}}{\Delta S_{fus}}$$

In other words, the melting point of a solid, T_{mp}, is equal to molar enthalpy of fusion, ΔH_{fus}, divided by molar entropy of fusion, ΔS_{fus}. Boiling takes place when the drive toward disorder overcomes the tendency to lose energy. Condensation, shown in **Figure 18,** takes place when the tendency to lose energy overcomes the drive to increase disorder. In other words, when $\Delta H_{vap} > T\Delta S_{vap}$, the liquid state is favored. The gas state is preferred when $\Delta H_{vap} < T\Delta S_{vap}$.

The same situation happens at the boiling point. ΔG is zero when liquid and gas are in equilibrium, so $\Delta H_{vap} = T\Delta S_{vap}$. Thus, given that *bp* stands for boiling point and *vap* stands for vaporization, the following equation is true.

$$T_{bp} = \frac{\Delta H_{vap}}{\Delta S_{vap}}$$

In other words, the boiling point of a liquid, T_{bp}, is equal to molar enthalpy of vaporization, ΔH_{vap}, divided by molar entropy of vaporization, ΔS_{vap}.

Figure 18
a Water condenses on the wings of the dragonfly when $\Delta H_{vap} > T\Delta S_{vap}$.
b Water freezes on the flower when $\Delta H_{fus} > T\Delta S_{fus}$.

Calculating Melting and Boiling Points

The enthalpy of fusion of mercury is 11.42 J/g, and the molar entropy of fusion is 9.79 J/mol•K. The enthalpy of vaporization at the boiling point is 294.7 J/g, and the molar entropy of vaporization is 93.8 J/mol•K. Calculate the melting point and the boiling point.

1 Gather information.

- molar mass of Hg = 200.59 g/mol
- enthalpy of fusion = 11.42 J/g
- molar entropy of fusion = 9.79 J/mol•K
- enthalpy of vaporization = 294.7 J/g
- molar entropy of vaporization = 93.8 J/mol•K
- melting point, T_{mp} = ?
- boiling point, T_{bp} = ?

2 Plan your work.

First calculate the molar enthalpy of fusion and molar enthalpy of vaporization, which have units of J/mol. Use the molar mass of mercury to convert from J/g to J/mol.

$$\Delta H_{fus} = 11.42 \text{ J/g} \times 200.59 \text{ g/mol} = 2291 \text{ J/mol}$$
$$\Delta H_{vap} = 294.7 \text{ J/g} \times 200.59 \text{ g/mol} = 59\,110 \text{ J/mol}$$

Set up the equations for determining T_{mp} and T_{bp}.

3 Calculate.

$$T_{mp} = \Delta H_{fus}/\Delta S_{fus} = \frac{2291 \text{ J/mol}}{9.79 \text{ J/mol•K}} = 234 \text{ K}$$

$$T_{bp} = \Delta H_{vap}/\Delta S_{vap} = \frac{59\,110 \text{ J/mol}}{93.8 \text{ J/mol•K}} = 630 \text{ K}$$

4 Verify your result.

Mercury is a liquid at room temperature, so the melting point must be below 298 K (25°C). Mercury boils well above room temperature, so the boiling point must be well above 298 K. These facts fit the calculation.

> **PRACTICE HINT**
>
> When setting up your equations, use the correct conversion factors so that you get the desired units when canceling. But be careful to keep values for vaporization together and separate from those for fusion. Keep track of your units! You may have to convert from joules to kilojoules or vice versa.

PRACTICE

Calculate the freezing and boiling points for each substance.

1 For ethyl alcohol, C_2H_5OH, the enthalpy of fusion is 108.9 J/g, and the entropy of fusion is 31.6 J/mol•K. The enthalpy of vaporization at the boiling point is 837 J/g, and the molar entropy of vaporization is 109.9 J/mol•K.

2 For sulfur dioxide, the molar enthalpy of fusion is 8.62 kJ/mol, and the molar entropy of fusion is 43.1 J/mol•K. ΔH_{vap} is 24.9 kJ/mol, and the molar entropy of vaporization at the boiling point is 94.5 J/mol•K.

3 For ammonia, ΔH_{fus} is 5.66 kJ/mol, and ΔS_{fus} is 29.0 J/mol•K. ΔH_{vap} is 23.33 kJ/mol, and ΔS_{vap} is 97.2 J/mol•K.

Pressure Can Affect Change-of-State Processes

Boiling points are pressure dependent because pressure has a large effect on the entropy of a gas. When a gas is expanded (pressure is decreased), its entropy increases because the degree of disorder of the molecules increases. At sea level, water boils at 100°C. In Denver, Colorado, where the elevation is 1.6 km, atmospheric pressure is about 0.84 times the pressure at sea level. At that elevation, water boils at about 95°C. On Pike's Peak, where the elevation is 4.3 km, water boils at about 85°C. People often use pressure cookers at that altitude to increase the boiling point of water.

Liquids and solids are almost incompressible. Therefore, changes of atmospheric pressure have little effect on the entropy of substances in liquid or solid states. Ordinary changes in pressure have essentially no effect on melting and freezing. Although the elevation is high and atmospheric pressure is very low, water on Pike's Peak still freezes at 273.15 K. You will learn more about pressure effects on state changes in the next section.

Section Review

UNDERSTANDING KEY IDEAS

1. What is the molar enthalpy of fusion?

2. What is the molar enthalpy of vaporization?

3. Compare the sizes of the entropy of fusion and entropy of vaporization of a substance.

4. Explain why liquid water at 273.3 K will not freeze in terms of Gibbs energy.

5. The following process has a ΔG equal to zero at 77 K and standard pressure. In how many states can nitrogen be present at this temperature and pressure?

$$N_2(l) \longrightarrow N_2(g)$$

6. **a.** How does atmospheric pressure affect the boiling point of a liquid?

   b. How does atmospheric pressure affect the melting point of a liquid?

PRACTICE PROBLEMS

7. The enthalpy of fusion of bromine is 10.57 kJ/mol. The entropy of fusion is 39.8 J/mol·K. Calculate the freezing point.

8. Calculate the boiling point of bromine given the following information:

   $Br_2(l) \longrightarrow Br_2(g)$
   $\Delta H_{vap} = 30.0$ kJ/mol
   $\Delta S_{vap} = 90.4$ J/mol·K

9. The enthalpy of fusion of nitric acid, HNO_3, is 167 J/g. The entropy of fusion is 45.3 J/mol·K. Calculate the melting point.

CRITICAL THINKING

10. In terms of enthalpy and entropy, when does melting take place?

11. Why is the enthalpy of vaporization of a substance always much greater than the enthalpy of fusion?

12. Why is the gas state favored when

$$T > \frac{\Delta H_{vap}}{\Delta S_{vap}} ?$$

13. Determine the change-of-state process described by each of the following:

   a. $\Delta H_{vap} > T\Delta S_{vap}$ **c.** $\Delta H_{vap} < T\Delta S_{vap}$

   b. $\Delta H_{fus} < T\Delta S_{fus}$ **d.** $\Delta H_{fus} > T\Delta S_{fus}$

Phase Equilibrium

KEY TERMS

- **phase**
- **equilibrium**
- **vapor pressure**
- **phase diagram**
- **triple point**
- **critical point**

OBJECTIVES

1. **Identify** systems that have multiple phases, and determine whether they are at equilibrium.

2. **Understand** the role of vapor pressure in changes of state between a liquid and a gas.

3. **Interpret** a phase diagram to identify melting points and boiling points.

Two-Phase Systems

A *system* is a set of components that are being studied. Within a system, a **phase** is a region that has the same composition and properties throughout. The lava lamp in **Figure 19** is a system that has two phases, each of which is liquid. The two phases in a lava lamp are different from each other because their chemical compositions are different. A glass of water and ice cubes is also a system that has two phases. This system has a solid phase and a liquid phase. However, the two phases have the same chemical composition. What makes the two phases in ice water different from each other is that they are different states of the same substance, water.

Phases do not need to be pure substances. If some salt is dissolved in the glass of water with ice cubes, there are still two phases: a liquid phase (the solution) and a solid phase (the pure ice). In this chapter, we will consider only systems like the ice water, that is, systems that contain one pure substance whose phases are different only by state.

phase

a part of matter that is uniform

Figure 19
The lava lamp is a system that has two liquid phases, and the ice water is a system that has a solid phase and a liquid phase.

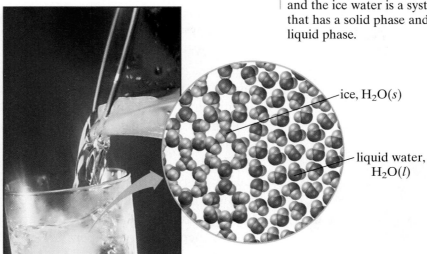

ice, $H_2O(s)$

liquid water, $H_2O(l)$

Figure 20
Iodine sublimes even at room temperature. Molecules escape from the solid and go into the gas phase, which is in equilibrium with the solid.

I$_2$(g)

I$_2$(s)

equilibrium

the state in which a chemical process and the reverse chemical process occur at the same rate such that the concentrations of reactants and products do not change

vapor pressure

the partial pressure exerted by a vapor that is in equilibrium with its liquid state at a given temperature

Figure 21
A few molecules of a liquid have enough energy to overcome intermolecular forces and escape from the surface into the gas phase, which is in equilibrium with the liquid.

Equilibrium Involves Constant Interchange of Particles

If you open a bottle of rubbing alcohol, you can smell the alcohol. Some molecules of alcohol have escaped into the gas phase. When you put the cap back on, an equilibrium is quickly reached. A dynamic **equilibrium** exists when particles are constantly moving between two or more phases yet no net change in the amount of substance in either phase takes place. Molecules are escaping from the liquid phase into a gas at the same rate that other molecules are returning to the liquid from the gas phase. That is, the rate of evaporation equals the rate of condensation.

Similarly, if you keep a glass of ice water outside on a 0°C day, a constant interchange of water molecules between the solid ice and the liquid water will take place. The system is in a state of equilibrium. **Figure 20** shows a system that has a solid and a gas at equilibrium.

Vapor Pressure Increases with Temperature

A closed container of water is a two-phase system in which molecules of water are in a gas phase in the space above the liquid phase. Moving randomly above the liquid, some of these molecules strike the walls and some go back into the liquid, as shown in **Figure 21.** An equilibrium, in which the rate of evaporation equals the rate of condensation, is soon created. The molecules in the gas exert pressure when they strike the walls of the container. The pressure exerted by the molecules of a gas, or vapor, phase in equilibrium with a liquid is called the **vapor pressure.** You can define boiling point as the temperature at which the vapor pressure equals the external pressure.

As the temperature of the water increases, the molecules have more kinetic energy, so more of them can escape into the gas phase. Thus, as temperature increases, the vapor pressure increases. This relationship is shown for water in **Table 6.** At 40°C, the vapor pressure of water is 55.3 mm Hg. If you increase the temperature to 80°C, the vapor pressure will be 355.1 mm Hg.

Table 6	Water-Vapor Pressure	
Temp. (°C)	Pressure (mm Hg)	Pressure (kPa)
0.0	4.6	0.61
10.0	9.2	1.23
20.0	17.5	2.34
30.0	31.8	4.25
40.0	55.3	7.38
50.0	92.5	12.34
60.0	149.5	19.93
70.0	233.7	31.18
80.0	355.1	47.37
90.0	525.8	70.12
100.0	760.0	101.32

Refer to Appendix A to find more values for water-vapor pressure.

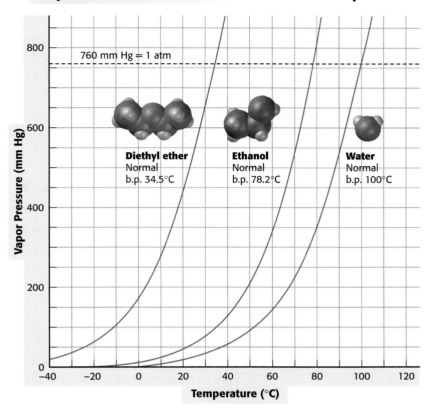

Vapor Pressures of Three Substances at Various Temperatures

760 mm Hg = 1 atm

Diethyl ether
Normal
b.p. 34.5°C

Ethanol
Normal
b.p. 78.2°C

Water
Normal
b.p. 100°C

Vapor Pressure (mm Hg)

Temperature (°C)

At 100°C, the vapor pressure has risen to 760.0 mm Hg, which is standard atmospheric pressure, 1 atm (101.32 kPa). The vapor pressure equals the external pressure, and water boils at 100°C. When you increase the temperature of a system to the point at which the vapor pressure of a substance is equal to standard atmospheric pressure—shown as a dotted line in **Figure 22**—you have reached the substance's *normal boiling point*.

The average kinetic energy of molecules increases about 3% for a 10°C increase in temperature, yet the vapor pressure about doubles or triples. The reason is that the fraction of very energetic molecules that can escape about doubles or triples for a 10°C increase in temperature. You can see this relationship at the high energy part of the curves in **Figure 23.**

Figure 22
The dotted line shows standard atmospheric pressure. The point at which the red line crosses the dotted line is the normal boiling point for each substance.

Energy Distribution of Gas Molecules at Different Temperatures

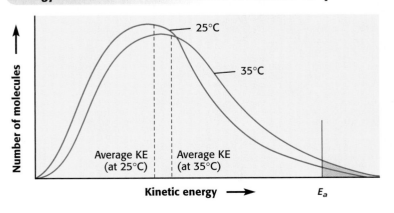

25°C

35°C

Number of molecules

Average KE (at 25°C)

Average KE (at 35°C)

Kinetic energy

E_a

Figure 23
For a 10°C rise in temperature, the average random kinetic energy of molecules increases slightly, but the fraction of molecules that have very high energy (>E_a) increases greatly, as shown by the shaded areas to the right.

Phase Diagrams

You know that a substance's state depends on temperature and that pressure affects state changes. To get a complete picture of how temperature, pressure, and states are related for a particular substance, you can look at a **phase diagram.** A phase diagram has three lines. One line is a vapor pressure curve for the liquid-gas equilibrium. A second line is for the liquid-solid equilibrium, and a third line is for the solid-gas equilibrium. All three lines meet at the **triple point.** The triple point is the only temperature and pressure at which three states of a substance can be in equilibrium.

Phase Diagrams Relate State, Temperature, and Pressure

The x-axis of **Figure 24** shows temperature, and the y-axis shows pressure. For any given point (x, y), you can see in what state water will be. For example, at 363 K (x = 90°C) and standard pressure (y = 101.3 kPa), you know that water is a liquid. If you look at the point for these coordinates, it in fact falls in the region labeled "Liquid."

Gas-Liquid Equilibrium

Look at point E on the line AD, where gas and liquid are in equilibrium at 101.3 kPa. If you increase the temperature slightly, liquid will evaporate and only vapor will remain. If you decrease the temperature slightly, vapor condenses and only water remains. Liquid exists to the left of line AD, and vapor exists to the right of AD. Along line AD, the vapor pressure is increasing, so the density of the vapor increases. The liquid decreases in density. At a temperature and pressure called the **critical point,** the liquid and vapor phases of a substance are identical. Above this point, the substance is called a *supercritical fluid.* A supercritical fluid is the state that a substance is in when the liquid and vapor phases are indistinguishable.

phase diagram

a graph of the relationship between the physical state of a substance and the temperature and pressure of the substance

triple point

the temperature and pressure conditions at which the solid, liquid, and gaseous phases of a substance coexist at equilibrium

critical point

the temperature and pressure at which the gas and liquid states of a substance become identical and form one phase

Figure 24
The phase diagram for water shows the physical states of water at different temperatures and pressures. (Note that the diagram is not drawn to scale.)

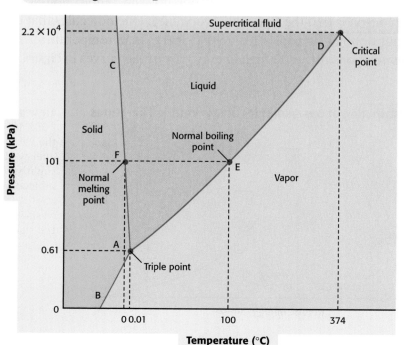

Phase Diagram for H₂O

www.scilinks.org
Topic: Phase Diagrams
SciLinks code: HW4130

Solid-Liquid Equilibrium

If you move to the left (at constant pressure) along the line EF, you will find a temperature at which the liquid freezes. The line AC shows the temperatures and pressures along which solid and liquid are in equilibrium but no vapor is present. If the temperature is decreased further, all of the liquid freezes. Therefore, only solid is present to the left of AC. Water is an unusual substance: the solid is less dense than the liquid. If the pressure is increased at point F, at constant temperature, water will melt. The line AC has a slightly negative slope, which is very rare in phase diagrams of other substances. If the pressure on this system is increased and you move up the line AC, you can see that pressure has very little effect on the melting point, so the decrease in temperature is very small.

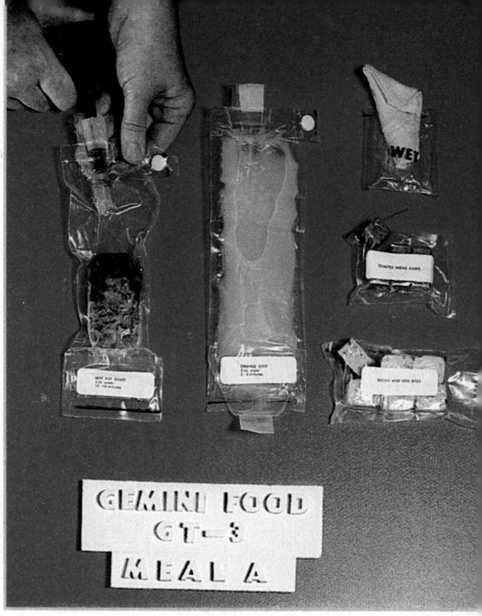

Solid-Gas Equilibrium

Along the line AB, solid is in equilibrium with vapor. If the pressure is decreased below the line AB, the solid will sublime. This relationship is the basis of freeze-drying foods, such as those shown in **Figure 25**. The food is frozen, and then a vacuum is applied. Water sublimes, which dehydrates the food very quickly. The food breaks down less when water is removed at the low temperature than when water evaporates at normal temperatures.

Phase Diagrams Are Unique to a Particular Substance

Each pure substance has a unique phase diagram, although the general structure is the same. Each phase diagram has three lines and shows the liquid-solid, liquid-gas, and solid-gas equilibria. These three lines will intersect at the triple point. The triple point is characteristic for each substance and serves to distinguish the substance from other substances.

Figure 25
Freeze-drying uses the process of sublimation of ice below the freezing point to dry foods. Many meals prepared for astronauts include freeze-dried foods.

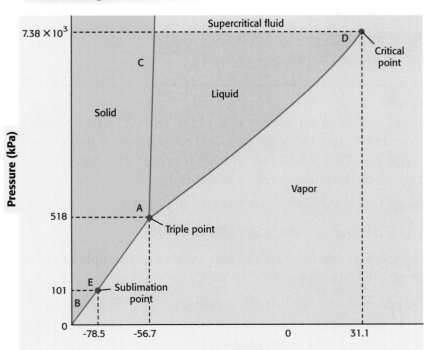

Phase Diagram for CO₂

Figure 26
The phase diagram for carbon dioxide shows the physical states of CO_2 at different temperatures and pressures. (Note that the diagram is not drawn to scale.)

The temperature at which the solid and liquid are in equilibrium—the melting point—is affected little by changes in pressure. Therefore, this line is very nearly vertical when pressure is plotted on the *y*-axis and temperature is plotted on the *x*-axis. Again, water is different in that the solid is less dense than the liquid. Therefore, an increase in pressure decreases the melting point. Most substances, such as carbon dioxide, shown in **Figure 26,** experience a slight increase in melting point when the pressure increases. However, the effect of pressure on boiling point is much greater.

SAMPLE PROBLEM B

How to Draw a Phase Diagram

The triple point of carbon dioxide is at −56.7°C and 518 kPa. The critical point is at 31.1°C and 7.38×10^3 kPa. Vapor pressure above solid carbon dioxide is 101.3 kPa at −78.5°C. Solid carbon dioxide is denser than liquid carbon dioxide. Sketch the phase diagram.

1 Gather information.

- triple point of CO_2 = −56.7°C, 518 kPa
- critical point of CO_2 = 31.1°C, 7.38×10^3 kPa
- The vapor pressure of the solid is 101.3 kPa (1 atm) at −78.5°C.

2 Plan your work.

- Label the *x*-axis "Temperature" and the *y*-axis "Pressure."
- The vapor pressure curve of the liquid goes from the triple point to the critical point.
- The vapor pressure curve of the solid goes from the triple point through −78.5°C and 101.3 kPa.
- The line for the equilibrium between solid and liquid begins at the triple point, goes upward almost vertically, and has a slightly positive slope.

3 Draw the graph.

The graph that results is shown in **Figure 26.**

PRACTICE HINT

In each phase diagram, the necessary data are triple point, critical point, vapor pressure of the solid or liquid at 1 atm (101.3 kPa), and relative densities of the solid and the liquid. The rough graphs that you draw will be helpful in making predictions but will lack accuracy for most of the data.

PRACTICE

1 **a.** The triple point of sulfur dioxide is at −73°C and 0.17 kPa. The critical point is at 158°C and 7.87×10^3 kPa. The normal boiling point of sulfur dioxide is −10°C. Solid sulfur dioxide is denser than liquid sulfur dioxide. Sketch the phase diagram of sulfur dioxide.

b. What state is sulfur dioxide in at 200 kPa and −100°C?

c. What state is sulfur dioxide in at 1 kPa and 80°C?

d. What happens as you increase the temperature of a sample of sulfur dioxide at 101.3 kPa from −20°C to 20°C?

e. What happens as you increase the pressure on a sample of sulfur dioxide at −11°C from 150 kPa to 300 kPa?

The phase diagram for carbon dioxide is similar to that for water, but there are differences. In the phase diagram for carbon dioxide, the horizontal line at 101.3 kPa does not intersect the solid-liquid line. Thus, carbon dioxide is never a liquid at standard pressure. In fact, if you set dry ice, which is solid carbon dioxide, in a room temperature environment, you can see that it sublimes, or changes directly from a solid to a gas.

The horizontal line at 101.3 kPa intersects the vapor pressure curve for the solid at −78.5°C. Therefore, solid carbon dioxide sublimes at this temperature. This sublimation point is equivalent to the normal boiling point of a liquid such as water. Because dry ice is at equilibrium with carbon dioxide gas at −78.5°C, it is frequently used to provide this low temperature in the laboratory.

 # Section Review

UNDERSTANDING KEY IDEAS

1. A glass of ice water has several ice cubes. Describe the contents of the glass in terms of phase?

2. How is the melting point of a substance defined?

3. What is the connection between vapor pressure and boiling point?

4. What is a supercritical fluid?

5. What happens when dry ice is warmed at 1 atm of pressure?

PRACTICE PROBLEMS

6. Describe what happens if you start with water vapor at 0°C and 0.001 kPa and gradually increase the pressure. Assume constant temperature.

7. **a.** The triple point of benzene is at 5.5°C and 4.8 kPa. The critical point is at 289°C and 4.29×10^3 kPa. Vapor pressure above solid benzene is 101.3 kPa at 80.1°C. Solid benzene is denser than liquid benzene. Sketch the phase diagram of benzene.

   b. In what state is benzene at 200 kPa and 80°C?

   c. In what state is benzene at 10 kPa and 100°C?

   d. What happens as you increase the temperature of a sample of benzene at 101.3 kPa from 0°C to 20°C?

   e. What happens as you decrease the pressure on a sample of benzene at 80°C from 150 kPa to 100 kPa?

CRITICAL THINKING

8. Look at the normal boiling point of diethyl ether in **Figure 22**. What do you think would happen if you warmed a flask of diethyl ether with your hand? (Hint: Normal body temperature is 37°C.)

9. Most rubbing alcohol is isopropyl alcohol, which boils at 82°C. Why does rubbing alcohol have a cooling effect on the skin?

10. **a.** Atmospheric pressure on Mount Everest is 224 mm Hg. What is the boiling point of water there?

    b. What is the freezing point of water on Mount Everest?

11. Why is the triple point near the normal freezing point of a substance?

12. You place an ice cube in a pot of boiling water. The water immediately stops boiling. For a moment, there are three phases of water present: the melting ice cube, the hot liquid water, and the water vapor that formed just before you added the ice. Is this three-phase system in equilibrium? Explain.

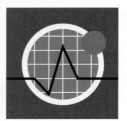
$C_8H_{10}N_4O_2$

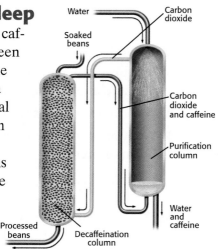

Caffeine gives coffee its bitter taste and some people a feeling of restlessness.

Food Scientist

Food science is the study of the chemistry, microbiology, and processing of foods. Food technicians are responsible for testing foods for quality and acceptability in carefully controlled taste tests. Microbiologists in the food industry monitor the safety of food products. Food analysts work in laboratories to monitor the composition of foods and the presence of pesticides. Some food scientists create new food products or food ingredients, such as artificial sweeteners. During their course of study, college students in the field of food science can gain valuable experience by working for food manufacturers and government agencies, such as the U.S. Food and Drug Administration. This experience can help students find jobs after graduation.

internet connect

www.scilinks.org
Topic: Supercritical Fluids
SciLinks code: HW4122

SCiLINKS Maintained by the National Science Teachers Association

Supercritical Fluids

New Uses for Carbon Dioxide

If the temperature and pressure of a substance are above the critical point, that substance is a *supercritical fluid*. Many supercritical fluids are used for their very effective and selective ability to dissolve other substances. This is very true of carbon dioxide. CO_2 can be made into a supercritical fluid at a relatively low temperature and pressure, so little energy is used in preparing it. Supercritical CO_2 is cheap, nontoxic, and nonflammable and is easy to remove.

Getting a Good Night's Sleep

Supercritical CO_2 is used to remove caffeine from coffee beans. First, the green coffee beans are soaked in water. The beans are then placed in the top of a column that is 70 ft high. Supercritical CO_2 fluid at about 93°C and 250 atm enters at the bottom of the column. The caffeine diffuses out of the beans and into the CO_2. The beans near the bottom of the column mix with almost pure CO_2, which dissolves the last caffeine from the beans. It takes about five hours for fresh beans to move out of the column.

The decaffeinated beans are removed from the bottom, dried, and roasted as usual. The caffeine-rich CO_2 is removed at the top and passed upward through another column. Drops of water fall through the supercritical CO_2 and dissolve the caffeine. The water solution of caffeine is removed and sold to make soft-drinks. The pure CO_2, is recirculated to be used again.

One process used to remove the caffeine from coffee dissolves the caffeine in supercritical CO_2.

Questions

1. Research advantages of using supercritical CO_2 as a solvent.
2. Research and report on other uses of supercritical fluids.

CHAPTER HIGHLIGHTS  11

KEY TERMS

surface tension
evaporation
boiling point
condensation
melting
melting point
freezing
freezing point
sublimation

intermolecular forces
dipole-dipole forces
hydrogen bond
London dispersion force

phase
equilibrium
vapor pressure
phase diagram
triple point
critical point

KEY IDEAS

SECTION ONE   States and State Changes

- Solid particles vibrate in fixed positions. Thus, solids have a definite shape, volume, and density.
- Liquid particles can move past each other. Thus, liquids change shape and have a definite volume and density.
- Gas particles are far apart from each other. Thus, gases can change shape, volume, and density.
- Solids, liquids, and gases convert from one state to another through freezing, melting, evaporation, condensation, sublimation, and deposition.

SECTION TWO   Intermolecular Forces

- The strongest force attracting particles together is the ionic force.
- All ions, atoms, and molecules are attracted by London dispersion forces.
- Polar molecules experience the dipole-dipole force. The dipole-dipole force is usually significant only when the molecules are quite polar.
- Hydrogen bonds are stronger dipole-dipole forces.
- Water's unique properties are due to the combination of the shape of a water molecule and the ability of water to form multiple hydrogen bonds.

SECTION THREE   Energy of State Changes

- Energy is needed to change solid to liquid, solid to gas, and liquid to gas. Thus, melting, sublimation, and evaporation are endothermic processes.
- For a given substance, the endothermic state change with the greatest increase in energy is sublimation. Evaporation has a slightly smaller increase in energy, and melting has a much smaller increase in energy.
- The molar enthalpy of fusion of a substance is the energy required to melt 1 mol of the substance at the melting point. The molar enthalpy of vaporization of a substance is the energy required to vaporize 1 mol of the substance.

SECTION FOUR   Phase Equilibrium

- A phase diagram shows all of the equilibria between the three states of a substance at various temperatures and pressures.
- On a phase diagram, the triple point is where three phases are in equilibrium. Above the critical point, a substance is a supercritical fluid.
- Water is unique in that increased pressure lowers the freezing point.

KEY SKILLS

Calculating Melting and Boiling Points
Sample Problem A p. 397

How to Draw a Phase Diagram
Sample Problem B p. 404

11 CHAPTER REVIEW

USING KEY TERMS

1. Most substances can be in three states. What are they?

2. Explain how solid naphthalene in mothballs is distributed evenly through clothes in a drawer.

3. What is the freezing point of a substance?

4. Clouds form when water vapor in the air collects as liquid water on dust particles in the atmosphere. What physical process takes place?

5. Carbon tetrachloride, CCl_4, is nonpolar. What forces hold the molecules together?

6. Compare dipole-dipole forces and hydrogen bonds.

7. What is the difference between the terms *state* and *phase*?

8. Define *boiling point* in terms of vapor pressure.

9. What is a triple point?

UNDERSTANDING KEY IDEAS

States and State Changes

10. Compare the arrangement and movement of particles in the solid, liquid, and gas states of matter.

11. What is surface tension?

12. A small drop of water assumes an almost spherical form on a Teflon™ surface. Explain why.

13. Compare the temperatures of a substance's melting point and freezing point.

14. What is happening when water is heated from 25°C to 155°C?

15. Give an example of deposition.

16. Give an example of sublimation.

Intermolecular Forces

17. Contrast ionic and molecular substances in terms of the types of attractive forces that govern their behavior.

18. Describe dipole-dipole forces.

19. Why are water molecules polar?

20. Is the melting point of $CaCl_2$ higher than that of NaCl or lower? Explain your answer.

21. A fellow student says, "All substances experience London dispersion forces of attraction between particles." Is this statement true? Explain your answer.

22. Which has larger London dispersion forces between its molecules, CF_4 or CCl_4?

23. Of the three forces, ionic, dipole-dipole, and London dispersion forces, which is the strongest?

24. Hydrogen peroxide, H_2O_2, is shown below. Why is hydrogen peroxide very soluble in water?

$$\begin{array}{c} H \\ \quad \backslash \\ \quad O{-}O \\ \qquad \quad \backslash \\ \qquad \quad H \end{array}$$

25. Why does ice float in water even though most solids sink in the pure liquid?

26. What accounts for the difference in boiling points between HCl (−85°C) and RbCl (1390°C)?

27. Why does CBr_4 boil at a higher temperature than CCl_4 does?

28. "All forces between particles are basically polar in nature." Is this statement true? Explain your answer.

Energy of State Changes

29. The molar enthalpy of fusion of water is 6.009 kJ/mol at 0°C. Explain what this statement means.

30. Why is the molar enthalpy of vaporization of a substance much higher than the molar enthalpy of fusion?

31. How do you calculate the entropy change during a change of state at equilibrium?

32. Why is the entropy of a substance higher in the liquid state than in the solid state?

33. During melting, energy is added, the temperature does not change, yet enthalpy increases. How can this be?

34. At 100°C, the enthalpy change for condensation of water vapor to liquid is negative. Is the entropy change positive, or is it negative?

35. ΔH for a process is positive, and ΔS is negative. What can you conclude about the process?

36. What is the enthalpy of fusion for 6 mol of benzene? (Hint: Use **Table 5** to determine the molar enthalpy of fusion of benzene.)

37. Explain why liquid water at 273.0 K will not melt in terms of Gibbs energy.

38. What thermodynamic values do you need to know to calculate a substance's melting point?

39. How does pressure affect the entropy of a gas?

40. How does pressure affect changes between the liquid and vapor states?

Phase Equilibrium

41. You have sweetened iced tea with sugar, and ice cubes are present. How many phases are present?

42. The term *vapor pressure* almost always means the equilibrium vapor pressure. What physical arrangement is needed to measure vapor pressure?

43. What is meant by the statement that a liquid and its vapor in a closed container are in a state of equilibrium?

44. Can solids have vapor pressure?

45. As the temperature of a liquid increases, what happens to the vapor pressure?

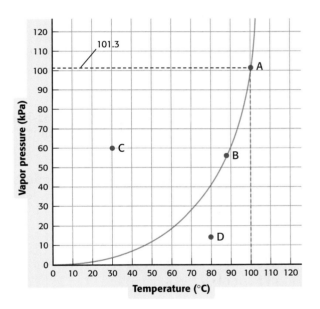

46. Use the above graph of the vapor pressure of water versus temperature to answer the following questions.
   a. At what point(s) does water boil at standard atmospheric pressure?
   b. At what point(s) is water only in the liquid phase?
   c. At what point(s) is water only in the vapor phase?
   d. At what point(s) is liquid water in equilibrium with water vapor?

47. What do the lines of a phase diagram tell you?

48. What two fixed points are on all phase diagrams?

49. Using **Figure 22,** determine the pressure on Pike's Peak, where the boiling point of water is 85°C.

50. Starting at the bottom of **Figure 24,** at a temperature of 100°C, describe what happens as pressure is increased at constant temperature.

51. Starting at the far left of the line *FE* in **Figure 24,** describe what happens as heat is added at constant pressure of 101 kPa.

PRACTICE PROBLEMS

Sample Problem A Calculating Melting and Boiling Points

Calculate the temperatures for the following phase changes. (Liquids are at the normal boiling point.)

52. The enthalpy of fusion of chlorine, Cl_2, is 6.40 kJ/mol, and the entropy of fusion is 37.2 J/mol·K.

53. The enthalpy of fusion of sulfur trioxide is 8.60 kJ/mol, and the entropy of fusion is 29.7 J/mol·K.

54. The enthalpy of fusion of potassium is 2.33 kJ/mol, and the entropy of fusion is 6.91 J/mol·K.

55. The enthalpy of vaporization of butane, C_4H_{10}, is 22.44 kJ/mol, and the entropy of vaporization is 82.2 J/mol·K.

56. ΔH_{vap} for silicon tetrachloride is 28.7 kJ/mol, and ΔS_{vap} is 86.7 J/mol·K.

57. ΔH_{vap} for acetic acid is 23.70 kJ/mol, and ΔS_{vap} is 60.6 J/mol·K.

58. ΔH_{vap} for hydrogen iodide is 19.76 kJ/mol, and ΔS_{vap} is 83.0 J/mol·K.

59. ΔH_{vap} for chloroform, $CHCl_3$, is 29.24 kJ/mol, and ΔS_{vap} is 87.5 J/mol·K.

Sample Problem B How to Draw a Phase Diagram

60. The critical point for krypton is at −64°C and a pressure of 5.5×10^3 kPa. The triple point is at −157.4°C and a pressure of 73.2 kPa. At −172°C, the vapor pressure is 13 kPa. The normal boiling point is −152°C. Sketch the phase diagram.

61. The critical point for ammonia is at 132°C and a pressure of 1.14×10^4 kPa. The triple point is at −77.7°C and a pressure of 6 kPa. Normal boiling point is −33.4°C. Sketch the phase diagram.

62. The critical point for carbon tetrachloride is at 283°C and 4.5×10^3 kPa. The triple point is at −87.0°C and 28.9 kPa. The normal boiling point is 76.7°C. Sketch the phase diagram.

MIXED REVIEW

63. The critical point for iodine is at 512°C and 1.13×10^4 kPa. The triple point is at 112.9°C and 11 kPa. The normal boiling point is 183°C. Sketch the phase diagram.

64. Calculate the melting point of acetic acid at standard pressure. The enthalpy of fusion of acetic acid is 11.54 kJ/mol, and the entropy of fusion is 39.8 J/mol·K.

65. ΔH_{vap} for gold is 324 kJ/mol, and ΔS_{vap} is 103.5 J/mol·K. Calculate the boiling point of gold.

66. The critical point for HBr is at 90°C and 8.56×10^3 kPa. The triple point is at −87.0°C and 29 kPa. The normal boiling point is −66.5°C. Sketch the phase diagram.

CRITICAL THINKING

67. How can water be made to evaporate rapidly at room temperature?

68. There is a saying, "Like water off a duck's back." Why does water run off a duck's back?

69. How does a pressure cooker work?

70. If deuterium—an isotope of hydrogen that has one neutron and one proton—were to replace hydrogen in water, would the resulting compound, D_2O, exhibit hydrogen bonds?

71. Which would have the higher boiling point: chloroform, $CHCl_3$, or bromoform, $CHBr_3$?

72. You know that the enthalpy change for vaporizing water is $\Delta H_{vap} = H_{gas} - H_{liq}$. What is the Gibbs energy change for this process?

73. Explain why steam produces much more severe burns than the same amount of boiling water does.

74. Chloroethane ($T_{bp} = -13°C$) has been used as a local anesthetic. When the liquid is sprayed onto the skin, it cools the skin enough to freeze and numb the skin. Explain the cooling effect of this liquid.

75. Is it possible to have *only* liquid water present in a container at 0.00°C? Explain.

76. Consider a system composed of water vapor and liquid at equilibrium at 100°C. Do the molecules of H_2O in the vapor have more kinetic energy than molecules in the liquid do? Explain.

77. Look at the phase diagram for carbon dioxide. How can CO_2 be made to boil?

ALTERNATIVE ASSESSMENT

78. Liquid crystals are substances that have properties of both liquids and crystals. Write a report on these substances and their various uses.

79. Consult reference materials at the library, and prepare a report that discusses freeze-drying, including its history and applications.

80. Some liquids lose all viscosity when cooled to extremely low temperatures—a phenomenon called *superfluidity*. Find out more about the properties of superfluid substances.

81. Many scientists think that more than 99% of the known matter in the universe is made of a fourth state of matter called plasma. Research plasmas, and report your findings to the class.

82. Prepare a report about the adjustments that must be made when cooking and baking at high elevations. Collect instructions for high-elevation adjustments from packages of prepared food mixes. Explain why changes must be made in recipes that will be prepared at high elevations. Check your library for cookbooks that contain information about food preparation at high elevations.

CONCEPT MAPPING

83. Use the following terms to complete the concept map below: *boiling point, liquids, vapor pressure, gases, melting point, states,* and *equilibrium*.

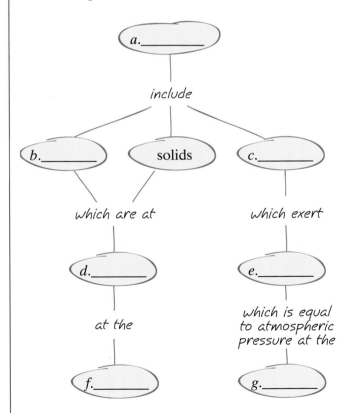

Study the graph below, and answer the questions that follow.
For help in interpreting graphs, see Appendix B, "Study Skills for Chemistry."

84. What is the normal boiling point of water?

85. Give the coordinates for a point at which only liquid water is present.

86. Give the coordinates for a point at which liquid water is in equilibrium with vapor.

87. Give the coordinates for a point at which only vapor is present.

88. What is the vapor pressure of water at point E?

89. What will happen if the vapor at point D is cooled at constant pressure?

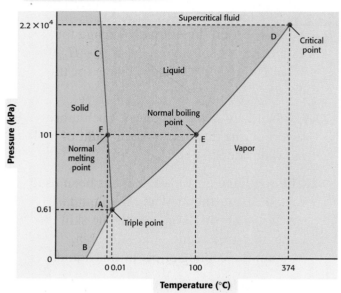

Phase Diagram for H₂O

TECHNOLOGY AND LEARNING

90. Graphing Calculator

Calculating Vapor Pressure by Using a Table

The graphing calculator can run a program that calculates a table for the vapor pressure in atmospheres at different temperatures (K) given the number of moles of a gas and the volume of the gas (V). Given a 0.50 mol gas sample with a volume of 10 L, you can calculate the pressure at 290 K by using a table. Use this program to make the table. Next, use the table to perform the calculations.

Go to Appendix C. If you are using a TI-83 Plus, you can download the program **VAPOR** and data and run the application as

directed. If you are using another calculator, your teacher will provide you with keystrokes and data sets to use. After you have run the program, answer the questions.

a. What is the pressure for 1.3 mol of a gas with a volume of 8.0 L and a temperature of 320 K?

b. What is the pressure for 1.5 mol of a gas with a volume of 10.0 L and a temperature of 340 K?

c. Two gases are measured at 300 K. One has an amount of 1.3 mol and a volume of 7.5 L, and the other has an amount of 0.5 mol and a volume of 10.0 L. Which gas has the lesser pressure?

STANDARDIZED TEST PREP

1. Surface tension is
   a. a skin on the surface of a liquid.
   b. the tendency of the surface of liquids to decrease the area.
   c. a measure of the entropy of a liquid.
   d. the same as vapor pressure.

2. Potatoes cook faster at sea level than at high altitudes because the water
   a. boils more rapidly.
   b. boils at a lower temperature.
   c. increases in temperature while boiling.
   d. boils at a higher temperature.

3. Pure liquids boil at high temperatures under high pressures because
   a. the molecules of liquid are closer together under high pressures.
   b. it takes a high temperature for the vapor pressure to equal the higher external pressure.
   c. the molecules of vapor are farther apart under high pressures.
   d. the entropy change is small at high temperatures and high pressures.

4. In comparing 1 mol of $H_2O(l)$ and $H_2O(g)$ at the same temperature, we find that the
   a. enthalpy of the liquid is higher.
   b. enthalpy of the vapor is higher.
   c. enthalpies are the same and not zero.
   d. enthalpies are both zero.

5. You increase the pressure on a gas. Which of the following would *not* take place?
   a. entropy decrease   **c.** evaporation
   b. enthalpy increase   **d.** sublimation

6. The formation of frost is an example of
   a. condensation.   **c.** deposition.
   b. evaporation.   **d.** melting point.

7. The vapor pressure of alcohol is higher than that of water at room temperature. Thus, the boiling point of alcohol is
   a. higher than that of water.
   b. lower than that of water.
   c. the same as the boiling point of water.
   d. indeterminable.

8. Of solids, liquids, and gases, _____ have the highest degree of disorder.
   a. solids   **c.** gases
   b. liquids   **d.** solids and gases

9. Which of the following substances has the strongest hydrogen bonds?
   a. CH_4 **c.** H_2Se
   b. C_2H_6 **d.** H_2O

10. Which of the following has the highest London dispersion forces between the molecules?
   a. H_2O **c.** H_2Se
   b. H_2S **d.** H_2Te

11. Which of the following has the greatest force between the particles?
   a. NaCl   **c.** Cl_2
   b. HCl   **d.** HOCl

12. Water boils at 100°C. Ethyl alcohol boils at 78.5°C. Which of the following statements are true?
   a. Water has the higher vapor pressure at 78.5°C.
   b. Ethyl alcohol has the higher vapor pressure at 78.5°C.
   c. Both have the same vapor pressure at 78.5°C.
   d. Vapor pressure is not related to boiling point.

CHAPTER 12

GASES

The hot-air balloon pictured here is being filled with hot air. Hot air expands, so the air in the balloon will be less dense. The balloon will be lifted up by the cooler, denser air outside it. The hot-air balloon demonstrates some important facts about gases: gases expand when heated, they have weight, and they have mass.

START-UP ACTIVITY

Pressure Relief

SAFETY PRECAUTIONS

PROCEDURE

1. Blow up a **round latex balloon,** and let the air out.

2. Put the round part of the balloon inside a **PET bottle.** Roll the neck of the balloon over the mouth of the bottle, and secure the neck of the balloon with a **rubber band** so that it dangles into the bottle.

3. Try to blow up the balloon. Record the results.

4. Answer the first two analysis questions below.

5. Design a modification to the balloon-in-a-bottle apparatus that will allow the balloon to inflate. Answer the third analysis question below.

6. If your teacher approves of your design, try it out.

ANALYSIS

1. What causes the balloon to expand?

2. What happens to air that is trapped in the bottle when you blow into the balloon?

3. Describe your design modification, and include a sketch, if applicable. Explain why your design works.

CONTENTS 12

Pre-Reading Questions

① **Among the states of matter, what is unique about gases?**

② **Do gases have mass and weight? How can you tell?**

③ **Gases are considered fluids. Why?**

internet connect

www.scilinks.org
Topic: Gases
SciLinks code: HW4152

SCLINKS Maintained by the National Science Teachers Association

415

Characteristics of Gases

- pressure
- newton
- pascal
- standard temperature and pressure
- kinetic-molecular theory

OBJECTIVES

1. **Describe** the general properties of gases.

2. **Define** pressure, give the SI unit for pressure, and convert between standard units of pressure.

3. **Relate** the kinetic-molecular theory to the properties of an ideal gas.

Properties of Gases

Each state of matter has its own properties. Gases have unique properties because the distance between the particles of a gas is much greater than the distance between the particles of a liquid or a solid. Although liquids and solids seem very different from each other, both have small intermolecular distances. Gas particles, however, are much farther apart from each other than liquid and solid particles are. In some ways, gases behave like liquids; in other ways, they have unique properties.

Gases Are Fluids

Gases are considered *fluids*. People often use the word *fluid* to mean "liquid." However, the word *fluid* actually means "any substance that can flow." Gases are fluids because they are able to flow. Gas particles can flow because they are relatively far apart and therefore are able to move past each other easily. In **Figure 1,** a strip of copper is reacting with nitric acid to form nitrogen dioxide, a brown gas. Like all gases, nitrogen dioxide is a fluid. The gas flows over the sides of the beaker.

Figure 1
The reaction in the beaker in this photo has formed NO_2, a brown gas, which flows out of the container.

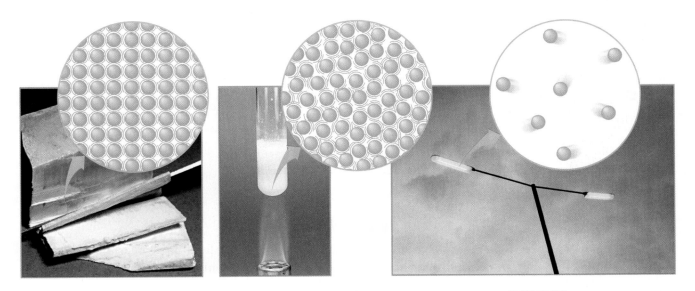

Gases Have Low Density

Gases have much lower densities than liquids and solids do. Because of the relatively large distances between gas particles, most of the volume occupied by a gas is empty space. As shown in **Figure 2,** particles in solids and liquids are almost in contact with each other, but gas particles are much farther apart. This distance between particles shows why a substance in the liquid or solid state always has a much greater density than the same substance in the gaseous state does. The low density of gases also means that gas particles travel relatively long distances before colliding with each other.

Gases Are Highly Compressible

Suppose you completely fill a syringe with liquid and try to push the plunger in when the opening is plugged. You cannot make the space the liquid takes up become smaller. It takes very great pressure to reduce the volume of a liquid or solid. However, if only gas is in the syringe, with a little effort you can move the plunger down and compress the gas. As shown in **Figure 3,** gas particles can be pushed closer together. The space occupied by the gas particles themselves is very small compared with the total volume of the gas. Therefore, applying a small pressure will move the gas particles closer together and will decrease the volume.

Figure 2
Particles of sodium in the solid, liquid, and gas phases are shown. The atoms of gaseous sodium are much farther apart, as in a sodium vapor street lamp.

Figure 3
The volume occupied by a gas can be reduced because gas molecules can move closer together.

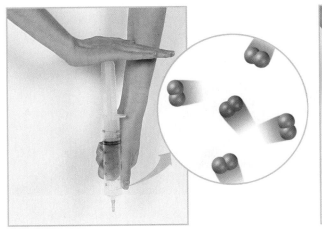
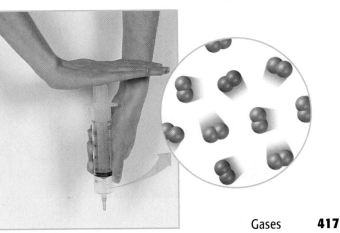

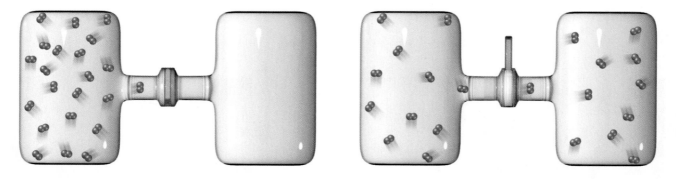

Figure 4
When the partition between the gas and the empty container (vacuum) is removed, the gas flows into the empty container and fills the entire volume of both containers.

Gases Completely Fill a Container

A solid has a certain shape and volume. A liquid has a certain volume but takes the shape of the lower part of its container. In contrast, a gas completely fills its container. This principle is shown in **Figure 4.** One of the containers contains a gas, and the other container is empty. When the barrier between the two containers is removed, the gas rushes into the empty container until it fills both containers equally. Gas particles are constantly moving at high speeds and are far apart enough that they do not attract each other as much as particles of solids and liquids do. Therefore, a gas expands to fill the entire volume available.

Gas Pressure

Gases are all around you. Earth's atmosphere, commonly known as *air,* is a mixture of gases: mainly nitrogen and oxygen. Because you cannot always feel air, you may have thought of gases as being weightless, but all gases have mass; therefore, they have weight in a gravitational field. As gas molecules are pulled toward the surface of Earth, they collide with each other and with the surface of Earth more often, as shown in **Figure 5.** Collisions of gas molecules are what cause *air pressure.*

Figure 5
Gas molecules in the atmosphere collide with Earth's surface, creating atmospheric pressure.

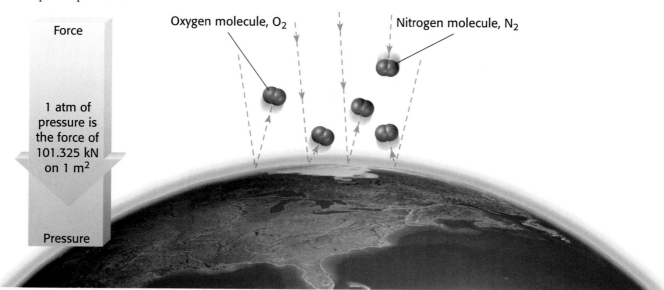

Force

1 atm of pressure is the force of 101.325 kN on 1 m²

Pressure

Oxygen molecule, O₂

Nitrogen molecule, N₂

You may have noticed that your ears sometimes "pop" when you ascend to high altitudes or fly in an airplane. This popping happens because the density of the air changes when you change altitudes. The atmosphere is denser as you move closer to Earth's surface because the weight of atmospheric gases at any elevation compresses the gases below. Less-dense air exerts less pressure. Your ears pop when the air inside your ears changes to the same pressure as the outside air.

Measuring Pressure

The scientific definition of **pressure** is "force divided by area." To find pressure, you need to know the force and the area over which that force is exerted.

The unit of force in SI units is the **newton.** One newton is the force that gives an acceleration of 1 m/s^2 to an object whose mass is 1 kg.

$$1 \text{ newton} = 1 \text{ kg} \times 1 \text{ m/s}^2 = 1 \text{ N}$$

The SI unit of pressure is the **pascal,** Pa, which is the force of one newton applied over an area of one square meter.

$$1 \text{ Pa} = 1 \text{ N/1 m}^2$$

One pascal is a small unit of pressure. It is the pressure exerted by a layer of water that is 0.102 mm deep over an area of one square meter.

Atmospheric pressure can be measured by a barometer, as shown in **Figure 6.** The atmosphere exerts pressure on the surface of mercury in the dish. This pressure goes through the fluid and up the column of mercury. The mercury settles at a point where the pressure exerted downward by its weight equals the pressure exerted by the atmosphere.

pressure

the amount of force exerted per unit area of surface

newton

the SI unit for force; the force that will increase the speed of a 1 kg mass by 1 m/s each second that the force is applied (abbreviation, N)

pascal

the SI unit of pressure; equal to the force of 1 N exerted over an area of 1 m^2 (abbreviation, Pa)

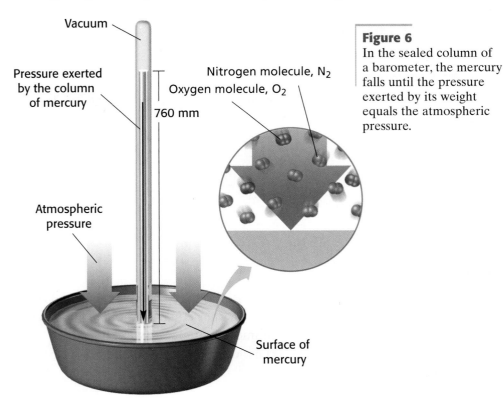

Vacuum

Pressure exerted by the column of mercury

Nitrogen molecule, N$_2$

Oxygen molecule, O$_2$

760 mm

Atmospheric pressure

Surface of mercury

Figure 6
In the sealed column of a barometer, the mercury falls until the pressure exerted by its weight equals the atmospheric pressure.

Table 1   Pressure Units

Unit	Abbreviation	Equivalent number of pascals
Atmosphere	atm	1 atm = 101 325 Pa
Bar	bar	1 bar = 100 025 Pa
Millimeter of mercury	mm Hg	1 mm Hg = 133.322 Pa
Pascal	Pa	1
Pounds per square inch	psi	1 psi = $6.892\ 86 \times 10^3$ Pa
Torr	torr	1 torr = 133.322 Pa

At sea level, the atmosphere keeps the mercury in a barometer at an average height of 760 mm, which is 1 atmosphere. One millimeter of mercury is also called a *torr*, after Evangelista Torricelli, the Italian physicist who invented the barometer. Other units of pressure are listed in **Table 1.**

In studying the effects of changing temperature and pressure on a gas, one will find a standard for comparison useful. Scientists have specified a set of standard conditions called **standard temperature and pressure,** or STP, which is equal to 0°C and 1 atm.

standard temperature and pressure

for a gas, the temperature of 0°C and the pressure 1.00 atmosphere

---

SAMPLE PROBLEM A

Converting Pressure Units

Convert the pressure of 1.000 atm to millimeters of mercury.

1 Gather information.

From **Table 1,** 1 atmosphere = 101 325 Pa, 1 mm Hg = 133.322 Pa

2 Plan your work.

Both units of pressure can be converted to pascals, so use pascals to convert atmospheres to millimeters of mercury.

The conversion factors are $\dfrac{101\ 325\ \text{Pa}}{1\ \text{atm}}$ and $\dfrac{1\ \text{mm Hg}}{133.322\ \text{Pa}}$.

3 Calculate.

$$1.000\ \text{atm} \times \frac{101\ 325\ \text{Pa}}{1\ \text{atm}} \times \frac{1\ \text{mm Hg}}{133.322\ \text{Pa}} = 760.0\ \text{mm Hg}$$

4 Verify your results.

By looking at **Table 1,** you can see that about 100 000 pascals equal one atmosphere, and the number of pascals that equal 1 mm Hg is just over 100. Therefore, the number of millimeters of mercury equivalent to 1 atm is just below 1000 (100 000 divided by 100). The answer is therefore reasonable.

PRACTICE HINT

Remember that millimeters of mercury is actually a unit of pressure, so it cannot be canceled by a length measurement.

1 The critical pressure of carbon dioxide is 72.7 atm. What is this value in units of pascals?

2 The vapor pressure of water at 50.0°C is 12.33 kPa. What is this value in millimeters of mercury?

3 In thermodynamics, the standard pressure is 100.0 kPa. What is this value in units of atmospheres?

4 A tire is inflated to a gauge pressure of 30.0 psi (Which must be added to the atmospheric pressure of 14.7 psi to find the total pressure in the tire). Calculate the total pressure in the tire in pascals.

The Kinetic-Molecular Theory

The properties of gases stated earlier are explained on the molecular level in terms of the **kinetic-molecular theory.** The kinetic-molecular theory is a model that is used to predict gas behavior.

The kinetic-molecular theory states that gas particles are in constant rapid, random motion. The theory also states that the particles of a gas are very far apart relative to their size. This idea explains the fluidity and compressibility of gases. Gas particles can easily move past one another or move closer together because they are farther apart than liquid or solid particles.

Gas particles in constant motion collide with each other and with the walls of their container. The kinetic-molecular theory states that the pressure exerted by a gas is a result of collisions of the molecules against the walls of the container, as shown in **Figure 7.** The kinetic-molecular theory considers collisions of gas particles to be perfectly elastic; that is, energy is completely transferred during collisions. The total energy of the system, however, remains constant.

kinetic-molecular theory

a theory that explains that the behavior of physical systems depends on the combined actions of the molecules constituting the system

www.scilinks.org
Topic: Kinetic Theory
SciLinks code: HW4157

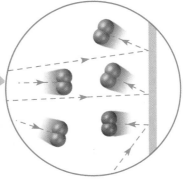

Figure 7
Gas molecules travel far, relative to their size, in straight lines until they collide with other molecules or the walls of the container.

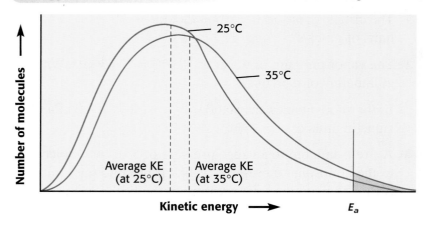

Figure 8
Increasing the temperature of a gas shifts the energy distribution in the direction of greater average kinetic energy.

Energy Distribution of Gas Molecules at Different Temperatures

Gas Temperature Is Proportional to Average Kinetic Energy

Molecules are always in random motion. The average kinetic energy of random motion is proportional to the absolute temperature, or temperature in kelvins. Heat increases the energy of random motion of a gas.

But not all molecules are traveling at the same speed. As a result of multiple collisions, the molecules have a range of speeds. **Figure 8** shows the numbers of molecules at various energies. For a 10°C rise in temperature from STP, the average energy increases about 3%, while the number of very high-energy molecules approximately doubles or triples.

Section Review

UNDERSTANDING KEY IDEAS

1. What characteristic of gases makes them different from liquids or solids?

2. Why are gases considered fluids?

3. What happens to gas particles when a gas is compressed?

4. What is the difference between force and pressure?

5. What is the SI unit of pressure, and how is it defined?

6. How does the kinetic-molecular theory explain the pressure exerted by gases?

7. How is a gas's ability to fill a container different from that of a liquid or a solid?

PRACTICE PROBLEMS

8. The atmospheric pressure on top of Mount Everest is 58 kPa. What is this pressure in atmospheres?

9. The vapor pressure of water at 0°C is 4.579 mm Hg. What is this pressure in pascals?

10. A laboratory high-vacuum system may operate at 1.0×10^{-5} mm Hg. What is this pressure in pascals?

CRITICAL THINKING

11. How does the kinetic-molecular theory explain why atmospheric pressure is greater at lower altitudes than at higher altitudes?

12. Molecules of hydrogen escape from Earth, but molecules of oxygen and nitrogen are held to the surface. Why?

The Gas Laws

KEY TERMS

- **Boyle's law**
- **Charles's law**
- **Gay-Lussac's law**
- **Avogadro's law**

OBJECTIVES

1. **State** Boyle's law, and use it to solve problems involving pressure and volume.

2. **State** Charles's law, and use it to solve problems involving volume and temperature.

3. **State** Gay-Lussac's law, and use it to solve problems involving pressure and temperature.

4. **State** Avogadro's law, and explain its importance in determining the formulas of chemical compounds.

Measurable Properties of Gases

In this section, you will study the relationships between the measurable properties of a gas, represented by the variables shown below.

P = pressure exerted by the gas

T = temperature in kelvins of the gas

V = total volume occupied by the gas

n = number of moles of the gas

Pressure-Volume Relationships

As you read in the last section, gases have pressure, and the space that they take up can be made smaller. In 1662, the English scientist Robert Boyle studied the relationship between the volume and the pressure of a gas. He found that as pressure on a gas increases in a closed container, the volume of the gas decreases. In fact, the product of the pressure and volume, PV, remains almost constant if the temperature remains the same. **Table 2** shows data for experiments similar to Boyle's.

Table 2	Pressure-Volume Data for a Sample of Gas at Constant Temperature	
Pressure (kPa)	**Volume (L)**	**PV (kPa × L)**
150	0.334	50.1
200	0.250	50.0
250	0.200	50.0
300	0.167	50.1

Figure 9

a Gas molecules in a car-engine cylinder spread apart to fill the cylinder.

b As the cylinder's volume decreases, there are more molecular collisions, and the pressure of the gas increases.

Boyle's Law

Figure 9 shows what happens to the gas molecules in a car cylinder as it compresses. As the volume decreases, the concentration, and therefore pressure, increases. This concept is shown in graphical form in **Figure 10.** The inverse relationship between pressure and volume is known as **Boyle's law.** The third column of **Table 2** shows that at constant temperature, the product of the pressure and volume of a gas is constant.

$$PV = k$$

If the temperature and number of particles are not changed, the PV product remains the same, as shown in the equation below.

$$P_1V_1 = P_2V_2$$

Boyle's law

the law that states that for a fixed amount of gas at a constant temperature, the volume of the gas increases as the pressure of the gas decreases and the volume of the gas decreases as the pressure of the gas increases

Figure 10
This pressure-volume graph shows an inverse relationship: as pressure increases, volume decreases.

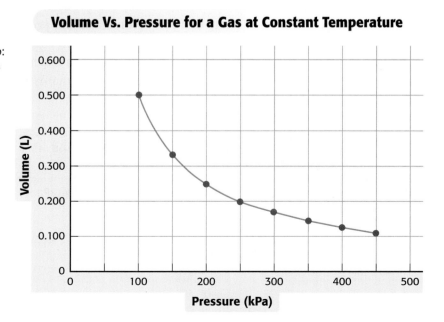

Volume Vs. Pressure for a Gas at Constant Temperature

SAMPLE PROBLEM B

Solving Pressure-Volume Problems

A given sample of gas occupies 523 mL at 1.00 atm. The pressure is increased to 1.97 atm, while the temperature remains the same. What is the new volume of the gas?

1 Gather information.

The initial volume and pressure and the final pressure are given. Determine the final volume.

$$P_1 = 1.00 \text{ atm} \qquad V_1 = 523 \text{ mL}$$

$$P_2 = 1.97 \text{ atm} \qquad V_2 = ?$$

2 Plan your work.

Place the known quantities into the correct places in the equation relating pressure and volume.

$$P_1V_1 = P_2V_2$$

$$(1.00 \text{ atm})(523 \text{ mL}) = (1.97 \text{ atm})V_2$$

3 Calculate.

$$V_2 = \frac{(1.00 \text{ atm})(523 \text{ mL})}{1.97 \text{ atm}} = 265 \text{ mL}$$

4 Verify your results.

The pressure was almost doubled, so the new volume should be about one-half the initial volume. The answer is therefore reasonable.

> **PRACTICE HINT**
>
> It does not matter what units you use for pressure and volume when using Boyle's law as long as they are the same on both sides of the equation.

PRACTICE

1 A sample of oxygen gas has a volume of 150.0 mL at a pressure of 0.947 atm. What will the volume of the gas be at a pressure of 1.000 atm if the temperature remains constant?

2 A sample of gas in a syringe has a volume of 9.66 mL at a pressure of 64.4 kPa. The plunger is depressed until the pressure is 94.6 kPa. What is the new volume, assuming constant temperature?

3 An air mass of volume 6.5×10^5 L starts at sea level, where the pressure is 775 mm Hg. It rises up a mountain where the pressure is 622 mm Hg. Assuming no change in temperature, what is the volume of the air mass?

4 A balloon has a volume of 456 mL at a pressure of 1.0 atm. It is taken under water in a submarine to a depth where the air pressure in the submarine is 3.3 atm. What is the volume of the balloon? Assume constant temperature.

PROBLEM SOLVING SKILL

Temperature-Volume Relationships

Heating a gas makes it expand. Cooling a gas makes it contract. This principle is shown in **Figure 11.** As balloons are dipped into liquid nitrogen, the great decrease in temperature makes them shrink. When they are removed from the liquid nitrogen, the gas inside the balloons warms up, and the balloons expand to their original volume.

In 1787, the French physicist Jacques Charles discovered that a gas's volume is directly proportional to the temperature on the Kelvin scale if the pressure remains the same.

Charles's Law

The direct relationship between temperature and volume is known as **Charles's law.** The kinetic-molecular theory states that gas particles move faster on average at higher temperatures, causing them to hit the walls of their container with more force. Repeated strong collisions cause the volume of a flexible container, such as a balloon, to increase. Likewise, gas volume decreases when the gas is cooled, because of the lower average kinetic energy of the gas particles at the lower temperature. If the absolute temperature is reduced by half, then the average kinetic energy is reduced by half, and the particles will strike the walls with half of the energy they had at the higher temperature. In that case, the volume of the gas will be reduced to half of the original volume if the pressure remains the same.

The direct relationship between volume and temperature is shown in **Figure 12,** in which volume-temperature data are graphed using the Kelvin scale. If you read the line in **Figure 12** all the way down to 0 K, it looks as though the gas's volume becomes zero. Does a gas volume really become zero at absolute zero? No. Before this temperature is reached, the gas becomes a liquid and then freezes to a solid, each of which has a certain volume.

Charles's law

the law that states that for a fixed amount of gas at a constant pressure, the volume of the gas increases as the temperature of the gas increases and the volume of the gas decreases as the temperature of the gas decreases

Figure 11

a Air-filled balloons are dipped into liquid nitrogen.

b The extremely low temperature of the nitrogen causes them to shrink in volume.

c When the balloons are removed from the liquid nitrogen, the air inside them quickly warms, and the balloons expand to their original volume.

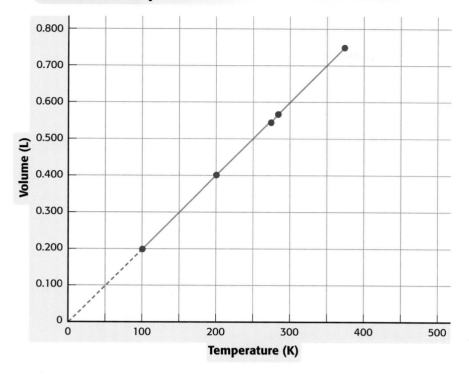

Volume Vs. Temperature for a Gas at Constant Pressure

Figure 12
When the temperature scale is in kelvins, the graph shows a direct proportion between volume of a sample of gas and the temperature.

Data for an experiment of the type carried out by Charles are given in **Table 3.** At constant pressure, the volume of a sample of gas divided by its absolute temperature is a constant, *k.* Charles's law can be stated as the following equation.

$$\frac{V}{T} = k$$

If all other conditions are kept constant, *V*/*T* will remain the same. Therefore, Charles's law can also be expressed as follows.

$$\frac{V_1}{T_1} = \frac{V_2}{T_2}$$

Table 3   Volume-Temperature Data for a Sample of Gas at a Constant Pressure

Volume (mL)	Temperature (K)	*V*/*T* (mL/K)
748	373	2.01
567	283	2.00
545	274	1.99
545	273	2.00
546	272	2.01
402	200	2.01
199	100	1.99

Solving Volume-Temperature Problems

A balloon is inflated to 665 mL volume at 27°C. It is immersed in a dry-ice bath at −78.5°C. What is its volume, assuming the pressure remains constant?

1 Gather information.

The initial volume and temperature and the final temperature are given. Determine the final volume.

$$V_1 = 665 \text{ mL} \qquad\qquad T_1 = 27°C$$

$$V_2 = ? \qquad\qquad T_2 = -78.5°C$$

2 Plan your work.

PRACTICE HINT

In gas law problems, always convert temperatures to kelvins. The gas law equations do not work for temperatures expressed in the Celsius or Fahrenheit scales.

Convert the temperatures from degrees Celsius to kelvins:

$$T_1 = 27°C + 273 = 300 \text{ K} \qquad T_2 = -78.5°C + 273 = 194.5 \text{ K}$$

Place the known quantities into the correct places in the equation relating volume and temperature.

$$\frac{V_1}{T_1} = \frac{V_2}{T_2}$$

$$\frac{665 \text{ mL}}{300 \text{ K}} = \frac{V_2}{194.5 \text{ K}}$$

3 Calculate.

$$V_2 = \frac{(665 \text{ mL})(194.5 \text{ K})}{300 \text{ K}} = 431 \text{ mL}$$

4 Verify your results.

Charles's law tells you that volume decreases as temperature decreases. The temperature decreased by about one-third, and according to the calculation, so did the volume. The answer is therefore reasonable.

PRACTICE

1 Helium gas in a balloon occupies 2.5 L at 300.0 K. The balloon is dipped into liquid nitrogen that is at a temperature of 80.0 K. What will the volume of the helium in the balloon at the lower temperature be?

2 A sample of neon gas has a volume of 752 mL at 25.0°C. What will the volume at 50.0°C be if pressure is constant?

3 A helium-filled balloon has a volume of 2.75 L at 20.0°C. The volume of the balloon changes to 2.46 L when placed outside on a cold day. What is the temperature outside in degrees Celsius?

4 When 1.50×10^3 L of air at 5.00°C is injected into a household furnace, it comes out at 30.0°C. Assuming the pressure is constant, what is the volume of the heated air?

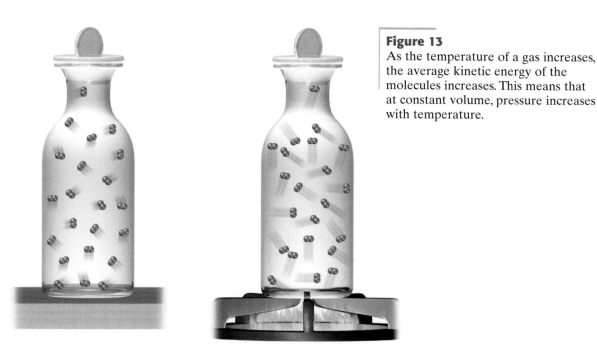

Figure 13
As the temperature of a gas increases, the average kinetic energy of the molecules increases. This means that at constant volume, pressure increases with temperature.

Temperature-Pressure Relationships

As you have learned, pressure is the result of collisions of particles with the walls of the container. You also know that the average kinetic energy of particles is proportional to the sample's average absolute temperature. Therefore, if the absolute temperature of the gas particles is doubled, their average kinetic energy is doubled. If there is a fixed amount of gas in a container of fixed volume, the collisions will have twice the energy, so the pressure will double, as shown in **Figure 13.** The French scientist Joseph Gay-Lussac is given credit for discovering this relationship in 1802. **Figure 14** shows a graph of data for the change of pressure with temperature in a gas of constant volume. The graph is a straight line with a positive slope, which indicates that temperature and pressure have a directly proportional relationship.

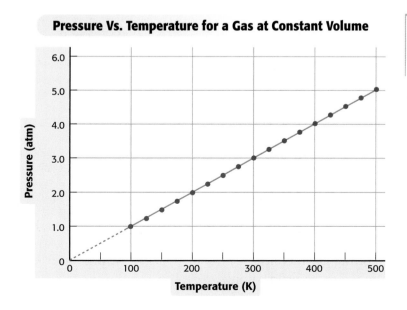

Pressure Vs. Temperature for a Gas at Constant Volume

Figure 14
This graph shows that gas pressure is directly proportional to Kelvin temperature, at constant volume.

Gay-Lussac's Law

Gay-Lussac's law

the law that states that the pressure of a gas at a constant volume is directly proportional to the absolute temperature

The direct relationship between temperature and pressure is known as **Gay-Lussac's law.** Because the pressure of a gas is proportional to its absolute temperature, the following equation is true for a sample of constant volume.

$$P = kT$$

This equation can be rearranged to the following form.

$$\frac{P}{T} = k$$

At constant volume, the following equation applies.

$$\frac{P_1}{T_1} = \frac{P_2}{T_2}$$

If any three of the variables in the above equation are known, then the unknown fourth variable can be calculated.

SAMPLE PROBLEM D

Solving Pressure-Temperature Problems

An aerosol can containing gas at 101 kPa and 22°C is heated to 55°C. Calculate the pressure in the heated can.

1 Gather information.

The initial pressure and temperature and the final temperature are given. Determine the final pressure.

$P_1 = 101$ kPa	$T_1 = 22$°C
$P_2 = ?$	$T_2 = 55$°C

PRACTICE HINT

When solving gas problems, be sure to identify which variables (P, V, n, T) remain constant and which change. That way you can pick the right equation to use.

2 Plan your work.

Convert the temperatures from degrees Celsius to kelvins:

$$T_1 = 22°C + 273 = 295 \text{ K} \qquad T_2 = 55°C + 273 = 328 \text{ K}$$

Use the equation relating temperature and pressure.

$$\frac{P_1}{T_1} = \frac{P_2}{T_2}$$

$$\frac{101 \text{ kPa}}{295 \text{ K}} = \frac{P_2}{328 \text{ K}}$$

3 Calculate.

$$P_2 = \frac{(101 \text{ kPa})(328 \text{ K})}{295 \text{ K}} = 112 \text{ kPa}$$

4 Verify your results.

The temperature of the gas increases by about 10%, so the pressure should increase by the same proportion.

1. At 122°C the pressure of a sample of nitrogen is 1.07 atm. What will the pressure be at 205°C, assuming constant volume?

2. The same sample of nitrogen as in item 1 starts at 122°C and 1.07 atm. After cooling, the pressure is measured to be 0.880 atm. What is the new temperature?

3. A sample of helium gas is at 122 kPa and 22°C. Assuming constant volume, what will the temperature be when the pressure is 203 kPa?

4. The air in a steel-belted tire is at a gauge pressure of 29.8 psi at a temperature of 20°C. After the tire is driven fast on a hot road, the temperature in the tire is 48°C. What is the tire's new gauge pressure?

Volume-Molar Relationships

In 1811, the Italian scientist Amadeo Avogadro proposed the idea that equal volumes of all gases, under the same conditions, have the same number of particles. This idea is shown in **Figure 15,** which shows equal numbers of molecules of the gases H_2, O_2, and CO_2, each having the same volume. A result of this relationship is that molecular masses can be easily determined. Unfortunately, Avogadro's insight was not recognized right away, mainly because scientists at the time did not know the difference between atoms and molecules. Later, the Italian chemist Stanislao Cannizzaro used Avogadro's principle to determine the true formulas of several gaseous compounds.

Figure 15
At the same temperature and pressure, balloons of equal volume contain equal numbers of molecules, regardless of which gas they contain.

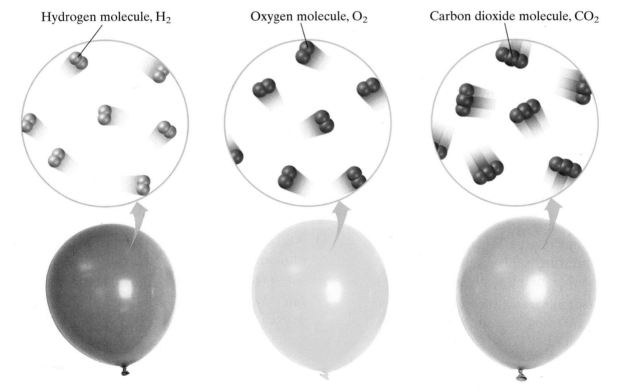

Hydrogen molecule, H_2

Oxygen molecule, O_2

Carbon dioxide molecule, CO_2

Avogadro's Law

Avogadro's idea turned out to be correct and is now known as **Avogadro's law.** With this knowledge, chemists gained insight into the formulas of chemical compounds for the first time. In 1858, Cannizzaro used Avogadro's law to deduce that the correct formula for water is H_2O. This important discovery will be discussed in more detail in the next section.

Avogadro's law also means that gas volume is directly proportional to the number of moles of gas at the same temperature and pressure. This relationship is expressed by the equation below, in which k is a proportionality constant.

$$V = kn$$

www.scilinks.org
Topic : Avogadro's Law
SciLinks code: HW4137

But volumes of gases change with changes in temperature and pressure. A set of conditions has been defined for measuring volumes of gases. For example, we know that argon exists as single atoms, and that its atomic mass is 39.95 g/mol. It has been determined that 22.41 L of argon at 0°C and 1 atm have a mass of 39.95 g. *Therefore, 22.41 L is the volume of 1 mol of any gas at STP.* The mass of 22.41 L of a gas at 0°C and a pressure of 1 atm will be equal to the gas's molecular mass.

Section Review

UNDERSTANDING KEY IDEAS

1. What is the name of the gas law relating pressure and volume, and what does it state?

2. What is the name of the gas law relating volume and absolute temperature, and what does it state?

3. What is the name of the gas law relating pressure and absolute temperature, and what does it state?

4. What relationship does Avogadro's law express?

PRACTICE PROBLEMS

5. A sample of gas occupies 1.55 L at 27.0°C and 1.00 atm pressure. What will the volume be if the pressure is increased to 50.0 atm, but the temperature is kept constant?

6. A sample of nitrogen gas occupies 1.55 L at 27°C and 1.00 atm pressure. What will the volume be at −100°C and the same pressure?

7. A 1.0 L volume of gas at 27.0°C exerts a pressure of 85.5 kPa. What will the pressure be at 127°C? Assume constant volume.

8. A sample of nitrogen has a volume of 275 mL at 273 K. The sample is heated and the volume becomes 325 mL. What is the new temperature in kelvins?

9. A small cylinder of oxygen contains 300.0 mL of gas at 15 atm. What will the volume of this gas be when released into the atmosphere at 0.900 atm?

CRITICAL THINKING

10. A student has the following data: $V_1 = 822$ mL, $T_1 = 75$°C, $T_2 = -25$°C. He calculates V_2 and gets −274 mL. Is this correct? Explain why or why not.

11. Aerosol cans have a warning not to dispose of them in fires. Why?

12. What volume of carbon dioxide contains the same number of molecules as 20.0 mL of oxygen at the same conditions?

Molecular Composition of Gases

KEY TERMS

- **ideal gas**
- **ideal gas law**
- **diffusion**
- **effusion**
- **Graham's law of diffusion**
- **Gay-Lussac's law of combining volumes**
- **partial pressure**
- **Dalton's law of partial pressure**

OBJECTIVES

① **Solve** problems using the ideal gas law.

② **Describe** the relationships between gas behavior and chemical formulas, such as those expressed by Graham's law of diffusion, Gay-Lussac's law of combining volumes, and Dalton's law of partial pressures.

③ **Apply** your knowledge of reaction stoichiometry to solve gas stoichiometry problems.

The Ideal Gas Law

You have studied four different gas laws, which are summarized in **Table 4.** Boyle's law states the relationship between the pressure and the volume of a sample of gas. Charles's law states the relationship between the volume and the absolute temperature of a gas. Gay-Lussac's law states the relationship between the pressure and the temperature of a gas. Avogadro's law relates volume to the number of moles of gas.

No gas perfectly obeys all four of these laws under all conditions. Nevertheless, these assumptions work well for most gases and most conditions. As a result, one way to model a gas's behavior is to assume that the gas is an **ideal gas** that perfectly follows these laws. An ideal gas, unlike a real gas, does not condense to a liquid at low temperatures, does not have forces of attraction or repulsion between the particles, and is composed of particles that have no volume.

ideal gas

an imaginary gas whose particles are infinitely small and do not interact with each other

> **STUDY** ▶ **TIP**
>
> **ORGANIZING THE GAS LAWS**
>
> You can use **Table 4** as a reference to help you learn and understand the gas laws.
>
> - You may want to make a more elaborate table in which you write down each law's name, a brief explanation of its meaning, its mathematical representation, and the variable that is kept constant.

Table 4	**Summary of the Basic Gas Laws**
Boyle's law	$P_1V_1 = P_2V_2$
Charles's law	$\dfrac{V_1}{T_1} = \dfrac{V_2}{T_2}$
Gay-Lussac's law	$\dfrac{P_1}{T_1} = \dfrac{P_2}{T_2}$
Avogadro's law	$V = kn$

The Ideal Gas Law Relates All Four Gas Variables

ideal gas law

the law that states the mathematical relationship of pressure (*P*), volume (*V*), temperature (*T*), the gas constant (*R*), and the number of moles of a gas (*n*); $PV = nRT$

In using the basic gas laws, we have made use of four variables: pressure, *P*, volume, *V*, absolute temperature, *T*, and number of moles, *n*. Boyle's law, Charles's law, Gay-Lussac's law, and Avogadro's law can be combined into one equation that gives the relationship between all four variables, *P, V, T,* and *n,* for any sample of gas. This relationship is called the **ideal gas law.** When any three variables are given, the fourth can be calculated. The ideal gas law is represented mathematically below.

$$PV = nRT$$

R is a proportionality constant. The value for *R* used in any calculation depends on the units used for pressure and volume. In this text, we will normally use units of kilopascals and liters when using the ideal gas law, so the value you will use for *R* is as follows.

$$R = \frac{8.314 \text{ L} \cdot \text{kPa}}{\text{mol} \cdot \text{K}}$$

If the pressure is expressed in atmospheres, then the value of *R* is 0.0821 (L•atm)/(mol•K).

The ideal gas law describes the behavior of real gases quite well at room temperature and atmospheric pressure. Under those conditions, the volume of the particles themselves and the forces of attraction between them can be ignored. The particles behave in ways that are similar enough to an ideal gas that the ideal gas law gives useful results. However, as the volume of a real gas decreases, the particles attract one another more strongly, and the volume is less than the ideal gas law would predict. At extremely high pressures, the volume of the particles themselves is close to the total volume, so the actual volume will be higher than calculated. This relationship is shown in **Figure 16.**

Figure 16
For an ideal gas, the ratio of *PV/nRT* is 1, which is represented by the dashed line. Real gases deviate somewhat from the ideal gas law and more at very high pressures.

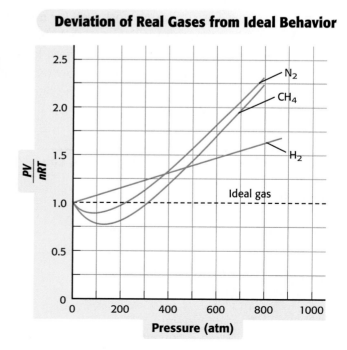

Deviation of Real Gases from Ideal Behavior

Using the Ideal Gas Law

How many moles of gas are contained in 22.41 liters at 101.325 kPa and 0°C?

1 Gather information.

Three variables for a gas are given, so you can use the ideal gas law to compute the value of the fourth.

$V = 22.41$ L $\qquad\qquad$ $T = 0°C$

$P = 101.325$ kPa $\qquad\qquad$ $n = ?$

2 Plan your work.

Convert the temperature to kelvins.

$$0°C + 273 = 273 \text{ K}$$

Place the known quantities into the correct places in the equation relating the four gas variables, $PV = nRT$.

$$(101.325 \text{ kPa})(22.41 \text{ L}) = n\left(\frac{8.314 \text{ L}\bullet\text{kPa}}{\text{mol}\bullet\text{K}}\right)(273 \text{ K})$$

PRACTICE HINT

When using the ideal gas law, be sure that the units of P and V match the units of the value of R that is used.

3 Calculate.

Solve for the unknown, n.

$$n = \frac{(101.325 \text{ kPa})(22.41 \text{ L})}{\left(\dfrac{8.314 \text{ L}\bullet\text{kPa}}{\text{mol}\bullet\text{K}}\right)(273 \text{ K})} = 1.00 \text{ mol}$$

4 Verify your results.

The product of pressure and volume is a little over 2000, and the product of R and the temperature is a little over 2000, so one divided by the other should be about 1. Also, recall that at STP the volume of 1 mol of gas is 22.41 L. The calculated result agrees with this value.

PRACTICE

1 How many moles of air molecules are contained in a 2.00 L flask at 98.8 kPa and 25.0°C?

2 How many moles of gases are contained in a can with a volume of 555 mL and a pressure of 600.0 kPa at 20°C?

3 Calculate the pressure exerted by 43 mol of nitrogen in a 65 L cylinder at 5°C.

4 What will be the volume of 111 mol of nitrogen in the stratosphere, where the temperature is −57°C and the pressure is 7.30 kPa?

Gas Behavior and Chemical Formulas

So far we have dealt only with the general behavior of gases. No calculation has depended on knowing which gas was involved. For example, in **Sample Problem E,** we determined the number of moles of a gas. The calculation would have been the same no matter which gas or mixture of gases was involved. Now we will consider situations in which it is necessary to know more about a gas, such as its molar mass.

Diffusion

diffusion

the movement of particles from regions of higher density to regions of lower density

Household ammonia is a solution of ammonia gas, NH_3, in water. When you open a bottle of household ammonia, the odor of ammonia gas doesn't take long to fill the room. Gaseous molecules, including molecules of the compounds responsible for the smell, travel at high speeds in all directions and mix quickly with molecules of gases in the air in a process called **diffusion.** Gases diffuse through each other at a very rapid rate. Even if the air in the room was very still, it would only be a matter of minutes before ammonia molecules were evenly distributed throughout the room. During diffusion, a substance moves from a region of high concentration to a region of lower concentration. Eventually, the mixture becomes homogeneous, as seen in the closed bottle of bromine gas in **Figure 17.** Particles of low mass diffuse faster than particles of higher mass.

The process of diffusion involves an increase in entropy. Entropy can be thought of as a measure of randomness. One way of thinking of randomness is to consider the probability of finding a particular particle at a particular location. This probability is much lower in a mixture of gases than in a pure gas. Diffusion of gases is a way in which entropy is increased.

Figure 17
When liquid bromine evaporates, gaseous bromine diffuses into the air above the surface of the liquid.

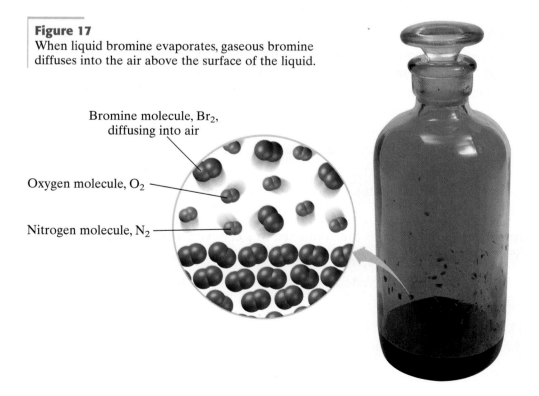

Bromine molecule, Br_2, diffusing into air

Oxygen molecule, O_2

Nitrogen molecule, N_2

Figure 18
Molecules of oxygen and nitrogen from inside the bicycle tire effuse out through a small nail hole.

Oxygen molecule, O_2

Nitrogen molecule, N_2

Effusion

The passage of gas particles through a small opening is called **effusion.** An example of effusion is shown in **Figure 18.** In a tire with a very small leak, the air in the tire effuses through the hole. The Scottish scientist Thomas Graham studied effusion in detail. He found that at constant temperature and pressure, the rate of effusion of a gas is inversely proportional to the square root of the gas's molar mass, M. This law also holds when one compares rates of diffusion, and molecular speeds in general. The molecular speeds, v_A and v_B, of gases A and B can be compared according to **Graham's law of diffusion,** shown below.

$$\frac{v_A}{v_B} = \sqrt{\frac{M_B}{M_A}}$$

For example, compare the speed of effusion of H_2 with that of O_2.

$$\frac{v_{H_2}}{v_{O_2}} = \sqrt{\frac{M_{O_2}}{M_{H_2}}} = \sqrt{\frac{32 \text{ g/mol}}{2 \text{ g/mol}}} = \sqrt{16} = 4$$

Hydrogen gas effuses four times as fast as oxygen. Particles of low molar mass travel faster than heavier particles.

According to the kinetic-molecular theory, gas particles that are at the same temperature have the same average kinetic energy. Therefore, the kinetic energy of two gases that are at the same temperature is

$$\frac{1}{2}M_A v_A{}^2 = \frac{1}{2}M_B v_B{}^2$$

Solving this equation for the ratio of speeds between v_A and v_B gives Graham's law of diffusion.

effusion

the passage of a gas under pressure through a tiny opening

Graham's law of diffusion

the law that states that the rate of diffusion of a gas is inversely proportional to the square root of the gas's density

internet connect

www.scilinks.org
Topic: Effusion/Diffusion
SciLinks code: HW4041

SCi
LINKS. Maintained by the National Science Teachers Association

Comparing Molecular Speeds

Oxygen molecules have an average speed of about 480 m/s at room temperature. At the same temperature, what is the average speed of molecules of sulfur hexafluoride, SF_6?

1 Gather information.

$$v_{O_2} = 480 \text{ m/s}$$

$$v_{SF_6} = ? \text{ m/s}$$

2 Plan your work.

This problem can be solved using Graham's law of diffusion, which compares molecular speeds.

$$\frac{v_{SF_6}}{v_{O_2}} = \sqrt{\frac{M_{O_2}}{M_{SF_6}}}$$

You need molar masses of O_2 and SF_6.
Molar mass of $O_2 = 2(16.00 \text{ g/mol}) = 32.00 \text{ g/mol}$
Molar mass of $SF_6 = 32.07 \text{ g/mol} + 6(19.00 \text{ g/mol}) = 146 \text{ g/mol}$

3 Calculate.

Place the known quantities into the correct places in Graham's law of diffusion.

$$\frac{v_{SF_6}}{480 \text{ m/s}} = \sqrt{\frac{32 \text{ g/mol}}{146 \text{ g/mol}}}$$

Solve for the unknown, v_{SF_6}.

$$v_{SF_6} = (480 \text{ m/s}) \sqrt{\frac{32 \text{ g/mol}}{146 \text{ g/mol}}} = 480 \text{ m/s} \times 0.47 = 220 \text{ m/s}$$

4 Verify your results.

SF_6 has a mass about 4 times that of O_2. The square root of 4 is 2, and the inverse of 2 is $\frac{1}{2}$. SF_6 should travel about half as fast as O_2. The answer is therefore reasonable.

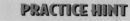

PRACTICE HINT

If you get confused, remember that molecular speeds and molar masses are inversely related, so the more massive particle will move at a slower speed.

PRACTICE

1 At the same temperature, which molecule travels faster, O_2 or N_2? How much faster?

2 At room temperature, Xe atoms have an average speed of 240 m/s. At the same temperature, what is the speed of H_2 molecules?

3 What is the molar mass of a gas if it diffuses 0.907 times the speed of argon gas?

4 Uranium isotopes are separated by effusion. What is the relative rate of effusion for $^{235}UF_6$ ($M = 349.03$ g/mol) and $^{238}UF_6$ ($M = 352.04$ g/mol)?

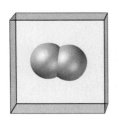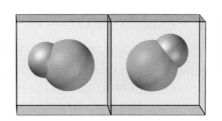

Figure 19
Hydrogen molecules combine with chlorine molecules in a 1:1 volume ratio to produce twice the volume of hydrogen chloride.

Gas Reactions Allow Chemical Formulas to Be Deduced

In 1808, Joseph Gay-Lussac made an important discovery: if the pressure and temperature are kept constant, gases react in volume proportions that are whole numbers. This is called **Gay-Lussac's law of combining volumes.** Consider the formation of gaseous hydrogen chloride from the reaction of hydrogen gas and chlorine gas. Gay-Lussac showed in an experiment that one volume of hydrogen gas reacts with one volume of chlorine gas to form two volumes of hydrogen chloride gas. **Figure 19** illustrates the volume ratios in the reaction in the form of a model. Let us use Avogadro's law and assume that two molecules of hydrogen chloride are formed, one in each of the two boxes on the right in **Figure 19.** Therefore, we must start with two atoms of hydrogen and two atoms of chlorine. Each box must contain one molecule of hydrogen and one molecule of chlorine, so there must be two atoms in each of these molecules.

Using several other reactions, such as the reaction of gaseous hydrogen and gaseous oxygen to form water vapor, the Italian chemist Stanislao Cannizzaro was able to deduce that oxygen is also diatomic and that the formula for water is H_2O. Dalton had guessed that the formula for water was HO, because this seemed the most likely combination of atoms for such a common compound. Before knowing atomic masses, chemists had only a set of relative weights. For example, it was known that 1 g of hydrogen can react with 8 g of oxygen to form water, so it was assumed that an oxygen atom was eight times as heavy as a hydrogen atom. It was not until the mid 1800s, just before the Civil War, that chemists knew the correct formula for water.

Dalton's Law of Partial Pressure

In 1805, John Dalton showed that in a mixture of gases, each gas exerts a certain pressure as if it were alone with no other gases mixed with it. The pressure of each gas in a mixture is called the **partial pressure.** The total pressure of a mixture of gases is the sum of the partial pressures of the gases. This principle is known as **Dalton's law of partial pressure.**

$$P_{total} = P_A + P_B + P_C$$

P_{total} is the total pressure, and P_A, P_B, and P_C are the partial pressures of each gas.

How is Dalton's law of partial pressure explained by the kinetic-molecular theory? All the gas molecules are moving randomly, and each has an equal chance to collide with the container wall. Each gas exerts a pressure proportional to its number of molecules in the container. The presence of other gas molecules does not change this fact.

Gay-Lussac's law of combining volumes

the law that states that the volumes of gases involved in a chemical change can be represented by the ratio of small whole numbers

partial pressure

the pressure of each gas in a mixture

Dalton's law of partial pressure

the law that states that the total pressure of a mixture of gases is equal to the sum of the partial pressures of the component gases

Gas Stoichiometry

The ideal gas law relates amount of gaseous substance in moles, *n,* with the other gas variables: pressure, volume, and temperature. Now that you have learned how to use the ideal gas law, an equation that relates the number of moles of gas to its volume, you can use it in calculations involving gases that react.

Gas Volumes Correspond to Mole Ratios

Topic Link

Refer to the "Stoichiometry" chapter for a discussion of stoichiometry.

Ratios of gas volumes will be the same as mole ratios of gases in balanced equations. Avogadro's law shows that the mole ratio of two gases at the same temperature and pressure is the same as the volume ratio of the two gases. This greatly simplifies the calculation of the volume of products or reactants in a chemical reaction involving gases. For example, consider the following equation for the production of ammonia.

$$3H_2(g) + N_2(g) \longrightarrow 2NH_3(g)$$

Consequently, 3 L of H_2 react with 1 L of N_2 to form 2 L of NH_3, and no H_2 or N_2 is left over (assuming an ideal situation of 100% yield).

Consider the electrolysis of water, a reaction expressed by the following chemical equation.

$$2H_2O(l) + \text{electricity} \longrightarrow 2H_2(g) + O_2(g)$$

The volume of hydrogen gas produced will be twice the volume of oxygen gas, because there are twice as many moles of hydrogen as there are moles of oxygen. As you can see in **Figure 20,** the volume of hydrogen gas produced is, in fact, twice as large as the volume of oxygen produced.

Furthermore, if we know the number of moles of a gaseous substance, we can use the ideal gas law to calculate the volume of that gas. **Skills Toolkit 1** on the following page shows how to find the volume of product given the mass of one of the reactants.

Figure 20
In the electrolysis of water, the volume of hydrogen is twice the volume of oxygen, because the mole ratio between the two gases is two to one.

Finding Volume of Unknown

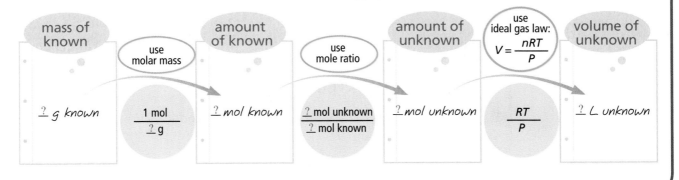

SAMPLE PROBLEM G

Using the Ideal Gas Law to Solve Stoichiometry Problems

How many liters of hydrogen gas will be produced at 280.0 K and 96.0 kPa if 1.74 mol of sodium react with excess water according to the following equation?

$$2Na(s) + 2H_2O(l) \longrightarrow 2NaOH(aq) + H_2(g)$$

1 Gather information.

$T = 280.0$ K $P = 96.0$ kPa $R = 8.314$ L·kPa/mol·K

$n = ?$ mol $V = ?$ L

2 Plan your work.

Use the mole ratio from the equation to compute moles of H_2.

$$1.74 \text{ mol Na} \times \frac{1 \text{ mol } H_2}{2 \text{ mol Na}} = 0.870 \text{ mol } H_2$$

Rearrange the ideal gas law to solve for the volume of hydrogen gas.

$$V = \frac{nRT}{P}$$

3 Calculate.

Substitute the three known values into the rearranged equation.

$$V = \frac{(0.870 \text{ mol } H_2)\left(\dfrac{8.314 \text{ L·kPa}}{\text{mol·K}}\right)(280.0 \text{ K})}{(96.0 \text{ kPa})} = 21.1 \text{ L } H_2$$

4 Verify your results.

Recall that 1 mol of gas at 0°C and 1 atm occupies 22.4 L. The conditions in this problem are close to this, so the volume should be near 22.4 L.

Practice problems on next page

PRACTICE HINT

Whenever you can relate the given information to moles, you can solve stoichiometry problems. In this case, the ideal gas law is the bridge that you need to get from moles to the answer.

PRACTICE

1 What volume of oxygen, collected at 25°C and 101 kPa, can be prepared by decomposition of 37.9 g of potassium chlorate?

$$2KClO_3(s) \longrightarrow 2KCl(s) + 3O_2(g)$$

2 Liquid hydrogen and oxygen are burned in a rocket. What volume of water vapor, at 555°C and 76.4 kPa, can be produced from 4.67 kg of H_2?

$$2H_2(l) + O_2(l) \longrightarrow 2H_2O(g)$$

3 How many grams of sodium are needed to produce 2.24 L of hydrogen, collected at 23°C and 92.5 kPa?

$$2Na(s) + 2H_2O(l) \longrightarrow 2NaOH(aq) + H_2(g)$$

③ Section Review

UNDERSTANDING KEY IDEAS

1. What gas laws are combined in the ideal gas law?

2. State the ideal gas law.

3. What is the relationship between a gas's molecular weight and speed of effusion?

4. How did Gay-Lussac's experiment allow the chemical formulas of hydrogen gas and chlorine gas to be deduced?

5. How does the total pressure of a mixture of gases relate to the partial pressures of the individual gases in a mixture?

6. In gas stoichiometry problems, what is the "bridge" between amount in moles and volume?

PRACTICE PROBLEMS

7. How many moles of argon are there in 20.0 L, at 25°C and 96.8 kPa?

8. A sample of carbon dioxide has a mass of 35.0 g and occupies 2.5 L at 400.0 K. What pressure does the gas exert?

9. How many moles of SO_2 gas are contained in a 4.0 L container at 450 K and 5.0 kPa?

10. Two gases effuse through a hole. Gas A has nine times the molecular mass of gas B. What is the ratio of the two molecular speeds?

11. What volume of ammonia can be produced from the reaction of 22.5 L of hydrogen with nitrogen?

$$N_2(g) + 3H_2(g) \longrightarrow 2NH_3(g)$$

12. What will be the volume, at 115 kPa and 355 K, of the nitrogen from the decomposition of 35.8 g of sodium azide, NaN_3?

$$2NaN_3(s) \longrightarrow 2Na(s) + 3N_2(g)$$

CRITICAL THINKING

13. Explain why helium-filled balloons deflate over time faster than air-filled balloons do.

14. Nitrous oxide is sometimes used as a source of oxygen gas:

$$2N_2O(g) \longrightarrow 2N_2(g) + O_2(g)$$

What volume of each product will be formed from 2.22 L N_2O? At STP, what is the density of the product gases when mixed? (Hint: Keep in mind Avogadro's law.)

Element Spotlight

N
Nitrogen
14.00674
$[He]2s^2 2p^3$

Where Is N?

Earth's crust:
<0.01% by mass

Air: 75% by mass

Sea water:
0.0016% by mass

Nitrogen

Nitrogen gas is the most abundant gas in the atmosphere. Nitrogen is important for making the proteins, nucleic acids, vitamins, enzymes, and hormones needed by plants and animals to live. However, nitrogen gas is too unreactive for plants and animals to use directly. Nitrogen-fixing bacteria convert atmospheric nitrogen into substances that green plants absorb from the soil. Animals then eat these plants or eat other animals that feed on these plants. When the animals and plants die and decay, the nitrogen in the decomposed organic matter returns as nitrogen gas to the atmosphere and as compounds to the soil. The nitrogen cycle then starts all over again.

Industrial Uses

- Nitrogen is used in the synthesis of ammonia.
- Ammonia, NH_3, is used to produce fertilizer, explosives, nitric acid, urea, hydrazine, and amines.
- Liquid nitrogen is used in superconductor research and as a cryogenic supercoolant for storing biological tissues.
- Nitrogen gas is used as an inert atmosphere for storing and processing reactive substances.
- Nitrogen gas, usually mixed with argon, is used for filling incandescent light bulbs.

Legumes, such as this soybean plant, have nitrogen-fixing bacteria on their roots. These bacteria can convert atmospheric nitrogen into compounds that can be used by plants in the formation of proteins.

Real-World Connection The atmosphere contains almost 4×10^{18} kg of N_2. That's more than the mass of 10 000 typical mountains!

A Brief History

1774–1777: Antoine Lavoisier determines that nitrogen is an element.

1909: Fritz Haber, a German chemist, discovers a method for synthesizing ammonia from hydrogen gas and nitrogen gas. The method is still used today and is called the Haber-Bosch process.

| 1600 | 1700 | 1800 | 1900 |

1772: Nitrogen is discovered by Daniel Rutherford in Scotland, Joseph Priestley and Henry Cavendish in England, and Carl Scheele in Sweden.

Questions

1. Name three nitrogen-containing products you come into contact with regularly.

2. Research the historical impacts of the invention of the Haber-Bosch process, which is mentioned in the timeline above.

internet connect

www.scilinks.org
Topic: Nitrogen Cycle
SciLinks code: HW4082

SCI LINKS. Maintained by the National Science Teachers Association

SECTION ONE Characteristics of Gases

- Gases are fluids, have low density, and are compressible, because of the relatively large intermolecular distances between gas particles.
- Gases expand to fill their entire container.
- Gases exert pressure in all directions.
- The kinetic-molecular theory states that gas particles are in constant random motion, are relatively far apart, and have volumes that are negligible when compared with the total volume of a gas.
- The average kinetic energy of a gas is proportional to its absolute temperature.

SECTION TWO The Gas Laws

- Pressure and volume of a gas at constant temperature are inversely proportional.
- The volume of a gas at constant pressure is proportional to the absolute temperature.
- The pressure of a gas at constant volume is proportional to the absolute temperature.
- Equal volumes of gas under the same conditions contain an equal number of moles of gas.

SECTION THREE Molecular Composition of Gases

- The ideal gas law is the complete statement of the relations between P, V, T, and n in a quantity of gas.
- Gases diffuse rapidly into each other.
- The rate of diffusion of a gas is inversely proportional to the square root of its molar mass.
- Each gas in a mixture produces a pressure as if it were alone in a container.
- Gay-Lussac's law of combining volumes can be used to deduce the chemical formula of a gas through observation of volume changes in a chemical reaction.

pressure
newton
pascal
standard temperature and pressure
kinetic-molecular theory

Boyle's law
Charles's law
Gay-Lussac's law
Avogadro's law

ideal gas
ideal gas law
diffusion
effusion
Graham's law of diffusion
Gay-Lussac's law of combining volumes
partial pressure
Dalton's law of partial pressure

Converting Pressure Units
Sample Problem A p. 420

Solving Pressure-Volume Problems
Sample Problem B p. 425

Solving Volume-Temperature Problems
Sample Problem C p..428

Solving Pressure-Temperature Problems
Sample Problem D p. 430

Using the Ideal Gas Law
Sample Problem E p. 435

Comparing Molecular Speeds
Sample Problem F p. 438

Using the Ideal Gas Law to Solve Stoichiometry Problems
Skills Toolkit 1 p. 441
Sample Problem G p. 441

CHAPTER REVIEW 12

USING KEY TERMS

1. What is the definition of *pressure?*

2. What is a newton?

3. Write a paragraph that describes how the kinetic-molecular theory explains the following properties of a gas: fluidity, compressibility, and pressure.

   WRITING SKILLS

4. How do pascals relate to newtons?

5. What relationship does Boyle's law express?

6. What relationship does Charles's law express?

7. What relationship does Gay-Lussac's law express?

8. What gas law combines the basic gas laws?

9. What are two characteristics of an ideal gas?

10. Describe in your own words the process of diffusion.

UNDERSTANDING KEY IDEAS

Characteristics of Gases

11. How does wind illustrate that gases are fluids?

12. Using air, water, and a syringe, how can you show the difference in compressibility of liquids and gases?

13. When you drive a car on hot roads, the pressure in the tires increases. Explain.

14. As you put more air in a car tire, the pressure increases. Explain.

15. When you expand your lungs, air flows in. Explain.

16. Even when there is air in bicycle tires, you can still push down on the handle of the pump rather easily. Explain.

17. When you put air in a completely flat bicycle tire, the entire tire expands. Explain.

18. What assumptions does the kinetic-molecular theory make about the nature of a gas?

19. How does the average kinetic energy of a gas relate to its temperature?

The Gas Laws

20. How are the volume and pressure of a gas related, if its temperature is kept constant?

21. Explain why pressure increases as a gas is compressed into a smaller volume.

22. How are the absolute temperature and volume of a gas related, at constant pressure?

23. Explain Charles's law in terms of the kinetic-molecular theory.

24. How is a gas's pressure related to its temperature, at constant volume?

25. Explain Gay-Lussac's law in terms of the kinetic-molecular theory.

26. What does Avogadro's law state about the relationship between gas volumes and amounts in moles?

Molecular Composition of Gases

27. When using the ideal gas law, what is the proportionality constant, and in what units is it usually expressed?

28. Ammonia, NH_3, and alcohol, C_2H_6O, are released together across a room. Which will you smell first?

29. How can Gay-Lussac's law of combining volumes be used to deduce chemical formulas?

30. Write the equation that expresses Dalton's law of partial pressures.

PRACTICE PROBLEMS

Sample Problem A Converting Pressure Units

31. The standard pressure at sea level is 101 325 pascals. What force is being exerted on each square meter of Earth's surface?

32. The vapor pressure of hydrogen peroxide is 100.0 torr at 97.9°C. What is this pressure in kPa?

33. The gauge pressure in a tire is 28 psi, which adds to atmospheric pressure of 14.0 psi. What is the internal tire pressure in kPa?

34. The weather bureau reports the atmospheric pressure as 925 millibars. What is this pressure in kPa?

Sample Problem B Solving Pressure-Volume Problems

35. A gas sample has a volume of 125 mL at 91.0 kPa. What will its volume be at 101 kPa?

36. A 125 mL sample of gas at 105 kPa has its volume reduced to 75.0 mL. What is the new pressure?

37. A diver at a depth of 1.0×10^2 m, where the pressure is 11.0 atm, releases a bubble with a volume of 100.0 mL. What is the volume of the bubble when it reaches the surface? Assume a pressure of 1.00 atm at the surface.

38. In a deep-sea station 2.0×10^2 m below the surface, the pressure in the module is 20.0 atm. How many liters of air at sea level are needed to fill the module with 2.00×10^7 L of air?

39. The pressure on a 240.0 mL sample of helium gas is increased from 0.428 atm to 1.55 atm. What is the new volume, assuming constant temperature?

40. A sample of air with volume 6.6×10^7 L changes pressure from 99.4 kPa to 88.8 kPa. Assuming constant temperature, what is the new volume?

Sample Problem C Solving Volume-Temperature Problems

41. Use Charles's law to solve for the missing value in the following. $V_1 = 80.0$ mL, $T_1 = 27°C$, $T_2 = 77°C$, $V_2 = ?$

42. A balloon filled with helium has a volume of 2.30 L on a warm day at 311 K. It is brought into an air-conditioned room where the temperature is 295 K. What is its new volume?

43. The balloon in item 42 is dipped into liquid nitrogen at −196°C. What is its new volume?

44. A gas at 65°C occupies 4.22 L. At what Celsius temperature will the volume be 3.87 L, at the same pressure?

45. A person breathes 2.6 L of air at −11°C into her lungs, where it is warmed to 37°. What is its new volume?

46. A scientist warms 26 mL of gas at 0.0°C until its volume is 32 mL. What is its new temperature in degrees Celsius?

Sample Problem D Solving Pressure-Temperature Problems

47. Use Gay-Lussac's law to solve for the unknown. $P_1 = 111$ kPa, $T_1 = 273$ K, $T_2 = 373$ K, $P_2 = ?$

48. A sample of hydrogen exerts a pressure of 0.329 atm at 47°C. What will the pressure be at 77°C, assuming constant volume?

49. A sample of helium exerts a pressure of 101 kPa at 25°C. Assuming constant volume, what will its pressure be at liquid nitrogen temperature, −196°C?

50. The pressure inside a tire is 39 psi at 20°C. What will the pressure be after the tire is driven at high speed on a hot highway, when the temperature in the tire is 48°C?

51. A tank of oxygen for welding is at 31°C and 11 atm. What is the pressure when it is taken to the South Pole, where the temperature is −41°C?

52. A cylinder of gas at 555 kPa and 22°C is heated until the pressure is 655 kPa. What is the new temperature?

Sample Problem E Using the Ideal Gas Law

53. How many moles of argon are there in 20.0 L, at 25°C and 96.8 kPa?

54. How many moles of air are in 1.00 L at −23°C and 101 kPa?

55. A 4.44 L container holds 15.4 g of oxygen at 22.55°C. What is the pressure?

56. A polyethylene plastic weather balloon contains 65 L of helium, which is at 20.0°C and 94.0 kPa. How many moles of helium are in the balloon?

57. What will be the volume of the balloon in item 56 in the stratosphere at −61°C and 1.1×10^3 Pa?

58. A polyethylene weather balloon is inflated with 12.0 g of helium at −23°C and 100.0 kPa. What is its volume?

Sample Problem F Comparing Molecular Speeds

59. An unknown gas effuses at a speed one-quarter of that of helium. What is the molar mass of the unknown gas? It is either sulfur dioxide or sulfur trioxide. Which gas is it?

60. An unknown gas effuses at one half the speed of oxygen. What is the molar mass of the unknown? It is either HBr or HI. Which gas is it?

61. Oxygen molecules have an average speed of 4.80×10^2 m/s at 25°C. What is the average speed of H_2 molecules at the same temperature?

62. Oxygen molecules have an average speed of 480 m/s at 25°C. What is the average speed of HI molecules at that temperature?

Sample Problem G Using the Ideal Gas Law to Solve Stoichiometry Problems

63. How many liters of hydrogen gas can be produced at 300.0 K and 104 kPa if 20.0 g of sodium metal is reacted with water according to the following equation?

$$2Na(s) + 2H_2O(l) \longrightarrow 2NaOH(aq) + H_2(g)$$

64. Magnesium will burn in oxygen to form magnesium oxide as represented by the following equation.

$$2Mg(s) + O_2(g) \longrightarrow 2MgO(s)$$

What mass of magnesium will react with 500.0 mL of oxygen at 150.0°C and 70.0 kPa?

65. Suppose a certain automobile engine has a cylinder with a volume of 500.0 mL that is filled with air (21% oxygen) at a temperature of 55°C and a pressure of 101.0 kPa. What mass of octane must be injected to react with all of the oxygen in the cylinder?

$$2C_8H_{18}(l) + 25O_2(g) \longrightarrow 16CO_2(g) + 18H_2O(g)$$

66. Methanol, CH_3OH, is made by using a catalyst to react carbon monoxide with hydrogen at high temperature and pressure. Assuming that 450.0 mL of CO and 825 mL of H_2 are allowed to react, answer the following questions. (Hint: First write the balanced chemical equation for this reaction.)
  a. Which reactant is in excess?
  b. How much of that reactant remains when the reaction is complete?
  c. What volume of $CH_3OH(g)$ is produced?

67. What volume of oxygen, measured at 27°C and 101.325 kPa, is needed for the combustion of 1.11 kg of coal? (Assume coal is 100% carbon.)

$$C(s) + O_2(g) \longrightarrow CO_2(g)$$

68. How many liters of hydrogen are obtained from the reaction of 4.00 g calcium with excess water, at 37°C and 0.962 atm?

69. How many grams of carbon dioxide are contained in 1.000 L of the gas, at 25.0°C and 101.325 kPa? What is the density of the gas at these conditions? What would the density of oxygen be at these conditions?

70. Below is a diagram showing the effects of pressure on a column of mercury and on a column of water. Which system is under a higher pressure? Explain your choice.

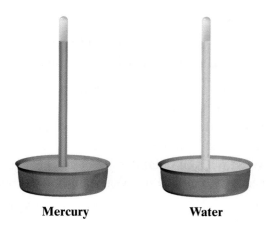

Mercury	**Water**

71. Solid LiOH can be used in spacecraft to remove CO_2 from the air.

$$2LiOH(s) + CO_2(g) \longrightarrow Li_2CO_3(s) + H_2O(l)$$

What mass of LiOH must be used to absorb the carbon dioxide that exerts a partial pressure of 5.0 kPa at 15°C in a spacecraft with volume of 8.0×10^4 L?

72. What is the density of helium gas at 0°C and 101.325 kPa?

73. What is the density of HI gas under the same conditions as in item 71?

74. The density of mercury is 13.6 times that of water. What would be the height of a water barometer at sea level?

75. A 6.55 L container has oxygen at 25°C and 95.5 kPa. What will the pressure be after 0.200 mol of nitrogen is added?

76. A sample of carbon dioxide occupies 638 mL at 0.893 atm and 12°C. What will the pressure be at a volume of 881 mL and temperature of 18°C?

77. How many grams of oxygen gas must be in a 10.0 L container to exert a pressure of 97.0 kPa at a temperature of 25°C?

78. A 3.00 L sample of air exerts a pressure of 101 kPa at 300.0 K. What pressure will the air exert if the temperature is increased to 400.0 K?

79. How many liters of hydrogen can be obtained from the reaction of 4.00 g Ca with excess water, at 37°C and 0.962 atm?

80. Explain in terms of the kinetic-molecular theory why increasing the temperature of a gas at constant volume increases the pressure of the gas.

81. A sample of air has a pressure of 80.0 kPa at 20.0°C. The temperature is changed, and the pressure becomes 120.0 kPa. What is the new temperature in degrees Celsius? Assume constant volume.

82. A sample of nitrogen has a volume of 421 mL at 27°C. To what temperature must it be heated so that its volume becomes 656 mL, assuming constant pressure?

83. A 10 L tank is filled with helium until its pressure is 2.7×10^3 kPa at 23°C. What mass of helium is contained in the tank?

84. Above 100°C and at constant pressure, two volumes of hydrogen react with one volume of oxygen to form two volumes of gaseous water. Set up a diagram for this reaction that is similar to **Figure 19,** and that shows how the molecular formulas for oxygen and water can be determined.

CRITICAL THINKING

85. Clean rooms, used for sterile biological research, are sealed tightly and operate under high air pressure. Explain why.

86. The partial pressure of oxygen in the air is 0.21 atm at sea level, where the total pressure is 1.00 atm. What is the partial pressure of oxygen when this air rises to where the total pressure is 0.86 atm?

87. A sample of gas has a volume of 100 mL. Suppose the temperature is raised from 200 to 400 K, and the pressure changed from 100 to 200 kPa. What is the final volume?

88. A 1.00 L sample of chlorine gas is at 0°C and 101 kPa. This is liquefied, and the liquid has a density of 1.37 g/mL. What is the volume of the liquid chlorine?

89. You have a 1 L container of a gas at 20°C and 1 atm. Without opening the container, how could you tell whether the gas is chlorine or fluorine?

90. The density of oxygen gas is 1.31 g/L at 25°C and 101.325 kPa. What is the density of SO_2 under the same conditions?

91. If a breath of air is 2.4 L at 28°C and 101.3 kPa, how many grams of oxygen are in it? (Air is 21% oxygen by volume.)

92. Gas companies often store their fuel supplies in liquid form and in large storage tanks. Liquid nitrogen is used to keep the temperature low enough for the fuel to remain condensed in liquid form. Although continuous cooling is expensive, storing a condensed fuel is more economical than storing the fuel as gas. Suggest a reason that storing a liquid is more economical than storing a gas.

93. How would the shape of a curve showing the kinetic-energy distribution of gas molecules at 50°C compare with the blue and red curves in **Figure 8?**

ALTERNATIVE ASSESSMENT

94. The air pressure of car tires should be checked regularly for safety reasons and to prevent uneven tire wear. Find out the units of measurement on a typical tire gauge, and determine how gauge pressure relates to atmospheric pressure.

95. Find a local hot-air balloon group, and discuss with group members how they use the gas laws to fly their balloons. The group may be willing to give a demonstration. Report your experience to the class.

96. Qualitatively compare the molecular masses of various gases by noting how long it takes you to smell them from a fixed distance. Work only with materials that are not dangerous, such as flavor extracts, fruit peels, and onions.

CONCEPT MAPPING

97. Use the following terms to complete the concept map below: *amount in moles, ideal gas law, pressure, temperature,* and *volume.*

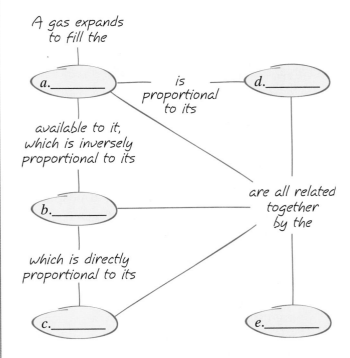

Study the graph below, and answer the questions that follow.
For help in interpreting graphs, see Appendix B, "Study Skills for Chemistry."

98. What variable decreases as you go down the *y*-axis of the graph?

99. What is the volume of the gas at a pressure of 150 kPa?

100. What is the pressure of the gas at a volume of 0.200 L?

101. Describe in your own words the relationship this graph illustrates.

102. List the values of volume and pressure that correspond to any two points on the graph, and show why they demonstrate Boyle's law.

103. Explain why you feel resistance if you try to compress a sample of gas inside a plugged syringe.

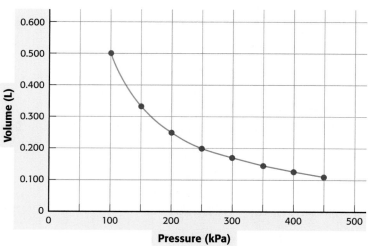

Volume Vs. Pressure for a Gas at Constant Temperature

TECHNOLOGY AND LEARNING

104. Graphing Calculator

Calculating Pressure Using the Ideal Gas Law

The graphing calculator can run a program that calculates the pressure in atmospheres, given the number of moles of a gas (*n*), volume (*V*), and temperature (*T*).

Go to Appendix C. If you are using a TI-83 Plus, you can download the program **IDEAL** and run the application as directed. If you are using another calculator, your teacher will provide you with keystrokes and data sets to use. Use the program on your calculator to answer these questions.

a. What is the pressure for a gas with an amount of 1.3 mol, volume of 8.0 L, and temperature of 293 K?

b. What is the pressure for a gas with an amount of 2.7 mol, volume of 8.5 L, and temperature of 310 K?

c. A gas with an amount of 0.75 mol and a volume of 6.0 L is measured at two different temperatures: 300 K and 275 K. At which temperature is the pressure greater?

1. The kinetic-molecular theory states that ideal gas molecules
  a. have weight and take up space.
  b. are in constant, rapid, random motion.
  c. exert forces of attraction and repulsion on each other.
  d. have high densities compared with liquids and solids.

2. Pressure can be measured in
  a. grams.
  c. pascals.
  b. meters.
  d. liters.

3. A sample of oxygen gas has a volume of 150 mL when its pressure is 0.947 atm. If the pressure is increased to 0.987 atm and the temperature remains constant, what will the new volume be?
  a. 140 mL
  c. 200 mL
  b. 160 mL
  d. 240 mL

4. A sample of neon gas occupies a volume of 752 mL at 25.0°C. If the pressure remains constant, what will be the volume of the gas at 50.0°C?
  a. 694 mL
  c. 815 mL
  b. 752 mL
  d. 955 mL

5. Which gas will effuse through a small opening fastest?
  a. the gas with the greatest molar mass
  b. the gas with the smallest molar mass
  c. the gas with the most polar molecules
  d. all will effuse at the same rate if temperature is constant

6. What is the pressure exerted by a 0.500 mol sample of nitrogen in a 10.0 L container at 25°C?
  a. 1.2 kPa
  c. 0.10 kPa
  b. 10 kPa
  d. 120 kPa

7. What volume of chlorine gas at 38°C and 1.63 atm is needed to react completely with 12.4 g of sodium to form NaCl?
  a. 4.22 L
  c. 7.90 L
  b. 4.45 L
  d. 10.2 L

8. A sample of gas in a closed container at a temperature of 100.0°C and 3.0 atm is heated to 300.0°C. What is the pressure of the gas at the higher temperature?
  a. 3.5 atm
  c. 5.9 atm
  b. 4.6 atm
  d. 9.0 atm

9. At room temperature, oxygen molecules have an average speed near 480 m/s. What is the speed of helium atoms at the same conditions?
  a. 3840 m/s
  c. 170 m/s
  b. 1360 m/s
  d. 60 m/s

10. An unknown gas effuses twice as fast as CH_4. What is the molar mass of the gas?
  a. 64 g/mol
  c. 8 g/mol
  b. 32 g/mol
  d. 4 g/mol

11. If 3 L N_2 is reacted completely with 3 L H_2, how many liters of *unreacted* gas remain?

$$N_2 + 3H_2 \longrightarrow 2NH_3$$

  a. 4 L
  c. 2 L
  b. 3 L
  d. 1 L

12. Avogadro's law states that
  a. equal amounts of gases at the same conditions occupy equal volumes, regardless of the identity of the gases.
  b. at constant pressure, gas volume is directly proportional to absolute temperature.
  c. the volume of a gas is inversely proportional to its amount in moles.
  d. at constant temperature, gas volume is inversely proportional to pressure.

$\mathbf{O}$cean water is an excellent example of a solution. A solution is a homogeneous mixture of two or more substances. *Homogeneous* means that the solution looks the same throughout, even under a microscope. In this case, water is the solvent—the substance in excess. Many substances are dissolved in the water. The most abundant substance is ordinary table salt, $NaCl$, which is present as the ions Na^+ and Cl^-. Other ions present include Mg^{2+}, Ca^{2+}, and Br^-. Oxygen gas is dissolved in the water. Without oxygen, the fish could not live. Because humans do not have gills like fish do, the diver needs scuba gear.

START-UP*ACTIVITY*

SAFETY PRECAUTIONS

Exploring Types of Mixtures

PROCEDURE

1. Prepare five mixtures in **five different 250 mL beakers.** Each should contain about 200 mL of water and one of the following substances: **12 g of sucrose, 3 g of soluble starch, 5 g of clay, 2 mL of food coloring,** and **20 mL of cooking oil.**

2. Observe the five mixtures and their characteristics. Record the appearance of each mixture after stirring.

3. Note which mixtures do not separate after standing. Transfer 10 mL of each of these mixtures to an individual **test tube.** Shine a **flashlight** through each mixture in a darkened room. Make a note of the mixture(s) in which the path of the light beam is visible.

ANALYSIS

1. What characteristic did the mixture(s) that separated after stirring share?

2. How do you think the mixture(s) in which the light's path was visible differed from those in which it was not?

Pre-Reading Questions

1. Give three examples of solutions you find in everyday life.

2. What main components do these solutions consist of?

3. How do you know that each of these examples is actually a solution?

internet connect

www.scilinks.org
Topic : Solutions
SciLinks code: HW4118

SCiLINKS. Maintained by the National Science Teachers Association

What Is a Solution?

KEY TERMS

- **solution**
- **suspension**
- **solvent**
- **solute**
- **colloid**

solution

a homogeneous mixture of two or more substances uniformly dispersed throughout a single phase

suspension

a mixture in which particles of a material are more or less evenly dispersed throughout a liquid or gas

> **Topic Link**
>
> *Refer to the "The Science of Chemistry" chapter for a discussion of the classification of mixtures.*

OBJECTIVES

1. **Distinguish** between solutions, suspensions, and colloids.
2. **Describe** some techniques chemists use to separate mixtures.

Mixtures

You have learned about the difference between pure substances and mixtures. Mixtures can either be *heterogeneous* or *homogeneous*. The particles of a heterogeneous mixture are large enough to see under a microscope. In a homogeneous mixture, however, the particles are molecule-sized, so the mixture appears uniform, even under a microscope. A homogenous mixture is also known as a **solution.**

Suspensions Are Temporary Heterogeneous Mixtures

The potter shown in **Figure 1** uses water to help sculpt the sides of a clay pot. As he dips his clay-covered fingers into a container of water, the water turns brown. The clay-water mixture appears uniform. However, if the container sits overnight, the potter will see a layer of mud on the bottom and clear water on top. The clay does not dissolve in water. The clay breaks up into small pieces that are of such low mass that, for a while, they remain suspended in the water. This type of mixture, in which the different parts spontaneously separate over time, is called a **suspension.** In a suspension, the particles may remain mixed with the liquid while the liquid is being stirred, but later they settle to the bottom.

Figure 1
Clay particles suspended in water form a suspension.

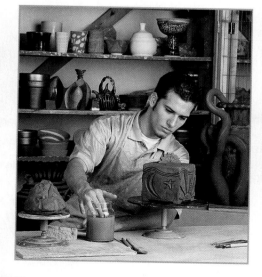

Before settling

After settling

Water molecule, H₂O

Fresh water

Water molecule, H₂O

Chloride ion, Cl⁻

Sodium ion, Na⁺

Salt water solution

Solutions Are Stable Homogeneous Mixtures

A student working in a pet shop is asked to prepare some water for a salt water aquarium. She prepares a bucket of fresh water. Then she adds a carefully measured quantity of salt crystals to the water, as shown in **Figure 2,** and stirs the water. These crystals consist of a mixture of salts which, when dissolved in water, will produce a solution with the same composition as sea water. After stirring, the student can no longer see any grains of salt in the water. No matter how long she waits, the salt will not separate from the water. The salt has dissolved in the water to form a stable homogeneous mixture. The particles are evenly distributed throughout the mixture, making it a true solution. The dissolved particles, which are ions in this case, are close to the size of the water molecules and are not clustered together.

Solution Is a Broad Term

Any mixture that is homogeneous on a microscopic level is a solution. According to that definition, air is a gaseous solution. However, when most people use the word *solution,* they are usually referring to a homogeneous *liquid* mixture. A homogeneous liquid mixture has one main component—a liquid—as well as one or more additional ingredients that are usually present in smaller amounts. The primary ingredient in a solution is called the **solvent,** and the other ingredients are the **solutes** and are said to be dissolved in the solvent. Water is the most common solvent. Although it is a very common substance, water is a unique solvent because so many substances can dissolve in it. Solutions in which water is the solvent are called *aqueous* solutions.

Figure 2
Fresh water is stable and homogeneous. The saltwater mixture is also stable and homogeneous because mixing occurs between molecules and ions.

solvent

in a solution, the substance in which the solute is dissolved

solute

in a solution, the substance that is dissolved in the solvent

Copper atom

Zinc atom

Brass

Figure 3
Brass is a solid solution of zinc atoms in copper atoms.

Another type of solution involves one solid mixed with another solid. Examples include solid alloys, such as brass, bronze, and steel. Brass, shown in **Figure 3,** is a mixture of copper and zinc. Brass is widely used in musical instruments because it is harder and more resistant to corrosion than pure copper.

Colloids Are Stable Heterogeneous Mixtures

Milk appears to be homogeneous. But as **Figure 4** shows, under a microscope you see that milk contains globules of fat and small lumps of the protein casein dispersed in a liquid called *whey*. Milk is actually a **colloid,** and not a solution. The particles of casein do not settle out after standing.

The particles in a colloid usually have an electric charge. These likecharged particles repel each other, so they do not collect into larger particles that would settle out. If you added acid to milk, the acid would neutralize the charge, and the particles would coagulate and settle to the bottom of the container.

colloid

a mixture consisting of tiny particles that are intermediate in size between those in solutions and those in suspensions and that are suspended in a liquid, solid, or gas

Figure 4
Milk is a colloidal suspension of proteins and fat globules in whey.

Separating Mixtures

There are many ways to separate mixtures into their components. The best method to use in a particular case depends on the kind of mixture and on the properties of the ingredients. The methods shown in **Figure 5** rely on the physical properties of the ingredients to separate them. Notice that some of these methods can be used outside a laboratory. In fact, you may use one or more of these methods in your own home. For example, you may use filtration to make coffee or evaporation when you cook. Centrifuges are used in dairies to separate the cream from the milk to make fat-free milk.

Figure 5

a Decanting separates a liquid from solids that have settled. To decant a mixture, carefully pour off the liquid to leave the solids behind.

b A centrifuge is used to separate substances of different densities. The centrifuge spins rapidly, and the denser substances collect at the bottom of the tube.

c Ground coffee is separated from liquid coffee by filtration. The filtrate—the liquid and whatever passes through the filter—collects in the coffeepot. The solid grounds stay on the filter.

d In saltwater ponds such as this one, sea water evaporates, and salts, mainly sodium chloride, are left behind.

Chromatography Separates by Differences in Attraction

You may have noticed that stains stick to some fabrics more than other fabrics. Also, different processes are used to remove different stains. This illustrates the principles used in *chromatography*. Chromatography separates components of a mixture based on how quickly different molecules dissolved in a *mobile phase* solvent move along a *solid phase.*

Paper chromatography is a powerful technique for separating components of a solution. For example, it can be used to separate the dyes in ink. The dyes are blotted onto the paper (solid phase), and a solvent such as water (mobile phase) travels up the paper. The solvent dissolves the ink as it travels up the paper. Dyes that are attracted more strongly to the paper than other dyes travel more slowly along the paper. **Figure 6** shows the separation of dyes that make up different colors of ink.

Quick LAB

SAFETY PRECAUTIONS

Design Your Own Lab: The Colors of Candies

In this activity you will investigate whether the colors of candy-coated chocolates are single dyes or mixtures of dyes.

PROCEDURE
Design a paper chromatography experiment that uses **candy-coated chocolates,** **chromatography paper, small beakers,** and **0.1% NaCl developing solution** to determine whether the dyes in the candies are mixtures of dyes or are single dyes. For example, is the green color a result of mixing two primary colors? Are the other colors mixtures?

ANALYSIS

1. Prepare a report that includes your experimental procedure, a data table that summarizes your results, and the experimental evidence.

2. Be sure to answer the questions posed in the procedure.

Distillation Separates by Differences in Boiling Point

Sometimes mixtures of liquids need to be purified or have their components separated. If the boiling points of the components are different, *distillation* can separate them based on their boiling points. As one component reaches its boiling point, it evaporates from the mixture and is allowed to cool and condense. This process continues until all the desired components have been separated from the mixture. For example, fermentation produces a solution of alcohol in water. If this is placed in a pot and boiled, the alcohol boils first. This alcohol-rich *distillate* can be collected by a distilling column. The distilling column is a cooler surface upon which the distillate recondenses, and can be collected as a liquid.

In some places where fresh water is scarce, distillation is used to obtain drinking water from sea water. However, because distillation requires a lot of energy, the process is expensive. Distillation is also used in the petroleum industry to separate crude oil into fractions according to their boiling points. The first fractions to distill are fluids with low boiling points, used as raw material in the plastics industry. Next comes gasoline, then at higher temperatures diesel fuel, then heating oil, then kerosene distill. What remains is the basis for lubricating greases.

Section Review

UNDERSTANDING KEY IDEAS

1. Explain why a suspension is considered a heterogeneous mixture.

2. Classify the following mixtures as homogeneous or heterogeneous:
   a. lemon juice   **c.** blood
   b. tap water   **d.** house paint

3. In a solution, which component is considered the solvent? Which is the solute?

4. Name the solvent and solute(s) in the following solutions:
   a. carbonated water   **c.** coffee
   b. apple juice   **d.** salt water

5. Does a solution have to involve a liquid? Explain your answer.

6. How is a colloid distinguished from a solution or a suspension?

7. What is the basic physical principle that chromatography is based upon?

8. How can distillation be used to prepare pure water from tap water?

CRITICAL THINKING

9. Explain how you could determine that brass is a solution, and not a colloid or suspension.

10. Explain why fog is a colloid.

11. You get a stain on a table cloth. Soapy water will not take the stain out, but rubbing alcohol will. How does this relate to chromatography?

12. If you allow a container of sea water to sit in the sun, the liquid level gets lower and lower, and finally crystals appear. What is happening?

SECTION 2

Concentration and Molarity

KEY TERMS

- concentration
- molarity

OBJECTIVES

① **Calculate** concentration using common units.

② **Define** molarity, and calculate the molarity of a solution.

③ **Describe** the procedure for preparing a solution of a certain molarity.

④ **Use** molarity in stoichiometric calculations.

Concentration

In a solution, the solute is distributed evenly throughout the solvent. This means that any part of a solution has the same ratio of solute to solvent as any other part of the solution. This ratio is the **concentration** of the solution. Some common ways of expressing concentration are given in **Table 1.**

concentration

the amount of a particular substance in a given quantity of a solution

Calculating Concentration

Concentrations can be expressed in many forms. One unit of concentration used in pollution measurements that involve very low concentrations is *parts per million,* or ppm. Parts per million is the number of grams of solute in 1 million grams of solution. For example, the concentration of lead in drinking water may be given in parts per million.

When you want to express concentration, you will begin with analytical data which may be expressed in units other than the units you want to use. In that case, each value must be converted into the appropriate units. Then, you must be sure to express the concentration in the correct ratio. **Sample Problem A** shows a typical calculation.

Table 1   Some Measures of Concentration

Name	Abbreviation or symbol	Units	Areas of application
Molarity	M	$\dfrac{\text{mol solute}}{\text{L solution}}$	in solution stoichiometry calculations
Molality	*m*	$\dfrac{\text{mol solute}}{\text{kg solvent}}$	with calculation of properties such as boiling-point elevation and freezing-point depression
Parts per million	ppm	$\dfrac{\text{g solute}}{1000\ 000\ \text{g solution}}$	to express small concentrations

See Appendix A for additional units of concentration.

Calculating Parts per Million

A chemical analysis shows that there are 2.2 mg of lead in exactly 500 g of a water sample. Convert this measurement to parts per million.

1 Gather information.

mass of solute: 2.2 mg
mass of solution: 500 g
parts per million = ?

2 Plan your work.

First, change 2.2 mg to grams:

$$2.2 \text{ mg} \times \frac{1 \text{ g}}{1000 \text{ mg}} = 2.2 \times 10^{-3} \text{ g.}$$

Divide this by 500 g to get the amount of lead in 1 g water, then multiply by 1 000 000 to get the amount of lead in 1 000 000 g water.

3 Calculate.

$$\frac{0.0022 \text{ g Pb}}{500 \text{ g H}_2\text{O}} \times \frac{1\,000\,000 \text{ parts}}{1 \text{ million}} =$$
4.4 ppm (parts Pb per million parts sample)

4 Verify your results.

Work backwards. If you divide 4.4 by 1 000 000 you get 4.4×10^{-6}. This result is the mass in grams of lead found per gram of the water sample. Multiply by 500 g to find the total amount of lead in the sample. The result is 2.2×10^{-3}, which is the given number of grams of lead in the sample.

PRACTICE HINT

Be sure to keep track of units in all concentration calculations. In the case of mass-to-mass ratios, such as parts per million, the masses of solute and solution must be expressed in the same units to obtain a correct ratio.

PRACTICE

1 Helium gas, 3.0×10^{-4} g, is dissolved in 200.0 g of water. Express this concentration in parts per million.

2 A sample of 300.0 g of drinking water is found to contain 38 mg Pb. What is this concentration in parts per million?

3 A solution of lead sulfate contains 0.425 g of lead sulfate in 100.0 g of water. What is this concentration in parts per million?

4 A 900.0 g sample of sea water is found to contain 6.7×10^{-3} g Zn. Express this concentration in parts per million.

5 A 365.0 g sample of water, contains 23 mg Au. How much gold is present in the sample in parts per million?

6 A 650.0 g hard-water sample contains 101 mg Ca. What is this concentration in parts per million?

7 An 870.0 g river water sample contains 2 mg of cadmium. Express the concentration of cadmium in parts per million.

Potassium ion, K⁺

Water molecule, H₂O

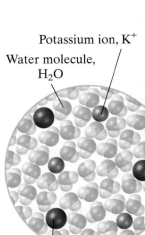

Bromine ion, Br⁻

Molarity

molarity

a concentration unit of a solution expressed as moles of solute dissolved per liter of solution

It is often convenient for chemists to discuss concentrations in terms of the number of solute particles in solution rather than the mass of particles in solution. Since the mole is the unit chemists use to measure the number of particles, they often specify concentrations using **molarity.** Molarity describes how many moles of solute are in each liter of solution.

Suppose that 0.30 moles of KBr are present in 0.40 L of solution. The molarity of the solution is calculated as follows:

$$\frac{0.30 \text{ mol KBr}}{0.40 \text{ L solution}} = 0.75 \text{ M KBr}$$

The symbol M is read as "molar" or as "moles per liter." Any amount of this solution has the same ratio of solute to solution, as shown in **Figure 7.**

Chemists often refer to the molarity of a solution by placing the formula for the solute in brackets. For example, $[CuSO_4]$ would be read as "the molarity of copper sulfate."

Preparing a Solution of Specified Molarity

Note that molarity describes concentration in terms of *volume of solution,* not volume of solvent. If you simply added 1.000 mol solute to 1.000 L solvent, the solution would not be 1.000 M. The added solute will change the volume, so the solution would not have a concentration of 1.000 M. The solution must be made to have exactly the specified volume of solution. The process of preparing a solution of a certain molarity is described in **Skills Toolkit 1.**

Topic Link

Refer to the "The Mole and Chemical Composition" chapter for a discussion of the use of the mole to express chemical amounts.

Preparing 1.000 L of a 0.5000 M Solution

Copper(II) sulfate, $CuSO_4$, is one of the compounds used to produce the chemiluminescence in light sticks. To make a 0.5000 M $CuSO_4$ solution, you need 0.5000 mol of the hydrate, $CuSO_4 \cdot 5H_2O$, for each liter of solution. To convert this amount of $CuSO_4 \cdot 5H_2O$ to a mass, multiply by the molar mass of $CuSO_4 \cdot 5H_2O$ (mass of $CuSO_4 \cdot 5H_2O$ = 0.5000 mol × 249.68 g/mol = 124.8 g).

Add some solvent (water) to the calculated mass of solute in the beaker to dissolve it, and then pour the solution into a 1.000 L volumetric flask.

Rinse the beaker with more water several times, and each time pour the rinse water into the flask until the solution almost reaches the neck of the flask.

Stopper the flask, and swirl thoroughly until all of the solid is dissolved.

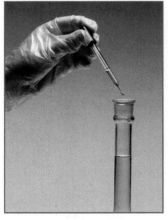

Carefully fill the flask with water to the 1.000 L mark.

Restopper the flask, and invert the flask at least 10 more times to ensure complete mixing.

The solution that results has 0.5000 mol of $CuSO_4$ dissolved in 1.000 L of solution—a 0.5000 M concentration.

Calculating with Molarity

In working with solutions in chemistry, you will find that numerical calculations often involve molarity. The key to all such calculations is the definition of molarity, which is stated as an equation below.

$$\text{molarity} = \frac{\text{moles of solute}}{\text{liters of solution}}$$

Skills Toolkit 2, below, shows how to use this equation in two common types of problems.

2   SKILLS Toolkit

1. Calculating the molarity of a solution when given the mass of solute and volume of solution

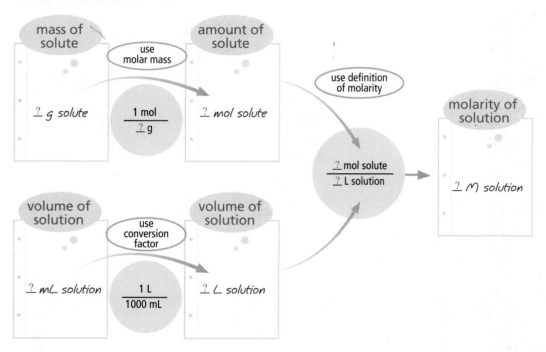

2. Calculating the mass of solute when given the molarity and volume of solution

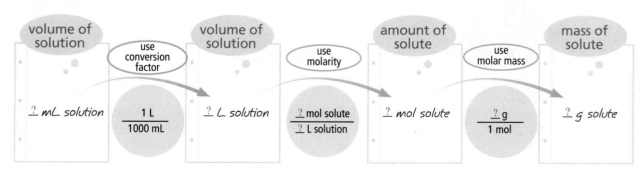

Calculating Molarity

What is the molarity of a potassium chloride solution that has a volume of 400.0 mL and contains 85.0 g KCl?

1 Gather information.

volume of solution = 400.0 mL
mass of solute = 85.0 g KCl
molarity of KCl solution = ?

2 Plan your work.

Convert the mass of KCl into moles of KCl by using the molar mass:

$$85.0 \ g \ KCl \times \frac{1 \ mol}{74.55 \ g \ KCl} = 1.14 \ mol \ KCl$$

Convert the volume in milliliters into volume in liters:

$$400.0 \ mL \times \frac{1 \ L}{1000 \ mL} = 0.4000 \ L$$

3 Calculate.

Molarity is moles of solute divided by volume of solution:

$$\frac{1.14 \ mol \ KCl}{0.4000 \ L} = 2.85 \ mol/L = 2.85 \ M \ KCl$$

4 Verify your results.

As a rough estimate, 85 g divided by 75 g/mol is about 1 mol. If you divide 1 mol by 0.4 L you get 2.5 M. This approximation agrees with the answer of 2.85 M.

PRACTICE HINT

Remember to check that any solution volumes are converted to liters before you begin calculations that involve molarity.

PRACTICE

1 Vinegar contains 5.0 g of acetic acid, CH_3COOH, in 100.0 mL of solution. Calculate the molarity of acetic acid in vinegar.

2 If 18.25 g HCl is dissolved in enough water to make 500.0 mL of solution, what is the molarity of the HCl solution?

3 If 20.0 g H_2SO_4 is dissolved in enough water to make 250.0 mL of solution, what is the molarity of the sulfuric acid solution?

4 A solution of $AgNO_3$ contains 29.66 g of solute in 100.0 mL of solution. What is the molarity of the solution?

5 A solution of barium hydroxide, $Ba(OH)_2$, contains 4.285 g of barium hydroxide in 100.0 mL of solution. What is the molarity of the solution?

6 What mass of KBr is present in 25 mL of a 0.85 M solution of potassium chloride?

7 If all the water in 430.0 mL of a 0.45 M NaCl solution evaporates, what mass of NaCl will remain?

PROBLEM SOLVING SKILL

Using Molarity in Stoichiometric Calculations

There are many instances in which solutions of known molarity are used in chemical reactions in the laboratory. Instead of starting with a known mass of reactants or with a desired mass of product, the process involves a solution of known molarity. The substances are measured out by volume, instead of being weighed on a balance. An example of such an application in stoichiometry is shown in **Sample Problem C.**

SAMPLE PROBLEM C

Solution Stoichiometry

What volume (in milliliters) of a 0.500 M solution of copper(II) sulfate, $CuSO_4$, is needed to react with an excess of aluminum to provide 11.0 g of copper?

1 Gather information.

$[CuSO_4] = 0.500$ M
mass of product = 11.0 g Cu
solution volume = ? L

2 Plan your work.

Write the balanced chemical equation for the reaction:

$$3CuSO_4(aq) + 2Al(s) \longrightarrow 3Cu(s) + Al_2(SO_4)_3(aq)$$

Look up the molar mass of Cu:

$$\text{molar mass of Cu} = 63.55 \text{ g/mol}$$

Convert the mass of Cu to moles, and then use the mole ratio of $CuSO_4$:Cu from the balanced chemical equation to determine the number of moles of $CuSO_4$ needed. The moles of $CuSO_4$ can be converted into volume of solution using the reciprocal of molarity.

$$\text{g Cu} \times \frac{1 \text{ mol Cu}}{\text{g Cu}} \times \frac{\text{mol } CuSO_4}{\text{mol Cu}} \times \frac{\text{L solution}}{\text{mol } CuSO_4} = \text{L } CuSO_4$$

PRACTICE HINT

As in all stoichiometry problems, the mole ratio is the key. In solution stoichoimetry, molarity provides the bridge between volume of solution and amount of solute.

3 Calculate.

Substitute the values given:

$$11.0 \text{ g Cu} \times \frac{1 \text{ mol Cu}}{63.55 \text{ g Cu}} \times \frac{3 \text{ mol } CuSO_4}{3 \text{ mol Cu}} \times \frac{1 \text{ L solution}}{0.500 \text{ mol } CuSO_4} \times$$

$$\frac{1000 \text{ mL solution}}{1 \text{ L solution}} = 346 \text{ mL } CuSO_4 \text{ solution}$$

4 Verify your results.

Work backwards. A volume of 0.346 L of a 0.500 M $CuSO_4$ solution contains 0.173 mol $CuSO_4$ (0.346 L × 0.500 M = 0.173 mol). A 0.173 mol sample of $CuSO_4$ contains 0.173 mol Cu, which has a mass of 11.0 g, so the answer is correct.

1. Commercial hydrochloric acid, HCl, is 12.0 molar. Calculate the mass of HCl in 250.0 mL of the solution.

2. An excess of zinc is added to 125 mL of 0.100 M HCl solution. What mass of zinc chloride is formed?

$$Zn + 2HCl \longrightarrow ZnCl_2 + H_2$$

3. Yellow CdS pigment is prepared by reacting ammonium sulfide with cadmium nitrate. What mass of CdS can be prepared by mixing 2.50 L of a 1.25 M $Cd(NO_3)_2$ solution with an excess of $(NH_4)_2S$?

$$Cd(NO_3)_2(aq) + (NH_4)_2S(aq) \longrightarrow CdS(s) + 2NH_4NO_3(aq)$$

 # Section Review

UNDERSTANDING KEY IDEAS

1. Why did chemists develop the concept of molarity?

2. In what units is molarity expressed?

3. Describe in your own words how to prepare 100.0 mL of a 0.85 M solution of sodium chloride.

4. If you dissolve 2.00 mol KI in 1.00 L of water, will you get a 2.00 M solution? Explain your answer.

PRACTICE PROBLEMS

5. A sample of 400.0 g of water is found to contain 175 mg Cd. What is this concentration in parts per million?

6. If 1.63×10^{-4} g of helium dissolves in 100.0 g of water, what is the concentration in parts per million?

7. A standard solution of NaOH is 1.000 M. What mass of NaOH is present in 100.0 mL of the solution?

8. A 32 g sample of LiCl is dissolved in water to form 655 mL of solution. What is the molarity of the solution?

9. Most household bleach contains sodium hypochlorite, NaOCl. A 2.84 L bottle contains 177 g NaOCl. What is the molarity of the solution?

10. What mass of $AgNO_3$ is needed to prepare 250.0 mL of a 0.125 M solution?

11. Calcium phosphate used in fertilizers can be made in the reaction described by the following equation:

$$2H_3PO_4(aq) + 3Ca(OH)_2(aq) \longrightarrow$$
$$Ca_3(PO_4)_2(s) + 6H_2O(aq)$$

What mass in grams of each product would be formed if 7.5 L of 5.00 M phosphoric acid reacted with an excess of calcium hydroxide?

CRITICAL THINKING

12. You have 1 L of 1 M NaCl, and 1 L of 1 M KCl. Which solution has the greater mass of solute?

13. Under what circumstances might it be easier to express solution concentrations in terms of molarity? in terms of parts per million?

14. One solution contains 55 g NaCl per liter, and another contains 55 g KCl per liter. Which solution has the higher molarity? How can you tell?

Solubility and the Dissolving Process

KEY TERMS

- **solubility**
- **miscible**
- **immiscible**
- **dissociation**
- **hydration**
- **saturated solution**
- **unsaturated solution**
- **supersaturated solution**
- **solubility equilibrium**
- **Henry's law**

solubility

the ability of one substance to dissolve into another at a given temperature and pressure; expressed in terms of the amount of solute that will dissolve in a given amount of solvent to produce a saturated solution

OBJECTIVES

1. **Identify** applications of solubility principles, and relate them to polarity and intermolecular forces.

2. **Explain** what happens at the particle level when a solid compound dissolves in a liquid.

3. **Predict** the solubility of an ionic compound by using a solubility table.

4. **Describe** solutions in terms of their degree of saturation.

5. **Describe** factors involved in the solubility of gases in liquids.

Solubility and Polarity

Some pairs of liquids form a solution when they are mixed. For example, any amount of ethylene glycol, a common antifreeze, mixes with any amount of water to form antifreeze solutions in radiators. These two compounds are both very polar and have 100% **solubility** with each other.

Oils, such as cooking oil, do not mix with water. An oil is nonpolar, and water is polar. However, paint thinner is soluble with the oil in oil-based paints. Both the paint and paint thinner are nonpolar. Polar compounds tend to dissolve in other polar compounds, and nonpolar compounds tend to dissolve in other nonpolar compounds.

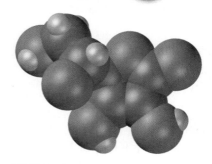

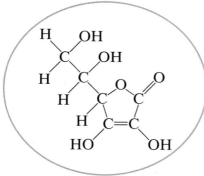

Figure 8

a The most common form of vitamin C is ascorbic acid, which is shown here.

b Because ascorbic acid is polar, it is very soluble in water but insoluble in fats and oils.

c Lemons, oranges, grapefruits, and limes are good sources of vitamin C.

Vitamin C Is a Water-Soluble Vitamin

The human body cannot make its own vitamin C; it must be obtained from external sources. Vitamin C also cannot be stored in the body. The disease scurvy, caused by a lack of vitamin C, has always been a threat to people with a limited diet. In 1747, Dr. James Lind studied the effect of diet on sailors who had scurvy. Those whose diet included citrus fruits recovered. In 1795, long before people knew that citrus fruits were rich in vitamin C, the British navy began to distribute lime juice during long sea voyages. For this reason, British sailors were often called "limeys."

Vitamin C was isolated and identified in the early 1930s by American chemists. The most important function of vitamin C is in the synthesis of collagen, a protein that makes up tendons and that enables muscle movements. **Figure 8** on the previous page shows that vitamin C has several –OH groups. These –OH groups form strong hydrogen bonds with the –OH groups in water, so vitamin C is very soluble in water. At room temperature, 33 g of vitamin C will dissolve in 100 g of water. Any excess in the diet is quickly eliminated by the kidneys. It is almost impossible to overdose on vitamin C.

Vitamin A Is a Fat-Soluble Vitamin

Vitamin A also must be obtained in food, especially yellow vegetables. It has many functions in the body, and is essential for good vision. Vitamin A is also needed for the respiratory tract, skin, and for normal growth of bones. Fortunately, vitamin A is fairly abundant in foods.

Vitamin A has a long, nonpolar carbon-hydrogen chain, as shown in **Figure 9.** Consequently, it has very low solubility in water. Its nonpolarity makes it very soluble in fats and oils, which are also nonpolar. Any excess of vitamin A in the diet builds up in body fat and is not easily eliminated from the body. So much can accumulate in fat that the amount of vitamin A may become toxic. So, as with other fat-soluble vitamins, it is possible to take too much vitamin A.

🖳 internet connect

www.scilinks.org
Topic : Vitamins
SciLinks code: HW4127

SCiLINKS. Maintained by the National Science Teachers Association

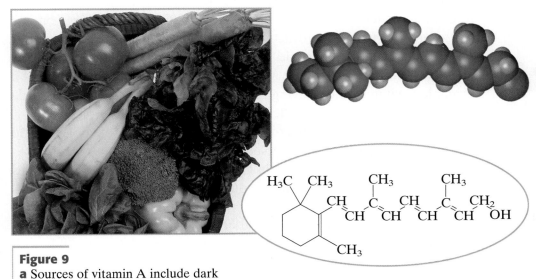

b Vitamin A is also known as retinol because it plays a vital role in helping the retina of your eye detect light.

c The vitamin A molecule is composed of mostly carbon and hydrogen, which makes the molecule nonpolar.

Figure 9
a Sources of vitamin A include dark green leafy vegetables, carrots, broccoli, tomatoes, and egg yolks.

Figure 10

a The nonpolar hydrocarbons in paint dissolve in the nonpolar oil in the paint thinner.

b Oil and water do not mix because oil is nonpolar and water is very polar.

The Rule Is "Like Dissolves Like"

In nonpolar molecules, such as vitamin A, London forces are the only forces of attraction between molecules. When nonpolar molecules are mixed with other nonpolar molecule, the intermolecular forces of the molecules easily match. Thus, nonpolar molecules are generally soluble with each other, as shown in **Figure 10a.** This is one part of the rule "like dissolves like": liquids that are completely soluble with each other are described as being **miscible** in each other.

If molecules are sufficiently polar, there is an additional electrical force pulling them toward each other. The negative partial charge on one side of a polar molecule attracts the positive partial charge on the other side of the next polar molecule. If you add polar molecules to other polar molecules, such as water, the attraction between the two is strong. An example is vitamin C dissolving in water. This is another part of the rule "like dissolves like": polar molecules dissolve other polar molecules.

However, if you try to mix oil and water, the nonpolar oil molecules do not mix with the polar water molecules. The two liquids are **immiscible.** They form two layers, as shown in **Figure 10b.** The polar water molecules attract each other, so they cannot be pushed apart by the nonpolar oil molecules to form a solution.

Miscibility can be difficult to determine for some substances. Ethyl alcohol, CH_3CH_2OH, is sufficiently polar to be completely miscible with water. But the alcohol octanol, $CH_3CH_2CH_2CH_2CH_2CH_2CH_2CH_2OH$, has a long nonpolar tail which causes it to be only slightly soluble in water.

miscible

describes two or more liquids that are able to dissolve into each other in various proportions

 Topic Link

Refer to the "States of Matter and Intermolecular Forces" chapter for a discussion of intermolecular forces.

immiscible

describes two or more liquids that do not mix with each other

Solubilities of Solid Compounds

Even two polar liquids placed in the same container may not dissolve in each other rapidly. Their strong intermolecular forces can only act on nearby molecules—not between molecules at the top of the container and those at the bottom.

The speed of the process can be increased by shaking the mixture. This action breaks the two liquids into small droplets and thereby increases the amount of contact between the surfaces of the liquids. This process works because the only place that dissolving can occur is at the surface between the two liquids, where the different molecules are near each other.

Similarly, in considering the solubility of solids in liquids, the only place where dissolution can occur is at the surface of the solid particles. The solid must be broken into smaller particles and then into molecules or ions, which can form a solution with the solvent molecules.

Greater Surface Area Speeds Up the Dissolving Process

As the discussion above indicated, the only place where dissolving can take place is at the surfaces where solute and solvent molecules are in contact. So if a solid has been broken into small particles, the surface area is much greater and the rate of the dissolving process is increased.

This is illustrated in **Figure 11.** The sugar granules dissolve more quickly than the sugar cubes. Because the sugar granules have more surface area, more of their molecules are directly exposed to the solvent and the dissolving process takes place faster. In the case of sugar cubes, most of the sugar molecules are inside the cubes and cannot dissolve until after the molecules at the outside of the cubes are dissolved.

Figure 11
Sugar granules dissolve in water more quickly than sugar cubes, because sugar granules have more surface area than sugar cubes.

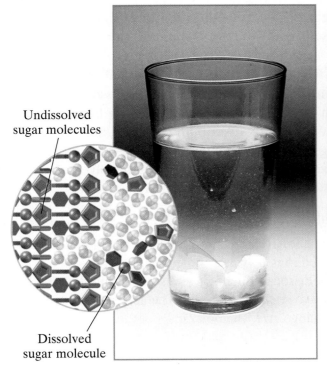

Undissolved sugar molecules

Dissolved sugar molecule

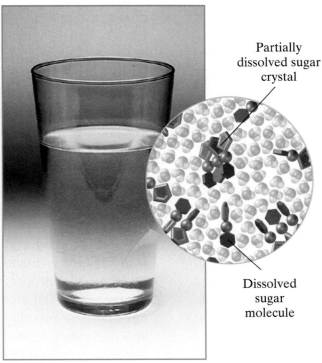

Partially dissolved sugar crystal

Dissolved sugar molecule

Figure 12
The effect of temperature on the solubility of some ionic solids is graphed here.

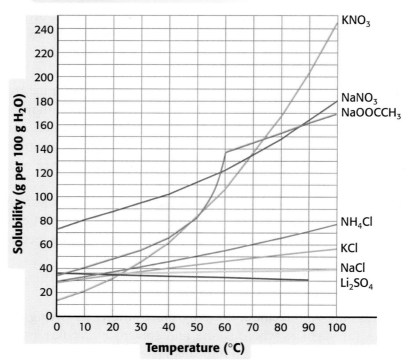

Solubility Vs. Temperature for Some Solid Solutes

Solubilities of Solids Generally Increase with Temperature

Another way to make most solids dissolve more and faster is to increase the temperature. Increasing the temperature is effective because, in general, solvent molecules with greater kinetic energy can dissolve more solute particles. **Figure 12** shows how an increase in temperature affects the solubility of several ionic compounds. In most cases, such as in the case of KNO_3, the solubility increases with temperature. However, temperature has little effect on the solubility of NaCl. The solubility of Li_2SO_4 actually decreases slightly as temperature increases.

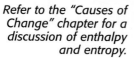

Refer to the "Causes of Change" chapter for a discussion of enthalpy and entropy.

dissociation

the separating of a molecule into simpler molecules, atoms, radicals, or ions

hydration

the strong affinity of water molecules for particles of dissolved or suspended substances that causes electrolytic dissociation

Both Enthalpy and Entropy Affect the Solubility of Salts

Until now, we have not made a distinction between the dissolving process of a covalent solid, such as sugar, and that of an ionic solid, such as table salt. Surface area and temperature affect both covalent and ionic solids. However, the dissolving of an ionic compound involves a unique factor: the separation of ions from the lattice into individual dissolved ions. This process, called **dissociation,** can be represented as an equation.

$$NaCl(s) \longrightarrow Na^+(aq) + Cl^-(aq)$$

If water is the solvent, as above, dissociation involves **hydration,** the surrounding of the dissociated ions by water molecules.

The actual process of dissociation, however, is more complex. It takes a large amount of energy to separate the ions. The separation requires a large positive enthalpy change, ΔH. The polar ends of the water molecules approach the ions and release energy, and this ΔH is very negative. The ΔH changes nearly cancel.

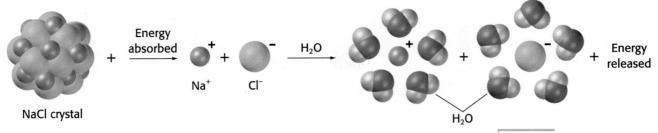

Energy absorbed

Na^+ Cl^-

H_2O

Energy released

NaCl crystal

H_2O

Entropy increases as the ions are scattered throughout the solution. On the other hand, there is a large decrease in entropy as the water molecules are structured around the ions. Smaller ions, such as the Na^+ ions in **Figure 13,** have a greater decrease in entropy than larger ions. The net result of all of the enthalpy and entropy changes that accompany the dissolving process determines the solubility of an ionic solid.

Figure 13
Individual ions are separated from the solid lattice by absorbed energy before they are hydrated by water molecules. Energy is released when hydrated ions are removed from the lattice.

Solubilities of Ionic Compounds

Solubilities are difficult to predict because of the many factors involved, so they must be measured experimentally. From experimental results of ionic solubilities in water, some patterns emerge, as shown in **Table 2.** Categories such as *soluble* and *insoluble* can be useful in many cases. The solubility of $Ba(OH)_2$ is 3.5 g per 100 g of water. It is described as slightly soluble. However, most substances are, at least to some extent, soluble in everything else. Even glass is very slightly soluble in water. In some delicate measurements, glass cannot be used as a container.

Table 2   Solubility Rules for Some Common Ionic Compounds
Compounds containing these ions are soluble in water:
Acetates, $CH_3CO_2^-$, except that of Fe^{3+}
Alkali metals (Group 1), except LiF
Ammonium, NH_4^+
Bromides, Br^-, except those of Ag^+, Pb^{2+}, and Hg_2^{2+}
Chlorides, Cl^-, except those of Ag^+, Pb^{2+}, and Hg_2^{2+}
Nitrates, NO_3^-
Sulfates, SO_4^{2-}, except those of Ca^{2+}, Sr^{2+}, Ba^{2+}, Pb^{2+}, and Hg_2^{2+}
Compounds containing these ions are insoluble in water:
Carbonates, CO_3^{2-}, except those of Group 1 and NH_4^+
Chromates, CrO_4^{2-}, except those of Group 1 and NH_4^+
Hydroxides, OH^-, except those of Group 1
Oxides, O^{2-}, except those of Group 1, Ca^{2+}, Sr^{2+}, and Ba^{2+} (which form hydroxides)
Phosphates, PO_4^{3-}, except those of Group 1 and NH_4^+
Sulfides, S^{2-}, except those of Group 1, Mg^{2+}, Ca^{2+}, Ba^{2+}, and NH_4^+

For a more detailed solubility table, see Appendix A.

Figure 14
a When a solution is saturated, additional solute added to the solvent will remain undissolved.

b As long as a solution is unsaturated, more solute can be added to the solvent and be dissolved.

Saturation

If you look in a chemistry handbook, or check the table in Appendix A, you will find that the solubility of potassium chloride, KCl, is 23.8 grams per 100 grams of water, at 20°C. This suggests some sort of limit.

When the maximum amount of solute, such as the 23.8 g of KCl mentioned above, is dissolved in a solution, the solution is said to be a **saturated solution.** As shown in **Figure 14a,** if a solution is saturated, any additional solute that is added collects at the bottom of the container. If more solute can be added to a solution and dissolve, as in **Figure 14b,** the solution is considered to be an **unsaturated solution. Figure 15** illustrates the relationship between solute added and solute dissolved.

The amount of solute that can dissolve depends on the forces between the solute particles and on forces between the solute particles and the solvent particles. When a solute is placed in contact with a solvent, molecules or ions of solute dissolve into the solvent. As soon as this happens, these same dissolved ions or molecules are capable of rejoining the undissolved solute. As the concentration of solute increases, the rate of return to the solute increases.

saturated solution

a solution that cannot dissolve any more solute under the given conditions

unsaturated solution

a solution that contains less solute than a saturated solution and that is able to dissolve additional solute

Figure 15
This graph shows the masses of solute that can be dissolved before saturation is reached. Any additional solute added beyond this point will remain undissolved.

Mass of Solute Added Vs. Mass of Solute Dissolved

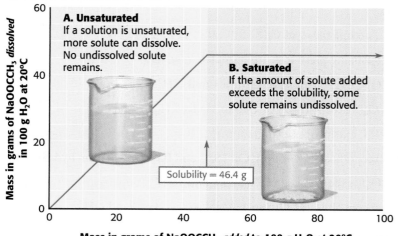

A. Unsaturated
If a solution is unsaturated, more solute can dissolve. No undissolved solute remains.

B. Saturated
If the amount of solute added exceeds the solubility, some solute remains undissolved.

Solubility = 46.4 g

Mass in grams of $NaOOCCH_3$ *dissolved* in 100 g H_2O at 20°C

Mass in grams of $NaOOCCH_3$ *added* to 100 g H_2O at 20°C

Solubility Can Be Exceeded

In a saturated solution, some excess solute remains undissolved, and the mass that dissolves is equal to the solubility value for that temperature. Under special conditions, **supersaturated solutions** can also exist. Supersaturated solutions have more solute dissolved than the solubility indicates would normally be possible, but only as long as there is no excess undissolved solute remaining.

Supersaturation is the reason why hand warmers, such as those shown in **Figure 16,** work. Inside the plastic pack, 60 g of sodium acetate, $NaOOCCH_3$, has been combined with 100 mL of water. This amount of sodium acetate is more than the amount that can dissolve in 100 mL of water at 20°C. As shown in **Figure 12,** only about 48 g $NaOOCCH_3$ can dissolve in 100 mL of water. When the solution is heated to 100°C, all of it dissolves, as shown in **Figure 16a.** When the solution is allowed to cool to 20°C, crystals of solute should form. **Figure 16b** shows that crystals do not form. Instead, the solution becomes supersaturated. However, if you disturb the cooled solution by clicking the disk in the center of the pack, crystallization immediately occurs, as shown in **Figure 16c.** The recrystallization of sodium acetate is exothermic, so its reappearance in **Figure 16c** releases heat.

supersaturated solution

a solution holding more dissolved solute than what is required to reach equilibrium at a given temperature

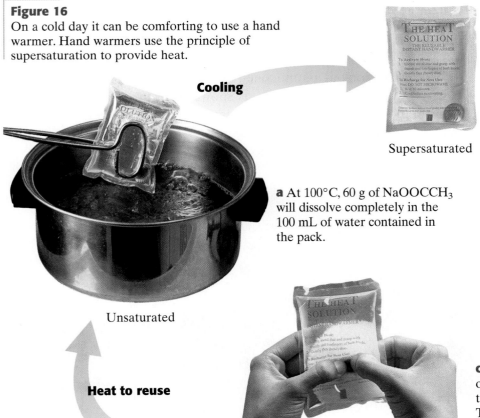

Figure 16
On a cold day it can be comforting to use a hand warmer. Hand warmers use the principle of supersaturation to provide heat.

Cooling

b When the solution is cooled to 20°C, $NaOOCCH_3$ does not crystallize unless the solution is disturbed.

Supersaturated

a At 100°C, 60 g of $NaOOCCH_3$ will dissolve completely in the 100 mL of water contained in the pack.

Crystallization

Unsaturated

Heat to reuse

Saturated

c Clicking the disk in the center of the pack triggers rapid recrystallization, and heat is released. The hand warmer can be reused if it is heated above the saturation point again.

Figure 17
In a saturated solution, the solute is recrystallizing at the same rate that it is dissolving.

Recrystallizing   Dissolving

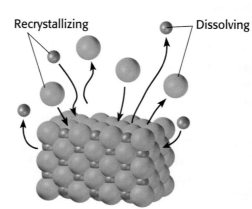

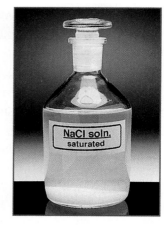

NaCl soln.
saturated

Saturation Occurs at a Point of Solubility equilibrium

In a saturated solution, solute particles are dissolving and recrystallizing at the same rate. It is a state of *dynamic equilibrium*. There is constant exchange, yet there is no net change.

When the amount of solute added to a solvent has reached its solubility limit, it is understood that the particles of solute in solution are in dynamic equilibrium with excess solute. This is illustrated in **Figure 17.** Na^+ ions and Cl^- ions are leaving the solid surface at the same rate as ions are also returning to the pile of excess solute at the bottom. These ions are considered to be in **solubility equilibrium.**

solubility equilibrium

the physical state in which the opposing processes of dissolution and crystallization of a solute occur at equal rates

Gases Can Dissolve in Liquids

When you first look at an unopened bottle of soda, you see very few bubbles. The liquid is homogeneous, as shown in **Figure 18a.** But when you open the bottle, you can hear gas escaping. Then you see many bubbles rising in the liquid, as pictured in **Figure 18b.** You probably know that the bubbles from soda are carbon dioxide, CO_2.

CO_2 under high pressure above solvent

Dissolved CO_2 molecules

Air at atmospheric pressure

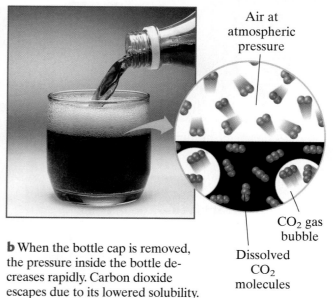

CO_2 gas bubble

Dissolved CO_2 molecules

Figure 18
a There are no bubbles in an unopened bottle of soda because carbon dioxide is dissolved in the liquid.

b When the bottle cap is removed, the pressure inside the bottle decreases rapidly. Carbon dioxide escapes due to its lowered solubility.

Gas Solubility Depends on Pressure and Temperature

Because a gas escapes when you open the bottle, you know that there is gaseous carbon dioxide present above the solution. Eventually, almost all the gas escapes when the bottle is opened to the atmosphere. Why?

In a gas, there is low attraction between the molecules. Likewise, there is usually little attraction between molecules of a gas and molecules of a liquid solvent. **Henry's law** states that the solubility of a gas increases as the partial pressure of the gas on the surface of the liquid increases.

At the high pressure of CO_2 in the unopened can, the gaseous CO_2 is in equilibrium with the dissolved gas. Therefore, the solution is saturated. When the bottle is opened and the pressure is released, the solubility decreases, and CO_2 bubbles escape. Finally, the dissolved CO_2 comes to equilibrium with the carbon dioxide in the air. The solution is saturated, but at this lower pressure, it is at a much lower concentration.

Temperature also affects gas solubility. After the soda bottle is open and becomes warm, the soda forms fewer bubbles and tastes flat. Even if you compare the taste of a newly opened warm soda with that of a newly opened cold soda, you will find that the warm soda will taste somewhat flat. Warm soda tastes flat because there is less CO_2 dissolved in it. Gases are less soluble in a liquid of higher temperature because the increased molecular motion in the solution allows gas molecules to escape their loose association with the solvent molecules.

Henry's law

the law that states that at constant temperature, the solubility of a gas in a liquid is directly proportional to the partial pressure of the gas on the surface of the liquid

Topic Link

Refer to the "Gases" chapter for a definition and discussion of partial pressure.

Section Review

UNDERSTANDING KEY IDEAS

1. Why is it possible to overdose on Vitamin A, but not on Vitamin C?

2. Why is ethanol miscible in water?

3. Why do sugar cubes dissolve more slowly than granulated sugar?

4. What factors are involved in determining the solubility of an ionic salt?

5. Would the compound $MgSO_4$ be considered soluble in water?

6. Would the compound PbS be considered soluble in water?

7. You keep adding sugar to a cold cup of coffee and stirring the coffee. Finally, solid sugar remains on the bottom of the cup. Explain why no more sugar dissolves.

8. What is the relation between supersaturation and the hand warmer in **Figure 16?**

CRITICAL THINKING

9. Ethylene glycol is represented by the formula $HOCH_2CH_2OH$. Is it likely to be soluble in water? in paint thinner? Explain your answer.

10. A solution of $BaCl_2$ is added to a solution of $AgNO_3$. Use **Table 2** to decide what reaction happens, and write a balanced equation.

11. Suppose a salt, when dissolving, has a positive ΔH and a negative ΔS. Is the solubility of the salt low or high?

12. A commercial "fizz saver" pumps helium under pressure into a soda bottle to keep gas from escaping. Will this keep CO_2 in the soda bottle? Explain your answer.

Physical Properties of Solutions

KEY TERMS

- **conductivity**
- **electrolyte**
- **nonelectrolyte**
- **hydronium ion**
- **colligative property**
- **surfactant**
- **detergent**
- **soap**
- **emulsion**

conductivity

the ability to conduct an electric current

electrolyte

a substance that dissolves in water to give a solution that conducts an electric current

OBJECTIVES

① **Distinguish** between nonelectrolytes, weak electrolytes, and strong electrolytes.

② **Describe** how a solute affects the freezing point and boiling point of a solution.

③ **Explain** how a surfactant stabilizes oil-in-water emulsions.

Electrical Conductivity in Solutions

Some substances conduct electricity and some cannot. The **conductivity** of a substance depends on whether it contains charged particles, and these particles must be able to move. Electrons move freely within a metal, thus allowing it to conduct electricity. Solid NaCl contains ions, but they cannot move, so solid NaCl is a nonconductor by itself. But an aqueous solution of ionic compounds such as NaCl contains charged ions, which can move about. Solutions of ionic compounds conduct electricity. Pure water does not conduct electricity.

Electrolytes Provide Ions in Solution

An **electrolyte** is a substance that dissolves in a liquid solvent and provides ions that conduct electricity. For example, sports drinks such as the one pictured in **Figure 19** contain electrolytes that your body needs replenished after strenuous physical activity. Electrolytes are considered to belong to one of two classes depending on their tendency to dissociate.

Figure 19
A sports drink not only supplies water but also supplies electrolytes.

Strong electrolytes completely dissociate into ions and conduct electricity well. *Weak electrolytes* provide few ions in solution. Therefore, even in high concentrations, solutions of weak electrolytes conduct electricity weakly. Ionic compounds are usually strong electrolytes. Covalent compounds may be strong electrolytes, weak electrolytes, or nonconductors.

Electrical Conductivities Span a Wide Range

As shown in **Figure 20,** the extent to which electrolytes dissociate into ions is indicated by the conductivity of their solutions. The apparatus shown has a light bulb attached to a battery, and there is a gap in the circuit between two electrodes. The electrodes are dipped in a solution. If the solution conducts electricity, the circuit is completed and the bulb lights. The amount of current that can be carried depends on the concentration of ions in the solution. A solution of a strong electrolyte has a high concentration of ions, so the bulb lights up brightly. A solution of a weak electrolyte has a low concentration of ions, so the bulb lights up dimly.

Virtually all of a strong electrolyte dissociates as it dissolves in a solvent. Sodium chloride, for example, ionizes completely in solution:

$$NaCl(s) \longrightarrow Na^+(aq) + Cl^-(aq)$$

The ions in solution can move about. NaCl is a strong electrolyte, and a solution of NaCl can conduct electricity. The sugar sucrose, on the other hand, does not ionize at all in solution. It is a **nonelectrolyte,** and does not conduct electricity.

internet connect

www.scilinks.org
Topic : Electrolytes and Nonelectrolytes
SciLinks code: HW4047

SC*i*LINKS. Maintained by the National Science Teachers Association

nonelectrolyte

a liquid or solid substance that does not alow the flow of an electric current, either in solution or in its pure state, such as water or sucrose

Figure 20
All four solutions have the same concentration. The brightness of the bulb indicates the degree of conduction and the degree of dissociation (ionization).

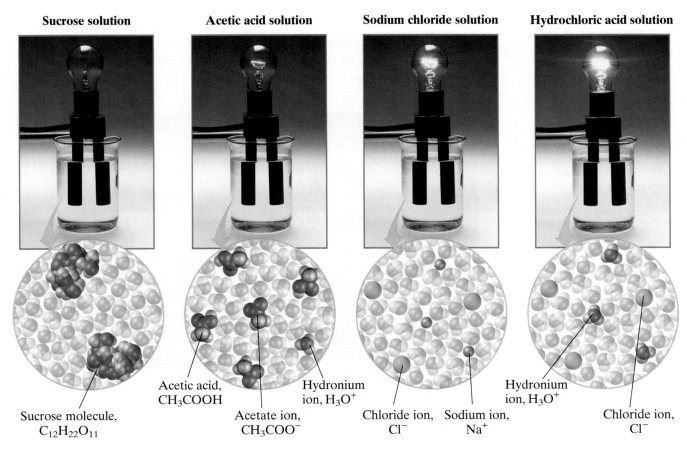

| Sucrose solution | Acetic acid solution | Sodium chloride solution | Hydrochloric acid solution |

Sucrose molecule, $C_{12}H_{22}O_{11}$

Acetic acid, CH_3COOH

Acetate ion, CH_3COO^-

Hydronium ion, H_3O^+

Chloride ion, Cl^- Sodium ion, Na^+

Hydronium ion, H_3O^+

Chloride ion, Cl^-

Figure 21
When acetic acid
dissolves in water,
very little of it is
changed into ions.

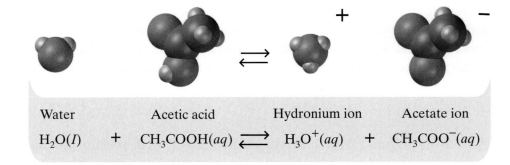

Water		Acetic acid		Hydronium ion		Acetate ion
$H_2O(l)$	+	$CH_3COOH(aq)$	$\rightleftharpoons$	$H_3O^+(aq)$	+	$CH_3COO^-(aq)$

hydronium ion

an ion consisting of a proton
combined with a molecule of
water; H_3O^+

Acids react with water to form the **hydronium ion,** H_3O^+. The reaction of acetic acid with water is shown in **Figure 21.** Acetic acid is a weak electrolyte. In water, only about 1% of acetic acid molecules ionize.

Hydrogen chloride dissolves in water to form a strongly conducting solution called *hydrochloric acid.* Hydrogen chloride is a strong electrolyte because it ionizes completely, as shown by the following equation:

$$HCl(g) + H_2O(l) \longrightarrow H_3O^+(aq) + Cl^-(aq)$$

Keep in mind that the use of the terms *strong* and *weak* have nothing to do with concentration. The term *strong* means that the substance provides a high *proportion* of ions in solution. Hydrochloric acid is a strong electrolyte at any concentration.

Tap Water Conducts Electricity

Have you wondered why you are warned not to use electrical appliances when you are near water? Have you wondered why you are also warned not to go swimming when a thunderstorm is near? The reason is because of the electrolytes in the water. Sea water, as shown in **Figure 22,** also conducts electricity.

Figure 22
The electricity from lightning
is conducted through groundwater, ponds, or ocean water,
which all contain electrolytes.

If you collect rainwater in a relatively unpolluted area, you will discover that the rainwater is essentially a nonconductor of electricity. A small concentration of carbonic acid from the carbon dioxide in the air added to the rainwater causes the rain water to be a weak conductor. Pure rainwater conducts almost as poorly as distilled water. However, most of the water we use comes from wells, lakes, or rivers. This water has been in contact with soil and rocks, which contain ionic compounds that dissolve in the water. Consequently, tap water conducts electricity. The conduction is not high, but the water can conduct enough current to stop a person's heart. So, for example, a person should not use an electrical appliance when in the bathtub or shower.

You should also not seek shelter under a tree during a thunderstorm. The tree not only sticks up like a lightning rod, but the sap in the tree also contains electrolytes, and conducts the electricity. Lightning finds a path to the ground through the trunk of the tree.

Colligative Properties

A solution made by dissolving a solute in a liquid, such as adding sulfuric acid to water, has particular chemical properties that the solvent alone did not have. The *physical properties* of water, such as how well the water mixes with other compounds, are also changed when substances dissolve in it.

As shown in **Figure 23,** salt can be added to icy sidewalks to melt the ice. The salt actually lowers the freezing point of water. Therefore, ice is able to melt at a lower temperature than it normally would. This change is called *freezing-point depression.* Nonvolatile solutes such as salt also increase the boiling point of a solvent. This change is called *boiling-point elevation.* For example, glycol in a car's radiator increases the boiling point of water in the radiator, which prevents overheating. It also lowers the freezing point, preventing freezing in cold weather.

Figure 23
Dissolved salt lowers the freezing point of water, causing it to melt at a lower temperature than it normally would.

Only the Concentration of Dissolved Particles Is Important

Any physical effect of the solute on the solvent is a **colligative property.** The lowering of the freezing point and the raising of the boiling point are examples of colligative properties. Only nonvolatile solutes have predictable effects on boiling point, but besides that requirement, the identity of the solute is relatively unimportant.

Any solute, whether an electrolyte or a nonelectrolyte, contributes to the colligative properties of the solvent. The degree of the effect depends on the concentration of solute particles (either molecules or ions) in a certain mass of solvent. The greater the particle concentration is, the greater the boiling-point elevation or the freezing-point depression is. For example, based on the number of moles of solute particles, 1 mol of sodium chloride, $NaCl$, is expected to give twice the amount of change as 1 mol of sucrose, $C_{12}H_{22}O_{11}$. This result occurs because $NaCl$ dissolves to give two moles of particles per mole, and sucrose dissolves to give only one. Likewise, 1 mol of calcium chloride, $CaCl_2$, has about three times the effect as 1 mol of sucrose because $CaCl_2$ dissolves to give three dissolved particles per mole. The following equations illustrate the logic.

$$C_{12}H_{22}O_{11}(s) \longrightarrow C_{12}H_{22}O_{11}(aq) \qquad \text{(1 dissolved particle)}$$

$$NaCl(s) \longrightarrow Na^+(aq) + Cl^-(aq) \qquad \text{(2 dissolved particles)}$$

$$CaCl_2(s) \longrightarrow Ca^{2+}(aq) + 2Cl^-(aq) \qquad \text{(3 dissolved particles)}$$

Figure 24, below, illustrates the differing numbers of solute particles generated by equal concentrations of each compound in solution.

Figure 24
Colligative properties depend on the concentration of solute particles. Equal amounts in moles of sugar, table salt, and calcium chloride affect the solvent in different degrees because of the different numbers of particles they form when dissolved.

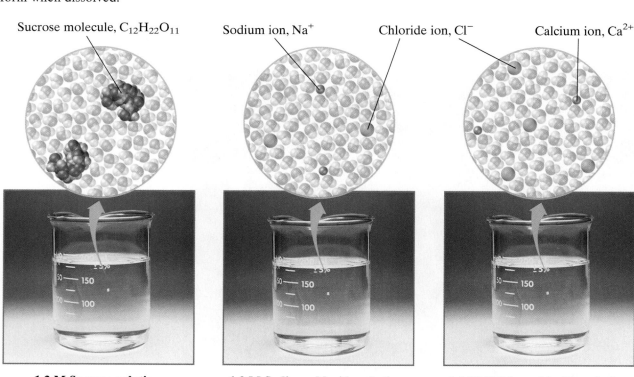

Sucrose molecule, $C_{12}H_{22}O_{11}$ Sodium ion, Na^+ Chloride ion, Cl^- Calcium ion, Ca^{2+}

1.3 M Sucrose solution   **1.3 M Sodium chloride solution**   **1.3 M Calcium chloride solution**

Dissolved Solutes Lower the Vapor Pressure of the Solvent

Colligative properties are all caused by a decrease in the vapor pressure of the solvent. Recall that all gases exert pressure, and vapor pressure is the pressure exerted by the vapor in equilibrium with its liquid state at a given temperature. The effect of a dissolved solute on the vapor pressure of a solvent can be understood when you consider the number of solvent particles in a solution. A solution has fewer solvent particles per volume than the pure solvent has, so fewer solvent particles are available to vaporize. Vapor pressure will therefore be decreased in proportion to the number of solute particles.

Figure 25 illustrates the difference between water's boiling point and freezing point (red lines), and the boiling point and freezing point of an aqueous solution (blue lines). Recall that the boiling point of a liquid is the temperature at which the liquid's vapor pressure is equal to the atmospheric pressure above the liquid. Because the vapor pressure of the water is lowered by the addition of a nonvolatile solute, the solution must be heated to a higher temperature for its vapor pressure to reach atmospheric pressure, at which point the solution boils.

The freezing point is the temperature at which water and ice are in equilibrium. Ice has a vapor pressure that is indicated by the line down to the left in **Figure 25.** The freezing point of water is the temperature at which the vapor pressure of pure water and ice are equal. Because the vapor pressure of the solution is lower, the vapor pressure of the solution intersects the line for the vapor pressure of ice at a lower temperature. Ice and water in the solution are in equilibrium at a lower temperature. The freezing point of the solution is therefore lower than that of pure water.

Topic Link

Refer to the "States of Matter and Intermolecular Forces" chapter for a discussion of vapor pressure.

Solute Effects on the Vapor Pressure of a Pure Solvent

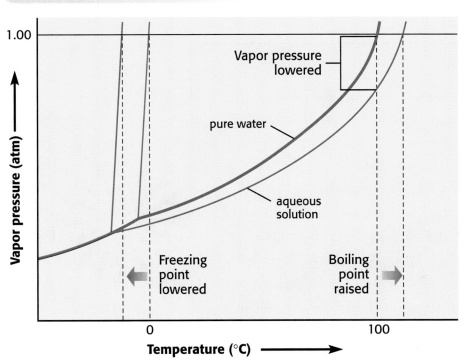

Figure 25
This is a modified phase diagram for pure water (red lines) and for an aqueous solution (blue lines). The addition of a solute has the effect of extending the range of the liquid phase.

Surfactants

Have you ever tried to wash your hands when they were very dirty without using soap? You were probably not very successful. Perspiration contains water and oils. The water evaporates, but the oil remains behind and coats the dirt particles. Oil and water do not mix, so washing without soap does not clean very well. However, if you use soap, the cleaning process is much more successful. Why?

The action of scrubbing your skin breaks the oil into tiny droplets. Soap molecules contain long nonpolar hydrocarbon chains, which are soluble in the nonpolar oil. As shown in **Figure 26,** soap molecules also have negatively charged ends, which are soluble in the water just outside the oil droplet. The negatively charged droplets repel each other and are carried away from the skin, along with any dirt that was on your skin.

Soap belongs to a general class of substances called **surfactants.** A surfactant is a substance that concentrates at the interface between two phases, either the solid-liquid, liquid-liquid, or gas-liquid phase. A **detergent** is a surfactant that is used for cleaning purposes. Usually, when we speak of detergents, we are talking about *synthetic detergents,* substances that are not natural products. A **soap** is a particular type of detergent and one that is a natural product. Soaps are sodium or potassium salts of fatty acids with long hydrocarbon chains. The formula for a typical soap, sodium palmitate, is shown below.

$$CH_3CH_2CH_2CH_2CH_2CH_2CH_2CH_2CH_2CH_2CH_2CH_2COO^-Na^+$$

Soap is an emulsifying agent. An **emulsion** is made of colloid-sized droplets suspended in a liquid in which they would ordinarily be insoluble, unless stabilized by an *emulsifying agent,* such as a soap. Without an emulsifying agent, polar and nonpolar molecules remain separate, as pictured in **Figure 27** on the next page.

surfactant

a compound that concentrates at the boundary surface between two immiscible phases, solid-liquid, liquid-liquid, or liquid-gas

detergent

a water-soluble cleaner that can emulsify dirt and oil

soap

a substance that is used as a cleaner and dissolves in water

emulsion

any mixture of two or more immiscible liquids in which one liquid is dispersed in the other

Figure 26
When you wash with soap, you create an emulsion of oil droplets dispersed in water, and stabilized by the soap.

internet connect

www.scilinks.org
Topic : Emulsions
SciLinks code: HW4050

SC*LINKS* Maintained by the National Science Teachers Association

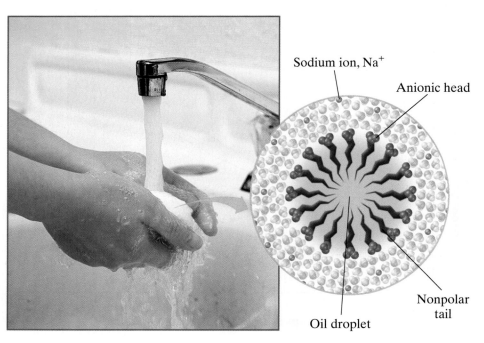

Sodium ion, Na⁺

Anionic head

Nonpolar tail

Oil droplet

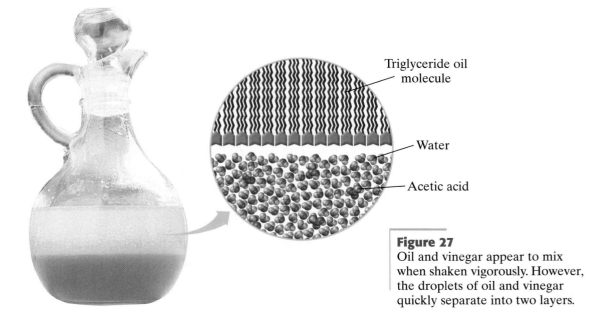

Triglyceride oil molecule

Water

Acetic acid

Figure 27
Oil and vinegar appear to mix when shaken vigorously. However, the droplets of oil and vinegar quickly separate into two layers.

Hard Water Limits Soap's Detergent Ability

Soaps are actually salts, although their physical properties are quite different from the salts you have studied so far. Like other salts, when dissolved, soaps form ions. Unlike other salts, the polyatomic anion of soap contains a long nonpolar part. It is this nonpolar hydrocarbon chain that is soluble with oils and dirt.

Soaps are not ideal cleansing agents because the salts of some of their anions are insoluble in water, especially salts of calcium, magnesium, and iron(II). Hard water has high concentrations of these cations, which react with anions such as the palmitate anion to form insoluble salts, such as the one shown in the following equation.

$$2C_{12}H_{25}COO^-(aq) + Ca^{2+}(aq) \longrightarrow (C_{12}H_{25}COO)_2Ca(s)$$

This is the type of substance responsible for bathtub rings. Before synthetic detergents were introduced for shampoos, some of the scum left over from soap would remain in people's hair after they washed it with soap. People would have to rinse their hair with vinegar to wash out the solid salts left over from the soap.

Synthetic Detergents Outperform Soaps in Hard Water

As noted above, soaps form precipitates when used in hard water. In the 1930s, chemists developed synthetic detergents as a substitute for soap to avoid this problem. Synthetic detergents can be used in hard water without forming precipitates. Today, almost all laundry products and shampoos contain synthetic detergents.

The early synthetic detergents were not biodegradable, gradually collected in the groundwater, and caused a serious pollution problem in some regions. Streams going over waterfalls developed huge mounds of soap suds. Most synthetic detergents are now made to biodegrade when they are disposed of.

internet connect

www.scilinks.org
Topic : Hard Water
SciLinks code: HW4154

SCiLINKS. Maintained by the National Science Teachers Association

Figure 28
The structure of sodium laurate (top), a typical soap, is shown here. Sodium dodecylbenzene sulfonate (bottom), is a synthetic detergent. It does not form an insoluble precipitate in hard water.

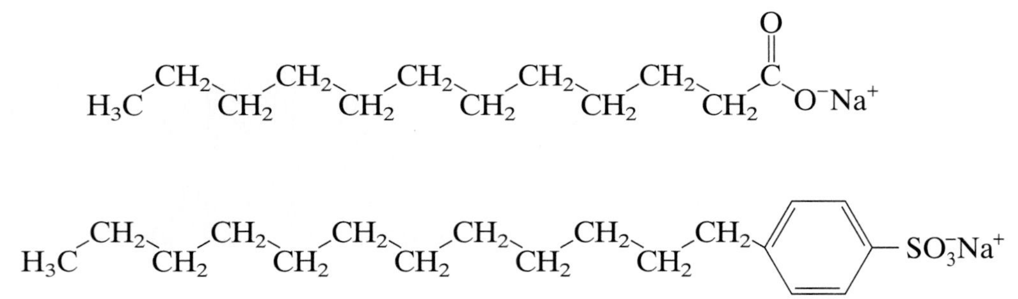

The basic structure of a synthetic detergent is the same as the structure of a soap. However, the long nonpolar tail of the detergent is connected to the salt of a sulfonic acid, –SOOOH, instead of the organic acid, –COOH. The difference between the structures of a synthetic detergent and soap is shown in **Figure 28.** The hexagon represents a six-carbon benzene ring, which is also nonpolar. The sulfonate group is negatively charged, just as the anionic group on a typical soap is. However, the sulfonate group does not form precipitates with magnesium, calcium, and iron(II) ions, which are found in hard water. The carboxylate anion, –COO⁻, reacts with hard water ions to form salts that come out of solution, but the sulfonate anion, –SOOO⁻, does not.

Section Review

UNDERSTANDING KEY IDEAS

1. What carries an electric current through a solution?

2. Is sugar an electrolyte? Why or why not?

3. How is a *weak electrolyte* different from a *strong electrolyte*?

4. Why does tap water conduct electricity, whereas distilled water does not?

5. What effect does a solute have on the boiling point of a solvent?

6. Why does spreading salt on an icy sidewalk cause the ice to melt?

7. What is the difference between the meaning of the terms *detergent* and *soap*?

8. What is hard water?

9. What is an emulsion?

10. What is an emulsifying agent?

CRITICAL THINKING

11. Will 1 mol of sugar have the same effect as 1 mol of table salt in lowering the freezing point of water? Explain.

12. Suppose you were taking a bath in distilled water but were using soap. Should you still worry about electric shock?

13. Are soap and synthetic detergents equally good as emulsifiers? Explain your answer.

14. A water softener removes calcium and magnesium ions from water. Why does a softening agent improve the cleansing ability of soap?

15. Why is soap described as a detergent? Why is it described as a surfactant?

CHAPTER HIGHLIGHTS

KEY TERMS

solution
suspension
solvent
solute
colloid

concentration
molarity

solubility
miscible
immiscible
dissociation
hydration
saturated solution
unsaturated solution
supersaturated solution
solubility equilibrium
Henry's law

conductivity
electrolyte
nonelectrolyte
hydronium ion
colligative property
surfactant
detergent
soap
emulsion

KEY IDEAS

SECTION ONE   What Is a Solution?

- A solution is a homogeneous mixture of a solute dissolved in a solvent.
- Several methods can be used to separate the components in a mixture.

SECTION TWO   Concentration and Molarity

- Units of concentration express the ratio of solute to solution, or solute to solvent, that is present throughout a solution.
- Molarity is moles of solute per liter of solution.

SECTION THREE   Solubility and the Dissolving Process

- Whether substances dissolve in each other depends on their chemical nature, on temperature, and on their ability to form hydrogen bonds.
- In general, polar dissolves in polar, and nonpolar dissolves in nonpolar.
- Ionic solubility can be roughly predicted using a table of ionic solubilities.
- A saturated solution has the solute in equilibrium with excess solute. A supersaturated solution has more dissolved solute than the equilibrium amount.
- Pressure and temperature affect the solubility of gases.

SECTION FOUR   Physical Properties of Solutions

- Ions are mobile in solution, so ionic solutions conduct electricity.
- Colligative properties involve the number of solute particles in solution.
- Surfactants make oil and water miscible.

KEY SKILLS

Calculating Parts per Million Sample Problem A p. 461	Preparing 1.000 L of a 0.5000 M Solution Skills Toolkit 1 p. 463	Calculating Molarity Skills Toolkit 2 p. 464 Sample Problem B p. 465	Solution Stoichiometry Sample Problem C p. 466

13 CHAPTER REVIEW

USING KEY TERMS

1. What happens to a suspension when it is allowed to stand over a period of time?

2. If sugar is dissolved in water, which component is the solute, and which component is the solvent?

3. Explain why milk is a colloid.

4. What ratio does molarity express?

5. Of the following three substances, which two are miscible with one another: oil, water, and ethanol?

6. One solution is made by dissolving sucrose in water. Another solution is made by dissolving NaCl in water. Which of these dissolving processes involves dissociation?

7. What mass of ammonium chloride can be added to 100 g of water at 20°C before the solution becomes saturated? (See **Figure 12**.)

8. If 20 g KCl is dissolved in 100 g of water at 20°C, is the solution unsaturated, saturated, or supersaturated? (See **Figure 12**.)

9. Write a paragraph explaining what happens to an ionic salt in the following steps: it is dissolved in water, more of it than its solubility amount is added to the solution, the solution is heated, the solution is cooled to room temperature, and the solution is disturbed by adding more solute. **WRITING SKILLS**

10. State Henry's law in your own words.

11. Why is NaCl described as a strong electrolyte?

12. Are all detergents surfactants? Explain.

13. Give an example of a nonelectrolyte.

14. What kind of mixture is soap able to form in order to make oil and water soluble?

15. Name two colligative properties.

16. What is the formula of a hydronium ion?

UNDERSTANDING KEY IDEAS

Solutions, Suspensions, and Colloids

17. What is a solution? How does it differ from a colloid?

18. What are the two components of a solution, and how do they relate to each other?

19. Explain how distillation can be used to obtain drinking water from sea water.

20. Explain how paper chromatography separates the components in a solution.

21. List these mixtures in order of increasing particle size: muddy water, sugar water, sand in water, and milk.

22. A few drops of milk are added to a glass of water, producing a cloudy mixture. The water is still cloudy after standing in the refrigerator for a week. What is this mixture called?

Concentration and Molarity

23. Name a unit of concentration commonly used to express small concentrations.

24. State the following expression in words: $[K_3PO_4]$

25. A solution of NaCl is 1 M. Why is the concentration of particles 2 M?

26. Describe how you would prepare 250.0 mL of a 0.500 M solution of NaCl by using apparatus found in a chemistry lab.

Solubility and the Dissolving Process

27. Explain why vitamin C is soluble in water.

28. Explain why gasoline is insoluble in water.

29. Why do small solid crystals dissolve in liquid more quickly than large crystals?

30. Would ammonium chloride be considered soluble in water?

31. Would the compound $BaSO_4$ be considered soluble in water?

32. Would the compound K_2O be considered soluble in water?

33. If a solution is saturated, how does the rate of dissolution of a solute compare with its rate of recrystallization?

34. Explain the effect of pressure on the solubility of a gas in a liquid.

35. Why does warmer liquid dissolve less gas than colder liquid?

Physical Properties of Solutions

36. A solution of salt in water conducts electricity, but a solution of sugar does not. Explain why.

37. A 1 M solution of NaCl in water has a freezing point that is 3.7°C lower than pure water. Estimate what the freezing point would be for a 1 M solution of $CaCl_2$.

38. Explain why soap is a surfactant, a detergent, and an emulsifying agent.

39. Explain why acetic acid is considered a weak electrolyte and why HCl is considered a strong electrolyte.

40. Draw a diagram of an oil droplet suspended in soapy water.

41. What is a colligative property?

Sample Problem A Calculating Parts per Million

42. A saturated solution of $PbCO_3$ contains 0.00011 g $PbCO_3$ in 100 g of water. What is this concentration in parts per million?

43. A 150 g sample of water contains 0.26 mg of cadmium ions. What is this concentration in parts per million?

44. Community water supplies usually contain 1.0 ppm of sodium fluoride. A particular water supply contains 0.0016 g of NaF in 1.60 L of water. Does it have enough NaF?

45. Most community water supplies have 0.5 ppm of chlorine added for purification. What mass of chlorine must be added to 100.0 L of water to achieve this level?

46. A 12.5 kg sample of shark meat contained 22 mg of methyl mercury, CH_3Hg^+. Is this amount within the legal limit of 1.00 ppm of methyl mercury in meat?

Sample Problem B Calculating Molarity

47. If 15.55 g NaOH are dissolved in enough water to make a 500.0 mL solution, what is the molarity of the solution?

48. A solution contains 32.7 g H_3PO_4 in 455 mL of solution. Calculate its molarity.

49. How many moles of $AgNO_3$ are needed to prepare 0.50 L of a 4.0 M solution?

50. What is the molarity of a solution that contains 20.0 g NaOH in 2.00 L of solution?

51. Calculate the molarity of a solution that contains 65.0 g $CuCl_2$ per 300.0 mL.

52. Calculate the molarity of a solution that contains 5.85 g KI in 0.125 L of solution.

53. Calculate the molarity of a H_3PO_4 solution of 6.66 g in 555 mL of solution.

54. What is the molarity of a solution that contains 0.0049 g NaCN in 1.55 L of solution?

55. Calculate the mass of NaOH in 65.0 mL of 2.25 M solution.

56. What mass of HCl is contained in 645 mL of 0.266 M solution?

57. What is the molarity of a hydrochloric acid solution that contains 18.3 g HCl in 100.0 mL of solution?

58. A saturated solution of NaCl contains 36 g NaCl in 114 mL of solution. What is the molarity of the solution?

59. Calculate the mass of LiF in 100.0 mL of 0.100 M solution.

60. How many grams of glucose, $C_6H_{12}O_6$, are in 255 mL of a 3.55 M solution?

61. Calculate the molarity of a solution of glucose, $C_6H_{12}O_6$, that contains 66 g in 500.0 mL of solution.

Sample Problem C Solution Stoichiometry

62. If 0.125 L of a 0.100 M solution of $Ba(OH)_2$ are mixed with 0.200 L of a 0.0750 M solution of H_3PO_4, the following reaction takes place.

$$3Ba(OH)_2 + 2H_3PO_4 \longrightarrow Ba_3(PO_4)_2 + 6H_2O$$

What mass of barium phosphate is formed?

63. You mix 1.00 L of 2.00 M $BaCl_2$ with 1.00 L of 2.00 M $AgNO_3$. What compounds remain in solution, and what are their concentrations?

$$BaCl_2(aq) + 2AgNO_3(aq) \longrightarrow$$
$$2AgCl(s) + Ba(NO_3)_2(aq)$$

64. How many milliliters of 18.0 M H_2SO_4 are required to react with 250 mL of 2.50 M $Al(OH)_3$ if the products are aluminum sulfate and water?

65. If 75.0 mL of an $AgNO_3$ solution reacts with enough Cu to produce 0.250 g Ag by single displacement, what is the molarity of the initial $AgNO_3$ solution if $Cu(NO_3)_2$ is the other product?

MIXED REVIEW

66. A 250 g sample of river water contained 0.0013 mg of selenium. What is this concentration in parts per million?

67. Commercial concentrated sulfuric acid solution is 18 M. How many grams of sulfuric acid are in 250 mL of the solution?

68. How many milliliters of 1.0 M $AgNO_3$ are needed to provide 168.88 g of pure $AgNO_3$?

69. What is the mass of potassium chromate, K_2CrO_4, in 20.0 mL of 6.0 M solution?

70. Sodium ions in blood serum normally are 0.145 M. How many grams of sodium ions are in 10.0 mL of serum?

71. A package of compounds used to achieve rehydration in sick patients contains 20.0 g of glucose, $C_6H_{12}O_6$. When this material is diluted to 1.00 L, what is the molarity of glucose?

72. The concentration of the solution in item 70 is also 0.020 M in KCl. How many grams of KCl are needed to prepare the 1.00 L of 0.020 M KCl?

73. A saturated solution contains 0.844 g of lead(II) bromide in 100.0 mL. What is the molarity of this solution?

74. Reagent nitric acid in the laboratory is usually 3.0 M. What mass of nitric acid is contained in 50.0 mL of this solution?

75. You are determining the concentration of a salt water solution by evaporating different samples and measuring the mass of salt that remains. Your data are shown below. Calculate the molar concentration of the original solution.

Sample volume (mL)	Mass of NaCl (g)
25.0	2.9
50.0	5.6
75.0	8.5
100.0	11.3

CRITICAL THINKING

76. Which will have the lower freezing point, 2.0 M NaCl or 1.0 M $AlCl_3$?

77. You evaporate 100.0 mL NaCl solution and obtain 11.3 g NaCl. What was the molarity of the original solution?

78. Calcium phosphate, $Ca_3(PO_4)_2$, is quite cheap and causes few pollution problems. Why is it not used to de-ice sidewalks? (Hint: See **Table 2** in Section 3.)

79. A calculation shows that a salt will have a negative ΔH and a positive ΔS when it dissolves. Is it actually soluble?

80. Imagine you are a sailor who must wash in sea water. Which is better to use, soap or synthetic detergent? Why?

81. You get a small amount of lubricating oil on your clothing. Which would work better to remove the oil, water or paint thinner? Explain your answer.

82. Air pressure in an airplane cabin while in flight is significantly lower than at sea level. Explain in terms of Henry's law how this affects the speed at which a carbonated beverage, after opening, loses its fizz.

83. Explain why oil and water do not mix.

84. Why would a substance that contains only ionic bonds not work as an emulsifying agent?

ALTERNATIVE ASSESSMENT

85. Design a solubility experiment that would identify an unknown substance that is either CsCl, RbCl, LiCl, NH_4Cl, KCl, or NaCl. (Hint: You will need a solubility versus temperature graph for each of the salts.) If your instructor approves your design, get a sample from the instructor, and perform your experiment.

86. Emergency-response teams working with chemical spills use solubility principles to control and clean up spills. Research the techniques they use, and explain why they work. Present your findings to the class.

87. Many reagent chemicals used in the lab are sold in the form of concentrated aqueous solutions, as shown in the table below. Different volumes are diluted to 1.00 L to make less-concentrated solutions. Create a computer spreadsheet that will calculate the volume of concentrated reagent needed to make 1.00 L solutions of any molar concentration that you enter.

Reagent	Concentration (M)
H_2SO_4	18.0
HCl	12.1
HNO_3	16.0
H_3PO_4	14.8
CH_3COOH	17.4
NH_3	15.0

CONCEPT MAPPING

88. Use the following terms to complete the concept map below: *concentration, dissociates, electrical conductivity, solute,* and *solvent.*

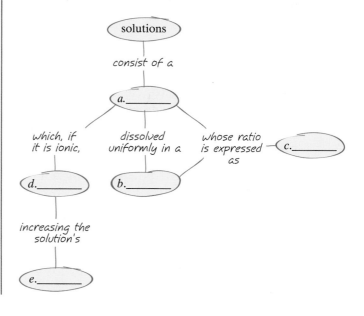

Study the graph below, and answer the questions that follow.
For help in interpreting graphs, see Appendix B, "Study Skills for Chemistry."

89. What do the numbers on the *y*-axis represent?

90. What does each curve on the graph represent?

91. Are most of the substances represented on the graph more or less soluble at higher temperatures?

92. Which salt is most soluble at 10°C? at 60°C? at 80°C?

93. If you heat water to 80°C, what amount of NaCl could you dissolve in it as compared to water that is at 20°C?

94. Which salt's solubility is most strongly affected by changes in temperature?

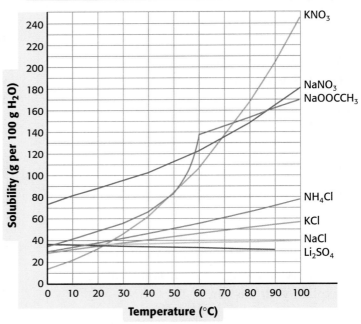

Solubility Vs. Temperature for Some Solid Solutes

TECHNOLOGY AND LEARNING

95. Graphing Calculator

Predicting Solubility from Tabular Data

The graphing calculator can run a program that graphs solubility data. Given solubility measurements for KCl, you will use the data to predict its solubility at various temperatures.

Go to Appendix C. If you are using a TI-83 Plus, you can download the program and data sets and run the application as directed. Press the **APPS** key on your calculator, then choose the application **CHEMAPPS**. Press **3**, then highlight **ALL** on the screen, press **1**, then highlight **LOAD** and press **2** to load the

data into your calculator. Press the keys **2nd** and then **QUIT**, and then run the program **SOLUBIL**. For L_1, press **2nd** and **LIST**, and choose **TMP21**. For L_2, press **2nd** and **LIST** and choose **SOL21**.

If you are using another calculator, your teacher will provide you with keystrokes and data sets to use.

a. At what temperature would you expect the solubility to be 48.9 g per 100 g H_2O?

b. At what temperature would you expect the solubility to be 35 g per 100 g H_2O?

c. What would you expect the solubility to be at a temperature of 100°C?

1. Two immiscible liquids can form a stable colloidal suspension by the addition of a(n)
 a. weak electrolyte.
 b. emulsifying agent.
 c. colloid.
 d. hydronium ion.

2. What mass of NaOH is contained in 2.5 L of a 0.010 M solution?
 a. 0.010 g
 b. 1.0 g
 c. 2.5 g
 d. 0.40 g

3. Water is an excellent solvent because
 a. it is a covalent compound.
 b. it is a nonconductor of electricity.
 c. of its high polarity.
 d. of its colligative properties.

4. Two liquids are likely to be immiscible if
 a. both have polar molecules.
 b. both have nonpolar molecules.
 c. one is polar and the other is nonpolar.
 d. one is water and the other is methyl alcohol, CH_3OH.

5. An _____ would increase the solubility of a gas in a liquid.
 a. addition of an electrolyte
 b. addition of an emulsifier
 c. agitation of the solution
 d. increase in its partial pressure

6. Acetic acid is considered a weak electrolyte because it
 a. is miscible with water.
 b. forms hydronium and hydroxide ions in aqueous solution.
 c. lowers the freezing point of water.
 d. ionizes only slightly in aqueous solution.

7. Because vitamin A is fat-soluble, it
 a. does not mix well with nonpolar substances.
 b. is stored in body fat.
 c. passes out of the body in the urine.
 d. cannot accumulate in the body.

8. Which of the following types of compounds is most likely to be a strong electrolyte?
 a. a polar compound
 b. a nonpolar compound
 c. a covalent compound
 d. an ionic compound

9. Which of the following will have the lowest freezing point?
 a. 1 M sugar solution
 b. 1 M NaCl solution
 c. 1 M $MgSO_4$ solution
 d. 1 M $CaCl_2$ solution

10. Soap, such as $CH_3(CH_2)_{10}COO^-Na^+$, is considered a detergent because it
 a. is an electrolyte.
 b. is nonpolar.
 c. is soluble in water.
 d. makes oil and water miscible.

11. A saturated solution can become supersaturated under which of the following conditions?
 a. it contains electrolytes
 b. the solution is heated and then allowed to cool
 c. more solvent is added
 d. more solute is added

12. Molarity is expressed in units of
 a. moles of solute per liter of solution.
 b. liters of solution per mole of solute.
 c. moles of solute per liter of solvent.
 d. liters of solvent per mole of solute.

CHAPTER

14

CHEMICAL EQUILIBRIUM

494

*S*ome marine animals use a salt called *calcium carbonate* to make their shells, such as the shell in the photo. You may think that these animals must constantly build up their shells so that their shells will not dissolve. Calcium carbonate, however, is not very soluble in water. So, even a shell that does not have an animal living in it will take a long time to dissolve. In this chapter, you will learn how to calculate the concentration of slightly soluble salts in water.

START-UP**ACTIVITY**

SAFETY PRECAUTIONS

Finding Equilibrium

PROCEDURE

1. Label one **rectangular plastic bucket** as "Reactants," and fill the bucket with **water.**

2. Label the other **rectangular plastic bucket** as "Products." Do not put any water in this bucket.

3. One student should be in charge of the "Reactants," and another student should be in charge of the "Products." Each student should use a **250 mL plastic beaker** to dip into the other student's bucket and get as much water as possible in the beaker. Then the students should pour the water from their beakers into their own bucket.

4. The students should dip and pour at equal rates.

ANALYSIS

1. What happens to the amount of "Reactants"? What happens to the amount of "Products"?

2. Does the system reach a state in which the system does not change?

3. How would the outcome of the experiment be different if the beakers had been of different sizes?

Pre-Reading Questions

① How do you know that a reaction has taken place?

② How do reactants differ from products?

③ What is a solution?

internet connect

www.scilinks.org
Topic: Chemical Reactions
SciLinks code: HW4029

*SC*LINKS. Maintained by the National Science Teachers Association

Reversible Reactions and Equilibrium

OBJECTIVES

1. **Contrast** reactions that go to completion with reversible ones.

2. **Describe** chemical equilibrium.

3. **Give** examples of chemical equilibria that involve complex ions.

Completion Reactions and Reversible Reactions

Topic Link

Refer to the "Chemical Equations and Reactions" chapter for a discussion of chemical reactions.

Do all reactants change into products during a reaction? Sometimes only a trace of reactants remains after the reaction is over. **Figure 1** shows an example of such a reaction. Oxygen gas reacts with sulfur to form sulfur dioxide, as shown in the following chemical equation:

$$S_8(s) + 8O_2(g) \longrightarrow 8SO_2(g)$$

If there is enough oxygen, almost all of the sulfur will react with it. In addition, the sulfur dioxide does not significantly break down into sulfur and oxygen. Reactions such as this one are called *completion reactions*. But in other reactions, the products can re-form reactants. Such reactions are called reversible reactions.

Figure 1
The burning of sulfur in oxygen gas to give sulfur dioxide gas, SO_2, is a completion reaction.

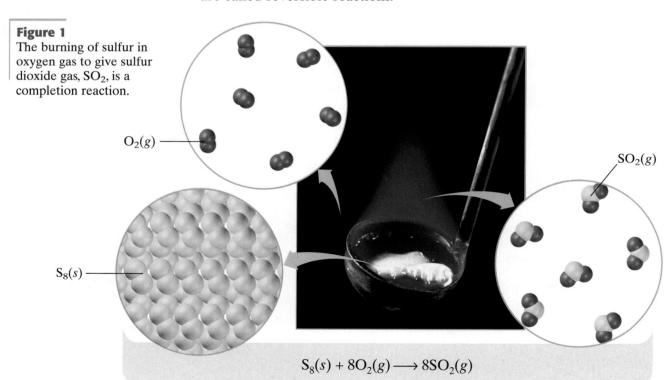

$O_2(g)$

$SO_2(g)$

$S_8(s)$

$$S_8(s) + 8O_2(g) \longrightarrow 8SO_2(g)$$

Figure 2
The precipitation of calcium sulfate, $CaSO_4$, is an example of equilibrium.

$Ca^{2+}(aq)$

$Cl^-(aq)$

$Na^+(aq)$

$SO_4^{2-}(aq)$

$CaSO_4(s)$

$$CaCl_2(aq) + Na_2SO_4(aq) \longrightarrow CaSO_4(s) + 2NaCl(aq)$$

Reversible Reactions Reach Equilibrium

An example of a **reversible reaction** is the formation of solid calcium sulfate by mixing solutions of calcium chloride and sodium sulfate, as shown in **Figure 2** and written below.

$$CaCl_2(aq) + Na_2SO_4(aq) \longrightarrow CaSO_4(s) + 2NaCl(aq)$$

Because the chloride and sodium ions are spectator ions, the net ionic equation better describes what is happening.

$$Ca^{2+}(aq) + SO_4^{2-}(aq) \longrightarrow CaSO_4(s)$$

Although solid calcium sulfate is the product, it can break down to make calcium ions and sulfate ions, as described below:

$$CaSO_4(s) \longrightarrow Ca^{2+}(aq) + SO_4^{2-}(aq)$$

This chemical equation is the reverse of the previous one. Another way to think about reversible reactions is that they form both products and reactants. Use arrows that point in opposite directions when writing a chemical equation for a reversible reaction.

$$Ca^{2+}(aq) + SO_4^{2-}(aq) \rightleftarrows CaSO_4(s)$$

The forward and reverse reactions take place at the same time. Some Ca^{2+} and SO_4^{2-} ions form $CaSO_4$, while some $CaSO_4$ is dissolving in water to form Ca^{2+} and SO_4^{2-} ions. The reactions occur at the same rate after the initial mixing of $CaCl_2$ and Na_2SO_4. As one unit of $CaSO_4$ forms, another unit of $CaSO_4$ dissolves. Therefore, the amounts of the products and reactants do not change. Reactions in which the forward and reverse reaction rates are equal are at **chemical equilibrium.**

reversible reaction

a chemical reaction in which the products re-form the original reactants

chemical equilibrium

a state of balance in which the rate of a forward reaction equals the rate of the reverse reaction and the concentrations of products and reactants remain unchanged

Opposing Reaction Rates Are Equal at Equilibrium

The reaction of hydrogen, H_2, and iodine, I_2, to form hydrogen iodide, HI, also reaches chemical equilibrium.

$$H_2(g) + I_2(g) \rightleftharpoons 2HI(g)$$

When equal amounts of H_2 and I_2 gases are mixed, only a very small fraction of the collisions between H_2 and I_2 result in the formation of HI.

$$H_2(g) + I_2(g) \longrightarrow 2HI(g)$$

After some time, the concentration of HI in the gas mixture goes up. As a result, fewer collisions occur between H_2 and I_2 molecules, and the rate of the forward reaction drops. This decline in the forward reaction rate is shown in the upper curve of **Figure 3.** Similarly, in the beginning, few HI molecules exist in the system, so they rarely collide with each other. As more HI molecules are made, they collide more often. A few of these collisions form H_2 and I_2 by the reverse reaction.

$$2HI(g) \longrightarrow H_2(g) + I_2(g)$$

The greater the number of HI molecules that form, the more often the reverse reaction occurs, as the lower curve in the figure shows. When the forward rate and the reverse rate are equal, the system is at chemical equilibrium.

If you repeated this experiment at the same temperature, starting with a similar amount of pure HI instead of the H_2 and I_2, the reaction would reach chemical equilibrium again. This equilibrium mixture would have exactly the same concentrations of each substance whether you started with HI or with a mixture of H_2 and I_2!

Figure 3
This graph shows how the forward and reverse rates of a reaction become equal at equilibrium.

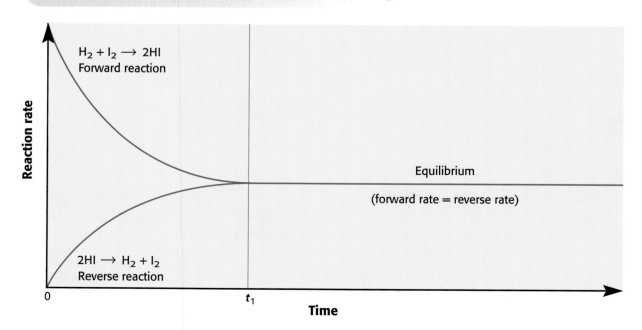

Rate Comparison for $H_2(g) + I_2(g) \rightleftharpoons 2HI(g)$

$H_2 + I_2 \longrightarrow 2HI$
Forward reaction

Reaction rate

Equilibrium
(forward rate = reverse rate)

$2HI \longrightarrow H_2 + I_2$
Reverse reaction

0   t_1

Time

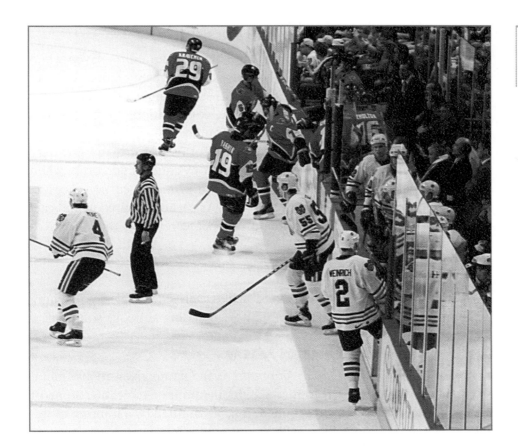

Figure 4
This moment during a hockey game, when some players leave the bench and others leave the ice, is an analogy for dynamic equilibrium.

Chemical Equilibria Are Dynamic

If you drop a small ball into a bowl, the ball will bounce around and then come to rest in the center of the bowl. The ball has reached static equilibrium. *Static equilibrium* is a state in which nothing changes.

Chemical equilibrium is different from static equilibrium because it is dynamic. In a *dynamic equilibrium,* there is no net change in the system. Two opposite changes occur at the same time in a dynamic equilibrium. A moment during a hockey game illustrates dynamic equilibrium shown in **Figure 4.** Even though 12 players are on the ice, the players are changing as some players enter the game and others return to the bench. In a chemical equilibrium, an atom may change from being a part of the products to being a part of the reactants many times. But the overall concentrations of products and reactants stay the same. For chemical equilibrium to be maintained, the rates of the forward and reverse reactions must be equal. Arrows of equal length also show equilibrium.

$$\text{reactants} \rightleftarrows \text{products}$$

In some cases, the forward reaction is nearly done before the rate of the reverse reaction becomes large enough to create equilibrium. So, this equilibrium has a higher concentration of products than reactants. This type of equilibrium, also shown by using two arrows, is shown below. Notice that the forward reaction has a longer arrow to show that the products are favored.

$$\text{reactants} \xrightleftharpoons{} \text{products}$$

Chemical Equilibrium　　**499**

Figure 5
In this complex ion, $[Cu(NH_3)_4]^{2+}$, ammonia molecules bond to the central copper(II) ion.

Topic Link

Refer to the "Periodic Table" chapter for a discussion of transition metals.

More Examples of Equilibria

Our environment has many examples of chemical equilibria. Even when systems are not in equilibrium, they are continuously changing in an effort to reach equilibrium. Equilibria control the composition of Earth's atmosphere, oceans, and lakes, as well as the composition of our body fluids. Even soot provides an example of an equilibrium.

Have you ever thought about how soot forms? Soot is the powdery form of carbon found in chimneys, around smoky candles, and often when combustion occurs. As you know, combustion produces carbon dioxide, CO_2, and poisonous carbon monoxide, CO. As carbon monoxide and carbon dioxide cool after combustion, a reversible reaction that produces soot occurs.

$$2CO(g) \rightleftharpoons C(s) + CO_2(g)$$

This reaction of gases and a solid will reach chemical equilibrium. Equilibria can involve any state of matter, including aqueous solutions.

Equilibria Involving Complex Ions

Complex ion, or *coordination compound,* is the name given to any metal atom or ion that is bonded to more than one atom or molecule. Some of the most interesting ions have a metal ion surrounded by a number of ligands. *Ligands* are molecules, such as ammonia, NH_3, or anions, such as cyanide, CN^-, that readily bond to metal ions. **Figure 5** shows a model of one complex ion, $[Cu(NH_3)_4]^{2+}$. Complex ions may be positively charged cations or negatively charged anions. Those formed from transition metals are often deeply colored, as **Table 1** shows. You can also see that the charge on the complex ion is the sum of the charges on the species from which the complex ion forms. For example, when the cobalt ion, Co^{2+}, bonds with four Cl^- ligands, the total charge is $(+2) + 4(-1) = -2$. Note that metal ions and ligands can form complexes that have no charge, such as $[Cu(NH_3)_2Cl_2]$, but are not complex ions because they are not charged.

Complex ions often form in systems that reach equilibrium. The following equation represents zinc nitrate dissolving in water.

$$Zn(NO_3)_2(s) \longrightarrow Zn^{2+}(aq) + 2NO_3^-(aq)$$

Table 1	**Examples of Complex Ions**		
Complex cation	**Color**	**Complex anion**	**Color**
$[Co(NH_3)_6]^{3+}$	yellow-orange	$[Co(CN)_6]^{3-}$	pale yellow
$[Co(NH_3)_5Cl]^{2+}$	violet	$[CoCl_4]^{2-}$	blue
$[Fe(H_2O)_5SCN]^{2+}$	deep red	$[Fe(CN)_6]^{3-}$	red
$[Cu(NH_3)_4]^{2+}$	blue-purple	$[Fe(CN)_6]^{4-}$	yellow

Refer to Appendix A for more examples of complex ions.

Notice from the table that water molecules can be ligands. In the absence of other ligands, water molecules bond with zinc ions. So, a more accurate description of this reaction is

$$Zn(NO_3)_2(s) + 4H_2O(l) \rightleftharpoons [Zn(H_2O)_4]^{2+}(aq) + 2NO_3^-(aq)$$

If another ligand, such as CN^-, is added, the new system will again reach chemical equilibrium. Both water molecules and cyanide ions "compete" to bond with zinc ions, as shown in the equation below.

$$[Zn(H_2O)_4]^{2+}(aq) + 4CN^-(aq) \rightleftharpoons [Zn(CN)_4]^{2-}(aq) + 4H_2O(l)$$

All of the ions in the previous equation are colorless, so you cannot see which complex ion has the greater concentration. However, in the following chemical equilibrium of nickel ions, ammonia, and water, the complex ions have different colors. Therefore, you can see which complex ion has the greater concentration.

$$[Ni(H_2O)_6]^{2+}(aq) + 6NH_3(aq) \rightleftharpoons [Ni(NH_3)_6]^{2+}(aq) + 6H_2O(l)$$
$$\text{green} \qquad\qquad\qquad\qquad \text{blue-violet}$$

The starting concentration of NH_3 will determine which one will have the greater concentration.

Section Review

UNDERSTANDING KEY IDEAS

1. Which of the following equations best represents a reaction that goes to completion?
   a. reactants $\underset{\longrightarrow}{\longleftarrow}$ products
   b. reactants $\rightleftharpoons$ products
   c. reactants $\longrightarrow$ products

2. At equilibrium, what is the relationship between the rates of the forward and reverse reactions?

3. Explain what each reaction below shows. Describe the relationship between the amounts of products and reactants for each case.
   a. reactants $\rightleftharpoons$ products
   b. reactants $\overset{\longrightarrow}{\longleftarrow}$ products

4. What is the difference between dynamic and static equilibria?

5. Write the formula and charge of the complex ion that forms when a copper ion, Cu^{2+}, bonds with four chloride ligands.

CRITICAL THINKING

6. What evidence might lead you to believe that a chemical reaction was *not* at equilibrium?

7. In what way are the number of players on the ice in a hockey game *not* like chemical equilibrium?

8. The final equation on this page describes a reaction in which all six water ligands are replaced by six ammonia molecules. Write the formula of a complex ion that would form if only two ligands were exchanged.

9. Which of the following nitrogen compounds or ions, NH_4^+, NH_3, or NH_2^-, cannot be a ligand? Why?

Systems at Equilibrium

- **equilibrium constant, K_{eq}**
- **solubility product constant, K_{sp}**

OBJECTIVES

1. **Write** K_{eq} expressions for reactions in equilibrium, and perform calculations with them.

2. **Write** K_{sp} expressions for the solubility of slightly soluble salts, and perform calculations with them.

The Equilibrium Constant, K_{eq}

Limestone caverns, shown in **Figure 6,** form over millions of years. They are made as rainwater, slightly acidified by H_3O^+, gradually dissolves rocks made of calcium carbonate. This reaction is still going today and is slowly enlarging caverns. The reverse reaction also takes place, and solid calcium carbonate is deposited as beautiful stalactites and stalagmites.

$$CaCO_3(s) + 2H_3O^+(aq) \rightleftharpoons Ca^{2+}(aq) + CO_2(g) + 3H_2O(l)$$

Figure 6
Stalactites and stalagmites in limestone caverns form because of a slight displacement of a reaction from equilibrium.

When some $CaCO_3$ first reacts with H_3O^+, the rate of the forward reaction is large. The rate of the reverse reaction is zero until some products form. As the reaction proceeds, the forward reaction rate slows as the reactant concentrations decrease. At the same time, the reverse rate increases as more products of the forward reaction form. When the two rates become equal, the reaction reaches chemical equilibrium. Because reaction rates depend on concentrations, there is a mathematical relationship between product and reactant concentrations at equilibrium. For the reaction of limestone and acidified water, the relationship is

$$K_{eq} = \frac{[Ca^{2+}][CO_2]}{[H_3O^+]^2}$$

where K_{eq} is a number called the **equilibrium constant** of the reaction. This expression has a specific numerical value that depends on temperature and must be found experimentally or from tables. The value of K_{eq} for this reaction is 1.4×10^{-9} at 25°C.

    Notice that equilibrium constants are unitless. In addition, equilibrium constants apply only to systems in equilibrium. **Skills Toolkit 1** contains the rules for writing the equilibrium constant expression for *any* reaction, including the one above.

equilibrium constant, K_{eq}

a number that relates the concentrations of starting materials and products of a reversible chemical reaction to one another at a given temperature

---

SKILLS Toolkit 1

Determining K_{eq} for Reactions at Chemical Equilibrium

1. Write a balanced chemical equation.

- Make sure that the reaction is at equilibrium before you write a chemical equation.

2. Write an equilibrium expression.

- To write the expression, place the product concentrations in the numerator and the reactant concentrations in the denominator.

- The concentration of any solid or a pure liquid that takes part in the reaction is left out because these concentrations never change.

- For a reaction occurring in aqueous solution, water is omitted because the concentration of water is almost constant during the reaction.

3. Complete the equilibrium expression.
- To complete the expression, raise each substance's concentration to the power equal to the substance's coefficient in the balanced chemical equation.

Calculating K_{eq} from Concentrations of Reactants and Products

An aqueous solution of carbonic acid reacts to reach equilibrium as described below.

$$H_2CO_3(aq) + H_2O(l) \rightleftharpoons HCO_3^-(aq) + H_3O^+(aq)$$

The solution contains the following solute concentrations: carbonic acid, 3.3×10^{-2} mol/L; bicarbonate ion, 1.19×10^{-4} mol/L; and hydronium ion, 1.19×10^{-4} mol/L. Determine the K_{eq}.

1 Gather information.

$$[H_2CO_3] = 3.3 \times 10^{-2}, [HCO_3^-] = [H_3O^+] = 1.19 \times 10^{-4}$$

2 Plan your work.

Write the equilibrium constant expression. For this reaction, the equilibrium constant expression is

$$K_{eq} = \frac{[HCO_3^-][H_3O^+]}{[H_2CO_3]}$$

3 Calculate.

Substitute the concentrations into the expression.

$$K_{eq} = \frac{[HCO_3^-][H_3O^+]}{[H_2CO_3]} = \frac{(1.19 \times 10^{-4}) \times (1.19 \times 10^{-4})}{(3.3 \times 10^{-2})} = 4.3 \times 10^{-7}$$

4 Verify your results.

The numerator of the expression is approximately 1×10^{-8}. The denominator is approximately 3×10^{-2}. The rough calculation below supports the value 4.3×10^{-7} for the K_{eq}.

$$K_{eq} \approx \frac{1 \times 10^{-8}}{3 \times 10^{-2}} = 3 \times 10^{-7}$$

PRACTICE HINT

Remember that $[H_2CO_3]$ means "the concentration of H_2CO_3 in mol/L." So, these quantities do not carry units because the use of square brackets around the formula implies a concentration in moles per liter.

Note that the chemical equation has coefficients of 1, so no concentration used in this expression is raised to a power.

PRACTICE

1 For the system involving N_2O_4 and NO_2 at equilibrium at a temperature of 100°C, the product concentration of N_2O_4 is 4.0×10^{-2} mol/L and the reactant concentration of NO_2 is 1.4×10^{-1} mol/L. What is the K_{eq} value for this reaction?

2 An equilibrium mixture at 852 K is found to contain 3.61×10^{-3} mol/L of SO_2, 6.11×10^{-4} mol/L of O_2, and 1.01×10^{-2} mol/L of SO_3. Calculate the equilibrium constant, K_{eq}, for the reaction where SO_2 and O_2 are reactants and SO_3 is the product.

Table 2 Equilibrium Constants at 25°C

Equation	K_{eq} expression and value
$N_2(g) + 3H_2(g) \rightleftharpoons 2NH_3(g)$	$\dfrac{[NH_3]^2}{[N_2][H_2]^3} =$ $\quad 3.3 \times 10^8$
$2NO_2(g) \rightleftharpoons N_2O_4(g)$	$\dfrac{[N_2O_4]}{[NO_2]^2} = 165$
$Hg^{2+}(aq) + Hg(l) \rightleftharpoons Hg_2^{2+}(aq)$	$\dfrac{[Hg_2^{2+}]}{[Hg^{2+}]} = 81$
$HCO_3^-(aq) + H_2O(l) \rightleftharpoons CO_3^{2-}(aq) + H_3O^+(aq)$	$\dfrac{[CO_3^{2-}][H_3O^+]}{[HCO_3^-]} =$ $\quad 4.7 \times 10^{-11}$
$N_2(g) + O_2(g) \rightleftharpoons 2NO(g)$	$\dfrac{[NO]^2}{[N_2][O_2]} =$ $\quad 4.5 \times 10^{-31}$

Refer to Appendix A for more K_{eq} values.

Figure 7
These pie charts show the relative amounts of reactants and products for three K_{eq} values of a reaction.

$K_{eq} = 0.02$

$K_{eq} = 1$

$K_{eq} = 50$

Reactants Products

K_{eq} Shows If the Reaction Is Favorable

When K_{eq} is large, the numerator of the equilibrium constant expression is larger than the denominator. Thus, the concentrations of the products will usually be greater than those of the reactants. In other words, when a reaction that has a large K_{eq} reaches equilibrium, the system's contents may be mostly products. Reactions in which more products form than reactants form are said to be "favorable." Look at the first entry in **Table 2**, where the reaction has a large K_{eq} value. The synthesis of ammonia is very favorable at 25°C.

When K_{eq} is small, the denominator of the equilibrium constant expression is larger than the numerator. The larger denominator shows that the concentrations of reactants at chemical equilibrium may be greater than those of products. A reaction that has larger concentrations of reactants than concentrations of products is an "unfavorable" reaction. Our air would be unbreathable if the reaction of oxygen and nitrogen to give nitrogen monoxide was favorable at 25°C! (See **Table 2** to know the value for this reaction.)

Figure 7 shows what the composition of a reaction mixture ($R \rightleftharpoons P$) would be at equilibrium for three different K_{eq} values. Notice that for $K_{eq} = 1$, there would be a 50:50 mixture of reactants (R) and products (P).

Calculating Concentrations of Products from K_{eq} and Concentrations of Reactants

K_{eq} for the equilibrium below is 1.8×10^{-5} at a temperature of 25°C. Calculate $[NH_4^+]$ when $[NH_3] = 6.82 \times 10^{-3}$.

$$NH_3(aq) + H_2O(l) \rightleftharpoons NH_4^+(aq) + OH^-(aq)$$

1 Gather information.

Determine the equilibrium expression first.

$$K_{eq} = \frac{[NH_4^+][OH^-]}{[NH_3]}$$

PRACTICE HINT

Remember to keep track of significant figures and to give no more significant figures than are needed.

NH_4^+ and OH^- ions are produced in equal numbers, so $[OH^-] = [NH_4^+]$.

2 Plan your work.

The numerator can be written as x^2 because $[OH^-] = [NH_4^+]$. Both K_{eq} and $[NH_3]$ are known and can be put into the expression.

3 Calculate.

Substitute known quantities.

$$1.8 \times 10^{-5} = K_{eq} = \frac{[NH_4^+][OH^-]}{[NH_3]} = \frac{x^2}{6.82 \times 10^{-3}}$$

Rearrange the equation.

$$x^2 = (1.8 \times 10^{-5}) \times (6.82 \times 10^{-3}) = 1.2 \times 10^{-7}$$

Take the square root of x^2.

$$[NH_4^+] = \sqrt{1.2 \times 10^{-7}} = 3.5 \times 10^{-4}$$

4 Verify your results.

Test to see whether the answer gives the equilibrium constant.

$$K_{eq} = \frac{[NH_4^+][OH^-]}{[NH_3]} = \frac{(3.5 \times 10^{-4})(3.5 \times 10^{-4})}{6.82 \times 10^{-3}} = 1.8 \times 10^{-5}$$

PRACTICE

1. If K_{eq} is 1.65×10^{-3} at 2027°C for the reaction below, what is the equilibrium concentration of NO when $[N_2] = 1.8 \times 10^{-3}$ and $[O_2] = 4.2 \times 10^{-4}$?

$$N_2(g) + O_2(g) \rightleftharpoons 2NO(g)$$

2. At 600°C, the K_{eq} for the reaction below is 4.32 when $[SO_3] = 0.260$ and $[O_2] = 0.045$. Calculate the equilibrium concentration for SO_2.

$$2SO_2(g) + O_2(g) \rightleftharpoons 2SO_3(g)$$

The Solubility Product Constant, K_{sp}

The maximum concentration of a salt in an aqueous solution is called the solubility of the salt in water. Solubilities can be expressed in moles of solute per liter of solution (mol/L or M). For example, the solubility of calcium fluoride in water is 3.4×10^{-4} mol/L. So, 0.00034 mol (or less than 0.03 g) of CaF_2 will dissolve in 1 L of water to give a saturated solution. If you try to dissolve 0.00100 mol of CaF_2 in 1 L of water, 0.00066 mol of CaF_2 will remain undissolved.

Like most salts, calcium fluoride is an ionic compound that dissociates into ions when it dissolves in water. Calcium fluoride is also one of a large class of salts that are said to be slightly soluble in water. The ions in solution and any solid salt are at equilibrium, as the following equation for CaF_2 and water shows.

$$CaF_2(s) \rightleftharpoons Ca^{2+}(aq) + 2F^-(aq)$$

Recall that solids are not a part of equilibrium constant expressions, so K_{eq} for this reaction is the product of $[Ca^{2+}]$ and $[F^-]^2$, which is equal to a constant. Equilibrium constants for the dissolution of slightly soluble salts are given a special name and symbol. They are called **solubility product constants, K_{sp},** and have no units. The K_{sp} value for calcium fluoride at 25°C is 1.6×10^{-10}.

$$K_{sp} = [Ca^{2+}][F^-]^2 = 1.6 \times 10^{-10}$$

This relationship is true whenever calcium ions and fluoride ions are in equilibrium with calcium fluoride, not just when the salt dissolves. For example, if you mix solutions of calcium nitrate and sodium fluoride, calcium fluoride precipitates. The net ionic equation for this precipitation is the reverse of the dissolution.

$$Ca^{2+}(aq) + 2F^-(aq) \rightleftharpoons CaF_2(s)$$

This equation is the same equilibrium. So, the K_{sp} for the dissolution of CaF_2 in this system is the same and is 1.6×10^{-10}.

Topic Link

Refer to the "Solutions" chapter for a discussion of solubility and concentrations.

solubility product constant, K_{sp}

the equilibrium constant for a solid that is in equilibrium with the solid's dissolved ions

Figure 8
Seashells are made mostly of calcium carbonate, which is only slightly soluble in water. The calcium carbonate is in equilibrium with ions present in sea water.

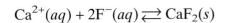

Table 3 **Solubility Product Constants at 25°C**

Salt	K_{sp}	Salt	K_{sp}
Ag_2CO_3	8.4×10^{-12}	CuS	1.3×10^{-36}
Ag_2CrO_4	1.1×10^{-12}	FeS	1.6×10^{-19}
Ag_2S	1.1×10^{-49}	$MgCO_3$	6.8×10^{-6}
AgBr	5.4×10^{-13}	$MnCO_3$	2.2×10^{-11}
AgCl	1.8×10^{-10}	PbS	9.0×10^{-29}
AgI	8.5×10^{-17}	$PbSO_4$	1.8×10^{-8}
$BaSO_4$	1.1×10^{-10}	$SrSO_4$	3.4×10^{-7}
$Ca_3(PO_4)_2$	2.1×10^{-33}	$ZnCO_3$	1.2×10^{-10}
$CaSO_4$	7.1×10^{-5}	ZnS	2.9×10^{-25}

Refer to Appendix A for more K_{sp} values.

2 SKILLS Toolkit

Determining K_{sp} for Reactions at Chemical Equilibrium

1. Write a balanced chemical equation.

- Remember that the solubility product is only for those salts that have low solubility. Soluble salts such as sodium chloride and ammonium nitrate do not have K_{sp} values.

- Make sure that the reaction is at equilibrium.

- Chemical equations should always be written so that the solid salt is the reactant and the ions are products.

2. Write a solubility product expression.

- To write the expression, write the product of the ion concentrations.

- Concentrations of any solid or pure liquid that take part in the reaction are left out because they never change.

3. Complete the solubility product expression.

- To complete the expression, raise each concentration to a power equal to the substance's coefficient in the balanced chemical equation.

- Remember that K_{sp} values depend on temperature.

Calculating K_{sp} from Solubility

Most parts of the oceans are nearly saturated with CaF_2. The mineral fluorite, CaF_2, may precipitate when ocean water evaporates. A saturated solution of CaF_2 at 25°C has a solubility of 3.4×10^{-4} M. Calculate the solubility product constant for CaF_2.

1 Gather information.

$$CaF_2(s) \rightleftharpoons Ca^{2+}(aq) + 2F^-(aq)$$

$$[CaF_2] = 3.4 \times 10^{-4}, [F^-] = 2[Ca^+]$$

$$K_{sp} = [Ca^{2+}][F^-]^2$$

2 Plan your work.

Because 3.4×10^{-4} mol CaF_2 dissolves in each liter of solution, you know from the balanced equation that every liter of solution will contain 3.4×10^{-4} mol Ca^{2+} and 6.8×10^{-4} mol F^-.

3 Calculate.

$$K_{sp} = [Ca^{2+}][F^-]^2 = (3.4 \times 10^{-4})(6.8 \times 10^{-4})^2 = 1.6 \times 10^{-10}$$

4 Verify your result.

To verify that the K_{sp} value is correct, determine the concentration of the F^- ion using $K_{sp} = 1.6 \times 10^{-10}$ and $[Ca^{2+}] = 3.4 \times 10^{-4}$.

$$[F^-]^2 = \frac{K_{sp}}{[Ca^{2+}]} = \frac{1.6 \times 10^{-10}}{3.4 \times 10^{-4}} = 4.7 \times 10^{-7}$$

$$[F^-] = 6.9 \times 10^{-4}$$

This calculation provides the concentration of the fluorine ions and confirms that the solubility product constant is 1.6×10^{-10}.

> **PRACTICE HINT**
>
> Do not forget to balance the chemical equation before using it to solve solubility problems.

PRACTICE

1. Copper(I) bromide is dissolved in water to saturation at 25°C. The concentration of Cu^+ ions in solution is 7.9×10^{-5} mol/L. Calculate the K_{sp} for copper(I) bromide at this temperature.

2. What is the K_{sp} value for $Ca_3(PO_4)_2$ at 298 K if the concentrations in a solution at equilibrium with excess solid are 3.42×10^{-7} M for Ca^{2+} ions and 2.28×10^{-7} M for PO_4^{3-} ions?

3. If a saturated solution of silver chloride contains an AgCl concentration of 1.34×10^{-5} M, confirm that the solubility product constant of this salt has the value shown in **Table 3.**

PROBLEM SOLVING SKILL

Calculating Ionic Concentrations Using K_{sp}

Copper(I) chloride has a solubility product constant of 1.2×10^{-6} and dissolves according to the equation below. Calculate the solubility of this salt in ocean water in which the $[Cl^-] = 0.55$.

$$CuCl(s) \rightleftarrows Cu^+(aq) + Cl^-(aq)$$

1 Gather information.

The given data and the chemical equation show that $[Cu^+][Cl^-] = K_{sp} = 1.2 \times 10^{-6}$. Additional information reveals that $[Cl^-] = 0.55$.

2 Plan your work.

The product of $[Cu^+][Cl^-]$ must equal $K_{sp} = 1.2 \times 10^{-6}$.

$$K_{sp} = [Cu^+][Cl^-] = 1.2 \times 10^{-6}$$

PRACTICE HINT

Recall that the K_{sp} expression holds even if the ions are from a source other than the salt.

3 Calculate.

Using the following equation, you can determine the concentration of copper ions.

$$[Cu^+] = \frac{K_{sp}}{[Cl^-]} = \frac{1.2 \times 10^{-6}}{0.55} = 2.2 \times 10^{-6}$$

The quantity is also the solubility of copper(I) chloride in ocean water because the dissolution of 1 mol of CuCl produces 1 mol of Cu^+. Therefore, the solubility of CuCl is 2.2×10^{-6} mol/L.

4 Verify your results.

You can recalculate the solubility product constant to check the answers.

$$K_{sp} = [Cu^+][Cl^-] = (0.55)(2.2 \times 10^{-6}) = 1.2 \times 10^{-6}$$

PRACTICE

PROBLEM SOLVING SKILL

1 The K_{sp} for silver carbonate is 8.4×10^{-12} at 298 K. The concentration of carbonate ions in a saturated solution is 1.28×10^{-4} M. What is the concentration of silver ions?

2 Lead-acid batteries employ lead(II) sulfate plates in a solution of sulfuric acid. Use data from **Table 3** to calculate the solubility of $PbSO_4$ in a battery acid that has an SO_4^{2-} concentration of 1.0 M.

3 Calculate the concentration of Pb^{2+} ions in solution when $PbCl_2$ is dissolved in water. The concentration of Cl^- ions in this solution is found to be 2.86×10^{-2} mol/L. At 25°C, the K_{sp} of $PbCl_2$ is 1.17×10^{-5}.

4 What is the concentration of Cu^+ ions in a saturated solution of copper(I) chloride given that the K_{sp} of CuCl is 1.72×10^{-7} at 25°C?

Using K_{sp} to Make Magnesium

Though slightly soluble hydroxides are not salts, they have solubility product constants. Magnesium hydroxide is an example.

$$Mg(OH)_2(s) \rightleftharpoons Mg^{2+}(aq) + 2OH^-(aq)$$

$$[Mg^{2+}][OH^-]^2 = K_{sp} = 1.8 \times 10^{-11}$$

This equilibrium is the basis for obtaining magnesium.

Table 4 lists the most abundant ions in ocean water and their concentrations. Notice that Mg^{2+} is the third most abundant ion in the ocean. Magnesium is so abundant that ocean water is used as the raw material from which magnesium is gotten. To get magnesium, calcium hydroxide is added to sea water. This raises the hydroxide ion concentration to a large value so that $[Mg^{2+}][OH^-]^2$ would be greater than 1.8×10^{-11}. As a result, magnesium hydroxide precipitates and can be collected. The next step in the process is to treat magnesium hydroxide with hydrochloric acid to make magnesium chloride. Finally, magnesium is obtained by the electrolysis of $MgCl_2$ in the molten state. One cubic meter of sea water yields 1 kg of magnesium metal.

Magnesium has a density of 1.7 g/cm^3, so it is one of the lightest metals. Because of magnesium's low density and rigidity, alloys of magnesium are used when light weight and strength are needed. Magnesium is found in ladders, cameras, cars, and airplanes.

Table 4	Ions in the Ocean
Ions	**Concentration in the ocean (mol/L)**
Cl^-	0.554
Na^+	0.470
Mg^{2+}	0.047
SO_4^{2-}	0.015
K^+	0.010
Ca^{2+}	0.009

Refer to Appendix A for more information on ocean water.

Section Review

UNDERSTANDING KEY IDEAS

1. Giving an example, explain how to write an expression for K_{eq} from a chemical equation.

2. Which species are left out from the K_{eq} expression, and why?

3. To which chemical systems can a K_{sp} be assigned?

4. When does K_{sp} not apply?

PRACTICE PROBLEMS

5. For the reaction in which hydrogen iodide is made at 425°C in the gas phase from its elements, calculate [HI], given that $[H_2] = [I_2] = 4.79 \times 10^{-4}$ and $K_{eq} = 54.3$.

6. Given that the K_{sp} value of CuS is 1.3×10^{-36}, what is $[Cu^{2+}]$ in a saturated solution?

7. Write the equation for the reaction in which solid carbon reacts with gaseous carbon dioxide to form gaseous carbon monoxide. At equilibrium, a 2.0 L reaction vessel is found to contain 0.40 mol of C, 0.20 mol of CO_2, and 0.10 mol of CO. Find the K_{eq}.

CRITICAL THINKING

8. A reaction has a single product and a single reactant, and both the product and the reactant are gases. If the concentration of each one was 0.1 M at equilibrium, what would the equilibrium constant be?

9. Write the solubility product expression for the slightly soluble compound aluminum hydroxide, $Al(OH)_3$.

③ Equilibrium Systems and Stress

KEY TERMS

- **Le Châtelier's principle**
- **common-ion effect**

OBJECTIVES

① **State** Le Châtelier's principle.

② **Apply** Le Châtelier's principle to determine whether the forward or reverse reaction is favored when a stress such as concentration, temperature, or pressure is applied to an equilibrium system.

③ **Discuss** the common-ion effect in the context of Le Châtelier's principle.

④ **Discuss** the practical uses of Le Châtelier's principle.

Le Châtelier's Principle

Figure 9
Henri Louis Le Châtelier (1850–1936) was particularly interested in applying science to industry and in getting maximum yield from chemical reactions.

Le Châtelier's principle

the principle that states that a system in equilibrium will oppose a change in a way that helps eliminate the change

If you hang a coat hanger by a rubber band, you create a system in mechanical equilibrium. The force of gravity, which pulls the hanger downward, is exactly balanced by the rubber band trying to pull the hanger upward. Now take a second coat hanger, and hook it onto the same rubber band beside the first one. The extra weight is a *stress* applied to this system. Stress is another word for something that causes a change in a system at equilibrium. The equilibrium is disturbed by the stress, and the rubber band lengthens. However, the system soon reaches a new equilibrium.

Chemical and mechanical equilibria are similar in that both respond to stresses by adjusting until new equilibria are reached. Henri Le Châtelier, the chemist pictured in **Figure 9,** studied the way in which chemical equilibria respond to changes. His findings are known as **Le Châtelier's principle.** This principle states that *when a system at equilibrium is disturbed, the system adjusts in a way to reduce the change.* Some situations in our lives are analogies for Le Châtelier's principle. For example, if you are disturbed by a loud noise, you may move to a quieter location.

Chemical equilibria respond to three kinds of stress: changes in the concentrations of reactants or products, changes in temperature, and changes in pressure. When a stress is first applied to a system, any chemical equilibrium is disturbed. As a result, the rates of the forward and backward reactions in the system are no longer equal. The system responds to the stress by forming more products or by forming more reactants. A new chemical equilibrium is reached when enough reactants or products form. At this point, the rates of the forward and backward reactions are equal again.

Changes in Concentration Alter Equilibrium Composition

What happens if you add more reactant to a system in chemical equilibrium? This increase in the reactant's concentration is a stress on the system. The system will respond to decrease the concentration of the reactant by changing some of the reactant into product. Therefore, the rate of the forward reaction must be greater than the rate of the reverse reaction. Because the forward reaction is increasing, the equilibrium is said to *shift right*. The reactant concentration will continue to drop until the reaction reaches equilibrium. Then, the forward and reverse reaction rates will be equal. Remember that changes in the amounts of solids and pure liquids do not affect K_{eq} values.

A reaction of two colored complex ions is described by the equation

$$[Cu(H_2O)_4]^{2+}(aq) + 4NH_3(aq) \rightleftharpoons [Cu(NH_3)_4]^{2+}(aq) + 4H_2O(l)$$
pale blue   blue-purple

internet connect

www.scilinks.org
Topic: Factors Affecting
    Equilibrium
SciLinks code: HW4057

SC*LINKS*. Maintained by the
    National Science
    Teachers Association

The beaker on the left in **Figure 10** contains this copper complex ion reaction in a chemical equilibrium that favors the formation of reactants. We know that the reverse reaction is favored, because the reaction mixture in the beaker is pale blue. But if additional ammonia is added to this beaker, the system responds to offset the increase by forming more of the product. This increase in the presence of product can be seen in the beaker to the right in **Figure 10,** which contains a blue-purple solution.

The equilibrium below occurs in a closed bottle of a carbonated liquid.

$$H_3O^+(aq) + HCO_3^-(aq) \rightleftharpoons 2H_2O(l) + CO_2(aq)$$

After you uncap the bottle, the dissolved carbon dioxide leaves the solution and enters the air. The forward reaction rate of this system will increase to produce more CO_2. This increase in the rate of the forward reaction decreases the concentration of H_3O^+ (H_3O^+ ions make soda taste sharp). As a result, the drink gets "flat." What would happen if you could increase the concentration of CO_2 in the bottle? The reverse reaction rate would increase, and $[H_3O^+]$ and $[HCO_3^-]$ would increase.

Figure 10
The color of the solution makes it easy to see which complex ion is dominant. Adding extra ammonia shifts the equilibrium in favor of the ammoniated complex ion.

$NH_3(aq)$

$[Cu(H_2O)_4]^{2+}(aq)$

$H_2O(l)$

$[Cu(NH_3)_4]^{2+}(aq)$

$$[Cu(H_2O)_4]^{2+}(aq) + 4NH_3(aq) \rightleftharpoons [Cu(NH_3)_4]^{2+}(aq) + 4H_2O(l)$$

Temperature Affects Equilibrium Systems

The effect of temperature on the gas-phase equilibrium of nitrogen dioxide, NO_2, and dinitrogen tetroxide, N_2O_4, can be seen because of the difference in color of NO_2 and N_2O_4. The intense brown NO_2 gas is the pollution that is responsible for the colored haze that you sometimes see on smoggy days.

$$2NO_2(g) \rightleftharpoons N_2O_4(g)$$
$$\text{brown} \qquad \text{colorless}$$

To understand how the nitrogen dioxide–dinitrogen tetroxide equilibrium is affected by temperature, we need to review endothermic and exothermic reactions. Recall that endothermic reactions absorb energy and have positive ΔH values. Exothermic reactions release energy and have negative ΔH values. The forward reaction is an exothermic process, as the equation below shows.

$$2NO_2(g) \longrightarrow N_2O_4(g) \qquad\qquad \Delta H = -55.3 \text{ kJ}$$

Consider that you are given the flask shown in the middle in **Figure 11.** The flask contains an equilibrium mixture of NO_2 and N_2O_4 and has a temperature of 25°C. Now suppose that you heat the flask to 100°C. The heated flask will look like the flask at the far right. The mixture becomes dark brown because the reverse reaction rate increased to remove some of the energy that you added to the system. The equilibrium shifts to the left, toward the formation of NO_2. Because this reaction is endothermic, the temperature of the flask drops as energy is absorbed. This equilibrium shift is true for all exothermic forward reactions: *Increasing the temperature of an equilibrium mixture usually leads to a shift in favor of the reactants.*

The opposite statement is true for endothermic forward reactions: *Increasing the temperature of an equilibrium mixture usually leads to a shift in favor of the products.*

Figure 11
The NO_2-N_2O_4 equilibrium system is shown at three different temperatures. Temperature changes put stress on equilibrium systems and cause either the forward or reverse reaction to be favored.

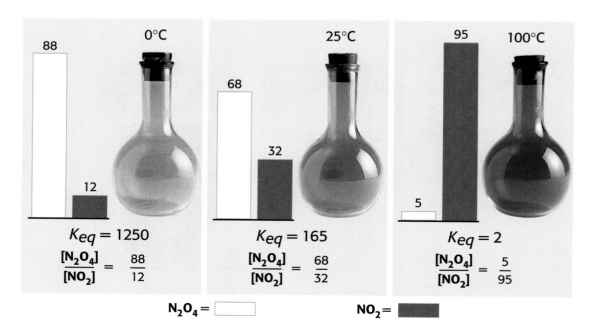

$$[Co(H_2O)_6]^{2+}(aq) + 4Cl^-(aq) \rightleftharpoons [CoCl_4]^{2-}(aq) + 6H_2O(l)$$

An example of the effect of temperature on an endothermic reaction is illustrated in **Figure 12**. The following equation describes an equilibrium that involves the two colored cobalt complex ions.

$$[Co(H_2O)_6]^{2+}(aq) + 4Cl^-(aq) \rightleftharpoons [CoCl_4]^{2-}(aq) + 6H_2O(l)$$
pink   blue

This particular equilibrium has an endothermic forward reaction. Therefore, this forward reaction is favored as the temperature of the reaction rises. The reverse reaction is favored at lower temperatures. The experiment illustrated in **Figure 12** confirms these predictions. The solution is pink at 0°C, which shows that the reactants are favored at low temperatures. At 100°C, the solution is blue, which shows that the products are favored at high temperatures.

Temperature changes affect not only systems at equilibrium but also the value of equilibrium constants. In fact, equilibrium constants changing with temperature is the reason that equilibria change with temperature. For example, consider K_{eq} for the ammonia synthesis equilibrium.

$$N_2(g) + 3H_2(g) \rightleftharpoons 2NH_3(g)$$

The forward reaction is exothermic ($\Delta H = -91.8$ kJ), so the equilibrium constant decreases a lot as temperature increases.

Figure 12
Depending on whether energy is removed or added, the reverse or forward reaction will be favored and a pink solution or a blue solution, respectively, will form.

Topic Link

Refer to the "Causes of Change" chapter for a discussion of enthalpy.

Pressure Changes May Alter Systems in Equilibrium

Pressure has almost no effect on equilibrium reactions that are in solution. Gases in equilibrium, however, may be affected by changes in pressure.

The NO_2 and N_2O_4 equilibrium can show the effect of a pressure stress on a chemical equilibrium. In **Figure 13a,** you see an equilibrium mixture of the gases in a syringe at low pressure. The gas has a larger concentration of N_2O_4 than of NO_2, which can be seen from the pale color of the gas. The gas is suddenly compressed to about half its former volume, which, as a consequence of Boyle's law, doubles the pressure. Before the system has time to adjust to the pressure stress, the concentration of each gas doubles. This effect is seen in the darker color of the gas in **Figure 13b.** Le Châtelier's principle predicts that the system will adjust in an attempt to reduce the pressure. According to the equation, 2 mol of NO_2 produce 1 mol of N_2O_4. The gas laws tell us that at constant volume and temperature, pressure is proportional to the number of moles. So the pressure reduces when there are fewer moles of gas. Thus, the equilibrium shifts to the right, and more N_2O_4 is produced. In **Figure 13c,** the color of the gas shows that the NO_2 concentration has fallen almost to its original level.

In an equilibrium, a pressure increase favors the reaction that produces fewer gas molecules. This change does not always favor the forward direction. The reverse reaction is favored for the equilibrium

$$2NOCl(g) \rightleftharpoons 2NO(g) + Cl_2(g)$$

There is no change in the number of molecules in the following reaction:

$$H_2O(g) + CO(g) \rightleftharpoons H_2(g) + CO_2(g)$$

In such cases, a change in pressure will not affect equilibrium.

Topic Link

Refer to the "Gases" chapter for a discussion of gas laws.

Figure 13
A sudden decrease in the volume of an NO_2-N_2O_4 mixture increases the pressure and makes the mixture darker at first. The color fades as more N_2O_4 forms to reduce the pressure and reach a new equilibrium.

a   b   c

The Common-Ion Effect

In pure water, the solubility of CuCl is 1.1×10^{-3} mol/L. Remember that you calculated the solubility of CuCl in ocean water to be 2.2×10^{-6} mol/L in **Sample Problem D.** So, CuCl is 500 times less soluble in sea water than it is in pure water. This dramatic reduction in solubility demonstrates Le Châtelier's principle and the dissolution of CuCl is shown below.

$$CuCl(s) \rightleftharpoons Cu^+(aq) + Cl^-(aq) \qquad [Cu^+][Cl^-] = K_{sp} = 1.2 \times 10^{-6}$$

If you have a saturated solution of copper(I) chloride in water and then add chloride-rich ocean water, you will increase $[Cl^-]$. However, the K_{sp}, which is the mathematical product of $[Cl^-]$ and $[Cu^+]$, must remain constant. Hence $[Cu^+]$ must decrease. This decrease can occur only by the precipitation of the CuCl salt. The ion Cl^- is the *common-ion* in this case because the Cl^- ion comes from two sources. The reduction of the solubility of a salt in the solution due to the addition of a common ion is called the **common-ion effect.**

Doctors use barium sulfate solutions to diagnose problems in the digestive tract. A patient swallows it, $BaSO_4$, so that the target organ becomes "visible" in X-ray film, as shown in **Figure 14.** $BaSO_4$ is a not a very soluble salt and is not absorbed by the body. But $BaSO_4$ is in equilibrium with a small concentration of $Ba^{2+}(aq)$, a poison. To reduce the Ba^{2+} concentration to a safe level, a common ion is added by mixing sodium sulfate, Na_2SO_4, into the $BaSO_4$ solution that a patient must swallow.

Another example of the common-ion effect emerges when solutions of potassium sulfate and calcium sulfate are mixed and either solution is saturated. Immediately after the two solutions are mixed, the product of the $[Ca^{2+}]$ and $[SO_4^{2-}]$ is greater than the K_{sp} of calcium sulfate. So, $CaSO_4$ precipitates to establish the equilibrium shown below.

$$CaSO_4(s) \rightleftharpoons Ca^{2+}(aq) + SO_4^{2-}(aq)$$

common-ion effect

the phenomenon in which the addition of an ion common to two solutes brings about precipitation or reduces ionization

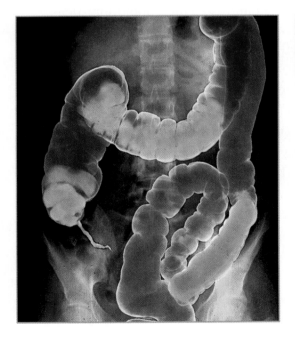

Figure 14
The light areas on this X ray of the digestive tract show insoluble barium sulfate.

www.scilinks.org
Topic: Common-Ion Effect
SciLinks code: HW4034

SC*i*LINKS. Maintained by the National Science Teachers Association

Practical Uses of Le Châtelier's Principle

The chemical industry, such as the factory in **Figure 15,** makes use of Le Châtelier's principle in many ways. One way is in the synthesis of ammonia by the Haber Process. High pressure is used to drive the following equilibrium to the right.

$$N_2(g) + 3H_2(g) \rightleftarrows 2NH_3(g)$$

Notice that the forward reaction converts 4 mol of gas into 2 mol of another gas. Therefore, the forward reaction is favored during an increase in pressure. The manufacturing process employs pressures as high as 400 times atmospheric pressure.

The ammonia synthesis is an exothermic reaction. Therefore, the forward reaction is favored at low temperatures.

$$0°C \qquad K_{eq} = 6.5 \times 10^8$$
$$250°C \qquad K_{eq} = 52$$
$$500°C \qquad K_{eq} = 5.8 \times 10^{-2}$$

The Haber Process is operated at temperatures of 500°C, which may seem odd because the K_{eq} is so small at that temperature. The reason that this process must be run at such a high temperature is that the reaction proceeds too slowly at lower temperatures.

Figure 15
Today ammonia is produced on a large scale in industrial plants such as this one.

internet connect

www.scilinks.org
Topic: Haber Process
SciLinks code: HW4153

SCiLINKS. Maintained by the National Science Teachers Association

---

③ Section Review

UNDERSTANDING KEY IDEAS

1. State Le Châtelier's principle.

2. To what stresses does a chemical equilibrium respond?

3. How does an equilibrium reaction respond to the addition of extra reactant? extra product?

4. What role does ΔH play in determining how temperature influences K_{eq}?

5. Which stress affects the value of an equilibrium constant?

6. Which equilibria are affected by pressure change?

7. Describe the common-ion effect, and give an example.

CRITICAL THINKING

8. If a reaction could exist that has $\Delta H = 0$, what effect would temperature have on this reaction at equilibrium?

9. For a purely gaseous equilibrium, does doubling the pressure have the same effect on equilibrium as doubling the concentration?

10. Do "spectator ions" affect K_{sp}?

11. Which of the two common ions, Ca^{2+} or F^-, is more effective in precipitating CaF_2?

12. In the Haber Process, how would removing NH_3 as it forms affect the equilibrium?

13. How does adding extra reactant affect K_{eq}?

Element Spotlight

Where Is Cl?

Earth's crust
0.045% by mass

Sea water
1.9% by mass
30.61% of dissolved materials

Chlorine Gives Us Clean Drinking Water

The practice of chlorinating drinking water began in the early 20th century. Chlorinating water greatly reduced the number of people who are affected by infectious diseases. No other method of disinfecting drinking water is as inexpensive and reliable as chlorination. Today, chlorine is used as a disinfectant in almost every drinking-water treatment plant in the United States and Canada. This method will probably continue to be used well into the 21st century.

Chlorine gas and chlorine compounds are used to purify swimming pools.

Industrial Uses

- Chlorine is a strong oxidizing agent and is used as a bleaching agent for paper.

- Chlorine gas is used to produce bromine compounds for use as a fire retardant.

- Table salt (sodium chloride) is one of the most important chlorine-containing compounds.

- Plastics, such as the vinyl used in sporting goods, contain chlorine.

- Many pharmaceuticals, including antibiotics and allergy medications, require chlorine-containing compounds for their manufacture.

Real-World Connection Many Third World regions have outbreaks of deadly diseases, such as cholera and typhoid fever, that could easily be prevented by disinfecting water supplies with chlorine.

A Brief History

1801: W. Cruickshank recommends the use of Cl_2 as a disinfectant.

1902: J. C. Downs patents the Downs cell for the production of $Cl_2(g)$ and $Na(s)$.

1600	1700	1800	1900

1648: J. R. Glauber prepares concentrated hydrochloric acid.

1774: C. W. Scheele isolates chlorine gas, which was called *dephlogisticated marine acid air* at the time.

1810: H. Davy proves that chlorine is an element and names it.

1908: P. Sommerfeld shows that HCl is present in the gastric juices of animals.

Questions

1. Research environmentally hazardous materials that contain chlorine. Also research ways that people are trying to reduce these hazards.

2. Research how chlorine was discovered.

internet connect

www.scilinks.org
Topic: Chlorine
SciLinks code: HW4031

SCiLINKS. Maintained by the National Science Teachers Association

CHAPTER HIGHLIGHTS

SECTION ONE Reversible Reactions and Equilibrium

- During completion reactions, products do not significantly re-form reactants.
- During reversible reactions, products re-form the original reactants.
- Reversible reactions can reach equilibrium.
- The forward and reverse reaction rates are equal at chemical equilibrium.
- At chemical equilibrium, reactant and product concentrations remain unchanged.
- Chemical equilibria are dynamic equilibria.
- Complex ions are metal ions or atoms that are bonded to more than one atom or molecule.
- Complex ion formations are often examples of equilibria.

SECTION TWO Systems at Equilibrium

- The constant K_{eq} is equal to an expression that has the concentrations of species in equilibrium.
- The equilibrium constant for the dissolution of a slightly soluble salt is the solubility product constant K_{sp}.
- Equilibrium constants and solubility product constants have no units.

SECTION THREE Equilibrium Systems and Stress

- Le Châtelier's principle states that chemical equilibria adjust to relieve applied stresses.
- Stresses due to changes in concentration, temperature, and pressure are subject to Le Châtelier's principle.
- Temperature changes affect the values of equilibrium constants.
- Pressure changes have almost no affect on equilibrium reactions in solution. Pressure changes can affect equilibrium reactions in the gas phase.
- The common-ion effect reduces the solubility of slightly soluble salts.

KEY TERMS

**reversible reaction
chemical equilibrium**

**equilibrium constant, K_{eq}
solubility product
  constant, K_{sp}**

**Le Châtelier's principle
common-ion effect**

Determining K_{eq} for Reactions at Chemical Equilibrium
Skills Toolkit 1, p. 503
Sample Problem A, p. 504

Calculating Concentrations from K_{eq}
Sample Problem B, p. 506

Determining K_{sp} for Reactions at Chemical Equilibrium
Skills Toolkit 2, p. 508
Sample Problem C, p. 509

Calculating Concentrations from K_{sp}
Sample Problem D, p. 510

USING KEY TERMS

1. Explain the meaning of the terms *reversible reaction*, *completion reaction*, and *reaction at equilibrium*.

2. What is the distinction between a static equilibrium and a dynamic equilibrium? In which class do chemical equilibria fit?

3. Define *complex ion* and give two examples.

4. What are ligands?

5. What is an equilibrium constant?

6. Distinguish between solubility and solubility product constant. Explain how one may be calculated from the other.

7. In the context of Le Châtelier's principle, what is a "stress"? List the stresses that affect chemical equilibria.

8. Give one example of a stress on a reaction in aqueous solution and at equilibrium.

9. In the precipitation of cadmium iodate, $Cd(IO_3)_2$, which of the following salts would exert a common-ion effect: $NaIO_3$, KI, $Cd(NO_3)_2$, NH_4IO_4, $CdCl_2$, or $CsNO_3$?

UNDERSTANDING KEY IDEAS

Reversible Reactions and Equlibrium

10. Use **Figure 3** to answer the following questions:
  a. What is happening to the rate of formation of $HI(g)$ before the system reaches equilibrium?
  b. When is the rate of the forward reaction the greatest?

11. Give two examples of static equilibrium and two examples of dynamic equilibrium. Your examples do not have to be chemical examples.

12. Identify the ligands in the following complex ions.

  a. $\left[\begin{array}{ccc} F & F & F \\ & Al & \\ F & F & F \end{array}\right]^{3-}$ **b.** $\left[\begin{array}{ccc} CN & & CN \\ & Ni & \\ CN & & CN \end{array}\right]^{2-}$

Systems at Equilibrium

13. Why must a balanced chemical equation be used when determining K_{eq}?

14. Describe and explain how the concentrations of A, B, C, and D change from the time when A and B first combine to the point at which equilibrium is established for the reaction

$$A + B \rightleftarrows C + D.$$

15. In general, which reaction (forward, reverse, or neither) is favored if the value of K at a specified temperature is
  a. equal to 1?
  b. very small?
  c. very large?

16. When nitrogen monoxide reacts with oxygen to produce nitrogen dioxide, an equilibrium is established.

  a. Write the balanced equation.

  b. Write the equilibrium constant expression.

17. Write equilibrium constant expressions for the following reactions:

  a. $2NO_2(g) \rightleftarrows N_2O_4(g)$

Chemical Equilibrium   **521**

b. $CO(g) + Cl_2(g) \rightleftarrows COCl_2(g)$

c. $AgCl(s) \rightleftarrows Ag^+(aq) + Cl^-(aq)$

d. $CH_3COOH(aq) + H_2O(l) \rightleftarrows$
$\qquad H_3O^+(aq) + CH_3COO^-(aq)$

18. The equilibrium constant for the reaction below is 1.8×10^{-5}.

$$NH_3(aq) + H_2O(l) \rightleftarrows NH_4^+(aq) + OH^-(aq)$$

  a. Write the equilibrium constant expression.

  b. What is the numerical value of equilibrium constant for the following reaction?

$$NH_4^+(aq) + OH^-(aq) \rightleftarrows NH_3(aq) + H_2O(l)$$

19. Write the solubility product expressions for the following slightly soluble salts: AgI, $SrSO_4$, Ag_2CO_3, Ag_2S, PbI_2, $AgIO_3$, $Mg_3(PO_4)_2$, and Hg_2Cl_2.

Equilibrum Systems and Stress

20. How would you explain Le Châtelier's principle in your own words to someone who finds the concept difficult to understand?

21. Predict whether each of the following pressure changes would favor the forward reaction or the reverse reaction.

$$2NO(g) + O_2(g) \rightleftarrows 2NO_2(g)$$

  a. increased pressure
  b. decreased pressure

22. Predict the effect of each of the following on the indicated equilibrium system in terms of which reaction (forward, reverse, or neither) will be favored.

$$H_2(g) + Cl_2(g) \rightleftarrows 2HCl(g) + 184 \text{ kJ}$$

  a. addition of Cl_2
  b. removal of HCl
  c. increased pressure
  d. decreased temperature
  e. removal of H_2
  f. decreased pressure

  g. increased temperature
  h. decreased system volume

23. What relative pressure (high or low) would result in the production of the maximum level of CO_2 according to the following equation? Explain.

$$2CO(g) + O_2(g) \rightleftarrows 2CO_2(g)$$

24. What relative conditions (reactant concentrations, pressure, and temperature) would favor a high equilibrium concentration of the substance in bold in each of the following equilibrium systems?

  a. $2CO(g) + O_2(g) \rightleftarrows \textbf{2CO}_\textbf{2}\textbf{(g)} + 167 \text{ kJ}$

  b. $Cu^{2+}(aq) + 4NH_3(aq) \rightleftarrows$
  $\qquad \textbf{[Cu(NH}_\textbf{3}\textbf{)}_\textbf{4}\textbf{]}^{\textbf{2+}}\textbf{(aq)} + 42 \text{ kJ}$

  c. $2HI(g) + 12.6 \text{ kJ} \rightleftarrows H_2(g) + \textbf{I}_\textbf{2}\textbf{(g)}$

  d. $4HCl(g) + O_2(g) \rightleftarrows$
  $\qquad 2H_2O(g) + \textbf{2Cl}_\textbf{2}\textbf{(g)} + 113 \text{ kJ}$

25. What changes in conditions would favor the reactants in the following equilibrium?

$$2SO_2(g) + O_2(g) \rightleftarrows 2SO_3(g) \qquad \Delta H = -198 \text{ kJ}$$

26. Write the equilibrium constant expression for the reaction in problem 25.

27. What changes in conditions would favor the products in the following equilibrium?

$$PCl_5(g) \rightleftarrows Cl_2(g) + PCl_3(g) \qquad \Delta H = 88 \text{ kJ}$$

28. Write the equilibrium constant expression for the reaction in problem 27.

29. Relate Le Châtelier's principle to the common-ion effect.

PRACTICE PROBLEMS

Sample Problem A Calculating K_{eq} from Concentrations of Reactants and Products

30. Vinegar—a solution of acetic acid, CH_3COOH, in water—is used in varying concentrations for different household tasks.

The following equilibrum exists in vinegar.

$$CH_3COOH(aq) + H_2O(l) \rightleftharpoons$$
$$H_3O^+(aq) + CH_3COO^-(aq)$$

If the concentration of the acetic acid solution at equilibrium is 3.00 M and the $[H_3O^+] = [CH_3COO^-] = 7.22 \times 10^{-3}$, what is the K_{eq} value for acetic acid?

31. Write the K_{eq} expressions for the following reactions:

a. $4H_3O^+(aq) + 2Cl^-(aq) + MnO_2(s) \rightleftharpoons$
$$Mn^{2+}(aq) + 6H_2O(l) + Cl_2(g)$$

b. $As_4O_6(s) + 6H_2O(l) \rightleftharpoons 4H_3AsO_3(aq)$

32. The following equation shows an equilibrium reaction between hydrogen and carbon dioxide.

$$H_2(g) + CO_2(g) \rightleftharpoons H_2O(g) + CO(g)$$

At 986°C, the following data were obtained at equilibrium in a 2.0 L reaction vessel.

Substance	Amount (mol)
H_2	0.4693
CO_2	0.0715
H_2O	0.2296
CO	0.2296

Calculate the K_{eq} for this reaction.

33. Determine the value of the equilibrium constant for each reaction below assuming that the equilibrium concentrations are as specified.
a. $A + B \rightleftharpoons C$; $[A] = 2.0$; $[B] = 3.0$; $[C] = 4.0$
b. $D + 2E \rightleftharpoons F + 3G$; $[D] = 1.5$; $[E] = 2.0$; $[F] = 1.8$; $[G] = 1.2$
c. $N_2(g) + 3H_2(g) \rightleftharpoons 2NH_3(g)$; $[N_2] = 0.45$; $[H_2] = 0.14$; $[NH_3] = 0.62$

34. An equilibrium mixture at a specific temperature is found to consist of 1.2×10^{-3} mol/L HCl, 3.8×10^{-4} mol/L O_2, 5.8×10^{-2} mol/L H_2O, and 5.8×10^{-2} mol/L Cl_2 according to the following:

$$4HCl(g) + O_2(g) \rightleftharpoons 2H_2O(g) + 2Cl_2(g).$$

Determine the value of the equilibrium constant for this system.

Sample Problem B Calculating Concentrations of Products from K_{eq} and Concentrations of Reactants

35. Analysis of an equilibrium mixture in which the following equilibrium exists gave $[OH^-] = [HCO_3^-] = 3.2 \times 10^{-3}$.

$$HCO_3^-(aq) + OH^-(aq) \rightleftharpoons CO_3^{2-}(aq) + H_2O(l)$$

The equilibrium constant is 4.7×10^3. What is the concentration of the carbonate ion?

36. When a sample of NO_2 gas equilibrated in a closed container at 25°C, a concentration of 0.0187 mol/L of N_2O_4 was found to be present. Use the data in **Table 2** to calculate the NO_2 concentration.

37. Methanol, CH_3OH, can be prepared in the presence of a catalyst by the reaction of H_2 and CO at high temperatures according to the following equation:

$$CO(g) + 2H_2(g) \rightleftharpoons CH_3OH(g)$$

What is the concentration of $CH_3OH(g)$ in moles per liter if the concentration of $H_2 = 0.080$ mol/L, the concentration of $CO = 0.025$ mol/L, and $K_{eq} = 290$ at 700 K?

38. At 450°C, the value of the equilibrium constant for the following system is 6.59×10^{-3}. If $[NH_3] = 1.23 \times 10^{-4}$ and $[H_2] = 2.75 \times 10^{-3}$ at equilibrium, determine the concentration of N_2 at that point.

$$N_2(g) + 3H_2(g) \rightleftharpoons 2NH_3(g)$$

Sample Problem C Calculating K_{sp} from Solubility

39. The solubility of cobalt(II) sulfide, CoS, is 1.7×10^{-13} M. Calculate the solubility product constant for CoS.

40. What is the solubility product for copper(I) sulfide, Cu_2S, given that the solubility of Cu_2S is 8.5×10^{-17} M?

41. The ionic substance T_3U_2 ionizes to form T^{2+} and U^{3-} ions. The solubility of T_3U_2 is 3.77×10^{-20} mol/L. What is the value of the solubility-product constant?

42. The ionic substance EJ dissociates to form E^{2+} and J^{2-} ions. The solubility of EJ is 8.45×10^{-6} mol/L. What is the value of the solubility-product constant?

Sample Problem D Calculating Ionic Concentrations Using K_{sp}

43. Four lead salts and their solubility products are as follows: PbS, 3.0×10^{-28}; $PbCrO_4$, 1.8×10^{-14}; $PbCO_3$, 7.4×10^{-14}; and $PbSO_4$, 2.5×10^{-8}. Calculate the Pb^{2+} concentration in a saturated solution of each of these salts.

44. Review problem 43, and note that all of the lead salts listed have anions whose charges are 2−. Why is the calculation more difficult for salts such as $PbCl_2$ or PbI_2?

45. What is the concentration of F^- ions in a saturated solution that is 0.10 M in Ca^{2+}? The K_{sp} of CaF_2 is 1.6×10^{-10}.

46. Silver bromide, AgBr, is used to make photographic black-and-white film. Calculate the concentration of Ag^+ and Br^- ions in a saturated solution at 25°C using **Table 3.**

47. The figure below shows the results of adding three different chemicals to distilled water and stirring well.

   a. Which substance(s) are completely soluble in water?

   b. Is it correct to say that AgCl is completely insoluble? Explain your answer. Is $Ba(OH)_2$ completely insoluble?

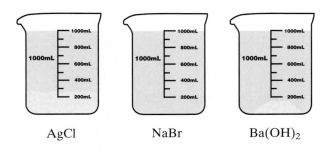

AgCl   NaBr $Ba(OH)_2$

MIXED REVIEW

48. At 1400°C, the following reaction attained equilibrium in a 1.0 L container.

$$H_2(g) + CO_2(g) \rightleftarrows H_2O(g) + CO(g)$$

An analysis of the equilibrium mixture found the following amounts: hydrogen, 0.060 mol; carbon dioxide, 0.060 mol; water vapor, 0.040 mol; and carbon monoxide, 0.040 mol. Calculate K_{eq} for the reaction.

49. Was it necessary to know the volume of the container in problem 48? Why or why not?

50. Why might an industrial process be operated at high temperature even though the reaction is more favorable at lower temperatures?

51. How does temperature affect equilibrium constants?

52. The reaction below has an equilibrium constant of 4.9×10^{11}.

$$Fe(OH)_2(s) + 2H_3O^+(aq) \rightleftarrows Fe^{2+}(aq) + 4H_2O(l)$$

Write the equilibrium constant expression, and determine the concentration of Fe^{2+} ions in equilibrium when the hydronium ion concentration is 1.0×10^{-7} mol/L.

53. When does a pressure change affect a chemical equilibrium?

54. What information may be conveyed by the knowledge that the equilibrium constant of a reaction is very small?

55. The K_{sp} of potassium periodate, KIO_4, is 3.7×10^{-4}. Determine whether a precipitate will form when 2.00 g of KCl and 2.00 g of $NaIO_4$ are dissolved in 1.00 L of water.

56. What is the concentration of Cu^+ ions in a saturated solution of copper(I) chloride given that the K_{sp} of CuCl is 1.72×10^{-7} at 25°C?

57. Calculate the solubility of a substance MN that ionizes to form M^{2+} and N^{2-} ions given that $K_{sp} = 8.1 \times 10^{-6}$.

CRITICAL THINKING

58. A student wrote, "The larger the equilibrium constant, the greater the rate at which reactants convert to products." How was he wrong?

59. Write the balanced equation for the reaction to which the following equilibrium constant applies.

$$\frac{[CO]^2}{[CO_2]} = K_{eq}$$

60. Write the equilibrium constant expression for the reaction below.

$$CaCO_3(s) \rightleftharpoons CaO(s) + CO_2(g)$$

Demonstrate that at a given temperature, the pressure of carbon dioxide in equilibrium with a mixture of calcium carbonate and calcium oxide cannot vary.

61. Imagine the following hypothetical reaction, taking place in a sealed, rigid container, to be neither exothermic nor endothermic.

$$A(g) + B(g) \rightleftharpoons C(g) \qquad \Delta H = 0$$

Would an increase in temperature favor the forward reaction or the reverse reaction? (Hint: Recall the gas laws.)

62. A very dilute solution of silver nitrate is added dropwise to a solution that contains equal concentrations of sodium chloride and potassium bromide. What salt will precipitate first?

63. Changes in the concentrations of the reactants and products at equilibrium have no effect on the value of the equilibrium constant. Explain this statement.

ALTERNATIVE ASSESSMENT

64. Your instructor will give you an index card with a specific equilibrium reaction on it. Describe how you would alter the reaction to produce either more of the products or more of the reactants. Show your method to your instructor. If your method is approved, obtain the necessary materials from your instructor and perform the experiment.

65. Research the practical uses of Le Châtelier's principle. Present your results to your class.

66. Develop a model that shows the concept of equilibrium. Be sure that your model includes the impact of Le Châtelier's principle on equilibrium.

CONCEPT MAPPING

67. Use the following terms to complete the concept map below: *chemical equilibrium, equilibrium constant, solubility product constant, reversible reactions, and Le Châtelier's principle.*

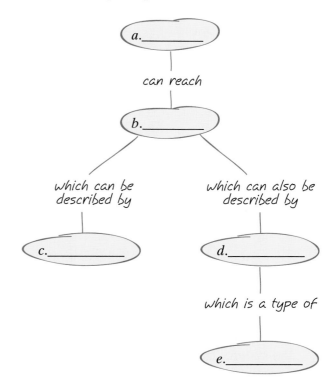

Study the graph below, and answer the questions that follow.
For help in interpreting graphs, see Appendix B, "Study Skills for Chemistry."

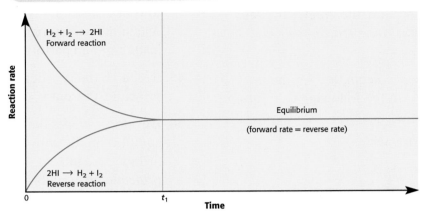

Rate Comparison for $H_2(g) + I_2(g) \rightleftharpoons 2HI(g)$

$H_2 + I_2 \longrightarrow 2HI$
Forward reaction

Reaction rate

Equilibrium

(forward rate = reverse rate)

$2HI \longrightarrow H_2 + I_2$
Reverse reaction

0   t_1

Time

68. What is the change in the rate of the forward reaction after the reaction is at equilibrium?

69. Does the reverse reaction rate ever equal zero? Why or why not?

70. Does the forward reaction rate ever equal zero? Why or why not?

TECHNOLOGY AND LEARNING

71. Graphing Calculator

Calculating the Equilibrium Constant, K, for a System

The graphing calculator can run a program that calculates K for a system, given the concentrations of the products and the concentrations of the reactants.

Given the balanced chemical equation

$$H_2(g) + I_2(g) \longrightarrow 2HI(g)$$

and the equilibrium mixture at 425°C, you can calculate the equilibrium constant for the system. Then you can use the program to make calculations.

Go to Appendix C. If you are using a TI-83 Plus, you can download the program

CONSTANT and data and run the application as directed. If you are using another calculator, your teacher will provide you with keystrokes and data sets to use. After you have graphed the data, answer the questions below.

a. What is the equilibrium constant given the following equilibrium concentrations: 0.012840 mol/L of H_2, 0.006437 mol/L of I_2, and 0.066807 mol/L of HI?

b. What is the equilibrium constant given the following equilibrium concentrations: 0.000105 mol/L of H_2, 0.000107 mol/L of I_2, and 0.000779 mol/L of HI?

c. What is the equilibrium constant given the following equilibrium concentrations: 0.000527 mol/L of H_2, 0.000496 mol/L of I_2, and 0.003757 mol/L of HI?

1. A chemical reaction is in equilibrium when
   a. forward and reverse reactions have ceased.
   b. the equilibrium constant equals 1.
   c. forward and reverse reaction rates are equal.
   d. no reactants remain.

2. A ligand is
   a. a component of wood.
   b. the metal atom in the center of a complex ion.
   c. a molecule or ion that bonds to the central metal ion in a complex ion.
   d. a colored transition metal ion.

3. Consider the following reaction:

$$2C(s) + O_2(g) \rightleftarrows 2CO(g)$$

The equilibrium constant expression for this reaction is

   a. $\dfrac{[CO]^2}{[O_2]}$.
   c. $\dfrac{2[CO]}{[O_2][2C]}$.

   b. $\dfrac{[CO]^2}{[O_2][C]^2}$.
   d. $\dfrac{[CO]}{[O_2]^2}$.

4. The solubility product of cadmium carbonate, $CdCO_3$, is 1.0×10^{-12}. In a saturated solution of this salt, the concentration of $Cd^{2+}(aq)$ ions is

   a. 5.0×10^{-13} mol/L.   **c.** 1.0×10^{-6} mol/L.

   b. 1.0×10^{-12} mol/L.   **d.** 5.0×10^{-7} mol/L.

5. Consider the following equation for an equilibrium system:

$$2PbS(s) + 3O_2(g) + C(s) \rightleftarrows 2Pb(l) + CO_2(g) + 2SO_2(g)$$

The concentration(s) that would be included in the denominator of the equilibrium constant expression are

   a. $Pb(l)$, $CO_2(g)$, and $SO_2(g)$.

   b. $PbS(s)$, $O_2(g)$, and $C(s)$.

   c. $O_2(g)$, $Pb(l)$, $CO_2(g)$, and $SO_2(g)$.

   d. $O_2(g)$.

6. Le Châtelier's principle states that
   a. at equilibrium, the forward and reverse reaction rates are equal.
   b. stresses include changes in concentrations, pressure, and temperature.
   c. to relieve stress, solids and solvents are omitted from equilibrium constant expressions.
   d. chemical equilibria respond to reduce applied stress.

7. If an exothermic reaction has reached equilibrium, then increasing the temperature will
   a. favor the forward reaction.
   b. favor the reverse reaction.
   c. favor both the forward and reverse reactions.
   d. have no effect on the equilibrium.

8. For the reaction _____, products are favored by increasing the pressure.

   a. $2NO(g) + O_2(g) \longrightarrow 2NO_2(g)$

   b. $CO_2(aq) + 2H_2O(l) \longrightarrow H_3O^+(aq) + HCO_3^-(aq)$

   c. $C(s) + O_2(g) \longrightarrow CO_2(g)$

   d. $2O_3(g) \longrightarrow 3O_2(g)$

ACIDS AND BASES

For health reasons, swimming pools are regularly treated with chemicals that are usually called "chlorine." However, the chemicals are more likely to be hypochlorites or similar compounds. The effectiveness of these disinfectants depends on how acidic or basic the water is. Pool owners must therefore measure and, if necessary, adjust, the acidity of their pool. A pH test kit, such as the one shown at left, can measure the balance of acids and bases dissolved in the pool water to ensure that it is in a healthy range.

START-UP ACTIVITY

What Does an Antacid Do?

PROCEDURE

1. Pour **100 mL of water** into a **150 mL beaker.** Use a **plastic pipet** to add **vinegar** one drop at a time while stirring with a **glass stirrer.** Keep **a piece of blue litmus paper** dipped in the solution, and record the number of drops needed to make the solution turn the litmus paper bright red.

2. Use a **mortar and pestle** to crush an **antacid tablet** until the tablet is powdered. Pour **100 mL of water** into another **150 mL beaker,** add the powdered tablet, and stir until the tablet is at least mostly dissolved.

3. Add vinegar dropwise to the antacid solution and monitor the solution by using **another piece of blue litmus paper.** Record the number of drops needed to turn the litmus paper bright red.

ANALYSIS

1. Which required more acid to turn the blue litmus paper red: the water or the antacid solution?

2. How does an antacid work to counteract excess stomach acid?

Pre-Reading Questions

① Give examples of acids and bases that you encounter in your everyday life.

② What polyatomic ion does the hydrogen ion, H^+, form in aqueous solution?

③ Acids are said to "neutralize" bases, and vice versa. How would you define the term *neutralize*?

internet connect

www.scilinks.org
Topic: Acids and Bases
SciLinks code: HW4002

SC*L*INKS. Maintained by the National Science Teachers Association

What Are Acids and Bases?

- **strong acid**
- **weak acid**
- **strong base**
- **weak base**
- **Brønsted-Lowry acid**
- **Brønsted-Lowry base**
- **conjugate acid**
- **conjugate base**
- **amphoteric**

OBJECTIVES

1. **Describe** the distinctive properties of strong and weak acids, and relate their properties to the Arrhenius definition of an acid.

2. **Describe** the distinctive properties of strong and weak bases, and relate their properties to the Arrhenius definition of a base.

3. **Compare** the Brønsted-Lowry definitions of acids and bases with the Arrhenius definitions of acids and bases.

4. **Identify** conjugate acid-base pairs.

5. **Write** chemical equations that show how an amphoteric species can behave as either an acid or a base.

Acids

Vinegar is acidic. So are the juices of the fruits that you see in **Figure 1.** Colas and some other soft drinks are also acidic. You can recognize these liquids as acidic by their tart, sour, or sharp taste. What they have in common is that they contain dissolved compounds that chemists describe as *acids*. Many other acids are highly caustic and should not be put to the taste test. One example of a hazardous acid is sulfuric acid, H_2SO_4, which is important in car batteries. Another example is hydrochloric acid, HCl, which is used to treat the water in swimming pools.

Figure 1
Fruits and fruit juices contain a variety of acids, including those shown here. Carbonic acid is found in cola. Vinegar contains acetic acid.

Carbonic acid

Acetic acid

Citric acid

Ascorbic acid

Acid Solutions Conduct Electricity Well

Acids are electrolytes, so their solutions in water are conductors of electric current, as **Figure 2** demonstrates. To understand why, consider what happens as hydrogen chloride, HCl, dissolves in water. Like other electrolytes, hydrogen chloride dissociates to produce ions. The hydrogen ion immediately reacts with a water molecule to form a hydronium ion, as shown in the equation below.

$$HCl(g) + H_2O(l) \longrightarrow H_3O^+(aq) + Cl^-(aq)$$

The resulting solution is called *hydrochloric acid.*

The hydronium ion, H_3O^+, is able to transfer charge through aqueous solutions much faster than other ions do. The positive charge is simply passed from water molecule to water molecule. The result is that acid solutions are excellent conductors of electricity.

Acids React with Many Metals

Another property shared by aqueous solutions of acids is that they react with many metals. All metals that are above hydrogen in the activity series react with acids to produce hydrogen gas. The reaction is caused by the hydronium ion present in the solution. The presence of the hydronium ion explains why all acids behave in this way. An example is the reaction of hydrochloric acid with zinc, which is shown in **Figure 3** and is represented by the following net ionic equation.

$$2H_3O^+(aq) + Zn(s) \longrightarrow 2H_2O(l) + H_2(g) + Zn^{2+}(aq)$$

Notice that even though hydrochloric acid is often represented by the formula HCl(*aq*), in aqueous solution hydrochloric acid actually consists of dissolved $H_3O^+(aq)$ and $Cl^-(aq)$ ions. The chloride ions do not play a part in the reaction and so do not appear in the net ionic equation.

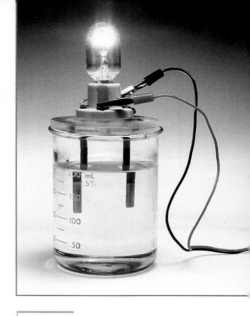

Figure 2
The beaker of hydrochloric acid solution contains $H_3O^+(aq)$ and $Cl^-(aq)$ ions. Because the solution is a good conductor, electricity can pass through the solution to cause the bulb to be brightly lit.

Topic Link

Refer to the "Solutions" chapter for a discussion of strong and weak electrolytes.

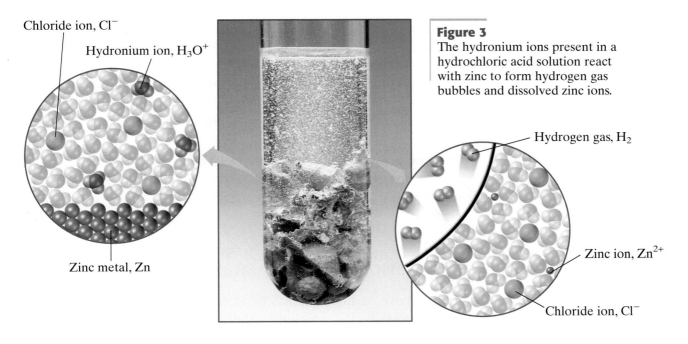

Chloride ion, Cl^-

Hydronium ion, H_3O^+

Zinc metal, Zn

Figure 3
The hydronium ions present in a hydrochloric acid solution react with zinc to form hydrogen gas bubbles and dissolved zinc ions.

Hydrogen gas, H_2

Zinc ion, Zn^{2+}

Chloride ion, Cl^-

Table 1 Some Strong Acids and Some Weak Acids	
Strong acids	**Weak acids**
hydrochloric acid, HCl	acetic acid, CH_3COOH
hydrobromic acid, HBr	hydrocyanic acid, HCN
hydriodic acid, HI	hydrofluoric acid, HF
nitric acid, HNO_3	nitrous acid, HNO_2
sulfuric acid, H_2SO_4	sulfurous acid, H_2SO_3
perchloric acid, $HClO_4$	hypochlorous acid, HOCl
periodic acid, HIO_4	phosphoric acid, H_3PO_4

Acids Generate Hydronium Ions

Recall that some electrolytes are *strong* and others are *weak,* depending on whether they dissociate completely or partially. Because acids are also electrolytes, they can be classified as strong or weak. **Table 1** lists the names and formulas of several strong acids and several weak acids.

Nitric acid, HNO_3, is an example of a **strong acid.** Its reaction with water is shown in **Figure 4** and is represented by the following equation:

$$HNO_3(l) + H_2O(l) \longrightarrow H_3O^+(aq) + NO_3^-(aq)$$

No HNO_3 molecules are present in a solution of nitric acid.

When a **weak acid** is dissolved in water, only a small fraction of its molecules are ionized at any given time. Hypochlorous acid, HOCl, is a weak acid. Its reaction with water is described by the equation below.

$$HOCl(l) + H_2O(l) \rightleftharpoons H_3O^+(aq) + ClO^-(aq)$$

Recall that the opposed arrows in this equation indicate equilibrium. Hypochlorite ions, ClO^-, react with hydronium ions to form HOCl at exactly the same rate that HOCl molecules react with water to form ions.

The presence of a considerable number of hydronium ions identifies an aqueous solution as acidic. A Swedish chemist, Svante Arrhenius, was among the first to recognize this fact. In 1890, he proposed that an acid be defined as any substance that, when added to water, increases the hydronium ion concentration. Later, you will learn about another way to define acids that goes beyond aqueous solutions.

In some acids, a single molecule can react to form more than one hydronium ion. This happens when sulfuric acid dissolves in water, as described by the following equations:

$$H_2SO_4(l) + H_2O(l) \longrightarrow H_3O^+(aq) + HSO_4^-(aq)$$

$$HSO_4^-(aq) + H_2O(l) \rightleftharpoons H_3O^+(aq) + SO_4^{2-}(aq)$$

As shown above, sulfuric acid has two ionizable hydrogens. One of them ionizes completely, after which the other ionizes partially as a weak acid.

strong acid

an acid that ionizes completely in a solvent

weak acid

an acid that releases few hydrogen ions in aqueous solution

Figure 4
According to Arrhenius, an acid in aqueous solution ionizes to form hydronium ions, H_3O^+, as shown here for nitric acid, HNO_3.

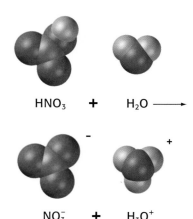

HNO_3 + H_2O $\longrightarrow$

NO_3^- + H_3O^+

Table 2   Some Strong Bases and Some Weak Bases

Strong bases	Weak bases
sodium hydroxide, NaOH	ammonia, NH_3
potassium hydroxide, KOH	sodium carbonate, Na_2CO_3
calcium hydroxide, $Ca(OH)_2$	potassium carbonate, K_2CO_3
barium hydroxide, $Ba(OH)_2$	aniline, $C_6H_5NH_2$
sodium phosphate, Na_3PO_4	trimethylamine, $(CH_3)_3N$

internet connect

www.scilinks.org
Topic: Bases
SciLinks code: HW4020

SciLINKS Maintained by the National Science Teachers Association

Bases

Bases are another class of electrolytes. Unlike acids, which are usually liquids or gases, many common bases are solids. Solutions of bases are slippery to the touch, but touching bases is an unsafe way to identify them. The slippery feel comes about because bases react with oils in your skin, converting them into soaps. This property of attacking oils and greases makes bases useful in cleaning agents, such as those in **Figure 5.**

Some bases, such as magnesium hydroxide, $Mg(OH)_2$, are almost insoluble in water. Other bases, such as potassium hydroxide, are so soluble that they will absorb water vapor from the air and dissolve in the water. A base that is very soluble in water is called an *alkali,* a term that you also know describes the Group 1 metals of the periodic table. These metals react with water to form hydroxides that are water-soluble alkalis. The hydroxide-rich solutions that form when bases dissolve in water are said to be *basic* or *alkaline.*

Just as acids may be strong or weak depending on whether they ionize completely or reach an equilibrium between ionized and un-ionized forms, bases are also classified as strong or weak. **Table 2** lists several bases of each class.

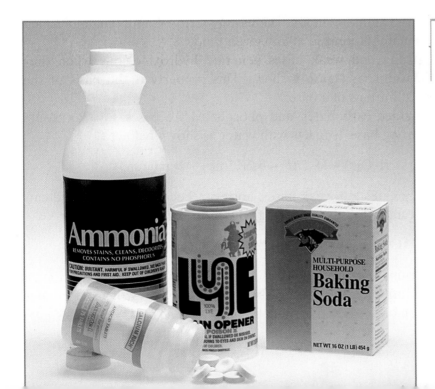

Figure 5
A variety of bases, including ammonia, sodium hydroxide, and sodium bicarbonate, can be found in products used around the home.

Figure 6

a Sodium hydroxide is a strong base. It dissociates completely in aqueous solution.

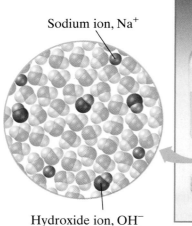

Sodium ion, Na⁺

Hydroxide ion, OH⁻

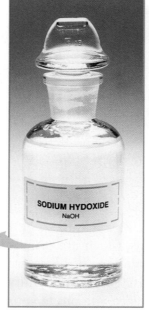

b Only a small portion of the dissolved molecules of ammonia, a weak base, react with water. Most of the ammonia remains as neutral molecules.

Ammonium ion, NH_4^+

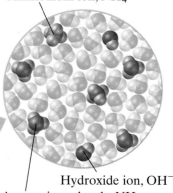

Hydroxide ion, OH⁻
Ammonia molecule, NH_3

strong base

a base that ionizes completely in a solvent

weak base

a base that releases few hydroxide ions in aqueous solution

Bases Generate Hydroxide Ions

The following equation can be used to describe the dissolving of sodium hydroxide, a **strong base,** in water.

$$NaOH(s) \longrightarrow Na^+(aq) + OH^-(aq)$$

The ions dissociate in solution as they become surrounded by water molecules and float away independently.

Ammonia is a typical **weak base.** At room temperature, ammonia is a gas, $NH_3(g)$, but it is very soluble in water, forming $NH_3(aq)$. A few of the ammonia molecules in solution react with water to reach the equilibrium system described by the equation below.

$$NH_3(aq) + H_2O(l) \rightleftharpoons NH_4^+(aq) + OH^-(aq)$$

$NH_4^+(aq)$ is the ammonium ion. The vast majority of ammonia molecules, however, remain un-ionized at any given time.

Both strong and weak bases generate hydroxide ions when they dissolve in water, as **Figure 6** shows. This property is the basis of the Arrhenius definitions of a base.

Many oxides, carbonates, and phosphates are bases, too. Potassium oxide is a strong base. It reacts with water as shown below:

$$K_2O(s) + H_2O(l) \longrightarrow 2K^+(aq) + 2OH^-(aq)$$

Soluble carbonates are weak bases, however, because the dissolved carbonate ion establishes the following equilibrium in water:

$$CO_3^{2-}(aq) + H_2O(l) \rightleftharpoons HCO_3^-(aq) + OH^-(aq)$$

The equilibrium system contains a low concentration of hydroxide ions, a low concentration of hydrogen carbonate ions, $HCO_3^-(aq)$, and a greater concentration of unreacted carbonate ions, $CO_3^{2-}(aq)$.

Quick LAB

Acids and Bases in the Home

Because taste and feel are not safe ways to determine whether a substance is an acid or a base, you should use an indicator to recognize acids and bases. As its name suggests, an *indicator* indicates whether a solution is acidic or basic. It does this by changing color.

PROCEDURE

1. Using a **blender**, grind **red cabbage** with **water**.

2. Strain the liquid into a **large beaker**, and dilute it with water. You now have an indicator.

3. Using your indicator, test **various household products** by adding each product to a separate **small beaker** containing a sample of the indicator. Start with an item that you know is an acid and an item that you know is a base.

ANALYSIS

1. What color is the indicator in acidic solution? in basic solution?

2. Are cleaning products more likely to be acidic or basic?

3. Are food products more likely to be acidic or basic?

Brønsted-Lowry Classification

The definitions of *Arrhenius acid* and *Arrhenius base* given earlier in this book are variants of the definitions of *acid* and *base* originally proposed by Arrhenius in the late 19th century. One drawback that the Arrhenius definitions have is that they are limited to aqueous solutions: HCl, for instance, should be considered an acid whether it is in the form of a pure gas or in aqueous solution. Another limitation is that the Arrhenius definition cannot classify substances that sometimes act as acids and sometimes act as bases.

Brønsted-Lowry Acids Donate Protons

In 1923, the Danish chemist Johannes Brønsted proposed a broader definition of *acid*. Surprisingly, the same year, the British scientist Thomas Lowry happened to make exactly the same proposal independently. Their idea was to apply the name *acid* to any species that can donate a proton. Recall that a proton is a hydrogen atom that has lost its electron; it is a hydrogen ion and can be represented as H^+. Such molecules or ions are now called **Brønsted-Lowry acids.** A reaction showing hydrochloric acid, a representative Brønsted-Lowry acid, is depicted in **Figure 7.**

Brønsted-Lowry acid

a substance that donates a proton to another substance

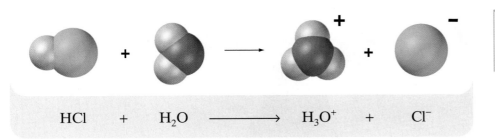

$$HCl + H_2O \longrightarrow H_3O^+ + Cl^-$$

Figure 7
When hydrogen chloride dissolves in water, a proton is transferred from the HCl molecule (leaving behind Cl^-) to an H_2O molecule (forming H_3O^+).

We might also think of the reaction occurring when HCl dissolves in water as proceeding in two steps:

$$HCl \longrightarrow H^+ + Cl^-$$

followed by $H^+ + H_2O \longrightarrow H_3O^+$

The reaction shows HCl acting as an Arrhenius acid, forming a hydronium ion. Because the reaction also involves a proton transfer, hydrochloric acid is also a Brønsted-Lowry acid. All Arrhenius acids are, by definition, also Brønsted-Lowry acids.

Brønsted-Lowry Bases Accept Protons

As you might expect based on their definition of an acid, Brønsted and Lowry defined a base as a proton acceptor. In the reaction taking place in **Figure 8,** ammonia, NH_3, serves as the proton acceptor and is therefore a **Brønsted-Lowry base.** Ammonia also functions as a proton acceptor when it dissolves in water, as in the equation below.

Brønsted-Lowry base

a substance that accepts a proton

$$NH_3(aq) + H_2O(l) \rightleftharpoons NH_4^+(aq) + OH^-(aq)$$

Again, notice that ammonia, a Brønsted-Lowry base, is also an Arrhenius base. All Arrhenius bases are also Brønsted-Lowry bases. Ammonia does not have to react in aqueous solution to be considered a Brønsted-Lowry base. Even as a gas, ammonia accepts a proton from hydrogen chloride, as **Figure 8** shows.

Figure 8
HCl(*g*) and NH$_3$(*g*) that have each escaped from aqueous solution combine to form a cloud of solid ammonium chloride, NH$_4$Cl(*s*)

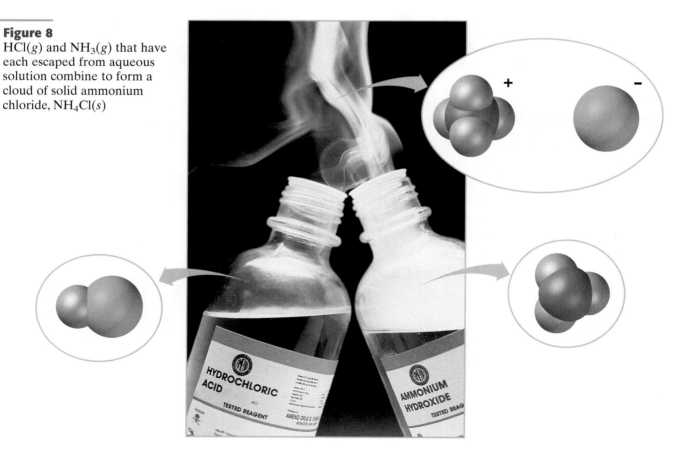

Conjugate Acids and Bases

In the language of Brønsted and Lowry, an acid-base reaction is very simple: one molecule or ion passes a proton to another molecule or ion. Whatever loses the proton is an acid, and whatever accepts the proton is a base.

Look again at the equation for the reversible reaction of ammonia, NH_3, with water:

$$\underset{\text{base}}{NH_3(aq)} + \underset{\text{acid}}{H_2O(l)} \rightleftharpoons NH_4^+(aq) + OH^-(aq)$$

Water donates a proton to ammonia, so it is an acid. Ammonia accepts the proton, so it is a base. The ammonium ion forms when NH_3 receives the proton, and the hydroxide ion forms when H_2O has lost the proton.

Notice that the reaction is reversible, so it can be written with the products as reactants. When the reaction is written that way, we can identify another acid and another base.

$$\underset{\text{acid}}{NH_4^+(aq)} + \underset{\text{base}}{OH^-(aq)} \rightleftharpoons NH_3(aq) + H_2O(l)$$

In this reaction, the ammonium ion donates a proton to the hydroxide ion. NH_4^+ is the acid, and OH^- is the base. The ammonium ion is called the **conjugate acid** of the base, ammonia. The hydroxide ion is called the **conjugate base** of the acid, water.

$$\underset{\text{base}}{NH_3(aq)} + \underset{\text{acid}}{H_2O(l)} \rightleftharpoons \underset{\text{conjugate acid}}{NH_4^+(aq)} + \underset{\text{conjugate base}}{OH^-(aq)}$$

Every Brønsted-Lowry acid has a conjugate base, and every Brønsted-Lowry base has a conjugate acid. **Table 3** lists many such acid-base pairs.

conjugate acid

an acid that forms when a base gains a proton

conjugate base

a base that forms when an acid loses a proton

Table 3 Conjugate Acid-Base Pairs	
Acid	**Conjugate base**
hydrochloric acid, HCl	chloride ion, Cl^-
sulfuric acid, H_2SO_4	hydrogen sulfate ion, HSO_4^-
hydronium ion, H_3O^+	water, H_2O
hydrogen sulfate ion, HSO_4^-	sulfate ion, SO_4^{2-}
hypochlorous acid, $HOCl$	hypochlorite ion, ClO^-
dihydrogen phosphate ion, $H_2PO_4^-$	monohydrogen phosphate ion, HPO_4^{2-}
ammonium ion, NH_4^+	ammonia, NH_3
hydrogen carbonate ion, HCO_3^-	carbonate ion, CO_3^{2-}
water, H_2O	hydroxide ion, OH^-
Conjugate acid	**Base**

Amphoteric Species Are Both Acids and Bases

Did you notice that some species appear in both the "Acid" and the "Base" columns in **Table 3?** Several species can both donate and accept protons. Such species are described as **amphoteric.** The hydrogen carbonate ion, HCO_3^-, is amphoteric in aqueous solution, for example. It can act as an acid by donating a proton to an ammonia molecule, as represented by the following equation:

$$\underset{\text{acid}}{HCO_3^-(aq)} + \underset{\text{base}}{NH_3(aq)} \rightleftharpoons CO_3^{2-}(aq) + NH_4^+(aq)$$

amphoteric

describes a substance, such as water, that has the properties of an acid and the properties of a base

Alternatively, the hydrogen carbonate ion can act as a base when, for example, it accepts a proton from a hydronium ion in the reaction described by the equation below.

$$\underset{\text{base}}{HCO_3^-(aq)} + \underset{\text{acid}}{H_3O^+(aq)} \rightleftharpoons H_2CO_3(aq) + H_2O(l)$$

Water itself is amphoteric. A water molecule can donate a proton and become a hydroxide ion in the process. Or it can accept a proton to become a hydronium ion. The consequences of this fact are described in the next section.

1 Section Review

UNDERSTANDING KEY IDEAS

1. List the observable properties of an acid.

2. How did Arrhenius define a base?

3. Giving examples, explain how strong acids and weak acids differ.

4. How does the Brønsted-Lowry definition of an acid differ from the Arrhenius definition of an acid?

5. Write an equation that demonstrates the properties of acids and bases, as defined by Brønsted and Lowry.

6. Define a conjugate acid-base pair, and give an example.

7. Show chemical equations for the reaction of water with (a) an acid of your choosing and (b) a base of your choosing.

CRITICAL THINKING

8. Could a Brønsted-Lowry acid *not* be an Arrhenius acid? Explain.

9. How would $[OH^-]$ in an ammonia solution compare with $[OH^-]$ in a sodium hydroxide solution of similar concentration?

10. Write the formulas of the conjugate acid and the conjugate base of the $H_2PO_4^-$ ion.

11. Write an equation describing a proton transfer between $H_2SO_4(aq)$ and $SO_4^{2-}(aq)$.

12. Why can magnesium hydroxide be described as a strong base even though it is only slightly soluble in water?

13. Identify two acids and their conjugate bases in the following reaction.

$$H_2SO_4(aq) + SO_3^{2-}(aq) \rightleftharpoons$$
$$HSO_4^-(aq) + HSO_3^-(aq)$$

Acidity, Basicity, and pH

KEY TERMS

- **self-ionization constant of water, K_w**
- **neutral**
- **pH**
- **indicator**

OBJECTIVES

1. **Use** K_w in calculations.

2. **Explain** the relationship between pH and H_3O^+ concentration.

3. **Perform** calculations using pH, $[H_3O^+]$, $[OH^-]$, and K_w.

4. **Describe** two methods of measuring pH.

The Self-Ionization of Water

You have just learned that water is both an acid and a base. This means that a water molecule can either give or receive a proton. So what happens when one molecule of water donates a proton to another molecule of water? The reaction is described by the equation below.

$$H_2O(l) + H_2O(l) \rightleftharpoons H_3O^+(aq) + OH^-(aq)$$
$$\text{base} \qquad \text{acid}$$

As also shown in **Figure 9,** a pair of water molecules are in equilibrium with two ions—a hydronium ion and a hydroxide ion—in a reaction known as the *self-ionization of water.*

Thus, even pure water contains ions. The chemical equation shows that the two ions are produced in equal numbers. Therefore, in pure water, the two ions must share the same concentration. Experiments show that this concentration is 1.00×10^{-7} M at 25°C.

$$[H_3O^+] = [OH^-] = 1.00 \times 10^{-7} \text{ M}$$

Figure 9
In water and aqueous solutions, an equilibrium exists between H_2O and the two ions shown.

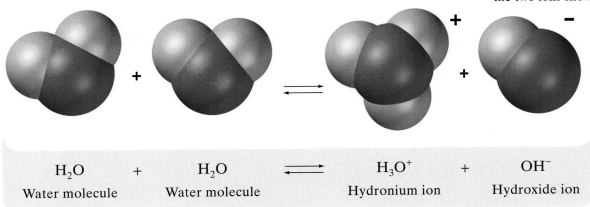

| H_2O | + | H_2O | $\rightleftharpoons$ | H_3O^+ | + | OH^- |
| Water molecule | | Water molecule | | Hydronium ion | | Hydroxide ion |

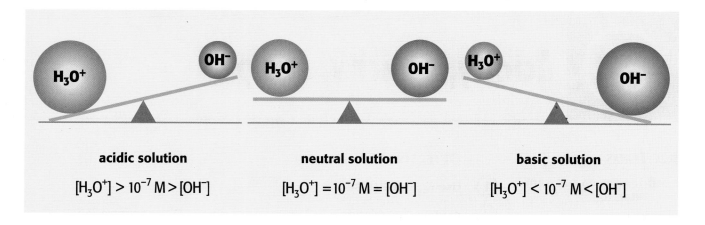

acidic solution

$$[H_3O^+] > 10^{-7}\,M > [OH^-]$$

neutral solution

$$[H_3O^+] = 10^{-7}\,M = [OH^-]$$

basic solution

$$[H_3O^+] < 10^{-7}\,M < [OH^-]$$

Figure 10
When the concentration of H_3O^+ goes up, the concentration of OH^- goes down, and vice versa.

The Self-Ionization Constant of Water

The equilibrium between water and the ions it forms is described by the following equation:

$$2H_2O(l) \rightleftharpoons H_3O^+(aq) + OH^-(aq)$$

Recall that an equilibrium-constant expression relates the concentrations of species involved in an equilibrium. The relationship for the water equilibrium is simply $[H_3O^+][OH^-] = K_{eq}$. This equilibrium constant, called the **self-ionization constant of water,** is so important that it has a special symbol, K_w. Its value can be found from the known concentrations of the hydronium and hydroxide ions in pure water, as follows:

the self-ionization constant of water, K_w

the product of the concentrations of the two ions that are in equilibrium with water; $[H_3O^+][OH^-]$

$$[H_3O^+][OH^-] = K_w = (1.00 \times 10^{-7})(1.00 \times 10^{-7}) = 1.00 \times 10^{-14}$$

The product of these two ion concentrations is always a constant. Thus, anything that increases one of the ion concentrations decreases the other, as elaborated in **Table 4** and illustrated in **Figure 10**. Likewise, if you know one of the ion concentrations, you can calculate the other. The concentration of hydronium ions in a solution expresses its *acidity*. The concentration of hydroxide ions in a solution expresses its *basicity*.

Topic Link

Refer to the "Chemical Equilibrium" chapter for a discussion of equilibrium constants.

Table 4   Concentrations and K_w

Solution	$[H_3O^+]$ (M)	$[OH^-]$ (M)	$K_w = [H_3O^+][OH^-]$
Pure water	1.0×10^{-7}	1.0×10^{-7}	1.0×10^{-14}
0.10 M strong acid	1.0×10^{-1}	1.0×10^{-13}	1.0×10^{-14}
0.010 M strong acid	1.0×10^{-2}	1.0×10^{-12}	1.0×10^{-14}
0.10 M strong base	1.0×10^{-13}	1.0×10^{-1}	1.0×10^{-14}
0.010 M strong base	1.0×10^{-12}	1.0×10^{-2}	1.0×10^{-14}
0.025 M strong acid	2.5×10^{-2}	4.0×10^{-13}	1.0×10^{-14}
0.025 M strong base	4.0×10^{-13}	2.5×10^{-2}	1.0×10^{-14}

Determining [OH$^-$] or [H$_3$O$^+$] Using K_w

What is [OH$^-$] in a 3.00×10^{-5} M solution of HCl?

1 Gather information.

Because HCl is a strong acid, all HCl in an aqueous solution ionizes according to the equation below.

$$HCl(g) + H_2O(l) \longrightarrow H_3O^+(aq) + Cl^-(aq)$$

Therefore, a 3.00×10^{-5} M solution of HCl has

$$[H_3O^+] = 3.00 \times 10^{-5} \text{ M}.$$

The self-ionization constant of water is

$$K_w = [H_3O^+][OH^-] = 1.00 \times 10^{-14}.$$

2 Plan your work.

$$K_w = 1.00 \times 10^{-14} = [H_3O^+][OH^-] = (3.00 \times 10^{-5})[OH^-]$$

Values of [H$_3$O$^+$] and of [H$_3$O$^+$][OH$^-$], K_w, are known. Therefore, [OH$^-$] can be found by division.

3 Calculate.

$$[OH^-] = \frac{K_w}{[H_3O^+]} = \frac{1.00 \times 10^{-14}}{3.00 \times 10^{-5}} = 3.33 \times 10^{-10} \text{ M}$$

4 Verify your results.

Multiplying the values for [H$_3$O$^+$] and [OH$^-$] gives the known K_w and confirms that the concentration of hydroxide ion in the solution is 3.33×10^{-10} M.

PRACTICE HINT

Remember that [H$_3$O$^+$] is equivalent to the concentration of the acid itself only if the acid is a strong acid. The same is true for [OH$^-$] and strong bases.

PRACTICE

1 Calculate the hydronium ion concentration in an aqueous solution that has a hydroxide ion concentration of 7.24×10^{-4} M.

2 What is [OH$^-$] in a 0.450 M solution of HNO$_3$?

3 What is [H$_3$O$^+$] in a solution of NaOH whose concentration is 3.75×10^{-2} M?

4 Calculate the hydroxide ion concentration of a 0.200 M solution of HClO$_4$.

5 If 1.2×10^{-4} moles of magnesium hydroxide, Mg(OH)$_2$, are dissolved in 1.0 L of aqueous solution, what are [OH$^-$] and [H$_3$O$^+$]?

PROBLEM SOLVING SKILL

The Meaning of pH

You have probably seen commercials in which products, such as that pictured in **Figure 11** on the next page, are described as "pH balanced." Perhaps you know that pH has to do with how basic or acidic something is. You may have learned that the pH of pure water is 7 and that acid rain has a lower pH. But what does pH actually mean?

pH and Acidity

neutral

describes an aqueous solution that contains equal concentrations of hydronium ions and hydroxide ions

When acidity and basicity are exactly balanced such that the numbers of H_3O^+ and OH^- ions are equal, we say that the solution is **neutral.** Pure water is neutral because it contains equal amounts of the two ions.

Two of the solutions listed in **Table 5** are neutral: both have a hydronium ion concentration of 1.00×10^{-7} M. The other solutions are either acidic or basic, depending on whether a strong acid or a strong base was dissolved in water. The solution listed last in **Table 5** was made by dissolving 0.100 mol of NaOH in 1.00 L of water, so it has a hydroxide ion concentration of 0.100 M. Its hydronium ion concentration can be calculated using K_w, as shown below.

$$[H_3O^+] = \frac{[H_3O^+][OH^-]}{[OH^-]} = \frac{1.00 \times 10^{-14}}{0.100} = 1.00 \times 10^{-13}$$

Notice that the hydronium ion concentrations in the listed solutions span a very wide range—in fact a trillionfold range. You can see that working with $[H_3O^+]$ can involve awkward negative exponents. In part to avoid this inconvenience, scientists adopted the suggestion, made by the Danish chemist Søren Sørensen in 1909, to focus not on the value of $[H_3O^+]$ but on the power of 10 that arises when $[H_3O^+]$ is expressed in scientific notation. Sørensen proposed using the negative of the power of 10 (that is, the negative logarithm) of $[H_3O^+]$ as the index of basicity and acidity. He called this measure the **pH.** The letters *p* and *H* represent **p**ower of **h**ydrogen. Keep in mind that because pH is a negative logarithmic scale, a *lower* pH reflects a *higher* hydronium ion concentration.

pH

a value used to express the acidity or alkalinity of a solution; it is defined as the logarithm of the reciprocal of the concentration of hydronium ions; a pH of 7 is neutral, a pH of less than 7 is acidic, and a pH of greater than 7 is basic

Table 5 pH Values at Specified $[H_3O^+]$		
Solution	**$[H_3O^+]$ (M)**	**pH**
1.00 L of H_2O	1.00×10^{-7}	7.00
0.100 mol HCl in 1.00 L of H_2O	1.00×10^{-1}	1.00
0.0100 mol HCl in 1.00 L of H_2O	1.00×10^{-2}	2.00
0.100 mol NaCl in 1.00 L of H_2O	1.00×10^{-7}	7.00
0.0100 mol NaOH in 1.00 L of H_2O	1.00×10^{-12}	12.00
0.100 mol NaOH in 1.00 L of H_2O	1.00×10^{-13}	13.00

Calculating pH from [H₃O⁺]

Based on Sørensen's definition, pH can be calculated by the following mathematical equation:

$$pH = -\log\,[H_3O^+]$$

Because of the negative sign, as the hydronium ion concentration increases, the pH will *decrease*. A solution of pH 0 is very acidic. A solution of pH 14 is very alkaline. A solution of pH 7 is neutral.

The equation above may be rearranged to calculate the hydronium ion concentration from the pH. In that form, the equation is as follows:

$$[H_3O^+] = 10^{-pH}$$

When the pH is a whole number, you can do this calculation in your head. For example, if a solution has a pH of 3, its $[H_3O^+]$ is 10^{-3} M, or 0.001 M.

Because pH is related to powers of 10, a change in one pH unit corresponds to a tenfold change in the concentrations of the hydroxide and hydronium ions. Therefore, a solution whose pH is 2.0 has a $[H_3O^+]$ that is ten times greater than a solution whose pH is 3.0 and 100 times greater than a solution whose pH is 4.0.

Figure 11
Soap manufacturers sometimes make claims about the pH of their products.

SKILLS Toolkit

Using Logarithms in pH Calculations

It is easy to find the pH or the $[H_3O^+]$ of a solution by using a scientific calculator. Because calculators differ, check your manual to find out which keys are used for log and antilog functions and how to use these functions.

1. Calculating pH from [H₃O⁺]
(see Sample Problem B for an example)
Use the definition of pH:

$$pH = -\log\,[H_3O^+]$$

- Take the logarithm of the hydronium ion concentration.
- Change the sign (+/−).
- The result is the pH.

2. Calculating [H₃O⁺] from pH
(see Sample Problem C for an example)
If you rearrange $pH = -\log\,[H_3O^+]$ to solve for $[H_3O^+]$, the equation becomes

$$[H_3O^+] = 10^{-pH}$$

- Change the sign of the pH (+/−)
- Raise 10 to the negative pH power (take the antilog).
- The result is $[H_3O^+]$.

Calculating pH for an Acidic or Basic Solution

What is the pH of (a) a 0.000 10 M solution of HNO_3, a strong acid, and (b) a 0.0136 M solution of KOH, a strong base?

1 Gather information.

(a) Concentration of HNO_3 solution = 0.000 10 M = 1.0×10^{-4} M; pH = ?
(b) Concentration of KOH solution = 0.0136 M = 1.36×10^{-2} M; pH = ?
$K_w = 1.00 \times 10^{-14}$

2 Plan your work.

Because HNO_3 and KOH are strong electrolytes, their aqueous solutions are completely ionized.
(a) Therefore, for HNO_3 solution, $[H_3O^+] = 1.0 \times 10^{-4}$ M.
(b) Therefore, for KOH solution, $[OH^-] = 1.36 \times 10^{-2}$ M.
The equation relating pH to $[H_3O^+]$ is pH = $-\log [H_3O^+]$. This equation alone is adequate for (a). For (b), you must first use K_w to calculate $[H_3O^+]$ from $[OH^-]$.

3 Plan your work.

Using a scientific calculator and following the instructions under item 1 of **Skills Toolkit 1,** one calculates for (a)

$$pH = -\log [H_3O^+] = -\log (1.0 \times 10^{-4}) = -(-4.00) = 4.00$$

For (b), $[H_3O^+] = \dfrac{K_w}{[OH^-]} = \dfrac{1.00 \times 10^{-14}}{1.36 \times 10^{-2}} = 7.35 \times 10^{-13}$ and then

$$pH = -\log [H_3O^+] = -\log (7.35 \times 10^{-13}) = -(-12.13) = 12.13.$$

4 Verify your results.

Because the solution in (a) is acidic, a pH between 0 and 7 is expected, so the calculated value of 4.00 is reasonable. The solution in (b) will be basic; therefore, a pH between 7 and 14 is expected. Therefore, the answer pH = 12.13 is reasonable.

PRACTICE HINT

Because it is easy to make calculator errors when dealing with logarithms, always check to see that answers to pH problems are reasonable. For example, negative pH values and pH values greater than 14 are unlikely.

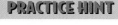

PRACTICE

1 Calculate pH if $[H_3O^+] = 5.0 \times 10^{-3}$ M.

2 What is the pH of a 0.2 M solution of a strong acid?

3 Calculate pH if $[OH^-] = 2.0 \times 10^{-3}$ M.

4 What is the pH of a solution that contains 0.35 mol/L of the strong base NaOH?

Calculating [H₃O⁺] and [OH⁻] from pH

What are the concentrations of the hydronium and hydroxide ions in a sample of rain that has a pH of 5.05?

1 Gather information.

pH = 5.05
$K_w = 1.00 \times 10^{-14}$
$[H_3O^+] = ?$
$[OH^-] = ?$

2 Plan your work.

Because you know the rain's pH, the equation $[H_3O^+] = 10^{-pH}$ can be used to find the hydronium ion concentration.
Then the equation $[H_3O^+][OH^-] = K_w$ can be rearranged and used to find $[OH^-]$.

PRACTICE HINT

For further verification, you could recalculate the pH from the found concentrations by using the methods in Sample Problems B and C.

3 Calculate.

Using a scientific calculator and following the instruction under item 2 of **Skills Toolkit 1,** one finds

$$[H_3O^+] = 10^{-pH} = 10^{-5.05} = 8.9 \times 10^{-6} \text{ M.}$$

$$\text{Next, } [OH^-] = \frac{K_w}{[H_3O^+]} = \frac{1.00 \times 10^{-14}}{8.9 \times 10^{-6}} = 1.1 \times 10^{-9} \text{ M.}$$

The concentrations of the hydronium and hydroxide ions are 8.9×10^{-6} M and 1.1×10^{-9} M, respectively.

4 Verify your results.

The rain's pH is mildly acidic, so the hydronium ion concentration should be more than 1.0×10^{-7} M, and the hydroxide concentration should be less than 1.0×10^{-7} M. Thus, the answers are reasonable.

PRACTICE

1 What is the hydronium ion concentration in a fruit juice that has a pH of 3.3?

2 A commercial window-cleaning liquid has a pH of 11.7. What is the hydroxide ion concentration?

3 If the pH of a solution is 8.1, what is $[H_3O^+]$ in the solution? What is $[OH^-]$ in the solution?

4 Normal human blood has a hydroxide ion concentration that ranges from 1.7×10^{-7} M to 3.5×10^{-7} M, but diabetics often have readings outside this range. A patient's blood has a pH of 7.67. Is there cause for concern?

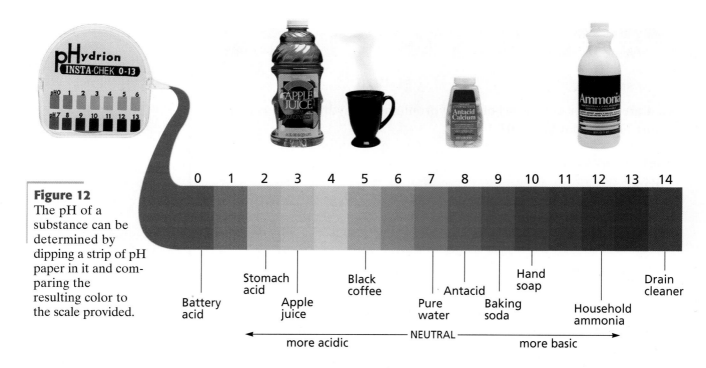

Figure 12
The pH of a substance can be determined by dipping a strip of pH paper in it and comparing the resulting color to the scale provided.

pHydrion
INSTA-CHEK 0-13

| 0 | 1 | 2 | 3 | 4 | 5 | 6 | 7 | 8 | 9 | 10 | 11 | 12 | 13 | 14 |

Battery acid
Stomach acid
Apple juice
Black coffee
Pure water
Antacid
Baking soda
Hand soap
Household ammonia
Drain cleaner

← more acidic — NEUTRAL — more basic →

Measuring pH

Measuring pH is an operation that is carried out frequently, for a variety of reasons, in chemical laboratories. There are two ways to measure pH. The first method, which uses indicators, is quick and convenient but does not give very precise results. The second method, which uses a pH meter, is very precise but is more complicated and expensive.

Indicators

indicator

a compound that can reversibly change color depending on the pH of the solution or other chemical change

Certain dyes, known as **indicators,** turn different colors in solutions of different pH. The pH paper pictured in **Figure 12** contains a variety of indicators and can develop a rainbow of colors, each of which corresponds to a particular pH value.

Thymol blue is an example of an indicator. It is yellow in solutions whose pH is between 3 and 8 but blue in solutions whose pH is 10 or higher. **Figure 13** shows the structure of the organic ion responsible for the yellow color—it is a weak acid. The blue form is the conjugate base.

Figure 13
The indicator thymol blue is yellow in neutral and acidic solutions. As $[OH^-]$ in the solution increases, the indicator turns blue.

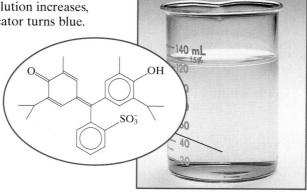

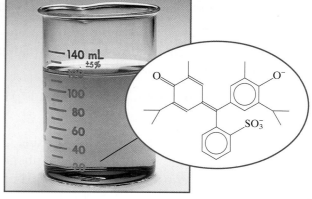

pH 4   pH 11

Dozens of indicator dyes are available. Some indicators, such as litmus, are natural products, but most are synthetic. Each indicator has its own colors and its individual range of pH over which it changes shade. By suitably blending several indicators, chemists have prepared "universal indicators," which turn different colors throughout the entire pH range. One such universal indicator is incorporated into the "pH paper" shown in **Figure 12.** By matching the color the paper develops to a standard chart, one can easily measure the solution's approximate pH.

pH Meters

The pH of a solution is being measured by a pH meter in **Figure 14.** A pH meter is an electronic instrument equipped with a probe that can be dipped into a solution of unknown pH. The probe has two electrodes, one of which is sensitive to the hydronium ion. An electrical voltage develops between the two electrodes, and the circuitry measures this voltage very precisely. The instrument converts the measurement into a pH reading, which is displayed on the meter's screen.

After calibration with standard solutions of known pH, a pH meter can measure pH with a precision of 0.01 pH units, which is much greater that the precision of measurements with indicators.

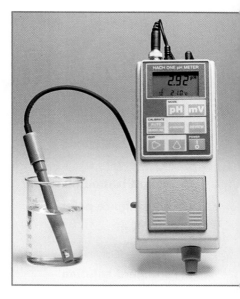

Figure 14
A pH meter is an electrochemical instrument that can measure pH accurately.

 Section Review

UNDERSTANDING KEY IDEAS

1. Describe the relationship between hydronium and hydroxide ion concentrations in an aqueous solution.

2. What does pH measure? How is pH defined?

3. What is a *neutral* solution? What is its pH?

4. Write equations linking the terms K_w, pH, $[H_3O^+]$, and $[OH^-]$.

5. The pH of pancreatic juice is 7.9. Is pancreatic juice acidic or basic?

6. What methods are used to measure pH? Briefly describe how each method works.

PRACTICE PROBLEMS

7. The hydronium ion concentration in a solution is 3.16×10^{-3} mol/L. What is $[OH^-]$? What is the pH?

8. The pH of vinegar is 2.9. Calculate the concentrations of H_3O^+ and OH.

9. If 5.3 g of the strong base NaOH is dissolved in water to form 1500 mL of solution, what are the pH, $[H_3O^+]$, and $[OH^-]$?

10. The $[OH^-]$ of a fruit juice is 3.2×10^{-11} M. What is the pH?

11. What amount in moles of a strong acid such as HBr must be dissolved in 1.00 L of water to prepare a solution whose pH is 2.00?

12. What volume of solution is needed to dissolve 1.0 mol of a strong base such as KOH to make a solution whose pH is 12.5?

CRITICAL THINKING

13. Why is "deionized water" *not* an entirely accurate description of pure water?

14. Can pH be negative? Why or why not?

15. Why would pH paper be unsuitable for measuring blood pH?

SECTION

3 Neutralization and Titrations

KEY TERMS

- **neutralization reaction**
- **equivalence point**
- **titration**
- **titrant**
- **standard solution**
- **transition range**
- **end point**

OBJECTIVES

① **Predict** the product of an acid-base reaction.

② **Describe** the conditions at the equivalence point in a titration.

③ **Explain** how you would select an indicator for an acid-base titration.

④ **Describe** the procedure for carrying out a titration to determine the concentration of an acid or base solution.

Neutralization

The solution of strong acid in the beaker on the left in **Figure 15** contains a high H_3O^+ concentration: high enough to react with and dissolve metals. The solution of strong base on the right is concentrated enough in OH^- to free a grease-clogged drain. Yet when these acidic and basic solutions are mixed in equal amounts, the solution formed has little effect on metal or grease. What has occurred? Because the relationship $[H_3O^+][OH^-] = 1.0 \times 10^{-14}$ must always be true, high concentrations of $H_3O^+(aq)$ and $OH^-(aq)$ cannot coexist. Most of these ions have reacted with each other in a process known as a **neutralization reaction.**

neutralization reaction

the reaction of the ions that characterize acids (hydronium ions) and the ions that characterize bases (hydroxide ions) to form water molecules and a salt

Figure 15
a This beaker contains a solution of nitric acid, a strong acid. This solution turns pH paper red.

b This beaker contains a solution of sodium hydroxide, a strong base. This solution turns pH paper blue.

c The neutralization reaction produces a sodium nitrate solution, which has a neutral pH of 7.

All Neutralizations Are the Same Reaction

When solutions of a strong acid and a strong base, having exactly equal amounts of $H_3O^+(aq)$ and $OH^-(aq)$ ions, are mixed, almost all of the hydronium and hydroxide ions react to form water. The reaction is described by the equation below.

$$H_3O^+(aq) + OH^-(aq) \longrightarrow 2H_2O(l)$$

This same reaction happens regardless of the identities of the strong acid and strong base.

Suppose, as in **Figure 16,** that the acid was hydrochloric acid, HCl, and the base was sodium hydroxide, NaOH. When these solutions are mixed, the result will be a solution of only water and the spectator ions sodium and chloride. This is just a solution of sodium chloride. You can prepare the same solution by dissolving common salt, NaCl(s), in water.

You may sometimes see this reaction described as follows:

$$HCl + NaOH \longrightarrow NaCl + H_2O$$

Arrhenius might have said "an acid plus a base produces a salt plus water." This representation can be misleading because the only reactants are $H_3O^+(aq)$ and $OH^-(aq)$ ions and the only product is H_2O.

Figure 16
After hydrochloric acid neutralizes a solution of sodium hydroxide, the only solutes remaining are $Na^+(aq)$ and $Cl^-(aq)$. When the water is evaporated, a small amount of sodium chloride crystals, which will be just like the ones shown, will be left.

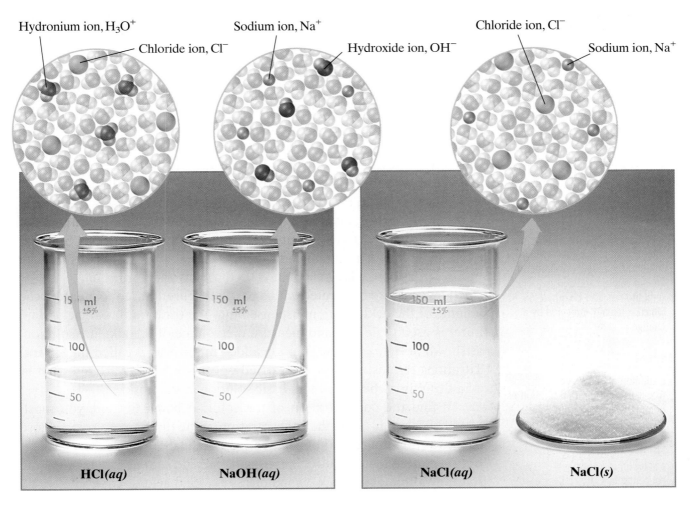

Hydronium ion, H_3O^+

Chloride ion, Cl^-

Sodium ion, Na^+

Hydroxide ion, OH^-

Chloride ion, Cl^-

Sodium ion, Na^+

HCl(aq) NaOH(aq) NaCl(aq) NaCl(s)

Figure 17

a A titration is done by using a buret, as shown here, to deliver a measured volume of titrant into a solution of unknown concentration.

b When reading the liquid level in a buret, you must read the level at the bottom of the meniscus. Here, the reading is 0.42 mL.

equivalence point

the point at which the two solutions used in a titration are present in chemically equivalent amounts

titration

a method to determine the concentration of a substance in solution by adding a solution of known volume and concentration until the reaction is completed, which is usually indicated by a change in color

titrant

a solution of known concentration that is used to titrate a solution of unknown concentration

standard solution

a solution of known concentration

Titrations

If an acidic solution is added gradually to a basic solution, at some point the neutralization reaction ends because the hydroxide ions become used up. Likewise, if a basic solution is added to an acid, eventually all of the hydronium ions will be used up. The point at which a neutralization reaction is complete is known as the **equivalence point.**

When a solution of a strong base is added to a solution of a strong acid, the equivalence point occurs when the amount of added hydroxide ions equals the amount of hydronium ions originally present. As you have learned, at 25°C this is the point at which both $H_3O^+(aq)$ and $OH^-(aq)$ ions have concentrations of 1.0×10^{-7} M, and the pH is 7.

The gradual addition of one solution to another to reach an equivalence point is called a **titration.** The purpose of a titration is to determine the concentration of an acid or a base. In addition to the two solutions, the equipment needed to carry out a titration usually includes two burets, a titration flask, and a suitable indicator. **Skills Toolkit 2,** later in this section, will describe how this equipment is used to perform a titration.

If an acid is to be titrated with a base, one buret is used to measure the volume of the acid solution dispensed into the titration flask. The second buret is used to deliver and measure the volume of the alkaline solution, as shown in **Figure 17.** The solution added in this way is called the **titrant.** Titrations can just as easily be carried out the other way around. That is, acid titrant may be added to a basic solution in the flask.

To find the concentration of the solution being titrated, you must, of course, already know the concentration of the titrant. A solution whose concentration is already known is called a **standard solution.** The concentration of a standard solution has usually been determined by reacting the solution with a precisely weighed mass of a solid acid or base.

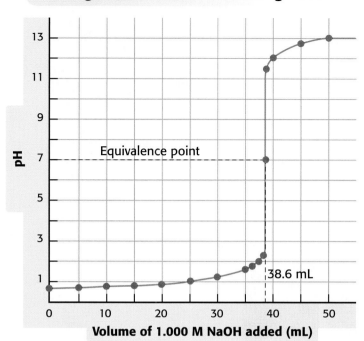

Strong Acid Titrated with Strong Base

Figure 18
This graph of pH versus the volume of 1.000 M NaOH added to an HCl solution indicates that the equivalence point occurred after 38.6 mL of titrant was added.

(Graph: y-axis labeled "pH" with values 1, 3, 5, 7, 9, 11, 13; x-axis labeled "Volume of 1.000 M NaOH added (mL)" with values 0, 10, 20, 30, 40, 50. Horizontal dashed line at pH 7 labeled "Equivalence point." Vertical line at 38.6 mL labeled "38.6 mL.")

The Equivalence Point

As titrant is added to the flask containing the solution of unknown concentration, pH is measured. A distinctively shaped graph, called a *titration curve*, results when pH is plotted against titrant volume. **Figure 18** shows a typical example. Because the curve is steep at the equivalence point, it is easy to locate the exact volume that corresponds to a pH of 7.00.

A titration is exact only if the equivalence point can be accurately detected. A pH meter can be used to monitor the pH during the titration, and indicators are also commonly used to detect the equivalence point.

Carrying Out a Titration

Skills Toolkit 2, on the next two pages, has step-by-step instructions to help you carry out an acid-base titration. Study and understand all of the steps *before* you start to perform a titration experiment. If your attention alternates between book and buret, you're likely to make mistakes. Experience helps, and your second titration should be much better than your first.

With each addition of titrant, the indicator will begin to change color but then will go back to its original color as you swirl the flask. The color will fade ever more slowly as the end point gets near. Immediately slow down to a drop-by-drop flow rate. Otherwise, you may miss the end point.

If you do miss the end point by adding too much titrant, however, you do not have to start all over. You can "back-titrate" by adding more unknown solution to the flask until the indicator turns back to its original color. Measure the volume of unknown solution that you added, then slowly add titrant again until the equivalence point is reached. You can then use the total volumes of unknown and titrant in your calculations.

Performing a Titration

The following procedure is used to determine the unknown concentration of an acid solution by titrating the solution with a standardized base solution.

Decide which buret will be used for the acid and which will be used for the base. Label each buret to avoid confusion. Rinse the acid buret three times with the acid to be used in the titration. Use the base solution to rinse the other buret in a similar manner.

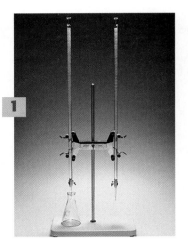

1

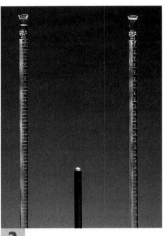

2 Fill the acid buret with the acid solution to a point above the 0 mL mark.

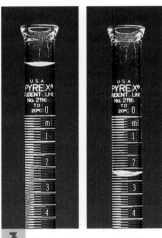

3 Release some acid into a waste flask to lower the volume into the calibrated portion of the buret.

4 Record the volume of the acid in the buret to the nearest 0.01 mL as your starting point.

5 Release a volume of acid (determined by your lab procedure) into a clean Erlenmeyer flask.

6 Record the new volume reading, and subtract the starting volume to find the volume of acid added.

7 Add three drops of an appropriate indicator (phenolphthalein in this case) to the flask.

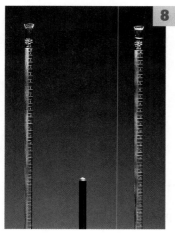

8

9

Fill the other buret with standardized base solution to a point above the 0 mL mark. Record the concentration of the standardized solution.

Release some base from the buret into a waste flask so that the top of the liquid is in the calibrated portion of the buret.

10 Record the volume of the base to the nearest 0.01 mL as your starting point.

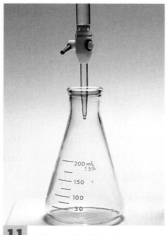

11 Place the flask containing the acid under the base buret. Notice that the tip of the buret extends into the mouth of the flask.

12 Slowly release base from the buret into the flask while constantly swirling the flask. The pink color should fade with swirling.

13 Near the end point, add base drop by drop.

14 The end point is reached when a very light pink color remains after 30 s of swirling.

15 Record the new volume, and determine the volume of base added.

pH 5   pH 7   pH 9
bromthymol blue

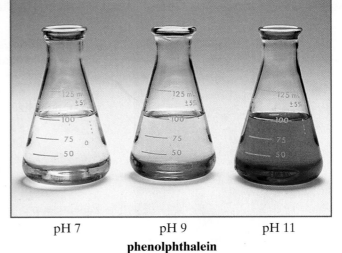

pH 7   pH 9   pH 11
phenolphthalein

Figure 19
Bromthymol blue changes color between a pH of 6.0 and 7.6, as in the neutralization of a strong acid and a strong base. Phenolphthalein changes color between a pH of 8.0 and 9.6, as in the neutralization of a weak acid with a strong base.

transition range

the pH range through which an indicator changes color

end point

the point in a titration at which a marked color change takes place

Selecting a Suitable Indicator

All indicators have a **transition range.** In this range, the indicator is partly in its acidic form and partly in its basic form. Thus, the indicator's color is intermediate between those of its acid and base colors. **Figure 19** illustrates the transition range for two typical indicators, bromthymol blue and phenolphthalein.

The instant at which the indicator changes color is the **end point** of the titration. If an appropriate indicator is chosen, the end point and the equivalence point will be the same. In order to determine the concentration of the titrated solution, you must determine the titrant volume at which the indicator changes color.

In titrations of a strong acid by a strong base, the equivalence point occurs at pH 7, and bromthymol blue would be an appropriate indicator, as **Table 6** confirms. However, when a weak acid is titrated by a strong base, the equivalence point is at a pH greater than 7 and thymol blue or phenolphthalein would be a better choice. On the other hand, methyl orange could be the best choice if your titration uses a weak base and a strong acid, because the equivalence point might be at pH 4.

Table 6	**Transition Ranges of Some Indicators**		
Indicator name	**Acid color**	**Transition range (pH)**	**Base color**
Thymol blue	red	1.2–2.8	yellow
Methyl orange	red	3.1–4.4	orange
Litmus	red	5.0–8.0	blue
Bromthymol blue	yellow	6.0–7.6	blue
Thymol blue	yellow	8.0–9.6	blue
Phenolphthalein	colorless	8.0–9.6	red
Alizarin yellow	yellow	10.1–12.0	red

Titration Calculations: From Volume to Amount in Moles

The goal of a titration is to determine either the original concentration of the solution in the titration flask or the original amount of acid or base.

Recall the simple equation, given below, that relates the amount n (in moles) of a solute to the concentration and volume.

$$n = cV$$

Here c is the concentration (in moles per liter) and V is the volume (in liters) of the solution. At the equivalence point in a titration of a strong acid by a strong base, the amount of hydroxide ion added equals the initial amount of hydronium ion. This relationship may be represented as $n_{H_3O^+} = n_{OH^-}$. If each of these amounts is replaced by the corresponding product of concentration and volume, the following equation is the result.

$$(c_{H_3O^+})(V_{H_3O^+}) = (c_{OH^-})(V_{OH^-})$$

This relationship is the one that you will need for most titration calculations. The equation applies whether the titrant is an acid or a base.

SAMPLE PROBLEM D

Calculating Concentration from Titration Data

A student titrates 40.00 mL of an HCl solution of unknown concentration with a 0.5500 M NaOH solution. The volume of base solution needed to reach the equivalence point is 24.64 mL. What is the concentration of the HCl solution in moles per liter?

1 Gather information.

$V_{H_3O^+} = 40.00 \text{ mL} = 0.040\ 00 \text{ L}$ $V_{OH^-} = 24.64 \text{ mL} = 0.024\ 64 \text{ L}$

$c_{OH^-} = 0.5500 \text{ mol/L}$ $c_{H_3O^+} = ?$

2 Plan your work.

The general equation $(c_{H_3O^+})(V_{H_3O^+}) = (c_{OH^-})(V_{OH^-})$ can be rearranged into the following equation:

$$c_{H_3O^+} = \frac{(c_{OH^-})(V_{OH^-})}{V_{H_3O^+}}$$

3 Calculate.

$$c_{H_3O^+} = \frac{(c_{OH^-})(V_{OH^-})}{V_{H_3O^+}} = \frac{(0.5500 \text{ mol/L})(0.024\ 64 \text{ L})}{0.040\ 00 \text{ L}} = 0.3388 \text{ mol/L}$$

4 Verify your results.

Amounts of hydronium and hydroxide ions should be the same.

$n_{H_3O^+} = c_{H_3O^+}V_{H_3O^+} = (0.3388 \text{ mol/L})(0.040\ 00 \text{ L}) = 0.013\ 55 \text{ mol}$

$n_{OH^-} = c_{OH^-}V_{OH^-} = (0.5500 \text{ mol/L})(0.024\ 64 \text{ L}) = 0.013\ 55 \text{ mol}$

Practice problems on next page

PRACTICE HINT

If you get confused, remember to keep track of amounts of acid and base. The rest is just a matter of converting molarity to moles by multiplying molarity by volume.

1. If 20.6 mL of 0.010 M aqueous HCl is required to titrate 30.0 mL of an aqueous solution of NaOH to the equivalence point, what is the molarity of the NaOH solution?

2. In the titration of 35.0 mL of drain cleaner that contains NaOH, 50.08 mL of 0.409 M HCl must be added to reach the equivalence point. What is the concentration of the base in the cleaner?

3. Titrating a sludge sample of unknown origin required 41.55 mL of 0.1125 M NaOH. How many moles of H_3O^+ did the sample contain?

4. Neutralizing 5.00 L of an acid rain sample required 11.3 mL of 0.0102 M KOH. Calculate the hydronium ion concentration in the rain sample.

③ Section Review

UNDERSTANDING KEY IDEAS

1. What are the reactants and the product common to all neutralization reactions?

2. Define *equivalence point.* How does the equivalence point differ from the *end point* of a titration?

3. What are standard solutions, and how are they standardized?

4. How would you choose an indicator for titrating a strong acid with a strong base?

5. What titration data are needed to calculate an unknown acid concentration?

6. What are the roles of the two burets in a titration experiment?

7. At the equivalence point of a titration, what is present in the solution?

PRACTICE PROBLEMS

8. If 29.5 mL of 0.150 M HCl neutralizes 25.0 mL of a basic solution, what was $[OH^-]$ in the basic solution?

9. What volume of 0.250 M nitric acid is needed to neutralize 17.35 mL of 0.195 M KOH solution?

10. In a titration of 30.00 mL of 0.0987 M HBr solution with a strong base of unknown concentration, the pH reached 7 after the addition of 37.43 mL of titrant. What was the concentration of the base?

11. If it took 72 mL of 0.55 NaOH titrant to neutralize 220 mL of an acidic solution, what was the hydronium ion concentration in the acidic solution?

12. In a titration of a sample of 0.31 M HNO_3, it took 75 mL of a 0.24 M KOH solution to reach a pH of 7. What was the volume of the sample?

CRITICAL THINKING

13. What indicator would you choose for the titration of acetic acid with potassium hydroxide?

14. Why is the steepness of a titration curve helpful in locating the equivalence point?

15. Explain why the titration of a strong acid with a weak base ends at a pH lower than 7.

Equilibria of Weak Acids and Bases

KEY TERMS

- acid-ionization constant, K_a
- buffer solution

OBJECTIVES

1. **Write** an equilibrium equation that shows how a weak acid is in equilibrium with its conjugate base.

2. **Calculate** K_a from the hydronium ion concentration of a weak acid solution.

3. **Describe** the components of a buffer solution, and explain how a buffer solution resists changes in pH.

Weak Acids and Bases

Consider the reaction represented by the following equation, in which one arrow is longer than the other:

$$A(aq) + B(aq) \rightleftharpoons C(aq) + D(aq)$$

Chemists use this notation to indicate that the forward reaction is *favored*. In other words, when the reaction has reached equilibrium, there will be more products than reactants.

Some Acids are Better Proton Donors Than Others

Some aspects of formic acid, HCOOH, are illustrated in **Figure 20.** Formic acid is a typical Brønsted-Lowry acid, able to donate a proton to a base, such as the acetate ion, CH_3COO^-. Thus, in a solution prepared by dissolving formic acid and sodium acetate in water, a reaction will occur.

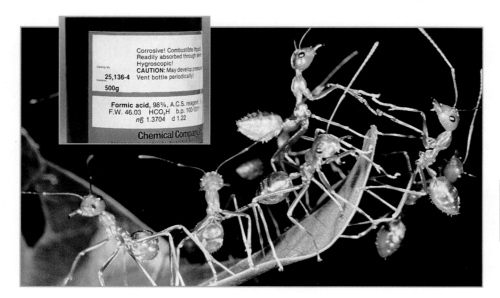

Figure 20
The name *formic acid* comes from *formica*, the Latin word for "ant." Formic acid was first isolated by distillation from ants in 1670. A molecular model of formic acid is shown.

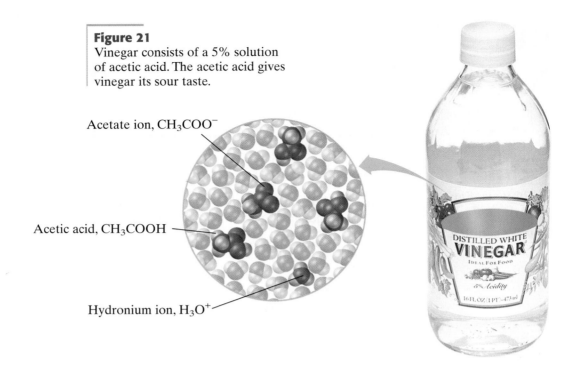

Figure 21
Vinegar consists of a 5% solution of acetic acid. The acetic acid gives vinegar its sour taste.

Acetate ion, CH_3COO^-

Acetic acid, CH_3COOH

Hydronium ion, H_3O^+

The reaction will produce the conjugate base of the formic acid, the formate ion, $HCOO^-$, and the conjugate acid of the acetate ion, acetic acid, CH_3COOH, as shown below.

$$HCOOH(aq) + CH_3COO^-(aq) \rightleftharpoons HCOO^-(aq) + CH_3COOH(aq)$$
$$\text{acid} \hspace{9.5cm} \text{acid}$$

Acetic acid is the active ingredient in vinegar, as shown in **Figure 21.** The unequal arrows in the equation above indicate that if you dissolved equal amounts of all four substances in water, the concentrations of $HCOO^-$ and CH_3COOH would be greater than the concentrations of $HCOOH$ and CH_3COO^- at equilibrium. If you think of this as a contest between the two acids to see which is better able to donate protons, formic acid would be the winner. $HCOOH$ is more willing to lose a proton than CH_3COOH is. Therefore, formic acid is considered a *stronger* acid than acetic acid is.

Some Bases Accept Protons More Readily Than Others

Now look at the same reaction from the standpoint of the two bases:

$$HCOOH(aq) + CH_3COO^-(aq) \rightleftharpoons HCOO^-(aq) + CH_3COOH(aq)$$
$$\text{base} \hspace{9.5cm} \text{base}$$

Both bases can accept protons, but the acetate ion, CH_3COO^-, has been more successful in accepting protons than the formate ion, $HCOO^-$, has. The formate ion is a weaker base than the acetate ion is. At equilibrium, there is more formate ion than acetate ion in solution.

In this example, formic acid is the stronger acid, but its conjugate base, the formate ion, is the weaker base. This example illustrates a general principle: *In an acid-base reaction, the conjugate base of the stronger acid is the weaker base, and the conjugate acid of the stronger base is the weaker acid.*

Table 7 Relative Strengths of Acids and Bases

Acid	Formula	K_a of acid	Conjugate base	Formula
Hydronium ion	H_3O^+	5.53×10^1	water	H_2O
Hydrogen sulfate ion	HSO_4^-	1.23×10^{-2}	sulfate ion	SO_4^{2-}
Phosphoric acid	H_3PO_4	7.52×10^{-3}	dihydrogen phosphate ion	$H_2PO_4^-$
Formic acid	HCOOH	1.82×10^{-4}	formate ion	$HCOO^-$
Benzoic acid	C_6H_5COOH	6.46×10^{-5}	benzoate ion	$C_6H_5COO^-$
Acetic acid	CH_3COOH	1.75×10^{-5}	acetate ion	CH_3COO^-
Carbonic acid	H_2CO_3	4.30×10^{-7}	hydrogen carbonate ion	HCO_3^-
Dihydrogen phosphate ion	$H_2PO_4^-$	6.31×10^{-8}	monohydrogen phosphate ion	HPO_4^{2-}
Hypochlorous acid	HOCl	2.95×10^{-9}	hypochlorite ion	ClO^-
Ammonium ion	NH_4^+	5.75×10^{-10}	ammonia	NH_3
Hydrogen carbonate ion	HCO_3^-	4.68×10^{-11}	carbonate ion	CO_3^{2-}
Monohydrogen phosphate ion	HPO_4^{2-}	4.47×10^{-13}	phosphate ion	PO_4^{3-}
Water	H_2O	1.81×10^{-16}	hydroxide ion	OH^-
Conjugate acid	**Formula**	**K_a of acid**	**Base**	**Formula**

Increasing acid strength (left side, upward arrow)

Increasing base strength (right side, downward arrow)

The Acid-Ionization Constant

The strengths of acids may be described in relative terms of *stronger* or *weaker,* but the strength of an acid may also be expressed quantitatively by its **acid-ionization constant, K_a.** This is just the equilibrium constant, K_{eq}, that describes the ionization of an acid in water.

Consider the following equation, which describes the equilibrium established when acetic acid dissolves in water.

$$CH_3COOH(aq) + H_2O(l) \rightleftharpoons H_3O^+(aq) + CH_3COO^-(aq)$$

The equilibrium expression for this reaction is written as follows:

$$\frac{[H_3O^+][CH_3COO^-]}{[CH_3COOH]} = K_a = 1.75 \times 10^{-5}$$

Recall that only solutes appear in equilibrium expressions. When water is a solvent, it is omitted. Remember, too, that K_a is unitless.

Table 7 lists many acid-ionization constants. Note that the stronger the acid is, the weaker its conjugate base is. Accordingly, the stronger the base is, the weaker its conjugate acid is.

acid-ionization constant, K_a

the equilibrium constant for a reaction in which an acid donates a proton to water

Topic Link

Refer to the "Chemical Equilibrium" chapter for a discussion of equilibrium constants.

Calculating K_a of a Weak Acid

A vinegar sample is found to have 0.837 M CH_3COOH. Its hydronium ion concentration is found to be 3.86×10^{-3} mol/L. Calculate K_a for acetic acid.

1 Gather information.

$[CH_3COOH] = 0.837$ M
$[H_3O^+] = 3.86 \times 10^{-3}$ M
$K_a = ?$

2 Plan your work.

The equation for the equilibrium is

$$CH_3COOH(aq) + H_2O(l) \rightleftharpoons H_3O^+(aq) + CH_3COO^-(aq)$$

which establishes that the expression for K_a is

$$K_a = \frac{[H_3O^+][CH_3COO^-]}{[CH_3COOH]}$$

The equation also shows that hydronium and acetate ions are produced in equal amounts, so $[CH_3COO^-] = [H_3O^+]$. Hence, all of the necessary concentration data are known.

3 Calculate.

$$K_a = \frac{[H_3O^+][CH_3COO^-]}{[CH_3COOH]} = \frac{[H_3O^+][H_3O^+]}{[CH_3COOH]} =$$

$$\frac{(3.86 \times 10^{-3})(3.86 \times 10^{-3})}{0.837} = 1.78 \times 10^{-5}$$

4 Verify your results.

The calculated acid-ionization constant is very close to the value listed in **Table 7,** so the answer seems reasonable.

> ### PRACTICE HINT
>
> Earlier in this chapter, a sample problem demonstrated how to calculate $[H_3O^+]$ from pH. In some K_a problems, you may need to perform this step first.

PRACTICE

1 Calculate $[H_3O^+]$ of a 0.150 M acetic acid solution.

2 Find K_a if a 0.50 M solution of a weak acid has a hydronium ion concentration of 1.3×10^{-4} M.

3 A solution prepared by dissolving 1.0 mol of benzoic acid in water to form 1.0 L of solution has a pH of 2.1. Calculate the acid-ionization constant.

4 Use **Table 7** to calculate the concentration of formate ion in 0.085 M formic acid.

Table 8 **Typical pH Values of Human Body Fluids**

Solution	pH	Solution	pH
Gastric juice	1.5	Blood	7.4
Urine	6.0	Tears	7.4
Saliva	6.5	Pancreatic juice	7.9
Milk	6.6	Bile	8.2

internet connect

www.scilinks.org
Topic: Buffers
SciLinks code: HW4023

SCiLINKS Maintained by the National Science Teachers Association

Buffer Solutions

You can see in **Table 8** that the pH of blood is 7.4. Keeping your blood pH between 7.35 and 7.45 is vital to your health. If your blood's pH goes outside this very narrow range, you will become ill. If your blood pH is lower than 7.35, you suffer *acidosis*. If your blood's pH rises above 7.45, symptoms of *alkalosis* appear. How does your body control the pH of blood within such narrow bounds? Your body relies on the properties of **buffer solutions**—solutions that resist changes in pH that would otherwise be caused by the addition of acids or bases. These solutions are said to be "buffered" against pH changes.

buffer solution

a solution made from a weak acid and its conjugate base that neutralizes small amounts of acids or bases added to it

A Buffer Has Two Ingredients

A buffer solution, often simply called a *buffer,* is a solution that contains approximately equal amounts of a weak acid and its conjugate base.

Imagine preparing two solutions. In the first, you dissolve one mole of sodium acetate in one liter of water. Sodium acetate is a strong electrolyte and ionizes completely in solution.

$$CH_3COONa(s) \longrightarrow CH_3COO^-(aq) + Na^+(aq)$$

For the second solution, you prepare one mole of acetic acid in one liter of water. Acetic acid is a weak acid that ionizes very little in water. The following equilibrium equation describes the solution:

$$CH_3COOH(aq) + H_2O(l) \Longleftarrow H_3O^+(aq) + CH_3COO^-(aq)$$

As the unequal arrows suggest, this equilibrium favors the reactants on the left side. About 99.6% of the acetic acid is un-ionized. Its pH is 2.4.

Now mix the two solutions. Both contain the acetate ion, so the common ion effect comes into play. Recall that Le Châtelier's principle predicts that the equilibrium will adjust to reduce the stress imposed by the increase in the $CH_3COO^-(aq)$ concentration. It does this by shifting even more heavily toward the left. In fact, now 99.996% of the acetic acid is un-ionized. The pH has doubled to 4.8.

The mixture is a buffer solution. It contains nearly equal amounts of the weak acid acetic acid, $CH_3COOH(aq)$, and its conjugate base, the acetate ion, $CH_3COO^-(aq)$. It is not necessary that the acid and its conjugate base be present in equal amounts to act as a buffer, but there must be a substantial concentration of each.

Topic Link

Refer to the "Chemical Equilibrium" chapter for a discussion of Le Châtelier's principle.

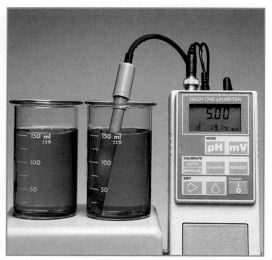

Figure 22
The left-hand beaker in each photo contains a neutral solution. The right-hand beaker in each photo contains the same solution plus hydrochloric acid.

a When a small amount of HCl is added to an unbuffered solution, the solution's pH drops significantly.

b When the same amount of HCl is added to a buffered solution, the pH of the solution does not change very much.

Buffer Solutions Stabilize pH

How do buffer solutions prevent large changes in pH when small amounts of acid or base are added, as demonstrated in **Figure 22?** Le Châtelier's principle can help us understand the effect. If HX is a weak acid and X^- is its conjugate base, then in a buffer solution composed of the two, the following equilibrium is established:

$$HX(aq) + H_2O(l) \rightleftharpoons H_3O^+(aq) + X^-(aq)$$

If a base is added to the buffer solution, the base will react with the H_3O^+ and remove some of this ion from solution. According to Le Châtelier's principle, the equilibrium will adjust by shifting to the right to make more H_3O^+, preventing too great a pH change. It is a similar story if an acid is added. The tendency for $[H_3O^+]$ to increase is countered by a shift of the equilibrium to the left and the formation of more HX molecules.

The greater the concentrations of the two buffer components, the greater the ability of the buffer to resist changes in pH. The efficiency of the buffer is greatest when the concentrations of the two components are equal, but this condition is not necessary for the buffer to work.

The equilibrium-constant expression for the reaction above is simply $[H_3O^+][X^-]/[HX] = K_a$. From this expression, it is easy to see that when the concentration of each member of the conjugate pair is equal, $[H_3O^+] = K_a$. Thus the pH of such a buffer solution is $-\log(K_a)$.

Buffers Are All Around

Now you understand what manufacturers of shampoos and antacids mean when they say that their products are buffered: the products have ingredients that resist pH changes. The pH of foods affects their taste and texture, so many packaged foods are buffered, too. Check ingredient labels for phosphates. The presence of phosphates probably means that the product contains the acid-base pair $H_2PO_4^-/HPO_4^{2-}$ to control the pH.

The liquid portion of blood is an example of a buffer solution. To keep the blood's pH very close to 7.40, the body uses a buffer in which the weak acid H_2CO_3, carbonic acid, is paired with its conjugate base, the hydrogen carbonate ion HCO_3^-. The equation below describes the equilibrium that is established.

$$H_2CO_3(aq) + H_2O(l) \rightleftharpoons H_3O^+(aq) + HCO_3^-(aq)$$

There are many medical conditions that can disrupt the equilibrium of this system. Uncontrolled diabetes can cause acidosis, in which the equilibrium is displaced too far to the right. Alcoholic intoxication causes alkalosis, in which the equilibrium lies too far to the left. Hyperventilation removes CO_2, which is also in equilibrium with H_2CO_3. The equilibrium will shift to the left, causing alkalosis.

Section Review

UNDERSTANDING KEY IDEAS

1. Identify the stronger acid and the stronger base in the reaction described by the following equation:

$$HOCl(aq) + NH_3(aq) \rightleftharpoons NH_4^+(aq) + ClO^-(aq)$$

2. Write the acid-ionization constant expression for the weak acid H_2SO_3.

3. The hydrogen sulfite ion, HSO_3^-, is a weak acid in aqueous solution. Write an equation showing the equilibrium established when hydrogen sulfite is dissolved in aqueous solution, using unequal arrows to show the equilibrium.

4. What is a buffer solution?

5. Give two examples of the practical uses of buffers.

PRACTICE PROBLEMS

6. Use **Table 7** to determine which direction is favored in the following reaction. Explain your answer.

$$H_2CO_3 + H_2O \rightleftharpoons HCO_3^- + H_3O^+$$

7. A 0.105 M solution of HOCl has a pH of 4.19. What is the acid-ionization constant?

8. A buffer solution, prepared from equal amounts of an acid and its conjugate base, has a pH of 10.1. What is the K_a of the acid?

9. Calculate the K_a of nitrous acid, given that a 1.00 M solution of the acid contains 0.026 mol of NO_2^- per liter of solution.

CRITICAL THINKING

10. Ammonia is a weak base. A 0.0123 M solution of ammonia has a hydroxide ion concentration of 4.63×10^{-4} M. Calculate the K_a of NH_4^+.

11. What would be the value of the acid-ionization constant for an acid that was so strong that not a single molecule remained un-ionized?

12. What would be a good acid-base pair from which to prepare a buffer solution whose pH is 10.3?

13. If 99.0% of the weak acid HX stays un-ionized in 1.0 M aqueous solution, what is the K_a?

14. Write all three K_a expressions for H_3PO_4. Which will have the smallest value?

15. Calculate K_{eq} for the following reaction:

$$H_2CO_3(aq) + CO_3^{2-}(aq) \rightleftharpoons 2HCO_3^-(aq)$$

Consumer Focus

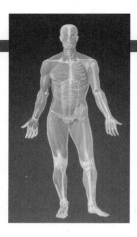

Antacids

The pH of gastric juice in the human stomach is 1.5. This strongly acidic environment activates digestive enzymes, such as pepsin, that work in the stomach.

Stomach acids and antacids

Acidity in the stomach is provided by 0.03 M hydrochloric acid, HCl(aq). Sometimes, a person's stomach generates too much acid. The discomfort known as *heartburn* results when the acid solution is forced into the esophagus. Heartburn can be temporarily relieved by taking an antacid to neutralize the excess stomach acid.

Although antacids contain other ingredients, all antacids contain a base that counteracts stomach acid. The base is either sodium hydrogen carbonate, $NaHCO_3$, calcium carbonate, $CaCO_3$, aluminum hydroxide, $Al(OH)_3$, or magnesium hydroxide, $Mg(OH)_2$.

Dangers of excess metals from antacids

In any antacid, the anion is the base that neutralizes the stomach acid. However, the cation in the antacid is also important. Antacids containing $NaHCO_3$ work fastest because $NaHCO_3$ is much more soluble than other antacid substances are. Overusing these antacids, however, can raise the level of positive ions in the body, just as salt does. Overuse can also seriously disrupt the acid-base balance in your blood.

Because of the risks associated with an excess of sodium, some antacid manufacturers have substituted calcium carbonate, $CaCO_3$, for $NaHCO_3$. But if calcium is taken in large amounts, it can promote kidney stones. Ingesting too much aluminum from antacid products, such as $Al(OH)_3$, can interfere with the body's absorption of phosphorus, which is needed for healthy bones. Excess magnesium from antacids that contains $Mg(OH)_2$ may pose problems for people who have kidney disorders.

You should know the active ingredient in any antacid product before you ingest the product, and you should never use an antacid product for more than a few days without consulting a doctor. It is best to avoid the need for an antacid in the first place. You can minimize the production of excess stomach acid by following a healthy diet, avoiding stress, and limiting your consumption of coffee, fatty foods, and chocolate.

Questions

1. What class of compound is common to all antacids?
2. Why should you pay attention to which ions an antacid contains?

internet connect

www.scilinks.org
Topic: Antacids
SciLinks code: HW4010

SCiLINKS. Maintained by the National Science Teachers Association

CHAPTER HIGHLIGHTS

KEY TERMS

strong acid
weak acid
strong base
weak base
Brønsted-Lowry acid
Brønsted-Lowry base
conjugate acid
conjugate base
amphoteric

self-ionization
  constant of water, K_w
neutral
pH
indicator

neutralization
  reaction
equivalence point
titration
titrant
standard solution
transition range
end point

acid-ionization
  constant, K_a
buffer solution

KEY IDEAS

SECTION ONE What Are Acids and Bases?
- Acid solutions have distinctive properties attributable to the H_3O^+ ion.
- Bases have distinctive properties attributable to the OH^- ion.
- Brønsted and Lowry defined an acid as donating a proton, and a base as accepting a proton.
- Every acid has a conjugate base, and every base has a conjugate acid.
- An amphoteric species, such as water, can behave as an acid or a base.

SECTION TWO Acidity, Basicity, and pH
- In aqueous solutions, $[H_3O^+]$ and $[OH^-]$ are interrelated by K_w.
- pH, which is a quantitative measure of acidity and basicity, is the negative logarithm of $[H_3O^+]$.

SECTION THREE Neutralization and Titrations
- A neutralization reaction between an acid and a base produces water.
- In a titration, a solution of unknown concentration is neutralized by a standard solution of known concentration.
- An indicator has a transition range of pH, within which lies its end point pH.

SECTION FOUR Equilibria of Weak Acids and Bases
- The acid-ionization constant reflects the strength of a weak acid and the strength of the acid's conjugate base.
- K_a can be used to calculate $[H_3O^+]$ in a solution of a weak acid.
- Buffer solutions are mixtures of a weak acid and its conjugate base, and resist pH changes.

KEY SKILLS

Determining [OH⁻] using K_w
Sample Problem A p. 541

Calculating pH for an Acidic or Basic Solution
Skills Toolkit 1 p. 543

Sample Problem B p. 544

Calculating [H₃O⁺] and [OH⁻] from pH
Skills Toolkit 1 p. 543
Sample Problem C p. 545

Performing a Titration
Skills Toolkit 2 p. 552

Calculating Concentration from Titration Data
Sample Problem D p. 555

Calculating K_a of a Weak Acid
Sample Problem E p. 560

15 CHAPTER REVIEW

USING KEY TERMS

1. How does a strong acid differ from a weak acid?

2. What kind of an electrolyte is a weak base?

3. How does Brønsted and Lowry's definition of an acid differ from Arrhenius's definition of an acid? Explain which definition is broader.

4. What is the conjugate acid of the base ammonia, NH_3?

5. Why is water considered amphoteric?

6. What is the concentration of hydroxide ions in pure water?

7. What is the value of K_w at 25°C?

8. What is the pH of a neutral solution?

9. Give the equation that relates pH to hydronium ion concentration.

10. How does the strength of an acid relate to the concentration of the acid? How does the strength of an acid relate to the pH of an aqueous solution of the acid? How does the concentration of an acid solution relate to the solution's pH?
WRITING SKILLS

11. What product do all neutralization reactions have in common?

12. At what point in a titration are the amounts of hydronium ions and hydroxide ions equal?

13. Group the following four terms into two pairs according to how the terms are related, and explain how they are related: *end point, standard solution, titrant,* and *transition range.*

14. What is the equilibrium constant that is applied to a weak acid?

15. How does the addition of a small amount of acid or base affect a buffered solution?

UNDERSTANDING KEY IDEAS

What Are Acids and Bases?

16. Compare the properties of an acid with those of a base.

17. What is a base according to Arrhenius? according to Brønsted and Lowry?

18. Why are weak acids and weak bases poor electrical conductors?

19. What is the difference between the strength and the concentration of an acid?

20. Identify each of the following compounds as an acid or a base according to the Brønsted-Lowry classification. For each species, write the formula and the name of its conjugate.
a. CH_3COO^-
b. HCN
c. $HOOCCOOH$
d. $C_6H_5NH_3^+$

21. Write an equation for the reaction between hydrocyanic acid, HCN, and water. Label the acid, base, conjugate acid, and conjugate base.

22. Write chemical equations that show how the hydrogen carbonate ion, HCO_3^-, acts as an amphoteric ion.

Acidity, Basicity, and pH

23. Explain the relationship between the self-ionization of water and K_w.

24. Write an equation that shows the self-ionization of water.

25. Three solutions have pHs of 3, 7, and 11. Which solution is basic? Which is acidic? Which is neutral?

26. By what factor does $[OH^-]$ change when the pH increases by 3? by 2? by 1? by 0.5?

27. Explain how you can calculate pH from $[H_3O^+]$ by using your calculator.

28. Describe two methods of measuring pH, and explain the advantages and disadvantages of each method.

Neutralization and Titrations

29. What is a neutralization reaction?

30. Describe two precautions that should be taken to ensure an accurate titration.

31. Explain what a titration curve is, and sketch its shape.

32. How would you select an indicator for a particular acid-base titration?

33. Would the pH at the equivalence point of a titration of a weak acid with a strong base be less than, equal to, or greater than 7.0?

34. Name an indicator you might use to titrate ammonia with hydrochloric acid.

Equilibria of Weak Acids and Bases

35. The K_a of nitrous acid, HNO_2, is 6.76×10^{-4}. Write the equation describing the equilibrium established when HNO_2 reacts with NH_3. Use unequal arrows to indicate whether reactants or products are favored.

36. a. What is the relationship between the strength of an acid and the strength of its conjugate base?

b. What is the realtionship between the strength of a base and the strength of its conjugate acid?

37. Propanoic acid, C_2H_5COOH, is a weak acid. Write the expression defining its acid-ionization constant.

38. Place the following acids in order of increasing strength:
  a. valeric acid, $K_a = 1.5 \times 10^{-5}$
  b. glutaric acid, $K_a = 3.4 \times 10^{-4}$
  c. hypobromous acid, $K_a = 2.5 \times 10^{-9}$
  d. acetylsalicylic acid (aspirin),
  $$K_a = 3.3 \times 10^{-4}$$

39. What are the components of a buffer solution? Give an example.

PRACTICE PROBLEMS

Sample Problem A Determining $[OH^-]$ or $[H_3O^+]$ Using K_w

40. If the hydronium ion concentration of a solution is 1.63×10^{-8} M, what is the hydroxide ion concentration?

41. Calculate the hydronium ion concentration in a solution of 0.365 mol/L of NaOH.

42. How much HCl would you need to dissolve in 1.0 L of water so that $[OH^-] = 6.0 \times 10^{-12}$ M?

43. The hydronium ion concentration in a solution is 1.87×10^{-3} mol/L. What is $[OH^-]$?

44. If 0.150 mol of KOH is dissolved in 500 mL of water, what are $[OH^-]$ and $[H_3O^+]$?

45. If a solution contains twice the concentration of hydronium ions as hydroxide ions, what is the hydronium ion concentration?

Sample Problem B Calculating pH for an Acidic or Basic Solution

46. Stomach acid contains HCl, whose concentration is about 0.03 mol/L. What is the pH of stomach acid?

47. If $[OH^-]$ of an aqueous solution is 0.0134 mol/L, what is the pH?

48. What is the pH of a 0.15 M solution of $HClO_4$, a strong acid?

49. LiOH is a strong base. What is the pH of a 0.082 M LiOH solution?

50. Find the pH of a solution consisting of 0.29 mol of HBr in 1.0 L of water.

51. What is the pH of aqueous solutions of the strong acid HNO_3, nitric acid, if the concentrations of the solutions are as follows: (a) 0.005 M, (b) 0.05 M, (c) 0.5 M, (d) 5 M?

52. Find the pH of a solution prepared by dissolving 0.65 mol of the strong base NaOH in 1.0 L of water.

53. What is the pH of a solution prepared by dissolving 0.15 mol of the strong base $Ba(OH)_2$ in one liter of water? (Hint: How much hydroxide ion does barium hydroxide generate per mole in solution?)

54. A solution has a hydronium ion concentration of 1.0×10^{-9} M. What is its pH?

55. If a solution has a hydronium ion concentration of 6.7×10^{-1} M, what is its pH?

56. What is the pH of a solution whose hydronium ion concentration is 2.2×10^{-12} M?

57. What is the pH of a solution whose H_3O^+ concentration is 1.9×10^{-6} M?

58. Calculate the pH of a 0.0316 M solution of the strong base RbOH.

Sample Problem C Calculating $[H_3O^+]$ and $[OH^-]$ from pH

59. The pH of a solution is 9.5. What is $[H_3O^+]$? What is $[OH^-]$?

60. A solution of a weak acid has a pH of 4.7. What is the hydronium ion concentration?

61. A 50 mL sample of apple juice has a pH of 3.2. What amount, in moles, of H_3O^+ is present?

62. Find $[H_3O^+]$ in a solution of pH 4.

63. What is the hydroxide ion concentration in a solution of pH 8.72?

64. Calculate the concentration of the H_3O^+ and OH^- ions in an aqueous solution of pH 5.0.

65. A solution has a pH of 10.1. Calculate the hydronium ion concentration and the hydroxide ion concentration.

66. What is the hydronium ion concentration in a solution of pH 5.5?

67. If the pH of a solution is 4.3, what is the hydroxide ion concentration?

68. What is the hydronium ion concentration in a solution whose pH is 10.0?

69. The pH of a solution is 3.0. What is $[H_3O^+]$?

70. What is $[H_3O^+]$ in a solution whose pH is 1.9?

71. If a solution has a pH of 13.3, what is its hydronium ion concentration?

Skills Toolkit Performing a Titration

72. To what volumetric mark should a buret be filled?

73. Why is it important to slow down the drop rate of the buret near the end of a titration?

74. What two buret readings need to be recorded in order to determine the volume of solution dispensed by the buret?

Sample Problem D Calculating Concentration from Titration Data

75. What volume of 0.100 M NaOH is required to neutralize 25.00 mL of 0.110 M H_2SO_4?

76. What volume of 0.100 M NaOH is required to neutralize 25.00 mL of 0.150 M HCl?

77. If 35.40 mL of 1.000 M HCl is neutralized by 67.30 mL of NaOH, what is the molarity of the NaOH solution?

78. If 50.00 mL of 1.000 M HI is neutralized by 35.41 mL of KOH, what is the molarity of the KOH solution?

79. If 133.73 mL of a standard solution of KOH, of concentration 0.298 M, exactly neutralized 50.0 mL of an acidic solution, what was the acid concentration?

80. To standardize a hydrochloric acid solution, it was used as titrant with a solid sample of sodium hydrogen carbonate, $NaHCO_3$. The sample had a mass of 0.3967 g, and 41.77 mL of acid was required to reach the equivalence point. Calculate the concentration of the standard solution.

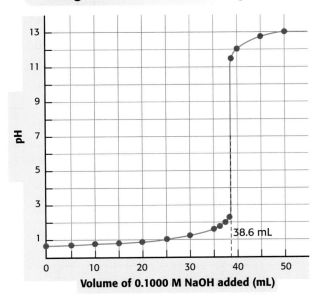

Strong Acid Titrated with Strong Base

pH vs. Volume of 0.1000 M NaOH added (mL)

38.6 mL

81. The graph above shows a titration curve obtained during the titration of a 25.00 mL sample of an acid with 0.1000 M NaOH. Calculate the concentration of the acid.

82. An HNO_3 solution has a pH of 3.06. What volume of 0.015 M LiOH will be required to titrate 65.0 mL of the HNO_3 solution to reach the equivalence point?

Sample Problem E Calculating K_a of a Weak Acid

83. The hydronium ion concentration in a 0.100 M solution of formic acid is 0.0043 M. Calculate K_a for formic acid.

84. $[NO_2^-] = 9.1 \times 10^{-3}$ mol/L in a nitrous acid solution of concentration 0.123 mol/L. What is K_a for HNO_2?

85. A solution of acetic acid had the following solute concentrations: $[CH_3COOH] = 0.035$ M, $[H_3O^+] = 7.4 \times 10^{-4}$ M, and $[CH_3COO^-] = 7.4 \times 10^{-4}$ M. Calculate the K_a of acetic acid based on these data.

86. Hydrazoic acid, HN_3, is a weak acid. A 0.01 M solution of hydrazoic acid contained a concentration of 0.0005 M of the N_3^- ion. Find the acid-ionization constant of hydrazoic acid.

87. Find K_a for a weak acid that contains 0.050 M H_3O^+ in a solution of 1.00 M concentration.

88. The pH of a solution prepared by dissolving 1.0 mol of a weak acid in water to form 1.0 L of solution was 3.1. What is the acid-ionization constant of the weak acid?

89. An aqueous solution of periodic acid had the following concentrations: $[HIO_4] = 1.1 \times 10^{-4}$ M and $[IO_4^-] = [H_3O^+] = 1.2 \times 10^{-2}$ M. What is the K_a of HIO_4?

MIXED REVIEW

90. If 25 mL of 1.00 M HCl is mixed with 75 mL of 1.00 M NaOH, what are the final amounts and concentrations of all ions present?

91. If 0.30 mol of HI is dissolved in 750 mL of water, what are $[H_3O^+]$, pH, and $[OH^-]$?

92. An aqueous solution is prepared by dissolving 0.321 g of $Ca(OH)_2$ in 555 mL of water. Find $[H_3O^+]$ and $[OH^-]$.

93. When 1.0 mol of a weak acid was dissolved in 10.0 L of water, the pH was found to be 3.90. What is K_a for the acid?

94. At the end point of a titration of 25 mL of 0.300 M NaOH with 0.200 M HNO_3, what would the concentration of sodium nitrate in the titration flask be?

95. What are the concentrations of all of the components of a benzoic acid solution if K_a is 6.5×10^{-5}, pH is 2.96, and C_6H_5COOH has a concentration of 0.020 M?

96. What is $[H_3O^+]$ in a buffer solution containing equal concentrations of acetic acid and sodium acetate?

97. Make a table listing the ionic concentrations in solutions of the following pH values: 14.25, 14.00, 13.75, 13.25, 13.00, 7.25, 7.00, 6.75, 1.00, 0.75, 0.50, 0.25, 0.00, and −0.25.

98. Write the equilibrium equation and the equilibrium constant expression for an ammonia–ammonium ion buffer solution.

99. If 18.5 mL of a 0.0350 M H_2SO_4 solution neutralizes 12.5 mL of aqueous LiOH, what mass of LiOH was used to make 1.00 L of the LiOH solution?

100. Use **Table 7** to calculate the pH of a buffer solution made from equal amounts of sodium monohydrogen phosphate and potassium dihydrogen phosphate.

101. A solution of 5% acetic acid ($K_a = 1.76 \times 10^{-5}$) has a concentration of 0.83 M. What is the pH of the solution?

CRITICAL THINKING

102. A solution of HCl has a pH of 1.00. You add NaOH to the solution, and the pH rises to 1.18. What happened? What does this outcome have to do with the relationship of acids to bases along the pH scale? Can you estimate what amounts of HCl and NaOH were used?

103. Why is a buret, rather than a graduated cylinder, used in titrations?

104. A small volume of indicator solution is usually added to the titration flask right before the titration. As a result, the sample is diluted slightly. Does this matter? Why or why not?

105. A student passes an end point in a titration. Is it possible to add an additional measured amount of the unknown and continue the titration? Explain how this process might work. How would the answer for the calculation of the molar concentration of the unknown differ from the answer the student would have gotten if the titration had been performed properly?

106. Design a buffer solution suitable for controlling the pH of a solution at a value close to 4.0.

Indicator	Acid color	pH transition range	Base color
Thymol blue	red	1.2–2.8	yellow
Bromphenol blue	yellow	3.0–4.6	blue
Bromcresol green	yellow	2.0–5.6	blue
Bromthymol blue	yellow	6.0–7.6	blue
Phenol red	yellow	6.6–8.0	red
Alizarin yellow	yellow	10.1–12.0	red

107. Refer to the table above to answer the following questions:
  a. Which indicator would be the best choice for a titration with an end point at a pH of 4.0?
  b. Which indicators would work best for a titration of a weak base with a strong acid?

108. Why does an indicator need to be a weak acid or a weak base?

109. A 0.0200 M solution of NaOH is being used to titrate 20.00 mL of 0.0400 M HCl using bromthymol blue as indicator. The end point, corresponding to a pH of 7.0, occurs when 40.00 mL of base has been added. Calculate the volumes of titrant that correspond to the beginning and to the end of bromthymol blue's transition range, 6.0–7.6.

110. What would be a suitable titrant (compound and concentration) with which to titrate 20.00 mL of a strong acid that has a concentration of about 0.015 M?

111. Explain the difference between *end point* and *equivalence point*. Why is it important that both occur at approximately the same pH in a titration?

112. Can you neutralize a strong acid solution by adding an equal volume of a weak base having the same molarity as the acid? Support your position.

113. In the 18th century, Antoine Lavoisier experimented with oxides such as CO_2 and SO_2. He observed that they formed acidic solutions. His observations led him to infer that for a substance to exhibit acidic behavior, it must contain oxygen. However, today that is known to be incorrect. Provide evidence to refute Lavoisier's conclusion.

ALTERNATIVE ASSESSMENT

114. Design an experiment to measure the pH of four types of hair shampoo: baby shampoo, shampoo for extra body, shampoo for oily hair, and shampoo with conditioner. Also compare two brands of "pH-balanced" shampoo. If your teacher approves your plan, carry it out.

115. Describe exactly how you would test the pH of an unknown solution. If your teacher approves your plan and has a solution available, complete the test.

116. Design an experiment to test the neutralization effectiveness of various brands of antacid. Show your procedure, including all safety procedures and cautions, to your teacher for approval. If your teacher approves your plan, carry it out. After experimenting, write an advertisement for the antacid you judge to be the most effective. Cite data from your experiments as part of your advertising claims.

117. Describe how you would prepare one or more buffer solutions, including which compounds to use. Predict the pH of each solution. If your teacher provides the needed materials, measure the pH to test your prediction.

118. Design an experiment to extract possible acid-base indicators from sources such as red cabbage, berries, and flower petals, using known acidic, basic, and neutral solutions to test the action of each indicator that you are able to isolate. If your teacher approves your design, carry it out.

CONCEPT MAPPING

119. Use the following terms to complete the concept map below: *hydronium ions (H_3O), hydroxide ions (OH^-), neutralization reaction, pH,* and *titration.*

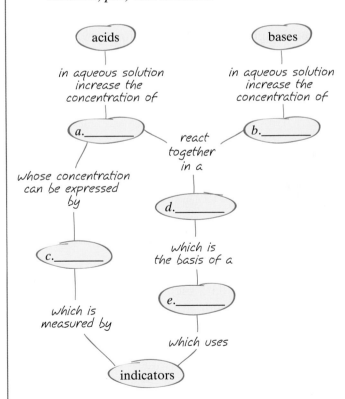

Study the graph below, and answer the questions that follow.
For help in interpreting graphs, see Appendix B, "Study Skills for Chemistry."

120. What variable is being measured along the *x*-axis?

121. What is the pH at the beginning of the titration?

122. What was the pH after 25 mL of titrant had been added?

123. What volume of titrant was needed to reach a pH of 2.0?

124. Where on the graph do you find the single most important data point?

125. If the titration continued beyond what the graph shows, how would you expect the pH to change past the end of the graph?

126. Roughly sketch the titration curve (pH versus volume) that you would expect if you titrated a weak base with a strong acid. Mark the equivalence point.

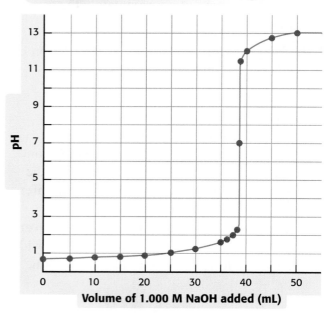

Strong Acid Titrated with Strong Base

Volume of 1.000 M NaOH added (mL)

TECHNOLOGY AND LEARNING

127. Graphing Calculator

Graphing Titration Data

The graphing calculator can run a program that graphs data such as pH versus volume of base. Graphing the titration data will allow you to determine which combination of acid and base is represented by the shape of the graph.

Go to Appendix C.. If you are using a TI-83 Plus, you can download the program and data and run the application as directed. Press the **APPS** key on your calculator, then choose the application **CHEMAPPS**. Press **5**,

then highlight **ALL** on the screen, press **1**, then highlight **LOAD** and press **2** to load the data into your calculator. Quit the application, and then run the program **TITRATN**. For L_1, press **2nd** and **LIST**, and choose **VOL1**. For L_2, press **2nd** and **LIST** and choose **PH1**. If you are using another calculator, your teacher will provide you with keystrokes and data sets to use.

a. At what approximate volume does the pH change from acidic to basic?

b. If the titrant was 0.24 M NaOH, and the volume of unknown was 230 mL, what was $[H_3O^+]$ in the unknown solution?

1. An aqueous solution of which of the following would have the highest concentration of hydronium ions?
   a. formic acid
   b. acetic acid
   c. potassium carbonate
   d. sodium formate

2. When an acid reacts with a metal,
   a. the hydronium ion concentration increases.
   b. the metal forms anions.
   c. hydrogen gas is produced.
   d. the K_w value changes.

3. The K_w value can be affected by
   a. dissolving a salt in the solution.
   b. changes in temperature.
   c. changes in the hydroxide ion concentration.
   d. the presence of a strong acid.

4. A solution of an Arrhenius acid would have
   a. no measurable pH.
   b. a pH of less than 7.
   c. a pH of 7.
   d. a pH of greater than 7.

5. The stronger an acid is,
   a. the smaller its K_a value.
   b. the stronger its conjugate base.
   c. the more hydronium ions produced in aqueous solution.
   d. the more hydroxide ions produced in aqueous solution.

6. Which of the following species is the conjugate acid of another species in the list?
   a. PO_4^{3-}
   b. H_3PO_4
   c. H_2O
   d. $H_2PO_4^-$

7. Which of the following solutions would have a pH value greater than 7?
   a. $[OH^-] = 2.4 \times 10^{-2}$ M
   b. $[H_3O^+] = 1.53 \times 10^{-2}$ M
   c. 0.0001 M HCl
   d. $[OH^-] = 4.4 \times 10^{-9}$ M

8. If the pH of a solution of a strong base is known, the _____ of the solution can be calculated.
   a. molar concentration
   b. $[OH^-]$
   c. $[H_3O^+]$
   d. All of the above

9. A neutral aqueous solution
   a. has a 7 M H_3O^+ concentration.
   b. contains neither hydronium ions nor hydroxide ions.
   c. has an equal number of hydronium ions and hydroxide ions.
   d. contains buffers that resist changes in pH.

10. Identify the salt that remains when a solution of H_2SO_4 is titrated with a solution of $Ca(OH)_2$.
    a. calcium sulfate
    b. calcium hydroxide
    c. calcium oxide
    d. calcium phosphate

11. Which of the following could not be titrated by 0.100 M HCl solution no matter which indicator you used?
    a. $OH^-(aq)$
    b. $NaHCO_3(s)$
    c. $H_2SO_4(aq)$
    d. $Na_3PO_4(s)$

REACTION RATES

A forest fire is an enormous combustion reaction that can go on as long as it has fuel, oxygen, and heat. The air tanker in the photograph is dropping a fire-retardant mixture to slow the spread of one of these fires. Fire retardants, which usually contain chemicals such as water, ammonium sulfate, and ammonium phosphate, work by forming a barrier between the fuel (brush and trees) and the oxygen. These chemicals help firefighters slow and eventually stop the combustion reaction. In this chapter, you will learn about the many factors that affect how fast a chemical reaction takes place.

START-UP ACTIVITY

Temperature and Reaction Rates

PROCEDURE

1. Submerge one **light stick** in a **bath of cold water** (about 10°C).

2. Submerge a second **light stick** in a **bath of hot water** (about 50°C).

3. Allow each light stick to reach the same temperature as its bath.

4. Remove the light sticks, and activate them.

5. In a dark corner of the room, observe and compare the light intensities of the two sticks.

ANALYSIS

1. Which stick was brighter?

2. Light is emitted from the stick because of a chemical reaction. What can you conclude about how temperature affects this reaction?

Pre-Reading Questions

①　Give two examples of units that could be used to measure a car's rate of motion.

②　What can you do to slow the rate at which milk spoils?

③　What is a catalytic converter in an automobile?

internet connect

www.scilinks.org
Topic: Inhibitors
SciLinks code: HW4172

SC LINKS. Maintained by the National Science Teachers Association

What Affects the Rate of a Reaction?

KEY TERMS

- **chemical kinetics**
- **reaction rate**

OBJECTIVES

① **Define** the rate of a chemical reaction in terms of concentration and time.

② **Calculate** the rate of a reaction from concentration-versus-time data.

③ **Explain** how concentration, pressure, and temperature may affect the rate of a reaction.

④ **Explain** why, for surface reactions, the surface area is an important factor.

Rates of Chemical Change

A *rate* indicates how fast something changes with time. In a savings account, the rate of interest tells how your money is growing over time. Speed is also a rate. From the speed of one of the race cars shown in **Figure 1,** you can tell the distance that the car travels in a certain time. If a car's speed is 67 m/s (150 mi/h), you know that it travels a distance of 67 meters every second. Rates are always measured in a unit of something per time interval. The rate at which the car's wheels turn would be measured in revolutions per second. The rate at which the car burns gasoline could be measured in liters per minute.

The rate of a chemical reaction measures how quickly reactants are changed into products. Some reactions are over in as little as 10^{-15} s; others may take hundreds of years. The study of reaction rates is called **chemical kinetics.**

chemical kinetics

the area of chemistry that is the study of reaction rates and reaction mechanisms

Figure 1
The winner of the race is the car that has the highest rate of travel.

Rate Describes Change over Time

At 500°C, the compound dimethyl ether slowly decomposes according to the equation below to give three products—methane, carbon monoxide, and hydrogen gas.

$$CH_3OCH_3(g) \longrightarrow CH_4(g) + CO(g) + H_2(g)$$

The concentration of dimethyl ether will keep decreasing during the reaction. Recall that the symbol Δ represents a change in some quantity. If the concentration of dimethyl ether changes by $\Delta[CH_3OCH_3]$ during a small time interval Δt, then the rate of the reaction is defined as

$$rate = \frac{-\Delta[CH_3OCH_3]}{\Delta t}$$

The sign is negative because, while $\Delta[CH_3OCH_3]$ is negative, the rate during the reaction must be a positive number.

The chemical equation shows that for every mole of dimethyl ether that decomposes, 1 mol each of methane, carbon monoxide, and hydrogen is produced. Thus, the concentrations of CH_4, CO, and H_2 will *increase* at the same rate that $[CH_3OCH_3]$ *decreases*. This means that the rate for this reaction can be defined in terms of the changes in concentration of any one of the products, as shown below.

$$rate = \frac{-\Delta[CH_3OCH_3]}{\Delta t} = \frac{\Delta[CH_4]}{\Delta t} = \frac{\Delta[CO]}{\Delta t} = \frac{\Delta[H_2]}{\Delta t}$$

The concentrations of the products are all increasing, so the signs of their rate expressions are positive.

Quick LAB

Concentration Affects Reaction Rate

PROCEDURE

1. Prepare two labeled beakers, one containing 0.001 M hydrochloric acid and the second containing 0.1 M hydrochloric acid.

2. Start a **stopwatch** at the moment you drop an **effervescent tablet** into the first beaker.

3. Stop the stopwatch when the tablet has finished dissolving.

4. Repeat steps 2–3 with a second **effervescent tablet,** using the second beaker.

ANALYSIS

1. What evidence is there that a chemical reaction occurred?

2. Were the dissolution times different? Did the tablet dissolve faster or slower in the more concentrated solution?

3. What conclusion can you draw about how the rate of a chemical reaction depends on the concentration of the reactants?

Balanced Coefficients Appear in the Rate Definition

Now consider the following reaction, which is the one illustrated in **Figure 2** below.

$$2N_2O_5(s) \longrightarrow 4NO_2(g) + O_2(g)$$

The stoichiometry is more complicated here because 2 mol of dinitrogen pentoxide produce 4 mol of nitrogen dioxide and 1 mol of oxygen. So, it is no longer true that the rate of decrease of the reactant concentration equals the rates of increase of the product concentrations. However, this difficulty can be overcome if, in order to define the reaction rate, we divide by the coefficients from the balanced equation. For this reaction, we get the following.

$$\text{rate} = \frac{-\Delta[N_2O_5]}{2\Delta t} = \frac{\Delta[NO_2]}{4\Delta t} = \frac{\Delta[O_2]}{\Delta t}$$

reaction rate

the rate at which a chemical reaction takes place; measured by the rate of formation of the product or the rate of disappearance of the reactants

The definition of **reaction rate** developed in these two examples may be generalized to cover any reaction.

It is important to realize that a reaction does not have a single, specific rate. Reaction rates depend on conditions such as temperature and pressure. Also, the rate of a reaction changes *during* the reaction. Usually, the rate decreases gradually as the reaction proceeds. The rate becomes zero when the reaction is complete.

Figure 2
Dinitrogen pentoxide decomposes to form oxygen and the orange-brown gas nitrogen dioxide.

Table 1	Concentration Data and Calculations for the Decomposition of N_2O_5				
t (s)	$[NO_2]$ (M)	$\Delta[NO_2]$ (M)	Δt (s)	$\Delta[NO_2]/\Delta t$ (M/s)	Rate (M/s)
0	0				
		4.68×10^{-3}	20.0	2.34×10^{-4}	5.85×10^{-5}
20.0	0.00468				
		4.22×10^{-3}	20.0	2.11×10^{-4}	5.28×10^{-5}
40.0	0.00890				
		3.82×10^{-3}	20.0	1.91×10^{-4}	4.78×10^{-5}
60.0	0.01272				
		3.44×10^{-3}	20.0	1.72×10^{-4}	4.30×10^{-5}
80.0	0.01616				

Reaction Rates Can Be Measured

To measure a reaction rate, you need to be able to keep track of how the concentration of one or more reactants or products changes over time. There are many ways of tracking these changes depending on the reaction you are studying.

For the reaction in **Figure 2,** you could measure how quickly the concentration of one product changes by measuring a change in color. Because nitrogen dioxide is the only gas in the reaction that has a color, you could use the red-brown color of the gas mixture to calculate $[NO_2]$. On the other hand, because the pressure of the system changes during the reaction, you could measure this change and, with help from the gas laws, calculate the concentrations.

internet connect

www.scilinks.org
Topic: Factors Affecting Equilibrium
SciLinks code: HW04057

SC*LINKS* Maintained by the National Science Teachers Association

Concentrations Must Be Measured Often

Remember that the Δt that occurs in the equations defining reaction rate is a *small* time interval. This means that studies of chemical kinetics require that concentrations be measured frequently. **Table 1** shows the results from a study of the following reaction.

$$2N_2O_5(g) \longrightarrow 4NO_2(g) + O_2(g)$$

The NO_2 concentrations were used to calculate the reaction rate in this example, but $[N_2O_5]$ or $[O_2]$ data could also have been used. As expected, the reaction rate decreases with time. It takes about 900 s before the reaction is 99% complete, and at that point, the rate is only 6.2×10^{-7} M/s. Reaction rates are generally expressed, as they are here, in moles per liter-second or M/s.

Notice in the table how the rate is calculated from pairs of data points—two different time readings and two different concentrations of NO_2. For example, the last rate in the table comes from the calculation shown below.

$$\text{rate} = \frac{\Delta[NO_2]}{4\Delta t} = \frac{0.01616 \text{ M} - 0.01272 \text{ M}}{4(80.0 \text{ s} - 60.0 \text{ s})} = 4.30 \times 10^{-5} \text{ M/s}$$

This result shows the rate of the reaction after it has been going on for about 70 s.

Figure 3
The graph shows the changes in concentration with time during the decomposition of dinitrogen pentoxide. The points represent the data used in **Table 1.**

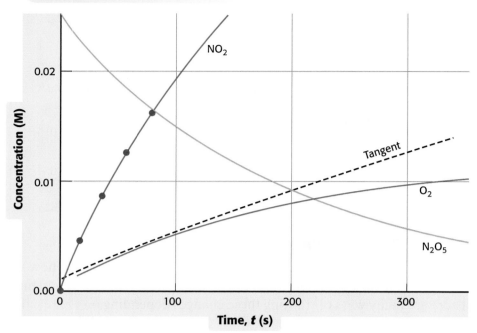

Reaction Rates Can Be Represented Graphically

Chemists often use graphs to help them think about chemical changes. Graphs are especially helpful in the field of chemical kinetics. For one example of how a graph can be useful, we can take another look at the decomposition reaction $2N_2O_5(g) \longrightarrow 4NO_2(g) + O_2(g)$. **Figure 3** is a graph that keeps track of this reaction with three curves, which show how the concentrations of the reactant and the products change with time. Notice that the concentration of dinitrogen pentoxide steadily falls. Also note that the concentration of oxygen and the concentration of nitrogen dioxide steadily increase.

Finally, notice that the graph also shows that the concentration of nitrogen dioxide increases four times faster than the concentration of oxygen increases. This result agrees with the 4:1 ratio of nitrogen dioxide to oxygen in the balanced equation.

Now, when some quantity is plotted versus time, the slope of the line tells you how fast that quantity is changing with time. So the slopes of the three curves in **Figure 3** measure the rates of change of each concentration. The slope of a curve at a particular point is just the slope of a straight line drawn as a tangent to the curve at that point. Because oxygen is a product and its coefficient in the equation is 1, the slope of the O_2 curve is simply the reaction rate.

$$\text{slope of } O_2 \text{ curve} = \frac{\Delta[O_2]}{\Delta t} = \text{rate of the reaction}$$

A line has been drawn as a tangent to the O_2 curve at $t = 70$ s. Its slope was measured in the usual way as rise/run and is 4.30×10^{-5} M/s. This value agrees with the rate calculated in **Table 1** at the same instant.

Calculating a Reaction Rate

The data below were collected during a study of the following reaction.

$$2Br^-(aq) + H_2O_2(aq) + 2H_3O^+(aq) \longrightarrow Br_2(aq) + 4H_2O(l)$$

Time t (s)	[H₃O⁺] (M)	[Br₂] (M)
0	0.0500	0
85	0.0298	0.0101
95	0.0280	0.0110
105	0.0263	0.0118

Use two methods to calculate what the reaction rate was after 100 s.

1 Gather information.

During the interval $\Delta t = 10$ s between $t = 95$ s and $t = 105$ s, the changes in the concentrations of hydronium ion and bromine were

$$\Delta[H_3O^+] = (0.0263\ M) - (0.0280\ M) = -0.0017\ M$$
$$\Delta[Br_2] = (0.0118\ M) - (0.0110) = 0.0008\ M$$

2 Plan your work.

For this reaction, two definitions of the reaction rate are as follows.

$$\text{rate} = \frac{-\Delta[H_3O^+]}{2\Delta t} = \frac{\Delta[Br_2]}{\Delta t}$$

3 Calculate.

From the change in hydronium ion concentration,

$$\text{rate} = \frac{-\Delta[H_3O^+]}{2\Delta t} = \frac{-(-0.0017\ M)}{2(10\ s)} = 8.5 \times 10^{-5}\ M/s$$

From the change in bromine concentration,

$$\text{rate} = \frac{\Delta[Br_2]}{\Delta t} = \frac{0.0008\ M}{10\ s} = 8 \times 10^{-5}\ M/s$$

4 Verify your results.

The two ways of solving the problem provide approximately the same answer.

PRACTICE HINT

The coefficient from the chemical equation, unless it is 1, must be included when calculating a reaction rate.

PRACTICE

1. For the reaction in **Sample Problem A,** write the expressions that define the rate in terms of the hydrogen peroxide and bromide ion concentrations.

2. The initial rate of the $N_2O_4(g) \longrightarrow 2NO_2(g)$ reaction is 7.3×10^{-6} M/s. What are the rates of concentration change for the two gases?

3. Use the data from **Sample Problem A** to calculate the reaction rate after 90 s.

PROBLEM SOLVING SKILL

Factors Affecting Rate

Concentration, pressure, temperature, and surface area are the most important factors on which the rate of a chemical reaction depends. Consider each of these effects for a type of reaction that is already familiar to you—combustion.

You know that the more fuel and oxygen there is, the faster a fire burns. This is an example of the general principle that the rate of a chemical reaction increases as the concentration of a reactant increases.

Many combustion processes, such as those of sulfur or wood, take place at a surface. The larger the surface area, the greater the chances that each particle will be involved in a reaction.

Concentration Affects Reaction Rate

Topic Link

Refer to the "Chemical Reactions and Equations" chapter for a discussion of collisions between molecules.

Though there are exceptions, almost all reactions, including the one shown in **Figure 4,** increase in rate when the concentrations of the reactants are increased.

It is easy to understand why reaction rates increase as the concentrations of the reactants increase. Think about the following reaction taking place within a container.

$$NO_2(g) + CO(g) \longrightarrow NO(g) + CO_2(g)$$

Clearly, the reaction can take place only when a nitrogen dioxide molecule collides with a carbon monoxide molecule. If the concentration of NO_2 is doubled, there are twice as many nitrogen dioxide molecules, and so the number of collisions with CO molecules will double. Only a very small fraction of those collisions will actually result in a reaction. Even so, the possibility that each reaction will take place is twice as much when the NO_2 concentration is doubled.

Reaction rates decrease with time because the reaction rate depends on the concentration of the reactants. As the reaction proceeds, the reactant is consumed and its concentration declines. This change in concentration, in turn, decreases the reaction rate.

Figure 4
Carbon burns faster in pure oxygen **a** than in air **b** because the concentration of the reacting species, O_2, is greater.

a

b

Concentration Affects Noncollision Reaction Rates

Not all reactions require a collision. The gas cyclopropane has a molecule in which three bonded carbon atoms form a triangle, with two hydrogen atoms attached to each carbon atom. Above room temperature, cyclopropane slowly changes into propene.

$$(CH_2)_3(g) \longrightarrow CH_2{=}CH{-}CH_3(g)$$

A collision is not necessary for this reaction, but the rate of the reaction still increases as the concentration of cyclopropane increases. In fact, the rate doubles if the $(CH_2)_3$ concentration doubles. This is not surprising. Because there are twice as many molecules, their reaction is twice as likely, and so the reaction rate doubles.

Pressure Affects the Rates of Gas Reactions

Pressure has almost no effect on reactions taking place in the liquid or solid states. However, it does change the rate of reactions taking place in the gas phase, such as the reaction shown in **Figure 5.**

As the gas laws confirm, doubling the pressure of a gas doubles its concentration. So changing the pressure of a gas or gas mixture is just another way of changing the concentration.

Temperature Greatly Influences the Reaction Rate

All chemical reactions are affected by temperature. In almost every case, the rate of a chemical reaction *increases* with increasing temperature. The increase in rate is often very large. A temperature rise of only 10%, say from 273 K to 300 K, will frequently increase the reaction rate tenfold. Our bodies work best at around 37°C or 310 K. Even a 1°C change in body temperature affects the rates of the body's chemical reactions enough that we may become ill as a result.

■ internet connect

www.scilinks.org
Topic: Factors Affecting
Reaction Rate
SciLinks code: HW4058

SCi
LINKS. Maintained by the
National Science
Teachers Association

Figure 5
This reaction between two gases, ammonia and hydrogen chloride, forms solid ammonium chloride in a white ring near the center of the glass tube.

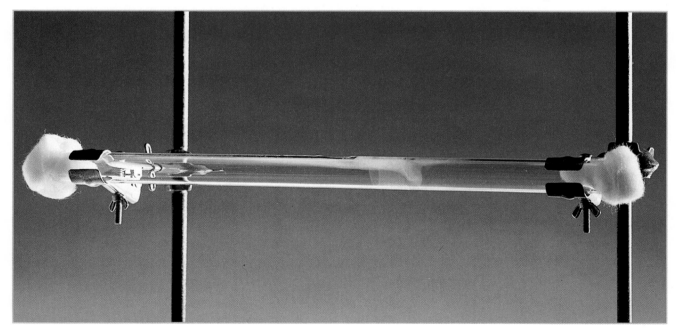

Temperature Affects Reactions in Everyday Life

The fact that reaction rates respond to temperature changes is part of everyday life. In the kitchen, we increase the temperature to speed up the chemical processes of cooking food, and we lower the temperature to slow down the chemical processes of food spoilage. When you put food in a refrigerator, you slow down the chemical reactions that cause food, such as the grapes shown in **Figure 6** to decompose. Most manufacturing operations use either heating or cooling to control their processes for optimal performance.

Why do chemical reactions increase in rate so greatly when the temperature rises? You have seen, in discussing reactions such as $NO_2(g)$ + $CO(g) \longrightarrow NO(g) + CO_2(g)$, that a collision between molecules (or other particles, such as ions or atoms) is necessary for a reaction to occur. A common misconception is that a rise in temperature increases the number of collisions and thereby boosts the reaction rate. It is true that a temperature rise does increase the collision frequency somewhat, but that effect is small. The main reason for the increase in reaction rate is that a temperature rise increases the fraction of molecules that have an energy great enough for collision to lead to reaction. If they are to react, molecules must collide with enough energy to rearrange bonds. A rise in temperature means that many more molecules have the required energy.

Surface Area Can Be an Important Factor

Most of the reactions that we have considered so far happen uniformly in three-dimensional space. However, many important reactions—such as precipitations, corrosions, and many combustions—take place at surfaces. The definition of *rate* given earlier does not apply to surface reactions. Even so, these reactions respond to changes in concentration, pressure, and temperature in much the same way as do other reactions.

A feature of surface reactions is that the amount of matter that reacts is proportional to the surface area. As **Figure 7** shows, you get a bigger blaze with small pieces of wood, because the surface area of many small pieces is greater than that of one larger piece of wood.

Figure 7

a Division of a solid makes the exposed surface of the solid larger.

b More divisions mean more exposed surface.

c Hence, more surface is available for other reactant molecules to come together.

Section Review

UNDERSTANDING KEY IDEAS

1. What does the word *rate* mean in everyday life, and what do chemists mean by *reaction rate*?

2. What is the name given to the branch of chemistry dealing with reaction rates? Why are such studies important?

3. Why is a collision between molecules necessary in many reactions?

4. How may reaction rates be measured?

5. Explain why reactant concentration influences the rate of a chemical reaction.

6. Give examples of the strong effect that temperature has on chemical reactions.

7. What is unique about surface reactions?

CRITICAL THINKING

8. Why must coefficients be included in the definition of reaction rate?

9. Calculating the reaction rate from a product appeared to give an answer different from that calculated from a reactant. Suggest a possible explanation.

10. The usual unit for reaction rate is M/s. Suggest a different unit that could be used for reaction rate, and explain why this unit would be appropriate.

11. Explain why an increase in the frequency of collisions is not an adequate explanation of the effect of temperature on reaction rate.

12. Would the factors that affect the rate of a chemical reaction influence a physical change in the same way? Explain, and give an example.

13. Why does pressure affect the rates of gas reactions?

How Can Reaction Rates Be Explained?

rate law

the expression that shows how the rate of formation of product depends on the concentration of all species other than the solvent that take part in a reaction

reaction mechanism

the way in which a chemical reaction takes place; expressed in a series of chemical equations

order

in chemistry, a classification of chemical reactions that depends on the number of molecules that appear to enter into the reaction

OBJECTIVES

(1) **Write** a rate law using experimental rate-versus-concentration data from a chemical reaction.

(2) **Explain** the role of activation energy and collision orientation in a chemical reaction.

(3) **Describe** the effect that catalysts can have on reaction rate and how this effect occurs.

(4) **Describe** the role of enzymes as catalysts in living systems, and give examples.

Rate Laws

You have learned that the rate of a chemical reaction is affected by the concentration of the reactant or reactants. The **rate law** describes the way in which reactant concentration affects reaction rate. A rate law may be simple or very complicated, depending on the reaction.

By studying rate laws, chemists learn *how* a reaction takes place. Researchers in chemical kinetics can often make an informed guess about the **reaction mechanism.** In other words, they can create a model to explain how atoms move in rearranging themselves from reactants into products.

Determining a General Rate Law Equation

For a reaction that involves a single reactant, the rate is often proportional to the concentration of the reactant raised to some power. That is, the rate law takes the following form.

$$\text{rate} = k[\text{reactant}]^n$$

This is a general expression for the rate law. The exponent, n, is called the **order** of the reaction. It is usually a whole number, often 1 or 2, but it could be a fraction. Occasionally, n equals 0, which means that the reaction rate is independent of the reactant concentration. The term k is the *rate constant,* a proportionality constant that varies with temperature.

Reaction orders cannot be determined from a chemical equation. They must be found by experiment. For example, you might guess that $n = 1$ for the following reaction.

$$CH_3CHO(g) \longrightarrow CH_4(g) + CO(g)$$

However, experiments have shown that the reaction order is 1.5.

Determining a Rate Law

Three experiments were performed to measure the initial rate of the reaction $2HI(g) \longrightarrow H_2(g) + I_2(g)$. Conditions were identical in the three experiments, except that the hydrogen iodide concentrations varied. The results are shown below.

Experiment	[HI] (M)	Rate (M/s)
1	0.015	1.1×10^{-3}
2	0.030	4.4×10^{-3}
3	0.045	9.9×10^{-3}

1 Gather information.

The general rate law for this reaction is as follows: $rate = k[HI]^n$
$n = ?$

2 Plan your work.

Find the ratio of the reactant concentrations between experiments 1 and 2, $\dfrac{[HI]_2}{[HI]_1}$

Then see how this affects the ratio $\dfrac{(rate)_2}{(rate)_1}$ of the reaction rates.

3 Calculate.

$$\dfrac{[HI]_2}{[HI]_1} = \dfrac{0.030 \text{ M}}{0.015 \text{ M}} = 2.0 \qquad \dfrac{(rate)_2}{(rate)_1} = \dfrac{4.4 \times 10^{-3} \text{ M/s}}{1.1 \times 10^{-3} \text{ M/s}} = 4.0$$

Thus, when the concentration changes by a factor of 2, the rate changes by 4, or 2^2. Hence n, the reaction order, is 2.

4 Verify your results.

On inspecting items 1 and 3 in the table, one sees that when the concentration triples, the rate changes by a factor of 9, or 3^2. This confirms that the order is 2.

PRACTICE HINT

To find a reaction order, compare a rate ratio with a concentration ratio.

PRACTICE

1 In a study of the $2NH_3(g) \longrightarrow N_2(g) + 3H_2(g)$ reaction, when the ammonia concentration was changed from 3.57×10^{-3} M to 5.37×10^{-3} M, the rate increased from 2.91×10^{-5} M/s to 4.38×10^{-5} M/s. Find the reaction order.

2 What is the order of a reaction if its rate increases by a factor of 13 when the reactant concentration increases by a factor of 3.6?

3 What concentration increase would cause a tenfold increase in the rate of a reaction of order 2?

4 When the CH_3CHO concentration was doubled in a study of the $CH_3CHO(g) \longrightarrow CH_4(g) + CO(g)$ reaction, the rate changed from 7.9×10^{-5} M/s to 2.2×10^{-4} M/s. Confirm that the order is 3/2.

PROBLEM SOLVING SKILL

Reaction Mixtures for NO + O₃ ⟶ NO₂ + O₂

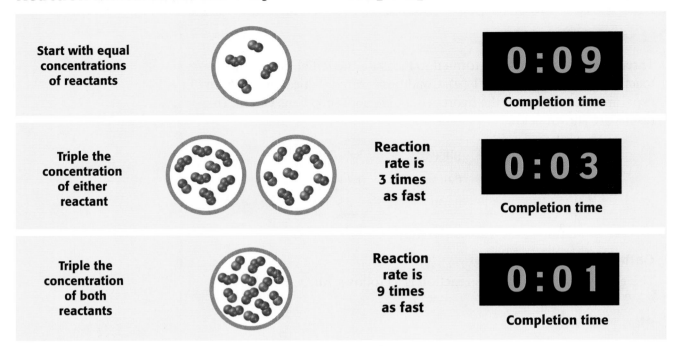

Start with equal concentrations of reactants			**0:09** Completion time
Triple the concentration of either reactant		Reaction rate is 3 times as fast	**0:03** Completion time
Triple the concentration of both reactants		Reaction rate is 9 times as fast	**0:01** Completion time

Figure 8
Nitrogen monoxide reacts with ozone. Increasing the concentration of either NO or O₃ will increase the reaction rate.

Rate Laws for Several Reactants

When a reaction has more than one reactant, a term in the rate law corresponds to each. There are three concentration terms in the rate law for the following reaction.

$$2Br^-(aq) + H_2O_2(aq) + 2H_3O^+(aq) \longrightarrow Br_2(aq) + 4H_2O(l)$$

There is an order associated with each term:

$$\text{rate} = k[Br^-]^{n_1}[H_2O_2]^{n_2}[H_3O^+]^{n_3}$$

For example, n_1 is the reaction order with respect to Br^-.

To be sure of the orders of reactions that have several reactants, one must perform many experiments. Often the concentration of only a single reactant is varied during a series of experiments. Then a new series is begun and a second reactant is varied, and so on.

Figure 8 shows the results of changing conditions during a study of the reaction represented by the equation below.

$$NO(g) + O_3(g) \longrightarrow NO_2(g) + O_2(g)$$

This is an important reaction because it participates in the destruction of the ozone layer high in the atmosphere. There are two terms in the rate law for this reaction, which is shown below.

$$\text{rate} = k[NO]^{n_1}[O_3]^{n_2}$$

In this case, it turns out that $n_1 = n_2 = 1$. The fact that the orders for each reactant are equal to one suggests that this reaction has a simple one-step mechanism in which an oxygen atom is transferred when the two reactant molecules collide.

Rate-Determining Step Controls Reaction Rate

Although a chemical equation can be written for the overall reaction, it does not usually show how the reaction actually takes place. For example, the reaction shown below is believed to take place in four steps, in the mechanism that follows.

www.scilinks.org
Topic: Reaction Mechanisms
SciLinks code: HW4162

SCiLINKS. Maintained by the National Science Teachers Association

$$2Br^-(aq) + H_2O_2(aq) + 2H_3O^+(aq) \longrightarrow Br_2(aq) + 4H_2O(l)$$

The order with respect to each of the three reactants was found to be 1.

(1) $\qquad Br^-(aq) + H_3O^+(aq) \rightleftarrows HBr(aq) + H_2O(l)(1)$

(2) $\qquad HBr(aq) + H_2O_2(aq) \longrightarrow HOBr(aq) + H_2O(l)$

(3) $\qquad Br^-(aq) + HOBr(aq) \rightleftarrows Br_2(aq) + OH^-(aq)$

(4) $\qquad OH^-(aq) + H_3O^+(aq) \rightleftarrows 2H_2O(l)$

These four steps add up to the overall reaction that was shown above. Three of the steps are shown as equilibria; these are fast reactions. Step 2, however, is slow. If one step is slower than the others in a sequence of steps, it will control the overall reaction rate, because a reaction cannot go faster than its slowest step. Such a step is known as the **rate-determining step.** Step 2 is the rate-determining step of the mechanism shown by steps 1–4. Species such as HOBr that form during a reaction but are then consumed are called **intermediates.**

rate-determining step

in a multistep chemical reaction, the step that has the lowest velocity, which determines the rate of the overall reaction

intermediate

a substance that forms in a middle stage of a chemical reaction and is considered a stepping stone between the parent substance and the final product

Quick LAB

SAFETY PRECAUTIONS

Modeling a Rate-Determining Step

PROCEDURE

1. Attach a **large-bore funnel** above a **small-bore funnel** onto a **ring stand.** Set a large **bowl** on the table, directly below the funnels.

2. Pour one cup of **sand** into the top funnel, and start a **stopwatch.**

3. When the last of the sand has fallen into the bowl, stop the stopwatch.

4. Write down the elapsed time.

5. Repeat steps 1 through 4 using the large-bore funnel above a **medium-bore funnel.**

6. Repeat steps 1 through 4 using the medium-bore funnel above the small-bore funnel.

7. Repeat steps 1 through 4 using the small-bore funnel above the large-bore funnel.

ANALYSIS

1. Which combination of funnels made the process go the fastest?

2. Which funnel controlled the rate of the process?

3. Does reversing the order of the two funnels in a trial change the results? Explain.

4. What strengths does this process have as a model for a chemical reaction? What weaknesses does it have?

Reaction Pathways and Activation Energy

Topic Link

Refer to the "Gases" chapter for a discussion of the energy distribution of gas molecules.

If two molecules approach each other, the outer electrons of each molecule repel the outer electrons of the other. So, ordinarily, the molecules just bounce off each other. For two molecules to react, they must collide violently enough to overcome the mutual repulsion, so that the electron clouds of the two molecules merge to some extent. This merging may lead to a distortion of the shapes of the colliding molecules and, ultimately, to the creation of new bonds.

Violent collisions happen only when the colliding pair of molecules have an unusually large amount of energy. The kinetic energies of individual gas molecules vary over a wide range. Only the molecules with especially high kinetic energy are likely to react. The other molecules must wait until a succession of "lucky" collisions brings their kinetic energies up to the necessary amount.

The minimum energy that a pair of colliding molecules (or atoms or ions) need to have before a chemical change becomes a possibility is called the **activation energy** of the reaction. It is represented by the symbol E_a. No reaction is possible if the colliding pair has less energy than E_a.

activation energy

the minimum energy required to start a chemical reaction

Activation-Energy Diagrams Model Reaction Progress

Imagine rolling a ball toward a speed bump in a parking lot. If you do not give the ball enough kinetic energy, it will roll partway up the bump, stop, reverse its direction, and come back toward you. If you give it enough energy, the ball will make it just to the top of the bump and stay there for a moment. After that, it may go either way. Given plenty of energy, the ball will pass easily over the bump. Then, gaining more kinetic energy as it descends, it will roll away down the far side of the speed bump.

The model of the ball and speed bump provides a good analogy of the reaction between two colliding molecules. Without enough kinetic energy, the two molecules will not change chemically. With a combined kinetic energy equal to the activation energy, the molecules reach a state where there is a 50:50 chance of either returning to the initial state without reacting, or of being rearranged and becoming products. This point, similar to the top of the speed bump, is called the **activated complex** or *transition state* of the reaction.

activated complex

a molecule in an unstable state intermediate to the reactants and the products in the chemical reaction.

Figure 9a is a graph of how the energy changes as a pair of hydrogen iodide molecules collide, form an activated complex, and then go on to become hydrogen and iodine molecules. As a chemical equation, the process could be written as follows.

$$2HI \quad \longrightarrow \quad H_2I_2 \quad \longrightarrow \quad H_2 + I_2$$

initial state	activated	final state
(reactant)	complex	**(products)**

In the initial state, the bonds are between the hydrogen and iodine atoms, H−I. In the activated complex, four weak bonds link the four atoms into a deformed square. In the final state the bonds link hydrogen to hydrogen, H−H, and iodine to iodine, I−I.

Activation Energies for the Decomposition of HI and HBr

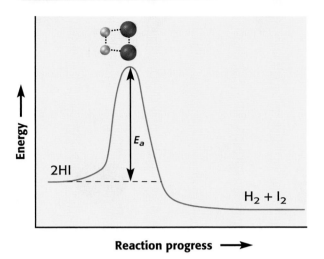

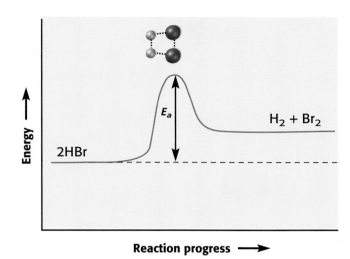

Figure 9

a The difference in energy between the bottom of this curve and the peak is the energy of activation for the decomposition of HI.

b The decomposition of HBr occurs at a faster rate than the decomposition of HI because this reaction has a lower activation energy.

Hydrogen Bromide Requires a Different Diagram

Figure 8b similarly represents how potential energy changes with reaction progress for the reaction below.

$$2HBr \longrightarrow H_2Br_2 \longrightarrow H_2 + Br_2$$

initial state   activated   final state

(reactant)   complex   **(products)**

One difference between the two graphs is that the activation energy is lower in the case of hydrogen bromide. Because the activation energy of HBr is lower than that of HI, a larger fraction of the HBr molecules have enough energy to clear the activation energy barrier than in the HI case. As a result, hydrogen bromide decomposes more quickly than hydrogen iodide does.

Notice in both **Figure 9** graphs that the initial states are not at the same energy as the final states. Note also that the products have a lower energy than the reactants in the case of the HI decomposition reaction in **Figure 9a,** while the opposite is true for hydrogen bromide decomposition in **Figure 9b.** This distinction reflects the fact that hydrogen iodide decomposition is exothermic,

$$2HI(g) \longrightarrow H_2(g) + I_2(g) \qquad \Delta H = -53 \text{ kJ}$$

while the decomposition of hydrogen bromide is endothermic.

$$2HBr(g) \longrightarrow H_2(g) + Br_2(g) \qquad \Delta H = 73 \text{ kJ}$$

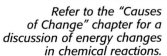

Refer to the "Causes of Change" chapter for a discussion of energy changes in chemical reactions.

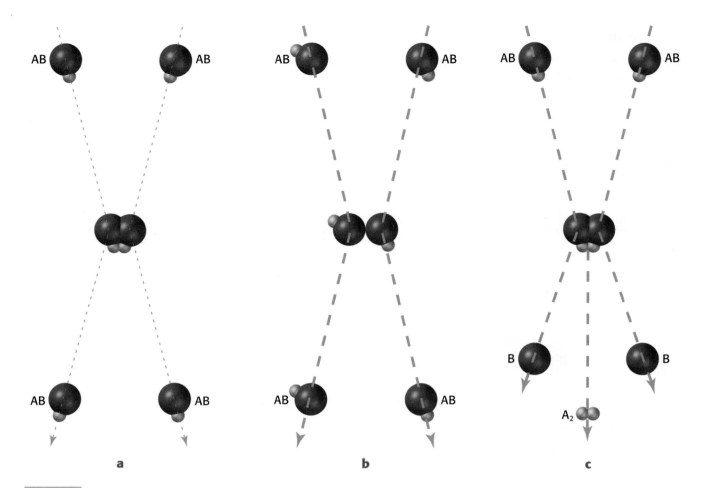

a   b   c

Figure 10
A reaction will not occur if the collision occurs too gently, as in **a,** or with the wrong orientation as in **b.** An effective collision, as in **c,** must deliver sufficient energy and bring together the atoms that bond in the products.

Not All Collisions Result in Reaction

Much of what we know about the collisions of molecules (and atoms) has come from studies of reactions between gases. However, it is believed that collisions happen similarly in solution. The principles of rate laws and activation energies apply in reactions that occur in solutions as well as in gas-phase reactions.

Collision between the reacting molecules is necessary for almost all reactions. Collision is not enough, though. The molecules must collide with enough energy to overcome the activation energy barrier. But another factor is also important. **Figure 10** illustrates the need for adequate energy and correct orientation in a collision.

A chemical reaction produces new bonds, and those bonds are formed between specific atoms in the colliding molecules. Unless the collision brings the correct atoms close together and in the proper orientation, the molecules will not react, no matter how much kinetic energy they have. For example, if a chlorine molecule collides with the oxygen end of the nitrogen monoxide molecule, the following reaction may occur.

$$NO(g) + Cl_2(g) \longrightarrow NOCl(g) + Cl(g)$$

This reaction will not occur if the chlorine molecule strikes the nitrogen end of the molecule.

Catalysts Increase Reaction Rate

Adding more reactant will usually increase the rate of a reaction. Adding extra product will sometimes cause the rate to decrease. Often, adding substances called **catalysts** to a reaction mixture will increase the reaction rate, even though the catalyst is still present and unchanged at the end of the reaction. The process, which is called **catalysis,** is shown in **Figure 11.**

Hydrogen peroxide solution, commonly used as a mild antiseptic and as a bleaching agent, decomposes only very slowly when stored in a bottle, forming oxygen as shown in the following equation.

$$2H_2O_2(aq) \longrightarrow 2H_2O(l) + O_2(g)$$

Adding a drop of potassium iodide solution speeds up the reaction. On the other hand, adding a few crystals of insoluble manganese dioxide, $MnO_2(s)$, causes a violent decomposition to occur. The iodide ion, $I^-(aq)$, and manganese dioxide are two of many catalysts for the decomposition of hydrogen peroxide.

Catalysis is widely used in the chemical industry, particularly in the making of gasoline and other petrochemicals. Catalysts save enormous amounts of energy. As you probably know, carbon monoxide is a poisonous gas that is found in automobile exhaust. The following oxidation reaction could remove the health hazard, but this reaction is very slow.

$$2CO(g) + O_2(g) \longrightarrow 2CO_2(g)$$

It is the job of the catalytic converter, built into the exhaust system of all recent models of cars, to catalyze this reaction.

Catalysis does not change the *overall* reaction at all. The stoichiometry and thermodynamics of the reaction are not changed. The changes affect only the path the reaction takes from reactant to product.

catalyst

a substance that changes the rate of a chemical reaction without being consumed or changed significantly

catalysis

the acceleration of a chemical reaction by a catalyst

Figure 11
Some catalysts work better than others. For example, MnO_2 is a better catalyst for the decomposition of H_2O_2 than I^-.

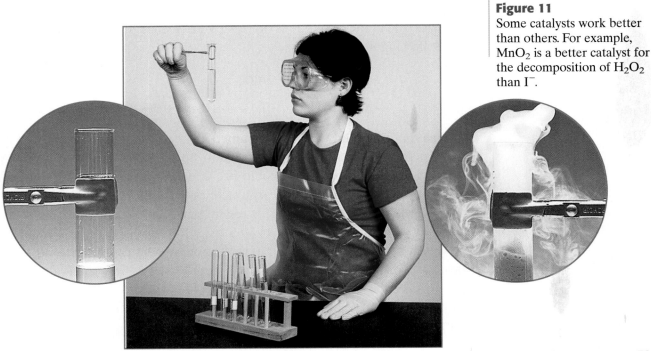

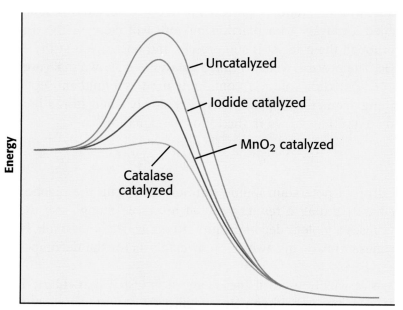

Comparison of Pathways for the Decomposition of H₂O₂

Figure 12
The four curves show that various catalysts reduce the activation energy for the hydrogen peroxide decomposition reaction, but by different amounts. Notice that the enzyme catalase almost cancels the activation energy.

Catalysts Lower the Activation Energy Barrier

Catalysis works by making a different pathway available between the reactants and the products. This new pathway has a different mechanism and a different rate law from that of the uncatalyzed reaction. The catalyzed pathway may involve a surface reaction, as in the decomposition of hydrogen peroxide catalyzed by manganese dioxide, and in biological reactions catalyzed by enzymes. Or, the catalytic mechanism may take place in the same phase as the uncatalyzed reaction.

The iodide-catalyzed decomposition of hydrogen peroxide is an example of catalysis that does not involve a surface. It probably works by the following mechanism.

(1) $$I^-(aq) + H_2O_2(aq) \longrightarrow IO^-(aq) + H_2O(l)$$

(2) $$IO^-(aq) + H_2O_2(aq) \longrightarrow I^-(aq) + O_2(g) + H_2O(l)$$

Notice that the iodide ion, I^-, consumed in step 1 is regenerated in step 2, and the hypoiodite ion, IO^-, generated in step 1 is consumed in step 2. In principle, a single iodide ion could break down an unlimited amount of hydrogen peroxide. This is the characteristic of all catalytic pathways—the catalyst is never used up. It is regenerated and so becomes available for use again and again.

Each pathway corresponds to a different mechanism, a different rate law, and a different activation energy. **Figure 12** shows the potential energy profiles for the uncatalyzed reaction and for catalysis by three different catalysts. Because the catalyzed pathways have lower activation energy barriers, the catalysts speed up the rate of the reaction.

Enzymes Are Catalysts Found in Nature

The most efficient of the three catalysts compared in **Figure 12** is an **enzyme.** Enzymes are large protein molecules. Their biological role is to catalyze metabolic processes that otherwise would happen too slowly to help the organism. For example, the enzyme lactase catalyzes the reaction of water with the sugar lactose, present in milk. People whose bodies lack the ability to produce lactase have what is known as *lactose intolerance.*

Enzymes are very specific and catalyze only one reaction. This is because the surface of an enzyme molecule has a detailed arrangement of atoms that interacts with the target molecule (lactose, for instance). The enzyme site and the target molecule are often said to have a "lock and key" relationship to each other.

Hydrogen peroxide is a toxic metabolic product in higher animals, and the enzyme catalase is present in their blood and other tissues to destroy H_2O_2. On the other hand, the bombardier beetle stores a supply of hydrogen peroxide for use as a defense mechanism. When threatened by a predator, the beetle injects catalase into its hydrogen peroxide store. The rapidly released oxygen gas provides pressure for a spray of irritating liquid that the beetle can squirt at its enemy, as shown in **Figure 13.**

Figure 13
Bombardier beetles can repel predators such as frogs with a chemical defense mechanism powered by the catalytic decomposition of hydrogen peroxide.

enzyme

a type of protein that speeds up metabolic reactions in plants and animals without being permanently changed or destroyed

② Section Review

[A](M)	[B] (M)	Rate (M/s)
0.08	0.06	0.012
0.08	0.03	0.006
0.04	0.06	0.003

UNDERSTANDING KEY IDEAS

1. How can reaction orders be measured?

2. What can be learned from reaction orders?

3. Explain why not all collisions between reactant molecules lead to reaction.

4. What are catalysts and how do they function?

5. Give an example of an enzyme-catalyzed reaction.

PRACTICE PROBLEMS

6. What is the order of a reaction if its rate triples when the reactant concentration triples?

7. The reaction $CH_3NC(g) \longrightarrow CH_3CN(g)$ is of order 1, with a rate of 1.3×10^{-4} M/s when the reactant concentration was 0.040 M. Predict the rate when $[CH_3NC] = 0.025$ M.

8. The following data relate to the reaction $A + B \longrightarrow C$. Find the order with respect to each reactant.

CRITICAL THINKING

9. Which corresponds to the faster rate: a mechanism with a small activation energy or one with a large activation energy?

10. If the reaction $NO_2(g) + CO(g) \longrightarrow NO(g) + CO_2(g)$ proceeds by a one-step mechanism, what is the rate law?

11. What happens if a pair of colliding molecules possesses less energy than E_a?

12. Why is the phrase "lock and key" used to describe enzyme catalysis?

13. How are a catalyst and an intermediate similar? How are they different?

14. Draw a diagram similar to **Figure 10** to show (a) an unsuccessful and (b) a successful collision between $H_2(g)$ and $Br_2(g)$.

KEY IDEAS

SECTION ONE What Affects the Rate of a Reaction?

- The rate of a chemical reaction is calculated from changes in reactant or product concentration during a small time interval.
- Reaction rates generally increase with reactant concentration or, in the case of gases, pressure.
- Rate increases with temperature because at a higher temperature a greater fraction of collisions have enough energy to cause a reaction.

SECTION TWO How Can Reaction Rates Be Explained?

- Rate laws, which are used to suggest mechanisms, are determined by studying how reaction rate depends on concentration.
- An activated complex occupies the energy high point on the route from reactant to product.
- Catalysts provide a pathway of lower activation energy.
- Enzymes are biological catalysts that increase the rates of reactions important to an organism.

KEY TERMS

chemical kinetics
reaction rate

rate law
reaction mechanism
order
rate-determining step
intermediate
activation energy
activated complex
catalyst
catalysis
enzyme

KEY SKILLS

Calculating a Reaction Rate
Sample Problem A p. 581

Determining a Rate Law
Sample Problem B p. 587

USING KEY TERMS

1. Explain why you must use a negative sign when you write the rate expression for a chemical reaction in terms of the change in concentration of a reactant.

2. Define *reaction rate*.

3. Explain the difference between a reaction rate and a rate law.

4. What is a mechanism, and what is its rate-determining step?

5. Explain why the names *activated complex* and *transition state* are suitable for describing the highest energy point on a reaction's route from reactant to product.

6. Explain the role of an intermediate in a reaction mechanism.

7. What are enzymes, and what common features do they all share?

UNDERSTANDING KEY IDEAS

What Affects the Rate of a Reaction?

8. What unit is most commonly used to express reaction rate?

9. Explain how to calculate a reaction rate from concentration-versus-time data.

10. Mention uses of the term *rate* other than in chemical kinetics.

11. Explain how a graph can be useful in defining and measuring the rate of a chemical reaction.

12. Suggest ways of measuring concentration in a reaction mixture.

13. Why is it necessary to divide by the coefficient in the balanced chemical equation when calculating a reaction rate? When can that step be omitted?

14. What does $\Delta[A]$ mean if A is the reactant in a chemical reaction?

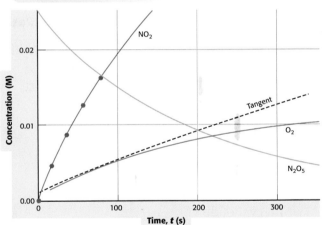

Figure 14

N_2O_5 Decomposition Data

15. In a graph like the one in **Figure 14,** what are the signs of the slopes for reactants and for products?

16. Which would have the greater effect on the rate of a chemical reaction in most cases—doubling the concentration of the reactant(s) or doubling the temperature?

17. Why does pressure affect the rate of the following reaction?

$$(CH_2)_3(g) \longrightarrow CH_2{=}CH{-}CH_3(g)$$

18. Explain the effect that area has on reactions that occur on surfaces.

19. Why are reaction orders not always equal to the coefficients in a chemical equation?

20. Write the general expression for the rate law of a reaction with three reactants A, B, and C.

21. Explain what a catalyst is and how it works.

22. Sketch a diagram showing how the potential energy changes with the progress of an endothermic reaction. Label the curve "Initial state," "Final state," and "Transition state." Then, draw a second curve to show the change brought about by a catalyst.

23. How do enzymes differ from other catalysts?

PRACTICE PROBLEMS

Sample Problem A Calculating a Reaction Rate

24. The concentration of a reaction product with a coefficient of 3 in the chemical equation increases by 1.5×10^{-4} M in 11 s. Calculate the reaction rate.

25. What is the rate of the reaction

$$2NO(g) + Br_2(g) \longrightarrow 2NOBr$$

given that the bromine concentration decreased by 5.3×10^{-5} M during an interval of 38 s?

26. During the same 38 s interval cited in problem 25, the nitric oxide concentration decreased by 1.04×10^{-4} M. Recalculate the rate.

27. Calculate the rate of a reaction, knowing that a graph of the concentration of a product versus time had a slope of 3.6×10^{-6} M/s. The product had a coefficient of 2.

28. Exposed to the air, a solution of potassium iodide slowly oxidizes in the following reaction.

$$6I^-(aq) + O_2(aq) + 2H_2O(l) \longrightarrow$$
$$2I_3^-(aq) + 4OH^-(aq)$$

A 250-mL volume of such a solution was found to contain 2.0×10^{-5} mol of I_3^- ions after 15 days. What was the reaction rate?

Sample Problem B Determining a Rate Law

29. In the reaction

$$2NO(g) + Br_2(g) \longrightarrow 2NOBr(g)$$

doubling the Br_2 concentration doubles the rate, but doubling the NO concentration quadruples the rate. Write the rate law.

30. What is the reaction order if the reaction rate triples when the concentration of a reactant is increased by a factor of 3?

31. The following reaction is first order.

$$(CH_2)_3(g) \longrightarrow CH_2= CH-CH_3(g)$$

What change in reaction rate would you expect if the pressure of $(CH_2)_3$ doubled?

32. The rate of the reaction

$$2NOCl(g) \longrightarrow 2NO(g) + Cl_2(g)$$

increased by a factor of 3.14 when the concentration of NOCl was changed by a factor of 1.77. What is the order of this reaction?

MIXED REVIEW

33. Explain why, even though a collision may have energy in excess of the activation energy, reaction may not occur.

34. Describe the procedure for measuring the slope of a curve.

35. What is the sign of $\Delta[A]$ if A is the reactant in a chemical reaction?

36. The names of most enzymes, for instance protease and carbonic anhydrase, end with *−ase*. Which one of these enzymes do you think would be the catalyst for the reaction shown below?

$$H_2CO_3(aq) \longrightarrow H_2O(l) + CO_2(aq)$$

(Hint: The other one breaks proteins into amino acids.)

37. $2SO_2(g) + O_2(g) \longrightarrow 2SO_3(g)$ is a slow reaction, but nitrogen dioxide acts as a catalyst. The first step in the proposed two-step mechanism is shown below.

$$SO_2(g) + NO_2(g) \longrightarrow SO_3(g) + NO(g).$$

Suggest the second step.

38. What is meant by the *rate-determining step* in a reaction mechanism?

39. When hydrogen peroxide solution, used as an antiseptic, is applied to a wound, it often bubbles. Explain why.

40. Using chemical terminology, explain the purpose of food refrigeration.

41. Why is a negative sign present in the equation defining the rate of a reaction in terms of the concentration of a reactant?

42. What is lactose? What is lactase?

43. Discuss whether or not an activated complex is an intermediate in a chemical reaction.

44. Why do reptiles move more sluggishly in cold weather?

45. Explain how to measure a reaction rate from a graph of reactant concentration versus time.

46. Write the equation for the decomposition of gaseous ammonia into its elements. Also write the general rate law. Are you able to assign a reaction order? Explain.

CRITICAL THINKING

47. Sketch a picture of the shape and bonding in the activated complex during the following reaction.

$$NO(g) + Cl_2(g) \longrightarrow NOCl(g) + Cl(g)$$

48. Why is it necessary, in defining the rate of a reaction, to require that Δt be small?

49. Point out any shortcomings of the "ball over a speed bump" analogy of a reaction path.

50. In a 1 L flask filled with NO_2 at 1 atm pressure, the content of dinitrogen tetroxide was 0.00027 mol after 15 s. Calculate the reaction rate. In a second experiment, at 2 atm of pressure, the N_2O_4 content after 15 s was 0.00108 mol. Write the rate law.

51. Explain why, unlike gas-phase reactions, a reaction in solution is hardly affected at all by pressure.

52. Could a catalyzed reaction pathway have an activation energy higher than the uncatalyzed reaction? Explain.

53. Would you expect the concentration of a catalyst to appear in the rate law of a catalyzed reaction? Explain.

ALTERNATIVE ASSESSMENT

54. Boilers are sometimes used to heat large buildings. Deposits of $CaCO_3$, $MgCO_3$, and $FeCO_3$ can hinder the boiler operation. Aqueous solutions of hydrochloric acid are commonly used to remove these deposits. The general equation for the reaction is written below.

$$MCO_3(s) + 2H_3O^+(aq) \longrightarrow$$
$$M^{2+}(aq) + 3H_2O(l) + CO_2(g)$$

In the equation, M stands for Ca, Mg, or Fe. Design an experiment to determine the effect of various HCl concentrations on the rates of this reaction. Present your design to the class.

CONCEPT MAPPING

55. Use the following terms to complete the concept map below: *activation energy, alternative reaction pathway, catalysts, enzymes,* and *reaction rate.*

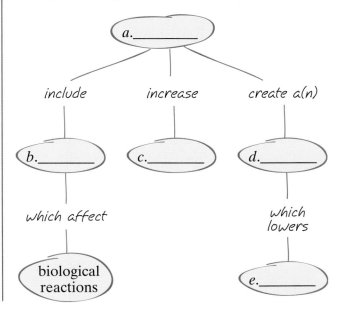

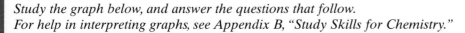

FOCUS ON GRAPHING

Study the graph below, and answer the questions that follow.
For help in interpreting graphs, see Appendix B, "Study Skills for Chemistry."

The graph relates to an experiment in which the concentrations of bromide ion, hydrogen peroxide, and bromine were monitored as the following reaction took place.

$$2Br^-(aq) + H_2O_2(aq) + 2H_3O^+(aq) \longrightarrow$$
$$Br_2(aq) + 4H_2O(l)$$

56. The three curves are lettered **a, b,** and **c.** Which curves have positive slopes and which have negative slopes?

57. Associate each curve with one of the species being monitored.

58. What were the initial concentrations of bromine and hydrogen peroxide?

59. Measure the slope of each of the three curves at $t = 500$ s.

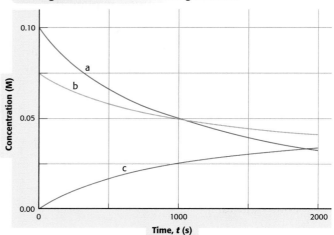

Changes in Concentration During Reaction

60. From each slope calculate a reaction rate. Do your three values agree?

TECHNOLOGY AND LEARNING

61. Graphing Calculator

Reaction Order

The graphing calculator can run a program that can tell you the order of a chemical reaction, provided you indicate the reactant concentrations and reaction rates for two experiments involving the same reaction.

Go to Appendix C. If you are using a TI-83 Plus, you can download the program **RXNORDER** and run the application as directed. If you are using another calculator, your teacher will provide you with keystrokes and data sets to use. At the prompts, enter the reactant concentrations and reaction rates. Run the program as needed to find the order of the following reactions. (All rates are given in M/s.)

a. $2N_2O_5(g) \longrightarrow 4NO_2(g) + O_2(g)$
N_2O_5: conc. 1 = 0.025 M; conc. 2 = 0.040 M
rate 1 = 8.1×10^{-5}; rate 2 = 1.3×10^{-4}

b. $2NO_2(g) \longrightarrow 2NO(g) + O_2(g)$
NO_2: conc. 1 = 0.040 M; conc. 2 = 0.080 M
rate 1 = 0.0030; rate 2 = 0.012

c. $2H_2O_2(g) \longrightarrow 2H_2O(g) + O_2(g)$
H_2O_2: conc. 1 = 0.522 M; conc. 2 = 0.887 M
rate 1 = 1.90×10^{-4}; rate 2 = 3.23×10^{-4}

d. $2NOBr(g) \longrightarrow 2NO(g) + Br_2(g)$
NOBr: conc. 1 = 1.27×10^{-4} M; conc. 2 = 4.04×10^{-4} M
rate 1 = 6.26×10^{-5}; rate 2 = 6.33×10^{-4}

e. $2HI(g) \longrightarrow H_2(g) + I_2(g)$
HI: conc. 1 = 4.18×10^{-4} M; conc. 2 = 8.36×10^{-4} M
rate 1 = 3.86×10^{-5}; rate 2 = 1.54×10^{-4}

1. The sequence of steps that occurs in a reaction process is called the
   a. heterogeneous reaction.
   b. rate law.
   c. overall reaction.
   d. reaction mechanism.

2. To be effective, a collision requires
   a. enough energy only.
   b. favorable orientation only.
   c. enough energy and a favorable orientation.
   d. a reaction mechanism.

3. The usual unit of reaction rate is
   a. mol/s.
   c. L/s.
   b. mol/L•s
   d. mol s/L.

4. How does the potential energy of the activated complex compare with the potential energies of the reactants and products?
   a. It is lower than both the potential energy of the reactants and the potential energy of the products.
   b. It is lower than the potential energy of the reactants but higher than the potential energy of the products.
   c. It is higher than the potential energy of the reactants but lower than the potential energy of the products.
   d. It is higher than both the potential energy of the reactants and the potential energy of the products.

5. If a collision between molecules is very gentle, the molecules are
   a. more likely to be orientated favorably.
   b. less likely to be orientated favorably.
   c. likely to react.
   d. likely to rebound without reacting.

6. A species that changes the rate of a reaction but is neither consumed nor changed is

   a. a catalyst.
   b. an activated complex.
   c. an intermediate.
   d. an enzyme.

7. Chemical kinetics is useful because
   a. the rate law enables the equation of the reaction to be derived.
   b. the rate law suggests possible reaction mechanisms.
   c. thermodynamic data can be obtained from activation energies.
   d. catalysts save money.

8. Which of the following is not a catalyst for the decomposition of hydrogen peroxide?
   a. $I^-(aq)$.
   c. $MnO_2(s)$.
   b. catalase.
   d. lactase.

9. A rate law relates
   a. reaction rate and temperature.
   b. reaction rate and concentration.
   c. temperature and concentration.
   d. energy and concentration.

10. The human body uses which of the following to speed up vital reactions?
   a. reactants
   c. collisions
   b. enzymes
   d. temperature

11. In a graph of how potential energy changes with reaction progress, the activated complex appears at the
   a. left end of the curve.
   b. right end of the curve.
   c. bottom of the curve.
   d. peak of the curve.

12. The slowest step in a mechanism is called
   a. the rate-determining step.
   b. the uncatalyzed reaction.
   c. the activation step.
   d. None of the above

CHAPTER

17

OXIDATION, REDUCTION, AND ELECTROCHEMISTRY

If you run out of gasoline in a car, you might have to walk to the nearest gas station for more fuel. But running out of energy when you are millions of kilometers from Earth is a different story! A robot designed to collect data on other bodies in our solar system needs a reliable, portable source of energy that can work in the absence of an atmosphere to carry out its mission. Many of the power sources for these explorer robots are batteries. In this chapter, you will learn about the processes of oxidation and reduction and how they are used in batteries to provide energy. You will also learn how these processes are used to purify metals and protect objects from corrosion.

START-UP ACTIVITY

Lights On

SAFETY PRECAUTIONS

PROCEDURE

1. Assemble **batteries,** a **light-emitting diode,** and **wires** so that the diode lights. Make a diagram of your construction.

2. Remove the batteries, and reconnect them in the opposite direction. Record the results.

ANALYSIS

1. A light-emitting diode allows electrons to move through it in only one direction—into the short leg and out of the longer leg. Based on your results, from which end of the battery must electrons leave?

2. How are the atoms changing in the end of the battery from which electrons leave?

3. What must happen to the electrons as they enter into the other end of the battery?

4. Why does a battery eventually run down?

Pre-Reading Questions

① What type of charge results from losing electrons? from gaining electrons?

② Name a device that converts chemical energy into electrical energy.

③ Why are batteries marked with positive and negative terminals?

SECTION

1

Oxidation-Reduction Reactions

KEY TERMS

- oxidation
- reduction
- oxidation-reduction reaction
- oxidation number
- half-reaction
- oxidizing agent
- reducing agent

Topic Link

Refer to the "Ions and Ionic Compounds" and "Covalent Compounds" chapters for more information about chemical bonding.

OBJECTIVES

(1) Identify atoms that are oxidized or reduced through electron transfer.

(2) Assign oxidation numbers to atoms in compounds and ions.

(3) Identify redox reactions by analyzing changes in oxidation numbers for different atoms in the reaction.

(4) Balance equations for oxidation-reduction reactions through the half-reaction method.

Electron Transfer and Chemical Reactions

You already know that atoms with very different electronegativities bond by an electron transfer. For example, sodium chloride is formed by the transfer of electrons from sodium atoms to chlorine atoms in the reaction shown in **Figure 1** and described by the following equation:

$$2Na(s) + Cl_2(g) \longrightarrow 2NaCl(s)$$

Though NaCl is the way the formula of sodium chloride is usually written, the compound is made up of ions. Therefore, it might be helpful to think of sodium chloride as if its formula were written Na^+Cl^- so that you remember the ions.

When the electronegativity difference between the atoms is smaller, a polar covalent bond can form when the atoms join, as shown below.

$$2C(s) + O_2(g) \longrightarrow 2CO(g)$$

The C–O bond has some ionic character because there is an unequal sharing of electrons between the carbon atom and the oxygen atom. The oxygen atom attracts the shared electrons more strongly than the carbon atom does.

Oxidation Involves a Loss of Electrons

oxidation

a reaction that removes one or more electrons from a substance such that the substance's valence or oxidation state increases

In the examples above, electrons were transferred at least in part from one atom to another. The sodium atom lost an electron to the chlorine atom. The carbon atom lost some of its control over its electrons to the oxygen atom. The loss, wholly or in part, of one or more electrons is called **oxidation.**

Thus, in making NaCl, the sodium atom is oxidized from Na to Na^+. Likewise, the carbon atom is oxidized when CO forms, even though the carbon atom does not become an ion.

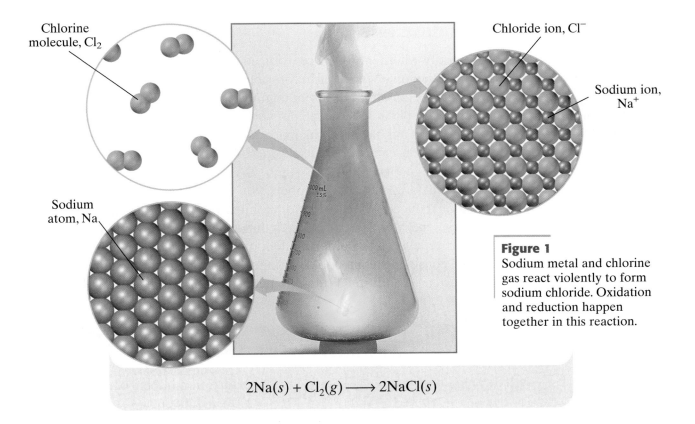

Figure 1
Sodium metal and chlorine gas react violently to form sodium chloride. Oxidation and reduction happen together in this reaction.

$$2Na(s) + Cl_2(g) \longrightarrow 2NaCl(s)$$

Reduction Involves a Gain of Electrons

In making NaCl, the electrons lost by the sodium atoms do not just disappear. They are gained by the chlorine atoms. The gain of electrons is described as **reduction.** The chlorine atoms are reduced as they change from Cl_2 to $2Cl^-$.

When joining with carbon atoms to make CO, oxygen atoms do not gain electrons but gain only a partial negative charge. But because the electrons in the C—O bonds spend more time near the oxygen atoms, the change is still a reduction.

More than one electron may be gained in a reduction. In the formation of Li_3N, described by the equation below, three electrons are gained by each nitrogen atom.

$$6Li(s) + N_2(g) \longrightarrow 2Li_3N(s)$$

Oxidation and Reduction Occur Together

When oxidation happens there must also be reduction taking place. You will learn later in this chapter that oxidation and reduction can happen at different places. In most situations, however, oxidation and reduction happen in a single place. Consider HgO being broken down into its elements, as described by the following equation:

$$2HgO(s) \longrightarrow 2Hg(l) + O_2(g)$$

In this reaction, mercury atoms are reduced, while oxygen atoms are oxidized. A single reaction in which an oxidation and a reduction happen is called an **oxidation-reduction reaction** or *redox reaction.*

reduction

a chemical change in which electrons are gained, either by the removal of oxygen, the addition of hydrogen, or the addition of electrons

www.scilinks.org
Topic: Redox Reactions
SciLinks code: HW4110

SC*LINKS* Maintained by the National Science Teachers Association

oxidation-reduction reaction

any chemical change in which one species is oxidized (loses electrons) and another species is reduced (gains electrons); also called *redox reaction*

Oxidation, Reduction, and Electrochemistry   **605**

Oxidation Numbers

oxidation number

the number of electrons that must be added to or removed from an atom in a combined state to convert the atom into the elemental form

To identify whether atoms are oxidized or reduced, chemists use a model of **oxidation numbers,** which can help them identify differences in an atom of an element in different compounds. By following the set of rules described in **Skills Toolkit 1** below, you can assign an oxidation number to each atom in a molecule or in an ion. **Sample Problem A** shows how to use the rules. You can see three different oxidation numbers for atoms of manganese in **Figure 2.**

By tracking oxidation numbers, you can tell whether an atom is oxidized or reduced. If the oxidation number of an atom increases during a reaction, the atom is oxidized. If the oxidation number decreases, the atom is reduced. Like other models, oxidation numbers have limits. You should consider them a bookkeeping tool to help keep track of electrons. In some cases, additional rules are needed to find values that make sense.

1 SKILLS Toolkit

Assigning Oxidation Numbers

1. Identify the formula.
- If no formula is provided, write the formula of the molecule or ion.

2. Assign known oxidation numbers.
- Place an oxidation number above each element's symbol according to the following rules.
  - **a.** The oxidation number of an atom of any free (uncombined) element in atomic or molecular form is zero.
  - **b.** The oxidation number of a monatomic ion is equal to the charge on the ion.
  - **c.** The oxidation number of an atom of fluorine in a compound is always -1 because it is the most electronegative element.
  - **d.** An atom of the more electronegative element in a binary compound is assigned the number equal to the charge it would have if it were an ion.
  - **e.** In compounds, atoms of the elements of Group 1, Group 2, and aluminum have positive oxidation numbers of $+1$, $+2$, and $+3$, respectively.
  - **f.** The oxidation number of each hydrogen atom in a compound is $+1$, unless it is combined with a metal atom; then it is -1.
  - **g.** The oxidation number of each oxygen atom in compounds is usually -2. When combined with fluorine atoms, oxygen becomes $+2$. In peroxides, such as H_2O_2, an oxygen atom has an oxidation number of -1.

3. Calculate remaining oxidation numbers, and verify the results.
- Use the total oxidation number of each element's atoms (the oxidation number for an atom of the element multiplied by the subscript for the element) and the following rules to calculate missing oxidation numbers.
  - **h.** The sum of the oxidation numbers for all the atoms in a molecule is zero.
  - **i.** The sum of the oxidation numbers for all atoms in a polyatomic ion is equal to the charge on that ion.

Figure 2
An atom of manganese in its elemental form has an oxidation number of 0. In MnO_2, the oxidation number of the manganese atom is +4. In the permanganate ion, MnO_4^-, the oxidation number of the manganese atom is +7.

In the figure: $\overset{0}{Mn}$, $\overset{+4\ -2}{MnO_2}$, $\overset{+7\ -2}{MnO_4^-}$, $\overset{+1}{K^+}$

SAMPLE PROBLEM A

Determining Oxidation Numbers

Assign oxidation numbers to the sulfur and oxygen atoms in the pyrosulfate ion, $S_2O_7^{2-}$.

1 Identify the formula.

The pyrosulfate ion has the formula $S_2O_7^{2-}$.

2 Assign known oxidation numbers.

According to Rule **g,** the oxidation number of the O atoms is −2, so this number is written above the O symbol in the formula. Because the oxidation number of the sulfur atoms is unknown, x is written above the S symbol. Thus the formula is as follows:

$$\overset{x\ \ -2}{S_2O_7^{2-}}$$

3 Calculate remaining oxidation numbers, and verify the results.

• Multiplying the oxidation numbers by the subscripts, we see that the S atoms contribute $2x$ and the O atoms contribute $7(-2) = -14$ to the total oxidation number. To come up with the correct total charge, (Rule **i**), $2x + (-14) = -2$. Solve this equation to find $x = +6$.

• In $S_2O_7^{2-}$, the oxidation number of the S atoms is +6, and the oxidation number of the O atoms is −2. The sum of the total oxidation numbers for each element is $2(+6) + 7(-2) = -2$, which is the charge on the ion.

PRACTICE

Determine the oxidation number for each atom in each of the following.

1
a. NH_4^+
b. Al
c. H_2O
d. Pb^{2+}

e. H_2
f. $PbSO_4$
g. $KClO_3$
h. BF_3

i. $Ca(OH)_2$
j. $Fe_2(CO_3)_3$
k. $H_2PO_4^-$
l. NH_4NO_3

Figure 3
Zinc metal reacts with hydrochloric acid, making bubbles of hydrogen gas.

Identifying Redox Reactions

Figure 3 shows the reaction of Zn with HCl. Is this a redox reaction? Hydrochloric acid is a solution in water of Cl^-, which plays no part in the reaction, and H_3O^+. The net change in this reaction is

$$2H_3O^+(aq) + Zn(s) \longrightarrow H_2(g) + 2H_2O(l) + Zn^{2+}(aq)$$

Using rules **a, b, f, g, h,** and **i** from **Skills Toolkit 1,** you can give oxidation numbers to all atoms as follows:

$$2\overset{+1\ -2}{H_3O^+}(aq) + \overset{0}{Zn}(s) \longrightarrow \overset{0}{H_2}(g) + 2\overset{+1\ -2}{H_2O}(l) + \overset{+2}{Zn^{2+}}(aq)$$

Comparing oxidation numbers, you see that the zinc atom changes from 0 to +2 and that two hydrogen atoms change from +1 to 0. So, this is a redox reaction. In a redox reaction, the oxidation numbers of atoms that are oxidized increase, and those of atoms that are reduced decrease.

Half-Reactions

In the reaction shown in **Figure 3,** each zinc atom loses two electrons and is oxidized. One way to show only this half of the overall redox reaction is by writing a **half-reaction** for the change.

half-reaction

the part of a reaction that involves only oxidation or reduction

$$Zn(s) \longrightarrow Zn^{2+}(aq) + 2e^-$$

Note that electrons are a product. Of course, there is also a half-reaction for reduction in which electrons are a reactant.

$$2e^- + 2H_3O^+(aq) \longrightarrow H_2(g) + 2H_2O(l)$$

By adding the two half-reactions together, you get the overall redox reaction shown earlier. Notice that enough electrons are in each half-reaction to keep the charges balanced. Keep in mind that free electrons do not actually leave the zinc atoms and float around before being picked up by the hydronium ions. Instead, they are "handed off" directly from one to the other.

Balancing Oxidation-Reduction Equations

Equations for redox reactions are sometimes difficult to balance. Use the steps in **Skills Toolkit 2** below to balance redox equations for reactions in acidic aqueous solution. An important step is to identify the key ions or molecules that contain atoms whose oxidation numbers change. These atoms are the starting points of the unbalanced half-reactions. For the reaction of zinc and hydrochloric acid, the unbalanced oxidation and reduction half-reactions would be as follows:

$$Zn(s) \longrightarrow Zn^{2+}(aq) \quad \text{and} \quad H_3O^+(aq) \longrightarrow H_2(g)$$

These reactions are then separately balanced. Finally, the balanced equations of the two half-reactions are added together to cancel the electrons.

SKILLS Toolkit 2

Balancing Redox Equations Using the Half-Reaction Method

1. Identify reactants and products.
- Write the unbalanced equation in ionic form, excluding any spectator ions.
- Assign oxidation numbers, and identify the atoms that change their oxidation numbers. Ignore all species whose atoms do not change their oxidation number.

2. Write and balance the half-reactions.
- Separate the equation into its two half-reactions.
- For each half-reaction, do the following:
  - **a.** Balance atoms other than hydrogen and oxygen.
  - **b.** Balance oxygen atoms by adding water molecules as needed.
  - **c.** Balance hydrogen atoms by adding one hydronium ion for each hydrogen atom needed and then by adding the same number of water molecules to the other side of the equation.
  - **d.** Balance the overall charge by adding electrons as needed.

3. Make the electrons equal, and combine half-reactions.
- Multiply each half-reaction by an appropriate number so that both half-reactions have the same number of electrons. Now the electrons lost equal the electrons gained, so charge is conserved.
- Combine the half-reactions, and cancel anything that is common to both sides of the equation.

4. Verify your results.
- Double-check that all atoms and charge are balanced.

The Half-Reaction Method

Write and balance the equation for the reaction when an acidic solution of MnO_4^- reacts with a solution of Fe^{2+} to form a solution containing Mn^{2+} and Fe^{3+} ions.

1 **Identify reactants and products.**

- The unbalanced equation in ionic form is as follows:

$$H_3O^+(aq) + MnO_4^-(aq) + Fe^{2+}(aq) \longrightarrow Mn^{2+}(aq) + Fe^{3+}(aq)$$

- Oxidation numbers for the atoms are as follows:

$$\overset{+1 \; -2}{H_3O^+} + \overset{+7 \; -2}{MnO_4^-} + \overset{+2}{Fe^{2+}} \longrightarrow \overset{+2}{Mn^{2+}} + \overset{+3}{Fe^{3+}}$$

- Atoms of Mn and Fe change oxidation numbers.

2 **Write and balance the half-reactions.**

Unbalanced: $Fe^{2+} \longrightarrow Fe^{3+}$ $\qquad\qquad$ $MnO_4^- \longrightarrow Mn^{2+}$

Balance O: $Fe^{2+} \longrightarrow Fe^{3+}$ $\qquad\qquad$ $MnO_4^- \longrightarrow Mn^{2+} + 4H_2O$

Balance H: $Fe^{2+} \longrightarrow Fe^{3+}$ $\qquad$ $8H_3O^+ + MnO_4^- \longrightarrow Mn^{2+} + 12H_2O$

Balance e^-: $Fe^{2+} \longrightarrow Fe^{3+} + e^-$ $\quad$ $5e^- + 8H_3O^+ + MnO_4^- \longrightarrow Mn^{2+} + 12H_2O$

3 **Make the electrons equal, and combine half-reactions.**

Multiply the half-reaction that involves iron by 5 to make the numbers of electrons the same in each half-reaction. Add the half-reactions, and cancel the electrons to get the final balanced equation:

$$8H_3O^+(aq) + MnO_4^-(aq) + 5Fe^{2+}(aq) \longrightarrow Mn^{2+}(aq) + 5Fe^{3+}(aq) + 12H_2O(l)$$

4 **Verify your results.**

Note that there are equal numbers of all atoms and a net charge of +17 on each side of the equation.

PRACTICE HINT

To help avoid confusion between charges and oxidation numbers, the sign of a charge is written last and the sign of an oxidation number is written first. Thus the Fe^{2+} ion has a 2+ charge and a +2 oxidation number.

PRACTICE

Use the half-reaction method to write a balanced equation for each of the following reactions in acidic, aqueous solution.

1 The reactants are Fe(s) and $O_2(aq)$, and the products are $Fe^{3+}(aq)$ and $H_2O(l)$.

2 Al(s) is placed in the acidic solution and forms $H_2(g)$ and $Al^{3+}(aq)$.

3 The reactants are sodium bromide and hydrogen peroxide, and the products are bromine and water.

4 The reactants are manganese dioxide and a soluble copper(I) salt, and the products are soluble manganese(II) and copper(II) salts.

Identifying Agents in Redox Reactions

In **Sample Problem B,** the permanganate ion caused the oxidation of the iron(II) ion. Substances that cause the oxidation of other substances are called **oxidizing agents.** They accept electrons easily and so are reduced. Common oxidizing agents are oxygen, hydrogen peroxide, and halogens.

Reducing agents cause reduction to happen and are themselves oxidized. The iron(II) ion caused the reduction of the permanganate ion and was the reducing agent. Common ones are metals, hydrogen, and carbon.

oxidizing agent

the substance that gains electrons in an oxidation-reduction reaction and is reduced

reducing agent

a substance that has the potential to reduce another substance

Section Review

UNDERSTANDING KEY IDEAS

1. Explain oxidation and reduction in terms of electron transfer.

2. How can you identify a reaction as a redox reaction?

3. Describe how an oxidation-reduction reaction may be broken down into two half-reactions, and explain why the latter are useful in balancing redox equations.

4. Compare the number of electrons lost in an oxidation half-reaction with the number of electrons gained in the corresponding reduction half-reaction.

5. Describe what an oxidizing agent and a reducing agent are.

PRACTICE PROBLEMS

6. Assign oxidation numbers to the atoms in each of the following:

   a. H_2SO_3 **c.** SF_6
   b. Cl_2 **d.** NO_3^-

7. Assign oxidation numbers to the atoms in each of the following:

   a. CH_4 **c.** $NaHCO_3$
   b. HSO_3^- **d.** $NaBiO_3$

8. Identify the oxidation number of a Cr atom in each of the following: CrO_3, CrO, $Cr(s)$, CrO_2, Cr_2O_3, $Cr_2O_7^{2-}$, and CrO_4^{2-}.

9. Which of the following equations represent redox reactions? For each redox reaction, determine which atom is oxidized and which is reduced, and identify the oxidizing agent and the reducing agent.

   a. $MgO(s) + H_2CO_3(aq) \longrightarrow$
   $$MgCO_3(s) + H_2O(l)$$
   b. $2KNO_3(s) \longrightarrow 2KNO_2(s) + O_2(g)$
   c. $H_2(g) + CuO(s) \longrightarrow Cu(s) + H_2O(l)$
   d. $NaOH(aq) + HCl(aq) \longrightarrow$
   $$NaCl(aq) + H_2O(l)$$
   e. $H_2(g) + Cl_2(g) \longrightarrow 2HCl(g)$
   f. $SO_3(g) + H_2O(l) \longrightarrow H_2SO_4(aq)$

10. Use the half-reaction method in acidic, aqueous solution to balance each of the following redox reactions:

    a. $Cl^-(aq) + Cr_2O_7^{2-}(aq) \longrightarrow$
    $$Cl_2(g) + Cr^{3+}(aq)$$
    b. $Cu(s) + Ag^+(aq) \longrightarrow Cu^{2+}(aq) + Ag(s)$
    c. $Br_2(l) + I^-(aq) \longrightarrow I_2(s) + Br^-(aq)$
    d. $I^-(aq) + NO_2^-(aq) \longrightarrow NO(g) + I_2(s)$

11. Determine which atom is oxidized and which is reduced, and identify the oxidizing agent and the reducing agent for each reaction in item 10.

CRITICAL THINKING

12. How is it possible for hydrogen peroxide to be both an oxidizing agent and a reducing agent?

Introduction to Electrochemistry

electrochemistry

the branch of chemistry that is the study of the relationship between electric forces and chemical reactions

internet connect

www.scilinks.org
Topic: Electrical Energy
SciLinks code: HW4045

SCi
LINKS. Maintained by the National Science Teachers Association

OBJECTIVES

(1) **Describe** the relationship between voltage and the movement of electrons.

(2) **Identify** the parts of an electrochemical cell and their functions.

(3) **Write** electrode reactions for cathodes and anodes.

Chemistry Meets Electricity

The science of **electrochemistry** deals with the connections between chemistry and electricity. It is an important subject because it is involved in many of the things you use every day. Electrochemical devices change electrical energy into chemical energy and vice versa. A simple flashlight is an everyday example of something that converts chemical energy into electrical energy, which is then converted into light energy.

Figure 4 shows the components of a typical flashlight. The power source is two batteries or cells. Each cell has a metal terminal at each end. If you examine a battery, you will find a $\oplus$ or + symbol near the top identifying the positive terminal. The bottom is the negative terminal, and is possibly unmarked or marked with a $\ominus$ or – symbol. By closing the switch, you turn the flashlight on. Electrons move from the negative terminal of the lower battery, through a metal circuit that includes the light bulb, and continue to the positive terminal of the upper battery. The circuit is completed by electrons and ions moving charge through each of the batteries and electrons moving charge from the upper battery to the lower one. When you turn off the flashlight, you break the pathway, and the movement of electrons and ions stops.

Figure 4
When the switch of a flashlight is closed, electrons "pushed" by the battery are forced through the thin tungsten filament of the bulb. This flow of electrons makes the filament hot enough to emit light.

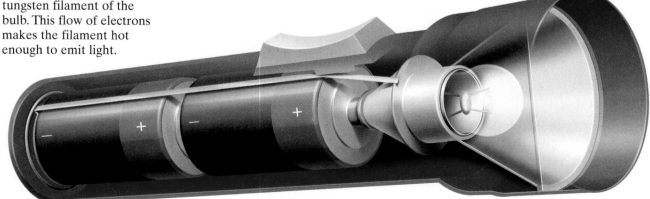

"Electrical Pressure" Is Expressed in Volts

Electrochemical reactions in a battery cause a greater electron density in the negative terminal than in the positive terminal. Electrons repel each other, so there is a higher "pressure" on the electrons in the negative terminal, which drives the electrons out of the battery and through the flashlight.

Electrical "pressure," often called *electric potential,* or **voltage,** is expressed in units of *volts.* The voltage of an ordinary flashlight cell is 1.5 volts or 1.5 V. When two cells are placed end-to-end, as in **Figure 4,** the voltages add together, so the overall voltage driving electrons is 3.0 V. The movement of electrons or other charged particles is described as *electric current* and is expressed in units of *amperes.*

voltage

the potential difference or electromotive force, measured in volts; it represents the amount of work that moving an electric charge between two points would take

Components of Electrochemical Cells

A flashlight battery is an electrochemical cell. An **electrochemical cell** consists of two electrodes separated by an electrolyte. An **electrode** is a conductor that connects with a nonmetallic part of a circuit. You have learned about two kinds of conductors. One kind includes metals, which conduct electric current through moving electrons. The second kind includes electrolyte solutions, which conduct through moving ions.

electrochemical cell

a system that contains two electrodes separated by an electrolyte phase

electrode

a conductor used to establish electrical contact with a nonmetallic part of a circuit, such as an electrolyte

The Cathode Is Where Reduction Occurs

Electrode reactions happen on the surfaces of electrodes. **Figure 5** shows half of an electrochemical cell. The copper strip is an electrode because it is a conductor in contact with the electrolyte solution. The reaction on this electrode is the reduction described by the following equation:

$$Cu^{2+}(aq) + 2e^- \longrightarrow Cu(s)$$

The copper electrode is a **cathode** because reduction happens on it.

cathode

the electrode on whose surface reduction takes place

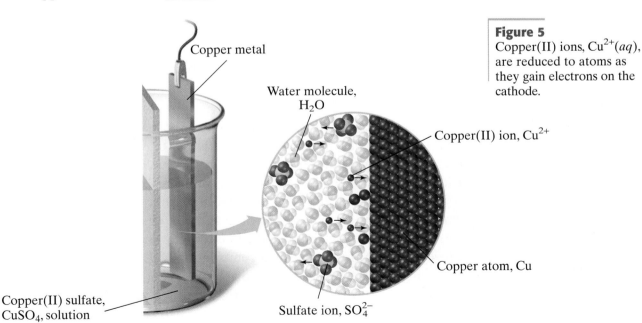

Figure 5
Copper(II) ions, $Cu^{2+}(aq)$, are reduced to atoms as they gain electrons on the cathode.

Copper metal

Water molecule, H_2O

Copper(II) ion, Cu^{2+}

Copper atom, Cu

Copper(II) sulfate, $CuSO_4$, solution

Sulfate ion, SO_4^{2-}

Cathodic reaction: $Cu^{2+}(aq) + 2e^- \longrightarrow Cu(s)$

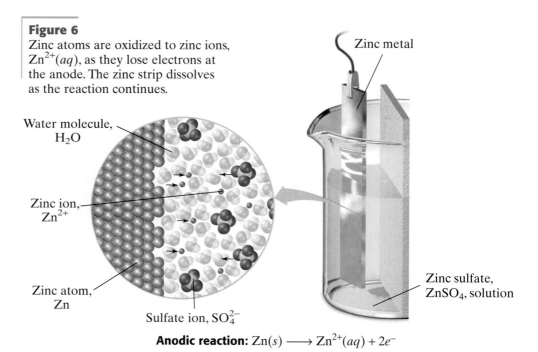

Figure 6
Zinc atoms are oxidized to zinc ions, $Zn^{2+}(aq)$, as they lose electrons at the anode. The zinc strip dissolves as the reaction continues.

Zinc metal

Water molecule, H_2O

Zinc ion, Zn^{2+}

Zinc atom, Zn

Sulfate ion, SO_4^{2-}

Zinc sulfate, $ZnSO_4$, solution

Anodic reaction: $Zn(s) \longrightarrow Zn^{2+}(aq) + 2e^-$

The Anode Is Where Oxidation Occurs

The electrons that cause reduction at the cathode are pushed there from a reaction at the second electrode of a cell. The **anode** is the electrode on which oxidation occurs. **Figure 6** shows the second half of the electrochemical cell. The zinc strip is an electrode because it is a conductor in contact with the solution. The reaction on this electrode is the oxidation described by the following equation:

$$Zn(s) \longrightarrow Zn^{2+}(aq) + 2e^-$$

The zinc electrode is an anode because oxidation happens on it.

The electrode reactions described here would have been called half-reactions in the last section. The difference is that a half-reaction is a helpful model, but the electrons shown are never really free. An electrode reaction describes reality because the electrons actually move from one electrode to another to continue the reaction.

Pathways for Moving Charges

The electrode reactions cannot happen unless the electrodes are part of a complete circuit. So, a cell must have pathways to move charges. Wires are often used to connect the electrodes through a meter or a light bulb. Electrons carry charges in the wires and electrodes.

Ions in solution carry charges between the electrolytes to complete the circuit. A porous barrier, as shown in **Figure 7,** or a salt bridge keeps the solutions from mixing, but lets the ions move. In the cell shown, charge is carried through the barrier by a combination of $Zn^{2+}(aq)$ ions moving to the right and $SO_4^{2-}(aq)$ ions moving to the left. Understand that positive charge may be carried from left to right through this cell either by negative particles (electrons or anions) moving from right to left or by cations moving from left to right.

anode

the electrode on whose surface oxidation takes place; anions migrate toward the anode, and electrons leave the system from the anode

Electrodes   Porous barrier

Figure 7
The light bulb is powered by the reaction in this cell.

The Complete Cell

Figure 7 shows a complete cell composed of the electrodes shown separately in earlier figures. The overall process when the anode reaction is added to the cathode reaction for this cell is the same as the following redox reaction:

$$Zn(s) + Cu^{2+}(aq) \longrightarrow Zn^{2+}(aq) + Cu(s)$$

Although the two electrode reactions occur at the same time, they occur at different places in the cell. This is an important distinction from the redox reactions discussed in the last section.

internet connect

www.scilinks.org

Topic: Electrochemical Cells
SciLinks code: HW4046

SCiLINKS. Maintained by the National Science Teachers Association

② Section Review

UNDERSTANDING KEY IDEAS

1. What is voltage?

2. How does voltage relate to the movement of electrons?

3. List the components of an electrochemical cell, and describe the function of each.

4. What are the names of the electrodes in an electrochemical cell? What type of reaction happens on each?

5. Describe the difference in how charge flows in wires and in electrolyte solutions.

6. A 12 V car battery has six cells connected end-to-end. What is the voltage of a single cell?

7. Will the reaction below happen at an anode or a cathode? Explain.

$$Fe(CN)_6^{3-}(aq) + e^- \longrightarrow Fe(CN)_6^{4-}(aq)$$

8. Describe the changes in oxidation number that happen in an anode reaction and in a cathode reaction.

CRITICAL THINKING

9. What would happen if you put one of the batteries in backward in a two-cell flashlight?

10. What would happen if you put both batteries in backward in a two-cell flashlight?

11. If an electrode reaction has dissolved oxygen, $O_2(aq)$, as a reactant, is the electrode an anode or a cathode? Explain.

12. Write an electrode reaction in which you change $Br^-(aq)$ to $Br_2(aq)$. Would this reaction happen at an anode or a cathode?

13. Write an electrode reaction in which $Sn^{4+}(aq)$ is changed to $Sn^{2+}(aq)$. Would this reaction happen at an anode or a cathode?

14. If you wanted to use an electrochemical cell to deposit a thin layer of silver metal onto a bracelet, which electrode would you make the bracelet? Explain using the equation for the electrode reaction that would occur.

15. What would happen at each electrode if batteries were connected to the cell in **Figure 7** so that electrons flowed in the opposite direction of the direction described in the text?

16. Compare the equations for electrode reactions with the equations for half-reactions.

17. Write the electrode reactions for a cell that involves only $Cu(s)$ and $Cu^{2+}(aq)$ in which the anode reaction is the reverse of the cathode reaction. What is the net result of operating this cell?

18. Is it correct to say that the net chemical result of an electrochemical cell is a redox reaction? Explain.

SECTION 3

Galvanic Cells

Key Terms

- corrosion
- standard electrode potential

OBJECTIVES

1. **Describe** the operation of galvanic cells, including dry cells, lead-acid batteries, and fuel cells.

2. **Identify** conditions that lead to corrosion and ways to prevent it.

3. **Calculate** cell voltage from a table of standard electrode potentials.

Types of Galvanic Cells

A battery is one kind of *galvanic cell,* a device that can change chemical energy into electrical energy. In these cells, a spontaneous reaction happens that causes electrons to move. **Figure 8** shows a kind of galvanic cell known as a Daniell cell. Daniell cells were used as energy sources in the early days of electrical research. Of course, Daniell cells would be impractical to use in a radio or portable computer today. There are many other kinds of galvanic cells. These include dry cells, lead-acid batteries, and fuel cells.

Figure 8
As a result of the reaction in a galvanic cell, the bulb lights up as electrons move in the wires from the anode to the cathode.

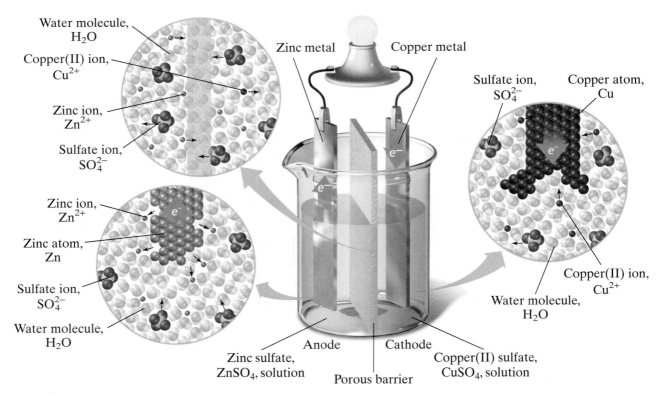

Water molecule, H_2O
Copper(II) ion, Cu^{2+}
Zinc ion, Zn^{2+}
Sulfate ion, SO_4^{2-}
Zinc ion, Zn^{2+}
Zinc atom, Zn
Sulfate ion, SO_4^{2-}
Water molecule, H_2O
Zinc metal
Copper metal
Sulfate ion, SO_4^{2-}
Copper atom, Cu
Copper(II) ion, Cu^{2+}
Water molecule, H_2O
Anode
Cathode
Zinc sulfate, $ZnSO_4$, solution
Porous barrier
Copper(II) sulfate, $CuSO_4$, solution

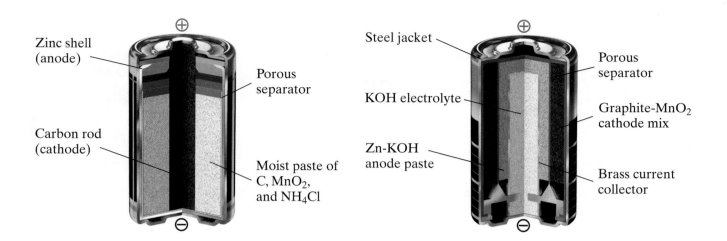

Zinc shell (anode)

Porous separator

Carbon rod (cathode)

Moist paste of C, MnO_2, and NH_4Cl

Steel jacket

Porous separator

KOH electrolyte

Graphite-MnO_2 cathode mix

Zn-KOH anode paste

Brass current collector

Figure 9
Two familiar kinds of dry cells use different electrolytes. The electrolytes make the cell on the left acidic and the cell on the right alkaline.

Dry Cells

Although the Daniell cell was useful for supplying energy in the lab, it wasn't very portable because it held solutions. The energy source that you know as a *battery* and use in radios and remote controls is a dry cell. In a dry cell, moist electrolyte pastes are used instead of solutions. This cell was invented over a century ago by Georges Leclanché, a French chemist. His original design was close to that shown on the left in **Figure 9.** A carbon rod, the battery's positive terminal, connects with a wet paste of carbon; ammonium chloride, NH_4Cl; manganese(IV) oxide, MnO_2; starch; and water. When the cell is used, the carbon rod is the cathode, and the following electrode reaction happens:

Cathode: $2MnO_2(s) + 2NH_4^+(aq) + 2e^- \longrightarrow Mn_2O_3(s) + 2NH_3(aq) + H_2O(l)$

The zinc case serves as the negative terminal of the battery. When the cell is used, zinc dissolves in the following electrode reaction:

Anode: $Zn(s) + 4NH_3(aq) \longrightarrow 2e^- + Zn(NH_3)_4^{2+}(aq)$

The white powder that you see on old corroded batteries is the chloride salt of this $Zn(NH_3)_4^{2+}(aq)$ complex ion.

The $NH_4^+(aq)$ ion is a weak acid, and for this reason the Leclanché cell is called the *acidic* version of the dry cell. The *alkaline cell,* shown in **Figure 9,** is a newer, better version. The ingredients of the alkaline cell are similar to the acidic version, but the carbon cathode is replaced by a piece of brass, and ammonium chloride is replaced by potassium hydroxide. The presence of this strong base gives the alkaline cell its name. The electrode reactions that occur when the cell is used are described by the equations below.

Cathode: $2MnO_2(s) + H_2O(l) + 2e^- \longrightarrow Mn_2O_3(s) + 2OH^-(aq)$

Anode: $Zn(s) + 2OH^-(aq) \longrightarrow 2e^- + Zn(OH)_2(s)$

A sturdy steel shell is needed to prevent the caustic contents from leaking out of the battery. Because of this extra packaging, alkaline cells are more expensive than cells of the older, acidic version.

Quick LAB

Listen Up

PROCEDURE

1. Press a **zinc strip** and a **copper strip** into a **raw potato.** The strips should be about 0.5 cm apart but should not touch one another.

2. While listening to the **earphone,** touch one wire from the earphone to one of the metal strips and the other wire from the earphone to the second metal strip using **alligator clips.** Record your observations.

3. While listening to the earphone, touch both wires to a single metal strip. Record your observations.

ANALYSIS

1. Compare your results from step 2 with your results from step 3. Suggest an explanation for any similarities or differences.

2. Suggest an explanation for the sound.

Lead-Acid Batteries

The batteries just discussed are called *dry cells* because the water is not free, but absorbed in pastes. In contrast, the Daniell cell and the lead-acid battery use aqueous solutions of electrolytes, so they should be used in an upright position.

Most car batteries are lead-acid storage batteries. Usually they have six cells mounted side-by-side in a single case, as shown in **Figure 10.** Though many attempts have been made to replace the heavy lead-acid battery by lighter alternatives, no other material has been found that can reliably and economically give the large surges of electrical energy needed to start a cold engine.

A fully charged lead-acid cell is made up of a stack of alternating lead and lead(IV) oxide plates isolated from each other by thin porous separators. All these components are in a concentrated solution of sulfuric acid. The positive terminal of one cell is linked to the negative terminal of the next cell in the same way that the batteries in the flashlight were connected. This arrangement of cells causes the outermost terminals of the lead-acid battery to have a voltage of 12.0 V. When the cell is used, it acts as a galvanic cell with the following reactions:

$$\text{Cathode: } PbO_2(s) + HSO_4^-(aq) + 3H_3O^+(aq) + 2e^- \longrightarrow PbSO_4(s) + 5H_2O(l)$$

$$\text{Anode: } Pb(s) + HSO_4^-(aq) + H_2O(l) \longrightarrow 2e^- + PbSO_4(s) + H_3O^+(aq)$$

Notice that $PbSO_4$ is produced at both electrodes.

Unlike the Daniell and Leclanché cells, the lead-acid cell is rechargeable. So, when the battery runs down, you do not need to replace it. Instead, an electric current is applied in a direction opposite to that discussed above. As a result of the input of energy, the reactions are reversed. The cell is eventually restored to its charged state. During recharge, the cell functions as an *electrolytic cell,* which you will learn about in the next section.

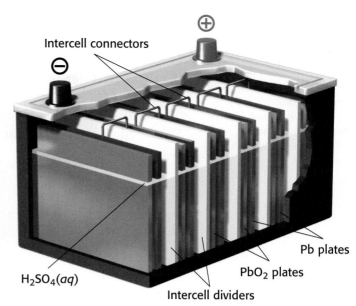

Intercell connectors

⊖

⊕

Pb plates

PbO₂ plates

$H_2SO_4(aq)$

Intercell dividers

Figure 10
The lead-acid battery is used to store energy in almost all vehicles. Although the cutaway view shows a single PbO_2 plate and a single Pb plate in each cell, there are actually several of each.

Fuel Cells

In a *fuel cell,* the oxidizing and reducing agents are brought in, often as gases, from outside of the cell, rather than being part of it. Unlike a dry cell, a fuel cell can work forever, in principle, changing chemical energy into electrical energy. **Figure 11** models a fuel cell that uses the reactions below.

$$\text{Cathode: } O_2(g) + 2H_2O(l) + 4e^- \longrightarrow 4OH^-(aq)$$

$$\text{Anode: } 2H_2(g) + 4OH^-(aq) \longrightarrow 4e^- + 4H_2O(l)$$

Because fuel cells directly change chemical energy into electrical energy, they are very efficient and are cleaner than the burning of fuels in power plants to generate electrical energy. Research into fuel cells continues, and fuel cells are used in a few experimental power plants.

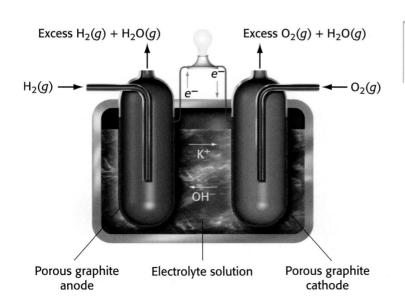

Excess $H_2(g) + H_2O(g)$ Excess $O_2(g) + H_2O(g)$

$H_2(g)$ → e^- ← $O_2(g)$

e^-

K^+

OH^-

Porous graphite anode   Electrolyte solution   Porous graphite cathode

Figure 11
The reactions in this fuel cell take place at carbon electrodes that contain metal catalysts. The water formed is removed as a gas.

Corrosion Cells

Oxygen is so reactive that many metals spontaneously oxidize in air. Fortunately, many of these reactions are slow. The disintegration of metals is called **corrosion.** Usually $O_2(g)$ is the oxidizing agent, but the direct reaction of O_2 with the metal is not usually how corrosion happens.

Water is usually involved in corrosion. Consider the corrosion of iron. Hydrated iron(III) oxide, or rust, forms by the following overall reaction:

$$4Fe(s) + 3O_2(aq) + 4H_2O(l) \longrightarrow 2Fe_2O_3 \cdot 2H_2O(s)$$

However, the reaction mechanism is more complicated than this equation suggests. An even simpler version of what happens is shown in **Figure 12.** The iron dissolves by the oxidation half-reaction

$$Fe(s) \longrightarrow Fe^{3+}(aq) + 3e^-$$

Any oxidation must be accompanied by a reduction taking place at the same time, but not necessarily at the same location. In fact, the electrons produced by the oxidation are consumed by the reduction of oxygen at a cathodic site elsewhere on the iron's surface. The reduction half-reaction is

$$O_2(aq) + 2H_2O(l) + 4e^- \longrightarrow 4OH^-(aq)$$

Electrons move in the metal and ions move in the water layer between the two reaction sites, as in an electrochemical cell. In fact, a corrosion cell is an unwanted galvanic cell. Chemical energy is converted into electrical energy, which heats the metal.

The three ingredients—oxygen, water, and ions—needed for the corrosion of metals are present almost everywhere on Earth. Even pure rainwater contains a few H_3O^+ and HCO_3^- ions from dissolved carbon dioxide. Higher ion concentrations—from airborne salt near the ocean, from acidic air pollutants, or from salts spread on icy roads—make corrosion worse in certain areas.

corrosion

the gradual destruction of a metal or alloy as a result of chemical processes such as oxidation or the action of a chemical agent

Figure 12
The cathodic reaction happens where the O_2 concentration is high. The anodic reaction happens in a region where the O_2 concentration is low, such as in a pit in the metal.

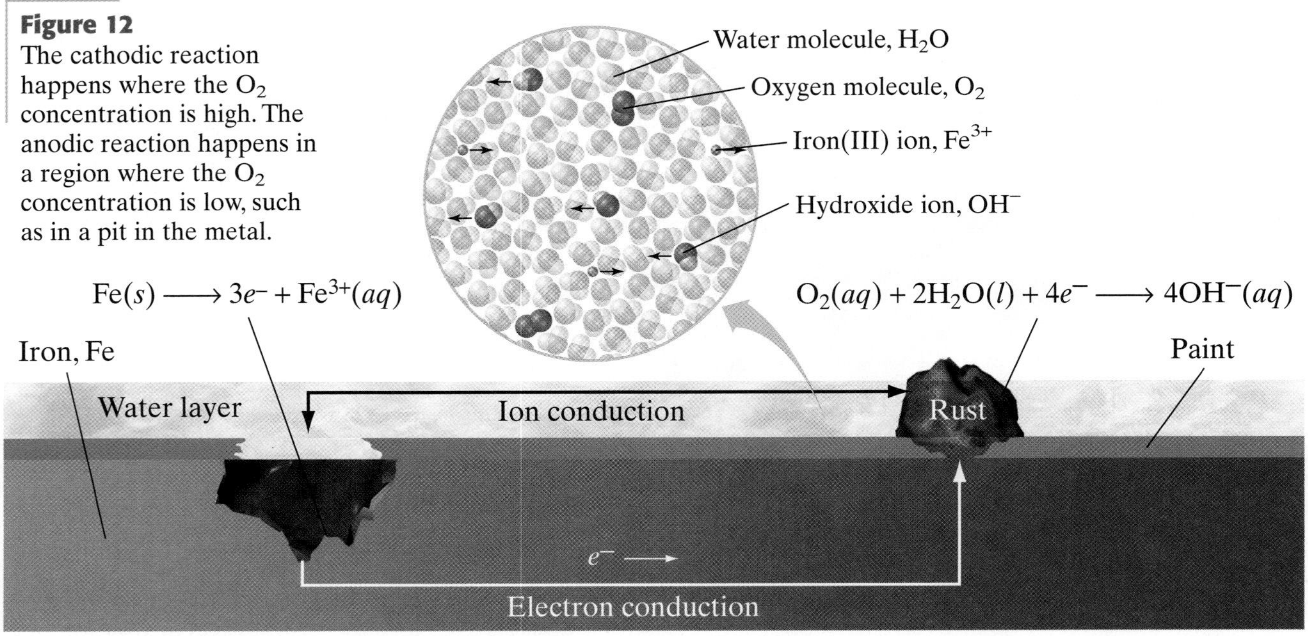

Water molecule, H_2O
Oxygen molecule, O_2
Iron(III) ion, Fe^{3+}
Hydroxide ion, OH^-

$Fe(s) \longrightarrow 3e^- + Fe^{3+}(aq)$

Iron, Fe

Water layer

Ion conduction

$O_2(aq) + 2H_2O(l) + 4e^- \longrightarrow 4OH^-(aq)$

Paint

Rust

$e^- \longrightarrow$

Electron conduction

Figure 13
The Alaskan oil pipeline is
cathodically protected by a
parallel zinc cable.

Methods to Prevent Corrosion

Corrosion is a major economic problem. About 20% of all the iron and steel produced is used to repair or replace corroded structures. That is why the prevention of corrosion is a major focus of research in materials science and electrochemistry. An obvious response to corrosion is to paint the metal or coat it with some other material that does not corrode. However, once a crack or scrape occurs in the coating, corrosion can begin and often spread even faster than on an uncoated surface.

Some metals corrode more easily than others do. Electronegativity is one factor. Gold, the metal with the highest electronegativity, is the most corrosion resistant. The alkali metals, with the lowest electronegativities, easily corrode. The properties of the metal oxides that form are also important. Despite having low electronegativities, aluminum, chromium, and titanium are corrosion-resistant metals. This is because the oxides of these metals form layers that cover the underlying metal, stopping oxidation. In contrast, rust is a porous powder that flakes off, so it does not protect the iron surface.

Surprisingly, it is better to coat steel with another metal that *does* corrode. Trash cans, for example, are made of zinc-coated steel. This coating does not stop corrosion. But the zinc corrodes first, making the steel underneath last much longer than it would without the zinc layer.

Whenever two metals are in electrical contact, a corrosion cell is likely to form. In fact, the metal that is the anode in the cell corrodes faster than it would if it were not connected to another metal. This idea explains the use of sacrificial anodes on ships and pipelines, as shown in **Figure 13.** As the anode corrodes, it gives electrons to the cathode. The corrosion of the anode slows or stops the corrosion of the important structural metal in a process called *cathodic protection*.

internet connect

www.scilinks.org
Topic: Corrosion
SciLinks code: HW4147

SCiLINKS. Maintained by the National Science Teachers Association

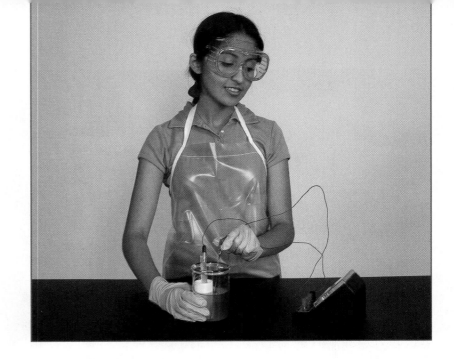

Figure 14
It would be difficult to measure every possible combination of electrodes using a technique like the one shown here. This is why the SHE is used as a reference for all other electrodes.

Determining the Voltage of a Cell

A voltmeter is an electronic instrument that measures the voltage between its two leads. The student in **Figure 14** is using the meter to measure the difference in potentials between the electrodes in a Daniell cell. With such a meter, measuring the voltage of a cell is easy. However, the voltage of a cell depends on such factors as temperature and concentration. And because there are so many combinations of electrode reactions, it would be very difficult to measure the voltage for each combination.

Standard Electrode Potentials

Picking one electrode as a standard and determining electrode potentials in reference to that standard is much easier than measuring the potential between every combination of electrodes is. The electrode that has been chosen as a standard is the *standard hydrogen electrode* (SHE). It consists of a platinum electrode in a 1.00 M H_3O^+ solution in the presence of H_2 gas at 1 atm pressure and 25°C. The SHE is assigned a potential of 0.0000 V and its reaction is

$$2H_3O^+(aq) + 2e^- \rightleftarrows H_2(g) + 2H_2O(l)$$

When measuring potentials, a salt bridge, a narrow tube filled with a concentrated solution of a salt, must be used to link the compartments. When a SHE is joined to another electrode by a salt bridge, a voltmeter can be used to determine the **standard electrode potential,** $E°$, for the electrode. Some standard electrode potentials are shown in **Table 1,** on the last page of this section.

The standard electrode potential is sometimes called the standard reduction potential because it is listed by the reduction half-reactions. However, a voltmeter allows no current in the cell during the measurement. Therefore, the conditions are neither galvanic nor electrolytic—the cell is at equilibrium. As a result, the half-reactions listed in the table are shown as reversible. If the reaction occurs in the opposite direction, as an oxidation half-reaction, $E°$ will have the opposite sign.

standard electrode potential

the potential developed by a metal or other material immersed in an electrolyte solution relative to the potential of the hydrogen electrode, which is set at zero

Calculating the Voltage of a Cell

Think of $E°$ as a measure of the ability of an electrode to gain electrons. A more positive value means the electrode is more likely to be a cathode. The standard cell voltage—the voltage of a cell under standard conditions—can be found by subtracting the standard potentials of the two electrodes, as follows:

$$E°_{cell} = E°_{cathode} - E°_{anode}$$

To determine the reaction that will happen naturally, use the electrode with the most positive $E°$ value as the cathode, as shown in **Sample Problem C.** Otherwise, a given reaction happens naturally if $E°_{cell}$ is positive. If $E°_{cell}$ is negative, the reaction could be made to happen if energy is added.

SAMPLE PROBLEM C

Calculating Cell Voltage

Calculate the voltage of a cell for the naturally occurring reaction between a liquid mercury electrode in a solution of mercury(I) nitrate and a cadmium metal electrode in a solution of cadmium nitrate.

1 Gather information.

From **Table 1** the standard electrode potentials are

$$Hg_2^{2+}(aq) + 2e^- \longrightarrow 2Hg(l) \qquad E° = +0.7973 \text{ V}$$
$$Cd^{2+}(aq) + 2e^- \longrightarrow Cd(s) \qquad E° = -0.4030 \text{ V}$$

2 Plan your work.

The mercury electrode has the more positive $E°$, so it is the cathode.

3 Calculate.

$$E°_{cell} = E°_{Hg} - E°_{Cd} = (+0.7973 \text{ V}) - (-0.4030 \text{ V}) = +1.2003 \text{ V}$$

4 Verify your results.

The positive value of $E°_{cell}$ shows that the reaction is spontaneous and will occur naturally with the mercury electrode as the cathode.

PRACTICE

Use data from **Table 1** to answer the following:

1 Calculate the voltage of a cell if the reactions are as follows:
$$Fe(s) \longrightarrow Fe^{3+}(aq) + 3e^-$$
$$O_2(g) + 2H_2O(l) + 4e^- \longrightarrow 4OH^-(aq)$$

2 Calculate the voltage of a cell for the naturally occurring reaction between a copper electrode in a copper(II) solution and a zinc electrode in a solution containing zinc ions.

3 Calculate the voltage of a cell for the naturally occurring reaction between a silver electrode in a solution containing silver ions and a copper electrode in a copper(II) solution.

Table 1 Standard Electrode Potentials

Electrode reaction	$E°(V)$	Electrode reaction	$E°(V)$
$Li^+(aq) + e^- \rightleftarrows Li(s)$	−3.0401	$Cu^{2+}(aq) + 2e^- \rightleftarrows Cu(s)$	+0.3419
$K^+(aq) + e^- \rightleftarrows K(s)$	−2.931	$O_2(g) + 2H_2O(l) + 4e^- \rightleftarrows 4OH^-(aq)$	+0.401
$Na^+(aq) + e^- \rightleftarrows Na(s)$	−2.71	$I_2(s) + 2e^- \rightleftarrows 2I^-(aq)$	+0.5355
$2H_2O(l) + 2e^- \rightleftarrows H_2(g) + 2OH^-(aq)$	−0.828	$Fe^{3+}(aq) + e^- \rightleftarrows Fe^{2+}(aq)$	+0.771
$Zn^{2+}(aq) + 2e^- \rightleftarrows Zn(s)$	−0.7618	$Hg_2^{2+}(aq) + 2e^- \rightleftarrows 2Hg(l)$	+0.7973
$Fe^{2+}(aq) + 2e^- \rightleftarrows Fe(s)$	−0.447	$Ag^+(aq) + e^- \rightleftarrows Ag(s)$	+0.7996
$PbSO_4(s) + H_3O^+(aq) + 2e^- \rightleftarrows$ $Pb(s) + HSO_4^-(aq) + H_2O(l)$	−0.42	$Br_2(l) + 2e^- \rightleftarrows 2Br^-(aq)$	+1.066
$Cd^{2+}(aq) + 2e^- \rightleftarrows Cd(s)$	−0.4030	$Cl_2(g) + 2e^- \rightleftarrows 2Cl^-(aq)$	+1.358
$Pb^{2+}(aq) + 2e^- \rightleftarrows Pb(s)$	−0.1262	$PbO_2(s) + 4H_3O^+(aq) + 2e^- \rightleftarrows$ $Pb^{2+}(aq) + 6H_2O(l)$	+1.455
$Fe^{3+}(aq) + 3e^- \rightleftarrows Fe(s)$	−0.037	$PbO_2(s) + HSO_4^-(aq) + 3H_3O^+(aq) +$ $2e^- \rightleftarrows PbSO_4(s) + 5H_2O(l)$	+1.691
$2H_3O^+(aq) + 2e^- \rightleftarrows H_2(g) + 2H_2O(l)$	0.0000	$Ce^{4+}(aq) + e^- \rightleftarrows Ce^{3+}(aq)$	+1.72
$AgCl(s) + e^- \rightleftarrows Ag(s) + Cl^-(aq)$	+0.222	$F_2(g) + 2e^- \rightleftarrows 2F^-(aq)$	+2.866

Refer to Appendix A for additional standard electrode potentials.

 # Section Review

UNDERSTANDING KEY IDEAS

1. How does a fuel cell differ from a battery?

2. How do an acidic dry cell and an alkaline dry cell differ?

3. Of the metals Zn, Fe, and Ag, which will corrode the easiest? Explain.

4. Describe how a standard electrode potential is measured.

PRACTICE PROBLEMS

5. Calculate the voltage and identify the anode for a cell in which the following electrode reactions take place:

$$Ag(s) + Cl^-(aq) \longrightarrow AgCl(s) + e^-$$
$$Cl_2(g) + 2e^- \longrightarrow 2Cl^-(aq)$$

6. Calculate the voltage and identify the cathode for a cell in which the natural reaction between the following electrodes happens:

$$Zn^{2+}(aq) + 2e^- \rightleftarrows Zn(s)$$
$$2H_2O(l) + 2e^- \rightleftarrows H_2(g) + 2OH^-(aq)$$

CRITICAL THINKING

7. Write the overall equation for the reaction occurring in a lead-acid cell during discharge. What happens to the sulfuric acid concentration during this process? Why is it possible to use a hydrometer, which measures the density of a liquid, to determine if a lead-acid cell is fully charged?

8. A sacrificial anode is allowed to corrode. Why is use of a sacrificial anode considered to be a way to prevent corrosion?

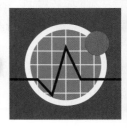

SCIENCE AND TECHNOLOGY

A small pump sends a precise sample size into the fuel cell inside this device.

Fuel Cells

Historical Perspective

In 1839, Sir William Robert Grove, a British lawyer and physicist, built the first fuel cell. More than 100 years later, fuel cells finally found a practical application—in space exploration. During short space missions, batteries can provide enough energy to keep the astronauts warm and to run electrical systems. But longer missions need energy for much longer periods of time, and fuel cells are better suited for this than batteries are. Today, fuel cells are critical to the space shuttle missions and to future missions on the international space station.

Blood Alcohol Testing

A more down-to-Earth use of fuel cells is found in traffic-law enforcement. Police officers need quick and simple ways to determine a person's blood alcohol level in the field. In the time it takes to bring a person to the station or to a hospital for a blood or urine test, the person's blood alcohol content (BAC) might change. Fuel cells, such as the one in the device shown above, provide a quick and accurate way to measure BAC from a breath sample. The alcohol ethanol from the person's breath is oxidized to acetic acid at the anode. At the cathode, gaseous oxygen is reduced and combined with hydronium ions (released from the anode) to form water. The reactions generate an electric current. The size of this current is related to the BAC.

Questions

1. Research at least two more uses for fuel cells. Identify the reactants and products for the cell in each use.
2. Research careers open to chemical engineers. Determine the level of education, the approximate salaries, and the areas of study needed to become a chemical engineer.

Electrolytic Cells

OBJECTIVES

① **Describe** how electrolytic cells work.

② **Describe** the process of electrolysis in the decomposition of water and in the production of metals.

③ **Describe** the process of electroplating.

Cells Requiring Energy

Galvanic cells generate electrical energy, but another kind of cell *consumes* electrical energy. This energy is used to drive a chemical reaction. The cell shown in **Figure 15** is a laboratory-scale version of the industrial process used to refine copper. The anode is impure copper, which includes such metals as zinc, silver, and gold. The oxidation reaction at the anode changes Cu atoms in the impure sample to $Cu^{2+}(aq)$. The opposite reaction happens at the cathode. The $Cu^{2+}(aq)$ ions are reduced to Cu atoms. Pure copper is formed, adding to the pure copper cathode.

Impurities such as Zn and other active metals also dissolve as cations. But they are not reduced at the cathode. Inactive metals, such as Au and Ag, fall to the bottom as an *anode sludge*. This sludge is a valuable source of these more-expensive metals in the industrial process.

Figure 15
The experiment shown here mirrors the industrial refining of copper.

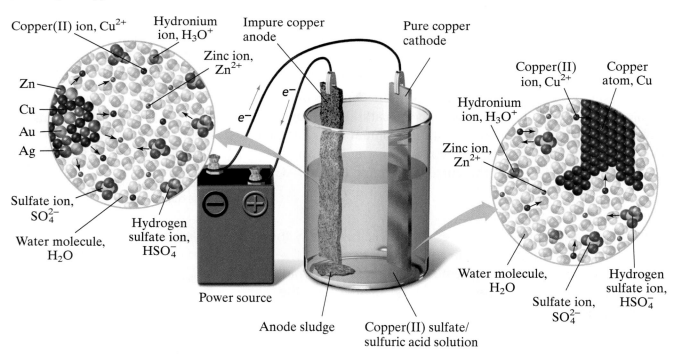

Copper(II) ion, Cu^{2+}
Hydronium ion, H_3O^+
Zinc ion, Zn^{2+}
Zn
Cu
Au
Ag
Sulfate ion, SO_4^{2-}
Water molecule, H_2O
Hydrogen sulfate ion, HSO_4^-

Impure copper anode
Pure copper cathode
e^-
e^-

Copper(II) ion, Cu^{2+}
Copper atom, Cu
Hydronium ion, H_3O^+
Zinc ion, Zn^{2+}
Water molecule, H_2O
Sulfate ion, SO_4^{2-}
Hydrogen sulfate ion, HSO_4^-

Power source
Anode sludge
Copper(II) sulfate/ sulfuric acid solution

Electrolysis

In an **electrolytic cell,** chemical changes are brought about by driving electrical energy through an electrochemical cell. In fact, the words *electrolysis* and *electrolytic* mean "splitting by electricity" and **electrolysis** does refer to the decomposition of a compound, usually into its elements. However, many changes other than decomposition can happen when you use an electrolytic cell. For example, adiponitrile, one of the raw materials used to make nylon, is synthesized in an electrolytic cell. What makes an electrolytic cell useful is that the overall reaction is a *nonspontaneous* process that is forced to happen by an input of energy.

Electrolysis of Water

The electrolysis of water, shown in **Figure 16,** leads to the overall reaction in which H_2O is broken down into its elements, H_2 and O_2. Pure water does not have enough ions in it and is not conductive enough for electrolysis. An electrolyte, such as sodium sulfate, must be added. The $Na^+(aq)$ and $SO_4^{2-}(aq)$ ions play no part in the electrode reactions, which are as follows:

$$\text{Anode: } 6H_2O(l) \longrightarrow 4e^- + O_2(g) + 4H_3O^+(aq)$$

$$\text{Cathode: } 4H_2O(l) + 4e^- \longrightarrow 2H_2(g) + 4OH^-(aq)$$

As always, oxidation happens on the anode, while reduction happens on the cathode. Note that hydronium ions form at the anode. Thus, the solution near the anode becomes acidic. But hydroxide ions form at the cathode. So, the solution near the cathode becomes basic.

electrolytic cell

an electrochemical device in which electrolysis takes place when an electric current is in the device

electrolysis

the process in which an electric current is used to produce a chemical reaction, such as the decomposition of water

Topic Link

Refer to the "Causes of Change" chapter for more information about spontaneous and nonspontaneous reactions.

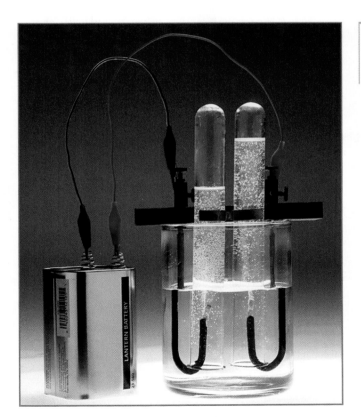

Figure 16
Electrical energy from the battery is used to break down water. Hydrogen forms at the cathode, and oxygen forms at the anode.

Sodium Production by Electrolysis

Sodium is such a reactive metal that preparing it through a chemical process can be dangerous. Sir Humphry Davy first isolated it in 1807 by the electrolysis of molten sodium hydroxide. Today, sodium is made by the electrolysis of molten sodium chloride in a Downs cell, as shown in **Figure 17.**

Pure sodium chloride melts at 801°C. The addition of calcium chloride, $CaCl_2$, to the NaCl lowers the melting point. The Downs cell can then work at 590°C, and less energy is needed to run the cell. The equations below describe the major reactions that occur.

$$\text{Anode: } 2Cl^-(l) \longrightarrow 2e^- + Cl_2(g)$$

$$\text{Cathode: } 2Na^+(l) + 2e^- \longrightarrow 2Na(s)$$

Because chlorine reacts with most metals, the anode is made of graphite. The cathode is steel. Because sodium melts at 98°C and is less dense than the molten salts, it floats to the top and can be removed.

In addition to sodium, a small amount of calcium also forms at the cathode by the reaction:

$$\text{Cathode: } Ca^{2+}(l) + 2e^- \longrightarrow Ca(s)$$

The calcium is more dense than the molten salts, so it falls to the bottom of the cell, where it slowly changes back to Ca^{2+} by the following reaction:

$$Ca(l) + 2Na^+(l) \longrightarrow 2Na(l) + Ca^{2+}(l)$$

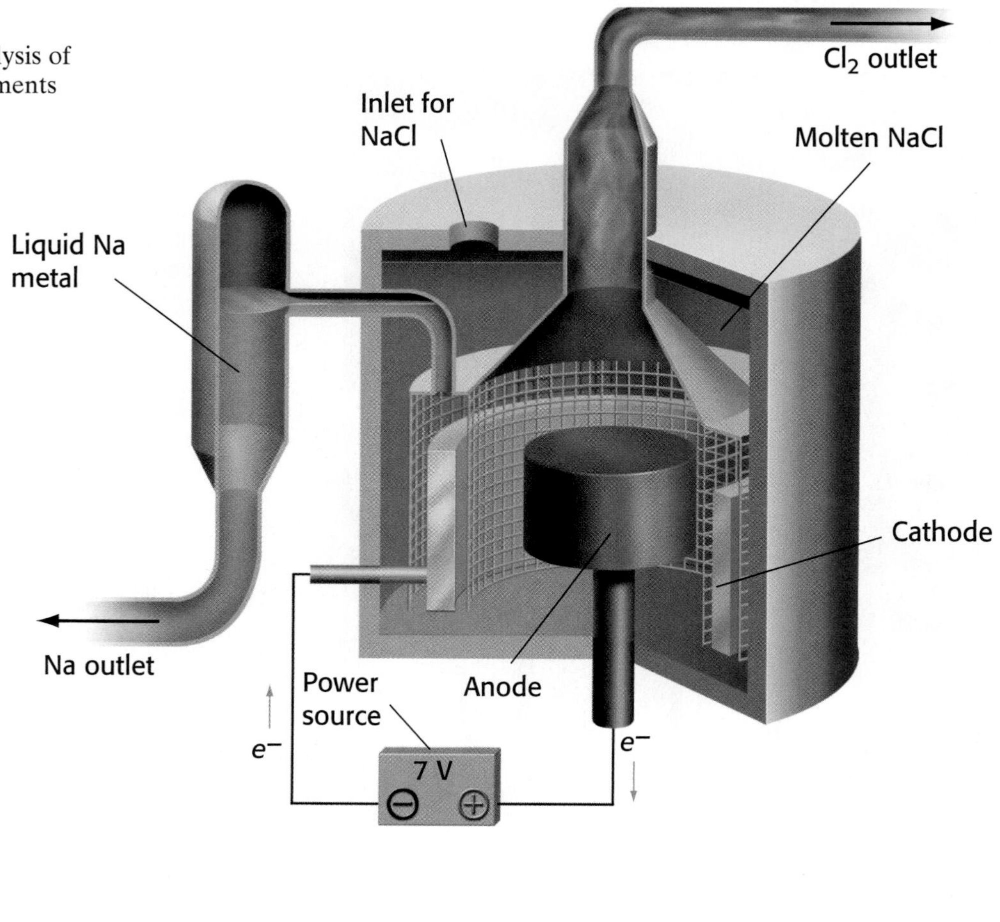

Figure 17
In a Downs cell the electrolysis of molten NaCl forms the elements sodium and chlorine.

Aluminum Production by Electrolysis

Aluminum is the most abundant metal in Earth's crust. Aluminum is light, weather resistant, and easily worked. You have seen aluminum in use in drink cans, in food packaging, and even in airplanes. However, aluminum is never found in nature as a pure metal. Instead, it is isolated from its ore through electrolysis.

The process used to get aluminum from its ore, bauxite, is the electrochemical Hall-Héroult process. This process is the largest single user of electrical energy in the United States—nearly 5% of the national total. This need for energy makes the manufacture of aluminum expensive. Recycling aluminum saves almost 95% of the cost! Aluminum recycling is one of the most economically worthwhile recycling programs that has been developed.

The bauxite is processed to extract and purify hydrated alumina, Al_2O_3. The alumina is fed into huge carbon-lined tanks, like the one in **Figure 18.** There the alumina dissolves in molten cryolite, Na_3AlF_6, at 970°C. Liquid aluminum forms at the cathode. Being more dense than the molten cryolite, aluminum sinks to the floor of the tank. As reduction continues, the level of aluminum rises. As needed, the liquid aluminum is drained and allowed to cool.

Carbon rods serve as the anode. The carbon is oxidized during the anodic reaction, forming CO_2. The rods are eaten away by this oxidation and must be replaced from time to time.

The Hall-Héroult process has been in use for more than a century. But scientists do not completely understand how alumina dissolves and what exactly the species are that participate in the electrode reactions. Although scientists still debate how the process works, they agree that the overall reaction is

$$2Al_2O_3(l) + 3C(s) \longrightarrow 4Al(l) + 3CO_2(g)$$

Figure 18
The Hall-Héroult process is used to make aluminum by the electrolysis of dissolved alumina, Al_2O_3.

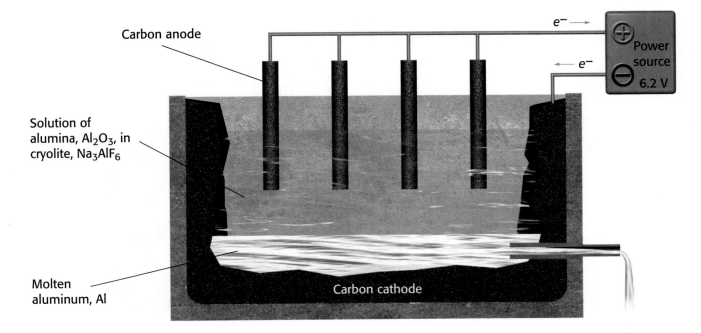

Carbon anode

Solution of alumina, Al_2O_3, in cryolite, Na_3AlF_6

Molten aluminum, Al

Carbon cathode

e^-

Power source

e^-

6.2 V

Electroplating

Many of the metal things that you use every day—forks and spoons, cans for food and drinks, plumbing fixtures, jewelry and decorative ornaments, automobile and appliance parts, nails, nuts, and bolts—have been treated to change their surfaces. Often this involves putting a layer of another metal on top of the main metal. **Electroplating** is one way of applying these finishes. Forks, spoons, and jewelry are often electroplated to give the objects the appearance of silver or gold while still keeping the cost of the objects low. The chrome parts on automobiles have been electroplated to improve the parts' appearance and protect them from corrosion. Electroplating is also used for many electronic and computer parts to give them a certain physical property or to make them last longer by protecting them against corrosion.

To electroplate a bracelet with silver, as shown in the simplified model in **Figure 19,** the bracelet is made the cathode of an electrolytic cell. The anode is a strip of pure silver metal. Both electrodes are placed in a solution of silver ions. The net reactions that happen are very simple. The anode slowly dissolves by the following oxidation reaction:

$$\text{Anode: } Ag(s) \longrightarrow e^- + Ag^+(aq)$$

The cathode reaction on the surface of the bracelet is the reverse of the anode reaction.

$$\text{Cathode: } Ag^+(aq) + e^- \longrightarrow Ag(s)$$

The result is that a thin coating of silver forms on the bracelet. The longer the plating is continued, the thicker the silver layer becomes.

electroplating

the electrolytic process of plating or coating an object with a metal

internet connect

www.scilinks.org
Topic: Electroplating
SciLinks code: HW4049

*SC*LINKS. Maintained by the National Science Teachers Association

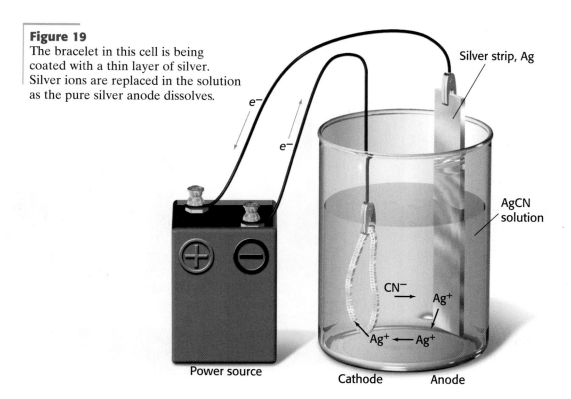

Figure 19
The bracelet in this cell is being coated with a thin layer of silver. Silver ions are replaced in the solution as the pure silver anode dissolves.

Silver strip, Ag

e^-

e^-

AgCN solution

CN^-

Ag^+

$Ag^+ \longleftarrow Ag^+$

Power source

Cathode   Anode

Benefits and Concerns About Electroplating

A major benefit of electroplating a metal is that it becomes more resistant to corrosion. Chrome-plated car parts and zinc-plated food cans are two uses of electroplating to reduce corrosion. Another benefit of electroplating is that it improves the appearance of an object. An item that is gold-plated or silver-plated is much cheaper than the same item in solid gold or silver but looks the same.

However, there are also drawbacks with the electroplating process. In practice, electroplating is not as simple as the description above suggests. It is difficult to get metal to deposit in depressions, so the metal layer often is not uniform. Sometimes, deposits are loose and powdery. Additives are used and conditions such as temperature and pH are carefully controlled to overcome these problems. Over time, impurities build up in the solutions used for electroplating. Eventually, the spent solutions must be discarded. They can contain high concentrations of such toxic metals as cadmium or chromium and require careful disposal to protect the environment.

4 Section Review

UNDERSTANDING KEY IDEAS

1. How does an electrolytic cell differ from a galvanic cell?

2. What chemical process occurs at the anode of an electrolytic cell?

3. What chemical process occurs at the cathode of an electrolytic cell?

4. What form of energy is used to drive an electrolytic cell?

5. List three commercial products made by using electrolytic cells.

6. What does *electrolysis* mean?

7. Write the equation for the cathodic reaction that occurs in the Downs cell used to make sodium.

8. How is aluminum manufactured?

9. Describe the electroplating process.

CRITICAL THINKING

10. In the copper refining process, why does zinc not also deposit on the cathode?

11. Elemental aluminum was first prepared in 1827 by the reaction of aluminum chloride with potassium.

   a. Write the balanced equation for this reaction.

   b. Determine if the reaction is a redox reaction.

   c. Would this reaction need to happen in an electrolytic cell? Explain.

12. Cryolite, Na_3AlF_6, is an ionic mineral used in the preparation of aluminum. Explain why the sodium ions are not reduced during the electrolytic process that produces aluminum.

13. The following reaction happens naturally:

   $$Zn(s) + Cu^{2+}(aq) \longrightarrow Cu(s) + Zn^{2+}(aq)$$

   Explain why it is necessary to use an electric current to deposit a layer of zinc on a copper bracelet.

14. Explain why a galvanic cell is often used in an electrolytic cell. What function does the galvanic cell serve?

15. Why is it so important to recycle rather than discard aluminum products?

KEY IDEAS

SECTION ONE Oxidation-Reduction Reactions

- The loss or gain of electrons in a chemical reaction is called *oxidation* or *reduction,* respectively.
- In a redox reaction, oxidation and reduction occur at the same time.
- An oxidation number may be assigned to each atom in a molecule or ion.
- Half-reactions, in which only the oxidation or the reduction is described, are useful in balancing redox equations.
- Reducing agents readily donate electrons; oxidizing agents readily accept electrons.

SECTION TWO Introduction to Electrochemistry

- An electrochemical cell is made up of two electrodes linked by one or more ionic conductors.
- When electric current is in a cell, electrode reactions take place.
- Oxidation happens at the anode. Reduction happens at the cathode.

SECTION THREE Galvanic Cells

- Many examples of galvanic cells are power sources that generate electrical energy from chemical energy.
- Fuel cells differ from batteries in that their oxidizing and reducing agents are gases introduced to the cell from outside.
- The corrosion of metals generally happens in a galvanic cell.
- Whether there is electric current or not, the electrodes of a cell have different potentials. The difference between these is the voltage of the cell.

SECTION FOUR Electrolytic Cells

- Electrical energy is used to power an electrolytic cell.
- In electrolysis, a compound is decomposed, usually to its elements; H_2, O_2, Al, Na, and Cl_2 are among the elements that can be isolated by electrolysis.
- A layer of a second metal is deposited cathodically in electroplating.

KEY TERMS

oxidation
reduction
oxidation-reduction reaction
oxidation number
half-reaction
oxidizing agent
reducing agent

electrochemistry
voltage
electrode
electrochemical cell
cathode
anode

corrosion
standard electrode potential

electrolytic cell
electrolysis
electroplating

KEY SKILLS

Assigning Oxidation Numbers
Skills Toolkit 1   p. 606
Sample Problem A   p. 607

Balancing Redox Equations Using the Half-Reaction Method
Skills Toolkit 2   p. 609
Sample Problem B   p. 610

Calculating Cell Voltage
Sample Problem C   p. 623

CHAPTER REVIEW

USING KEY TERMS

1. What is a redox reaction?

2. Explain the terms *oxidation* and *reduction* in terms of electrons.

3. Explain how oxidation numbers are used to identify redox reactions?

4. Explain what half-reactions are and why they are useful.

5. Define *electrochemistry*.

6. Define *electrode, anode,* and *cathode*.

7. Explain voltage and current in terms of electrons.

8. Distinguish between galvanic and electrolytic cells.

9. What is a fuel cell?

10. What is corrosion, and what is a corrosion cell?

11. Why does a sacrificial anode provide cathodic protection?

12. What is electroplating?

UNDERSTANDING KEY IDEAS

Oxidation-Reduction Reactions

13. Assign oxidation numbers to the atoms in the ionic compound $MgBr_2(s)$.

14. Assign oxidation numbers to the atoms in the ionic compound $NH_4NO_3(s)$.

15. Assign oxidation numbers to the atoms in the ion $PF_6^-(aq)$.

16. Identify each of the following half-reactions as oxidation or reduction reactions.
    a. $K(s) \longrightarrow e^- + K^+(aq)$
    b. $Cu^{2+}(aq) + e^- \longrightarrow Cu^+(aq)$
    c. $Br_2(l) + 2e^- \longrightarrow 2Br^-(aq)$

17. Is the reaction below a redox reaction? Explain your answer.
$$Ca(s) + Cl_2(g) \longrightarrow CaCl_2(s)$$

18. Describe how to identify the oxidizing agent and the reducing agent in a reaction.

19. Nitrogen monoxide, $NO(g)$, reacts with phosphorus, $P_4(s)$, to produce nitrogen, $N_2(g)$, and diphosphorus pentoxide, P_2O_5. Write the balanced equation for this reaction. Identify the atoms that have been oxidized and reduced, and identify the oxidizing and reducing agents.

Introduction to Electrochemistry

20. What is the distinction between a half-reaction and an electrode reaction?

21. Describe the components of an electrochemical cell.

22. Identify which of the following reactions (as written) is an anodic reaction and which is a cathodic reaction. Write the balanced overall ionic equation for the redox reaction of the cell.
$$Cd(s) \longrightarrow Cd^{2+}(aq) + 2e^-$$
$$Ag^+(aq) + e^- \longrightarrow Ag(s)$$

23. What is the significance of the $\oplus$ symbol on a dry cell or battery?

24. What reaction happens at the cathode of an electrochemical cell?

25. In an electrochemical cell, what role does the porous barrier play? What would happen without it?

26. Explain why the combination of the two electrode reactions of an electrochemical cell always gives the equation of a redox reaction.

Galvanic Cells

27. Describe a galvanic cell, and give an example.

28. Write the equations of the two electrode reactions that occur when a Daniell cell is in use. Identify the anode reaction and the cathode reaction.

29. What is the essential advantage of a fuel cell over other types of galvanic cells that are used to generate electrical energy?

30. Explain why a corrosion cell is a galvanic cell.

31. Discuss methods of reducing corrosion.

32. The standard electrode potential for the reduction of $Zn^{2+}(aq)$ to $Zn(s)$ is −0.762 V. What does this value indicate?

33. Which half-reaction would be more likely to be an oxidation: one with a standard electrode potential of −0.42 V, or one with a standard electrode potential of +0.42 V? Explain your answer.

Electrolytic Cells

34. Define *electrolytic cell,* and give an example.

35. Describe the apparatus used in the electrolysis of water.

36. Explain why sodium can be prepared by electrolysis.

37. Describe some benefits of electroplating.

38. What are some problems in the electroplating industry?

PRACTICE PROBLEMS

Determining Oxidation Numbers

39. Determine the oxidation number of each atom in CO_2.

40. Determine the oxidation number of each atom in CoO.

41. Determine the oxidation number of each atom in $BaCl_2$.

42. Determine the oxidation number of each atom in K_2SO_4.

43. Determine the oxidation number of the sulfur atom in S^{2-}.

44. Determine the oxidation number of the lanthanum atom in La^{3+}.

45. Determine the oxidation number of each atom in CH_4.

46. Determine the oxidation number of each atom in NH_4^+.

47. Determine the oxidation number of each atom in $CaCO_3$.

48. Determine the oxidation number of each atom in $PtCl_6^{2-}$.

49. Determine the oxidation number of each atom in $COCl_2$.

50. Determine the oxidation number of each atom in PO_4^{3-}.

The Half-Reaction Method

51. Write the balanced half-reaction for the conversion of $Fe(s)$ to $Fe^{2+}(aq)$.

52. Write the balanced half-reaction for the conversion of $Cl_2(g)$ to $Cl^-(aq)$.

53. Combine the half-reactions from items 51 and 52 into a single reaction.

54. Write the balanced half-reaction for the conversion of $HOBr(aq)$ to $Br_2(aq)$ in acidic solution.

55. Write the balanced half-reaction for the conversion of $H_2O(l)$ to $O_2(aq)$ in acidic solution.

56. Combine the half-reactions from items 54 and 55 into a single reaction.

57. Write the balanced half-reaction for the change of $O_2(aq)$ to $H_2O(l)$ in acidic solution.

58. Write the balanced half-reaction for the change of $SO_2(aq)$ to $HSO_4^-(aq)$ in acidic solution.

59. Combine the reactions from items 57 and 58 into a single reaction.

60. Using half-reactions, balance the redox equation of $Zn(s)$ and $Fe^{3+}(aq)$ reacting to form $Zn^{2+}(aq)$ and $Fe^{2+}(aq)$.

Calculating Cell Voltage

61. The standard electrode potentials of two electrodes in a cell are 1.30 V and 0.45 V. What is the voltage of the cell?

Use the information from Table 1 to answer the following items.

62. Calculate the voltage of a cell for the naturally occurring reaction between the following electrodes:

$$AgCl(s) + e^- \rightleftarrows Ag(s) + Cl^-(aq)$$

$$2H_3O^+(aq) + 2e^- \rightleftarrows 2H_2O(l) + H_2(g)$$

63. Calculate the voltage of a cell that has the following electrode reactions:

$$2H_3O^+(aq) + 2e^- \longrightarrow 2H_2O(l) + H_2(g)$$

$$Fe^{2+}(aq) \longrightarrow Fe^{3+}(aq) + e^-$$

64. Calculate the voltage of a cell in which the overall reaction is

$$2Fe^{3+}(aq) + Cd(s) \longrightarrow Cd^{2+}(aq) + 2Fe^{2+}(aq)$$

65. Calculate the voltage of a cell that has the following electrode reactions:

$$O_2(g) + 2H_2O(l) + 4e^- \longrightarrow 4OH^-(aq)$$

$$H_2(g) + 2OH^-(aq) \longrightarrow 2H_2O(l) + 2e^-$$

66. Calculate the voltage of a cell in which the two electrode reactions are the reduction of chlorine gas to chloride ions and the oxidation of copper metal to copper(II) ions.

67. Calculate the voltage of a cell in which the two electrode reactions are those of the lead-acid battery.

MIXED REVIEW

68. Assign an oxidation number to the N atom in each of the following oxides of nitrogen: N_2O, NO, NO_2, N_2O_3, N_2O_4, and N_2O_5.

69. Refer to the figure below to answer the following questions:

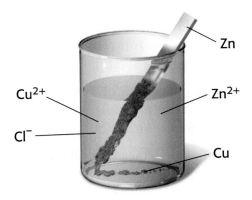

a. What observations suggest that a chemical reaction has taken place?

b. Write a balanced equation for the reaction taking place, and explain how you can identify it as a redox reaction.

c. Identify what is oxidized and what is reduced in the reaction.

d. Describe how the quantity of zinc metal and of copper metal will change as the reaction continues.

e. Describe the role of the chloride ions in the reaction.

70. What name is given to an electrode at which oxidation occurs? at which reduction occurs?

71. Explain how electric charges travel through the various parts of an electrochemical cell, including how they cross the electrode surfaces.

72. Why are dry cells so called, even though their chemistry involves water?

73. Explain how copper is refined.

74. Commercial aluminum smelters are located, not near bauxite mines, but near electrical power plants. Why?

75. Determine the oxidation number of each phosphorus atom in P_4.

76. Determine the oxidation number of each atom in H_2SO_3.

77. Write the half-reaction for the conversion of $Fe^{3+}(aq)$ to $Fe^{2+}(aq)$. Is this change an oxidation or a reduction?

78. Using half-reactions, balance the equation for the redox reaction when $Cr_2O_7^{2-}(aq)$ and $Fe^{2+}(aq)$ react to form $Cr^{3+}(aq)$ and $Fe^{3+}(aq)$ in acidic solution.

79. Calculate the voltage of a cell in which the overall reaction is the electrolysis of aqueous cadmium chloride into its elements.

80. Calculate the voltage of a cell in which the overall reaction is the electrolysis of solid AgCl into its elements.

CRITICAL THINKING

81. Suggest how the word *oxidation* might have come to be the general term for a loss of electrons.

82. Think of ammonium nitrite as $(NH_4^+)(NO_2^-)$, and assign oxidation numbers within each ion. Now think of ammonium nitrate as $N_2H_4O_2$, and assign oxidation numbers. Which assignment makes more sense?

83. Why is the negative battery terminal the one with the greater "electron pressure"?

84. What is different about electric current in metals compared with current in electrolyte solutions?

85. The activity of the halogens decreases as you move down the group on the periodic table. Use the information in **Table 1** to explain this trend.

86. A fuel cell uses methanol, CH_3OH, as fuel. Assuming complete oxidation of the fuel to CO_2 and H_2O in acidic aqueous solution, how many electrons are transferred per CH_3OH molecule?

87. Using **Table 1,** calculate the voltage of a cell for the naturally occurring reaction between the electrodes below. If a wire were connected between the Ag and Cu, which way would electrons travel? Which electrode would be the anode?

$$Ag^+(aq) + e^- \rightleftarrows Ag(s)$$
$$Cu^{2+}(aq) + 2e^- \rightleftarrows Cu(s)$$

88. Electrolytic cells can be considered the opposite of galvanic cells. Discuss whether they are closer to being the opposite of batteries or of fuel cells.

89. How does the oxidation number of an atom of Mn change in the reaction below?

$$4Mn^{2+}(aq) + MnO_4^-(aq) + 8H_3O^+(aq) +$$
$$15H_2P_4O_7^{2-}(aq) \longrightarrow 12H_2O(l) +$$
$$5Mn(H_2P_4O_7)_3^{3-}(aq)$$

ALTERNATIVE ASSESSMENT

90. Your teacher will assign you a known metal with an unknown reduction potential. Devise a method to determine the $E°$ value of the metal from a list of metals with known $E°$ values. Present your method to your teacher.

91. Investigate the development and operation of the sodium-sulfur battery that has been proposed to run electric cars. Choose a stand for or against its use, and present your findings in a persuasive speech to your classmates.

92. Consumer use of rechargeable batteries is growing. Nickel-cadmium batteries, a common type of rechargeable battery, are used in cellular phones, electric shavers, and portable video-game systems. Make a list of the items with which you come into contact that use nickel-cadmium batteries or other rechargeable batteries. Write a short essay about technology that was not and could not have been available before the development of the nickel-cadmium battery.

93. For one week, keep a record of how many times you use devices powered by batteries. Record what kind of device you used and the number and type of batteries it contained. Your teacher will provide you with various batteries and a balance. Record the mass of each type of battery you used during the week. Assuming that your battery usage is typical of everyone in the country, estimate the mass of waste material produced by battery usage in one year. Write a short report offering ways to reduce the amount of waste.

94. Read the book *Apollo 13* on which the movie of the same title was based, and report information about the power sources in the spacecraft.

95. Research metals other than copper, sodium, and aluminum that are manufactured or refined electrochemically.

96. Using a voltmeter from home or one borrowed from your teacher, devise a method of measuring the voltage of a flashlight battery *while it is delivering power to the bulb.* Compare that voltage with that of a new battery. Also record the voltage as the battery "runs down." Write a report on your results.

97. Manufacturers often claim that their batteries are "heavy duty" or "long lasting." Design an experiment to test the value and efficacy of AA batteries. If your teacher approves, carry out your procedure.

98. Carpentry nails are steel, but some are plated with a second metal. Obtain a variety of nails and evaluate their tendency to rust by laying them on a piece of cloth moistened by water containing salt and vinegar, so they are exposed both to the liquid and to air. After two weeks, report your findings.

99. Find at least five items that have been electroplated. Describe why each object was electroplated.

CONCEPT MAPPING

100. Use the following terms to complete the concept map below: *cathode, electrodes, electrochemical cell, anode, oxidation,* and *reduction.*

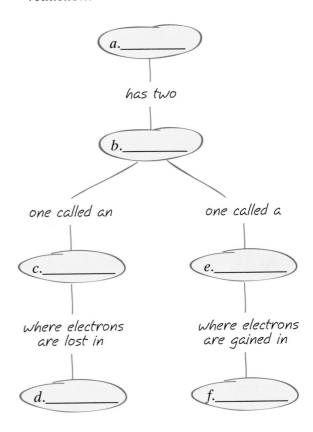

Study the graph below, and answer the questions that follow.
For help in interpreting graphs, see Appendix B, "Study Skills for Chemistry."

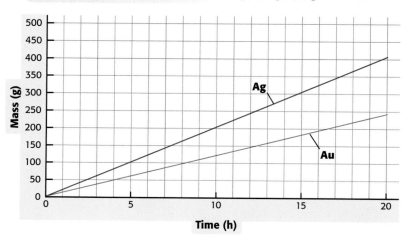

Mass of Metal Deposited by Electroplating Using a 5 A Current

Time (h)

101. What is the rate of production, in g/h, for each metal?

102. What is the rate of production, in mol/h, for each metal?

103. How much time is needed for 1.00 mol of electrons to be absorbed?

104. Write the half-reaction for the production of silver metal from silver ions. Write the half-reaction for the production of gold metal from gold(III) ions. Using these half-reactions, provide an explanation for the difference in the rates of production of the metals.

TECHNOLOGY AND LEARNING

105. Graphing Calculator

Calculate the Equilibrium Constant, using the Standard Cell Voltage

The graphing calculator can run a program that calculates the equilibrium constant for an electrochemical cell using an equation called the Nernst equation, given the standard potential and the number of electrons transferred. Given that the standard potential is 2.041 V and that two electrons are transferred, you will calculate the equilibrium constant. The program will be used to make the calculations.

Go to Appendix C. If you are using a TI-83 Plus, you can download the program NERNST and data and run the application as directed. If you are using another calculator, your teacher will provide you with keystrokes and data sets to use. After you have run the program, answer the following questions.

a. What is the equilibrium constant when the standard potential is 0.099?

b. What is the equilibrium constant when the standard potential is 1.125?

c. What is the equilibrium constant when the standard potential is 2.500?

1. In the following reaction, which species is reduced?

$$2K + Br_2 \longrightarrow 2K^+ + 2Br^-$$

  a. K only
  b. Br_2 only
  c. both K and Br_2
  d. neither K nor Br_2

2. The electrode at which reduction occurs is
  a. always the anode.
  b. always the cathode.
  c. either the anode or the cathode.
  d. always the half-cell.

3. In a galvanic cell, the anode
  a. can be either positive or negative.
  b. is positive.
  c. is not charged.
  d. is negative.

4. A cell contains two electrodes. The first electrode is a strip of zinc metal in a solution containing zinc ions. The second is a strip of copper metal in a solution containing copper ions. When this cell operates as a galvanic cell,
  a. Cu is oxidized and Zn^{2+} is reduced.
  b. Cu is reduced and Zn^{2+} is oxidized.
  c. Cu^{2+} is oxidized and Zn is reduced.
  d. Cu^{2+} is reduced and Zn is oxidized.

5. When water is electrolyzed, oxygen gas is formed at
  a. the anode only.
  b. the cathode only.
  c. the midpoint between the anode and the cathode.
  d. both the anode and the cathode.

6. When sulfuric acid, H_2SO_4, is used in the electrolysis of water, its purpose is to
  a. react with the water.
  b. keep the electrode clean.
  c. provide adequate conductivity.
  d. supply energy.

7. During the electrolysis of molten $CaBr_2$, which of the following would you expect to be made at the anode?
  a. H_2
  b. Ca
  c. Br_2
  d. O_2

8. When a rechargeable cell is being recharged, the cell acts as a(n)
  a. fuel cell.
  b. electrolytic cell.
  c. galvanic cell.
  d. Leclanché cell.

9. When silver is electroplated onto another metal, Ag^+ is
  a. oxidized at the anode.
  b. reduced at the anode.
  c. oxidized at the cathode.
  d. reduced at the cathode.

10. The oxidation number of the sulfur atom in the SO_4^{2-} ion is
  a. +2.
  b. −2.
  c. +6.
  d. +4.

11. A half-reaction
  a. involves electrons.
  b. is useful in balancing equations.
  c. could occur at an electrode.
  d. All of the above

NUCLEAR CHEMISTRY

Before nuclear power was used, submarines could stay submerged for only a brief period of time. A diesel-powered submarine had to surface regularly to recharge its batteries and refuel. But with a lump of nuclear fuel about the size of a golf ball, the first nuclear-powered submarine could remain underwater for months and travel about 97 000 km (about 60 000 mi). Today, nuclear power enables submarines to refuel only once every nine years.

START-UP ACTIVITY

Half-Lives and Pennies

SAFETY PRECAUTIONS

PROCEDURE

1. Make a data table with two columns. Label the first column "Trials." Label the second column "Number of pennies." Count the **pennies** your teacher has given you, and record this number in the table. Also, record "0" in the column labeled "Trials."

2. Place the pennies in a **plastic cup.** Cover the cup with one hand, and gently shake it for several seconds.

3. Pour the pennies on your desk or laboratory table. Remove all the pennies that are heads up. Count the remaining pennies, and record this number in column two. In the first column, record the number of times you performed step 2.

4. Repeat steps 2 and 3 until you have no pennies to place in your cup.

5. Plot your data on **graph paper.** Label the *x*-axis "Trial," and label the *y*-axis "Number of pennies."

ANALYSIS

1. What does your graph look like?

2. Describe any trend that your data display.

Pre-Reading Questions

1. What particles make up an atom?

2. Name some types of radiation that compose the electromagnetic spectrum.

3. Can energy be created? Explain.

4. What quantities are conserved in a chemical reaction?

internet connect

www.scilinks.org
Topic: Nuclear Power
SciLinks code: HW4087

SCILINKS® Maintained by the National Science Teachers Association

internet connect

www.scilinks.org
Topic : Nuclear Reactors
SciLinks code: HW4089

SCILINKS® Maintained by the National Science Teachers Association

Atomic Nuclei and Nuclear Stability

OBJECTIVES

(1) **Describe** how the strong force attracts nucleons.

(2) **Relate** binding energy and mass defect.

(3) **Predict** the stability of a nucleus by considering factors such as nuclear size, binding energy, and the ratio of neutrons to protons in the nucleus.

Nuclear Forces

Topic Link

Refer to the "Atoms and Moles" chapter for a discussion of Rutherford's experiment.

nucleon

a proton or a neutron

nuclide

an atom that is identified by the number of protons and neutrons in its nucleus

Figure 1
In this figure, *X* represents the element, *Z* represents the atom's atomic number, and *A* represents the element's mass number.

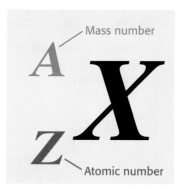

In 1911, Ernest Rutherford's famous gold-foil experiment determined the distribution of charge and mass in an atom. Rutherford's results showed that all of an atom's positive charge and almost all of its mass are contained in an extremely small nucleus.

Other scientists later determined more details about the nuclei of atoms. Atomic nuclei are composed of protons. The nuclei of all atoms except hydrogen also are composed of neutrons. The number of protons is the atomic number, *Z*, and the total number of protons and neutrons is the mass number, *A*. The general symbol for the nucleus of an atom of element *X* is shown in **Figure 1.**

The protons and neutrons of a nucleus are called **nucleons.** A **nuclide** is a general term applied to a specific nucleus with a given number of protons and neutrons. Nuclides can be represented in two ways. One way, shown in **Figure 1,** shows an element's symbol with its atomic number and mass number. A second way is to represent the nuclide by writing the element's name followed by its mass number, such as radium-228 or einsteinium-253. It is not essential to include the atomic number when showing a nuclide because all nuclides of an element have the same atomic number.

Recall that isotopes are atoms that have the same atomic number but different mass numbers. So, isotopes are nuclides that have the same number of protons but different numbers of neutrons. The following symbols represent nuclei of isotopes of tellurium.

$$^{122}_{52}\text{Te} \quad ^{124}_{52}\text{Te} \quad ^{128}_{52}\text{Te}$$

These three isotopes of tellurium are stable. So, their nuclei do not break down spontaneously. Yet, each of these nuclei are composed of 52 protons. How can these positive charges exist so close together? Protons repel each other because of their like charges. So, why don't nuclei fall apart? There must be some attraction in the nucleus that is stronger than the repulsion due to the positive charges on protons.

The Strong Force Holds the Nucleus Together

In 1935, the Japanese physicist Hideki Yukawa proposed that a force between protons that is stronger than the electrostatic repulsion can exist between protons. Later research showed a similar attraction between two neutrons and between a proton and a neutron. This force is called the **strong force** and is exerted by nucleons only when they are very close to each other. All the protons and neutrons of a stable nucleus are held together by this strong force.

Although the strong force is much stronger than electrostatic repulsion, the strong force acts only over very short distances. Examine the nuclei shown in **Figure 2.** The nucleons are close enough for each nucleon to attract all the others by the strong force. In larger nuclei, some nucleons are too far apart to attract each other by the strong force. Although forces due to charges are weaker, they can act over greater distances. If the repulsion due to charges is not balanced by the strong force in a nucleus, the nucleus will break apart.

Protons and Neutrons Are Made of Quarks

In the early 1800s, John Dalton suggested that atoms could not be broken down. However, the discovery of electrons, protons, and neutrons showed that this part of his atomic theory is not correct. So, scientists changed the atomic theory to state that these subatomic particles were indivisible and were the basic building blocks of all matter. However, the atomic theory had to change again when scientists discovered in the 1960s that protons and neutrons are made of even smaller particles called *quarks,* as shown in **Figure 3.**

Quarks were first identified by observing the products formed in high-energy nuclear collisions. Six types of quarks are recognized. Each quark type is known as a flavor. The six flavors are up, down, top, bottom, strange, and charm. Only two of these—the up and down quarks—compose protons and neutrons. A proton is made up of two up quarks and one down quark, while a neutron consists of one up quark and two down quarks. The other four types of quarks exist only in unstable particles that spontaneously break down during a fraction of a second.

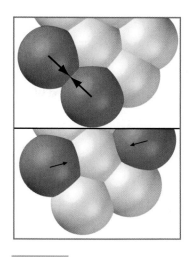

Figure 2
In the nucleus, the nuclear force acts only over a distance of a few nucleon diameters. Arrows describe magnitudes of the strong force acting on the protons.

strong force

the interaction that binds nucleons together in a nucleus

Topic Link

Refer to the "Atoms and Moles" chapter for a discussion of protons, neutrons, and Dalton's theory.

Figure 3
Protons and neutrons, which are made of quarks, make up nuclei.

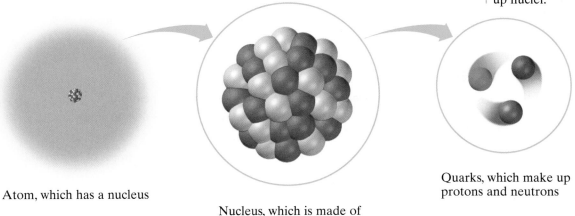

Atom, which has a nucleus

Nucleus, which is made of protons and neutrons

Quarks, which make up protons and neutrons

Binding Energy and Nuclear Stability

When protons and neutrons that are far apart come together and form a nucleus, energy is released. As a result, a nucleus is at a lower energy state than the separate nucleons were. A system is always more stable when it reaches a lower energy state. One way to describe this reaction is as follows:

$$\text{separate nucleons} \longrightarrow \text{nucleus} + \text{energy}$$

The energy released in this reaction is enormous compared with the energy changes that take place during chemical reactions. The energy released when nucleons come together is called *nuclear binding energy*. Where does this enormous quantity of energy come from? The answer can be found by comparing the total mass of the nucleons with the nucleus they form.

The mass of any atom is less than the combined masses of its separated parts. This difference in mass is known as the **mass defect,** also called *mass loss*. Electrons have masses so small that they can be left out of mass defect calculations. For helium, $^{4}_{2}\text{He}$, the mass of the nucleus is about 99.25% of the total mass of two protons and two neutrons. According to the equation $E = mc^2$, energy can be converted into mass, and mass can be converted into energy. So, a small quantity of mass is converted into an enormous quantity of energy when a nucleus forms.

mass defect

the difference between the mass of an atom and the sum of the masses of the atom's protons, neutrons, and electrons

Binding Energy Can Be Calculated for Each Nucleus

As **Figure 4** shows, the mass defect for one $^{4}_{2}\text{He}$ nucleus is 0.0304 amu. The equation $E = mc^2$ can be used to calculate the binding energy for this nucleus. Remember to first convert the mass defect, which has units of amu to kilograms, to match the unit for energy which is joules (kg·m²/s²).

$$0.0304 \ \text{amu} \times \frac{1.6605 \times 10^{-27} \ \text{kg}}{1 \ \text{amu}} = 5.05 \times 10^{-29} \ \text{kg}$$

The binding energy for one $^{4}_{2}\text{He}$ nucleus can now be calculated.

$$E = (5.05 \times 10^{-29} \ \text{kg})(3.00 \times 10^8 \ \text{m/s})^2 = 4.54 \times 10^{-12} \ \text{J}$$

This quantity of energy may seem rather small, but remember that 4.54×10^{-12} J is released for every $^{4}_{2}\text{He}$ nucleus that forms. The binding energy for 1 mol of $^{4}_{2}\text{He}$ nuclei is much more significant.

$$4.54 \times 10^{-12} \ \frac{\text{J}}{\text{He nucleus}} \times 6.022 \times 10^{23} \ \frac{\text{He nuclei}}{\text{mol}} =$$
$$2.73 \times 10^{12} \ \text{J/mol}$$

Figure 4
The mass defect represents the difference in mass between the helium nucleus and the total mass of the separated nucleons.

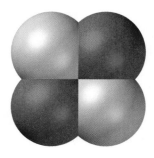

Helium nucleus

$^{4}_{2}\text{He}$ nucleus = 2(mass of proton) + 2(mass of neutron)
= 2(1.007 276 47 amu) + 2(1.008 664 90 amu)
= 4.031 882 74 amu
mass defect = (total mass of separate nucleons) − (mass of helium nucleus)
= 4.031 882 74 amu − 4.001 474 92 amu
= 0.030 407 82 amu per nucleus of $^{4}_{2}\text{He}$

Binding Energy Is One Indicator of Nuclear Stability

A system's stability depends on the amount of energy released as the system is established. When 16 g of oxygen nuclei is formed, 1.23×10^{13} J of binding energy is released. This amount of energy is about equal to the energy needed to heat 4.6×10^6 L of liquid water from 0°C to 100°C and to boil the water away completely.

The binding energy of a selenium nucleus, $^{80}_{34}\text{Se}$, is much greater than that of an $^{16}_{8}\text{O}$ nucleus. Does this difference in energy mean that the $^{80}_{34}\text{Se}$ nucleus is more stable than the $^{16}_{8}\text{O}$ nucleus? Not necessarily. After all, $^{80}_{34}\text{Se}$ contains 64 more nucleons than $^{16}_{8}\text{O}$ does. To make a good comparison of these nuclei, you must look at the binding energy per nucleon. Examine the graph in **Figure 5.** Notice that the binding energy per nucleon rises rapidly among the lighter nuclei. The greater the binding energy per nucleon is, the more stable the nucleus is.

In the graph, the binding energy per nucleon levels off when the mass number is approximately 60. The curve reaches a maximum when the mass number is around 55. Therefore, the most stable nuclei are $^{56}_{26}\text{Fe}$ and $^{58}_{28}\text{Ni}$. These isotopes are relatively abundant in the universe in comparison to other heavy metals, and they are the major components of Earth's core.

Atoms that have larger mass numbers than $^{56}_{26}\text{Fe}$ and $^{58}_{28}\text{Ni}$ have nuclei too large to have larger binding energies per nucleon than these iron and nickel isotopes. In these cases, the net attractive force on a proton is reduced because of the increased distance of the neighboring protons. So, the repulsion between protons results in a decrease in the binding energy per nucleon. Nuclei that have mass numbers greater than 209 and atomic numbers greater than 83 are never stable.

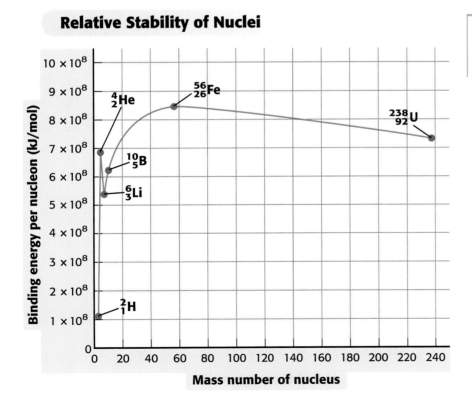

Relative Stability of Nuclei

Figure 5
This graph indicates the relative stability of nuclei. Isotopes that have a high binding energy per nucleon are more stable. The most stable nucleus is $^{56}_{26}\text{Fe}$.

Predicting Nuclear Stability

Finding the binding energy per nucleon for an atom is one way to predict a nucleus's stability. Another way is to compare the number of neutrons with the number of protons of a nucleus. Examine the graph in **Figure 6.** The number of neutrons, N, is plotted against the number of protons, Z, of each stable nucleus. All known stable nuclei are shown as red dots.

The maroon line shows where the data lie for $N/Z = 1$. For elements that have small atomic numbers, the most stable nuclei are those for which $N/Z = 1$. Notice in **Figure 6** that the dots that represent elements that have small atomic numbers are clustered near the line that represents $N/Z = 1$. The green line shows where the data would lie for $N/Z = 1.5$. For elements that have large atomic numbers, the most stable nuclei are those where $N/Z = 1.5$. The reason for the larger N/Z number is that neutrons are needed to stabilize the nuclei of heavier atoms. Notice in **Figure 6** that the dots that represent elements with large atomic numbers are clustered near the line $N/Z = 1.5$.

The dots representing 256 known stable nuclei cluster over a range of neutron-proton ratios, which are referred to as a *band of stability*. This band of stability is shown in yellow in **Figure 6.**

Figure 6
The graph shows the ratio of protons to neutrons for 256 of the known stable nuclei.

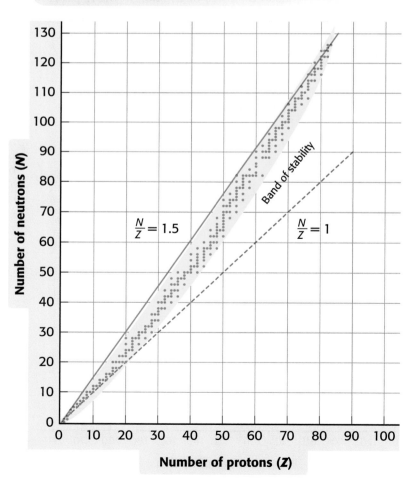

Neutron-Proton Ratios of Stable Nuclei

Some Rules to Help You Predict Nuclear Stability

You probably see that the graph in **Figure 6** shows several trends. The following rules for predicting nuclear stability are based on this graph.

1. Except for $_1^1H$ and $_2^3He$, all stable nuclei have a number of neutrons that is equal to or greater than the number of protons.

2. A nucleus that has an *N/Z* number that is too large or too small is unstable. For small atoms, *N/Z* is very close to 1. As the nuclei get larger, this number increases gradually until the number is near 1.5 for the largest nuclei.

3. Nuclei with even numbers of neutrons and protons are more stable. Almost 60% of all stable nuclei have even numbers of protons and even numbers of neutrons.

4. Nuclei that have so-called magic numbers of protons and neutrons tend to be more stable than others. These numbers—2, 8, 20, 28, 50, 82, and 126—apply to the number of protons or the number of neutrons. Notice in **Figure 5** the large binding energy of $_2^4He$. This nucleus is very small and has two protons and two neutrons. Such "extra stability" also is true of the element calcium, which has six stable isotopes that range from $_{20}^{40}Ca$ to $_{20}^{48}Ca$, all of which have 20 protons. Tin, having the magic number of 50 protons, has 10 stable isotopes, the largest number of any element. The heaviest stable element, bismuth, having only one stable isotope, has the magic number of 126 neutrons in $_{83}^{209}Bi$.

5. No atoms that have atomic numbers larger than 83 and mass numbers larger than 209 are stable. The nuclei of these atoms are too large to be stable.

Section Review

UNDERSTANDING KEY IDEAS

1. What are the nucleons of an atom?

2. What role does the strong force play in the structure of an atom?

3. What is the band of stability?

4. What is mass defect?

5. Explain what happens to the mass that is lost when a nucleus forms.

6. How do the nuclides $_8^{16}O$ and $_8^{15}O$ differ?

7. Why is bismuth, $_{83}^{209}Bi$, stable?

8. Which are more stable, nuclei that have an even number of nucleons or nuclei that have an odd number of nucleons?

CRITICAL THINKING

9. Which is generally more stable, a small nucleus or a large nucleus? Explain.

10. How does nuclear binding energy relate to the stability of an atom?

11. Which is expected to be more stable, $_3^6Li$ or $_3^9Li$? Explain.

12. Use **Figure 6** and the rules for predicting nuclear stability to determine which of the following isotopes are stable and which are unstable.

a. $_{15}^{32}P$

b. $_6^{14}C$

c. $_{23}^{51}V$

d. $_{12}^{24}Mg$

e. $_{43}^{97}Tc$

Nuclear Change

KEY TERMS

- **radioactivity**
- **beta particle**
- **gamma ray**
- **nuclear fission**
- **chain reaction**
- **critical mass**
- **nuclear fusion**

OBJECTIVES

1. **Predict** the particles and electromagnetic waves produced by different types of radioactive decay, and write equations for nuclear decays.

2. **Identify** examples of nuclear fission, and describe potential benefits and hazards of its use.

3. **Describe** nuclear fusion and its potential as an energy source.

Radioactive Decay

Nuclear changes can be easier to understand than chemical changes because only a few types of nuclear changes occur. One type is the spontaneous change of an unstable nucleus to form a more stable one. This change involves the release of particles, electromagnetic waves, or both and is generally called **radioactivity** or radioactive decay. Specifically, radioactivity is the spontaneous breakdown of unstable nuclei to produce particles or energy. **Table 1** summarizes the properties of both the particles and the energy released by radioactive decay.

radioactivity

the process by which an unstable nucleus emits one or more particles or energy in the form of electromagnetic radiation

Table 1   Characteristics of Nuclear Particles and Rays

Particle	Mass (amu)	Charge	Symbol	Stopped by
Proton	1.007 276 47	+1	$p, p^+, {}^1_{+1}p, {}^1_1H$	a few sheets of paper
Neutron	1.008 664 90	0	$n, n^0, {}^1_0n$	a few centimeters of lead
β particle (electron)	0.000 548 580	−1	$\beta, \beta^-, {}^{0}_{-1}e*$	a few sheets of aluminum foil
Positron†	0.000 548 580	+1	$\beta^+, {}^{0}_{+1}e*$	same as electron
α particle (He-4 nucleus)	4.001 474 92	+2	$\alpha, \alpha^{2+}, {}^4_2He$	skin or one sheet of paper
Gamma ray	0	0	γ	several centimeters of lead

*The superscript zero in the symbols for electron and positron does not mean that they have zero mass. It means their mass number is zero.

†The positron is the antiparticle of the electron. Each particle has an antiparticle, but only the positron is frequently involved in nuclear changes.

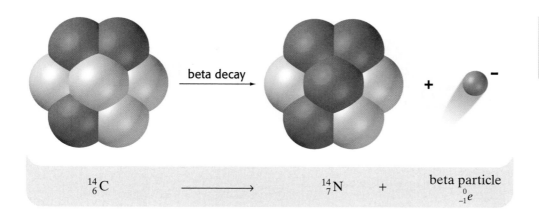

$$^{14}_{6}C \xrightarrow{\hspace{2cm}} \quad ^{14}_{7}N \quad + \quad \text{beta particle} \atop ^{0}_{-1}e$$

Figure 7
When the unstable carbon-14 nucleus emits a beta particle, the carbon-14 nucleus changes into a nitrogen-14 nucleus.

Stabilizing Nuclei by Converting Neutrons into Protons

Recall that the stability of a nucleus depends on the ratio of neutrons to protons, or the N/Z number. If a particular isotope has a large N/Z number or too many neutrons, the nucleus will decay and emit radiation.

A neutron in an unstable nucleus may emit a high-energy electron, called a **beta particle** (β particle), and change to a proton. This process is called *beta decay*. This process often occurs in unstable nuclei that have large N/Z numbers.

$$^{1}_{0}n \xrightarrow{\text{beta decay}} ^{1}_{+1}p + ^{0}_{-1}e$$

Because this process changes a neutron into a proton, the atomic number of the nucleus increases by one, as you can see in **Figure 7**. As a result of beta decay, carbon becomes a different element, nitrogen. However, the mass number does not change because the total number of nucleons does not change as shown by the following equation.

$$^{14}_{6}C \longrightarrow ^{14}_{7}N + ^{0}_{-1}e$$

Stabilizing Nuclei by Converting Protons into Neutrons

One way that a nucleus that has too many protons can become more stable is by a process called *electron capture*. In this process, the nucleus merely absorbs one of the atom's electrons, usually from the 1s orbital. This process changes a proton into a neutron and decreases the atomic number by one. The mass number stays the same.

$$^{1}_{+1}p + ^{0}_{-1}e \xrightarrow{\text{electron capture}} ^{1}_{0}n$$

A typical nucleus that decays by this process is chromium-51.

$$^{51}_{24}Cr + ^{0}_{-1}e \xrightarrow{\text{electron capture}} ^{51}_{23}V + \gamma$$

The final symbol in the equation, γ, indicates the release of **gamma rays.** Many nuclear changes leave a nucleus in an energetic or excited state. When the nucleus stabilizes, it releases energy in the form of gamma rays. **Figure 8** shows a thunderstorm during which gamma rays may also be produced.

beta particle

a charged electron emitted during a certain type of radioactive decay, such as beta decay

gamma ray

the high-energy photon emitted by a nucleus during fisson and radioactive decay

Figure 8
Thunderstorms may produce terrestrial gamma-ray flashes (TGFs).

Figure 9
Nuclei can release positrons
to form new nuclei. Matter is
then converted into energy
when positrons and electrons
collide and are converted
into gamma rays.

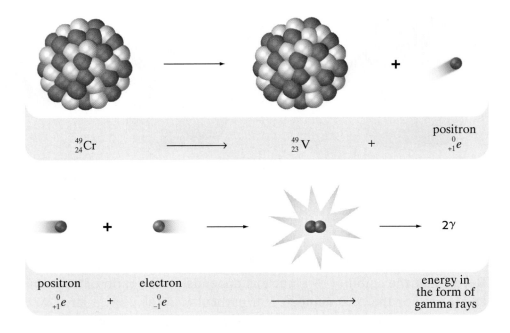

$$^{49}_{24}\text{Cr} \longrightarrow ^{49}_{23}\text{V} + \text{positron } ^{0}_{+1}e$$

$$\text{positron } ^{0}_{+1}e + \text{electron } ^{0}_{-1}e \longrightarrow \text{energy in the form of gamma rays } 2\gamma$$

Gamma Rays Are Also Emitted in Positron Emission

Some nuclei that have too many protons can become stable by emitting positrons, which are the antiparticles of electrons. The process is similar to electron capture in that a proton is changed into a neutron. However, in *positron emission,* a proton emits a positron.

$$^{1}_{+1}p \xrightarrow{\text{positron emission}} ^{1}_{0}n + ^{0}_{+1}e$$

Topic Link

Refer to the "Atoms and Moles" chapter for a discussion of electromagnetic waves.

Notice that when a proton changes into a neutron by emitting a positron, the mass number stays the same, but the atomic number decreases by one. The isotope chromium-49 decays by this process, as shown by the model in **Figure 9.**

$$^{49}_{24}\text{Cr} \longrightarrow ^{49}_{23}\text{V} + ^{0}_{+1}e$$

Another example of an unstable nucleus that emits a positron is potassium-38, which changes into argon-38.

$$^{38}_{19}\text{K} \longrightarrow ^{38}_{18}\text{Ar} + ^{0}_{+1}e$$

The positron is the opposite of an electron. Unlike a beta particle, a positron seldom makes it into the surroundings. Instead, the positron usually collides with an electron, its antiparticle. Any time a particle collides with its antiparticle, all of the masses of the two particles are converted entirely into electromagnetic energy or gamma rays. This process is called *annihilation of matter,* which is illustrated in **Figure 9.**

$$^{0}_{-1}e + ^{0}_{+1}e \xrightarrow{\text{annihilation}} 2\gamma$$

The gamma rays from electron-positron annihilation have a characteristic wavelength; therefore, these rays can be used to identify nuclei that decay by positron emission. Such gamma rays have been detected coming from the center of the Milky Way galaxy.

Stabilizing Nuclei by Losing Alpha Particles

An unstable nucleus that has an N/Z number that is much larger than 1 can decay by emitting an alpha particle. In addition, none of the elements that have atomic numbers greater than 83 and mass numbers greater than 209 have stable isotopes. So, many of these unstable isotopes decay by emitting alpha particles, as well as by electron capture or beta decay. Uranium-238 is one example.

$$^{238}_{92}U \xrightarrow{\text{alpha decay}} {}^{234}_{90}Th + {}^{4}_{2}He$$

Notice that the atomic number in the equation decreases by two while the mass number decreases by four. Alpha particles have very low penetrating ability because they are large and soon collide with other matter. Exposure to external sources of alpha radiation is usually harmless. However, if substances that undergo alpha decay are ingested or inhaled, the radiation can be quite damaging to the body's internal organs.

Many heavy nuclei go through a series of reactions called a decay series before they reach a stable state. The decay series for uranium-238 is shown in **Figure 10**. After the $^{238}_{92}U$ nucleus decays to $^{234}_{90}Th$, the nucleus is still unstable because it has a large N/Z number. This nucleus undergoes beta decay to produce $^{234}_{91}Pa$. By another beta decay, $^{234}_{91}Pa$ changes to $^{234}_{92}U$. After a number of other decays (taking millions of years), the nucleus finally becomes a stable isotope, $^{206}_{82}Pb$.

Topic Link

Refer to the "Atoms and Moles" chapter for a discussion of alpha particles.

Figure 10
Uranium-238 decays to lead-206 through a decay series.

Uranium-238 Decay Series

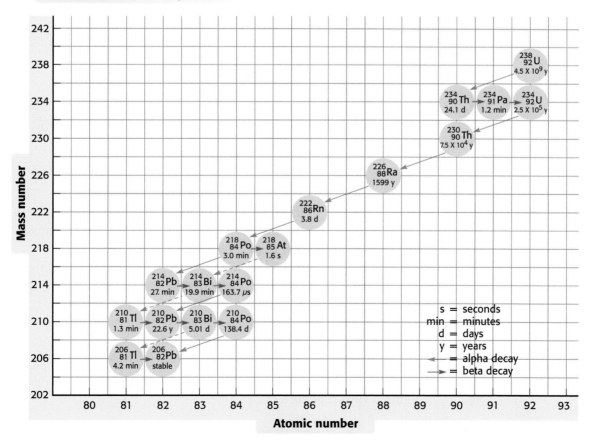

Nuclear Equations Must Be Balanced

Look back at all of the nuclear equations that have appeared so far in this chapter. Notice that the sum of the mass numbers (superscripts) on one side of the equation always equals the sum of the mass numbers on the other side of the equation. Likewise, the sums of the atomic numbers (subscripts) on each side of the equation are equal. Look at the following nuclear equations, and notice that they balance in terms of both mass and nuclear charge.

$$^{238}_{92}U \longrightarrow {}^{234}_{90}Th + {}^{4}_{2}He \quad \begin{array}{l}[238 = 234 + 4 \text{ mass balance}]\\ [92 = 90 + 2 \text{ charge balance}]\end{array}$$

$$^{234}_{90}Th \longrightarrow {}^{234}_{91}Pa + {}^{0}_{-1}e \quad \begin{array}{l}[234 = 234 + 0 \text{ mass balance}]\\ [90 = 91 + (-1) \text{ charge balance}]\end{array}$$

Remember that whenever the atomic number changes, the identity of the element changes. In the above examples, uranium changes into thorium, and thorium changes into protactinium.

SKILLS Toolkit

Balancing Nuclear Equations

The following rules are helpful for balancing a nuclear equation and for identifying a reactant or a product in a nuclear reaction.

1. Check mass and atomic numbers.
- The total of the mass numbers must be the same on both sides of the equation.
- The total of the atomic numbers must be the same on both sides of the equation. In other words, the nuclear charges must balance.
- If the atomic number of an element changes, the identity of the element also changes.

2. Determine how nuclear reactions change mass and atomic numbers.
- If a beta particle, $^{0}_{-1}e$, is released, the mass number does not change but the atomic number increases by one.
- If a positron, $^{0}_{+1}e$ is released, the mass number does not change but the atomic number decreases by one.
- If a neutron, $^{1}_{0}n$, is released, the mass number decreases by one and the atomic number does not change.
- Electron capture does not change the mass number but decreases the atomic number by one.
- Emission of an alpha particle, $^{4}_{2}He$, decreases the mass number by four and decreases the atomic number by two.
- When a positron and an electron collide, energy in the form of gamma rays is generated.

Balancing a Nuclear Equation

Identify the product formed when polonium-212 emits an alpha particle.

1 Gather information.

- Check the periodic table to write the symbol for polonium-212: $^{212}_{84}\text{Po}$.
- Write the symbol for an alpha particle: $^{4}_{2}\text{He}$.

2 Plan your work.

- Set up the nuclear equation.

$$^{212}_{84}\text{Po} \longrightarrow \ ^{4}_{2}\text{He} + \ ?$$

3 Calculate.

- The sums of the mass numbers must be the same on both sides of the equation: $212 = 4 + A$; $A = 212 - 4 = 208$

$$^{212}_{84}\text{Po} \longrightarrow \ ^{4}_{2}\text{He} + \ ^{208}?$$

- The sums of the atomic numbers must be the same on both sides of the equation: $84 = 2 + Z$; $Z = 84 - 2 = 82$

$$^{212}_{84}\text{Po} \longrightarrow \ ^{4}_{2}\text{He} + \ ^{208}_{82}?$$

- Check the periodic table to identify the element that has an atomic number of 82, and complete the nuclear equation.

$$^{212}_{84}\text{Po} \longrightarrow \ ^{4}_{2}\text{He} + \ ^{208}_{82}\text{Pb}$$

4 Verify your results.

- Emission of an alpha particle does decrease the atomic number by two (from 84 to 82) and does decrease the mass number by four (from 212 to 208).

PRACTICE HINT

Unlike a chemical equation, the elements are usually different on each side of a balanced nuclear equation.

PRACTICE

Write balanced equations for the following nuclear equations.

1. $^{218}_{84}\text{Po} \longrightarrow \ ^{4}_{2}\text{He} + \ ?$

2. $^{142}_{61}\text{Pm} + \ ? \longrightarrow \ ^{142}_{60}\text{Nd}$

3. $^{253}_{99}\text{Es} + \ ^{4}_{2}\text{He} \longrightarrow \ ^{1}_{0}n + \ ?$

4. Write the balanced nuclear equation that shows how sodium-22 changes into neon-22.

Nuclear Fission

nuclear fission

the splitting of the nucleus of a large atom into two or more fragments, a process that produces additional neutrons and a lot of energy

So far, you have learned about one class of nuclear change in which a nucleus decays by adding or losing particles. Another class of nuclear change is called **nuclear fission.** Nuclear fission occurs when a very heavy nucleus splits into two smaller nuclei, each more stable than the original nucleus. Some nuclei undergo fission without added energy. A very small fraction of naturally occurring uranium nuclei is of the isotope $^{235}_{92}U$, which undergoes spontaneous fission. However, most fission reactions happen artificially by bombarding nuclei with neutrons.

Figure 11 shows what happens when an atom of uranium-235 is bombarded with a neutron. The following equation represents the first reaction shown in **Figure 11.**

$$^{235}_{92}U + ^{1}_{0}n \xrightarrow{\text{nuclear fission}} ^{93}_{36}Kr + ^{140}_{56}Ba + 3\,^{1}_{0}n$$

Notice that the products include Kr-93, Ba-140, and three neutrons.

chain reaction

a reaction in which a change in a single molecule makes many molecules change until a stable compound forms

As shown in **Figure 11,** each of the three neutrons emitted by the fission of one nucleus can cause the fission of another uranium-235 nucleus. Again, more neutrons are emitted. These reactions continue one after another as long as enough uranium-235 remains. This process is called a **chain reaction.** One characteristic of a chain reaction is that the particle that starts the reaction, in this case a neutron, is also produced from the reaction. A minimum quantity of radioactive material, called **critical mass,** is needed to keep a chain reaction going.

critical mass

the minimum mass of a fissionable isotope that provides the number of neutrons needed to sustain a chain reaction

Figure 11
A neutron strikes a uranium-235 nucleus, which splits into a krypton nucleus and a barium nucleus. Three neutrons are also produced. Each neutron may cause another fission reaction.

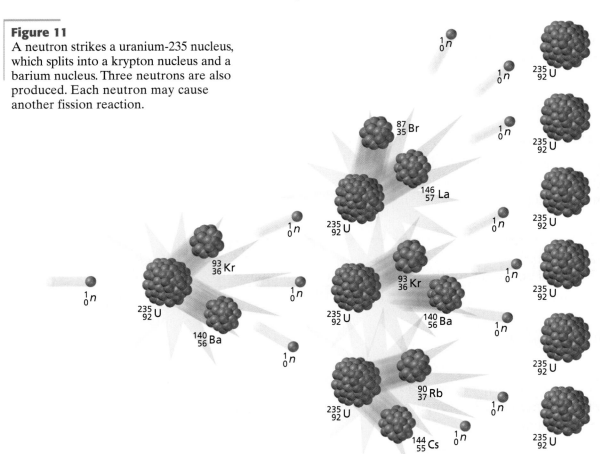

Chain Reactions Occur in Nuclear Reactors

Fission reactions can produce a large amount of energy. For example, the fission of 1 g of uranium-235 generates as much energy as the combustion of 2700 kg of coal. Fission reactions are used to generate electrical energy in nuclear power plants. Uranium-235 and plutonium-239 are the main radioactive isotopes used in these reactors.

In a nuclear reactor, represented in **Figure 12,** the fuel rods are surrounded by a moderator. The moderator is a substance that slows down neutrons. Control rods are used to adjust the rate of the chain reactions. These rods absorb some of the free neutrons produced by fission. Moving these rods into and out of the reactor can control the number of neutrons that are available to continue the chain reaction. Chain reactions that occur in reactors can be very dangerous if they are not controlled. An example of the danger that nuclear reactors can create is the accident that happened at the Chernobyl reactor in the Ukraine in 1986. This accident occurred when technicians briefly removed most of the reactor's control rods during a safety test. However, most nuclear reactors have mechanisms that can prevent most accidents.

As shown in **Figure 12,** water is heated by the energy released from the controlled fission of U-235 and changed into steam. The steam drives a turbine to produce electrical energy. The steam then passes into a condenser and is cooled by a river or lake's water. Notice that water heated by the reactor or changed into steam is isolated. Only water used to condense the steam is gotten from and is returned to the environment.

internet connect

www.scilinks.org
Topic: Fission
SciLinks code: HW4085

SCiLINKS. Maintained by the National Science Teachers Association

Figure 12
This model shows a pressurized, light-water nuclear reactor, the type most often used to generate electrical energy in the United States. Note that each of the three water systems is isolated from the others for safety reasons.

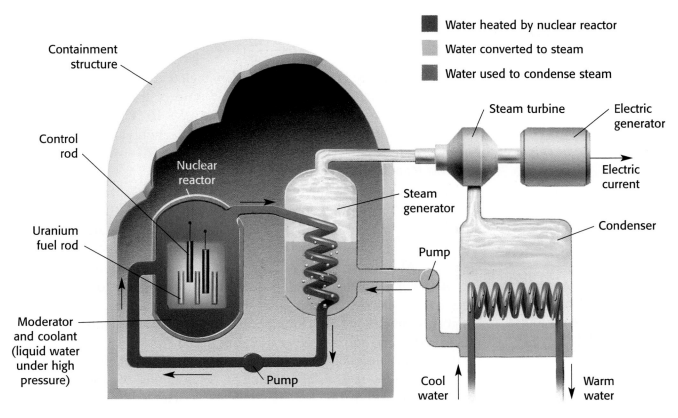

Water heated by nuclear reactor

Water converted to steam

Water used to condense steam

Containment structure

Control rod

Nuclear reactor

Uranium fuel rod

Moderator and coolant (liquid water under high pressure)

Pump

Steam generator

Steam turbine

Electric generator

Electric current

Condenser

Pump

Cool water

Warm water

Nuclear Chemistry   **655**

Nuclear Fusion

Nuclear fusion, which is when small nuclei combine, or fuse, to form a larger, more stable nucleus, is still another type of nuclear change. The new nucleus has a higher binding energy per nucleon than each of the smaller nuclei does, and energy is released as the new nucleus forms. In fact, fusion releases greater amounts of energy than fission for the same mass of starting material. Fusion is the process by which stars, including our sun, generate energy.

In the sun, four hydrogen nuclei fuse to form a single $^4_2$He nucleus

$$4^1_1\text{H} \longrightarrow {}^4_2\text{He} + 2^{\ 0}_{+1}e.$$

The reaction above is a net reaction. Very high temperatures are required to bring the nuclei together. The temperature of the sun's core, where some of the fusion reactions occur, is about $1.5 \times 10^7 °$C. When the hydrogen nuclei are fused, some mass is converted to energy.

Fusion Reactions Are Hard to Maintain

Scientists are investigating ways to control fusion reactions so that they may be used for both energy generation and research. One problem is that starting a fusion reaction takes a lot of energy. So far, researchers need just as much energy to start a fusion reaction as is released by the reaction. As a result, fusion is not a practical source of energy.

Topic Link

Refer to the "Periodic Table" chapter for a discussion of nuclear fusion.

Another challenge is finding a suitable place for a fusion reaction. In fusion reactions, the reactants are in the form of a plasma, a random mixture of positive nuclei and electrons. Because no form of solid matter can withstand the tremendous temperatures required for fusion to occur, this plasma is hard to contain. Scientists currently use extremely strong magnetic fields to suspend the charged plasma particles. In this way, the plasma can be kept from contacting the container walls. Scientists have also experimented with high-powered laser light to start the fusion process.

Nuclear Energy and Waste

The United States depends on nuclear power to generate electrical energy. In fact, about 100 nuclear reactors generate 20% of electrical energy needs in the United States. Nuclear power also generates waste like many other sources of energy, such as fossil fuels. Nuclear waste is "spent fuel" that can no longer be used to create energy. But this material is still radioactive and dangerous and must be disposed of with care.

Nuclear waste is often stored in "spent-fuel pools" that cover the spent fuel with at least 6 m of water. This amount of water prevents radiation from the waste from harming people. Nuclear waste can also be stored in a tightly sealed steel container. These containers have inert gases that surround the waste. These containers can also be surrounded by steel or concrete. Most of the nuclear waste that is put into a container has first been put in a spent-fuel pool to cool for about one year.

Some isotopes from the spent fuel can be extracted and used again as reactor fuel. However, this process is not currently done on a large scale in the United States.

internet connect

www.scilinks.org
Topic: Nuclear Energy
SciLinks code: HW4084

SCiLINKS. Maintained by the National Science Teachers Association

Section Review

UNDERSTANDING KEY IDEAS

1. What is the name of a high-energy electron that is emitted from an unstable nucleus?

2. How are nuclear fission and nuclear fusion similar? How are they different?

3. Describe what happens when a positron and an electron collide.

4. How is critical mass related to a chain reaction?

PRACTICE PROBLEMS

5. Write the balanced equations for the following nuclear reactions.

   a. Uranium-233 undergoes alpha decay.

   b. Copper-66 undergoes beta decay.

   c. Beryllium-9 and an alpha particle combine to form carbon-13. The carbon-13 nucleus then emits a neutron.

   d. Uranium-238 absorbs a neutron. The product then undergoes successive beta emissions to become plutonium-239.

6. A fusion reaction that takes place in the sun is the combination of two helium-3 nuclei to form two hydrogen nuclei and one other nucleus. Write the balanced nuclear equation for this fusion reaction. Be sure to include both products that are formed.

CRITICAL THINKING

7. In electron capture, why is the electron that is absorbed by the nucleus usually taken from the $1s$ orbital?

8. Can annihilation of matter occur between a positron and a neutron? Explain your answer.

9. Why do the nuclear reactions in a decay series eventually stop?

10. Cobalt-59 is bombarded with neutrons to produce cobalt-60, which is then used to treat certain cancers. The nuclear equation for this reaction shows the gamma rays that are released when cobalt-60 is produced.

$$^{59}_{27}\text{Co} + ^{1}_{0}n \longrightarrow ^{60}_{27}\text{Co} + \gamma$$

Is this an example of a nuclear change that involves the creation of a nucleus of another element? Explain your answer.

Uses of Nuclear Chemistry

OBJECTIVES

1. **Define** the half-life of a radioactive nuclide, and explain how it can be used to determine an object's age.

2. **Describe** some of the uses of nuclear chemistry.

3. **Compare** acute and chronic exposures to radiation.

Half-Life

The start-up activity for this chapter involved shaking pennies and then removing those that landed heads up after they were poured out of the cup. Each time you repeated this step, you should have found that about half the pennies were removed. Therefore, if you started with 100 pennies, about 50 should have been removed after the first shake. After the second shake, about 25 should have been removed, and so on. So, half of the amount of pennies remained after each step. This process is similar to what happens to radioactive materials that undergo nuclear decay. A radioactive sample decays at a constant rate. This rate of decay is measured in terms of its **half-life.**

half-life

the time required for half of a sample of a radioactive substance to disintegrate by radioactive decay or natural processes

Constant Rates of Decay Are Key to Radioactive Dating

The half-life of a radioactive isotope is a constant value and is not influenced by any external conditions, such as temperature and pressure. The use of radioactive isotopes to determine the age of an object, such as the one shown in **Figure 14,** is called *radioactive dating*. The radioactive isotope carbon-14 is often used in radioactive dating.

Nearly all of the carbon on Earth is present as the stable isotope carbon-12. A very small percentage of the carbon in Earth's crust is carbon-14. Carbon-14 undergoes decay to form nitrogen-14. Because carbon-12 and carbon-14 have the same electron configuration, they react chemically in the same way. Both of these carbon isotopes are in carbon dioxide, which is used by plants in photosynthesis.

As a result, all animals that eat plants contain the same ratio of carbon-14 to carbon-12 as the plants do. Other animals eat those animals, and so on up the food chain. So all animals and plants have the same ratio of carbon-14 to carbon-12 throughout their lives. Any carbon-14 that decays while the organism is alive is replaced through photosynthesis or eating. But when a plant or animal dies, it stops taking in carbon-containing substances, so the carbon-14 that decays is not replaced.

Figure 14
Using radioactive-dating techniques, scientists determined this Egyptian cat was made between 950–342 BCE.

TABLE 2

Half-Lives of Some Radioactive Isotopes

Isotope	Half-life	Radiation emitted	Isotope formed
Carbon-14	5.715×10^3 y	β^-, γ	nitrogen-14
Iodine-131	8.02 days	β^-, γ	xenon-131
Potassium-40	1.28×10^9 y	β^+, γ	argon-40
Radon-222	3.82 days	α, γ	polonium-218
Radium-226	1.60×10^3 y	α, γ	radon-222
Thorium-230	7.54×10^4 y	α, γ	radium-226
Thorium-234	24.10 days	β^-, γ	protactinium-234
Uranium-235	7.04×10^8 y	α, γ	thorium-231
Uranium-238	4.47×10^9 y	α, γ	thorium-234
Plutonium-239	2.41×10^4 y	α, γ	uranium-235

internet connect
www.scilinks.org
Topic: Radioactive Dating
SciLinks code: HW4105
SCiLINKS. Maintained by the National Science Teachers Association

Table 2 shows that the half-life of carbon-14 is 5715 years. After that interval, only half of the original amount of carbon-14 will remain. In another 5715 years, half of the remaining carbon-14 atoms will have decayed and leave one-fourth of the original amount.

Once amounts of carbon-12 and carbon-14 are measured in an object, the ratio of carbon-14 to carbon-12 is compared with the ratio of these isotopes in a sample of similar material whose age is known. Using radioactive dating, with carbon-14, scientists can estimate the age of the object.

A frozen body that was found in 1991 in the Alps between Austria and Italy was dated using C-14. The body is known as the Iceman. A small copper ax was found with the Iceman's body, which shows that the Iceman lived during the Age of copper (4000 to 2200 BCE). Radioactive dating with C-14 revealed that the Iceman lived between 3500 and 3000 BCE and is the oldest prehistoric human found in Europe.

Generally, the more unstable a nuclide is, the shorter its half-life is and the faster it decays. **Figure 15** shows the radioactive decay of iodine-131, which is a very unstable isotope that has a short half-life.

internet connect
www.scilinks.org
Topic: Discovering Radioactivity
SciLinks code: HW4150
SCiLINKS. Maintained by the National Science Teachers Association

Figure 15
The radioactive isotope $^{131}_{53}\text{I}$ has a half-life of 8.02 days. In each successive 8.02-day period, half the atoms of $^{131}_{53}\text{I}$ in the original sample decay to $^{131}_{54}\text{Xe}$.

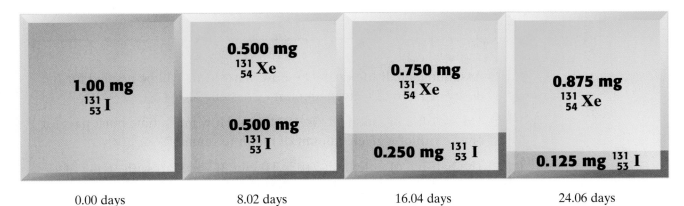

| 1.00 mg $^{131}_{53}\text{I}$ | 0.500 mg $^{131}_{54}\text{Xe}$ / 0.500 mg $^{131}_{53}\text{I}$ | 0.750 mg $^{131}_{54}\text{Xe}$ / 0.250 mg $^{131}_{53}\text{I}$ | 0.875 mg $^{131}_{54}\text{Xe}$ / 0.125 mg $^{131}_{53}\text{I}$ |
| 0.00 days | 8.02 days | 16.04 days | 24.06 days |

Determining the Age of an Artifact or Sample

An ancient artifact is found to have a ratio of carbon-14 to carbon-12 that is one-eighth of the ratio of carbon-14 to carbon-12 found in a similar object today. How old is this artifact?

1 Gather information.

- The half-life of carbon-14 is 5715 years.
- The artifact has a ratio of carbon-14 to carbon-12 that is one-eighth of the ratio of carbon-14 to carbon-12 found in a modern-day object.

2 Plan your work.

- First, determine the number of half-lives that the carbon-14 in the artifact has undergone.
- Next, find the age of the artifact by multiplying the number of half-lives by 5715 y.

3 Calculate.

- For an artifact to have one-eighth of the ratio of carbon-14 to carbon-12 found in a modern-day object, three half-lives must have passed.

$$\frac{1}{8} = \frac{1}{2} \times \frac{1}{2} \times \frac{1}{2}$$

- To find the age of the artifact, multiply the half-life of carbon-14 three times for the three half-lives that have elapsed.

$$3 \times 5715 \text{ y} = 17\ 145 \text{ y}$$

4 Verify your results.

- Start with your answer, and work backward through the solution to be sure you get the information found in the problem.

$$\frac{17\ 145 \text{ y}}{3} = 5715 \text{ y}$$

PRACTICE HINT

Make a diagram that shows how much of the original sample is left to solve half-life problems.

$$1 \longrightarrow 1/2 \longrightarrow 1/4 \longrightarrow 1/8 \longrightarrow 1/16 \longrightarrow 1/32 \longrightarrow \text{etc.}$$

Each arrow represents one half-life.

PRACTICE

1 Assuming a half-life of 1599 y, how many years will be needed for the decay of 15/16 of a given amount of radium-226?

2 The half-life of radon-222 is 3.824 days. How much time must pass for one-fourth of a given amount of radon to remain?

3 The half-life of polonium-218 is 3.0 min. If you start with 16 mg of polonium-218, how much time must pass for only 1.0 mg to remain?

Some Isotopes Are Used for Geologic Dating

By analyzing organic materials in the paints, scientists used carbon-14 to date the cave painting shown in **Figure 16.** Two factors limit dating with carbon-14. The first limitation is that C-14 cannot be used to date objects that are completely composed of materials that were never alive, such as rocks or clay. The second limitation is that after four half-lives, the amount of radioactive C-14 remaining in an object is often too small to give reliable data. Consequently, C-14 is not useful for dating specimens that are more than about 50 000 years old. Anything older must be dated on the basis of a radioactive isotope that has a half-life longer than that of carbon-14. One such isotope is potassium-40.

Potassium-40, which has a half-life of 1.28 billion years, represents only about 0.012% of the potassium present in Earth today. Potassium-40 is useful for dating ancient rocks and minerals. Potassium-40 produces two different isotopes in its radioactive decay. About 11% of the potassium-40 in a mineral decays to argon-40 by emitting a positron.

$$\ce{^{40}_{19}K} \longrightarrow \ce{^{40}_{18}Ar} + \ce{^{0}_{+1}}e$$

The argon-40 may remain in the sample. The remaining 89% of the potassium-40 decays to calcium-40 by emitting a beta particle.

$$\ce{^{40}_{19}K} \longrightarrow \ce{^{40}_{20}Ca} + \ce{^{0}_{-1}}e$$

The calcium-40 is not useful for radioactive dating because it cannot be distinguished from other calcium in the rock. The argon-40, however, can be measured. **Figure 17** shows the decay of potassium-40 through four half-lives.

Figure 16
Scientists determined that this cave painting at Lascaux, called *Chinese Horse*, was created approximately 13 000 BCE.

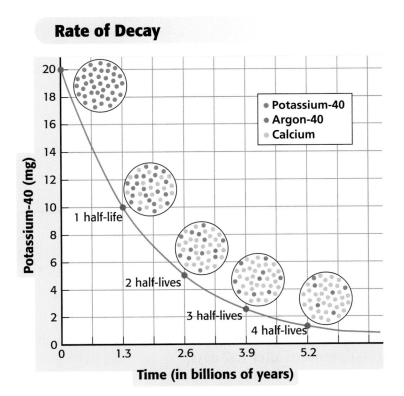

Rate of Decay

Legend:
- Potassium-40
- Argon-40
- Calcium

Y-axis: Potassium-40 (mg)
X-axis: Time (in billions of years)

1 half-life
2 half-lives
3 half-lives
4 half-lives

Figure 17
Potassium-40 decays to argon-40 and calcium-40, but scientists monitor only the ratio of potassium-40 to argon-40 to determine the age of the object.

Determining the Original Mass of a Sample

A rock is found to contain 4.3 mg of potassium-40. The rock is dated to be 3.84 billion years old. How much potassium-40 was originally present in this rock?

1 Gather information.

- The rock is 3.84 billion years old and contains 4.3 mg of $^{40}_{19}$K.
- The half-life of potassium-40 is 1.28 billion years.

2 Plan your work.

- Find the number of half-lives that the $^{40}_{19}$K in the rock has undergone.
- Next, find the mass of the $^{40}_{19}$K that was originally in the rock. Double the present amount for every half-life that the isotope has undergone.

3 Calculate.

- Divide the age of the rock by the half-life of the isotope to find the number of half-lives.

$$\frac{3.84 \text{ billion y}}{1.28 \text{ billion y}} = 3 \text{ half-lives have elapsed}$$

 PRACTICE HINT

Remember to double the amount of radioactive isotope each time you go back one half-life.

- The mass of the original potassium-40 sample is calculated by doubling 4.3 mg three times.

  $4.3 \text{ mg} \times 2 = 8.6 \text{ mg}$ were present in the rock 1 half-life ago

  $8.6 \text{ mg} \times 2 = 17 \text{ mg}$ were present in the rock 2 half-lives ago

  $17 \text{ mg} \times 2 = 34 \text{ mg}$ were present in the rock 3 half-lives ago

4 Verify your results.

After three half-lives, one-eighth of the original $^{40}_{19}$K remains. So, $8 \times 4.3 = 34$ mg.

PRACTICE

1 The half-life of polonium-210 is 138.4 days. How many milligrams of polonium-210 remain after 415.2 days if you start with 2.0 mg of the isotope?

2 After 4797 y, how much of an original 0.250 g sample of radium-226 remains? Its half-life is 1599 y.

3 The half-life of radium-224 is 3.66 days. What was the original mass of radium-224 if 0.0800 g remains after 7.32 days?

Other Uses of Nuclear Chemistry

Scientists create new elements by using nuclear reactions. But the use of nuclear reactions has extended beyond laboratories. Today, nuclear reactions have become part of our lives. Nuclear reactions that protect your life may be happening in your home.

Smoke Detectors Contain Sources of Alpha Particles

Smoke detectors depend on nuclear reactions to sound an alarm when a fire starts. Many smoke detectors contain a small amount of americium-241, which decays to form neptunium-237 and alpha particles.

$$^{241}_{95}Am \longrightarrow \, ^{237}_{93}Np + \, ^{4}_{2}He$$

The alpha particles cannot penetrate the plastic cover and can travel only a short distance. When alpha particles travel through the air, they ionize gas molecules in the air, which change the molecules into ions. These ions conduct an electric current. Smoke particles reduce this current when they mix with the ionized molecules. In response, the smoke detector sets off an alarm.

Detecting Art Forgeries with Neutron Activation Analysis

Nuclear reactions can be used to help museum directors detect whether an artwork, such as the one shown in **Figure 18,** is a fake. The process is called *neutron activation analysis*. A tiny sample from the suspected forgery is placed in a machine. A nuclear reactor in the machine bombards the sample with neutrons. Some of the atoms in the sample absorb neutrons and become radioactive isotopes. These isotopes emit gamma rays as they decay.

Scientists can identify each element in the sample by the characteristic pattern of gamma rays that each element emits.

Figure 18
Neutron activation analysis can be used to determine if this artwork is real.

Scientists can then determine the exact proportions of the elements present. This method gives scientists a "fingerprint" of the elements in the sample. If the fingerprint matches materials that were not available when the work was supposedly created, then the artwork is a fake.

Nuclear Reactions Are Used in Medicine

The use of nuclear reactions by doctors has grown to the point where a whole field known as nuclear medicine has developed. Nuclear medicine includes the use of nuclear reactions both to diagnose certain conditions and to treat a variety of diseases, especially certain types of cancer.

For years, doctors have used a variety of devices, such as X-ray imaging, to get a view inside a person's body. Nuclear reactions have enabled them to get a much more detailed view of the body. For example, doctors can take a close look at a person's heart by using a thallium stress test. The person is given an intravenous injection of thallium-201, which acts chemically like calcium and collects in the heart muscle. As the thallium-201 decays, low-energy gamma rays are emitted and are detected by a special camera that produces images, such as the one shown in **Figure 19.**

The radioactive isotope most widely used in nuclear medicine is technetium-99, which has a short half-life and emits low-energy gamma rays. This radioactive isotope is used in bone scans. Bone repairs occur when there is a fracture, infection, arthritis, or an invading cancer. Bones that are repairing themselves take in minerals and absorb the technetium at the same time. If an area of bone has an unusual amount of repair, the technetium will gather there. Cameras detect the gamma rays that result from its decay.

Another medical procedure that uses nuclear reactions is called *positron emission tomography* (PET), which is shown in **Figure 20.** PET uses radioactive isotopes that have short half-lives. An unstable isotope that contains too many protons is injected into the person.

Figure 19
This image reveals the size of the heart, how well the chambers are pumping, and whether there is any scarring of muscle from previous heart attacks.

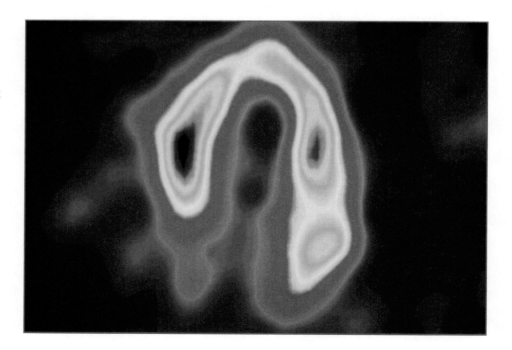

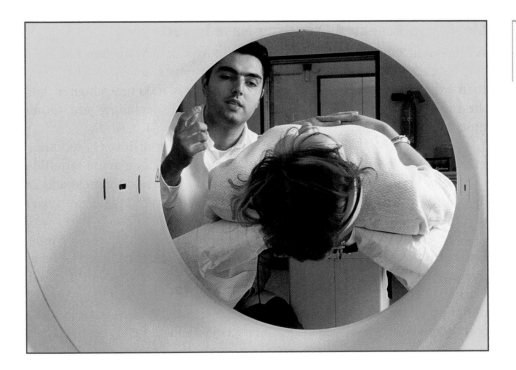

As this isotope decays, positrons are emitted. Recall that when a positron collides with an electron, both are annihilated, and two gamma rays are produced. These gamma rays leave the body and are detected by a scanner. A computer converts the images into a detailed three-dimensional picture of the person's organs.

Exposure to Radiation Must Be Checked

Table 3 shows how radiation can affect a person's health using the unit *rem,* which expresses the biological effect of an absorbed dose of radiation in humans. People who work with radioactivity wear a film badge to monitor the amount of radiation to which they are exposed. Radioactivity was discovered when sealed photographic plates exposed to radiation became fogged. A film badge works on the same principle. Any darkening of the film indicates that the badge wearer was exposed to radiation, and the degree of darkening indicates the total exposure.

Table 3	Effect of Whole-Body Exposure to a Single Dose of Radiation
Dose (rem)	**Probable effect**
0–25	no observable effect
25–50	slight decrease in white blood cell count
50–100	marked decrease in white blood cell count
100–200	nausea, loss of hair
200–500	ulcers, internal bleeding
> 500	death

Table 4 Units Used in Measurements of Radioactivity	
Units	**Measurements**
Curie (C)	radioactive decay
Becquerel (Bq)	radioactive decay
Roentgens (R)	exposure to ionizing radiation
Rad (rad)	energy absorption caused by ionizing radiation
Rem (rem)	biological effect of the absorbed dose in humans

Single and Repeated Exposures Have Impact

As shown in **Table 3,** the biological effect of exposure to nuclear radiation can be expressed in rem. Healthcare professionals are advised to limit their exposure to 5 rem per year. This exposure is 1000 times higher than the recommended exposure level for most people, including you. Other units of radiation measurement can be seen in **Table 4.**

People exposed to a single large dose or a few large doses of radiation in a short period of time are said to have experienced an acute radiation exposure. More than 230 people suffered acute radiation sickness and 28 died when a meltdown occurred in 1986 at the Chernobyl nuclear power plant in the Ukraine.

The effects of nuclear radiation on the body can add up over time. Exposure to small doses of radiation over a long period of time can be as dangerous as a single large dose if the total radiation received is equal. Chronic radiation exposure occurs when people get many low doses of radiation over a long period of time. Some scientific studies have shown a correlation between chronic radiation exposure and certain types of cancer.

Section Review

UNDERSTANDING KEY IDEAS

1. What is meant by the *half-life* of a radioactive nuclide?

2. Explain how carbon-14 dating is used to determine the age of an object.

3. Why is potassium-40 used to date objects older than 50 000 years old?

4. Identify three practical applications of nuclear chemistry.

PRACTICE PROBLEMS

5. What fraction of an original sample of a radioactive isotope remains after three half-lives have passed?

6. How many half-lives of radon-222 have passed in 11.46 days? If 5.2×10^{-8} g of radon-222 remain in a sealed box after 11.46 days, how much was present in the box initially? Refer to **Table 2.**

7. The half-life of protactinium-234 in its ground state is 6.69 h. What fraction of a given amount remains after 26.76 h?

8. The half-life of thorium-227 is 18.72 days. How many days are required for three-fourths of a given amount to decay?

CRITICAL THINKING

9. Someone tells you that neutron activation analysis can reveal whether a famous painter or a rival living at the same time created a painting. What is wrong with this reasoning?

10. Why are isotopes that have relatively short half-lives the only ones used in medical diagnostic tests?

11. A practical rule is that a radioactive nuclide is essentially gone after 10 half-lives. What percentage of the original radioactive nuclide is left after 10 half-lives? How long will it take for 10 half-lives to pass for plutonium-239? Refer to **Table 2.**

H
Hydrogen
1.007 94
$1s^1$

Element Spotlight

Where Is H?

Earth's crust
0.9 by mass

Universe
approximately 93% of all atoms

Hydrogen Is an Element unto Itself

Hydrogen is a unique element in many respects. Its scarcity as a free element on Earth is partially due to the low density of hydrogen gas. The low density permits hydrogen molecules to escape Earth's gravitational pull and drift into space.

Hydrogen does not fit precisely anywhere in the periodic table. It could be placed in Group 1 because it has a single valence electron. But it could also be placed with the halogens in Group 17 because it needs only one electron to get a full outer shell.

Industrial Uses

- Hydrogen gas is prepared industrially by the thermal decomposition of hydrocarbons, such as natural gas, oil-refinery gas, gasoline, fuel oil, and crude oil.

- Most of the hydrogen gas produced is used for synthesizing ammonia.

- Hydrogen is used in the hydrogenation of unsaturated vegetable oils to make solid fats.

- Liquid hydrogen is a clear, colorless liquid that has a boiling point of $-252.87°C$, the lowest boiling point of any known liquid other than liquid helium. Because of its low temperature, liquid hydrogen is used to cool superconducting materials.

- Liquid hydrogen is used to fuel rockets, satellites, and spacecraft.

Liquid hydrogen is used as fuel for some rockets.

Real-World Connection Nuclear fusion, in which hydrogen atoms form helium atoms, occurs in our sun.

A Brief History

1783: Jacques Charles fills a balloon with hydrogen and flies in a basket over the French countryside.

1931: Harold Urey discovers deuterium, an isotope of hydrogen, in water.

1937: The Hindenburg, a hydrogen-filled dirigible, explodes during a landing in Lakehurst, New Jersey.

1600	1700	1800	1900

1660: Robert Boyle prepares hydrogen from a reaction between iron and sulfuric acid.

1766: Henry Cavendish prepares a pure sample of hydrogen and distinguishes it from other gases. He names it "inflammable air."

1898: James Dewar produces liquid hydrogen and develops a glass vacuum flask to hold it.

1934: Ernest Rutherford, Marcus Oliphant, and Paul Harteck discover tritium.

1996: Scientists at Lawrence Livermore National Laboratory succeed in making solid, metallic hydrogen.

Questions

1. Research how hydrogen is used to fuel rockets and spacecrafts.

2. Write a paragraph about stars and fusion.

internet connect

www.scilinks.org
Topic : Hydrogen
SciLinks code: HW4155

SCiLINKS Maintained by the National Science Teachers Association

KEY IDEAS

SECTION ONE Atomic Nuclei and Nuclear Stability

- The strong force overcomes the repulsive force between protons to keep a nucleus intact.
- The mass that is converted to energy when nucleons form a nucleus is known as the mass defect.
- If the mass defect is known, the nuclear binding energy can be calculated by using the equation $E = mc^2$.
- The ratio of neutrons to protons defines a band of stability that includes the stable nuclei.

SECTION TWO Nuclear Change

- Unstable nuclei are radioactive and can emit radiation in the form of alpha particles, beta particles, and gamma rays.
- Unstable nuclei that have large N/Z usually emit beta particles.
- Unstable nuclei that have small N/Z or have too few neutrons can undergo either electron capture or positron emission, emitting gamma rays in the process.
- Large nuclei that have large N/Z frequently emit alpha particles.
- Nuclear equations are balanced in terms of mass and nuclear charge.
- In nuclear fission, a heavy nucleus splits into two smaller nuclei; in nuclear fusion, two or more smaller nuclei combine to form one larger nucleus.
- Nuclear fission reactions that cause other fissions are chain reactions. Chain reactions must be controlled to generate usable energy.

SECTION THREE Uses of Nuclear Chemistry

- Half-life is the time required for one half of the mass of a radioactive isotope to decay.
- The half-life of the carbon-14 isotope can be used to date organic material that is up to 50 000 years old. Other radioactive isotopes are used to date older rock and mineral formations.
- Radioactive isotopes have a number of practical applications in industry, medicine, and chemical analysis.

KEY TERMS

nucleons
nuclide
strong force
mass defect

radioactivity
beta particle
gamma ray
nuclear fission
chain reaction
critical mass
nuclear fusion

half-life

KEY SKILLS

Balancing a Nuclear Equation
Skills Toolkit 1 p. 652
Sample Problem A p. 653

Determining the Age of an Artifact or Sample
Sample Problem B p. 660

Determining the Original Mass of a Sample
Sample Problem C p. 662

CHAPTER REVIEW 18

USING KEY TERMS

1. What is the energy emitted when a nucleus forms?

2. What is a nucleon?

3. What is the high-energy electromagnetic radiation produced by decaying nuclei?

4. What nuclear reaction happens when two small nuclei combine?

5. Explain the difference between fission and fusion.

6. Name the process that describes an unstable nucleus that emits particles and energy.

7. Define *critical mass*.

8. Define *half-life*.

9. What is the combination of neutrons and protons in a nucleus known as?

10. Name two types of nuclear changes.

UNDERSTANDING KEY IDEAS

Atomic Nuclei and Nuclear Stability

11. Explain how the strong force holds a nucleus together despite the repulsive forces between protons.

12. Describe what happens to unstable nuclei.

13. a. What is the relationship among the number of protons, the number of neutrons, and the stability of the nucleus for small atoms?

b. What is the relationship among number of protons, the number of neutrons, and the stability of the nucleus for large atoms?

14. What is the relationship between binding energy and the formation of a nucleus from protons and neutrons?

15. What is the relationship between mass defect and binding energy?

16. Why is nuclear stability better indicated by binding energy per nucleon than by total binding energy per nucleus?

17. What is a quark?

Nuclear Change

18. What is the relationship between an alpha particle and a helium nucleus?

19. Compare the penetrating powers of alpha particles, beta particles, and gamma rays.

20. Is the decay of an unstable isotope into a stable isotope always a one-step process? Explain.

21. a. What role does a neutron serve in starting a nuclear chain reaction and in keeping it going?

b. Why must neutrons in a chain reaction be controlled?

c. Why must there be a minimum mass of material in order to sustain a chain reaction?

22. Under what conditions does fusion occur?

23. Why do positron emission and electron capture have the same effect on a nucleus?

Uses of Nuclear Chemistry

24. Explain why nuclei that emit alpha particles, such as americium-241, are safe to use in smoke detectors.

25. How does acute radiation exposure differ from chronic radiation exposure?

26. Why do animals contain the same ratio of carbon-14 to carbon-12 as plants do?

27. What type of radioactive nuclide is injected into a person who is about to undergo a PET scan?

28. Describe how nuclear chemistry can be used to detect an art forgery.

29. What does the unit *rem* describe?

PRACTICE PROBLEMS

Sample Problem A Balancing a Nuclear Equation

30. The decay of uranium-238 results in the spontaneous ejection of an alpha particle. Write the nuclear equation that describes this process.

31. What type of radiation is emitted in the decay described by the following equation?

$$^{43}_{19}K \longrightarrow \,^{43}_{20}Ca + \,?$$

32. When a radon-222 nucleus decays, an alpha particle is emitted. Write the nuclear equation to show what happens when a radon-222 nucleus decays. What is the other product that forms?

33. One radioactive decay series that begins with uranium-235 and ends with lead-207 shows the partial sequence of emissions: alpha, beta, alpha, beta, alpha, alpha, alpha, alpha, beta, beta, and alpha. Write an equation for each reaction in the series.

34. Balance the following nuclear reactions.

a. $^{239}_{93}Np \longrightarrow \,^{0}_{-1}e + \,?$

b. $^{9}_{4}Be + \,^{4}_{2}He \longrightarrow \,?$

c. $^{32}_{15}P + \,? \longrightarrow \,^{33}_{15}P$

d. $^{236}_{92}U \longrightarrow \,^{94}_{36}Kr + \,? + 3^{1}_{0}n$

35. Complete and balance the following nuclear equations:

a. $^{187}_{75}Re + \,? \longrightarrow \,^{188}_{75}Re + \,^{1}_{1}H$

b. $^{9}_{4}Be + \,^{4}_{2}He \longrightarrow \,? + \,^{1}_{0}n$

c. $^{22}_{11}Na + \,? \longrightarrow \,^{22}_{10}Ne$

36. Write the nuclear equation for the release of a positron by $^{117}_{54}Xe$.

Sample Problem B Determining the Age of an Artifact or Sample

37. Copper-64 is used to study brain tumors. Assume that the original mass of a sample of copper-64 is 26.00 g. After 64 hours, all that remains is 0.8125 g of copper-64. What is the half-life of this radioactive isotope?

38. The half-life of thorium-234 is 24.10 days. How many days until only one-sixteenth of a 52.0 g sample of thorium-234 remains?

39. The half-life of carbon-14 is 5715 y. How long will it be until only half of the carbon-14 in a sample remains?

40. The half-life of of element *Y* is 350 s. How many minutes are required for one-half of a given amount of *Y* to decay?

Sample Problem C Determining the Original Mass of a Sample

41. The half-life of one radon isotope is 3.8 days. If a sample of gas contains 4.38 g of radon-222, how much radon will remain in the sample after 15.2 days?

42. Phosphorus-32 is used to treat a certain form of leukemia. Starting with 10.0 mg, what mass of phosphorus-32 would remain after 57 days? The half-life of phosphorus-32 is 14.3 days.

43. After 4797 y, how much of an original 0.450 g of radium-226 remains? The half-life of radium-226 is 1599 y.

44. The half-life of cobalt-60 is 10.47 min. How many milligrams of Co-60 remain after 104.7 min if you start with 10.0 mg of Co-60?

MIXED REVIEW

45. Calculate the neutron-proton ratios for the following nuclides, and determine where they lie in relation to the band of stability.

a. $^{235}_{92}U$

b. $^{16}_{8}O$

c. $^{56}_{26}Fe$

d. $^{156}_{60}Nd$

46. Calculate the binding energy per nucleon of $^{238}_{92}U$ in joules. The atomic mass of a $^{238}_{92}U$ nucleus is 238.050 784 amu.

47. The energy released by the formation of a nucleus of $^{56}_{26}Fe$ is 7.89×10^{-11} J. Use Einstein's equation, $E = mc^2$, to determine how much mass is lost (in kilograms) in this process.

48. Calculate the binding energy for one mole of deuterium atoms. The measured mass of deuterium is 2.0140 amu.

49. What nuclear process is occuring in the sun shown? Also, write a nuclear reaction that describes this process.

50. The radiation given off by iodine-131 in the form of beta particles is used to treat cancer of the thyroid gland. Write the nuclear equation to describe the decay of an iodine-131 nucleus.

51. The plutonium isotope $^{239}_{94}Pu$ is sometimes detected in nuclear reactors. Consider that nuclear fuel contains a large portion of the common uranium isotope $^{238}_{92}U$ in addition to fissionable $^{235}_{92}U$. Describe a process by which $^{239}_{94}Pu$ might form in a nuclear reactor.

52. The parent nuclide of the thorium decay series is $^{232}_{90}Th$. The first four decays are as follows: alpha emission, beta emission, beta emission, and alpha emission. Write the nuclear equations for this series of emissions.

53. The half-life of radium-224 is 3.66 days. What was the original mass of radium-224 if 0.0500 g remains after 7.32 days?

54. How many milligrams remain of a 15.0 mg sample of radium-226 after 6396 y? The half-life of this isotope is 1599 y.

55. The mass of a $^{7}_{3}Li$ atom is 7.016 00 amu. Calculate its mass defect.

56. Determine whether each of the following nuclear reactions involves alpha decay, beta decay, positron emission, or electron capture.

a. $^{234}_{90}Th \longrightarrow\ ^{0}_{-1}e + ^{234}_{91}Pa$

b. $^{238}_{92}U \longrightarrow\ ^{4}_{2}He + ^{234}_{90}Th$

c. $^{15}_{8}O \longrightarrow\ ^{0}_{+1}e + ^{15}_{7}N$

57. Uranium-238 decays through alpha decay with a half-life of 4.46×10^9 y. How long would it take for seven-eighths of a sample of uranium-238 to decay?

58. Write the nuclear equation for the release of an alpha particle by $^{157}_{70}Yb$.

59. The half-life of iodine-131 is 8.02 days. What percentage of an iodine-131 sample will remain after 40.2 days?

60. The mass of a $^{20}_{10}Ne$ atom is 19.992 44 amu. Calculate its mass defect.

61. Calculate the nuclear binding energy of one lithium-6 atom. The measured atomic mass of lithium-6 is 6.015 amu.

62. Write the nuclear equation for the release of a beta particle by $^{210}_{82}Pb$.

63. The half-life of an element X is 5.25 y. How many days are required for one-fourth of a given amount of X to decay?

64. Complete the following nuclear reactions.

  a. $^{12}_{5}\text{B} \longrightarrow ^{12}_{6}\text{C} + ?$

  b. $^{225}_{89}\text{Ac} \longrightarrow ^{221}_{87}\text{Fr} + ?$

  c. $^{63}_{28}\text{Ni} \longrightarrow ? + ^{0}_{-1}e$

  d. $^{212}_{83}\text{Bi} \longrightarrow ? + ^{4}_{2}\text{He}$

65. Actinium-217 decays by releasing an alpha particle. Write an equation for this decay process, and determine what element is formed.

66. Indicate if the following equations represent fission reactions or fusion reactions.

  a. $^{1}_{1}\text{H} + ^{2}_{1}\text{H} \longrightarrow ^{3}_{2}\text{He} + \gamma$

  b. $^{1}_{0}n + ^{235}_{92}\text{U} \longrightarrow ^{146}_{57}\text{La} + ^{87}_{35}\text{Br} + 3^{1}_{0}n$

  c. $^{21}_{10}\text{Ne} + ^{4}_{2}\text{He} \longrightarrow ^{24}_{12}\text{Mg} + ^{1}_{0}n$

  d. $^{208}_{82}\text{Pb} + ^{58}_{26}\text{Fe} \longrightarrow ^{265}_{108}\text{Hs} + ^{1}_{0}n$

67. Predict whether the total mass of the 26 protons and neutrons that make up the iron nucleus will be more, less, or equal to 55.847 amu, the mass of an iron atom from the periodic table. If it is not equal, explain why not.

68. A sample of francium-212 will decay to one-sixteenth its original amount after 80 min. What is the half-life of francium-212?

69. A sample of strontium-90 is found to have decayed to one-eighth of its original amount after 87.3 y. What is the half-life of strontium-90?

70. Identify which of the four common types of nuclear radiation (beta, neutron, alpha, or gamma) correspond to the following descriptions:

  a. an electron

  b. uncharged particle

  c. can be stopped by a piece of paper

  d. high-energy light

71. Calculate the time required for three-fourths of a sample of cesium-138 to decay given that its half-life is 32.2 min.

72. Calculate that half-life of cesium-135 if seven-eighths of a sample decays in 6×10^{6} y.

73. An archaeologist discovers a wooden mask whose carbon-14 to carbon-12 ratio is one-sixteenth the ratio measured in a newly fallen tree. How old does the wooden mask seem to be, given this evidence?

74. The half-life of tritium, $^{3}_{1}\text{H}$, is 12.3 y. How long will it take for seven-eighths of the sample to decay?

75. The longest-lived radioactive isotope yet discovered is the beta-emitter tellurium-130. It has been determined that it would take 2.5×10^{21} y for 99.9% of this isotope to decay. Write the equation for this reaction, and identify the isotope into which tellurium-130 decays.

76. It takes about 10^{6} y for just half the samarium-149 in nature to decay by alpha-particle emission. Write the decay equation, and find the isotope that is produced by the reaction.

77. Describe some of the similarities and differences between atomic electrons and beta particles.

CRITICAL THINKING

78. Medium-mass nuclei have larger binding energies per nucleon than heavier nuclei do. What can you conclude from this fact?

79. Why are elevated temperatures necessary to initiate fusion reactions but not fission reactions?

80. The release of radioactive strontium into the atmosphere was once a major concern, especially for infants whose main food source was milk. Why were scientists concerned about radioactive strontium? (Hint: Check a periodic table for the members of the group that include strontium.)

81. Why do lighter elements undergo fusion more readily than heavier elements do?

82. Why is the constant rate of decay of radioactive nuclei so important in radioactive dating?

83. Why would someone working around radio-active waste in a landfill use a radiation monitor instead of a watch to determine when the workday is over? At what point would that person decide to stop working?

84. Radioactive isotopes are often used as "tracers" to follow the path of an element through a chemical reaction. For example, the oxygen atoms in O_2 produced by a plant during photosynthesis come from the oxygen in H_2O and not the oxygen in carbon dioxide, CO_2. Explain how you could use a radioactive isotope of oxygen to identify the source of the oxygen atoms in the O_2 produced during photosynthesis.

85. Suppose a nucleus captures two neutrons and decays to produce one neutron. Is this process likely to produce a chain reaction? Explain your reasoning.

86. Explain why charged particles do not penetrate matter deeply.

87. Many radioactive isotopes have half-lives of several billion years. Other radioactive isotopes have half-lives of billionths of a second. Suggest a way in which the half-lives of such isotopes are measured.

ALTERNATIVE ASSESSMENTS

88. Your local grocery store may sell perishable foods that have been irradiated. Find out what stores in your area sell irradiated foods, and determine the shelf life of these foods. What are the shelf lives of the same foods without irradiation? Do research to identify foods that are routinely irradiated even though their label may not indicate irradiation. Report your findings to the class.

89. Research some important historical findings that have been validated through radioactive dating. Report your findings to the class.

90. Design an experiment that illustrates the concept of half-life.

91. Research and evaluate environmental issues regarding the storage, containment, and disposal of nuclear wastes.

92. Suppose you are an energy consultant who has been asked to evaluate a proposal to build a power plant in a remote area of the desert. Research the requirements for each of the following types of power plant: nuclear-fission power plant, coal-burning power plant, solar-energy farm. Decide which of these power plants would be best for its surroundings, and write a paragraph supporting your decision.

CONCEPT MAPPING

93. Use the following terms to complete the concept map below: *critical mass, chain reaction, nuclear fission,* and *nucleon.*

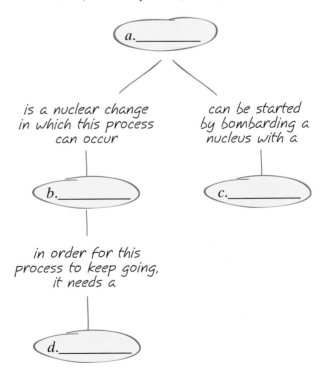

Study the graph below, and answer the questions that follow.
For help in interpreting graphs, see Appendix B, "Study Skills for Chemistry."

94. Do stable nuclei that have N/Z numbers approximately equal to 1 have small or large atomic numbers?

95. Do stable nuclei that have N/Z numbers approximately equal to 1.5 have small or large atomic numbers?

96. Calculate the N/Z number for a nucleus A that has 70 neutrons and 50 protons.

97. Calculate the N/Z number for a nucleus B that has 90 neutrons and 60 protons.

98. Does nucleus A or nucleus B have an N/Z number closer to 1.5?

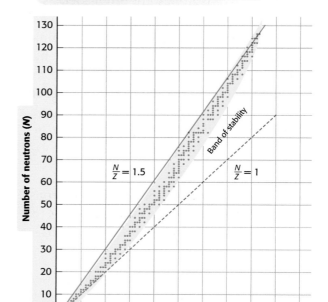

Neutron-Proton Ratios of Stable Nuclei

TECHNOLOGY AND LEARNING

99. Graphing Calculator

Calculating the Amount of Radioactive Material

The graphing calculator can run a program that graphs the relationship between the amount of radioactive material and elapsed time. Given the half-life of the radioactive material and the initial amount of material in grams, you will graph the relationship between the amount of radioactive material and the elapsed time. Then, with the elapsed time, you will trace the graph to calculate the amount of radioactive material.

Go to Appendix C. If you are using a TI-83 Plus, you can download the program **RADIOACT** and run the application as

directed. If you are using another calculator, your teacher will provide you with key-strokes and data sets to use. After you have run the program, answer these questions.

a. Determine the amount of neptunium-235 left after 2.0 years, given the half-life of neptunium-235 is 1.08 years and the initial amount was 8.00 g.

b. Determine the amount of neptunium-235 left after 5.0 years, given the half-life of neptunium-235 is 1.08 years and the initial amount was 8.00 g.

c. Determine the amount of uranium-232 left after 100 years, given the half-life of uranium-232 is 69 years and the initial amount was 10.0 g.

1. Which of the following pairs represent a nucleon combination?
   a. protons and electrons
   b. neutrons and electrons
   c. protons and neutrons
   d. alpha particle and beta particle

2. When a nucleus is formed,
   a. some mass is gained.
   b. energy is absorbed.
   c. some mass is converted to energy.
   d. electrons are captured by the nucleus that forms.

3. Which of the following combinations of proton number and neutron number yields a nucleus that is least stable?
   a. even/even   **c.** odd/even
   b. even/odd   **d.** odd/odd

4. When several smaller nuclei form a larger nucleus, _____ has occurred.
   a. nuclear fission
   b. nuclear fusion
   c. alpha emission
   d. gamma ray production

5. Which of the following nuclear equations is correctly balanced?
   a. $^{37}_{18}Ar + ^{0}_{-1}e \longrightarrow ^{37}_{17}Cl$
   b. $^{6}_{3}Li + 2^{1}_{0}n \longrightarrow ^{4}_{2}He + ^{3}_{1}H$
   c. $^{254}_{99}Es + ^{4}_{2}He \longrightarrow ^{258}_{101}Md + 2^{1}_{0}n$
   d. $^{14}_{7}N + ^{4}_{2}He \longrightarrow ^{17}_{8}O + ^{2}_{1}H$

6. Gamma rays
   a. have the same energy as beta particles do.
   b. are visible light.
   c. have no charge and no mass.
   d. are not a form of electromagnetic radiation.

7. Combining a nucleus of curium-246 with a nucleus of carbon-12 will produce a nucleus of
   a. einsteinium-254.
   b. nobelium-258.
   c. fermium-254.
   d. mendelevium-256.

8. The half-life of thorium-234 is 24 days. If you start with a 42.0 g sample of thorium-24, how much will remain after 72 days?
   a. 42.0 g   **c.** 10.5 g
   b. 21.0 g   **d.** 5.25 g

9. Alpha particles can safely be used in smoke detectors because they
   a. are not a form of radiation.
   b. cannot even penetrate paper.
   c. are harmless even if swallowed.
   d. combine with beta particles to produce a neutron.

10. Every radioactive isotope
   a. emits an alpha particle.
   b. has a characteristic half-life.
   c. has the same atomic mass.
   d. has an atomic number greater than 92.

11. A nucleus that is stable
   a. does not have a half-life.
   b. always emits an alpha particle.
   c. must have an equal number of protons and neutrons.
   d. has a low binding energy.

12. When a radioactive nuclide emits a beta particle, its
   a. atomic number decreases by one.
   b. mass number decreases by one.
   c. atomic number increases by one.
   d. atomic number remains the same.

CARBON AND ORGANIC COMPOUNDS

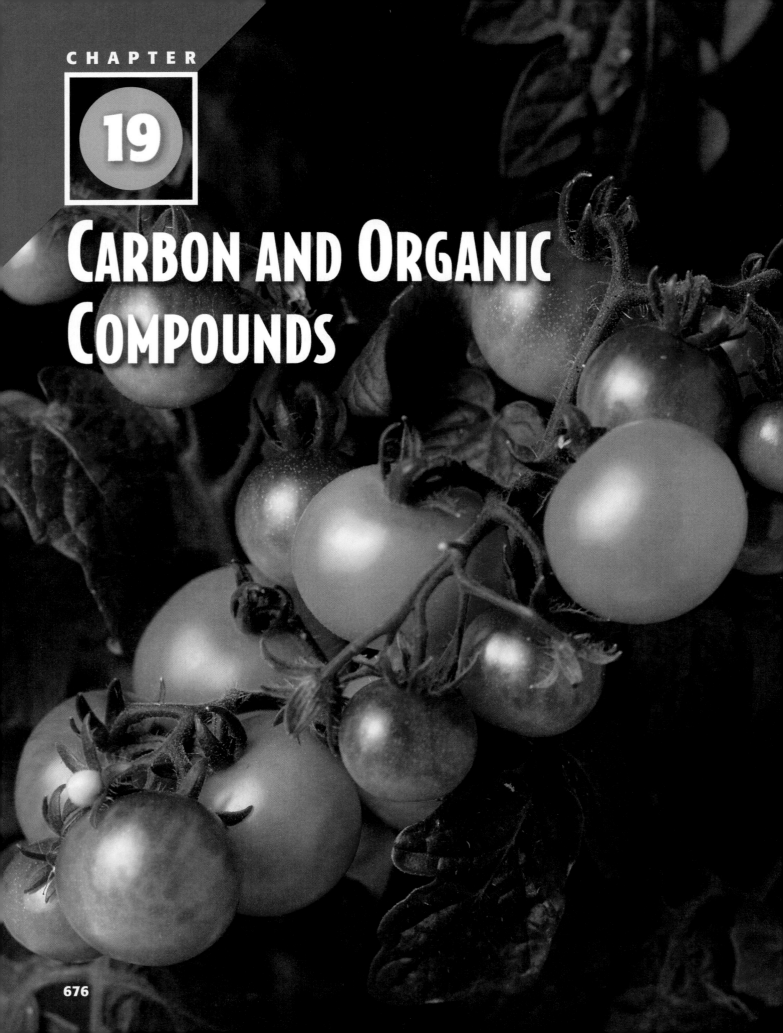

Tomatoes contain many compounds of carbon, including some that have properties that help people stay healthy. Two of these compounds are lycopene and beta-carotene. Lycopene gives tomatoes their red color and is believed to help prevent heart disease and some forms of cancer. In the human body, beta-carotene is converted to vitamin A, an essential nutrient.

Like the vine that supports the tomatoes in this picture, carbon forms the backbone for the chemicals that make up living organisms. In this chapter, you will learn about the nature of carbon and its many compounds.

START-UP ACTIVITY

Testing Plastics

PROCEDURE

1. Examine **two plastic samples** with a **magnifying lens** to look for any structural differences.

2. To test the rigidity of each sample, try to bend both pieces.

3. To test the hardness of each sample, press into each sample with your fingernail and try to make a permanent mark.

4. To test the strength of each sample, try tearing each plastic piece.

ANALYSIS

1. Which plastic sample would you use to hold liquids?

2. What physical differences did you observe between the two samples?

3. Why do you think most communities recycle only one of these plastics?

Pre-Reading Questions

① How many covalent bonds can a carbon atom form?

② How does the structure of a compound affect its chemical reactivity?

③ What are two possible ways to show the structure of CH_4?

internet connect

www.scilinks.org
Topic: Organic Compounds
SciLinks code: HW4092

SCLINKS. Maintained by the National Science Teachers Association

Compounds of Carbon

KEY TERMS

- **hydrocarbon**
- **alkane**
- **alkene**
- **alkyne**
- **aromatic hydrocarbon**
- **functional group**
- **isomer**

OBJECTIVES

(1) **Explain** the unique properties of carbon that make the formation of organic molecules possible.

(2) **Relate** the structures of diamond, graphite, and other allotropes of carbon to their properties.

(3) **Describe** the nature of the bonds formed by carbon in alkanes, alkenes, alkynes, aromatic compounds, and cyclic compounds.

(4) **Classify** organic compounds such as alcohols, esters, and ketones by their functional groups.

(5) **Explain** how the structural difference between isomers is related to the difference in their properties.

Properties of Carbon

internet connect

www.scilinks.org
Topic: Carbon
SciLinks code: HW4138

SC/LINKS. Maintained by the National Science Teachers Association

The water bottle shown in **Figure 1** is made of a strong but flexible plastic. These properties result from the bonds formed by the carbon atoms that make up the plastic. Carbon atoms nearly always form covalent bonds. Three factors make the bonds that carbon atoms form with each other unique.

First, even a single covalent bond between two carbon atoms is quite strong. In contrast, the single covalent bond that forms between two oxygen atoms, such as in hydrogen peroxide (HO—OH), is so weak that this compound decomposes at room temperature. Second, carbon compounds are not extremely reactive under ordinary conditions. Butane, C_4H_{10}, is stable in air, but tetrasilane, Si_4H_{10}, catches fire spontaneously in air. Third, because carbon can form up to four single covalent bonds, a wide variety of compounds is possible.

Figure 1
Carbon-carbon bonds within long-chained molecules, such as the polyethylene used to make water bottles, are very strong.

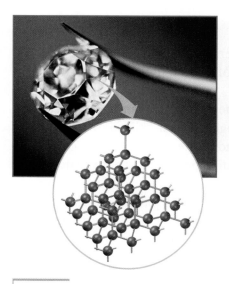

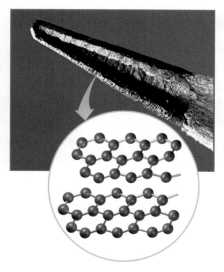

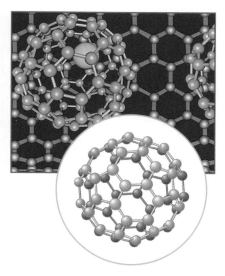

Figure 2
a Diamond is a carbon allotrope in which the atoms are densely packed in a tetrahedral arrangement.

b Graphite is a carbon allotrope in which the atoms form separate layers that can slide past one another.

c Buckminsterfullerene is a carbon allotrope in which 60 carbon atoms form a sphere.

Carbon Exists in Different Allotropes

As an element, carbon atoms can form different bonding arrangements, or *allotropes*. Three carbon allotropes are illustrated in **Figure 2**. As shown in **Figure 2a,** a diamond contains an enormous number of carbon atoms that form an extremely strong, tetrahedral network, which makes diamond the hardest known substance.

In contrast, graphite, another allotrope of carbon, is very soft. As illustrated in **Figure 2b,** the carbon atoms in graphite are bonded in a hexagonal pattern and lie in planes. The covalent bonds in each plane are very strong. However, weaker forces hold the planes together so that the planes can slip past each other. The sliding layers make graphite useful as a lubricant and as pencil lead. As you write with a pencil, the graphite layers slide apart, leaving a trail of graphite on the paper.

Other Carbon Allotropes Include Fullerenes and Nanotubes

In the mid-1980s, another type of carbon allotrope, the fullerene, was discovered. As illustrated in **Figure 2c,** fullerenes consist of near-spherical cages of carbon atoms. The most stable of these structures is C_{60}, which is formed by 60 carbon atoms arranged in interconnecting rings. The discoverers of these allotropes named C_{60} *buckminsterfullerene* in honor of the architect and designer Buckminster Fuller, whose geodesic domes had a similar shape. These allotropes can be found in the soot that forms when carbon-containing materials burn with limited oxygen.

In 1991, yet another carbon allotrope was discovered. Hexagons of carbon atoms were made to form a hollow cylinder known as a *nanotube*. A nanotube has a diameter about 10 000 times smaller than a human hair. Despite its thinness, a single nanotube is between 10 and 100 times stronger than steel by weight. Scientists are currently experimenting to find ways in technology and industry to use the unique properties of nanotubes.

Organic Compounds

Most compounds of carbon are referred to as *organic compounds*. Organic compounds contain carbon, of course, and most also contain atoms of hydrogen.

In addition to hydrogen, many other elements can bond to carbon. These elements include oxygen, nitrogen, sulfur, phosphorus, and the halogens. These bonded atoms are found in the different types of organic compounds found in living things, including proteins, carbohydrates, lipids (fats), and nucleic acids. In addition, these atoms are used to make a wide variety of synthetic organic compounds including plastics, fabrics, rubber, and pharmaceutical drugs. **Figure 3** shows examples of some natural and synthetic organic compounds.

More than 12 million organic compounds are known, and thousands of new ones are discovered or synthesized each year. There are more known compounds of carbon than compounds of all the other elements combined. To make the study of these many organic compounds easier, chemists group those with similar characteristics. The simplest class of organic compounds are those that contain only carbon and hydrogen and are known as **hydrocarbons.** Hydrocarbons can be classified into three categories based on the type of bonding between the carbon atoms.

hydrocarbon

an organic compound composed only of carbon and hydrogen

Figure 3

a This shirt and the paper are both made of cellulose. Cellulose is made from chains of glucose molecules.

b Your hair is made of proteins that are made from smaller organic compounds called amino acids. Serine is an example of an amino acid.

c Citrus fruits contain citric acid, an organic acid.

d Caffeine is an organic compound that contains nitrogen.

Alkanes Are the Simplest Hydrocarbons

The simplest hydrocarbons, **alkanes,** have carbon atoms that are connected only by single bonds. Three examples include methane, ethane, and propane. The structural formulas for each of these alkanes are drawn as follows.

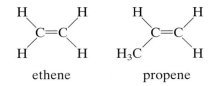

methane, CH_4 ethane, C_2H_6 propane, C_3H_8

alkane

a hydrocarbon characterized by a straight or branched carbon chain that contains only single bonds

If you examine the structural formulas for these three alkanes, you will notice that each member of the series differs from the one before by one carbon atom and two hydrogen atoms. This difference is more obvious when you compare the molecular formulas of each compound. The molecular formulas of the alkanes fit the general formula C_nH_{2n+2}, where n represents the number of carbon atoms. If the alkane contains 30 carbon atoms, then its formula is $C_{30}H_{62}$.

☐ internet connect

www.scilinks.org
Topic: Alkanes
SciLinks code: HW4134

SCiLINKS. Maintained by the National Science Teachers Association

Many Hydrocarbons Have Multiple Bonds

The second class of hydrocarbons is the **alkenes,** which contain at least one double bond between two carbon atoms. The structural formulas for two alkenes are drawn as follows.

alkene

a hydrocarbon that contains one or more double bonds

ethene propene

Because alkenes with one double bond have twice as many hydrogen atoms as carbon atoms, their general formula is written C_nH_{2n}.

The third class of hydrocarbons is the **alkynes,** which contain at least one triple bond between two carbon atoms. The simplest alkyne is ethyne, C_2H_2, which is shown in **Figure 4.** The general formula for an alkyne with one triple bond is C_nH_{2n-2}.

alkyne

a hydrocarbon that contains one or more triple bonds

Figure 4
Ethyne, commonly called *acetylene,* is one of the very few alkynes that are of practical importance. This welder is using an acetylene torch.

$$H—C≡C—H$$

Carbon Atoms Can Form Rings

Carbon atoms that form covalent bonds with one another can be arranged in a straight line or in a ring structure. They can also be branched. For example, 4 carbon atoms and 10 hydrogen atoms can be arranged to form butane, C_4H_{10}, which has a linear structure. Four carbon atoms can also form a compound called cyclobutane, C_4H_8, which has a ring structure.

butane   cyclobutane

Notice that the prefix *cyclo-* is added to the name of the alkane to indicate that it has a ring structure.

Benzene Is an Important Ring Compound

A most important organic ring compound is the hydrocarbon benzene, C_6H_6. Benzene is the simplest member of a class of organic compounds known as **aromatic hydrocarbons.** These compounds have a variety of practical uses from insecticides to artificial flavorings. Benzene can be drawn as a six-carbon ring with three double bonds, as shown below.

aromatic hydrocarbon

a member of the class of hydrocarbons (of which benzene is the first member) that consists of assemblages of cyclic conjugated carbon atoms and that is characterized by large resonance energies

However, experiments show that all the carbon-carbon bonds in benzene are the same. In other words, benzene is a molecule with resonance structures. **Figure 5** illustrates how the electron orbitals in benzene overlap to form continuous molecular orbitals known as delocalized clouds. The following structural formula is often used to show the ring structure of benzene.

The hexagon represents the six carbon atoms, while the circle represents the delocalized electron clouds. The hydrogen atoms are not shown in this simplified structural formula.

Topic Link

Refer to the "Covalent Compounds" chapter for a discussion of resonance structures.

Figure 5
Electron orbitals in benzene overlap to form continuous orbitals that allow the delocalized electrons to spread uniformly over the entire ring.

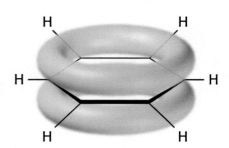

Other Organic Compounds

Hydrocarbons are only one class of organic compounds. The other classes of organic compounds include other atoms such as oxygen, nitrogen, sulfur, phosphorus, and the halogens along with carbon (and usually hydrogen).

Less than 200 years ago, scientists believed that organic compounds could be made only by living things. The word *organic* that is used to describe these compounds comes from this belief. Then in 1828 a German chemist named Friedrich Wöhler synthesized urea, an organic compound, from inorganic substances.

Many Compounds Contain Functional Groups

Like most organic compounds, urea contains a group of atoms that is responsible for its chemical properties. Such a group of atoms is known as a **functional group.** Many common organic functional groups can be seen in **Figure 6.** Because single bonds between carbon atoms are rarely involved in most chemical reactions, functional groups, which contain bonds between carbon atoms and atoms of other elements, are often responsible for how an organic compound reacts. Organic compounds are commonly classified by the functional groups they contain. **Table 1** on the next page provides an overview of some common classes of organic compounds and their functional groups.

functional group

the portion of a molecule that is active in a chemical reaction and that determines the properties of many organic compounds

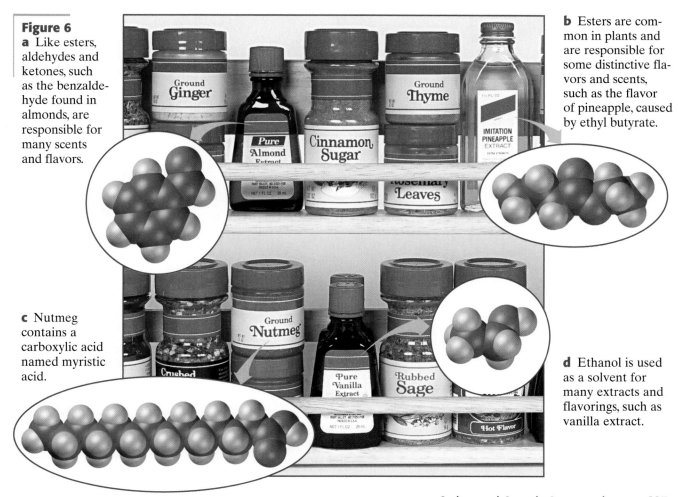

Figure 6
a Like esters, aldehydes and ketones, such as the benzaldehyde found in almonds, are responsible for many scents and flavors.

b Esters are common in plants and are responsible for some distinctive flavors and scents, such as the flavor of pineapple, caused by ethyl butyrate.

c Nutmeg contains a carboxylic acid named myristic acid.

d Ethanol is used as a solvent for many extracts and flavorings, such as vanilla extract.

Table 1   Classes of Organic Compounds

Class	Functional group	Example	Use
Alcohol	—OH	 2-propanol	disinfectant
Aldehyde	$\overset{O}{\underset{\|}{-C-H}}$	 benzaldehyde	almond flavor
Halide	—F, Cl, Br, I	 trichlorofluoromethane (Freon-11)	refrigerant
Amine	$-\overset{\|}{\underset{\|}{N}}$	 caffeine	beverage ingredient
Carboxylic acid	$\overset{O}{\underset{\|}{-C-OH}}$	 tetradecanoic acid (myristic acid)	soap-making ingredient
Ester	$\overset{O}{\underset{\|}{-C-O-}}$	 ethyl butanoate	perfume ingredient
Ether	—O—	 methyl phenyl ether (anisole)	perfume ingredient
Ketone	$\overset{O}{\underset{\|}{-C-}}$	 propanone (acetone)	solvent in nail-polish remover

Functional Groups Determine Properties

The presence of a functional group in an organic compound causes the compound to have properties that differ greatly from those of the corresponding hydrocarbon. In fact, while molecules of very different sizes with the same functional group will have similar properties, molecules of similar sizes with different functional groups will have very different properties.

Compare the structural formulas of the molecules shown in **Table 2.** Notice that each of these molecules consists of four carbon atoms joined to one another by a single bond and arranged in a linear fashion. Notice, however, that each molecule, with the exception of butane, has a different functional group attached to one or more of these carbon atoms. As a result, each molecule has properties that differ greatly from butane.

For example, compare the boiling point of butane with those of the other compounds in **Table 2.** Butane is a gas at room temperature. Because of the symmetrical arrangement of the atoms, butane is nonpolar. Because the intermolecular forces between butane molecules are weak, butane has very low boiling and melting points and a lower density than the other four-carbon molecules.

Next compare the structural formulas of butane and 1-butanol in **Table 2.** Notice that the only difference between these two molecules is the presence of the functional group —OH on one of the carbon atoms in 1-butanol. The presence of this functional group causes 1-butanol to exist as a liquid at room temperature with much higher melting and boiling points and a significantly greater density than butane.

Table 2   Comparing Classes of Organic Compounds

Name	Structural formula	Melting point (°C)	Boiling point (°C)	Density (g/mL)
Butane		−138.4	−0.5	0.5788
1-butanol		−89.5	117.2	0.8098
Butanoic acid		−4.5	163.5	0.9577
2-butanone		−86.3	79.6	0.8054
Diethyl ether		−116.2	34.5	0.7138

Figure 7

Both of these molecules are alcohols. They are isomers of each other because they both have the molecular formula $C_4H_{10}O$.

1-butanol

2-methyl-1-propanol
(isobutyl alcohol)

Different Isomers Have Different Properties

Examine the two molecules shown in **Figure 7.** Both have the same molecular formula: $C_4H_{10}O$. They differ, however, in the way in which their atoms are arranged. These two molecules are known as **isomers.** Isomers are compounds that have the same formula but differ in their chemical and physical properties because of the difference in the arrangement of their atoms. The greater the structural difference between two isomers, the more significant is the difference in their properties. Because the structural difference between the two isomers shown in **Figure 7** is minor, both molecules have similar boiling points and densities.

isomer

one of two or more compounds that have the same chemical composition but different structures

① Section Review

UNDERSTANDING KEY IDEAS

1. List the three factors that make the bonding of carbon atoms unique.

2. What are allotropes?

3. How are alkanes, alkenes, and alkynes similar? How are they different from each other?

4. Draw the simplified representation of the resonance structure for benzene.

5. List four elements other than carbon and hydrogen that can bond to carbon in organic compounds.

6. What is an aromatic compound?

7. What is a functional group?

8. What is an isomer? What do two molecules that are isomers of each other have in common?

CRITICAL THINKING

9. Draw a structural formula for the straight-chain hydrocarbon with the molecular formula C_3H_6. Is this an alkane, alkene, or alkyne?

10. Can molecules with molecular formulas C_4H_{10} and $C_4H_{10}O$ be isomers of one another? Why or why not?

11. Draw a structural formula for an alkyne that contains seven carbon atoms.

12. Draw the structural formulas for two isomers of C_4H_{10}.

13. Why is benzene not considered a cycloalkene even though double bonds exist between the carbon atoms that are arranged in a ring structure?

14. Write the molecular formulas for an alkane, alkene, and alkyne with 5 carbon atoms each. Why are these three hydrocarbons not considered isomers?

15. Draw C_4H_6 as a cycloalkene.

Names and Structures of Organic Compounds

KEY TERMS

- saturated hydrocarbon
- unsaturated hydrocarbon

OBJECTIVES

1 **Name** simple hydrocarbons from their structural formulas.

2 **Name** branched hydrocarbons from their structural formulas.

3 **Identify** functional groups from a structural formula, and assign names to compounds containing functional groups.

4 **Draw** and interpret structural formulas and skeletal structures for common organic compounds.

Naming Straight-Chain Hydrocarbons

Inorganic carbon compounds, such as carbon dioxide, are named by using a system of prefixes and suffixes. Organic compounds have their own naming scheme, which includes prefixes and suffixes that denote the class of organic compound. Learning just a few rules will help you decipher the names of most common organic compounds.

For example, the names of all alkanes end with the suffix *-ane*. The simplest alkane is methane, CH_4, the main component of natural gas. **Table 3** lists the names and formulas for the first 10 straight-chain alkanes. For alkanes that consist of five or more carbon atoms, the prefix comes from a Greek or Latin word that indicates the number of carbon atoms in the chain.

Table 3   Straight-Chain Alkane Nomenclature

Number of carbon atoms	Name	Formula
1	methane	CH_4
2	ethane	CH_3—CH_3
3	propane	CH_3—CH_2—CH_3
4	butane	CH_3—CH_2—CH_2—CH_3
5	pentane	CH_3—CH_2—CH_2—CH_2—CH_3
6	hexane	CH_3—CH_2—CH_2—CH_2—CH_2—CH_3
7	heptane	CH_3—CH_2—CH_2—CH_2—CH_2—CH_2—CH_3
8	octane	CH_3—CH_2—CH_2—CH_2—CH_2—CH_2—CH_2—CH_3
9	nonane	CH_3—CH_2—CH_2—CH_2—CH_2—CH_2—CH_2—CH_2—CH_3
10	decane	CH_3—CH_2—CH_2—CH_2—CH_2—CH_2—CH_2—CH_2—CH_2—CH_3

Naming Short-Chain Alkenes and Alkynes

The scheme used to name straight-chain hydrocarbons applies to both saturated and unsaturated compounds. A **saturated hydrocarbon** is a hydrocarbon in which each carbon atom forms four single covalent bonds with other atoms. The alkanes are saturated hydrocarbons. An **unsaturated hydrocarbon** is a hydrocarbon in which not all carbon atoms have four single covalent bonds. The alkenes and alkynes are unsaturated hydrocarbons.

The rules for naming an unsaturated hydrocarbon with fewer than four carbon atoms are similar to those for naming alkanes. A two-carbon alkene is named *ethene,* with the suffix *-ene* indicating that the molecule is an alkene. A three-carbon alkyne is named *propyne,* with the suffix *-yne* indicating that the molecule is an alkyne.

Naming Long-Chain Alkenes and Alkynes

The name for an unsaturated hydrocarbon containing four or more carbon atoms must indicate the position of the double or triple bond within the molecule. First number the carbon atoms in the chain so that the first carbon atom in the double bond has the lowest number. Examine **Figure 8,** which shows structural formulas for two alkenes with five carbon atoms.

The correct name for the alkene shown on the left in **Figure 8** is *1-pentene.* The molecule is correctly numbered from left to right because the first carbon atom with the double bond must have the lowest number. The name *1-pentene* indicates that the double bond is present between the first and second carbon atoms. The alkene shown on the right in **Figure 8** is correctly named *2-pentene,* indicating that the double bond is present between the second and third carbon atoms. Note that 1-pentene and 2-pentene are the only possible pentenes, because 3-pentene would be the same molecule as 2-pentene and the lower numbering is preferred.

If there is more than one multiple bond in a molecule, number the position of each multiple bond, and use a prefix to indicate the number of multiple bonds. For example, the following molecule is called *1,3-pentadiene.* (Note the placement of the prefix *di-*.)

H   H   H

H—C=C—C=C—C—H

   H   H H

---

saturated hydrocarbon

an organic compound formed only by carbon and hydrogen linked by single bonds

unsaturated hydrocarbon

a hydrocarbon that has available valence bonds, usually from double or triple bonds with carbon

---

Figure 8
Both the names and structural formulas indicate the position of the double bond in each alkene. Notice that you cannot tell from the space-filling models where the double bond is located.

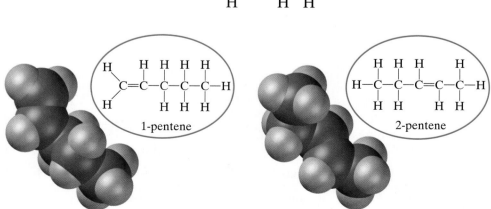

1-pentene

2-pentene

Naming Branched Hydrocarbons

When naming a hydrocarbon that is not a simple straight chain, first determine the number of carbon atoms in the longest chain. It can be named based on the corresponding alkane in **Table 3.** The longest chain may not appear straight in a structural formula, as in the example below.

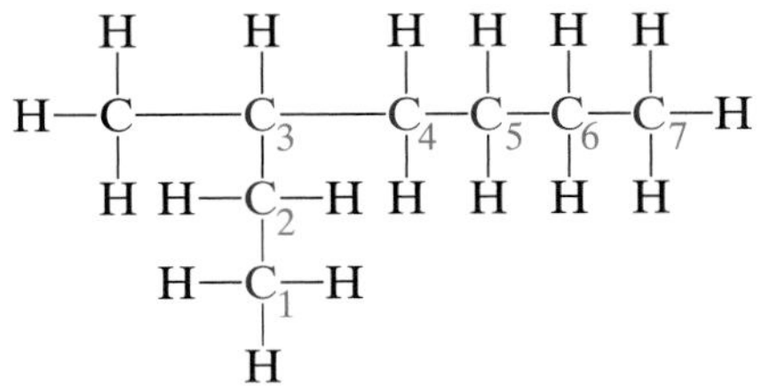

The "parent" chain in the compound shown above contains seven carbon atoms, so it is heptane. Next, number the carbon atoms on the parent chain so that any branches on the chain have the lowest numbers possible.

Name the Attached Groups and Indicate Their Positions

In the structural formula above, all the numbered carbon atoms, with one exception, are bonded to at least two hydrogen atoms. The exception is the third carbon atom, which has a $-CH_3$ group attached. This group is called a *methyl group,* because it is similar to a methane molecule, but with one less hydrogen atom. Because the methyl group is attached to the third carbon, the complete name for this branched alkane is *3-methylheptane.*

You can omit the numbers if there is no possibility of ambiguity. For example, a propane chain can have a methyl group only on its second carbon (if the methyl group were on the first or third carbon of propane, the molecule would be butane). So, what you might want to call *2-methylpropane* would be called *methylpropane.*

With unsaturated hydrocarbons that have attached groups, the longest chain containing the double bond is considered the parent compound. In addition, if more than one group is attached to the longest chain, the position of attachment of each group is given. Prefixes are used if the same group is attached more than once. Examine the following structural formula for a branched alkene.

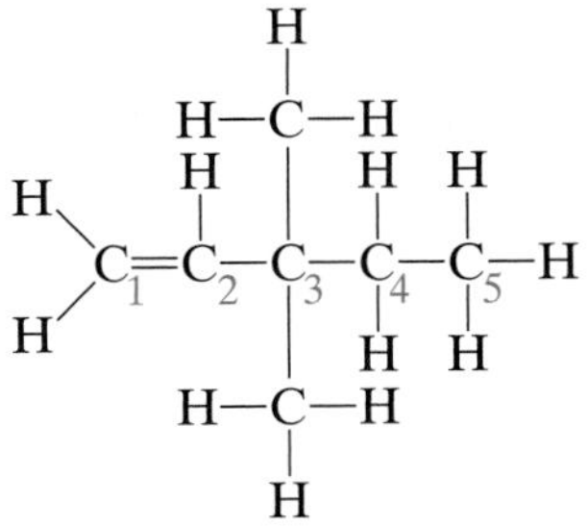

The chain containing the double bond has five carbon atoms. Therefore, the compound is a pentene. Notice that the first carbon atom has a double bond, making the chain 1-pentene. Because two methyl groups are attached to the third carbon atom, the correct name for this branched alkene is *3,3-dimethyl-1-pentene.*

Naming a Branched Hydrocarbon

Name the following hydrocarbon.

```
              H
              |
          H—C—H
              |
              H
          |
H—C≡C—C—C—H
          |
          H
      H—C—H
          |
          H
```

1 Gather information.

- The triple bond makes the branched hydrocarbon an alkyne.

2 Plan your work.

- Identify the longest continuous chain (the "parent" chain), and name it.
- Number the parent chain so that the triple bond is attached to the carbon atom with the lowest possible number.
- Name the groups that make up the branches.
- Identify the positions that the branches occupy on the longest chain.

PRACTICE HINT

Many organic structural formulas look quite confusing, but keep in mind that the name will be based on one of the simple alkane names listed in **Table 3**.

3 Name the structure.

- The longest continuous chain has four carbon atoms.
- The parent chain is butyne.
- The numbering begins with the triple bond.
- Two methyl, —CH₃, groups are present.
- Both methyl groups are attached to the third carbon atom.
- The name of this branched hydrocarbon is *3, 3-dimethyl-1-butyne*.

```
              H
              |
          H—C—H
              |
              H
          |
H—C≡C—C—C—H
          |
          H
      H—C—H
          |
          H
```

```
              H
              |
          H—C—H
              |
              H
          |
H—C≡C—C—C—H
  1   2  3  4
          |
          H
      H—C—H
          |
          H
```

4 Verify your results.

- The parent name *butyne* indicates that four carbon atoms are present in the longest chain. The *1-butyne* indicates that the first carbon atom has a triple bond. The *3,3-dimethyl-* indicates that two methyl groups, —CH₃, are attached to the third carbon atom in the longest chain.

PRACTICE

Name the following branched hydrocarbons.

1 **a.**

```
                H
                |
      H  H—C—H  H      H        H
      |    |    |      |        |
  H—C————C————C————C————C—H
      |    |    |      |        |
      H  H—C—H  H  H  H—C—H     H
           |              |
           H              H
```

b.

```
            H  H  H
            |  |  |
  H—C≡C—C——C——C—H
            |  |  |
            H  H  H
```

1 **c.**

```
            H           H
   H H—C—H H H—C—H  H  H  H  H  H
   |   |   | |   |  |  |  |  |  |
H—C   C   C   C   C—C—C—C—C—H
   |   |   | |   |  |  |  |  |
   H   H H—C—H H  H  H  H  H  H
            |
            H
```

d.

```
            H
   H H—C—H          H  H
   |   |            |  |
H—C   C   C=C—C—C—H
   |   |   |  |  |  |
   H   H   H  H  H  H
```

Names of Compounds Reflect Functional Groups

Names for organic compounds with functional groups are based on the same system used for hydrocarbons with branched chains. First, the longest chain is named. Then a prefix or suffix indicating the functional group is added to the hydrocarbon name. **Table 4** lists the prefixes and suffixes for various functional groups. When necessary, the position of the functional group is noted in the same way that the position of hydrocarbon branches is noted. Consider the following structural formula.

```
      H  H  H
      |  |  |
   H—C—C—C—H
      |  |  |
      H  O  H
         |
         H
```

Because the longest chain consists of three carbon atoms, the name for this compound is based on propane. From **Table 1,** you can see that the presence of the —OH functional group classifies this compound as an alcohol. Therefore, as indicated by **Table 4,** the name for this compound is *propanol,* whose suffix *-ol* indicates that this molecule is an alcohol. Because the functional group is attached to the second carbon atom, the correct name for this compound is *2-propanol.* A number of organic compounds are often referred to by their common names, even by chemists. The common name for 2-propanol is *isopropyl alcohol.*

Table 4	Naming Compounds with Functional Groups	
Class of compound	**Suffix or prefix**	**Example**
Alcohol	*-ol*	propanol
Aldehyde	*-al*	butanal
Amine	*-amine* or *amino-*	methylamine
Carboxylic acid	*-oic acid*	ethanoic acid
Ketone	*-one*	propanone

Naming a Compound with a Functional Group

Name the following organic compound.

$$\begin{array}{ccccccc} & H & H & H & O & H & H \\ & | & | & | & \| & | & | \\ H- & C- & C- & C- & C- & C- & C-H \\ & | & | & | & & | & | \\ & H & H & H & & H & H \end{array}$$

1 Gather information.

- Notice that the functional group indicates that this compound is a ketone.

2 Plan your work.

- Identify the longest continuous chain (the "parent" chain), and name it.
- Number the parent chain so that the functional group is attached to the carbon atom with the lowest possible number.
- Identify the position that the functional group occupies on the longest chain.
- Name the organic compound.

PRACTICE HINT

The steps to follow for naming organic compounds with functional groups are similar to those for naming branched hydrocarbons.

3 Name the structure.

- The longest continuous chain has six carbon atoms: the parent chain is hexane.
- The carbon atoms are numbered from right to left to give the ketone functional group the lowest number.

$$\begin{array}{ccccccc} & H & H & H & O & H & H \\ & | & | & | & \| & | & | \\ H- & C- & C- & C- & C- & C- & C-H \\ & _6 & _5 & _4 & _3 & _2 & _1 \\ & | & | & | & & | & | \\ & H & H & H & & H & H \end{array}$$

- The name of this organic compound is *3-hexanone*.

4 Verify your results.

- The name *3-hexanone* indicates that six carbon atoms are present in the parent chain. The suffix *-one* indicates that this compound is a ketone. The *3-* indicates that the functional group is attached to the third carbon atom in the parent chain.

PRACTICE

Name the following organic compounds.

1 a.

$$\begin{array}{ccc} & H & OH \\ & | & | \\ H- & C- & C-H \\ & | & | \\ & H & H \end{array}$$

c.

$$\begin{array}{cccc} & H & H & H & O \\ & | & | & | & \| \\ H- & C- & C- & C- & C-OH \\ & | & | & | & \\ & H & H & H & \end{array}$$

b.

$$\begin{array}{ccccc} & H & H & H & O & H \\ & | & | & | & \| & | \\ H- & C- & C- & C- & C- & C-H \\ & | & | & | & & \\ & H & H & H & & \end{array}$$

d.

$$\begin{array}{cccccc} & H & H & OH & H & H & H \\ & | & | & | & | & | & | \\ H- & C- & C- & C- & C- & C- & C-H \\ & | & | & | & | & | & \\ & H & H & H & H & H & \end{array}$$

Representing Organic Molecules

Table 5 shows four ways of representing the organic molecule cyclohexane. Each type of model used to represent an organic compound has both advantages and disadvantages. Each one highlights a different feature of the molecule, from the number and kinds of atoms in a chemical formula to the three-dimensional shape of the space-filling model. Keep in mind that a picture or model cannot fully convey the true three-dimensional shape of a molecule or show the motion within a molecule caused by the atoms' constant vibration.

Structural Formulas Can Be Simplified

Structural formulas are sometimes represented by what are called *skeletal structures*, which show bonds, but leave out some or even all of the carbon and hydrogen atoms. You have already seen the skeletal structure for benzene, which is a hexagon with a ring inside it.

A skeletal structure usually shows the carbon framework of a molecule only as lines representing bonds. These lines are often drawn in a zigzag pattern to indicate the tetrahedral arrangement of bonds between a carbon atom and other atoms. Carbon atoms are understood to be at each bond along with enough hydrogen atoms so that each carbon atom has four bonds. Atoms other than carbon and hydrogen are always shown, which highlights any functional groups present.

Table 5 Types of Molecular Models

Type of model	Example	Advantages	Disadvantages
Chemical formula	C_6H_{12}	shows number of atoms in a molecule	does not show bonds, atom sizes, or shape
Structural formula		shows arrangement of all atoms and bonds in a molecule	does not show actual shape of molecule or atom sizes; larger molecules can be too complicated to draw easily
Skeletal structure		shows arrangements of carbon atoms; is simple	does not show actual shape or atom sizes; does not show all atoms or bonds
Space-filling model		shows three-dimensional shape of molecule; shows most of the space taken by electrons	uses false colors to differentiate between elements; bonds are not clearly indicated; parts of large molecules may be hidden

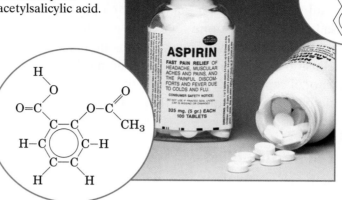

Figure 9
a The chemical name for aspirin is acetylsalicylic acid.

b Because the complete structural formula of acetyl-salicylic acid is complex ...

c ...chemists usually draw its skeletal structure instead. The presence of a benzene ring indicates that it is an aromatic compound.

SAMPLE PROBLEM C

Drawing Structural and Skeletal Formulas

Draw both the structural formula and the skeletal structure for 1,2,3-propanetriol.

1 Gather information.

- The name *propanetriol* indicates that the molecule is an alcohol that consists of three carbon atoms making up the parent chain.
- The suffix *-triol* indicates that three alcohol groups are present.
- The *1,2,3-* prefix indicates that an alcohol group is attached to the first, second, and third carbon atoms.

2 Plan your work.

- Draw the carbon framework showing the parent chain.
- Add the alcohol groups to the appropriate carbon atoms.
- Add enough hydrogen atoms so that each carbon atom has four bonds.
- Show the carbon framework as a zigzag line.
- Include the functional groups as part of the skeletal structure.

3 Draw the structures.

Structural formula:

$$C-C-C \longrightarrow HO-\overset{\overset{\displaystyle OH}{|}}{C}-C-OH \longrightarrow HO-\overset{\overset{\displaystyle H}{|}}{\underset{\underset{\displaystyle H}{|}}{C}}-\overset{\overset{\displaystyle OH}{|}}{\underset{\underset{\displaystyle H}{|}}{C}}-\overset{\overset{\displaystyle H}{|}}{\underset{\underset{\displaystyle H}{|}}{C}}-OH$$

Skeletal formula:

4 Verify your results.

- The structural formula should show all bonds and atoms in the compound 1, 2, 3-propanetriol.
- The skeletal formula should show only carbon-carbon bonds plus any functional groups present in the molecule.

> **PRACTICE HINT**
>
> Unless it is a part of a functional group, hydrogen is not shown in a skeletal structure. In the sample, the hydrogens shown are part of the alcohol functional group. The other hydrogen atoms bonded to carbon are not shown.

Draw both structural and skeletal formulas for each of the following compounds.

1 2-octanone

2 butanoic acid

3 1,1,1,2-tetrabromobutane
(Hint: *Bromo-* indicates that a Br atom is attached to the parent chain.)

4 2,2-dichloro-1,1-difluoropropane
(Hint: Both Cl and F atoms are attached to the parent chain.)

② Section Review

UNDERSTANDING KEY IDEAS

1. How does a saturated hydrocarbon differ from an unsaturated hydrocarbon?

2. What does the prefix *dec-* indicate about the composition of an organic compound?

3. What is the functional group for an aldehyde?

4. How are the carbon atoms in the parent chain numbered in a branched alkene or alkyne?

5. Which class of compounds forms the basis for naming most other carbon compounds?

PRACTICE PROBLEMS

6. Name the following branched hydrocarbon.
(Hint: The —CH$_2$—CH$_3$ group is an ethyl group.)

$$CH_3—CH—C{=}CH_2$$

with CH$_3$ above the CH and CH$_2$—CH$_3$ below the C

7. Name the following branched hydrocarbon.

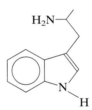

8. Draw the structural formula for dichloromethane.

9. Draw the structural and skeletal formulas for 2-bromo-4-chloroheptane.

10. Write the molecular formula for the compound with the following skeletal structure.

(Hint: Draw a full structural formula, and include all the carbon and hydrogen atoms in your count.)

CRITICAL THINKING

11. Why do the names of organic acids not contain any numbers to indicate the position of the functional group?

12. What is incorrect about the name *nonene*?

13. How is methanol different from methanal? How are they similar?

14. How many double bonds are present in 1,3-butadiene? Where are they located in the molecule?

Organic Reactions

OBJECTIVES

① **Describe** and distinguish between substitution and addition reactions.

② **Describe** and distinguish between condensation and elimination reactions.

substitution reaction

a reaction in which one or more atoms replace another atom or group of atoms in a molecule

addition reaction

a reaction in which an atom or molecule is added to an unsaturated molecule

Substitution and Addition Reactions

The single bonds between carbon and hydrogen atoms in organic compounds are not highly reactive. However, these compounds do participate in a variety of chemical reactions, one of which is called a *substitution reaction*. A **substitution reaction** is a reaction in which one or more atoms replace another atom or group of atoms in a molecule. Another type of reaction involving organic compounds is an **addition reaction** in which an atom or molecule is added to an unsaturated molecule and increases the saturation of the molecule.

Halogens Often Replace Hydrogen Atoms

As saturated hydrocarbons, the alkanes have the lowest chemical reactivity of organic compounds. However, under certain conditions these compounds can undergo substitution reactions, especially with the halogens. An example of such a reaction is that between an alkane, such as methane, and a halogen, such as chlorine. In this substitution reaction, a chlorine atom replaces a hydrogen atom on the methane molecule.

$$
\begin{array}{ccccccc}
& H & & & & H & \\
& | & & & & | & \\
H-\!\!&C&\!\!-H & + & Cl-Cl & \longrightarrow & H-\!\!&C&\!\!-Cl & + & H-Cl \\
& | & & & & | & \\
& H & & & & H & \\
\text{methane} & & \text{chlorine} & & & \text{chloromethane} & & \text{hydrogen chloride}
\end{array}
$$

The substitution reactions can continue, replacing the remaining hydrogen atoms in the methane molecule one at a time. The products are dichloromethane, trichloromethane, and tetrachloromethane. Trichloromethane is commonly known as *chloroform,* which was once used as an anesthetic. The common name for tetrachloromethane is carbon *tetrachloride,* which for many years was commonly used as a solvent.

Because the single covalent bonds are hard to break, catalysts are often added to the reaction mixture. For example, trichlorofluoromethane, CCl_3F, commonly known as *Freon-11,* was used as a refrigerant. It was made by a substitution reaction catalyzed by SbF_3.

Hydrogenation Is a Common Addition Reaction

A common type of addition reaction is hydrogenation, in which one or more hydrogen atoms are added to an unsaturated molecule. As a result of hydrogenation, the product of the reaction contains fewer double or triple bonds than the reactant. Hydrogenation is used to convert vegetable oils into fats. Vegetable oils are long chains of carbon atoms with many double bonds. When hydrogen gas is bubbled through an oil, double bonds between carbon atoms in the oil are broken and hydrogen atoms are added. Only a portion of the very long oil and fat molecules are shown in the following hydrogenation reaction.

$$\left(\begin{array}{c}H\;H\;H\;H\;H\;H\;H\\ |\;\;\;\;\;\;|\;\;\;\;\;\;|\\ -C-C=C-C-C=C-C-\\ |\;\;\;\;\;\;\;\;\;\;\;|\;\;\;\;\;\;\;\;\;\;\;|\\ H\;\;\;\;\;\;\;\;\;H\;\;\;\;\;\;\;\;\;H\end{array}\right) + H_2 \xrightarrow{\text{catalyst}} \left(\begin{array}{c}H\;H\;H\;H\;H\;H\;H\\ |\;\;\;\;\;\;|\\ -C-C=C-C-C-C-C-\\ |\;\;\;\;\;\;\;\;\;\;\;|\;\;\;|\;\;\;|\;\;\;|\\ H\;\;\;\;\;\;\;\;\;H\;H\;H\;H\end{array}\right)$$

oil   fat

Making Consumer Products by Hydrogenation

The margarine and vegetable shortening shown in **Figure 10** are two products made by the hydrogenation of oil. Although they contain double bonds, oils are still not very reactive. As a result, the hydrogenation of an oil requires the addition of a catalyst and temperatures of about 260°C.

Another application of hydrogenation is the manufacture of cyclohexane from benzene as shown by the following reaction.

$$\text{benzene} + 3H_2 \xrightarrow{\text{catalyst}} \text{cyclohexane}$$

benzene   cyclohexane

Over 90% of the cyclohexane that is made is used in the manufacture of nylon. The rest is used mostly as a solvent for paints, varnish, and oils.

Figure 10
Hydrogenation is used to turn vegetable oil into solid margarine and butter.

Some Addition Reactions Form Polymers

The addition reactions you have examined so far involve adding atoms to a molecule. Some addition reactions involve joining smaller molecules together to make larger ones. The smaller molecules are known as *monomers*. The larger molecule that is made by the addition reaction is called a **polymer.**

Consider how polyethylene is made. Polyethylene is a strong but flexible plastic used to make a variety of consumer products, including the water bottle shown at the beginning of this chapter. The monomer from which polyethylene is made is ethene, C_2H_4. Because ethene is commonly known as *ethylene,* the polymer it forms is often called *polyethylene.* The following equation shows how a portion of the polymer forms. Notice that these are condensed formulas that show all the atoms but not the bonds between the carbon and hydrogen atoms.

$$CH_2{=}CH_2 \ + \ CH_2{=}CH_2 \ \longrightarrow \ -CH_2-CH_2-CH_2-CH_2-$$

Monomers Can Be Added in Different Ways

Notice the open single bonds at each end of the product in the reaction shown above. An ethene molecule can be added at each end. The process of adding ethene molecules, one at each end, continues until polyethylene is eventually produced. Polyethylene is a very long alkane polymer chain. These chains form a product that is strong yet flexible.

Occasionally, monomers are added so that a chain branches. For example, an ethene monomer is sometimes added to form a side chain. A polymer with many side chains remains flexible. Such polymers are used to manufacture the plastic that wraps a variety of consumer products such as those shown in **Figure 11.**

polymer

a large molecule that is formed by more than five monomers, or small units

www.scilinks.org
Topic: Polymers
SciLinks code: HW4098

Figure 11
Plastic wrap is used to protect foods from spoiling. It is flexible because the side chains in the polymer prevent side-by-side molecules from packing together rigidly.

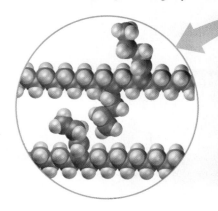

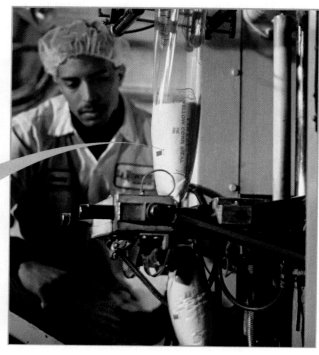

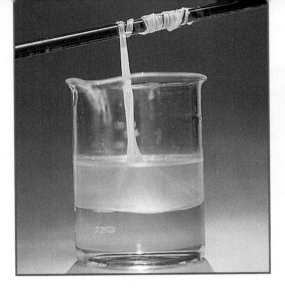

Condensation and Elimination

Polymers can also be formed by a **condensation reaction** in which two molecules combine to form a larger molecule. Water can be generated during this reaction. In some instances, hydrochloric acid is formed as a byproduct of a condensation reaction. You will see that polyesters and nylons are made by condensation reactions.

Another type of reaction that produces water is known as an *elimination reaction*. An **elimination reaction** is a reaction in which a simple molecule is removed from adjacent carbon atoms on the same organic molecule. Another simple molecule that can be a product of an elimination reaction is ammonia.

condensation reaction

a chemical reaction in which two or more molecules combine to produce water or another simple molecule

elimination reaction

a reaction in which a simple molecule, such as water or ammonia, is removed and a new compound is produced

Condensation Reactions Produce Nylon

Figure 12 shows a polymer being formed in a condensation reaction. The bottom layer in the beaker shown in **Figure 12** is hexanediamine, an organic molecule with an amine group at each end. The top layer in the beaker is adipic acid, an organic molecule with a carboxyl group at each end. The condensation reaction takes place between an amine group on hexanediamine and a carboxyl group on adipic acid as shown below.

$$\underset{\text{hexanediamine}}{\text{H}-\overset{\displaystyle\overset{\text{H}}{|}}{\text{N}}-\text{CH}_2-\text{CH}_2-\text{CH}_2-\text{CH}_2-\text{CH}_2-\text{CH}_2-\overset{\displaystyle\overset{\text{H}}{|}}{\text{N}}-\text{H}} \; + \; \underset{\text{adipic acid}}{\text{HO}-\overset{\displaystyle\overset{\text{O}}{\|}}{\text{C}}-\text{CH}_2-\text{CH}_2-\text{CH}_2-\text{CH}_2-\overset{\displaystyle\overset{\text{O}}{\|}}{\text{C}}-\text{OH}} \; \longrightarrow$$

$$\text{H}-\overset{\displaystyle\overset{\text{H}}{|}}{\text{N}}-\text{CH}_2-\text{CH}_2-\text{CH}_2-\text{CH}_2-\text{CH}_2-\text{CH}_2-\overset{\displaystyle\overset{\text{H}}{|}}{\text{N}}-\overset{\displaystyle\overset{\text{O}}{\|}}{\text{C}}-\text{CH}_2-\text{CH}_2-\text{CH}_2-\text{CH}_2-\overset{\displaystyle\overset{\text{O}}{\|}}{\text{C}}-\text{OH} \; + \; \underset{\text{water}}{\text{H}_2\text{O}}$$

Notice that a water molecule is eliminated when an H atom from the amine group and an —OH group from the carboxyl group are removed. Another adipic acid molecule is then added to the amine group shown on the left, while another hexanediamine molecule is added to the carboxyl group shown on the right. This process continues, linking hundreds of reactants to form a product called *nylon 66*.

Many Polymers Form by Condensation Reactions

In addition to nylon 66, many other polymers are made by condensation reactions. The polymer shown in **Figure 13** is polyethylene terephthalate, abbreviated PET, which is used to make permanent-press clothing and soda bottles. The following formulas show how two monomers are combined in this condensation reaction.

$$n\text{HO}-\text{CH}_2-\text{CH}_2-\text{OH} \ + \ n\text{HO}-\overset{\overset{\text{O}}{\|}}{\text{C}}-\bigcirc-\overset{\overset{\text{O}}{\|}}{\text{C}}-\text{OH} \longrightarrow$$

$$\left(\text{O}-\text{CH}_2-\text{CH}_2-\text{O}-\text{C}-\bigcirc-\overset{\overset{\text{O}}{\|}}{\text{C}}\right)_n \ + \ 2n\text{H}_2\text{O}$$

Notice that the first reactant shown is an alcohol because it contains two —OH groups. The second reactant shown is an organic acid because it contains two —COOH groups. When PET is made, water is formed from an —H from the alcohol and an —OH from the acid. The two monomers then bond. The functional group present in the product shown above classifies this molecule as an ester. Therefore, PET is a polyester.

Elimination Reactions Often Form Water

An elimination reaction involves the removal of a small molecule from two adjacent carbon atoms, as shown below.

$$\text{H}-\overset{\overset{\text{H}}{|}}{\underset{\underset{\text{H}}{|}}{\text{C}}}-\overset{\overset{\text{OH}}{|}}{\underset{\underset{\text{H}}{|}}{\text{C}}}-\text{H} \ \xrightarrow[\Delta]{\text{conc. H}_2\text{SO}_4} \ \text{H}-\overset{\overset{}{|}}{\underset{\underset{\text{H}}{|}}{\text{C}}}=\overset{\overset{}{|}}{\underset{\underset{\text{H}}{|}}{\text{C}}}-\text{H} \ + \ \text{H}_2\text{O}$$

ethanol　　　　　　　　　　　ethene　　　water

The acid catalyzes a reaction that eliminates water from the ethanol molecule, which leaves a double bond.

Figure 14
An elimination reaction occurs when sucrose and concentrated sulfuric acid are mixed. Water is formed, which leaves a product that is mostly carbon.

Figure 14 shows another example of an elimination reaction; one whose results can be seen easily. When sucrose reacts with concentrated sulfuric acid, water is eliminated, which leaves behind mostly carbon. Carbon is the black substance you can see forming in the photos and rising out of the beaker on the far right.

 Section Review

UNDERSTANDING KEY IDEAS

1. Explain why an addition reaction increases the saturation of a molecule.

2. What molecule is often a product of both condensation and elimination reactions?

3. What kind of organic reaction can form fluoromethane, CH_3F, from methane?

4. Give an example of a polymer, and tell what monomers it consists of.

5. How does a condensation reaction get its name?

6. Name the type of organic reaction that results in the formation of a double bond.

CRITICAL THINKING

7. Explain why alkanes do not undergo addition reactions.

8. Explain how an elimination reaction can be considered the opposite of an addition reaction.

9. Draw the skeletal structure of part of a polyethylene molecule consisting of eight monomers.

10. Can two different monomers be involved in an addition reaction? Why or why not?

11. Why is a molecule with only one functional group unable to undergo a condensation reaction to form a polymer?

12. Why does a substitution reaction involving an alkane and a halogen not increase the saturation of the organic compound?

CONSUMER FOCUS

Recycling Codes for Plastic Products

More than half the states in the United States have enacted laws that require plastic products to be labeled with numerical codes that identify the type of plastic used in them.

Sorting your plastics

Used plastic products can be sorted by the codes shown in **Table 6** and properly recycled or processed. Only Codes 1 and 2 are widely accepted for recycling. Codes 3 and 6 are rarely recycled. Find out what types of plastics are recycled in your area. If you know what the codes mean, you will have an idea of how successfully a given plastic product can be recycled. This information may affect your decision to buy or not buy particular items.

Questions

1. What do the recycling codes on plastic products indicate?

2. Why is it important to sort plastics before recycling them?

Table 6	**Recycling Codes for Plastic Products**			
Recycling code	**Type of plastic**	**Physical properties**	**Examples**	**Uses for recycled products**
1	polyethylene terephthalate (PET)	tough, rigid; can be a fiber or a plastic; solvent resistant; sinks in water	soda bottles, clothing, electrical insulation, automobile parts	backpacks, sleeping bags, carpet, new bottles, clothing
2	high density polyethylene (HDPE)	rough surface; stiff plastic; resistant to cracking	milk containers, bleach bottles, toys, grocery bags	furniture, toys, trash cans, picnic tables, park benches, fences
3	polyvinyl chloride (PVC)	elastomer or flexible plastic; tough; poor crystallization; unstable to light or heat; sinks in water	pipe, vinyl siding, automobile parts, clear bottles for cooking oil, bubble wrap	toys, playground equipment
4	low density polyethylene (LDPE)	moderately crystalline, flexible plastic; solvent resistant; floats on water	shrink wrapping, trash bags, dry-cleaning bags, frozen-food packaging, meat packaging	trash cans, trash bags, compost containers
5	polypropylene (PP)	rigid, very strong; fiber or flexible plastic; lightweight; heat-and-stress-resistant	heatproof containers, rope, appliance parts, outdoor carpet, luggage, diapers, automobile parts	brooms, brushes, ice scrapers, battery cable, insulation, rope
6	polystyrene (P/S, PS)	somewhat brittle, rigid plastic; resistant to acids and bases but not organic solvents; sinks in water, unless it is a foam	fast-food containers, toys, videotape reels, electrical insulation, plastic utensils, disposable drinking cups, CD jewel cases	insulated clothing, egg cartons, thermal insulation

CHAPTER HIGHLIGHTS

KEY TERMS

hydrocarbon
alkane
alkene
alkyne
aromatic hydrocarbon
functional group
isomer

saturated hydrocarbon
unsaturated
  hydrocarbon

substitution reaction
addition reaction
polymer
condensation reaction
elimination reaction

KEY IDEAS

SECTION ONE Compounds of Carbon

• The properties of carbon allotropes depend on the arrangement of the atoms and how they are bonded to each other.

• The simplest organic compounds are the hydrocarbons, which consist of only carbon and hydrogen atoms.

• Alkanes, alkenes, and alkynes are hydrocarbons. Organic compounds containing one or more rings with delocalized electrons are aromatic hydrocarbons.

• Organic compounds are classified by their functional groups.

SECTION TWO Names and Structures of Organic Compounds

• The names of the alkanes form the basis for naming most other organic compounds.

• When an organic compound is named, the parent chain is identified, and the carbon atoms are numbered so that any branches or multiple bonds have the lowest possible numbers.

• Organic molecules can be represented in various ways, and each model has advantages and disadvantages.

SECTION THREE Organic Reactions

• In a substitution reaction, an atom or group of atoms is replaced.

• In an addition reaction, an atom or group of atoms is added to replace a double or triple bond.

• Polymers are very long organic molecules formed by successive addition of monomers and are used in plastics.

• In a condensation reaction, two molecules or parts of the same molecule combine, which usually forms water.

• In an elimination reaction, a molecule, usually water, is formed by combining atoms from adjacent carbon atoms.

KEY SKILLS

Naming a Branched Hydrocarbon
Sample Problem A p. 690

Naming a Compound with a Functional Group
Sample Problem B p. 692

Drawing Structural and Skeletal Formulas
Sample Problem C p. 694

USING KEY TERMS

1. The benzene ring is the simplest member of what class of organic compounds?

2. Two compounds may have the same molecular formulas but different structural formulas. Each of these compounds is known as a(n) _____.

3. What class of organic compounds includes all saturated hydrocarbons?

4. If an element exists in more than one bonding pattern, what term is used for each of these forms?

5. What type of reaction involves the replacement of a hydrogen atom by a halogen atom?

6. The chemical and physical properties of an organic compound are largely determined by the presence of a(n) _____.

7. Which two types of organic reactions usually form small molecules such as water?

8. What type of molecule results when many smaller units are joined in addition reactions?

9. What type of hydrocarbon molecule contains a triple bond between two of its carbon atoms?

10. The hexagon and circle often used to depict a benzene molecule is an example of what kind of structure?

UNDERSTANDING KEY IDEAS

Compounds of Carbon

11. Explain why alkynes are more reactive than alkanes.

12. **a.** Why is diamond so hard and strong?
    b. Why is graphite so soft and easy to break apart?

13. How are fullerenes and nanotubes alike? How are they different?

14. What is the molecular formula for the alkane that contains 14 carbon atoms?

15. Draw the two possible resonance structures for benzene.

16. Explain the connection between the strength of the carbon-carbon single bond and the ability of carbon to be the basis of large molecules.

17. How does pentane differ from cyclopentane?

18. Explain why isomers have different chemical and physical properties.

19. Explain why the properties of butane differ from those of butanol.

Names and Structures of Organic Compounds

20. Use **Table 4** to identify the functional group from the name for each of the following organic compounds.
    a. propanol
    b. ethanoic acid
    c. propanal
    d. hexanone

21. What functional groups are present in a molecule of adrenaline, whose structural formula is shown below?

22. What group of organic compounds forms the basis for naming the other organic compounds?

23. What rule must be followed when the carbon atoms in an alkene or alkyne is numbered?

24. How does a skeletal structure differ from a structural formula? How are they the same?

25. Why is the name *pentyne* not completely correct?

26. What information does the name *1-aminobutane* provide about the structure of this organic compound?

27. List the main advantage and disadvantage of using a skeletal structure as a model.

Organic Reactions

28. Identify the functional groups that are the source of the water molecule produced in the condensation reaction between hexanediamine and adipic acid.

29. What are two reactions by which polymers can be formed?

30. Compare substitution and addition reactions.

31. What is the structural requirement for a molecule to be a monomer in an addition reaction?

32. Explain what hydrogenation is.

33. How does adding monomers as branches to a parent chain affect the properties of a polymer?

34. What is the difference between condensation and elimination?

35. Why are catalysts added to substitution reactions involving alkanes?

36. What is the chemical difference between an oil and a fat?

37. How are a nylon and polyethylene similar? How are they different?

38. Name the following compounds.

a.
$$CH_3-C=CH-C-CH_3$$
with CH_3 groups on the second carbon and the fourth carbon, and CH_3 below

b.
$$CH_3-CH_2-C-C=C-CH_3$$
with CH_3 above and CH_3 below the third carbon

c. $CH_3-CH_2-C=CH-CH_2-CH_3$
with CH_2-CH_3 below

d. $CH_3-CH=C-CH_2-CH_2-CH_2-CH_3$
with $CH_2-CH_2-CH_3$ below

39. Draw the structural formulas for each of the following compounds.
 a. 1,4-dichlorohexane
 b. 2-bromo-4-chloroheptane

40. Name the following organic compounds, and then write the molecular formula for each compound.

a.
b.

41. Draw structural formulas for the following compounds.
 a. butanoic acid
 b. 1-nonanol
 c. 2-pentanone

42. The skeletal structure for proline, an amino acid, is shown below. Draw its structural formula.

43. Name the following hydrocarbons.
 a. CH_3-I
 b. $Cl-CH_2-CH_2-Cl$

44. Name the following alcohols.

   a. CH₃—OH

   b.

$$CH_3—\overset{\displaystyle OH}{\underset{|}{CH}}—CH_2—OH$$

   c.

$$CH_3—\overset{\displaystyle OH}{\underset{|}{CH}}—CH_2—CH_3$$

   d.

$$CH_3—CH_2—CH_2—CH_2—\overset{\displaystyle OH}{\underset{|}{CH}}—CH_3$$

45. The skeletal structure for vitamin A is shown below. Draw its structural formula.

46. Draw skeletal structures for the following organic compounds.
   a. 2,3,4-trichloropentane
   b. 2,2 dichloro-1,1-difluoropropane

47. Name the following compounds.

   a.

$$CH_3—CH_2—CH_2—\overset{\displaystyle O}{\overset{\|}{C}}—CH_2—CH_3$$

   b.

$$CH_3—\underset{\underset{\displaystyle Br}{|}}{CH}—CH_2—\overset{\overset{\displaystyle Cl}{|}}{\underset{\underset{\displaystyle Cl}{|}}{CH}}$$

   c.

CRITICAL THINKING

48. Explain why some alcohols and organic acids are soluble in water, whereas hydrocarbons are virtually insoluble.

49. Identify all the functional groups in the following molecules.
   a. cinnamaldehyde
   b. salicylic acid

50. Correctly number the carbon atoms in the following hydrocarbon.

$$H—\overset{\overset{\displaystyle H}{|}}{\underset{\underset{\displaystyle H}{|}}{C}}—\overset{\overset{\displaystyle H}{|}}{\underset{\underset{\displaystyle H}{|}}{C}}—C≡C—\overset{\overset{\displaystyle H}{|}}{\underset{\underset{\displaystyle H}{|}}{C}}—\overset{\overset{\displaystyle H}{|}}{\underset{\underset{\displaystyle H}{|}}{C}}—\overset{\overset{\displaystyle H}{|}}{\underset{\underset{\displaystyle H}{|}}{C}}—\overset{\overset{\displaystyle H}{|}}{\underset{\underset{\displaystyle H}{|}}{C}}—H$$

51. Draw the structural formula for 2,3, 4-trimethylnonane.

52. Copolymers are made from two different monomers. For example, some plastic food wrap is an addition polymer made from 1,1-dichloroethene and chloroethene. Draw a possible structure for this copolymer showing a structure that is four monomers in length.

53. When propyne reacts with H₂ under the proper conditions, the triple bond is broken and hydrogen atoms are added to the alkyne to form an alkane.
   a. Draw the structural formula for the alkane product.
   b. What is the name of this alkane?

54. When 2-methylpropene is mixed with HI, 2-iodo-2-methylpropane is produced.
   a. Draw the structural formula for the organic reactant.
   b. Draw the structural formula of the product.

55. Convert the following structural formula to a skeletal structure.

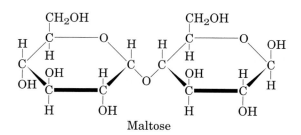

Maltose

56. Polymethylmethacrylate, sold under the trade names Plexiglas® and Lucite®, is an addition polymer made from the monomer methyl methacrylate, whose structural formula is shown below.

Draw a portion of the polymer showing four monomers that have combined.

57. The Kevlar™ that is used in bulletproof vests is a condensation polymer that can be made from the following monomer.

Draw a portion of a Kevlar™ polymer showing four molecules of a monomer that have combined.

58. Draw two structural formulas for an alcohol with the molecular formula C_3H_8O.

59. Classify the organic compounds shown below by their functional groups.

a.
$$\begin{array}{c} O \\ \parallel \\ H-C-OH \end{array}$$

b.
$$\begin{array}{c} H\ H\quad\ H\ H \\ |\ \ |\quad\ |\ \ | \\ H-C-C-O-C-C-H \\ |\ \ |\quad\ |\ \ | \\ H\ H\quad\ H\ H \end{array}$$

c.
$$\begin{array}{c} H\ O\ H\ O\ H \\ |\ \ \parallel\ \ |\ \ \parallel\ \ | \\ H-C-C-C-C-C-H \\ |\quad\ |\quad\ | \\ H\quad\ H\quad\ H \end{array}$$

d.
$$\begin{array}{c} H\quad\ H \\ |\quad\ / \\ H-C-N \\ |\quad\ \backslash \\ H\quad\ H \end{array}$$

e.
$$\begin{array}{c} H \\ | \\ H-C-H \\ | \\ Cl \end{array}$$

60. Ethylene glycol, $HO—CH_2—CH_2—OH$, is a monomer that forms a condensation polymer used as a car wax. Draw a portion of this wax polymer showing four molecules of the monomer that have combined.

ALTERNATIVE ASSESSMENT

61. Dimethyl mercury is an organic compound that poses a serious environmental threat to all living things. Research how this compound affects living things. Include information on whether dimethyl mercury poses a threat to your local environment. If so, determine what is being done to eliminate this problem.

62. Environmental concerns have led to the development of plastics that are labeled "biodegradable." Devise a set of experiments to study how well biodegradable plastics break down. If your teacher approves your plan, carry out your experiments on various consumer products labeled "biodegradable."

63. At one time, a group of compounds known as PCBs were very popular for a number of industrial applications. Find out the general structural formula for these compounds and the properties that made them so popular. Why were PCBs eventually banned for most industrial uses? Present your findings to the class.

CONCEPT MAPPING

concept map

64. Use the following terms to complete the concept map below: *organic reactions, substitution, addition, condensation, hydrogen, halogen,* and *water.*

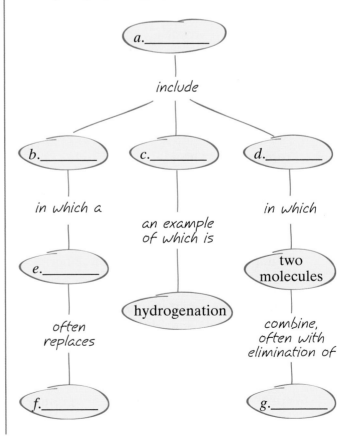

Study the graph below, and answer the questions that follow.
For help in interpreting graphs, see Appendix B, "Study Skills for Chemistry."

65. a. Determine the percentage composition by weight of hexane, C_6H_{14}.

b. Using the charts to the right as a model, make a pie chart for hexane using a protractor to draw the correct sizes of the pie slices. (Hint: A circle has 360°. To draw the correct angle for each slice, multiply each percentage by 360°.)

66. a. Compare the charts for methane, ethane, and hexane. In which of these three charts is the slice for carbon the largest?

b. In which of the three charts is the slice for carbon the smallest?

67. Based on your answers to the previous item, complete the following statement:
For saturated hydrocarbons, as the number of carbon atoms in the molecule increases, the percentage of carbon in the molecule will _____.

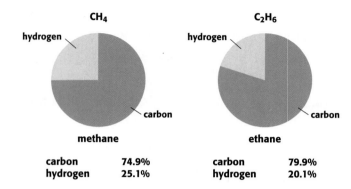

	CH$_4$		C$_2$H$_6$	
	methane		ethane	
carbon	74.9%		carbon	79.9%
hydrogen	25.1%		hydrogen	20.1%

68. a. Determine the percentage composition of hexene, C_6H_{12}.

b. Using the charts above as a model, make a pie chart for hexene.

c. Compare the charts for hexane and hexene. Which of these charts shows a larger slice for carbon?

TECHNOLOGY AND LEARNING

69. Graphing Calculator

Hydrocarbon formulas

The graphing calculator can run a program that can tell you the formula of any straight-chain hydrocarbon, provided you indicate the number of carbons and the number of double bonds in the compound.

Go to Appendix C. If you are using a TI-83 Plus, you can download the program **HYDROCAR** and run the application as directed. If you are using another calculator, your teacher will provide you with key-strokes and data sets to use. At the prompts, enter the number of carbon atoms and the

number of double bonds in the molecule. Run the program as needed to answer the following questions.

a. Dodecane is an alkane with 12 carbons and no double bonds. What is its formula?

b. The name 1,5-hexadiene describes a molecule with six carbons (hexa–) and two double bonds (–diene). What is its formula?

c. What is the formula for 1, 3, 5-hexatriene?

d. What is the formula for 3-nonene?

e. What is the formula for 1,3,5,7-octatetraene?

f. What is the formula for 2,4,6-octatriene?

1. Which of the following hydrocarbons must be an alkane?
   a. C_2H_2
   b. C_5H_{10}
   c. C_7H_{12}
   d. $C_{14}H_{30}$

2. A hydrocarbon with the formula C_8H_{18} is called
   a. octene.
   b. octyne.
   c. octane.
   d. propane.

3. During a condensation polymerization reaction,
   a. single bonds replace all double bonds that are present in the monomer.
   b. water is often produced.
   c. alcohol groups are formed.
   d. an aldehyde group is changed to a ketone group.

4. In naming an organic compound,
   a. remember that the locations of all functional groups are optional.
   b. do not consider the number of carbon atoms in the molecule as a factor.
   c. begin by identifying and naming the longest hydrocarbon chain.
   d. use side chains as the basis for naming the molecule.

5. Which of the following can have the greatest number of isomers?
   a. C_2H_6
   b. C_3H_8
   c. C_6H_6
   d. $C_{20}H_{42}$

6. Which of the following compounds cannot have different isomers?
   a. C_7H_{16}
   b. C_5H_{10}
   c. C_3H_8
   d. $C_6H_{12}O_6$

7. Examine the following structural formula.

   The correct name for this compound is
   a. 2,2-dimethylbutane.
   b. 1,1,1-trimethylpropane.
   c. 2-ethyl-2-methylpropane.
   d. 3,3-dimethylbutane.

8. Examine the following skeletal structure.

   The correct molecular formula for this compound is
   a. $C_2H_4O_2$.
   b. $C_5H_4O_2$.
   c. $C_5H_8O_2$.
   d. CHO.

9. Which of the following is an aromatic compound?
   a.
   b.
   c.
   d.

BIOLOGICAL CHEMISTRY

A spider web can stop an insect that is flying at top speed, and a single thread of spider silk can hold the weight of a spider that is large in size. Scientists have marveled that a material as lightweight as spider silk can be so strong. The silk that spiders use to form their webs is made up of a biological chemical—a protein—called fibroin. Scientists are searching for ways to use fibroin to make building materials that are strong and lightweight, like spider silk. The study of spider silk is just one example of how biological chemists are looking to nature to solve problems in the industrial world.

START-UP*ACTIVITY*

Exploring Carbohydrates

SAFETY PRECAUTIONS

PROCEDURE

1. Measure out one-half teaspoon of **sugar** into a **small beaker.**

2. Measure out one-half teaspoon of **cornstarch** into a **second small beaker.**

3. Your teacher will provide you with a **slice of apple,** a **slice of potato,** and a **slice of turkey.**

4. Add a drop of **iodine solution** to all five samples.

ANALYSIS

1. In the presence of starch, iodine turns dark blue-black. Note which samples test positive for starch.

2. Explain your observations.

Pre-Reading Questions

① Describe at least one way that the laws of chemistry apply to living systems.

② What biological molecule contains the information that determines your traits?

③ In chemical terms, what is the purpose of the food we eat?

internet connect

www.scilinks.org
Topic: Spider Proteins
SciLinks code: HW4132

SC*LINKS*. Maintained by the National Science Teachers Association

Carbohydrates and Lipids

KEY TERMS

- carbohydrate
- monosaccharide
- disaccharide
- polysaccharide
- condensation reaction
- hydrolysis
- lipid

carbohydrate

any organic compound that is made of carbon, hydrogen, and oxygen and that provides nutrients to the cells of living things

monosaccharide

a simple sugar that is the basic subunit of a carbohydrate

disaccharide

a sugar formed from two monosaccharides

polysaccharide

one of the carbohydrates made up of long chains of simple sugars; polysaccharides include starch, cellulose, and glycogen

OBJECTIVES

1. **Describe** the structure of carbohydrates.

2. **Relate** the structure of carbohydrates to their role in biological systems.

3. **Identify** the reactions that lead to the formation and breakdown of carbohydrate polymers.

4. **Describe** a property that all lipids share.

Carbohydrates in Living Systems

Most of the energy that you get from food comes in the form of **carbohydrates.** For most of us, starch, found in such foods as potatoes, bread, and rice, is our major carbohydrate source. Sugars—in fruit, honey, candy, and many packaged foods—are also carbohydrates. Plants make carbohydrates, such as the starch in potato tubers, shown in **Figure 1.**

Raw potato is difficult to digest because the starch is present in tight granules. Cooking bursts the granules, so that starch can be attacked by our digestive juices. During digestion, the starch is broken down into another carbohydrate called glucose, which—unlike starch—can be carried by the bloodstream.

Carbohydrates are compounds of carbon, hydrogen, and oxygen. They usually have the general formula $C_{6n}H_{10n+2}O_{5n+1}$. When $n = 1$ (6 C atoms), the carbohydrate is a **monosaccharide;** glucose is an example. A **disaccharide** is a carbohydrate with $n = 2$ (12 C atoms). Starch is an example of a **polysaccharide,** in which n can be many thousands.

Figure 1
Potatoes have a lot of starch, a polysaccharide.

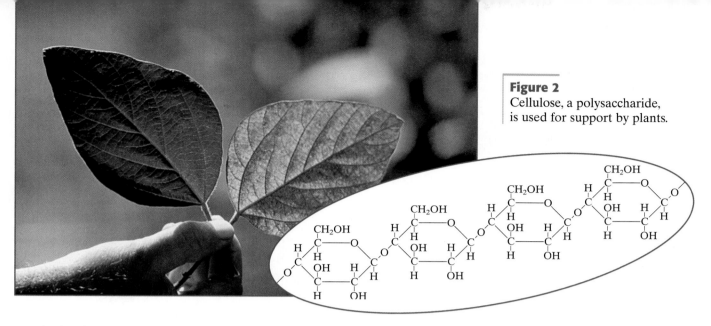

Figure 2
Cellulose, a polysaccharide, is used for support by plants.

Carbohydrates Have Many Functions

Starch is the polysaccharide that plants use for storing energy. Many animals make use of a similar energy-storage carbohydrate called *glycogen*. It is often stored in muscle tissue as an energy source.

Mammals rely on bones and muscles, which are made primarily of proteins, to give their bodies structure and support. However, insects and crustaceans, such as crabs and lobsters, rely on hard shells made of the polysaccharide chitin for structure.

The carbohydrate you come into contact with the most is the one you are looking at right now—cellulose, in paper, which comes from wood fiber. Cellulose is the most abundant organic compound on Earth. It is the polysaccharide that most plants use to give their structures rigidity. The leaves, stems, and roots of these plants are all made of cellulose, shown in **Figure 2.**

Structure of Simple Sugars

To a chemist, *sugar* is the name given to all monosaccharides and disaccharides. To a cook, *sugar* means one particular disaccharide, sucrose. The cyclic sugar glucose is important to the body because it is the chemical that the bloodstream uses to carry energy to every cell in the body. Shown below are the structures for glucose, $C_6H_{12}O_6$, and fructose, another sugar.

glucose   **fructose**

The glucose molecule has a ring made of six atoms—five carbon atoms and one oxygen atom. A sixth carbon atom is part of a $-CH_2OH$ side chain. Four other hydroxyl, $-OH$, groups connect to the carbons in the ring, as do four H atoms. The fructose molecule has a ring of five atoms, four carbon and one oxygen. Fructose has two $-CH_2OH$ side chains. Fructose and glucose have the same molecular formula, $C_6H_{12}O_6$, even though they have very different structures.

Figure 3
Three different disaccharides—sucrose, maltose, and lactose—are present in a malted milk shake.

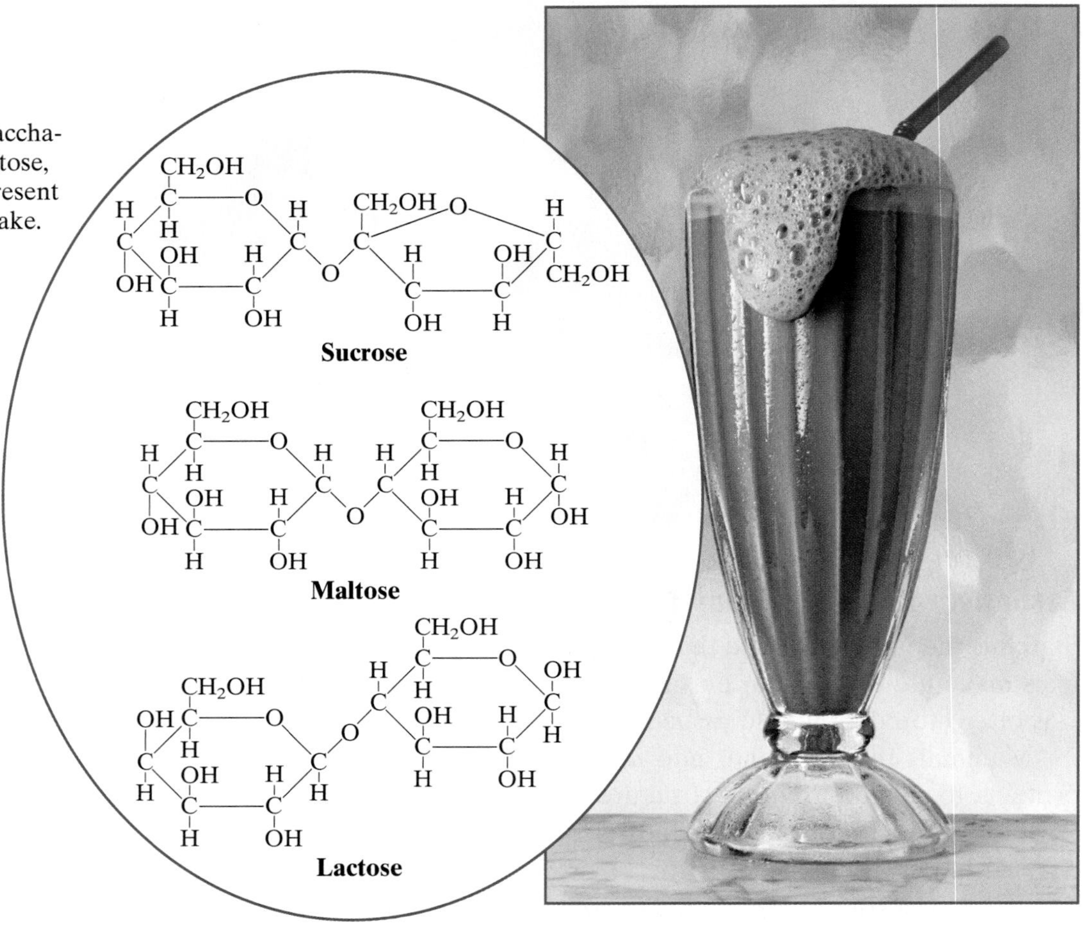

Sucrose

Maltose

Lactose

Sugars Combine to Make Disaccharides

Monosaccharides, such as glucose, have one ring. However, two can combine to form a double-ringed disaccharide. Three examples of disaccharides—lactose, maltose, and sucrose—are found in the malted milk shake shown in **Figure 3.** Notice that the disaccharides are each made up of two monosaccharides. Each molecule of maltose, the sugar that adds to the flavor of malted milk shakes, is made up of two glucose units. Each molecule of sucrose, the sugar you use to sweeten food, is made up of a glucose and a fructose unit.

Structure of Polysaccharides

Topic Link

Refer to the "Carbon and Organic Compounds" chapter for a discussion of polymers.

Just as two monosaccharides combine to form a disaccharide, many monosaccharides or disaccharides can combine to form a long chain called a polysaccharide. Polysaccharides may be represented by the general formula below or by structural models such as the ones shown in **Figures 1 and 2.**

$$\cdots O-(C_6H_{10}O_4)-O-(C_6H_{10}O_4)-O-(C_6H_{10}O_4)-O-(C_6H_{10}O_4)\cdots$$

Earlier, you learned about the linking together of small molecular units in a process known as *polymerization*. Polymerization is a series of synthesis reactions that link many monomers together to make a very large, chainlike molecule. The formation of polysaccharides is similar to polymerization. In fact, polysaccharides and other large, chainlike molecules found in living things are called *biological polymers*. Amylose, a biological polymer listed in **Table 1,** is a form of starch.

Table 1 Types of Carbohydrates

Type	Example	Role
Monosaccharides	fructose	sweetener found in fruits
	glucose	cell fuel
Disaccharides	sucrose	sweetener (table sugar)
Polysaccharides	chitin	insect exoskeleton, support, protection
	amylose	energy storage (plants)
	glycogen	energy storage (animals)

Carbohydrate Reactions

Photosynthesis and respiration, described later, are the main ways that carbohydrates are made and broken down in living systems. These processes are also the primary ways that living things capture and use energy. Thus, carbohydrate reactions play a major role in the chemistry of life.

Formation of Disaccharides and Polysaccharides

Because glucose and other sugars dissolve easily in water, they are not useful for long-term energy storage. This is why living things change sugars to starch or glycogen, neither of which is soluble in water.

Disaccharides and polysaccharides are formed from sugars during **condensation reactions,** in which water is a byproduct. Though there are many more steps that are not shown here, the net equation below describes the formation of the disaccharide sucrose.

condensation reaction

a chemical reaction in which two or more molecules combine to produce water or another simple molecule

glucose   fructose → sucrose   water

Breakdown of Carbohydrates

When an organism is ready to use energy that was previously stored as a polysaccharide, a different kind of reaction takes place. Polysaccharides are changed back to sugars during **hydrolysis** reactions. In these reactions, the decomposition of a biological polymer takes place along with the breakdown of a water molecule, as shown in the equation below.

hydrolysis

a chemical reaction between water and another substance to form two or more new substances

sucrose   water → glucose   fructose

The reaction is the reverse of the condensation reaction by which sucrose formed. In humans, polysaccharides, such as starch and glycogen, and disaccharides, such as sucrose, are broken down in this way to make glucose.

Lipids

lipid

a type of biochemical that does not dissolve in water, including fats and steroids; lipids store energy and make up cell membranes

Lipids are a class of biological molecules that do not dissolve in water. However, they generally can have a polar, hydrophilic region at one end of the molecule. For example, the lipid shown below is oleic acid, which is found in the fat of some animals.

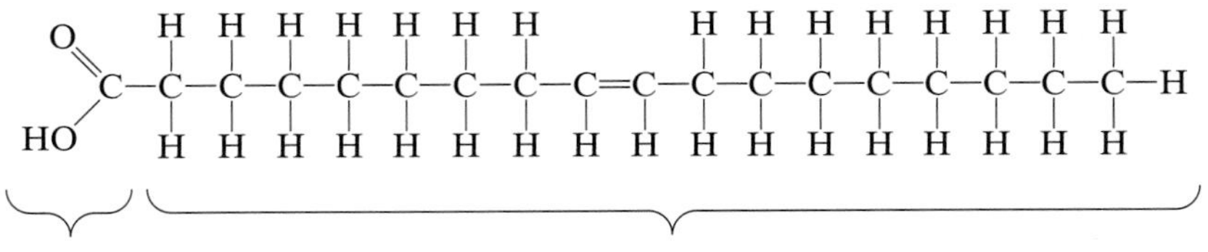

[structure of oleic acid]

hydrophilic region   hydrophobic region

The hydrophilic region on the right side of the molecule allows it to interact with polar molecules. The hydrophobic region on the left side of the molecule allows it to interact with nonpolar molecules.

Lipids have a variety of roles in living systems. They are used in animals for energy storage as *fats*. Cell membranes are made up of lipids called *phospholipids*. *Steroids*—such as cholesterol, shown in **Figure 4**—are lipids used for chemical signaling. Waxes, such as those found in candles and beeswax are also lipids.

Figure 4
Like all steroids, cholesterol has a structure with four connected rings.

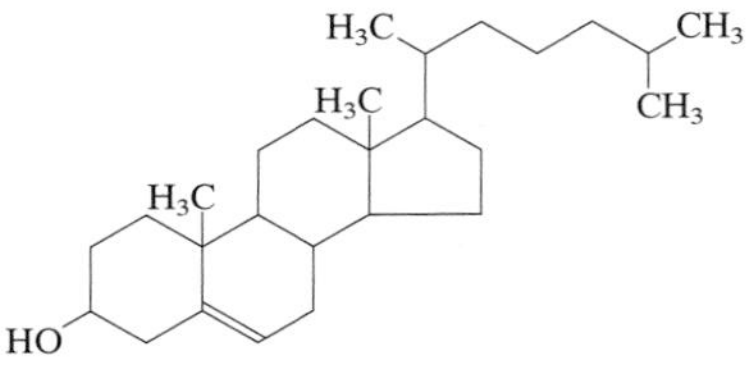

① Section Review

UNDERSTANDING KEY IDEAS

1. Describe the general chemical formula of carbohydrates.

2. What do chemists mean by a *sugar,* and what are the two principal classes of sugars?

3. What role do carbohydrates play in the survival of animals and plants?

4. Name several polysaccharides, and explain the biological role of each.

5. What is the molecular formula of glucose, and what is the role of this compound in human body systems?

6. What names are given to the reactions by which large carbohydrate molecules are built up and broken down?

7. How does the formation of a biological polymer compare to the formation of most manufactured polymers?

8. What property do all lipids share?

CRITICAL THINKING

9. What is the formula of the compound formed by the condensation of two disaccharides?

10. Why do we cook starchy foods?

11. Classify the following carbohydrates into monosaccharides, disaccharides, or polysaccharides: cellulose, glucose, lactose, starch, maltose, sucrose, chitin, and fructose.

12. Why is glycogen often called *animal starch*?

13. **a.** What type of reaction does the following equation describe?

   b. Name the reactants and the products.

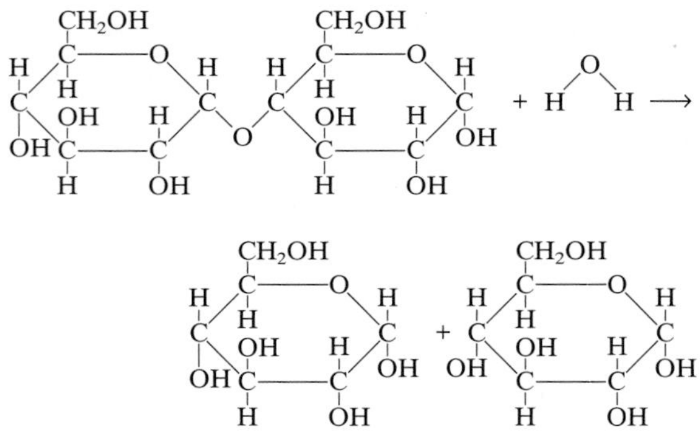

[chemical structures/equations of two sugar molecules]

Proteins

KEY TERMS

- **protein**
- **amino acid**
- **polypeptide**
- **peptide bond**
- **enzyme**
- **denature**

OBJECTIVES

1. **Describe** the general amino acid structure.

2. **Explain** how amino acids form proteins through condensation reactions.

3. **Explain** the significance of amino-acid side chains to the three-dimensional structure and function of a protein.

4. **Describe** how enzymes work and how the structure and function of an enzyme is affected by changes in temperature and pH.

Amino Acids and Proteins

A **protein** is a biological polymer that is made up of nitrogen, carbon, hydrogen, oxygen, and sometimes other elements. Our bodies are mostly made out of proteins. For example, the most abundant protein in your body is collagen, which is found in skin and bones. Your hair has structural proteins, such as keratin, shown in **Figure 5.** Proteins in muscles allow your muscles to contract, making body movement possible.

Different proteins have different physical properties. Some—such as casein in milk, ovalbumin in egg whites, and hemoglobin in blood—are water-soluble. Others—such as keratin in hair, fibroin in spider silk, and collagen in connective tissue—are flexible solids.

What do all these proteins have in common? They are all made up of **amino acids.** In the same way that sugars are the building blocks of carbohydrates, amino acids are the building blocks of proteins.

protein

an organic compound that is made of one or more chains of amino acids and that is a principal component of all cells

amino acid

any one of 20 different organic molecules that contain a carboxyl and an amino group and that combine to form proteins

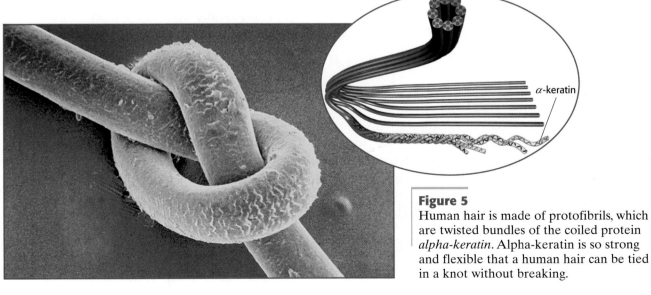

Figure 5
Human hair is made of protofibrils, which are twisted bundles of the coiled protein *alpha-keratin.* Alpha-keratin is so strong and flexible that a human hair can be tied in a knot without breaking.

α-keratin

Table 2 Structures and Roles of Several Amino Acids

Name	Structure	Role	Name	Structure	Role
Cysteine	SH CH$_2$ H$_2$N—C—COOH H	cross-links to other cysteine units	Valine	H$_3$C CH$_3$ CH H$_2$N—C—COOH H	contributes to hydrophobicity (nonpolar)
Glutamic acid	O OH C CH$_2$ CH$_2$ H$_2$N—C—COOH H	gives an acidic side chain	Asparagine	O NH$_2$ C CH$_2$ H$_2$N—C—COOH H	gives hydrogen-bonding sites (polar)
Glycine	H H$_2$N—C—COOH H	acts as a spacer	Histidine	H H N—C N H C CH$_2$ H$_2$N—C—COOH H	gives a basic side chain

Amino-Acid Structure and Protein Synthesis

Amino refers to the –NH$_2$ group of atoms. Generally, organic acids have the carboxylic acid group, –COOH. Thus, *amino acids* are compounds that have both the basic –NH$_2$ and the acidic –COOH groups. There are 20 amino acids from which natural proteins are made. All of them have the same basic structure shown below. The *R* represents a *side chain*.

A side chain is a chemical group that differs from one amino acid to another. **Table 2** shows the detailed structure of six of these amino acids.

The reaction by which proteins are made from amino acids is similar to the condensation of carbohydrates. A water molecule forms from the –OH of the carboxylic acid group of one amino acid and an –H of the amino group of another. The condensation of amino acids is shown below.

The biological polymer that forms is called a **polypeptide.** The link that joins the N and C atoms of two different amino acids in a protein is called a **peptide bond.** In protein synthesis, hundreds of peptide bonds are formed one after another. This process makes a long polypeptide chain. The chain's backbone has the pattern –N–C–C–N–C–C–N–C–C–. Half the C atoms have side chains (R), as shown below.

internet connect
www.scilinks.org
Topic: Proteins
SciLinks code: HW4104
SC/LINKS Maintained by the National Science Teachers Association

polypeptide

a long chain of several amino acids

peptide bond

the chemical bond that forms between the carboxyl group of one amino acid and the amino group of another amino acid

Properties and Interactions of Side Chains

The properties of a part of a polypeptide chain depend on the properties of the side chains present. For example, the side chain of glutamic acid is acidic. The side chain of histidine is basic. The side chains of asparagine and several other amino acids are strongly polar. On the other hand, amino acids with nonpolar side groups, such as valine, are nonpolar.

Some amino acid side chains can form ionic or covalent bonds with other side chains. Cysteine is a unique amino acid, because the –SH group in cysteine can form a covalent bond with other cysteine units. Two cysteine units, at different points on a protein molecule, can bond to form a *disulfide bridge,* shown in **Figure 6.** Such bonding can form a looped protein or link two separate polypeptides. In fact, curly hair is a result of the presence of disulfide bridges in hair protein. Some amino acid side chains can form ionic bonds with other amino acid side chains. These bonds also link different points on a protein. For example, glutamic acid can give up a proton to histidine. When this happens, an ionic bond will form between the two amino acids.

Also, weaker interactions can affect how segments of proteins interact with one another. You have read about these interactions in earlier chapters. Two are shown in **Figure 6.** One of these weak interactions is between the nonpolar hydrocarbon side chains present on many amino acids. These groups are hydrophobic and do not tend to be found in polar and ionic environments. Instead, nonpolar segments of a protein tend to be found with nonpolar molecules or with other nonpolar segments of the same protein.

The side chains of certain amino acids, such as asparagine, allow for another kind of interaction—hydrogen bonding. The hydrogen atoms on hydroxyl groups, –OH, and amino groups, $-NH_2$, are drawn to places where they can hydrogen bond to oxygen atoms, especially to carboxyl groups, –C=O, in the polypeptide backbone or in the side chains.

Topic Link

Refer to the "States of Matter and Intermolecular Forces" chapter for a discussion of intermolecular forces.

Figure 6
Four different kinds of interaction between side chains on a polypeptide molecule help to make the shape that a protein takes. Three are shown here.

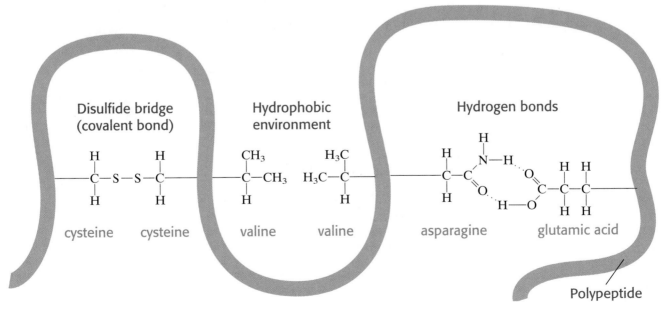

Four Levels of Protein Structure

Proteins are not just long polypeptide chains. Because of the interactions of the side chains and other forces, each protein usually folds up into a unique shape. The three-dimensional shape that the chain forms gives characteristic properties to each protein. If a polypeptide chain folds into the wrong shape, it can function differently. It may also be unable to carry out its biological role. The levels of protein structure are shown in **Table 3.**

The amino-acid sequence of the polypeptide chain is said to be the *primary structure* of a protein. Thus, the primary structure of a protein is simply the order in which the amino acids bonded together.

Most proteins have segments in which the polypeptide chain is coiled or folded. These coils and folds are often held in place by hydrogen bonding. They give the protein its *secondary structure*. Two common kinds of secondary structures are the *alpha helix* and the *beta pleated sheet,* both of which are shown in the table. The alpha (α) helix is shaped like a coil with hydrogen bonds that form along a single segment of a polypeptide. The beta (β) pleated sheet is shaped like an accordion with hydrogen bonds that form between adjacent polypeptide segments.

In alpha-keratin, shown in **Figure 5,** the entire length of the protein has an α-helix structure. However, other proteins will have only sections that are α-helixes. Different sections of the same protein may have a pleated sheet secondary structure. These different sections of a protein can fold in different directions. These factors, combined with the intermolecular forces acting between side chains give each protein a distinct three-dimensional shape. This shape is the *tertiary structure* of the protein.

A *quaternary structure* arises when different polypeptide chains that have their own three-dimensional structure come together to form a larger protein. For example, four separate polypeptides make up a single molecule of hemoglobin, the protein that carries O_2 within red blood cells.

Table 3	Levels of Protein Structure		
Primary structure	**Secondary structure**	**Tertiary structure**	**Quaternary structure**

valine   proline

α-helix

β-pleated sheet

Amino-Acid Substitution Can Affect Shape

The sequence of amino acids—the primary structure—helps dictate the protein's final shape. A substitution of just one amino acid in the polypeptide sequence can have major effects on the final shape of the protein.

A hereditary blood cell disease called *sickle cell anemia* gives one example of the importance of amino-acid sequence. As the blood circulates, hemoglobin proteins in red blood cells pick up oxygen in the lungs and deliver it to all regions of the body. Normal red blood cells have the dimpled disk shape shown on the left in **Figure 7.** However, people with sickle cell anemia have blood cells with a crescent, or "sickle," shape. These cells are less efficient at carrying oxygen, which can cause respiration difficulties. Worse, the sickled cells tend to clump together in narrow blood vessels, causing clotting and sometimes death.

The cause of the sickle cell shape lies in the amino-acid sequence of the polypeptide. In sickle cell hemoglobin, the sixth amino acid in one of the polypeptide chains is valine. The sixth amino acid in healthy hemoglobin is glutamic acid. Because of the difference in only one amino acid, the entire shape of the hemoglobin is different in the unhealthy blood cells. This tiny change in the primary structure of the protein is enough to affect the health and life of people who have this disease.

Figure 7

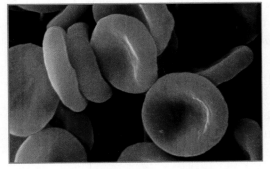

a The round, flat shape of healthy red blood cells shows they have normal hemoglobin molecules.

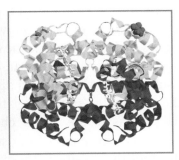

b Hemoglobin consists of four polypeptide chains; a fragment of one chain is shown in green.

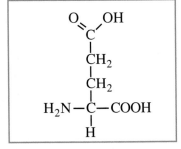

c Each of the chains is a polymer of 141 or 146 amino acid units, such as the glutamic acid monomer shown here.

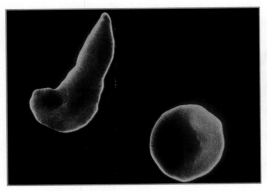

d Because of their shape, sickle cells clog small blood vessels.

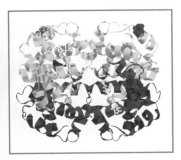

e A genetic mutation causes one glutamic acid to be replaced by valine in the hemoglobin molecules, as shown in red.

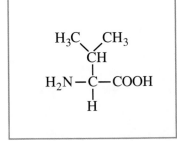

f The sickle shape of the cell comes from the different shape of the hemoglobin caused by the valine substitution.

Biological Chemistry   **721**

Enzymes

An **enzyme** is a protein that catalyzes a chemical reaction. Almost all of the chemical reactions in living systems take place with the help of enzymes. In fact, some biochemical processes would not take place at all without enzymes.

Enzymes have remarkable catalytic power. For example, blood cells change carbon dioxide, CO_2, to carbonic acid, H_2CO_3, which is easily carried to the lungs. Once in the lungs, carbonic acid decomposes back into carbon dioxide so that the CO_2 can be exhaled by the lungs. The reaction described by the equation below takes place in our lungs and tissues.

$$CO_2(aq) + H_2O(l) \rightleftharpoons H_2CO_3(aq)$$

Topic Link

Refer to the "Reaction Rates" chapter for a discussion of catalysis.

The enzyme *carbonic anhydrase* allows this reaction to take place 10 million times faster than it normally would. The forward and reverse processes are accelerated equally. Hence the reaction's equilibrium constant is unaffected by the enzyme's presence. Enzymes are very efficient. A single molecule of carbonic anhydrase can cause 600 000 carbon dioxide molecules to react each second.

How Enzymes Work

In the late 19th century, the German chemist Emil Fischer proposed that enzymes work like a lock and key. That is, only an enzyme of a specific shape can fit the reactants of the reaction that it is catalyzing. A model of an enzyme mechanism is shown in **Figure 8.** Only a small part of the enzyme's surface, known as the *active site,* is believed to make the enzyme active. In reactions that use an enzyme, the reactant is called a *substrate.* The substrate has bumps and dips that fit exactly into the dips and bumps of the active site, much like three-dimensional puzzle pieces. Also, the active site has groups of side chains that form hydrogen bonds and other interactions with parts of the substrate. While the enzyme and the substrate hold this position, the bond breaking (or bond formation) takes place and the products are released. Once the products are released, the enzyme is available for a new substrate.

ype of protein that speeds up metabolic reactions in plant and animals without being permanently changed or destroyed

Figure 8

a The enzyme reacts with the substrate in a fast, reversible reaction.

b The substrate-enzyme complex can either revert to the reactants or . . .

c . . . proceed to the products.

Scientists have added to Fischer's idea and suggested that some enzymes are flexible structures. An enzyme might wrap its active site around the substrate as the substrate approaches. Further flexing of the enzyme causes some bonds in the substrate to break and frees the products. Whatever the actual mechanism of an enzyme, its shape is very important to its ability to catalyze a reaction. Because protein function depends so much on the shape of the protein, changing a protein's shape can inactivate a protein.

Denaturing an Enzyme Destroys Its Function

You do not have to change the primary structure of an enzyme to inactivate it. You can **denature** a protein. To denature a protein means to cause it to lose its tertiary and quaternary structures so that the polypeptide becomes a random tangle. Mild changes, such as shifts in solvent, temperature, pH, or salinity, may be enough to denature the enzyme. For example, the enzymatic ability to decompose hydrogen peroxide is lost by plant and animal cells when they are heated.

Of course, many proteins other than enzymes can also be easily denatured. When you prepare protein foods for meals you are usually denaturing proteins. For example, when you cook an egg, the egg white changes from runny and clear to firm and white, because the proteins are denatured by the change in temperature. Denaturing is the reason you can "cook" some foods without heating them. For example, when you make a dish called *ceviche* (suh VEE CHAY), you denature the proteins in raw fish by changing the pH of the protein's environment. By marinating the fish in acidic lime juice, you are denaturing the proteins much in the same way as if you heated the fish. Some recipes for pickled herring work in the same way, using vinegar (acetic acid) to denature the raw fish proteins.

denature

to change irreversibly the structure or shape—and thus the solubility and other properties—of a protein by heating, shaking, or treating the protein with acid, alkali, or other species

Quick LAB

Denaturing an Enzyme

PROCEDURE

1. Get **15 potato cubes** from your teacher. Place one potato cube on a **paper plate.**

2. Using a **dropper,** drop **hydrogen peroxide solution** onto the potato cube. Note the amount of bubbling (the enzymatic activity). Let this

amount of bubbling count as a score of 10.

3. Place the remaining potato cubes in a **beaker of water** at room temperature. Place the beaker on a preheated **hot plate** that remains switched on.

4. Using **tongs,** remove one cube every 30 s, and test its enzymatic activity, assigning

a score between 0 and 10 based on the amount of bubbling.

ANALYSIS

1. Graph the enzymatic activity score versus heating time.

2. What happens to the enzymatic activity of a potato with heating? Explain.

Curbing Enzyme Action

Enzymes can be too strong by themselves. One example of an overly strong enzyme is a *proteolytic* (or protein-splitting) enzyme called *trypsin*, which plays a part in the digestion of protein food. Trypsin is used in the small intestine to help break down proteins into amino acids through hydrolysis. However, the small intestine is itself made of proteins, which can also be broken down by trypsin! Rather than producing trypsin that will destroy its own organs, the body makes an inactive form of trypsin, a protein called *trypsinogen*.

Trypsinogen is stored in the pancreas. It is added to semidigested food as it passes through the small intestine. Small amounts of another protein, *enteropeptidase,* which is enzymatically active, are also added. When an enteropeptidase molecule meets a molecule of trypsinogen, enteropeptidase attacks one of the bonds in trypsinogen. When this bond is broken, one of the products is trypsin. Thus, this strong enzyme is made only at a time and place when it can break down food with the fewest dangerous side effects.

Section Review

UNDERSTANDING KEY IDEAS

1. Describe the meaning of the two parts of the name *amino acid*.

2. Draw the general structure of an amino acid.

3. What is a peptide bond, and what name is given to enzymes that catalyze its hydrolysis?

4. **a.** Identify three side chains found in amino acids.
   b. Draw the three amino acids that have these side chains.
   c. What property does each of these chains give to a polypeptide chain?

5. What causes sickle cell anemia?

6. Describe the secondary structure of proteins.

7. What is meant by *denaturing* an enzyme, and what changes in conditions might bring it about?

8. Briefly describe how enzymes are believed to work to catalyze a reaction.

CRITICAL THINKING

9. What do condensation of sugars and condensation of amino acids have in common?

10. What different meanings do the words *polypeptide* and *protein* have?

11. List four different ways in which one part of a polypeptide chain may interact with another part. List them in the order that reflects *decreasing* strength of the interaction. (Hint: Apply what you have learned in previous chapters about the strength of different types of bonds and intermolecular forces.)

12. Proteolytic enzymes catalyze the hydrolysis of polypeptides. Predict the products if you carried out the hydrolysis of the following molecule, a dipeptide.

$$
\begin{array}{c}
H_3C \quad\; CH_3 \\
\backslash \;/ \\
CH \qquad\qquad H \\
| \qquad\qquad | \\
H-N-C-C-N-C-C-OH \\
\;\;\; | \;\; \|\;\; | \;\; | \;\; \| \\
\;\;\; H \;\; O \;\; H \;\; H \;\; O
\end{array}
$$

Nucleic Acids

KEY TERMS

- **nucleic acid**
- **DNA**
- **gene**
- **DNA fingerprint**
- **clone**
- **recombinant DNA**

OBJECTIVES

① **Relate** the structure of nucleic acids to their function as carriers of genetic information.

② **Describe** how DNA uses the genetic code to control the synthesis of proteins.

③ **Describe** important gene technologies and their significance.

Nucleic Acids and Information Storage

You are probably like one or both of your parents in personality or physical features. Some traits may be due to the environment you grew up in, but many traits you inherited from your parents. Before you were born, you began as a single cell that had equal amounts of information from your mother and father about *their* hereditary characteristics. As that cell divided and redivided, that information was duplicated and now resides in every cell of your body.

Hereditary information is not just about the shape and color of your eyes, but also about the very fact that you have eyes—and that you are a human and not a snail or a cabbage. All that information, including the "construction plans" for building your body, is stored chemically in compounds called **nucleic acids.**

Nucleic-Acid Structure

Like polysaccharides and polypeptides, nucleic acids are biological polymers. Nucleic acids are formed from equal numbers of three chemical units: a sugar, a phosphate group, and one of several nitrogenous bases. The "backbone" of the nucleic acid is a -sugar-phosphate-sugar-phosphate- chain, with various nitrogenous bases connected to the sugar units. **Figure 9** shows the structures of the four most common nitrogenous bases.

nucleic acid

an organic compound, either RNA or DNA, whose molecules are made up of one or two chains of nucleotides and carry genetic information

Thymine   Cytosine   Adenine   Guanine

Figure 9
There are four common nitrogenous bases of nucleic acids. Thymine and cytosine bases have a single six-membered ring. Adenine and guanine bases have connected six- and five-membered rings.

Deoxyribonucleic Acid, or DNA

Deoxyribonucleic acid is the full name of the most famous nucleic acid, which is usually known by the abbreviation **DNA.** DNA acts as the biochemical storehouse of genetic information in the cells of all living things.

The sugar in DNA is *deoxyribose,* which has a ring in which four of the atoms are carbon and the fifth atom is oxygen. The phosphate group comes from phosphoric acid, $(HO)_3PO$. Two of the −OH groups from the phosphoric acid condense with the −OH groups on two different sugar molecules, linking all three together as shown below.

The nitrogenous bases connect to the sugar units in the backbone. There is one base per sugar unit. Any one of the four bases—adenine, guanine, thymine, and cytosine—is connected along the strand at the sugar units. All genetic information is encoded in the sequence of the four bases, which are abbreviated to A, G, T, and C. Just as history is written in books using a 26-letter alphabet, heredity is written in DNA using a 4-letter alphabet.

Living things vary in the size and number of DNA molecules in their cells. Cells may have just one or many molecules of DNA. Some bacteria cells have a single molecule of DNA that has about 8 million bases. Human cells have 46 molecules of DNA that have a total of about 6 billion bases.

Figure 10
The three-dimensional structure of DNA is made stable by hydrogen bonding between base pairs.

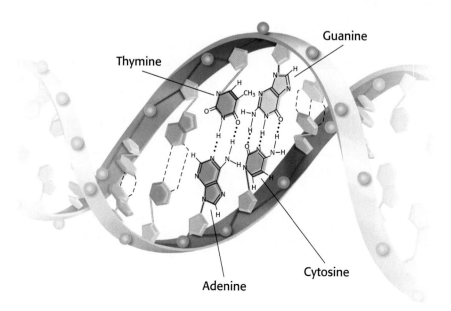

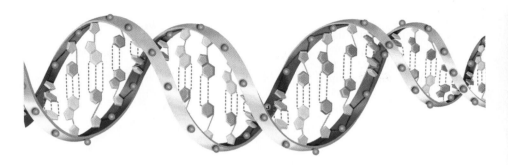

DNA's Three-Dimensional Structure

There are single strands of DNA, but the biological polymer is mostly found as a double helix in which two DNA strands spiral around each other as shown in **Figure 11.** The two strands are not duplicates of each other. Instead, they are complementary. This means that where an adenine (A) is found in one strand, thymine (T) is found in the other. Likewise, a guanine (G) in one strand is matched with a cytosine (C) in the other.

The reason for the complementary nature of DNA can be seen in **Figure 10.** When A and T are lined up opposite each other, the two bases are ideally placed for forming two hydrogen bonds, which bond the two strands together. Likewise, G and C can easily form three hydrogen bonds between themselves. No other pairing can form the right hydrogen bonds to keep the strands together. Thus, the three-dimensional configuration of DNA looks like a twisted ladder or spiral staircase, with A–T and G–C base pairs providing the rungs or steps.

Figure 11
The double helix of DNA can be seen by scanning-tunneling microscopy (above) or shown as a molecular model (above left).

Quick LAB

Isolation of Onion DNA

PROCEDURE

1. Place **5 mL of onion extract** in a **test tube.** The extract was taken from whole onions that were processed in a laboratory.

2. Hold the test tube at a 45° angle. Use a **pipet** to add **5 mL of ice-cold ethanol** to the tube one drop at a time. Note: Allow the ethanol to run slowly down the side of the tube so that it forms a distinct layer.

3. Let the test tube stand for 2–3 min.

4. Insert a **glass stirring rod** into the boundary between the onion extract and ethanol. Gently twirl the stirring rod by rolling the handle between your thumb and finger.

5. Remove the stirring rod from the liquids, and examine any material that has stuck to it. You are looking at onion DNA. Touch the

DNA to the lip of the test tube, and observe how it acts as you try to remove it.

ANALYSIS

1. Why do you think the DNA is now visible?

2. How has the DNA changed from when it was undisturbed in the onion's cells?

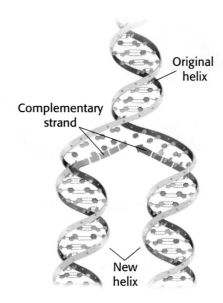

Figure 12
DNA replicates by building complementary strands on the single strands that form as the original helix unwinds.

Original helix

Complementary strand

New helix

gene

a segment of DNA that is located in a chromosome and that codes for a specific hereditary trait

DNA Replication

There is a copy of your DNA in each cell in your body, because DNA is able to replicate itself efficiently. To begin replication, a part of the double helix unwinds, providing two strands. Each strand acts as a template for the making of a new strand. New nucleic acid units made by the cell meet up one by one with their complementary bases on the template. Hydrogen bonds form between the correct base pairs: A to T, T to A, C to G, and G to C. As the nucleic acid units line up on the template strand, covalent bonds form between the sugars and phosphate groups of neighboring units or the complementary strand, as shown in **Figure 12.** Eventually, the original double helix is replaced by two perfect copies.

RNA and Protein Synthesis

Our proteins determine what our cells do. However, our DNA determines what these proteins are made of. A **gene** is a segment of DNA that has the code for the amino acid sequence to build a polypeptide. The way that the gene is translated into an amino-acid sequence is elaborate. It uses many proteins and another nucleic acid, *ribonucleic acid,* or RNA.

Protein synthesis begins with the cell making an RNA strand that codes for a specific protein. The DNA double helix unwinds and RNA units match up with the DNA bases. The process is similar to DNA replication. However, instead of using DNA units, the cell uses RNA units, which differ from DNA by an oxygen on the sugar unit and in one of the bases. RNA has the base *uracil,* shown in **Figure 13,** instead of thymine. The uracil bases hydrogen-bond with the adenine on the DNA strand, as in the following base sequence.

DNA strand: C C C C A C C C T A C G G T G
RNA strand: G G G G U G G G A U G C C A C

The cell then uses the RNA strand as instructions for building a protein. Amino acids line up according to the sequence of bases in the RNA. The polypeptide chain grows as bonds form between the amino acids.

The Genetic Code

There are 20 different amino acids but only four RNA bases. Thus, a single base cannot specify a single amino acid. In fact, a group of three, or a *triplet* of bases in RNA indicates a particular amino acid. For example, the sequence of bases GUC causes valine to be added to a growing polypeptide. The complete *genetic code* lists the RNA triplets and their corresponding amino acids. You can use **Skills Toolkit 1** to decode RNA sequences to their corresponding amino acid sequences, as shown below.

RNA strand:	GGG	GUG	GGA	UGC	CAC
amino acid:	glycine	valine	glycine	cysteine	histidine

Because there are $4^3 = 64$ triplet combinations of the four bases, most of the 20 amino acids are encoded by more than one triplet. Almost all living things use the same code to translate their proteins.

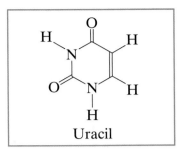

Figure 13
Uracil is a nitrogenous base that is unique to RNA. Uracil pairs with adenine.

SKILLS Toolkit 1

Using the Genetic Code

This table shows the triplet codes of RNA that specify each of the 20 amino acids. The triplets UAA, UAG, UGA, and AUG signal the end of the gene and the start of the next gene.

1. Find the first base of the RNA triplet along the left side of the table.
2. Follow that row to the right until you are beneath the second base of the triplet.
3. Move up or down in that section until you are even, on the right side of the chart, with the third base of the triplet.

The Genetic Code

First base	Second base				Third base
	U	C	A	G	
U	UUU, UUC Phenylalanine / UUA, UUG Leucine	UCU, UCC, UCA, UCG Serine	UAU, UAC Tyrosine / UAA, UAG Stop	UGU, UGC Cysteine / UGA—Stop / UGG—Tryptophan	U C A G
C	CUU, CUC, CUA, CUG Leucine	CCU, CCC, CCA, CCG Proline	CAU, CAC Histidine / CAA, CAG Glutamine	CGU, CGC, CGA, CGG Arginine	U C A G
A	AUU, AUC Isoleucine, AUA / AUG—Start	ACU, ACC, ACA, ACG Threonine	AAU, AAC Asparagine / AAA, AAG Lysine	AGU, AGC Serine / AGA, AGG Arginine	U C A G
G	GUU, GUC, GUA, GUG Valine	GCU, GCC, GCA, GCG Alanine	GAU, GAC Aspartic acid / GAA, GAG Glutamic acid	GGU, GGC, GGA, GGG Glycine	U C A G

DNA fingerprint

the pattern of bands that results when an individual's DNA sample is fragmented, replicated, and separated

Figure 14
Scientists study images called autoradiographs, which show the pattern of nitrogenous bases in the DNA of an organism.

Gene Technology

After learning the role that DNA plays in life, biological chemists have gone on to research ways of using DNA that differ from natural processes. These efforts have many benefits and promise many more to come. But at the same time, gene technology has raised fears about the possibilities of misuse or mistake, as well as ethical issues about the uniqueness and sanctity of life.

Mapping and Identifying DNA

There are thought to be about 30 000 genes in human DNA. However, genes are only a tiny part of our DNA. There are large parts of our DNA that either have no function or have functions that have not been found yet.

Both the coding and noncoding base sequences differ from person to person. Unless you have an identical twin, the chance that someone else shares your DNA pattern is next to zero. Because no one else has the same DNA as you, your DNA pattern gives a unique "fingerprint" of you and your cells. Scientists use a technique called **DNA fingerprinting** to identify where a sample of DNA comes from. In DNA fingerprinting, scientists compare *autoradiographs* of DNA samples, such as those shown in **Figure 14.** Autoradiographs are images that show the DNA's pattern of nitogenous bases.

You may have heard that DNA fingerprinting is used in forensics to prove whether a suspect can be linked to a crime. There are other applications. Two people who are closely related to each other have DNA patterns that are more similar than the DNA of two unrelated people, so DNA is useful in identifying a person's family members and tracing heredity. Likewise, because species that share a common extinct ancestor have similar DNA patterns, scientists can track presumed evolutionary links.

Identifying DNA from Small Samples

It takes a lot of DNA to make a DNA fingerprint. However, forensic applications of DNA fingerprinting can make use of a single hair, or the smallest trace of blood. Scientists can use small samples of DNA because they can rapidly copy, or "amplify," DNA strands. By making many copies of a tiny sample of DNA, a scientist can make enough DNA to see the pattern of bases.

Scientists use a method called *polymerase chain reaction,* or PCR, which replicates a short "targeted" sequence of double-stranded DNA. Large amounts of the four monomeric components of DNA are added to a solution that has the DNA, an enzyme, and *primers*. A primer is a short length of single-stranded DNA that has the complementary sequence of the first few bases of the target. The solution is then subjected to a number of heating-cooling cycles. Heating denatures the DNA and separates the double strands. Cooling causes the primer to connect to the end of the target. The enzyme then replicates the DNA using the primer as a starting point. In this way, the amount of DNA is doubled during each cooling. After 20 cycles, the amount of DNA increases by a factor of 2^{20}, or more than 1 million.

Figure 15

a Each of these identical twins has the same genetic information as her sister.

b Growers can produce many orchids by artificial cloning of the meristem tissue of a single orchid plant.

c The kitten at left is an artificial clone of an adult calico cat.

Cloning

Identical twins arise from the chance splitting of a group of embryonic cells early in the growth of a human baby. Each cell of a very young embryo can grow into a complete organism, but this ability is lost as an embryo grows larger and its cells become more specialized.

Undifferentiated cells are cells that have not yet specialized to become part of a specific tissue in the body. These cells include *stem cells* in animals and *meristem* cells in plants, which may be cultured artificially so they grow into complete organisms. These organisms are genetically identical to the organisms from which the cells were harvested and are **clones** of their "parent." Cloning a mammal is a difficult task. However, it was accomplished in 1997 by Scottish scientist Ian Wilmut. His work produced a sheep named Dolly. Dolly's genes were taken from the mammary cell of one sheep and placed in the enucleated, or empty, egg cell of another sheep. Dolly's embryo was then raised in the uterus of a third sheep. Scientists have artificially cloned many other living things—not only sheep, but plants, such as orchids, and other animals, such as the kitten shown in **Figure 15.**

clone

an organism that is produced by asexual reproduction and that is genetically identical to its parent; to make a genetic duplicate

Recombinant DNA

The greatest advances in gene technology have come from *recombinant DNA technology*. Making use of proteins that cut and reconnect DNA molecules, scientists have learned to insert genes from one species into the DNA of another. When this **recombinant DNA** is placed in a cell, the cell is able to make the protein coded by the foreign gene.

recombinant DNA

DNA molecules that are artificially created by combining DNA from different sources

The earliest success was in redesigning the DNA of bacteria to make human insulin, a protein that people with diabetes lack. Many proteins can be made in this way, and drug companies are rapidly finding ways to cure diseases and make life-saving drugs using recombinant DNA.

Bacteria are not the only living things that have been treated with recombinant DNA. Plants have been made more resistant to insects and frost damage. Spiders do not make large quantities of spider silk proteins, which may be used as strong building materials, so genetically changed goats with spider genes make milk that has these potentially useful proteins. This very active scientific field has grown much since the late 1900s.

Though genetically changed organisms offer new solutions to many difficult problems, many people worry about the drawbacks of using such technologies. For example, a genetically changed organism may thrive so well in an ecosystem that natural organisms cannot compete and are wiped out. Also, some people object to products that come from recombinant DNA because of ethical issues about the creation of new life forms for human use.

③ Section Review

UNDERSTANDING KEY IDEAS

1. From what three components is DNA made?

2. Describe the three-dimensional shape of DNA.

3. Describe how DNA uses the genetic code to control the synthesis of proteins.

4. Why is a very small trace of blood enough for DNA fingerprinting?

5. What was the first protein to be made commercially by recombinant DNA technology?

PRACTICE PROBLEMS

6. For what sequence of amino acids does the RNA base sequence AUGAAGUUUG-GCUAA code?

7. A segment of a DNA strand has the base sequence ACGTTGGCT.

   a. What is the base sequence in a complementary strand of RNA?

   b. What is the corresponding amino acid sequence?

   c. What is the base sequence in a complementary strand of DNA?

CRITICAL THINKING

8. Why might identical twins be called clones?

9. What features of the four base pairs make them ideal for holding DNA strands together?

10. Is it possible to specify the 20 amino acids using only two base pairs as the code? Explain.

SCIENCE AND TECHNOLOGY

Protease Inhibitors

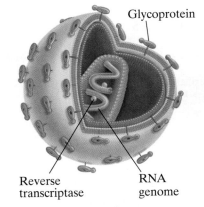

Glycoprotein

Reverse transcriptase

RNA genome

HIV Virus

HIV, or human immunodeficiency virus, is the virus that causes AIDS by severely weakening the human immune system. Since the discovery of HIV in 1983, scientists have searched for drugs that will combat the growth of the virus in human cells. Protease inhibitors are one of the newest classes of drugs to be developed.

Viruses are not living cells. They are bits of genetic material (RNA or DNA) combined with protein molecules. Viruses enter (infect) cells and release their genetic material. The cell uses this genetic material as a code to make more viruses.

The HIV virus is a *retrovirus*, a virus that contains RNA, which it carries into the cell along with an enzyme called *reverse transcriptase*. The HIV virus uses the reverse transcriptase enzyme to make a DNA copy of the RNA genetic pattern. The DNA segment enters the cell's nucleus, where it becomes a part of the cell's genes. There, it causes the cell to make all of the parts needed to make new viruses. The new viruses assemble and leave the cell to infect new cells. The cell is usually destroyed in the process.

Inhibiting Viral Reproduction

Most of the drugs that have been used to treat HIV infections are compounds that inhibit the reverse transcriptase enzyme, in turn preventing the RNA from forming a DNA copy. The new drugs, protease inhibitors, do their work after the parts of the virus have been made. The polypeptides that are needed to put together new viruses must be cut apart into the individual proteins. Protease is an enzyme that breaks the polypeptides in the right places. Inhibiting protease keeps many of the new viruses from forming.

Questions

1. Research to find out more about HIV. Identify the kind of cells the virus attacks, and describe how the viral infection leads to AIDS.
2. Find out more about other retroviruses. How are drugs used to combat infections caused by these retroviruses?

CAREER APPLICATION

Nurse Practitioner

A nurse practitioner does all of the things that registered nurses in hospitals or physicians' offices do. In fact, most nurse practitioners (NPs) begin as nurses and, after a few years of experience, study to become a nurse practitioner. Nurse practitioners have some of the same responsibilities as physicians. NPs can do extensive diagnoses of disease, carry out medical tests, counsel families, and in some cases, prescribe medicine. They often have specialties, such as pediatrics, mental health, or geriatrics. For some families, the NP is the primary health care provider.

internet connect

www.scilinks.org
Topic: Protease Inhibitors
SciLinks code: HW4102

SC*i*LINKS Maintained by the National Science Teachers Association

SECTION

4

Energy in Living Systems

KEY TERMS

- **photosynthesis**
- **respiration**
- **ATP**

OBJECTIVES

① **Explain** how plants use photosynthesis to gather energy.

② **Explain** how plants and animals use energy from respiration to carry out biological functions.

www.scilinks.org
Topic: Biochemical Processes
SciLinks code: HW4022

SC*LINKS.* Maintained by the National Science Teachers Association

photosynthesis

the process by which plants, algae, and some bacteria use sunlight, carbon dioxide, and water to produce carbohydrates and oxygen

Obtaining Energy

Moving our muscles is one way in which we use energy, but many other ways that we use energy are harder to see. We use energy in digesting our food, in pumping our blood, in keeping warm, and in making the many compounds that our bodies need to function and grow. Energy is needed for every action of every organ in our bodies. All living things need energy to build and repair themselves and to fuel their activities. With rare exceptions, all forms of life on Earth draw energy ultimately from sunlight. Green plants get energy directly from the sun's rays through the process of **photosynthesis.** Other living things rely on plants, directly or indirectly, as their source of energy.

The flow of energy throughout an ecosystem is related to the carbon cycle. The carbon cycle follows carbon atoms as they become part of one compound and then another. The reactions that involve these carbon compounds, shown in **Figure 16,** give plants and animals the energy that they need.

Figure 16
Plants use carbon dioxide, water, and sunlight to produce oxygen and glucose. Glucose is used by plants and animals to produce chemical energy in the form of a substance called *ATP*.

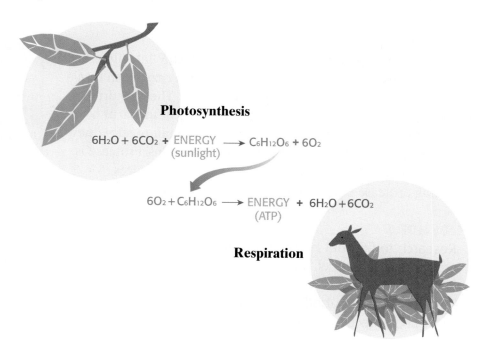

Photosynthesis

$$6H_2O + 6CO_2 + \text{ENERGY (sunlight)} \longrightarrow C_6H_{12}O_6 + 6O_2$$

$$6O_2 + C_6H_{12}O_6 \longrightarrow \text{ENERGY (ATP)} + 6H_2O + 6CO_2$$

Respiration

Plants Use Photosynthesis as a Source of Carbohydrates

Look at the diagram of the carbon cycle shown in **Figure 16.** Notice that the reactants needed for the second equation—glucose and oxygen—are produced in the first equation, photosynthesis. And the reactants and conditions needed for the first equation—carbon dioxide, water, and energy—are produced in the second equation, although the energy is in a different form.

Most plants use *chlorophyll,* a magnesium-containing organic molecule, to capture the energy of sunlight. The light absorbed by the chlorophyll is mostly from the red and the blue regions of the visible spectrum. What is reflected is green light, from the central region of the spectrum, which accounts for the color of most plants.

The overall chemistry of photosynthesis, which takes place in green plants and many other living things, such as the algae in **Figure 17,** is described by the following endothermic equation.

$$6\ O{=}C{=}O\ +\ 6\ H_2O\ \xrightarrow{\text{light}}\ C_6H_{12}O_6\ +\ 6\ O{=}O$$

carbon dioxide   **water**   **glucose**   **oxygen**

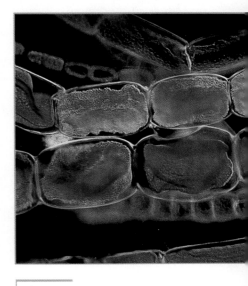

Figure 17
Green plants and algae have chlorophyll, a multiringed compound that contains magnesium.

🖸 internet connect

www.scilinks.org
Topic: Photosynthesis
SciLinks code: HW4096

SC*i*LINKS Maintained by the National Science Teachers Association

Animals Consume Carbohydrates as a Source of Energy

The carbohydrates that plants make by photosynthesis are used as a source of energy, not only by plants themselves, but also by animals.

Both plants and animals need carbohydrates for energy, and both plants and animals store simple carbohydrates by making them into larger carbohydrate polymers, such as starch and glycogen. Because animals cannot make carbohydrates directly from the sun's energy as plants do, animals eat plants or other animals to obtain the carbohydrates that plants have made. **Figure 18** shows one way that we get plant carbohydrates. Once an animal eats a plant, it breaks the plant's larger carbohydrates down into simpler carbohydrates, such as glucose. Glucose, which is soluble in blood, can be carried to the rest of the body for energy use.

Figure 18
Carbohydrates, such as the starch found in this baked potato, are the main energy source for most humans.

Using Energy

Glucose itself is changed into a more readily available source of energy through *respiration*. The equation for chemical respiration is shown in **Figure 16**. In everyday speech, *respiration* means getting gases into and out of the lungs. In biological chemistry, **respiration** refers to the entire process of getting oxygen into body tissues and allowing it to react with glucose to generate energy.

Respiration Requires Oxygen and Glucose

You may have noticed that you breathe more heavily when you exercise, as does the runner in **Figure 19**. This is because you need to get more oxygen into your system and you need to remove carbon dioxide more rapidly from your system.

The lungs move oxygen from the air into the blood as oxygen-carrying hemoglobin. The lungs also move carbon dioxide out of the blood—where it is present as $HCO_3^-(aq)$ and $H_2CO_3(aq)$—and into the air. The bloodstream carries oxygen and glucose to all the cells of your body for respiration. The bloodstream must also remove the products of respiration. That is, it takes carbon dioxide to the lungs, and it takes water to the kidneys.

Chemical respiration, or *cellular respiration,* takes place in the cells of a plant or animal and is fueled by glucose and oxygen. The overall process is the opposite of the photosynthesis reaction, as shown in the following equation.

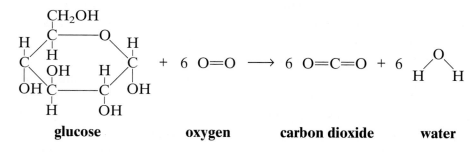

glucose   oxygen   carbon dioxide   water

For every molecule of glucose that is broken down by respiration, six molecules of oxygen, O_2, are consumed. The overall process produces six molecules of carbon dioxide, CO_2, and six molecules of water.

Respiration Is Exothermic

While photosynthesis takes in energy, respiration gives off energy. The thermodynamic values for the equation below show that the reaction is very exothermic ($\Delta H = -2803$ kJ) and highly spontaneous ($\Delta G = -2880$ kJ).

$$C_6H_{12}O_6(s) + 6O_2(g) \longrightarrow 6CO_2(g) + 6H_2O(l)$$

However, the goal of cellular respiration is not to liberate energy as heat or light but to produce chemical energy in the form of special polyatomic ions, as discussed next.

Figure 19
Muscular activity leads to an increase in respiration rate.

respiration

the process by which cells produce energy from carbohydrates; atmospheric oxygen combines with glucose to form water and carbon dioxide

☐ internet connect

www.scilinks.org
Topic: Respiration
SciLinks code: HW4111

SC/LINKS. Maintained by the National Science Teachers Association

The hydrolysis of ATP to ADP reaction with structures of ATP and ADP shown, and the hydrolysis equation below.

Adenosine triphosphate (ATP)   **Adenosine diphosphate (ADP)**

$$H_2O + ATP \longrightarrow {}^-O-\overset{\displaystyle O}{\underset{\displaystyle OH}{P}}-OH + ADP$$

Adenosine Triphosphate and Adenosine Diphosphate

Adenosine triphosphate, **ATP,** and adenosine diphosphate, ADP, are the high-energy and low-energy forms of a chemical that acts as energy "cash" in biological systems. The structures of ATP and ADP are shown in **Figure 20.** The main structural difference between them is that ATP has an extra phosphate group, $-PO_3^-$.

The hydrolysis of ATP to ADP is exothermic ($\Delta H = -20$ kJ) and spontaneous, as the following equation shows.

$$ATP(aq) + H_2O(l) \longrightarrow ADP(aq) + H_2PO_4^-(aq) \qquad \Delta G = -31 \text{ kJ}$$

Many reactions in a cell would not take place spontaneously if left alone. These reactions can "use" the spontaneity of ATP hydrolysis to take place by coupling with the ATP $\longrightarrow$ ADP reaction. ATP hydrolysis thus allows these other nonspontaneous reactions to take place.

The Two Stages of Cellular Respiration

Cellular respiration has two stages. Both stages produce ATP. The first stage of cellular respiration includes *glycolysis.* The name means "glucose-splitting," which makes sense because the six-carbon glucose is split into two molecules of pyruvic acid, $CH_3COCOOH$ or $C_3H_4O_3$. The glycolysis reaction has about a dozen steps. Other products react further to make more ATP. The net gain of eight ATP is shown in the following equation.

$$C_6H_{12}O_6 + O_2 + 8ADP + 8H_2PO_4^- \longrightarrow 2C_3H_4O_3 + 8ATP + 10H_2O$$

The second stage of cellular respiration, called the *Krebs cycle,* also has several steps. The overall result is the oxidation of pyruvic acid to form CO_2, as shown in the following equation.

$$2C_3H_4O_3 + 5O_2 + 30ADP + 30H_2PO_4^- \longrightarrow 6CO_2 + 30ATP + 34H_2O$$

The two stages together produce 38 ATP ions per glucose molecule. The reaction shown for glucose has an enthalpy change of -2803 kJ. Thus, $(38 \times -20 \text{ kJ})/(-2803 \text{ kJ})$ or 27% of the energy of glucose can be stored as ATP. The remaining energy helps to keep the body warm.

Figure 20
The hydrolysis of ATP produces ADP and releases energy.

ATP

adenosine **t**ri**p**hosphate, an organic molecule that acts as the main energy source for cell processes; composed of a nitrogenous base, a sugar, and three phosphate groups

internet connect

www.scilinks.org
Topic: ATP
SciLinks code: HW4018

SCiLINKS. Maintained by the National Science Teachers Association

Table 4	Approximate "Cost" of Daily Activities	
Activity (for 30 min)	Energy required (kJ)	ATP required (mol)
Running	1120	56
Swimming	840	42
Bicycling	1400	70
Walking	560	28

ATP Is Energy Currency

The conversion of ATP to ADP gives the energy needed for many cellular activities. So, ATP represents energy that is immediately available in the cell. Also, ATP is continuously resynthesized by cellular respiration as long as an organism is alive.

On the molecular level, there are three kinds of work that a cell does, and ATP gives the energy needed for them all. ATP gives the energy needed for *synthetic work,* making compounds that do not form spontaneously because they are accompanied by a positive ΔG. By coupling the reaction to the ATP $\longrightarrow$ ADP conversion, the overall process becomes spontaneous. ATP also gives the energy needed for *mechanical work.* The ATP $\longrightarrow$ ADP conversion changes the shape of muscle cells, which allows muscles to flex and move. Finally, the ATP $\longrightarrow$ ADP conversion fuels *transport work,* carrying solutes across a membrane. Again, the ATP $\longrightarrow$ ADP conversion is harnessed to allow specific proteins in the membrane to pump ions into or out of the cell. **Table 4** shows just how much ATP is needed for some daily activities.

Section Review

UNDERSTANDING KEY IDEAS

1. What are two reactions that involve carbon and together give plants energy?

2. Write the chemical reaction representing the photosynthesis of glucose.

3. What role does the bloodstream play in respiration?

4. Write the net equation for the reaction that makes ATP and pyruvic acid from glucose during cellular respiration.

5. Briefly, what biological role is played by ATP?

CRITICAL THINKING

6. To some small extent, plants make some ATP during photosynthesis. Why can't plants use photosynthesis as an energy source all the time instead of making carbohydrates?

7. In what sense is it true to say that sunlight fills the energy needs of a cheetah?

8. Explain the roles of glycogen, glucose, and ATP as energy sources in animals.

9. For chemical reactions, Gibbs energy is a more important quantity than enthalpy. Show that the efficiency of the glucose to ATP conversion in terms of ΔG is only 41%.

10. Explain how nonspontaneous biochemical reactions can take place with the help of ATP.

12
Mg
Magnesium
24.3050
[Ne]$3s^2$

Element Spotlight

Magnesium: An Unlimited Resource

Extracting magnesium from sea water is an efficient and economical process. Sea water is mixed with lime, CaO, from oyster shells to form insoluble magnesium hydroxide, $Mg(OH)_2$, which can be easily filtered out. Hydrochloric acid is added to the solid to form magnesium chloride. The electrolysis of molten magnesium chloride will produce pure magnesium metal.

Industrial Uses

- Magnesium oxide, MgO, is used in paper manufacturing, as well as in fertilizers, medicine, and household cleaners.

- Aqueous magnesium hydroxide, $Mg(OH)_2$, is known as *milk of magnesia,* an antacid.

- Magnesium alloys are used in aircraft fuselages, engine parts, missiles, luggage, optical and photo equipment, lawn mowers, and portable tools.

Spinach is a good source of dietary magnesium. Magnesium is the central atom in the green plant pigment chlorophyll.

Real-World Connection If 90 million metric tons of magnesium were extracted per year for 1 million years, the magnesium content of the oceans would drop by 0.01%.

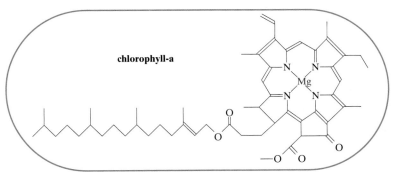

chlorophyll-a

A Brief History

1808: Humphry Davy discovers that the compound magnesia alba is the oxide of a new metal.

1828: A.A.B. Bussy obtains the first pure magnesium metal.

1944: L. M. Pidgeon discovers how to extract magnesium from its ore, dolomite.

1700	1800	1900

1833: Michael Faraday makes magnesium metal through the electrolysis of molten magnesium chloride.

1852: Robert Bunsen designs an electrolytic cell that allows molten Mg to be collected without burning when it makes contact with the air.

Questions

1. Find out more about chlorophyll. How is chlorophyll's structure important to its role in photosynthesis?

2. Magnesium is used to make fireworks. Find out what property makes this substance useful in fireworks.

internet connect

www.scilinks.org
Topic: Magnesium
SciLinks code: HW4077

SciLINKS Maintained by the National Science Teachers Association

KEY IDEAS

SECTION ONE Carbohydrates and Lipids

- Carbohydrates are compounds of carbon, hydrogen, and oxygen made by living things for energy storage and support. They can be ringed and have many −OH groups.
- Carbohydrates are classified into monosaccharides, disaccharides, or polysaccharides according to the number of rings present. The smaller carbohydrates are called *sugars*.
- Sugars combine by condensation, a reaction in which a water molecule is formed. The reverse reaction, hydrolysis, breaks down polysaccharides into smaller carbohydrate units.
- Lipids are nonpolar molecules that include fats, phospholipids, steroids, and waxes.

SECTION TWO Proteins

- The 20 amino acids from which proteins are formed all have the formula H_2N-CHR-COOH. They differ in the identity of R, which stands for different side chains.
- Proteins are formed by condensation of amino acids.
- The form and function of a protein depends on its three-dimensional shape, which itself depends on the amino acid sequence in the polypeptide chain.

SECTION THREE Nucleic Acids

- Nucleic acids are made of -phosphate-sugar-phosphate-sugar- chains with nitrogenous bases connected to the sugar units.
- DNA uses four bases and forms a double helix by specific A−T and G−C pairing. Replication can take place only when the helix splits apart.
- The arrangement of base triplets on DNA encodes genetic information by dictating the synthesis of proteins.
- Gene technologies involve working with DNA and include DNA fingerprinting, cloning, and recombinant DNA.

SECTION FOUR Energy in Living Systems

- Green plants use solar energy, carbon dioxide, and water to synthesize glucose during photosynthesis.
- The reverse of photosynthesis is respiration, in which glucose is broken down into carbon dioxide and water. Energy is harvested in the process by the production of about 38 ATP ions per glucose molecule.
- Through the release of energy during the breaking of its third phosphate bond, ATP fuels life's processes: motion, synthesis, and transport.

KEY TERMS

carbohydrate
monosaccharide
disaccharide
polysaccharide
condensation reaction
hydrolysis
lipid

protein
amino acid
polypeptide
peptide bond
enzyme
denature

nucleic acid
DNA
gene
DNA fingerprint
clone
recombinant DNA

photosynthesis
respiration
ATP

KEY SKILLS

Using the Genetic Code
Skills Toolkit 1 p. 729

CHAPTER REVIEW 20

USING KEY TERMS

1. What do all carbohydrates have in common?

2. How are carbohydrates classified?

3. What type of reaction changes sugars into polysaccharides?

4. Where are peptide bonds found?

5. Describe and name the four different levels of protein structure.

6. What is an enzyme, and what is a proteolytic enzyme?

7. Felicia adds vinegar (acetic acid) to milk, which causes casein (a protein in the milk) to curdle. The casein has the same amino acid sequence before and after the vinegar is added, so Felicia knows that no new substances have formed. How might she explain what has happened to the casein?

8. Contrast the terms nucleic acid, DNA, and gene.

9. Name three examples of gene technologies.

10. Describe how green plants use sunlight.

11. What does *respiration* mean in everyday language, and what larger meaning does it have in biological chemistry?

12. Describe the role of ATP and what the name *ATP* stands for.

UNDERSTANDING KEY IDEAS

Carbohydrates and Lipids

13. Name two examples from each of the following classes of carbohydrate: monosaccharide, disaccharide, and polysaccharide.

14. What different roles do the polysaccharides starch and cellulose play in plant systems?

15. Identify the reaction that changes sugars into polysaccharides and the reaction that changes polysaccharides into sugars.

16. Both cholesterol and oleic acid are lipids. What property do they have in common?

Proteins

17. List the four groups attached to the central carbon of an amino acid.

18. What are the products of protein synthesis?

19. Explain the cause of genetic disease sickle cell anemia.

20. Describe five properties that particular amino-acid side-chain groups give to the polypeptide formed by their condensation.

21. How is a disulfide bridge formed?

22. What is the *lock-and-key* model of enzyme action?

Nucleic Acids

23. Describe the structure of a DNA molecule and what the name *DNA* stands for.

24. In DNA replication, why is a G on the original strand partnered by a C on the complementary strand, and not by an A, a T, or a G?

25. What is the genetic code? Give an example of how it is used.

26. What is recombinant DNA technology?

27. What was significant about the sheep Dolly?

28. Describe the procedure for DNA amplification by polymerase chain reaction (PCR).

29. Identify the specialized molecule that absorbs light in photosynthesis.

30. Write the balanced chemical equation that describes the overall process in photosynthesis.

31. Explain why plants are generally green.

32. What is glycolysis?

33. How is ATP produced?

34. How are living things able to respond immediately to energy-demanding situations?

PRACTICE PROBLEMS

PROBLEM SOLVING SKILL

Skills Toolkit 1 Using the Genetic Code

35. What sequence of amino acids do the following RNA base sequences code?
  a. AAG AUU GGA CAC
  b. AUG UCU UCG AGU UCA UAG
  c. AAA GGG CCC UUU

36. A segment of a DNA strand has the base sequence TACACACGTTGGATT.
  a. What is the base sequence in a complementary strand of RNA?
  b. What is the corresponding amino acid sequence?
  c. What is the base sequence in a complementary strand of DNA?

37. a. Write one possible RNA sequence that codes for the following amino acids: aspartic acid-glutamine-tryptophan.
  b. What is the sequence in a complementary strand of DNA?

MIXED REVIEW

38. Imagine that you have created a very short polypeptide from the following RNA sequence: GACGAAGGAGAG.
  a. What is the amino acid sequence of the polypeptide?
  b. What property does the polypeptide have?

39. Write balanced chemical equations to describe the following metabolic processes: (a) starch $\longrightarrow$ glucose; (b) glucose $\longrightarrow$ carbon dioxide; (c) ATP $\longrightarrow$ ADP.

40. A nucleic acid has the following sequence: UCAUCGUGGAACUUG.
  a. Is the nucleic acid RNA or DNA?
  b. How can you tell?
  c. For what amino acid sequence does this nucleic acid code?

41. Hydrolysis of the disaccharide lactose produces glucose and galactose. What type of carbohydrate is galactose?

42. Identify each of the following structures as a carbohydrate, an amino acid, or a nitrogenous base.
  a.

  b.

  c.

CRITICAL THINKING

43. Explain how a similar reaction forms three kinds of biological polymers: polysaccharides, polypeptides, and nucleic acids.

44. How is it possible to denature a protein without breaking the polypeptide chain?

45. Why is a special molecule, hemoglobin, needed to move oxygen *from* the lungs, while no molecule is needed to move carbon dioxide *to* the lungs?

A

B

46. Study the image above. Identify whether respiration or photosynthesis take place in organisms A and B. Explain your answer.

47. Compare the advantages and disadvantages of DNA fingerprinting compared with literal fingerprints as a forensic tool.

48. A lab technician sweeps his hair back while he prepares a sample for polymerase chain reaction (PCR). Later, the DNA fingerprinting tests of the sample indicate that the lab technician was at the scene of the crime. What other explanation is there for the results of the DNA test?

49. How can spider silk proteins be present in the milk of goats?

50. Explain how all of the following statements can be true: "Many plants use *starch* to provide energy"; "Energy is supplied by *glucose* in both plants and animals"; and "*ATP* is the energy source in all living cells."

ALTERNATIVE ASSESSMENT

51. News reports about gene technology are sometimes one-sided, stressing the advantages but ignoring the dangers, or vice versa. Find such a report and write "the other side of the story."

52. Research to find out more about the structure of phospholipids and the properties that make them ideal for the construction of cell membranes.

CONCEPT MAPPING

53. Use the following terms to complete the concept map below: *DNA, polypeptides, amino acids, nucleic acids,* and *carbohydrates.*

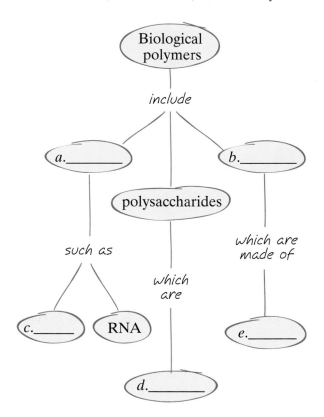

FOCUS ON GRAPHING

Study the graph below, and answer the questions that follow.
For help in interpreting graphs, see Appendix B, "Study Skills for Chemistry."

54. What characteristic of the two proteins in this bar graph is being compared?

55. The two proteins compared are α-keratin in wool and fibroin in spider silk. Which color represents the protein found in wool?

56. According to the graph, what is significant about spider fibroin protein?

57. What are the mole percentages of alanine in α-keratin and fibroin?

58. Why do you think the mole percentages of all of the amino acids are not shown?

59. Spider fibroin protein is a much stronger material than α-keratin in wool. Violet would like to create a strong protein for manufacturing fishing line. What amino acids might she decide to use to build the protein? Use the graph to support your answer.

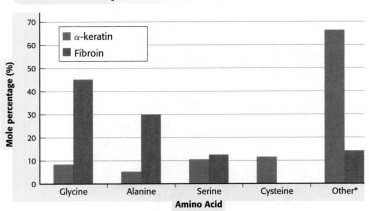

Amino Acid Composition of Proteins in Wool and Silk

TECHNOLOGY AND LEARNING

60. Graphing Calculator

Polypeptides and Amino Acids

Go to Appendix C. If you are using a TI-83 Plus, you can download the program **PEPTIDE** and run the application as directed. If you are using another calculator, your teacher will provide you with keystrokes to use. There are 20 amino acids that occur in proteins found in nature. The program will prompt you to input a number of amino acids. After you do, press **ENTER.** The program will respond with the number of different straight-chain polypeptides possible given that number of amino acid units.

a. Aspartame is an artificial sweetener that is a dipeptide, a protein made of two amino acids. How many possible dipeptides are there?

b. Enkephalins produced in the brain serve to help the body deal with pain. Several of them are pentapeptides. That is, they are polypeptides made of five amino acids. How many different pentapeptides are there?

c. The calculator uses the following equation:

number of polypeptides = $20^{\text{(number of amino acids)}}$

This equation can also be expressed as:

number of polypeptides =
$(2^{\text{(number of amino acids)}})(10^{\text{(number of amino acids)}})$

Given this equation, estimate how many possible polypeptides there are that are made of 100 amino acids? (Hint: The answer is too large for your calculator. However, you can use the graphing calculator to find the value of $2^{\text{(number of amino acids)}}$.)

1. _____ are compounds used by plants for energy storage and structural support.
a. Proteins
b. Disaccharides
c. Polypeptides
d. Polysaccharides

2. The mechanism of enzyme catalysis requires the substrate to bind
a. to DNA.
b. to an active site.
c. through a disulfide bridge.
d. permanently.

3. In forming a DNA double helix, _____ links with thymine.
a. adenine
b. cytosine
c. guanine
d. thymine

4. In the following list, _____ is _not_ an amino acid.
a. glutamic acid
b. histidine
c. nucleic acid
d. cysteine

5. The _____ group is present in all amino acids.
a. –COOH
b. –OH
c. –CO—NH—
d. –SH

6. The _____ bond is never employed in establishing the tertiary structure of proteins.
a. covalent
b. ionic
c. metallic
d. hydrogen

7. Glycolysis produces
a. glucose.
b. ethanol.
c. pyruvic acid.
d. carbon dioxide.

8. In the genetic code, groups of _____ bases specify an amino acid.
a. 3
b. 4
c. 20
d. 38

9. The products of photosynthesis that are the reactants of respiration are
a. carbon dioxide and water.
b. glucose and oxygen.
c. oxygen and water.
d. carbon dioxide and pyruvic acid.

10. In forensic work, polymerase chain reaction is used to
a. compare the DNA of a suspect with a crime-scene sample.
b. test if a DNA sample is of human origin.
c. establish the innocence of the victim.
d. increase the amount of DNA available.

11. _____ is not a protein.
a. Keratin
b. Trypsin
c. Chitin
d. Fibroin

12. _____ is a carbohydrate and a sugar, but not a disaccharide.
a. Fructose
b. Lactose
c. Maltose
d. Sucrose

LABORATORY PROGRAM

Skills Practice Labs and Inquiry Labs

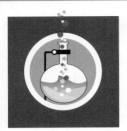

Working in the World of a Chemist

Meeting Today's Challenges

Even though you have already taken science classes with lab work, you will find the two types of laboratory experiments in this book organized differently from those you have done before. The first type of lab is called a Skills Practice Lab. Each Skills Practice Lab helps you gain skills in lab techniques that you will use to solve a real problem presented in the second type of lab, which is called an Investigation. The Skills Practice Lab serves as a Technique Builder, and the Investigation is presented as an exercise in Problem Solving.

Both types of labs refer to you as an employee of a professional company, and your teacher has the role of supervisor. Lab situations are given for real-life circumstances to show how chemistry fits into the world outside of the classroom. This will give you valuable practice with skills that you can use in chemistry and in other careers, such as creating a plan with available resources, developing and following a budget, and writing business letters.

As you work in these labs, you will better understand how the concepts you studied in the chapters are used by chemists to solve problems that affect life for everyone.

Skills Practice Labs

The Skills Practice Labs provide step-by-step procedures for you to follow, encouraging you to make careful observations and interpretations as you progress through the lab session. Each Skills Practice Lab gives you an opportunity to practice and perfect a specific lab technique or concept that will be needed later in an Investigation.

Skills Practice Lab

What Should You Do Before a Skills Practice Lab?

Preparation will help you work safely and efficiently. The evening before a lab, be sure to do the following:

◆ Read the lab procedure to make sure you understand what you will do.

◆ Read the safety information that begins on page 751, as well as any safety information provided in the lab procedure itself.

◆ Write down any questions you have in your lab notebook so that you can ask your teacher about them before the lab begins.

◆ Prepare all necessary data tables so that you will be able to concentrate on your work when you are in the lab.

What Should You Do After a Skills Practice Lab?

Most teachers require a lab report as a way of making sure that you understand what you are doing. Your teacher will give you specific details about how to organize your lab reports, but most lab reports will include the following:

◆ title of the lab

- summary paragraph(s) describing the purpose and procedure

- data tables and observations that are organized and comprehensive

- worked-out calculations with proper units

- answers that are, boxed, circled, or highlighted for items in the Analysis and Interpretation, Conclusion, and Extensions sections

Inquiry Labs

Inquiry Labs
Design Your Own Experiment

The Inquiry Labs differ from Skills Practice Labs because they do not provide step-by-step instructions. In each Inquiry Lab, you are required to develop your own procedure to solve a problem presented to your company by a client. You must decide how much money to spend on the project and what equipment to use. Although this may seem difficult. Inquiry Labs contain a number of clues about how to successfully solve the problem.

What Should You Do Before an Inquiry Lab?

Before you will be allowed to work on the lab, you must turn in a preliminary report. Usually, you must describe in detail the procedure you plan to use, provide complete data tables for the data and observations you will collect, and list exactly what equipment you will need and the costs. Only after your teacher, acting as your supervisor, approves your plans are you allowed to proceed. Before you begin writing a preliminary report, follow these steps.

- Read the Inquiry Lab thoroughly, and search for clues.

- Jot down notes in your lab notebook as you find clues.

- Consider what you must measure or observe to solve the problem.

- Think about Skills Practice Labs you have done that used a similar technique or reaction.

- Imagine working through a procedure, keeping track of each step, and determining what equipment you need.

- Carefully consider whether your approach is the best, most efficient one.

What Should You Do After an Inquiry Lab?

After you finish, organize a report of your data as described in the Memorandum. This is usually in the form of a one- or two-page letter to the client. Your teacher may have additional requirements for your report. Carefully consider how to convey the information the client needs to know. In some cases, a graph or diagram can communicate information better than words can.

If you need help with graphing or with using significant figures, ask your teacher.

Materials List for Inquiry Labs

Refer to the Equipment and Chemical lists below when planning your procedure for the Inquiry Labs. Include in your budget only the items you will need to solve the problem presented to your company by the client. Remember, you must always include the cost of lab space and the standard disposal fee in your budget.

Equipment

Aluminum foil

Balance

Beaker, 250 mL

Beaker, 400 mL

Beaker tongs

Büchner funnel

Bunsen burner/related equipment

Buret

Cobalt glass plate

Crucible and cover

Crucible tongs

Desiccator

Drying oven

Erlenmeyer flask, 250 mL

Evaporating dish

Filter flask with sink attachment

Filter paper

Flame-test wire

Glass funnel

Glass plate

Glass stirring rod

Graduated cylinder,100 mL

Hot plate

Index card (3 in. x 5 in.)

Lab space/fume hood/utilities

Litmus paper

Magnetic stirrer

Mortar and pestle

Paper clips

pH meter

Equipment (continued)

Plastic bags

Ring stand/ring/wiregauze or pipestem triangle

Ring stand with buretclamp

Rubber policeman

Spatula

Spectroscope

Standard disposal fee

Stopwatch

6 test tubes/holder/rack

Thermistor probe

Thermometer

Wash bottle

Watch glass

Weighing paper

Reagents and Additional Materials

Ring stand with buret clamp

Rubber policeman

Spatula

Spectroscope

Standard disposal fee

Stopwatch

Ring stand with buretclamp

Rubber policeman

Spatula

6 test tubes/holder/rack

Thermistor probe

Thermometer

Wash bottle

Watch glass

Weighing paper

Safety in the Chemistry Laboratory

Any chemical can be dangerous if it is misused. Always follow the instructions for the experiment. Pay close attention to the safety notes. Do not do anything differently unless you are instructed to do so by your teacher.

Chemicals, even water, can cause harm. The challenge is to know how to use chemicals correctly. If you follow the rules stated below, pay attention to your teacher's directions, and follow the precautions on chemical labels and in the experiments, then you will be using chemicals correctly.

These Safety Precautions Always Apply in the Lab

1. Always wear a lab apron and safety goggles.

Laboratories contain chemicals that can damage your clothing even if you aren't working on an experiment at the time. Keep the apron strings tied.

Some chemicals can cause eye damage and even blindness. If your safety goggles are uncomfortable or if they cloud up, ask your teacher for help. Try lengthening the strap, washing the goggles with soap and warm water, or using an anti-fog spray.

2. Do not wear contact lenses in the lab.

Even if you wear safety goggles, chemicals can get between contact lenses and your eyes and cause irreparable eye damage. If your

doctor requires you to wear contact lenses instead of glasses, then you should wear eye-cup safety goggles in the lab. Ask your doctor or your teacher how to use this very important and special eye protection.

3. NEVER WORK ALONE IN THE LABORATORY.

Do lab work only under the supervision of your teacher.

4. Wear the right clothing for lab work.

Necklaces, neckties, dangling jewelry, long hair, and loose clothing can knock things over or catch on fire. Tuck in neckties, or take them off. Do not wear a necklace or other dangling jewelry, including hanging earrings. It also might be a good idea to remove your wristwatch so that it is not damaged by a chemical splash.

Pull back long hair, and tie it in place. Wear cotton clothing if you can. Nylon and polyester fabrics burn and melt more readily than cotton does. It's best to wear fitted garments, but if your clothing is loose or baggy, tuck it in or tie it back so that it does not get in the way or catch on fire. It is also important to wear pants, not shorts or skirts.

Wear shoes that will protect your feet from chemical spills. Do not wear open-toed shoes or sandals or shoes with woven leather straps. Shoes made of solid leather or polymer are preferred over shoes made of cloth.

5. Only books and notebooks needed for the experiment should be in the lab.

Do not bring textbooks, purses, bookbags, backpacks, or other items into the lab; keep these things in your desk or locker.

6. Read the entire experiment before entering the lab.

Memorize the safety precautions. Be familiar with the instructions for the experiment. Only materials and equipment authorized by your teacher should be used. When you do your lab work, follow the instructions and safety precautions described in the experiment.

7. Read chemical labels.

Follow the instructions and safety precautions stated on the labels.

8. Walk with care in the lab.

Sometimes you will have to carry chemicals from the supply station to your lab station. Avoid bumping into other students and spilling the chemicals. Stay at your lab station at other times.

9. Food, beverages, chewing gum, cosmetics, and smoking are NEVER allowed in the lab.

(You should already know this.)

10. NEVER taste chemicals or touch them with your bare hands.

Keep your hands away from your face and mouth while working, even if you are wearing gloves.

11. Use a sparker to light a Bunsen burner.

Do not use matches. Be sure that all gas valves are turned off and that all hot plates are turned off and unplugged when you leave the lab.

12. Be careful with hot plates, Bunsen burners, and other heat sources.

Keep your body and clothing away from flames. Do not touch a hot plate after it has just been turned off because it is probably still hot. The same is true of glassware, crucibles, and other things that have been removed from the flame of a Bunsen burner or from a drying oven.

13. Do not use electrical equipment with frayed or twisted wires.

14. Be sure your hands are dry before you use electrical equipment.

Before plugging an electrical cord into a socket, be sure the equipment is turned off. When you are finished with the equipment, turn it off. Before you leave the lab, unplug the equipment, but be sure to turn it off FIRST.

15. Do not let electrical cords dangle from work stations.

Dangling cords can cause tripping or electrical shocks. The area under and around electrical equipment should be dry, and cords should not lie in puddles of spilled liquid.

16. Know fire-drill procedures and the locations of exits.

17. Know the location and operation of safety showers and eyewash stations.

18. If your clothes catch on fire, walk to the safety shower, stand under it, and turn it on.

19. If you get a chemical in your eyes, walk immediately to the eyewash station, turn it on, and lower your head so that your eyes are in the running water.

Hold your eyelids open with your thumbs and fingers, and roll your eyeballs around. Flush your eyes continuously for at least 15 minutes. Call out to your teacher as you do this.

20. If you spill anything on the floor or lab bench, call your teacher rather than trying to clean it up by yourself.

Your teacher will tell you if it is OK for you to do the cleanup; if not, your teacher will know how the spill should be cleaned up safely.

21. If you spill a chemical on your skin, wash the chemical off at the sink and call your teacher.

If you spill a solid chemical on your clothing, brush it off carefully without scattering it onto somebody else, and call your teacher. If you get liquid on your clothing, wash it off right away using the faucet at the sink, and call your teacher. If the spill is on your pants or somewhere else that will not fit under the sink faucet, use the safety shower. Remove the pants or other affected clothing while you are under the shower, and call your teacher. (It may be temporarily embarrassing to remove pants or other clothing in front of your class, but failing to flush that chemical off your skin could cause permanent damage.)

22. The best way to prevent an accident is to stop it before it happens.

If you have a close call, tell your teacher so that you and your teacher can find a way to prevent it from happening again. Otherwise, the next time, it could be a harmful accident instead of just a close call. If you get a headache, feel sick to your stomach, or feel dizzy, tell your teacher immediately.

23. All accidents, no matter how minor, should be reported to your teacher.

24. For all chemicals, take only what you need.

If you take too much and have some left over, DO NOT put it back in the bottle. If a chemical is accidently put into the wrong bottle, the next person to use it will have a contaminated sample. Ask your teacher what to do with leftover chemicals.

25. NEVER take any chemicals out of the lab.

26. Horseplay and fooling around in the lab are very dangerous.

NEVER be a clown in the laboratory.

27. Keep your work area clean and tidy.

After your work is done, clean your work area and all equipment.

28. Always wash your hands with soap and water before you leave the lab.

29. All of these rules apply all of the time you are in the lab.

CLOTHING PROTECTION

◆ Wear laboratory aprons in the laboratory. Keep the apron strings tied so that they do not dangle.

EYE SAFETY

◆ Wear safety goggles in the laboratory at all times. Know how to use the eyewash station.

CLEAN UP

◆ Keep your hands away from your face and mouth.

◆ Always wash your hands before leaving the laboratory.

CHEMICAL SAFETY

◆ Never taste, eat, or swallow any chemicals in the laboratory. Do not eat or drink any food from laboratory containers. Beakers are not cups, and evaporating dishes are not bowls.

◆ Never return unused chemicals to their original containers.

◆ It helps to label the beakers and test tubes containing chemicals. (This is not a new rule, just a good idea.)

◆ Never transfer substances by sucking on a pipet or straw; use a suction bulb.

WASTE DISPOSAL

◆ Some chemicals are harmful to our environment. You can help protect the environment by following the instructions for proper disposal.

GLASSWARE SAFETY

◆ Never place glassware, containers of chemicals, or anything else near the edges of a lab bench or table.

HAND SAFETY

◆ If a chemical gets on your skin or clothing or in your eyes, rinse it immediately, and alert your teacher.

CAUSTIC SAFETY

◆ If a chemical is spilled on the floor or lab bench, tell your teacher, but do not clean it up yourself unless your teacher says it is OK to do so.

HEATING SAFETY

◆ When heating a chemical in a test tube, always point the open end of the test tube away from yourself and other people.

Safety Quiz

Refer to the list of rules on p. 751–753, and identify whether a specific rule applies or whether the rule presented is a new rule.

1. Tie back long hair, and confine loose clothing. **(Rule ? applies)**

2. Never reach across an open flame. **(Rule ? applies)**

3. Use proper procedures when lighting Bunsen burners. Turn off hot plates, Bunsen burners, and other heat sources when they are not in use. **(Rule ? applies)**

4. Heat flasks or beakers on a ring stand with wire gauze between the glass and the flame. **(Rule ? applies)**

5. Use tongs when heating containers. Never hold or touch containers while heating them. Always allow heated materials to cool before handling them. **(Rule ? applies)**

6. Turn off gas valves when they are not in use. **(Rule ? applies)**

7. Use flammable liquids only in small amounts. **(Rule ? applies)**

8. When working with flammable liquids, be sure that no one else is using a lit Bunsen burner or plans to use one. **(Rule ? applies)**

9. What additional rules apply to every lab? **(Rule ? applies)**

10. Check the condition of glassware before and after using it. Inform your teacher of any broken, chipped, or cracked glassware because it should not be used. **(Rule ? applies)**

11. Do not pick up broken glass with your bare hands. Place broken glass in a specially designated disposal container. **(Rule ? applies)**

12. Never force glass tubing into rubber tubing, rubber stoppers,, or wooden corks. To pro-tect your hands, wear heavy cloth gloves or wrap toweling around the glass and the tubing, stopper, or cork, and gently push in the glass. **(Rule ? applies)**

13. Do not inhale fumes directly. When instructed to smell a substance, use your hand to wave the fumes toward your nose, and inhale gently. **(Rule ? applies)**

14. Keep your hands away from your face and mouth. **(Rule ? applies)**

15. Always wash your hands before leaving the laboratory.**(Rule ? applies)**

Finally, if you are wondering how to answer the question that asks what additional rules apply to every lab, here is the correct answer.

Any time you see any of the safety symbols, you should remember that all 29 of the numbered laboratory rules apply.

Skills Practice Lab

OBJECTIVES

- Demonstrate proficiency in using a Bunsen burner, a balance, and a graduated cylinder.

- Demonstrate proficiency in handling solid and liquid chemicals.

- Develop proper safety techniques for all lab work.

- Use neat and organized data-collecting techniques.

- Use graphing techniques to plot data.

MATERIALS

- balance
- beakers, 250 mL (2)
- Bunsen burner and related equipment
- copper wire
- crucible tongs
- evaporating dish
- graduated cylinder, 100 mL
- heat-resistant mat
- NaCl
- spatula
- test tube
- wax paper or weighing paper

1 Laboratory Techniques

Introduction

You have applied to work at a company that does research, development, and analysis work. Although the company does not require employees to have extensive chemical experience, all applicants are tested for their ability to follow directions, heed safety precautions, perform simple laboratory procedures, clearly and concisely communicate results, and make logical inferences.

The company will consider your performance on the test in deciding whether to hire you and determining what your initial salary will be. Pay close attention to the procedures and safety precautions because you will continue to use them throughout your work if you are hired by this company. In addition, you will need to pay attention to what is happening around you, make careful observations, and keep a clear and legible record of these observations in your lab notebook.

This laboratory orientation session will teach you some of the following techniques:
- how to use a Bunsen burner
- how to handle solids and liquids
- how to use a balance
- how to practice basic safety techniques in lab work

Safety Procedures

- Wear safety goggles when working around chemicals, acids, bases, flames, or heating devices. Contents under pressure may become projectiles and cause serious injury.
- Never look directly at the sun through any optical device or use direct sunlight to illuminate a microscope.
- Avoid wearing contact lenses in the lab.
- If any substance gets in your eyes, notify your instructor immediately and flush your eyes with running water for at least 15 min.

- Secure loose clothing, and remove dangling jewelry. Don't wear open-toed shoes or sandals in the lab.
- Wear an apron or lab coat to protect your clothing when working with chemicals.
- If a spill gets on your clothing, rinse it off immediately with water for at least 5 min while notifying your instructor.

- Always use caution when working with chemicals.
- Never mix chemicals unless specifically directed to do so.
- Never taste, touch, or smell chemicals unless specifically directed to do so.
- Add acid or base to water; never do the opposite.
- Never return unused chemicals to the original container.
- Never transfer substances by sucking on a pipette or straw; use a suction bulb.
- Follow instructions for proper disposal.

- Avoid wearing hair spray or hair gel on lab days.
- Whenever possible, use an electric hot plate as a heat source instead of an open flame.
- When heating materials in a test tube, always angle the test tube away from yourself and others.
- Glass containers used for heating should be made of heat-resistant glass.
- Know your school's fire-evacuation routes.

- Clean and decontaminate all work surfaces and personal protective equipment as directed by your instructor.
- Dispose of all sharps (broken glass and other contaminated sharp objects) and other contaminated materials (biological and chemical) in special containers as directed by your instructor.

Data Table 1

Material	Mass (g) step 11	Mass (g) step 12
empty beaker		
beaker and 50 mL of water		
50 mL of water		
beaker and 100 mL of water		
100 mL of water		
beaker and 150 mL of water		
150 mL of water		

Procedure

1. Copy Data Tables 1 and 2 in your lab notebook. Be sure that you have plenty of room for observations about each test.

Data Table 2

Material	Mass (g)
weighing paper	
weighing paper and NaCl	

2. Record in your lab notebook the location and use of the following emergency items: lab shower, eyewash station, and emergency telephone numbers.

3. Check to be certain that the gas valve at your lab station and at the neighboring lab stations are turned off. Notify your teacher immediately if a valve is on, because the fumes must be cleared before any work continues.

4. Compare the Bunsen burner in **Figure A** with your burner. Construction may vary, but the air and methane gas, CH_4, always mix in the barrel, the vertical tube in the center of the burner.

Figure A

5. Partially close the air ports at the base of the barrel, turn the gas on full, hold the sparker about 5 cm above the top of the barrel, and proceed to light. Adjust the gas valve until the flame extends about 8 cm above the barrel. Adjust the air supply until you have a quiet, steady flame with a sharply defined, light-blue inner cone. If an internal flame develops, turn off the gas valve, and let the burner cool down. Otherwise, the metal of the burner can get hot enough to set fire to anything nearby that is flammable. Before you relight the burner, partially close the air ports.

6. Using crucible tongs, hold a 10 cm piece of copper wire for 2–3 s in the part of the flame labeled "a" in **Figure B.** Repeat this step for the parts of the flame labeled "b" and "c." Record your observations in your lab notebook.

7. Experiment with the flame by completely closing the air ports at the base of the burner. Observe and record the color of the flame and the sounds made by the burner. Using crucible tongs, hold an evaporating dish in the tip of the flame for about 3 min. Place the dish on a heat-

resistant mat, and shut off the burner. After the dish cools, examine its un derside, and record your observations.

8. Before using the balance, make sure that it is level and showing a mass of zero. If necessary, adjust the calibration knob. To avoid discrepancies, use the same balance for all measurements during a lab activity. Never put chemicals directly on the balance pan.

9. Place a piece of weighing paper on the balance pan. Determine the mass of the paper, and record this mass to the nearest 0.01 g in your data table. Put a small quantity of NaCl on a separate piece of weighing paper. Then, transfer 13 g of the NaCl to the weighing paper on the balance pan. Record the exact mass to the nearest 0.01 g in your data table.

Figure B

10. Remove the weighing paper and NaCl from the balance pan. Lay the test tube flat on the table, and transfer the NaCl into the tube by rolling the weighing paper and sliding it into the test tube. As you lift the test tube to a vertical position, tap the paper gently, and the solid will slip into the test tube, as shown in **Figure C.**

11. Measure the mass of a dry 250 mL beaker, and record the mass in your data table. Add water up to the 50 mL mark, determine the new mass,

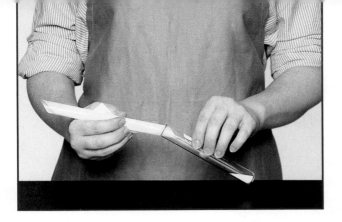

Figure C

and record the new mass in your data table. Repeat the procedure by filling the beaker to the 100 mL mark and then to the 150 mL mark, and record the mass each time. Subtract the mass of the empty beaker from the other measurements to determine the masses of the water.

12. Repeat step 11 with a second dry 250 mL beaker, but use a graduated cylinder to measure the volumes of water to the nearest 0.1 mL before pouring the water into the beaker. Read the volumes by using the bottom of the meniscus, the curve formed by the water's surface.

13. Clean all apparatus and your lab station. Put the wire, NaCl, and weighing paper in the containers designated by your teacher. Pour the water from the beakers into the sink. Scrub the cooled evaporating dish with soap, water, and a scrub brush. Be certain that the gas valves at your lab station and the nearest lab station are turned off. Be sure lab equipment is completely cool before storing it. Always wash your hands thoroughly after all lab work is finished and before you leave the lab.

Analysis

1. **Analyzing data** Based on your observations, which type of flame is hotter: the flame formed when the air ports are open or the flame formed when they are closed? What is the hottest part of the flame? (Hint: The melting point of copper is 1083°C.)

2. **Examining data** Which of the following measurements could have been made by your balance: 3.42 g of glass, 5.666 72 g of aspirin, or 0.000 017 g of paper?

3. **Constructing graphs** Make a graph of mass versus volume for data from steps 11 and 12. The mass of water (g) should be graphed along the y-axis as the dependent variable, and the volume of water (mL) should be graphed along the x-axis as the independent variable.

Conclusions

4. **Interpreting information** When methane is burned, it usually produces carbon dioxide and water. If there is a shortage of oxygen, the flame is not as hot and black carbon solid is formed. Which steps in the lab demonstrate these flames?

5. **Applying conclusions** Which is the most accurate method for measuring volumes of liquids, a beaker or a graduated cylinder? Explain why.

6. **Evaluating data** In Mandeville High School, Jarrold got only partway through step 7 of this experiment when he had to put everything away. Soon after Jarrold left, his lab drawer caught on fire. How did this happen?

7. **Drawing conclusions** The density of water is equal to its mass divided by its volume. Calculate the density of water by using your data from step 11. Then, calculate the density of water by using your data from step 12.

Extensions

8. **Designing experiments** You have been asked to design an experiment to find the density of sand. The density of sand is equal to its mass divided by its volume. Describe how you could measure the density of sand by using the equipment from this lab.

9. **Research and communications** Scientists use a number of different instruments to measure the mass of an object. Find information on different types of balances, and make a poster comparing at least three different kinds of balances. The poster should show the smallest amount of mass that could be measured on the balance and identify something appropriate to measure on the balance.

Inquiry LAB
Design Your Own Experiment

THE PROBLEM

January 9, 2004

Director of Research
CheMystery Labs, Inc.
52 Fulton Street
Springfield, VA 22150

Dear Director of Research:

Juliette Brand Foods is preparing to enter the rapidly expanding popcorn market with a new popcorn product. As you may know, the key to making popcorn pop is the amount of water contained within the kernel.

As of today, the product development division has created three different production techniques for the popcorn, each of which creates popcorn that contains differing amounts of water. We need an independent lab such as yours to measure the percentage of water contained in each sample and to determine which technique produces the best-popping popcorn.

I have enclosed samples from each of the three techniques, labeled "technique beta," "technique gamma," and "technique delta." Please send us the bill when the work is complete.

Sincerely,

Mary Biedenbecker

Mary Biedenbecker, Director
Product Development Division

References

Popcorn pops because of the natural moisture inside each kernel. When the internal water is heated above 100°C, the kernel expands rapidly and the liquid water changes to a gas, which takes up much more space than the liquid.

The percentage of water in popcorn can be determined by the following equation.

$$\frac{\text{initial mass} - \text{final mass}}{\text{initial mass}} \times 100 = \text{percent } H_2O$$

The popping process works best when the kernels are first coated with a small amount of vegetable oil. Make sure you account for the presence of this oil when measuring masses.

THE PLAN

CheMystery Labs, Inc. 52 Fulton Street, Springfield, VA 22150

CheMystery Labs, Inc.
52 Fulton Street
Springfield, VA 22150

Memorandum

Date: January 11, 2004
To: Leon Fuller
From: Martha Li-Hsien

Your team needs to design a procedure for determining the percentage of water in three samples of popcorn. Some of the popcorn was damaged in the mail, so each team will have only 80 kernels of popcorn per technique. Make sure to use your samples carefully!

Before you begin the lab work, I must approve your procedure. Give the following items to me ASAP:

- a detailed one-page plan for your procedure, including any necessary data tables
- a detailed list of the equipment and materials you will need

When you finish your experiment, prepare a report in the form of a two-page letter to Mary Biedenbecker that includes the following:

- a paragraph summarizing how you analyzed the samples
- your findings about the percentage of water in each sample, including calculations and a discussion of the multiple trials
- a detailed and organized data table
- a graph comparing your findings with those of the other teams
- suggestions for improving the analysis procedure

Required Precautions

- Wear safety goggles when working around chemicals, acids, bases, flames, or heating devices. Contents under pressure may become projectiles and cause serious injury.
- Avoid wearing contact lenses in the lab.
- If any substance gets in your eyes, notify your instructor immediately and flush your eyes with running water for at least 15 min.
- Secure loose clothing, and remove dangling jewelry. Don't wear open-toed shoes or sandals in the lab.

- Wear an apron or lab coat to pro-

tect your clothing when working with chemicals.
- If a spill gets on your clothing, rinse it off immediately with water for at least 5 min while notifying your instructor.
- Always use caution when working with chemicals.
- Never taste, touch, or smell chemicals unless specifically directed to do so.
- Follow instructions for proper disposal.
- Whenever possible, use an electric hot plate as a heat source instead of an open flame.

- When heating materials in a test tube, always angle the test tube away from yourself and others.
- Know your school's fire-evacuation routes.
- Clean and decontaminate all work surfaces and personal protective equipment as directed by your instructor.
- Dispose of all sharps (broken glass and other contaminated sharp objects) and other contaminated materials (biological and chemical) in special containers as directed by your instructor.

Skills Practice Lab

OBJECTIVES

♦ **Recognize** how the solubility of a salt varies with temperature.

♦ **Demonstrate** proficiency in fractional crystallization and in filtration.

♦ **Solve** the percentage of two salts recovered by fractional crystallization.

MATERIALS

♦ balance

♦ beaker tongs or hot mitt

♦ beakers, 150 mL (4)

♦ Bunsen burner or hot plate

♦ filter paper

♦ graduated cylinder, 100 mL

♦ ice and rock salt

♦ NaCl–KNO$_3$ solution (50 mL)

♦ nonmercury thermometer

♦ ring stand set up

♦ rubber policeman

♦ spatula

♦ stirring rod, glass

♦ tray, tub, or pneumatic trough

♦ vacuum filtration setup or gravity-filtration setup

OPTIONAL EQUIPMENT

♦ CBL unit

♦ graphing calculator with cable

♦ Vernier temperature probe

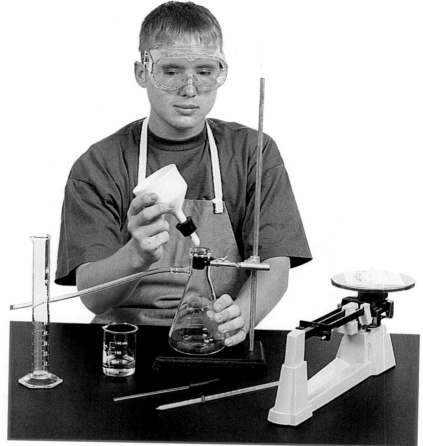

Introduction

Your company has been contacted by a fireworks factory. A mislabeled container of sodium chloride, NaCl, was accidentally mixed with potassium nitrate, KNO$_3$. KNO$_3$ is used as an oxidizer in fireworks to ensure that the fireworks burn thoroughly. The fireworks company wants your company to investigate ways they could separate the two compounds. They have provided an aqueous solution of the mixture for you to work with.

The substances in a mixture can be separated by physical means. For example, if one substance dissolves in a liquid solvent but another does not, the mixture can be filtered. The substance that dissolved will be carried through the filter by the solvent, but the other substance will not.

Because both NaCl and KNO$_3$ dissolve in water, filtering alone cannot separate them. However, there are differences in the way they dissolve. The graph in **Figure A** shows the same amount of NaCl dissolving in water regardless of the temperature of the water. On the other hand, KNO$_3$ is very soluble in warm water but much less soluble at 0°C.

Safety Procedures

- Wear safety goggles when working around chemicals, acids, bases, flames, or heating devices. Contents under pressure may become projectiles and cause serious injury.
- Never look directly at the sun through any optical device or use direct sunlight to illuminate a microscope.
- Avoid wearing contact lenses in the lab. If any substance gets in your eyes, notify your instructor immediately and flush your eyes with running water for at least 15 min.

- Secure loose clothing, and remove dangling jewelry. Don't wear open-toed shoes or sandals in the lab.
- Wear an apron or lab coat to protect your clothing when working with chemicals.
- If a spill gets on your clothing, rinse it off immediately with water for at least 5 min while notifying your instructor.

- Always use caution when working with chemicals.
- Never mix chemicals unless specifically directed to do so.
- Never taste, touch, or smell chemicals unless specifically directed to do so.

- Add acid or base to water; never do the opposite.
- Never return unused chemicals to the original container.
- Never transfer substances by sucking on a pipette or straw; use a suction bulb.
- Follow instructions for proper disposal.

- Avoid wearing hair spray or hair gel on lab days.
- Whenever possible, use an electric hot plate as a heat source instead of an open flame.
- When heating materials in a test tube, always angle the test tube away from yourself and others.
- Glass containers used for heating should be made of heat-resistant glass.
- Know your school's fire-evacuation routes.

- Clean and decontaminate all work surfaces and personal protective equipment as directed by your instructor.
- Dispose of all sharps (broken glass and other contaminated sharp objects) and other contaminated materials (biological and chemical) in special containers as directed by your instructor.

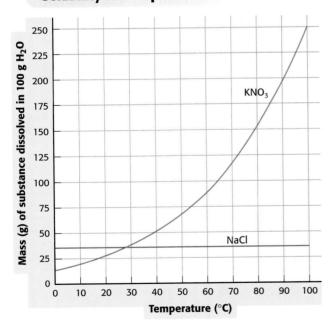

Solubility vs. Temperature for Two Salts

Figure A
This graph shows the relationship between temperature and the solubility of NaCl and KNO_3.

You will make use of the differences in solubility to separate the two salts. This technique is known as fractional crystallization. If the water solution of NaCl and KNO_3 is cooled from room temperature to a temperature near 0°C, some KNO_3 will crystallize. This KNO_3 residue can then be separated from the NaCl solution by filtration. The NaCl can be isolated from the filtrate by evaporation of the water. To determine whether this method is efficient, you will measure the mass of each of the recovered substances. Then, your client can decide whether this method is cost-effective.

Filtration-Technique Option

Figure B
Vacuum filtration

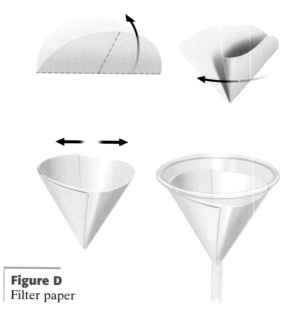

Figure C
Gravity filtration

Vacuum-Filtration Setup

1. To set up a vacuum filtration, screw an aspirator nozzle onto the faucet. Attach the other end of the plastic tubing to the side arm of the filter flask.

2. Place a one-hole rubber stopper on the stem of the funnel, and fit the stopper snugly in the neck of the filter flask, as shown in **Figure B.**

3. Place a piece of filter paper on the bottom of the funnel so that it is flat and covers all of the holes in the funnel.

4. When you are ready, turn on the water at the faucet that has the aspirator nozzle attached. This action creates a vacuum, which helps the filtering process go much faster. If the suction is working properly, the filter paper should be pulled against the bottom of the funnel, which results in covering all of the holes. If the filter paper appears to have bubbles of air under it or is not centered well, turn the water off, reposition the filter paper, and begin again.

Gravity-Filtration Setup

1. Set up a ring stand with a ring. Gently rest a glass funnel inside the ring, and place a beaker under the glass funnel, as shown in **Figure C.**

2. Fold a piece of filter paper in half along its diameter, and then fold it again to form a quadrant, as shown in **Figure D.** Separate the folds of the filter paper so that three thicknesses are on one side and one thickness is on the other.

3. Fit the filter paper in the funnel, and wet it with a little water so that it will adhere to the sides of the funnel. Gently but firmly press the paper against the sides of the funnel so that no air is between the funnel and the filter paper. Be certain that the filter paper does not extend above the sides of the funnel.

Figure D
Filter paper

Procedure ▮▮▮▮▮▮▮▮

Advance Preparation

1. Copy the data table below in your lab notebook. Be sure that you have plenty of room for observations about each test.

Data Table 1
Mass of beaker 1
Volume of NaCl–KNO$_3$ solution added to beaker 1
Temperature of mixture before cooling
Mass of filter paper
Mass of beaker 4
Mass of beaker 4 with NaCl
Mass of beaker 1 with filter paper and KNO$_3$
Temperature of mixture after cooling

2. Obtain four clean, dry 150 mL beakers, and label them 1, 2, 3, and 4.

Thermometer procedure continues on page 767.

CBL and Sensors

3. Connect the CBL to the graphing calculator with the unit-to-unit link cable using the I/O ports located on each unit. Connect the temperature probe to the CH1 port. Turn on the CBL and the graphing calculator. Start the program CHEMBIO on the graphing calculator.

   a. Select option *SET UP PROBES* from the MAIN MENU. Enter 1 for the number of probes. Select the temperature probe from the list. Enter 1 for the channel number.

   b. Select the *COLLECT DATA* option from the MAIN MENU. Select the *TRIGGER* option from the DATA COLLECTION menu.

4. Set up your filtering apparatus. If you are using a Büchner funnel for vacuum filtration or a glass funnel for gravity filtration, follow the setup procedure under "Filtration-Technique Option."

5. Measure the mass of beaker 1 to the nearest 0.01 g, and record the mass in your data table.

6. Measure about 50 mL of the NaCl–KNO$_3$ solution into a graduated cylinder. Record the exact volume in your data table. Pour this mixture into beaker 1.

7. Using the temperature probe, measure the temperature of the mixture. Press TRIGGER on the CBL to collect the temperature reading of the mixture. Record this temperature in your data table. Select *CONTINUE* from the TRIGGER menu on the graphing calculator.

8. Measure the mass of a piece of filter paper to the nearest 0.01 g, and record the mass in your data table.

9. Make an ice bath by filling a tray, tub, or trough half-full with ice. Add a handful of rock salt. The salt lowers the freezing point of water so that the ice bath can cool to a lower temperature. Fill the ice bath with water until it is three-quarters full.

10. Using a fresh supply of ice and distilled water, fill beaker 2 half-full with ice, and add water. Do not add rock salt to this ice-water mixture. You will use this water to wash your purified salt.

First Filtration

11. Put beaker 1 with your NaCl–KNO$_3$ solution into the ice bath. Place the temperature probe in the solution to monitor the temperature. Stir the solution with a stirring rod while it cools. (Do not stir the solution with the temperature probe.) The lower the temperature of the mixture is, the more KNO$_3$ that will crystallize out of solution. When the temperature nears 4°C, press TRIGGER on the CBL to collect the temperature reading of the mixture. Record this temperature in your data table. Select *STOP* from the TRIGGER menu on the graphing calculator. Proceed with step 11a if you are using the Büchner funnel or step 11b if you are using a glass funnel.

a. Vacuum filtration

Prepare the filtering apparatus by pouring approximately 50 mL of ice-cold distilled water from beaker 2 through the filter paper. After the water has gone through the funnel, empty the filter flask into the sink. Reconnect the filter flask, and pour the salt-and-water mixture in beaker 1 into the funnel. Use the rubber policeman to transfer all of the cooled mixture into the funnel, especially any crystals that are visible. It may be helpful to add small amounts of ice-cold water from beaker 2 to beaker 1 to wash any crystals onto the filter paper. After all of the solution has passed through the funnel, wash the KNO_3 residue by pouring a very small amount of ice-cold water from beaker 2 over it. When this water has passed through the filter paper, turn off the faucet and carefully remove the tubing from the aspirator. Empty the filtrate, which has passed through the filter paper and is now in the filter flask, into beaker 3. When finished, continue with step 12.

b. Gravity filtration

Place beaker 3 under the glass funnel. Prepare the filtering apparatus by pouring approximately 50 mL of ice-cold water from beaker 2 through the filter paper. The water will pass through the filter paper and drip into beaker 3. When the dripping stops, empty beaker 3 into the sink. Place beaker 3 back under the glass funnel so that it will collect the filtrate from the funnel. Pour the salt-water mixture into the funnel. Use the rubber policeman to transfer all of the cooled mixture into the funnel, especially any visible crystals. It may be helpful to add small amounts of ice-cold water from beaker 2 to beaker 1 to wash any crystals onto the filter paper. After all of the solution has passed through the funnel, wash the KNO_3 by pouring a very small amount of ice-cold water from beaker 2 over it.

12. After you have finished filtering, use either a hot plate or a Bunsen burner, ring stand, ring, and wire gauze to heat beaker 3. When the liquid in beaker 3 begins to boil, continue heating gently

until enough water has vaporized to decrease the volume to approximately 25–30 mL. Be sure to use beaker tongs. Remember that hot glassware does not always look hot.

Second Filtration

13. Allow the solution in beaker 3 to cool. Then set it in the ice bath and stir until the temperature is approximately 4°C.

14. Measure the mass of beaker 4, and record the mass in your data table.

15. Repeat step 11a or step 11b, pouring the solution from beaker 3 onto the filter paper and using beaker 4 to collect the filtrate that passes through the filter.

16. Wash and dry beaker 1. Carefully remove the filter paper with the KNO_3 from the funnel, and put it in the beaker. Avoid spilling the crystals. Place the beaker in a drying oven overnight.

Figure E
Use beaker tongs to move a beaker that has been heated, even if you believe that the beaker is cool.

Recovery of NaCl

17. Heat beaker 4 with a hot plate or Bunsen burner until the water begins to boil. Continue to heat the beaker gently until all of the water has vaporized and the salt appears dry. Turn off the hot plate or burner, and allow the beaker to cool. Use beaker tongs to move the beaker, as shown in **Figure E.** Measure the mass of beaker

4 with the NaCl to the nearest 0.01 g, and record the mass in your data table.

18. The next day, use beaker tongs to remove beaker 1 with the filter paper and KNO_3 from the drying oven. Allow the beaker to cool. Measure the mass using the same balance you used to measure the mass of the empty beaker. Record the new mass in your data table. Be sure to use beaker tongs. Remember that hot glassware does not always look hot.

19. Clean all apparatus and your lab station. Once the mass of the NaCl has been determined, add water to dissolve the NaCl, and rinse the solution down the drain. Do not wash KNO_3 down the drain. Dispose of the KNO_3 in the waste container designated by your teacher. Wash your hands thoroughly after all lab work is finished and before you leave the lab.

Thermometer

3. Set up your filtering apparatus. If you are using a Büchner funnel for vacuum filtration or a glass funnel for gravity filtration, follow the setup procedure under "Filtration-Technique Option."

4. Measure the mass of beaker 1 to the nearest 0.01 g, and record the mass in your data table.

5. Measure about 50 mL of the NaCl–KNO_3 solution into a graduated cylinder. Record the exact volume in your data table. Pour this mixture into beaker 1.

6. Using a thermometer, measure the temperature of the mixture. Record this temperature in your data table.

7. Measure the mass of a piece of filter paper to the nearest 0.01 g, and record the mass in your data table.

8. Make an ice bath by filling a tray, tub, or trough half-full with ice. Add a handful of rock salt. The salt lowers the freezing point of water so that the ice bath can cool to a lower temperature. Fill the ice bath with water until it is three-quarters full.

9. Using a fresh supply of ice and distilled water, fill beaker 2 half-full with ice, and add water. Do not add rock salt to this ice-water mixture. You will use this water to wash your purified salt.

First Filtration

10. Put beaker 1 with your NaCl–KNO_3 solution into the ice bath. Place a thermometer in the solution to monitor the temperature. Stir the solution with a stirring rod while it cools. The lower the temperature of the mixture is, the more KNO_3 that will crystallize out of solution. When the temperature nears 4°C, record the temperature in your data table. Proceed with step 10a if you are using the Büchner funnel or step 10b if you are using a glass funnel. Never stir a solution with a thermometer; the bulb is very fragile.

a. Vacuum filtration
Prepare the filtering apparatus by pouring approximately 50 mL of ice-cold distilled water from beaker 2 through the filter paper. After the water has gone through the funnel, empty the filter flask into the sink. Reconnect the filter flask, and pour the salt-and-water mixture in beaker 1 into the funnel. Use the rubber policeman to transfer all of the cooled mixture into the funnel, especially any crystals that are visible. It may be helpful to add small amounts of ice-cold water from beaker 2 to beaker 1 to wash any crystals onto the filter paper. After all of the solution has passed through the funnel, wash the KNO_3 residue by pouring a very small amount of ice-cold water from beaker 2 over it. When this water has passed through the filter paper, turn off the faucet and carefully remove the tubing from the aspirator. Empty the filtrate, which has passed through the filter paper and is now in the filter flask, into beaker 3. When finished, continue with step 11.

b. Gravity filtration
Place beaker 3 under the glass funnel. Prepare the filtering apparatus by pouring approximately 50 mL of ice-cold water from beaker 2 through the filter paper. The water

will pass through the filter paper and drip into beaker 3. When the dripping stops, empty beaker 3 into the sink. Place beaker 3 back under the glass funnel so that it will collect the filtrate from the funnel. Pour the salt-water mixture into the funnel. Use the rubber policeman to transfer all of the cooled mixture into the funnel, especially any visible crystals. It may be helpful to add small amounts of ice-cold water from beaker 2 to beaker 1 to wash any crystals onto the filter paper. After all of the solution has passed through the funnel, wash the KNO_3 by pouring a very small amount of ice-cold water from beaker 2 over it.

11. After you have finished filtering, use either a hot plate or a Bunsen burner, ring stand, ring, and wire gauze to heat beaker 3. When the liquid in beaker 3 begins to boil, continue heating gently until enough water has vaporized to decrease the volume to approximately 25–30 mL. Be sure to use beaker tongs. Remember that hot glassware does not always look hot.

Second Filtration

12. Allow the solution in beaker 3 to cool. Then set it in the ice bath and stir until the temperature is approximately 4°C.

13. Measure the mass of beaker 4, and record the mass in your data table.

14. Repeat step 10a or step 10b, pouring the solution from beaker 3 onto the filter paper and using beaker 4 to collect the filtrate that passes through the filter.

15. Wash and dry beaker 1. Carefully remove the filter paper with the KNO_3 from the funnel, and put it in the beaker. Avoid spilling the crystals. Place the beaker in a drying oven overnight.

Recovery of NaCl

16. Heat beaker 4 with a hot plate or Bunsen burner until the water begins to boil. Continue to heat the beaker gently until all of the water has vaporized and the salt appears dry. Turn off the hot plate or burner, and allow the beaker to cool. Use beaker tongs to move the beaker, as shown in Figure E. Measure the mass of beaker 4 with the NaCl to the nearest 0.01 g, and record the mass in your data table.

17. The next day, use beaker tongs to remove beaker 1 with the filter paper and KNO_3 from the drying oven. Allow the beaker to cool. Measure the mass using the same balance you used to measure the mass of the empty beaker. Record the new mass in your data table. Be sure to use beaker tongs. Remember that hot glassware does not always look hot.

18. Clean all apparatus and your lab station. Once the mass of the NaCl has been determined, add water to dissolve the NaCl, and rinse the solution down the drain. Do not wash KNO_3 down the drain. Dispose of the KNO_3 in the waste container designated by your teacher. Wash your hands thoroughly after all lab work is finished and before you leave the lab.

Analysis

1. **Analyzing results** Find the mass of NaCl in your 50 mL sample by subtracting the mass of the empty beaker 4 from the mass of beaker 4 with NaCl.

2. **Analyzing data** Find the mass of KNO_3 in your 50 mL sample by subtracting the mass of beaker 1 and the mass of the filter paper from the mass of beaker 1 with the filter paper and KNO_3.

3. **Analyzing data** Determine the total mass of the two salts.

Conclusions

4. Applying conclusions How many grams of KNO_3 and NaCl would be found in a 1.0 L sample of the solution? (Hint: For each substance, make a conversion factor by using the mass of the compound and the volume of the solution.)

5. Analyzing graphs Use the graph at the beginning of this exploration to determine how much of each compound would dissolve in 100 g of water at room temperature and at the temperature of your ice-water bath.

6. Drawing conclusions Calculate the percentage by mass of NaCl in the salt mixture. Calculate the percentage by mass of KNO_3 in the salt mixture. Assume that the density of your 50 mL solution is 1.0 g/mL.

7. Applying conclusions The fireworks company has another 55 L of the salt mixture dissolved in water just like the sample you worked with. How many kilograms of each compound can the company expect to recover from this sample? (Hint: Use your answer from item 4 to help you answer this question.)

8. Evaluating methods Use the graph shown at the beginning of this lab to estimate how much KNO_3 could still be contaminating the NaCl you recovered.

9. Relating ideas Use the graph shown at the beginning of this lab to explain why it is impossible to completely separate the two compounds by fractional crystallization.

10. Evaluating methods Why was it important to use ice-cold water to wash the KNO_3 after filtration?

11. Evaluating methods If it was important to use very cold water to wash the KNO_3, why was the salt-and-ice-water mixture from the bath not used? After all, it had a lower temperature than the ice and distilled water from beaker 2 did. (Hint: Consider what is contained in rock salt.)

12. Evaluating methods Why was it important to keep the amount of cold water used to wash the KNO_3 as small as possible?

13. Interpreting graphics Using the graph shown at the beginning of this lab, determine the minimum mass of water necessary to dissolve the amounts of each compound from Analysis items 1 and 2. Calculate the mass dissolved at room temperature and at 4°C. What volumes of water would be necessary? (Hint: The density of water is about 1.0 g/mL.)

Extensions

1. Designing experiments Describe how you could use the properties of the compounds to test the purity of your recovered samples. If your teacher approves your plan, use it to check your separation of the mixtures. (Hint: Check a chemical handbook for more information about the properties of NaCl and KNO_3.)

2. Designing experiments How could you improve the yield or the purity of the compounds you recovered? If you can think of ways to modify the procedure, ask your teacher to approve your plan and run the procedure again.

January 20, 2004

George Taylor
Director of Analytical Services
CheMystery Labs, Inc.
52 Fulton Street
Springfield, VA 22150

Dear George:

I thought of your new company when a problem came up here at Goldstake. I think I have some work for your company. While performing exploratory drilling for natural gas near Afton in western Wyoming, our engineers encountered a new subterranean, geothermal aquifer. We estimate the size of the aquifer to be 1×10^{12} L.

The Bureau of Land Management advised us to alert the Environmental Protection Agency. Preliminary qualitative tests of the water identified two dissolved salts: potassium nitrate and copper nitrate.

The EPA is concerned that a full-scale mining operation may harm the environment if the salts are present in large quantities. They are requiring us to halt all operations while we obtain more information for an environmental impact statement. We need your firm to separate the sample, purify the sample, and make a determination of the amounts of the two salts in the Afton Aquifer.

Sincerely,

Lynn L. Brown

Lynn L. Brown
Director of Operations
Goldstake Mining Corporation

References

The procedure for this Investigation is similar to one your team recently completed involving the separation of sodium chloride, NaCl, and potassium nitrate, KNO_3.

THE PLAN

CheMystery Labs, Inc. 52 Fulton Street, Springfield, VA 22150

CheMystery Labs, Inc.
52 Fulton Street
Springfield, VA 22150

Memorandum

Date: January 23, 2004
To: Andre Kalaviencz
From: George Taylor

Because this is our first mining-industry contract, we need to plan carefully to get good results at minimum cost. Each research team will receive only a 50.0 mL sample of the aquifer water.

I need the following information from each team before the work begins.
- a detailed, one-page plan for the procedure that you will use to accomplish the analysis, including all necessary data tables
- a list of the materials and supplies you will need

When you have completed your labwork, present the following information to Goldstake in a two page report:
- the mass of potassium nitrate, KNO_3, and copper nitrate, $Cu(NO_3)_2$, in the 50.0 mL sample
- the extrapolated mass of KNO_3 and $Cu(NO_3)_2$ in the Afton Aquifer
- a short paragraph that summarizes and describes the procedures you used
- detailed and organized data and analysis section that shows your calculations and explanations of any possible sources of error

---

Required Precautions

- Wear safety goggles when working around chemicals, acids, bases, flames, or heating devices. Contents under pressure may become projectiles and cause serious injury.
- Avoid wearing contact lenses in the lab.
- If any substance gets in your eyes, notify your instructor immediately and flush your eyes with running water for at least 15 min.
- Secure loose clothing, and remove dangling jewelry. Don't wear open-toed shoes or sandals in the lab.

- Wear an apron or lab coat to pro-

tect your clothing when working with chemicals.
- If a spill gets on your clothing, rinse it off immediately with water for at least 5 min while notifying your instructor.
- Always use caution when working with chemicals.
- Never taste, touch, or smell chemicals unless specifically directed to do so.
- Follow instructions for proper disposal.
- Whenever possible, use an electric hot plate as a heat source instead of an open flame.

- When heating materials in a test tube, always angle the test tube away from yourself and others.
- Know your school's fire-evacuation routes.
- Clean and decontaminate all work surfaces and personal protective equipment as directed by your instructor.
- Dispose of all sharps (broken glass and other contaminated sharp objects) and other contaminated materials (biological and chemical) in special containers as directed by your instructor.

OBJECTIVES

- Identify a set of flame-test color standards for selected metal ions.

- Relate the colors of a flame test to the behavior of excited electrons in a metal ion.

- Draw conclusions and identify an unknown metal ion by using a flame test.

USING SCIENTIFIC **METHODS**

- Demonstrate proficiency in performing a flame test and in using a spectroscope.

MATERIALS

- beaker, 250 mL
- Bunsen burner
- $CaCl_2$ solution
- cobalt glass plates
- crucible tongs
- distilled water
- flame-test wire
- glass test plate
- HCl solution (1.0 M)
- K_2SO_4 solution
- Li_2SO_4 solution
- Na_2SO_4 solution
- NaCl crystals
- NaCl solution
- spectroscope
- $SrCl_2$ solution
- unknown solution

3 Flame Tests

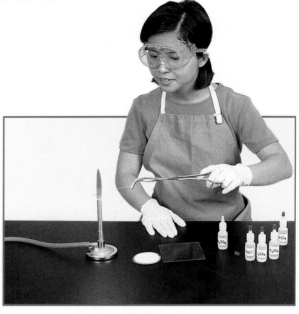

Introduction

Your company has been contacted by Julius and Annette Benetti. They are worried about some abandoned, rusted barrels of chemicals that their daughter found while playing in the vacant lot behind their home. The barrels have begun to leak a colored liquid that flows through their property before emptying into a local sewer. The Benettis want your company to identify the compound in the liquid. Earlier work indicates that it is a dissolved metal compound. Many metals, such as lead, have been determined to be hazardous to our health. Many compounds of these metals are often soluble in water and are therefore easily absorbed into the body.

Electrons in atoms jump from their ground state to excited states by absorbing energy. Eventually these electrons fall back to their ground state, re-emitting the absorbed energy in the form of light. Because each atom has a unique structure and arrangement of electrons, each atom emits a unique spectrum of light. This characteristic light is the basis for the chemical test known as a flame test. In this test the atoms are excited by being placed within a flame. As they re-emit the absorbed energy in the form of light, the color of the flame changes. For most metals, these changes are easily visible. However, even the presence of a tiny speck of another substance can interfere with the identification of the true color of a particular type of atom.

To determine what metal is contained in the barrels behind the Benettis' house, you must first perform flame tests with a variety of standard solutions of different metal compounds. Then you will perform a flame test with the unknown sample from the site to see if it matches any of the solutions you've used as standards. Be sure to keep your equipment very clean, and perform multiple trials to check your work.

772   Skills Practice Lab 3

Safety Procedures

- Wear safety goggles when working around chemicals, acids, bases, flames, or heating devices. Contents under pressure may become projectiles and cause serious injury.
- Never look directly at the sun through any optical device or use direct sunlight to illuminate a microscope.
- Avoid wearing contact lenses in the lab.
- If any substance gets in your eyes, notify your instructor immediately and flush your eyes with running water for at least 15 min.

- Secure loose clothing, and remove dangling jewelry. Don't wear open-toed shoes or sandals in the lab.
- Wear an apron or lab coat to protect your clothing when working with chemicals.
- If a spill gets on your clothing, rinse it off immediately with water for at least 5 min while notifying your instructor.

- If a chemical gets on your skin or clothing or in your eyes, rinse it immediately, and alert your instructor.
- If a chemical is spilled on the floor or lab bench, alert your instructor, but do not clean it up yourself unless your teacher says it is OK to do so.

- Always use caution when working with chemicals.

- Never mix chemicals unless specifically directed to do so.
- Never taste, touch, or smell chemicals unless specifically directed to do so.
- Add acid or base to water; never do the opposite.
- Never return unused chemicals to the original container.
- Never transfer substances by sucking on a pipette or straw; use a suction bulb.
- Follow instructions for proper disposal.

- Avoid wearing hair spray or hair gel on lab days.
- Whenever possible, use an electric hot plate as a heat source instead of an open flame.
- When heating materials in a test tube, always angle the test tube away from yourself and others.
- Glass containers used for heating should be made of heat-resistant glass.
- Know your school's fire-evacuation routes.

- Clean and decontaminate all work surfaces and personal protective equipment as directed by your instructor.
- Dispose of all sharps (broken glass and other contaminated sharp objects) and other contaminated materials (biological and chemical) in special containers as directed by your instructor.

Data Table 1

Metal Compound	Color of flame	Wavelengths (nm)
$CaCl_2$ solution		
K_2SO_4 solution		
Li_2SO_4 solution		
Na_2SO_4 solution		
$SrCl_2$ solution		
Na_2SO_4 (cobalt glass)		
K_2SO_4 (cobalt glass)		
Na_2SO_4 and K_2SO_4		
Na_2SO_4 and K_2SO_4 (cobalt glass)		
NaCl solution		
NaCl crystals		
Unknown solution		

Procedure

1. Copy the Data Table 1 in your lab notebook. Be sure that you have plenty of room for observations about each test.

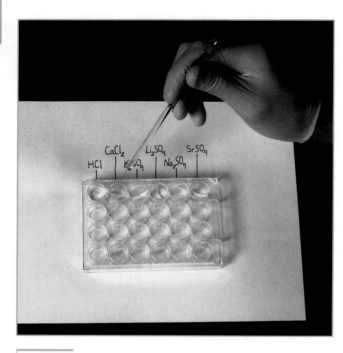

Figure A
Be sure that you record the position of the various metal ion solutions in each well of the well strip.

2. Label a beaker "Waste." Thoroughly clean and dry a well strip. Fill the first well one-fourth full with 1.0 M HCl. Clean the test wire by first dipping it in the HCl and then holding it in the flame of the Bunsen burner. Repeat this procedure until the flame is not colored by the wire. When the wire is ready, rinse the well with distilled water, and collect the rinse water in the waste beaker.

3. Put 10 drops of each metal ion solution listed in the materials list except NaCl in a row in each well of the well strip. Put a row of 1.0 M HCl drops on a glass plate across from the metal ion solutions. Record the position of all of the chemicals placed in the wells. The wire will need to be cleaned thoroughly with HCl between each test solution to avoid contamination from the previous test.

4. Dip the wire into the $CaCl_2$ solution, as shown in **Figure A,** and then hold it in the Bunsen burner flame. Observe the color of the flame, and record it in the data table. Repeat the procedure again, but this time look through the spectroscope to view the results. Record the wavelengths you see from the flame. Perform each test three times. Clean the wire with the HCl as you did in step 2.

5. Repeat step 4 with the K_2SO_4 and with each of the remaining solutions in the well strip. For each solution that you test, record the color of each flame and the wavelength observed with the spectroscope. After the solutions are tested, clean the wire thoroughly, rinse the well strip with distilled water, and collect the rinse water in the waste beaker.

6. Test another drop of Na_2SO_4, but this time view the flame through two pieces of cobalt glass. Clean the wire, and repeat the test by using the K_2SO_4. View the flame through the cobalt glass. Record in your data table the colors and wavelengths of the flames. Clean the wire and the well strip, and rinse the well strip with distilled water. Pour the rinse water into the waste beaker.

7. Put a drop of K_2SO_4 in a clean well. Add a drop of Na_2SO_4. Flame-test the mixture. Observe the flame without the cobalt glass. Repeat the test, this time observing the flame through the cobalt glass. Record the colors and wavelengths of the flames in the data table. Clean the wire, and rinse the well strip with distilled water. Pour the rinse water into the waste beaker.

8. Test a drop of the NaCl solution in the flame, and then view it through the spectroscope. (Do not use the cobalt glass.) Record your observations. Clean the wire, and rinse the well strip with distilled water. Pour the rinse water into the waste beaker. Place a few crystals of NaCl in a clean well, dip the wire in the crystals, and do the flame test once more. Record the color of the flame test. Clean the wire, and rinse the well strip with distilled water. Pour the rinse water into the waste beaker.

Figure B
Flame test

9. Dip the wire into the unknown solution; then hold it in the Bunsen burner flame, as shown in **Figure B.** Perform flame tests for the wire, both with and without the cobalt glass. Record your observations. Clean the wire, and rinse the well strip with distilled water. Pour the rinse water into the waste beaker.

10. Clean all apparatus and your lab station. Dispose of the contents of the waste beaker into the container designated by your teacher. Wash your hands thoroughly after cleaning up the lab area and equipment.

Analysis

1. **Organizing data** Examine your data table, and create a summary of the flame test for each metal ion.

2. **Analyzing data** Account for any differences in the individual trials for the flame tests for the metal ions.

3. **Explaining events** Explain how viewing the flame through cobalt glass can make analyzing the ions being tested easier.

4. **Explaining events** Explain how the lines seen in the spectroscope relate to the position of electrons in the metal atom.

5. **Identifying patterns** For three of the metal ions tested, explain how the flame color you saw relates to the lines of color you saw when you looked through the spectroscope.

Conclusions

6. **Evaluating results** What metal ions are in the unknown solution from the barrels on the vacant lot?

7. **Evaluating methods** How would you characterize the flame test with respect to its sensitivity? What difficulties could occur when identifying ions by the flame test?

8. **Evaluating methods** Explain how you can use a spectroscope to identify the components of solutions containing several different metal ions.

9. **Applying ideas** Some stores sell jars of "fireplace crystals." When sprinkled on a log, these crystals make the flames blue, red, green, and violet. Explain how these crystals can change the flame's color. What ingredients would you expect the crystals to contain?

Extensions

10. **Designing experiments** A student performed flame tests on several unknown substances and observed that they all were shades of red. What could the student do to correctly identify these substances? Explain your answer.

11. **Designing experiments** During a flood, the labels from three bottles of chemicals were lost. The three unlabeled bottles of white solids were known to contain the following substances: strontium nitrate, ammonium carbonate, and potassium sulfate. Explain how you could easily test the substances and relabel the three bottles. (Hint: Ammonium ions do not provide a distinctive flame color.)

3 Spectroscopy and Flame Tests
Identifying Materials

THE PROBLEM

January 27, 2004

Director of Investigations
CheMystery Labs, Inc.
52 Fulton Street
Springfield, VA 22150

Dear Director:

As you may have seen in news reports, one of our freelance pilots, David Matthews, was killed in a crash of an experimental airplane.

The reports did not mention that Matthews's airplane was a recently perfected design that he had developed for us. The notes he left behind indicate that the coating on the nose cone was the key to the plane's speed and maneuverability. Unfortunately, he did not reveal what substances he used, and we were able to recover only flakes of material from the nose cone after the accident.

We have sent you samples of these flakes dissolved in a solution. Please identify the material Matthews used so that we can duplicate his prototype. We will pay $200,000 for this work, provided that you can identify the material within three days.

Sincerely,

Jared MacLaren

Jared MacLaren
Experimental Testing Agency

References

Review information about spectroscopic analysis. The procedure is similar to one your team recently completed to identify an unknown metal in a solution. As before, use small amounts of metal, and clean equipment carefully to avoid contamination. Perform multiple trials for each sample.

The following information is the bright-line emission data (in nm) for the four possible metals.

- Lithium: 670, 612, 498, 462
- Potassium: 700, 695, 408, 405
- Strontium: 710, 685, 665, 500, 490, 485, 460, 420, 405
- Calcium: 650, 645, 610, 485, 460, 445, 420

THE PLAN

CheMystery Labs, Inc. 52 Fulton Street, Springfield, VA 22150

CheMystery Labs, Inc.
52 Fulton Street
Springfield, VA 22150

Memorandum

Date: January 28, 2004
To: Edwin Thien
From: Marissa Bellinghausen

We have narrowed down the material used to four possibilities. It is a compound of either lithium, potassium, strontium, or calcium. Using flame tests and the wavelengths of spectroscopic analysis, you should be able to identify which of these is in the sample.

Because our contract depends on timeliness, give me a preliminary report that includes the following as soon as possible:
- a detailed, one-page summary of your plan for the procedure
- an itemized list of equipment

After you complete your analysis, prepare a report in the form of a two-page letter to MacLaren. The report must include the following:
- the identity of the metal in the sample
- a summary of your procedure
- a detailed and organized analysis and data sections showing tests and results

Required Precautions

- Wear safety goggles when working around chemicals, acids, bases, flames, or heating devices. Contents under pressure may become projectiles and cause serious injury.
- Avoid wearing contact lenses in the lab.
- If any substance gets in your eyes, notify your instructor immediately and flush your eyes with running water for at least 15 min.
- Secure loose clothing, and remove dangling jewelry. Don't wear open-toed shoes or sandals in the lab.

- Wear an apron or lab coat to protect your clothing when working with chemicals.
- If a spill gets on your clothing, rinse it off immediately with water for at least 5 min while notifying your instructor.
- Always use caution when working with chemicals.
- Never taste, touch, or smell chemicals unless specifically directed to do so.
- Follow instructions for proper disposal.
- Whenever possible, use an electric hot plate as a heat source instead of an open flame.

- When heating materials in a test tube, always angle the test tube away from yourself and others.
- Know your school's fire-evacuation routes.
- Clean and decontaminate all work surfaces and personal protective equipment as directed by your instructor.
- Dispose of all sharps (broken glass and other contaminated sharp objects) and other contaminated materials (biological and chemical) in special containers as directed by your instructor.

OBJECTIVES

◆ Observe the physical properties of common elements.

◆ Observe the properties and trends in the elements on the periodic table.

◆ Draw conclusions and identify unknown elements based on observed trends in properties.

USING SCIENTIFIC **METHODS**

MATERIALS

◆ blank periodic table

◆ elemental samples of Ar, C, Cu, Sn, and Pb

◆ note cards, 3 × 5

◆ periodic table

4 The Mendeleev Lab of 1869

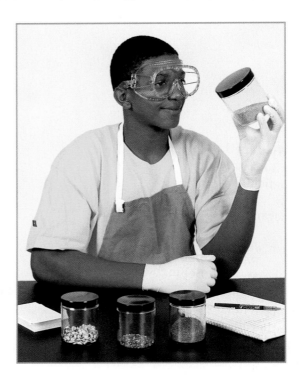

Introduction

Russian chemist Dmitri Mendeleev is generally credited as being the first chemist to observe that patterns emerge when the elements are arranged according to their properties. Mendeleev's arrangement of the elements was unique because he left blank spaces for elements that he claimed were undiscovered as of 1869. Mendeleev was so confident that he even predicted the properties of these undiscovered elements. His predictions were eventually proven to be quite accurate, and these new elements fill the spaces that originally were blank in his table.

Use your knowledge of the periodic table to determine the identity of each of the nine unknown elements in this activity. The unknown elements are from the groups in the periodic table that are listed below. Each group listed below contains at least one unknown element.

1   2   11   13   14   17   18

None of the known elements serves as one of the nine unknown elements.

No radioactive elements are used during this experiment. The relevant radioactive elements include Fr, Ra, At, and Rn. You may not use your textbook or other reference materials. You have been provided with enough information to determine each of the unknown elements.

Safety Procedures

- Wear safety goggles when working around chemicals, acids, bases, flames, or heating devices. Contents under pressure may become projectiles and cause serious injury.
- Never look directly at the sun through any optical device or use direct sunlight to illuminate a microscope.
- Avoid wearing contact lenses in the lab.

- If any substance gets in your eyes, notify your instructor immediately and flush your eyes with running water for at least 15 min.
- Secure loose clothing, and remove dangling jewelry. Don't wear open-toed shoes or sandals in the lab.
- Wear an apron or lab coat to protect your clothing when working with chemicals.
- If a spill gets on your clothing, rinse it off immediately with water for at least 5 minutes while notifying your instructor.

Data Table 1

Unknown	Element
1	
2	
3	
4	

Procedure

1. Copy the data table in your lab notebook. Be sure that you have plenty of room for observations about each test.

2. Use the note cards to copy the information listed on each of the sample cards in the worksheets that your teacher has given you. If the word *observe* is listed, you will need to visually inspect the sample and then write the observation in the appropriate space.

3. Arrange the note cards of the known elements in a crude representation of the periodic table. In other words, all of the known elements from Group 1 should be arranged in the appropriate order. Arrange all of the other cards accordingly.

4. Once the cards of the known elements are in place, inspect the properties of the unknowns to see where their properties would best "fit" the trends of the elements of each group.

5. Assign the proper element name to each of the unknowns. Add the symbol for each one of the unknown elements to your data table.

6. Clean up your lab station, and return the left-over note cards and samples of the elements to your teacher. Do not pour any of the samples down the drain or in the trash unless your teacher directs you to do so. Wash your hands thoroughly before you leave the lab and after all your work is finished.

Conclusions

1. **Interpreting information** Summarize your group's reasoning for the assignment of each unknown. Explain in a few sentences exactly how you predicted the identity of the nine unknown elements.

Skills Practice Lab

7 Percent Composition of Hydrates

Introduction

You are a research chemist working for a company that is developing a new chemical moisture absorber and indicator. The company plans to seal the moisture absorber into a transparent, porous pouch attached to a cellophane window on the

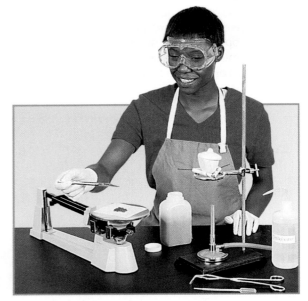

inside of packages for compact disc players. This way, moisture within the packages will be absorbed, and any package that has too much moisture can be quickly detected and dried out. Your company's efforts have focused on copper(II) sulfate, $CuSO_4$, which can absorb water to become a hydrate that shows a distinctive color change.

When many ionic compounds are crystallized from a water solution, they include individual water molecules as part of their crystalline structure. If the substances are heated, this water of crystallization may be driven off and leave behind the pure anhydrous form of the compound. Because the law of multiple proportions also applies to crystalline hydrates, the number of moles of water driven off per mole of the anhydrous compound should be a simple whole-number ratio. You can use this information to help you determine the formula of the hydrate.

To help your company decide whether $CuSO_4$ is the right substance for the moisture absorber and indicator, you will need to examine the hydrated and anhydrous forms of the compound and determine the following:

* the empirical formula of the hydrate, including its water of crystallization
* if the compound is useful as an indicator when it changes from the hydrated to the anhydrous form
* the mass of water absorbed by the 25 g of anhydrous compound, which the company proposes to use

Even if you can guess what the formula for the hydrate should be, carefully perform this lab so that you know how well your company's supply of $CuSO_4$ absorbs moisture.

OBJECTIVES

* Demonstrate proficiency in using the balance and the Bunsen burner.

* Determine that all the water has been driven from a hydrate by heating your sample to a constant mass.

* Relate results to the law of conservation of mass and the law of multiple proportions.

* Perform calculations by using the molar mass.

* Analyze the results and determine the empirical formula of the hydrate and its percentage by mass of water.

MATERIALS

* balance
* Bunsen burner
* crucible and cover
* crucible tongs
* $CuSO_4$, hydrated crystals
* desiccator
* distilled water
* dropper or micropipet
* ring and pipe-stem triangle
* ring stand
* spatula
* stirring rod, glass
* weighing paper

Safety Procedures

- Wear safety goggles when working around chemicals, acids, bases, flames, or heating devices. Contents under pressure may become projectiles and cause serious injury.
- Never look directly at the sun through any optical device or use direct sunlight to illuminate a microscope.
- Avoid wearing contact lenses in the lab.
- If any substance gets in your eyes, notify your instructor immediately and flush your eyes with running water for at least 15 min.

- Secure loose clothing, and remove dangling jewelry. Don't wear open-toed shoes or sandals in the lab.
- Wear an apron or lab coat to protect your clothing when working with chemicals.
- If a spill gets on your clothing, rinse it off immediately with water for at least 5 min while notifying your instructor.

- If a chemical gets on your skin or clothing or in your eyes, rinse it immediately, and alert your instructor.
- If a chemical is spilled on the floor or lab bench, alert your instructor, but do not clean it up yourself unless your teacher says it is OK to do so.

- Always use caution when working with chemicals.
- Never mix chemicals unless specifically directed to do so.
- Never taste, touch, or smell chemicals unless specifically directed to do so.

- Add acid or base to water; never do the opposite.
- Never return unused chemicals to the original container.
- Never transfer substances by sucking on a pipette or straw; use a suction bulb.
- Follow instructions for proper disposal.

- Avoid wearing hair spray or hair gel on lab days.
- Whenever possible, use an electric hot plate as a heat source instead of an open flame.
- When heating materials in a test tube, always angle the test tube away from yourself and others.
  Glass containers used for heating should be made of heat-resistant glass.
- Know your school's fire-evacuation routes.

- Check the condition of glassware before and after using it. Inform your teacher of any broken, chipped, or cracked glassware, because it should not be used.
- Do not pick up broken glass with your bare hands. Place broken glass in a specially designated disposal container.

- Clean and decontaminate all work surfaces and personal protective equipment as directed by your instructor.
- Dispose of all sharps (broken glass and other contaminated sharp objects) and other contaminated materials (biological and chemical) in special containers as directed by your instructor.

Data Table 1	
Mass of empty crucible and cover	
Initial mass of sample, crucible, and cover	
Mass of sample, crucible, and cover after first heating	
Mass of sample, crucible, and cover after second heating	
Constant mass of sample, crucible, and cover	

Procedure

1. Copy Data Table 1 in your lab notebook. Be sure that you have plenty of room for observations about each test.

2. Make sure that your equipment is very clean so that you will get the best possible results. Once you have heated the crucible and cover, do not touch them with your bare hands. Remember that you will need to cool the heated crucible in the desiccator before you measure its mass. Never put a hot crucible on a balance; it will damage the balance.

3. Place the crucible and cover on the triangle with the lid slightly tipped, as shown in **Figure A.** The small opening will allow gases to escape. Heat the crucible and cover until the crucible glows slightly red. Use the tongs to transfer the crucible and cover to the desiccator, and allow them to cool for 5 min. Determine the mass of the crucible and cover to the nearest 0.01 g, and record the mass in your data table.

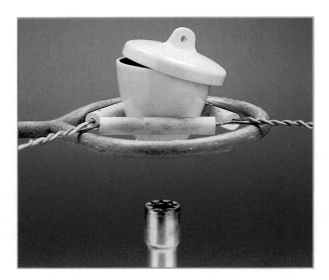

Figure A

4. Using a spatula, add approximately 5 g of copper sulfate hydrate crystals to the crucible. Break up any large crystals before placing them in the crucible. Determine the mass of the covered crucible and crystals to the nearest 0.01 g, and record the mass in your data table.

5. Place the crucible with the copper sulfate hydrate on the triangle, and again position the cover so there is only a small opening. If the opening is too large, the crystals may spatter as they are heated. Heat the crucible very gently on a low flame to avoid spattering. Increase the temperature gradually for 2 or 3 min, and then heat until the crucible glows red for at least 5 min. Be very careful not to raise the temperature of the crucible and its contents too suddenly. You will observe a color change, which is normal, but if the substance remains yellow after cooling, it was overheated and has begun to decompose. Allow the crucible, cover, and contents to cool for 5 min in the desiccator, and then measure their mass. Record the mass in your data table.

6. Heat the covered crucible and contents to redness again for 5 min. Allow the crucible, cover, and contents to cool in the desiccator, and then determine their mass and record it in the data table. If the two mass measurements differ by no more than 0.01 g, you may assume that all of the water has been driven off. Otherwise, repeat the process until the mass no longer changes, which indicates that all of the water has evaporated. Record this constant mass in your data table.

7. After recording the constant mass, set aside a part of your sample on a piece of weighing paper. Using the dropper or pipet, as shown in **Figure B,** put a few drops of water onto this sample to rehydrate the crystals. Record your observations in your lab notebook.

Figure B

8. Clean all apparatus and your lab station. Make sure to completely shut off the gas valve before leaving the laboratory. Remember to wash your hands thoroughly. Place the rehydrated and anhydrous chemicals in the disposal containers designated by your teacher.

Analysis

1. **Explaining events**
   Why do you need to heat the clean crucible before using it in this lab? Why do the tongs used throughout this lab need to be especially clean?

2. **Explaining events**
   Why do you need to use a cover for the crucible? Could you leave the cover off each time you measure the mass of the crucible and its contents and still get accurate results? Explain your answer.

3. **Examining data**
   Calculate the mass of anhydrous copper sulfate (the residue that remains after heating to constant mass) by subtracting the mass of the empty crucible and cover from the mass of the crucible, cover, and heated $CuSO_4$. Use the molar mass for $CuSO_4$, determined from the periodic table, to calculate the number of moles present.

4. **Analyzing data**
   Calculate the mass and moles of water originally present in the hydrate by using the molar mass determined from the periodic table.

Conclusions

5. **Interpreting information** Explain why the mass of the sample decreased after it was heated, despite the law of conservation of mass.

6. **Drawing conclusions** Using your answers from items 3 and 4, determine the empirical formula for the copper sulfate hydrate.

7. **Analyzing results** What is the percentage by mass of water in the original hydrated compound?

8. **Applying conclusions** How much water could 25 g of anhydrous $CuSO_4$ absorb?

9. **Applying conclusions** When you rehydrated the small amount of anhydrous copper sulfate, what were your observations? Explain whether this substance would make a good indicator of moisture.

10. **Applying conclusions** Some cracker tins include a glass vial of drying material in the lid. This is often a mixture of magnesium sulfate and cobalt chloride. As the mixture absorbs moisture to form hydrated compounds, the cobalt chloride changes from blue-violet $CoCl_2 \cdot 2H_2O$ to pink $CoCl_2 \cdot 6H_2O$. When this hydrated mixture becomes totally pink, it can be restored to the dihydrate form by being heated in the oven. Write equations for the reactions that occur when this mixture is heated.

11. **Drawing conclusions** Three pairs of students obtained the results in the table below when they heated a solid. In each case, the students observed that when they began to heat the solid, drops of a liquid formed on the sides of the test tube.

    a. Could the solid be a hydrate? Explain how you could find out.

    b. If the solid has a molar mass of 208 g/mol after being heated, how many formula units of water are there in one formula unit of the unheated compound?

Data Table 2

Sample number	Mass before heating (g)	Constant mass after heating (g)
1	1.92	1.26
2	2.14	1.40
3	2.68	1.78

Extensions

12. **Designing experiments** Some electronic equipment is packaged for shipping with a small packet of drying material. You are interested in finding out whether the electronic equipment was exposed to moisture during shipping. How could you determine this?

7 Hydrates
Gypsum and Plaster of Paris

Director of Research
CheMystery Labs, Inc.
52 Fulton Street
Springfield, VA 22150

Dear Director:

Lost Art Gypsum Mine previously sold its raw gypsum to a manufacturing company that used the gypsum to make anhydrous calcium sulfate, $CaSO_4$ (a desiccant), and plaster of Paris. That company has now gone out of business, and we are currently negotiating the purchase of the firm's equipment to process our own gypsum into $CaSO_4$ and plaster of Paris.

Your company has been recommended to plan the large-scale industrial process for our new plant. We will need a detailed report on the development of the process and formulas for these products. This report will be presented to the bank handling our loan for the new plant. As we discussed on the telephone today, we are willing to pay you $250,000 for the work, and the contract papers will arrive under separate cover today.

Sincerely,

Alex Farros

Alex Farros
Vice President
Lost Art Gypsum Mine

References

Review information about hydrates and water of crystallization. Gypsum and plaster of Paris are hydrated forms of calcium sulfate, $CaSO_4$. One of the largest gypsum mines in the world is located outside Paris, France. Plaster of Paris contains less water of crystallization than gypsum. Plaster of Paris is commonly used in plaster walls and art sculptures.

CheMystery Labs, Inc. 52 Fulton Street, Springfield, VA 22150

CheMystery Labs, Inc.
52 Fulton Street
Springfield, VA 22150

Memorandum

Date: February 10, 2004
To: Kenesha Smith
From: Martha Li-Hsien

Your team needs to develop a procedure to experimentally determine the correct empirical formulas for both hydrates of this anhydrous compound. You will use gypsum samples from the mine and samples of the plaster of Paris product.

As soon as possible, I need a preliminary report from you that includes the following:
- a detailed one-page summary of your plan for the procedure, including all necessary data tables
- an itemized list of equipment

After you complete the analysis, prepare a two-page report that includes the following information:
- formulas for anhydrous calcium sulfate, plaster of Paris, and gypsum
- a summary of your procedure
- detailed and organized data and analysis sections that show calculations, along with estimates and explanations of any possible sources of error

Required Precautions

- Wear safety goggles when working around chemicals, acids, bases, flames, or heating devices. Contents under pressure may become projectiles and cause serious injury.
- Avoid wearing contact lenses in the lab.
- If any substance gets in your eyes, notify your instructor immediately and flush your eyes with running water for at least 15 min.
- Secure loose clothing, and remove dangling jewelry. Don't wear open-toed shoes or sandals in the lab.

- Wear an apron or lab coat to protect your clothing when working with chemicals.
- If a spill gets on your clothing, rinse it off immediately with water for at least 5 min while notifying your instructor.
- Always use caution when working with chemicals.
- Never taste, touch, or smell chemicals unless specifically directed to do so.
- Follow instructions for proper disposal.
- Whenever possible, use an electric hot plate as a heat source instead of an open flame.

- When heating materials in a test tube, always angle the test tube away from yourself and others.
- Know your school's fire-evacuation routes.
- Clean and decontaminate all work surfaces and personal protective equipment as directed by your instructor.
- Dispose of all sharps (broken glass and other contaminated sharp objects) and other contaminated materials (biological and chemical) in special containers as directed by your instructor.

9 Stoichiometry and Gravimetric Analysis

OBJECTIVES

♦ Observe the reaction between strontium chloride and sodium carbonate, and write a balanced equation for the reaction.

♦ Demonstrate proficiency with gravimetric methods.

♦ Measure the mass of insoluble precipitate formed.

USING SCIENTIFIC **METHODS**

♦ Relate the mass of precipitate formed to the mass of reactants before the reaction.

♦ Calculate the mass of sodium carbonate in a solution of unknown concentration.

MATERIALS

♦ balance
♦ beaker tongs
♦ beakers, 250 mL (3)
♦ distilled water
♦ drying oven
♦ filter paper
♦ glass funnel or Büchner funnel
♦ glass stirring rod
♦ graduated cylinder, 100 mL
♦ Na_2CO_3 solution (15 mL)
♦ ring and ring stand
♦ rubber policeman
♦ spatula
♦ $SrCl_2$ solution, 0.30 M (45 mL)
♦ water bottle

Introduction

You are working for a company that makes water-softening agents for homes with hard water. Recently, there was a mix-up on the factory floor, and sodium carbonate solution was mistakenly mixed in a vat with an unknown quantity of distilled water. You must determine the amount of Na_2CO_3 in the vat in order to predict the percentage yield of the water-softening product.

When chemists are faced with problems that require them to determine the quantity of a substance by mass, they often use a technique called gravimetric analysis. In this technique, a small sample of the material undergoes a reaction with an excess of another reactant. The chosen reaction is one that almost always provides a yield near 100%. If the mass of the product is carefully measured, you can use stoichiometry calculations to determine how much of the reactant of unknown amount was involved in the reaction. Then by comparing the size of the analysis sample with the size of the original material, you can determine exactly how much of the substance is present.

This procedure involves a double-displacement reaction between strontium chloride, $SrCl_2$, and sodium carbonate, Na_2CO_3. In general, this reaction can be used to determine the amount of any carbonate compound in a solution.

You will react an unknown amount of sodium carbonate with an excess of strontium chloride. After purifying the product, you will determine the following:

• how much product is present
• how much Na_2CO_3 must have been present to produce that amount of product
• how much Na_2CO_3 is contained in the 575 L of solution

Safety Procedures

- Wear safety goggles when working around chemicals, acids, bases, flames, or heating devices. Contents under pressure may become projectiles and cause serious injury.
- Never look directly at the sun through any optical device or use direct sunlight to illuminate a microscope.
- Avoid wearing contact lenses in the lab.
- If any substance gets in your eyes, notify your instructor immediately and flush your eyes with running water for at least 15 min.

- Secure loose clothing, and remove dangling jewelry. Do not wear open-toed shoes or sandals in the lab.
- Wear an apron or lab coat to protect your clothing when working with chemicals.
- If a spill gets on your clothing, rinse it off immediately with water for at least 5 min while notifying your instructor.

- Always use caution when working with chemicals.
- Never mix chemicals unless specifically directed to do so.
- Never taste, touch, or smell chemicals unless specifically directed to do so.

- Add acid or base to water; never do the opposite.
- Never return unused chemicals to the original container.
- Never transfer substances by sucking on a pipette or straw; use a suction bulb.
- Follow instructions for proper disposal.

- Avoid wearing hair spray or hair gel on lab days.
- Whenever possible, use an electric hot plate as a heat source instead of an open flame.
- When heating materials in a test tube, always angle the test tube away from yourself and others.
- Glass containers used for heating should be made of heat-resistant glass.
- Know your school's fire-evacuation routes.

- Clean and decontaminate all work surfaces and personal protective equipment as directed by your instructor.
- Dispose of all sharps (broken glass and other contaminated sharp objects) and other contaminated materials (biological and chemical) in special containers as directed by your instructor.

Data Table 1
Volume of Na_2CO_3 solution added
Volume of $SrCl_2$ solution added
Mass of dry filter paper
Mass of beaker with paper towel
Mass of beaker with paper towel, filter paper, and precipitate
Mass of precipitate

Procedure

1. Organizing Data

Copy the data table in your lab notebook. Be sure that you have plenty of room for observations about each test.

2. Clean all of the necessary lab equipment with soap and water. Rinse each piece of equipment with distilled water.

3. Measure the mass of a piece of filter paper to the nearest 0.01 g, and record this value in your data table.

4. Refer to page 764 to set up a filtering apparatus, either a Büchner funnel or a gravity filtration, depending on what equipment is available.

5. Label a paper towel with your name, your class, and the date. Place the towel in a clean, dry 250 mL beaker, and measure and record the mass of the towel and beaker to the nearest 0.01 g.

6. Measure about 15 mL of the Na_2CO_3 solution into the graduated cylinder. Record this volume to the nearest 0.5 mL in your data table. Pour the Na_2CO_3 solution into a clean, empty 250 mL beaker. Carefully wash the graduated cylinder, and rinse it with distilled water.

7. Measure about 25 mL of the 0.30 M $SrCl_2$ solution into the graduated cylinder. Record this volume to the nearest 0.5 mL in your data table. Pour the $SrCl_2$ solution into the beaker with the Na_2CO_3 solution, as shown in **Figure A.** Gently stir the solution and precipitate with a glass stirring rod.

Figure A
Graduated cylinder pouring solution into beaker.

8. Carefully measure another 10 mL of $SrCl_2$ into the graduated cylinder. Record the volume to the nearest 0.5 mL in your data table. Slowly add it to the beaker. Repeat this step until no more precipitate forms.

9. Once the precipitate has settled, slowly pour the mixture into the funnel. Be careful not to overfill the funnel because some of the precipitate could be lost between the filter paper and the funnel. Use the rubber policeman to transfer as much of the precipitate into the funnel as possible.

10. Rinse the rubber policeman into the beaker with a small amount of distilled water, and pour this solution into the funnel. Rinse the beaker several more times with small amounts of distilled water, as shown in **Figure B.** Pour the rinse water into the funnel each time.

Figure B
Washing a beaker with water bottle

11. After all of the solution and rinses have drained through the funnel, slowly rinse the precipitate on the filter paper in the funnel with distilled water to remove any soluble impurities.

12. Carefully remove the filter paper from the funnel, and place it on the paper towel that you have labeled with your name. Unfold the filter paper, and place the paper towel, filter paper, and precipitate in the rinsed beaker. Then place the beaker in the drying oven. For best results, allow the precipitate to dry overnight.

13. Using beaker tongs, remove your sample from the drying oven, and allow it to cool. Measure and record the mass of the beaker with paper towel, filter paper, and precipitate to the nearest 0.01 g.

14. Dispose of the precipitate in a designated waste container. Pour the filtrate in the other 250 mL beaker into the designated waste container. Clean up the lab and all equipment after use, and dispose of substances according to your teacher's instructions. Wash your hands thoroughly after all lab work is finished and before you leave the lab.

Analysis

1. Organizing Data
Write a balanced equation for the reaction. What is the precipitate? Write its empirical formula. (Hint: It was a double-displacement reaction.)

2. Examining Data
Calculate the mass of the dry precipitate. Calculate the number of moles of precipitate produced in the reaction. (Hint: Use the results from step 13.)

3. Examining Data
How many moles of Na_2CO_3 were present in the 15 mL sample?

Conclusions

4. Evaluating Methods
There was 0.30 mol of $SrCl_2$ in every liter of solution. Calculate the number of moles of $SrCl_2$ that were added. Determine whether $SrCl_2$ or Na_2CO_3 was the limiting reactant. Would this experiment have worked if the other reactant had been chosen as the limiting reactant? Explain why or why not.

5. Evaluating Methods
Why was the precipitate rinsed in step 11? What soluble impurities could have been on the filter paper along with the precipitate? How would the calculated results vary if the precipitate had not been completely dry? Explain your answer.

6. Applying Conclusions
How many grams of Na_2CO_3 were present in the 15 mL sample?

7. Applying Conclusions
How many grams of Na_2CO_3 are present in the 575 L? (Hint: Create a conversion factor to convert from the sample, with a volume of 15 mL, to the entire solution, with a volume of 575 L.)

8. Evaluating Methods
Ask your teacher for the theoretical mass of Na_2CO_3 in the sample, and calculate your percentage error.

Extensions

1. Designing Experiments
What possible sources of error can you identify with your procedure? If you can think of ways to eliminate them, ask your teacher to approve your plan, and run the procedure again.

9 Gravimetric Analysis
Hard Water Testing

March 3, 2004

George Taylor, Director of Analysis
CheMystery Labs, Inc.
52 Fulton Street
Springfield, VA 22150

Dear Mr. Taylor:

The city's Public Works Department is investigating new sources of water. One proposal involves drilling wells into a nearby aquifer that is protected from brackish water by a unique geological formation. Unfortunately, this formation is made of calcium minerals. If the concentration of calcium ions in the water is too high, the water will be "hard," and treating it to meet local water standards would be too expensive for us.

Water containing more than 120 mg of calcium per liter is considered hard. I have enclosed a sample of water that has been distilled from 1.0 L to its present volume. Please determine whether the water is of suitable quality.

We are seeking a firm to be our consultant for the entire testing process. Interested firms will be evaluated based on this water analysis. We look forward to receiving your report.

Sincerely,

Dana Rubio

Dana Rubio
City Manager

References

Review the "Stoichiometry" chapter for information about mass-mass stoichiometry. In this investigation, you will use a double-displacement reaction, but Na_2CO_3 will be used as a reagent to identify how much calcium is present in a sample. Like strontium and other Group 2 metals, calcium salts react with carbonate-containing salts to produce an insoluble precipitate.

THE PLAN

CheMystery Labs, Inc. 52 Fulton Street, Springfield, VA 22150

CheMystery Labs, Inc.
52 Fulton Street
Springfield, VA 22150

Memorandum

Date: March 4, 2004
To: Shane Thompson
From: George Taylor

We can solve the city's problem by doing some careful gravimetric analysis, because calcium salts and carbonate compounds undergo double-displacement reactions to yield insoluble calcium carbonate as a precipitate.

Before you begin your work, I will need the following information from you so that I can create our bid:
- a detailed one-page summary of your plan for the procedure as well as all necessary data tables
- a description of necessary calculations
- an itemized list of equipment

After you complete the analysis, prepare a two-page report for Dana Rubio. Make sure to include the following items:
- a calculation of calcium concentration in mg/L for the water from the aquifer
- an explanation of how you determined the amount of calcium in the sample, including measurements and calculations
- a balanced chemical equation for the reaction
- explanations and estimations for any possible sources of error

Required Precautions

- Wear safety goggles when working around chemicals, acids, bases, flames, or heating devices. Contents under pressure may become projectiles and cause serious injury.
- Avoid wearing contact lenses in the lab.
- If any substance gets in your eyes, notify your instructor immediately and flush your eyes with running water for at least 15 min.
- Secure loose clothing, and remove dangling jewelry. Don't wear open-toed shoes or sandals in the lab.

- Wear an apron or lab coat to protect your clothing when working with chemicals.
- If a spill gets on your clothing, rinse it off immediately with water for at least 5 min while notifying your instructor.

- Always use caution when working with chemicals.
- Never taste, touch, or smell chemicals unless specifically directed to do so.
- Follow instructions for proper disposal.
- Whenever possible, use an electric hot plate as a heat source instead of an open flame.

- When heating materials in a test tube, always angle the test tube away from yourself and others.
- Know your school's fire-evacuation routes.
- Clean and decontaminate all work surfaces and personal protective equipment as directed by your instructor.

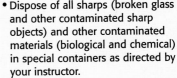

- Dispose of all sharps (broken glass and other contaminated sharp objects) and other contaminated materials (biological and chemical) in special containers as directed by your instructor.

Skills Practice Lab

OBJECTIVES

- **Demonstrate** proficiency in the use of calorimeters and related equipment.

- **Relate** temperature changes to enthalpy changes.

- **Determine** the heat of reaction for several reactions.

- **Demonstrate** that the heat of reaction can be additive.

MATERIALS

- balance
- distilled water
- glass stirring rod
- graduated cylinder, 100 mL
- HCl solution, 0.50 M (100 mL)
- HCl solution, 1.0 M (50 mL)
- NaOH pellets (4 g)
- NaOH solution, 1.0 M (50 mL)
- plastic-foam cups (or calorimeters)
- spatula
- thermometer
- watch glass

OPTIONAL EQUIPMENT

- CBL unit
- graphing calculator with link cable
- Vernier temperature probe

10 Calorimetry and Hess's Law

Introduction

A man working for a cleaning firm was told by his employer to pour some old cleaning supplies into a glass container for disposal. Some of the supplies included muriatic (hydrochloric) acid, HCl(*aq*), and a drain cleaner containing lye, NaOH(*s*). When the substances were mixed, the container shattered, spilling the contents onto the worker's arms and legs. The worker claims that the hot spill

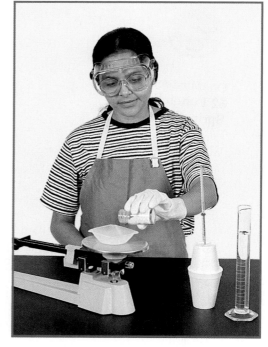

caused burns, and he is therefore suing his employer. The employer claims that the worker is lying because the solutions were at room temperature before they were mixed. The employer says that a chemical burn is unlikely because tests after the accident revealed that the mixture had a neutral pH, indicating that the HCl and NaOH were neutralized. The court has asked you to evaluate whether the worker's story is supported by scientific evidence.

Chemicals can be dangerous because of their special storage needs. Chemicals that are mixed and react are even more dangerous because many reactions release large amounts of heat. Glass is heat-sensitive and can shatter if there is a sudden change in temperature due to a reaction. Some glassware, such as Pyrex, is heat-conditioned but can still fracture under extreme heat conditions, especially if scratched.

You will measure the amount of heat released by mixing the chemicals in two ways. First you will break the reaction into steps and measure the heat change of each step. Then you will measure the heat change of the reaction when it takes place all at once. When you are finished, you will be able to use the calorimetry equation from the chapter "Causes of Change" to determine the following:

- the amount of heat evolved during the overall reaction
- the amount of heat for each step
- the amount of heat for the reaction in kilojoules per mole
- whether this heat could have raised the temperature of the water in the solution high enough to cause a burn

Safety Procedures

- Wear safety goggles when working around chemicals, acids, bases, flames, or heating devices. Contents under pressure may become projectiles and cause serious injury.
- Never look directly at the sun through any optical device or use direct sunlight to illuminate a microscope.
- Avoid wearing contact lenses in the lab.
- If any substance gets in your eyes, notify your instructor immediately and flush your eyes with running water for at least 15 min.

- Secure loose clothing, and remove dangling jewelry. Don't wear open-toed shoes or sandals in the lab.
- Wear an apron or lab coat to protect your clothing when working with chemicals.
- If a spill gets on your clothing, rinse it off immediately with water for at least 5 min while notifying your instructor.

- If a chemical gets on your skin or clothing or in your eyes, rinse it immediately, and alert your instructor.
  If a chemical is spilled on the floor or lab bench, alert your instructor, but do not clean it up yourself unless your teacher says it is OK to do so.

- Always use caution when working with chemicals.
- Never mix chemicals unless specifically directed to do so.
- Never taste, touch, or smell chemicals unless specifically directed to do so.
- Add acid or base to water; never do the opposite.
- Never return unused chemicals to the original container.
- Never transfer substances by sucking on a pipette or straw; use a suction bulb.
- Follow instructions for proper disposal.

- Check the condition of glassware before and after using it. Inform your teacher of any broken, chipped, or cracked glassware, because it should not be used.
- Do not pick up broken glass with your bare hands. Place broken glass in a specially designated disposal container.

- Clean and decontaminate all work surfaces and personal protective equipment as directed by your instructor.
- Dispose of all sharps (broken glass and other contaminated sharp objects) and other contaminated materials (biological and chemical) in special containers as directed by your instructor.

Procedure

1. Copy the data table below in your lab notebook. Reactions 1 and 3 will each require two additional spaces to record the mass of the empty watch glass and the mass of the watch glass with NaOH.

2. If you are not using a plastic-foam cup as a calorimeter, ask your teacher for instructions on using the calorimeter. At various points in the procedure, you will need to measure the temperature of the solution within the calorimeter.

Thermometer procedure continues on page 796.

Data Table 1

	Reaction 1	Reaction 2	Reaction 3
Total volumes of liquid(s)			
Initial temperature			
Final temperature			
Mass of empty watch glass			
Mass of watch glass with NaOH			

CBL and Sensors

3. Connect the CBL to the graphing calculator with the unit-to-unit link cable using the I/O ports located on each unit. Connect the temperature probe to the CH1 port. Turn on the CBL and the graphing calculator. Start the program CHEMBIO on the graphing calculator.

   a. Select option *SET UP PROBES* from the MAIN MENU. Enter 1 for the number of probes. Select the temperature probe from the list. Enter 1 for the channel number. Select *USE STORED* from the CALIBRATION menu.

   b. Select the *COLLECT DATA* option from the MAIN MENU. Select the *TRIGGER* option from the DATA COLLECTION menu.

4. Measure the temperature by gently inserting the Vernier temperature probe into the hole in the calorimeter lid.

Reaction 1: Dissolving NaOH

5. Pour about 100 mL of distilled water into a graduated cylinder. Measure and record the volume of the water to the nearest 0.1 mL. Pour the water into your calorimeter.

6. Using the temperature probe, measure the temperature of the water. Press TRIGGER on the CBL to collect the temperature reading. Record this temperature in your data table. Select *STOP* from the TRIGGER menu on the graphing calculator. Leave the probe in the calorimeter.

7. Select the *COLLECT DATA* option from the MAIN MENU. Select the *TIME GRAPH* option from the DATA COLLECTION menu. Enter 6 for the time between samples, in sec-

onds. Enter 99 for the number of samples (the CBL will collect data for 9.9 min). Press ENTER. Select *USE TIME SETUP* to continue. If you want to change the number of samples or the time between samples, select *MODIFY SETUP.* Enter 0 for *Ymin,* enter 100 for *Ymax,* and enter 5 for *Yscl.*

8. Determine and record the mass of a clean and dry watch glass to the nearest 0.01 g. Remove the watch glass from the balance. While wearing gloves, obtain about 2 g of NaOH pellets, and put them on the watch glass. Use forceps when handling NaOH pellets. Measure and record the mass of the watch glass and the pellets to the nearest 0.01 g. **It is important that this step be done quickly because NaOH is hygroscopic. It absorbs moisture from the air, and its mass increases as long as it remains exposed to the air.**

9. Press ENTER on the graphing calculator to begin collecting the temperature readings for the water in the calorimeter.

10. Immediately place the NaOH pellets in the calorimeter cup, and gently stir the solution with a stirring rod. Place the lid on the calorimeter.

11. When the CBL displays DONE, use the arrow keys to trace the graph. Time in seconds in graphed on the *x*-axis, and the temperature readings are graphed on the *y*-axis. Record the highest temperature reading from the CBL in your data table.

12. When the reaction is finished, pour the solution into the container designated by your teacher for disposal of basic solutions.

13. Be sure to clean all equipment and rinse it with distilled water before continuing with the next procedure.

Reaction 2: NaOH and HCl in solution

14. Pour about 50 mL of 1.0 M HCl into a graduated cylinder. Measure and record the volume of the HCl solution to the nearest 0.1 mL. Pour the HCl solution into your calorimeter.

15. Select the *COLLECT DATA* option from the MAIN MENU. Select the *TRIGGER* option from the DATA COLLECTION menu. Using the temperature probe, measure the temperature of the HCl solution. Press TRIGGER on the CBL to collect the temperature reading. Record this temperature in your data table.

16. Pour about 50 mL of 1.0 M NaOH into a graduated cylinder. Measure and record the volume of the NaOH solution to the nearest 0.1 mL. **For this step only, rinse the temperature probe in distilled water.** Using the temperature probe, measure the temperature of the NaOH solution. Press TRIGGER on the CBL to collect the temperature reading. Record this temperature in your data table. Select *STOP* from the TRIGGER menu on the graphing calculator. Put the probe in the calorimeter.

17. Select the *COLLECT DATA* option from the MAIN MENU. Select the *TIME GRAPH* option from the DATA COLLECTION menu. Enter 6 for the time between samples, in seconds. Enter 99 for the number of samples. Press ENTER. Select *USE TIME SETUP* to continue. If you want to change the number of samples or the time between samples, select *MODIFY SETUP.* Enter 0 for *Ymin*, enter 100 for *Ymax*, and enter 5 for *Yscl*. Press ENTER on the calculator to begin collecting temperature readings.

18. Pour the NaOH solution into the calorimeter cup, and stir gently. Place the lid on the calorimeter.

19. When the CBL displays DONE, use the arrow keys to trace the graph. Time in seconds in

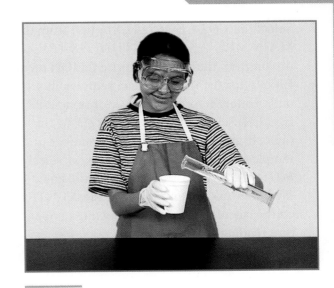

Figure A

graphed on the *x*-axis, and the temperature readings are graphed on the *y*-axis. Record the highest temperature reading from the CBL in your data table.

20. Pour the solution into the container designated by your teacher for disposal of mostly neutral solutions. Clean and rinse all equipment before continuing with the next procedure.

Reaction 3: Solid NaOH and HCl in solution

21. Pour about 100 mL of 0.50 M HCl into a graduated cylinder. Measure and record the volume to the nearest 0.1 mL. Pour the HCl solution into your calorimeter, as shown in **Figure A.**

22. Select the *COLLECT DATA* option from the MAIN MENU. Select the *TRIGGER* option from the DATA COLLECTION menu. Using the temperature probe, measure the temperature of the HCl solution. Press TRIGGER on the CBL to collect the temperature reading. Record this temperature in your data table. Select *STOP* from the TRIGGER menu on the graphing calculator.

23. Select the *COLLECT DATA* option from the MAIN MENU. Select the *TIME GRAPH* option from the DATA COLLECTION menu. Enter 6 for the time between samples, in seconds. Enter 99 for the number of samples. Press ENTER. Select *USE TIME SETUP* to continue. If you want to change the number of samples or the time between samples, select *MODIFY SETUP*. Enter 0 for *Ymin*, enter 100 for *Ymax*, and enter 5 for *Yscl*. Press ENTER on the calculator to begin collecting temperature readings.

24. Measure the mass of a clean and dry watch glass, and record it in your data table. Obtain approximately 2 g of NaOH. Place it on the watch glass, and record the total mass to the nearest 0.01 g. **It is important that this step be done quickly because NaOH is hygroscopic.**

25. Press ENTER on the graphing calculator to begin collecting the temperature readings for the water in the calorimeter.

26. Immediately place the NaOH pellets in the calorimeter, and gently stir the solution. Place the lid on the calorimeter.

27. When the CBL displays DONE, use the arrow keys to trace the graph. Time in seconds in graphed on the *x*-axis, and the temperature readings are graphed on the *y*-axis. Record the highest temperature reading from the CBL in your data table.

28. When the reaction is finished, pour the solution into the container designated by your teacher for disposal of basic solutions.

29. Clean all apparatus and your lab station. Check with your teacher for the proper disposal procedures. Any excess NaOH pellets should be disposed of in the designated container. Always wash your hands thoroughly after cleaning up the lab area and equipment.

Thermometer

3. Measure the temperature by gently inserting the thermometer into the hole in the calorimeter lid, as shown in **Figure B.** The thermometer takes time to reach the same temperature as the solution inside the calorimeter, so wait to be sure you have an accurate reading. **Thermometers break easily, so be careful with them, and do not use them to stir a solution.**

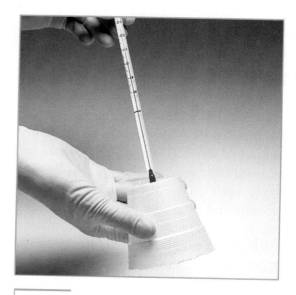

Figure B

Reaction 1: Dissolving NaOH

4. Pour about 100 mL of distilled water into a graduated cylinder. Measure and record the volume of the water to the nearest 0.1 mL. Pour the water into your calorimeter. Measure and record the water temperature to the nearest 0.1°C.

5. Determine and record the mass of a clean and dry watch glass to the nearest 0.01 g. Remove the watch glass from the balance. While wearing gloves, obtain about 2 g of NaOH pellets, and put them on the watch glass. Use forceps when handling NaOH pellets. Measure and record the mass of the watch glass and the pellets to the nearest 0.01 g. **It is important that this step be done quickly because NaOH is hygroscopic. It absorbs moisture from the air, and increases its mass as long as it remains exposed to the air.**

6. Immediately place the NaOH pellets in the calorimeter cup, and gently stir the solution with a stirring rod. **Do not stir with a thermometer.** Place the lid on the calorimeter. Watch the thermometer, and record the highest temperature in the data table. When the reaction is finished, pour the solution into the container designated by your teacher for disposal of basic solutions.

7. Be sure to clean all equipment and rinse it with distilled water before continuing with the next procedure.

Reaction 2: NaOH and HCl in solution

8. Pour about 50 mL of 1.0 M HCl into a graduated cylinder. Measure and record the volume of the HCl solution to the nearest 0.1 mL. Pour the HCl solution into your calorimeter. Measure and record the temperature of the HCl solution to the nearest 0.1°C.

9. Pour about 50 mL of 1.0 M NaOH into a graduated cylinder. Measure and record the volume of the NaOH solution to the nearest 0.1 mL. **For this step only, rinse the thermometer in distilled water, and measure the temperature of the NaOH solution in the graduated cylinder to the nearest 0.1°C. Record the temperature in your data table, and then replace the thermometer in the calorimeter.**

10. Pour the NaOH solution into the calorimeter cup, and stir gently. Place the lid on the calorimeter. Watch the thermometer, and record the highest temperature in the data table. When finished with this reaction, pour the solution into the container designated by your teacher for disposal of mostly neutral solutions.

11. Clean and rinse all equipment before continuing with the next procedure.

Reaction 3: Solid NaOH and HCl in solution

12. Pour about 100 mL of 0.50 M HCl into a graduated cylinder. Measure and record the volume to the nearest 0.1 mL. Pour the HCl solution into your calorimeter, as shown in **Figure C.** Measure and record the temperature of the HCl solution to the nearest 0.1°C.

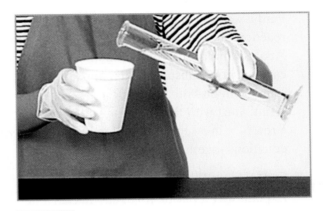

Figure C

13. Measure the mass of a clean and dry watch glass, and record it in your data table. Obtain approximately 2 g of NaOH. Place it on the watch glass, and record the total mass to the nearest 0.01 g. **It is important that this step be done quickly because NaOH is hygroscopic.**

14. Immediately place the NaOH pellets in the calorimeter, and gently stir the solution. Place the lid on the calorimeter. Watch the thermometer, and record the highest temperature in the data table. When finished with this reaction, pour the solution into the container designated by your teacher for disposal of mostly neutral solutions.

15. Clean all apparatus and your lab station. Check with your teacher for the proper disposal procedures. Any excess NaOH pellets should be disposed of in the designated container. Always wash your hands thoroughly after cleaning up the lab area and equipment.

Analysis

1. Organizing Data

Write a balanced chemical equation for each of the three reactions that you performed. (Hint: Be sure to include states of matter for all substances in each equation.)

2. Analyzing Results

Find a way to get the equation for the total reaction by adding two of the equations from Analysis and Interpretation item 1 and then canceling out substances that appear in the same form on both sides of the new equation. (Hint: Start with the equation whose product is a reactant in a second equation. Add those two equations together.)

3. Explaining Events

Explain why a plastic-foam cup makes a better calorimeter than a paper cup does.

4. Organizing Data

Calculate the change in temperature (Δt) for each of the reactions.

5. Organizing Data

Assuming that the density of the water and the solutions is 1.00 g/mL, calculate the mass, m, of liquid present for each of the reactions.

6. Analyzing Results

Using the calorimeter equation, calculate the heat released by each reaction. (Hint: Use the specific heat capacity of water in your calculations; $c_{p,H_2O} = 4.180$ J/g•°C.)

$$\text{Heat} = m \times \Delta t \times c_{p,H_2O}$$

7. Organizing Data

Calculate the moles of NaOH used in each of the reactions. (Hint: To find the number of moles in a solution, multiply the volume in liters by the molar concentration.)

8. Analyzing Results

Calculate the ΔH value in terms of kilojoules per mole of NaOH for each of the three reactions.

9. Analyzing Results

Using your answer to Analysis and Interpretation item 2 and your knowledge of Hess's law from the chapter "Causes of Change," explain how the enthalpies for the three reactions should be mathematically related.

10. Analyzing Results

Which of the following types of heat of reaction apply to the enthalpies calculated in Analysis and Interpretation item 8: heat of combustion, heat of solution, heat of reaction, heat of fusion, heat of vaporization, and heat of formation?

Conclusions

11. Evaluating Methods

Use your answers from Analysis and Interpretation items 7 and 8 to determine the ΔH value for the reaction of solid NaOH with HCl solution by direct measurement and by indirect calculation.

12. Drawing Conclusions

Third-degree burns can occur if skin comes into contact for more than 4 s with water that is hotter than 60°C (140°F). Suppose someone accidentally poured hydrochloric acid into a glass-disposal container that already contained the drain cleaner NaOH and the container shattered. The solution in the container was approximately 55 g of NaOH and 450 mL of hydrochloric acid solution containing 1.35 mol of HCl (a 3.0 M HCl solution). If the initial temperature of the solutions was 25°C, could a mixture hot enough to cause burns have resulted?

13. Applying Conclusions

For the reaction between the drain cleaner and HCl described in item 12, which chemical is the limiting reactant? How many moles of the other reactant remained unreacted?

14. Evaluating Results

When chemists make solutions from NaOH pellets, they often keep the solution in an ice bath. Explain why.

15. Evaluating Methods

You have worked with heats of solution for exothermic reactions. Could the same type of procedure be used to determine the temperature changes for endothermic reactions? How would the procedure stay the same? What would change about the procedure and the data?

16. Drawing Conclusions

Which is more stable, solid NaOH or NaOH solution? Explain your answer.

Extensions

1. Designing Experiments

You have worked with heats of solution for exothermic reactions. Could the same type of procedure be used to determine the temperature changes for endothermic reactions? How would the procedure stay the same? What would change about the procedure and the data?

2. Designing Experiments

A chemical supply company is going to ship NaOH pellets to a very humid place, and you have been asked to give advice on packaging. Design a package for the NaOH pellets. Explain the advantages of your package's design and materials. (Hint: Remember that the reaction in which NaOH absorbs moisture from the air is exothermic and that NaOH reacts exothermically with other compounds as well.)

OBJECTIVES

◆ Conduct a paper chromatography experiment with three different water-soluble colored markers.

◆ Design a successful method to ensure that the chromatography paper remains vertical throughout the experiment.

◆ Observe the dye components of three different water-soluble markers.

MATERIALS

◆ beaker, 250 mL

◆ chromatography paper

◆ developing solution: NaCl solution, 0.1% by mass

◆ graduated cylinder, 10 mL

◆ hot plate

◆ markers

◆ paper clips

◆ pencils

◆ ruler

◆ scissors

13 Paper Chromatography of Colored Markers

Introduction

There is a wide variety of marker products on the market today ranging in color and function. All of these markers contain different dye components that are responsible for their color.

Paper chromatography is an analytical technique that uses paper as a medium to separate the different dye components dissolved in a mixture. In this process, the mixture to be separated is placed on a piece of chromatography paper. A solvent is then allowed to soak up into the paper. As the solvent travels across the paper, some of the components of the mixture are carried with it. Particles of the same component group together. The components that are most soluble and least attracted to the paper travel farther than others. A color band is created and the different components can be seen separated on the paper. The success of chromatography hinges on the slight difference in the physical properties of the individual components.

In this activity you will use a paper chromatography to determine the components of the dyes found in water-soluble markers. Your goal is to use paper chromatography to determine the dye components of three different water-soluble markers. You will also need to design a simple method that will keep the chromatography paper vertical while it is in the developing solution.

Safety Procedures

- Wear safety goggles when working around chemicals, acids, bases, flames, or heating devices. Contents under pressure may become projectiles and cause serious injury.
- Never look directly at the sun through any optical device or use direct sunlight to illuminate a microscope.
- Avoid wearing contact lenses in the lab.
- If any substance gets in your eyes, notify your instructor immediately and flush your eyes with running water for at least 15 min.

- Secure loose clothing, and remove dangling jewelry. Don't wear open-toed shoes or sandals in the lab.
- Wear an apron or lab coat to protect your clothing when working with chemicals.
- If a spill gets on your clothing, rinse it off immediately with water for at least 5 min while notifying your instructor.

- Always use caution when working with chemicals.
- Never mix chemicals unless specifically directed to do so.
- Never taste, touch, or smell chemicals unless specifically directed to do so.
- Add acid or base to water; never do the opposite.
- Never return unused chemicals to the original container.
- Never transfer substances by sucking on a pipette or straw; use a suction bulb.
- Follow instructions for proper disposal.

- Clean and decontaminate all work surfaces and personal protective equipment as directed by your instructor.
- Dispose of all sharps (broken glass and other contaminated sharp objects) and other contaminated materials (biological and chemical) in special containers as directed by your instructor.

Procedure

1. Copy the data table below in your lab notebook.

Data Table 1

Marker Color	Dye Components

2. Obtain a clean 250 mL beaker and a 7.0 × 2.5 cm piece of chromatography paper.

3. Choose three different markers for this activity. Write the color of each marker in your data table.

4. Using a ruler, draw a horizontal line in pencil approximately 1.0 cm from one of the ends of the paper. Mark three small dots on this line, using a different marker for each dot.

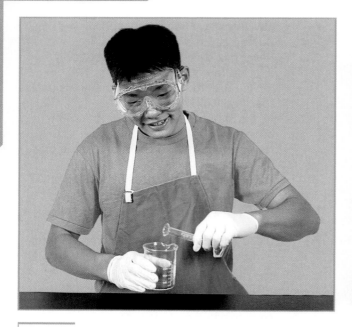

Figure A

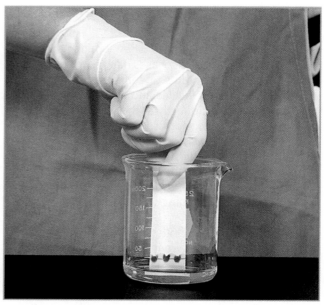

Figure B

5. Using a pencil, label each of the dots on the chromatography paper according to the color of the markers.

6. Measure out 7.0 mL of the developing solution in a 10 mL graduated cylinder.

7. Pour the 7.0 mL of solution in a 250 mL beaker, as shown in **Figure A.** Make sure the bottom of the beaker is completely covered. The level of the liquid must be below the marks on your chromatography paper.

8. You will need to design an experimental technique to ensure that your paper sample does not slide into the developing solution. The chromatography paper must remain vertical as the developing solution rises into the paper.

9. Carefully place your paper (with the dots at the bottom) into the liquid, as shown in **Figure B.**

10. When the level of the liquid has advanced through most of the paper, remove the paper from the developing solution. Hold up the paper and observe the colors.

11. The chromatography samples can be carefully dried on a hot plate.

12. You may repeat this process using overwrite or color-change markers.

13. Clean all apparatus and your lab station. Return equipment to its proper place. Dispose of chemicals and solutions in the containers designated by your teacher. Do not pour any chemicals down the drain or in the trash unless your teacher directs you to do so.

Analysis

1. Describing Events
What was the purpose of this experiment?

2. Explaining Events
Why were only water-soluble markers used in this experiment? Could permanent markers be used?

3. Explaining Events
Why must the spotted marks remain above the level of the liquid in the beaker?

Conclusions

4. Applying Conclusions
Why shouldn't you use a ballpoint pen when marking the initial line and spots on the chromatography paper? Explain.

5. Evaluating Results
Make observations about the dye components (colors) of each marker based on your results.

6. Applying Conclusions
Explain how law enforcement officials could use paper chromatography to identify a pen that was used in a ransom note.

7. Applying Conclusions
List some other applications for using paper chromatography.

8. Evaluating Methods
Compare your results with those of another lab group. Were the dye components found in other markers different from those found in yours?

Extensions

1. Research and Communications
Gasoline is a mixture of many different chemicals. Chemists can identify the different components of the mixture using chromatography. Research what gasoline is composed of and make a chart of the common components.

Skills Practice Lab

OBJECTIVES

- Translate word equations into chemical formulas.
- Count the number of drops of sodium hydroxide needed to completely react with different acid samples.
- Calculate the average number of drops of sodium hydroxide needed for each acid.
- Relate the number of drops to the coefficients in the balanced chemical equations.

MATERIALS

- buret clamps
- burets (2)
- H_2SO_4, 0.1 M
- HCl, 0.1 M
- NaOH, 0.3 M
- phenolphthalein indicator
- pipets
- ring stands
- test tubes
- test-tube rack

15A Drip-Drop Acid-Base Experiment

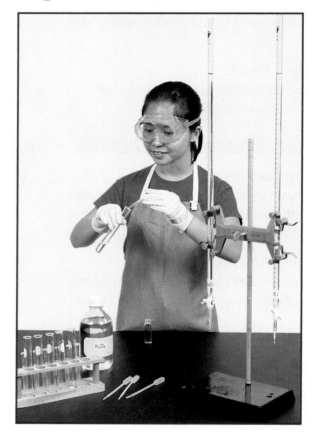

Introduction

The purpose of this lab is to investigate the simple reaction between two different acids and a base. We will be counting the number of drops of sodium hydroxide that are needed to react completely with all of the acid. The starting acid and base solutions are colorless and clear, and the final products are colorless and clear.

To monitor the progress of the chemical reaction, the acid-base indicator phenolphthalein will be used. Phenolphthalein is colorless when acidic and pink in color when neutral or basic. In this activity, we will know that all of the acid has been consumed by the base when the test-tube solution starts to turn pink. We can monitor the progress of the reaction so that a single drop of the base results in a sudden change from colorless to pink. At that point, we will know that all of the acid has reacted with the base.

You will need to count the number of drops of sodium hydroxide that are necessary to neutralize two different acids. Find the relationship between the sodium hydroxide drops necessary and the coefficients in the balanced chemical equation.

Safety Procedures

- Wear safety goggles when working around chemicals, acids, bases, flames, or heating devices. Contents under pressure may become projectiles and cause serious injury.
- Never look directly at the sun through any optical device or use direct sunlight to illuminate a microscope.
- Avoid wearing contact lenses in the lab.
- If any substance gets in your eyes, notify your instructor immediately and flush your eyes with running water for at least 15 min.

- Secure loose clothing, and remove dangling jewelry. Don't wear open-toed shoes or sandals in the lab.
- Wear an apron or lab coat to protect your clothing when working with chemicals.
- If a spill gets on your clothing, rinse it off immediately with water for at least 5 min while notifying your instructor.

- Always use caution when working with chemicals.
- Never mix chemicals unless specifically directed to do so.

- Never taste, touch, or smell chemicals unless specifically directed to do so.
- Add acid or base to water; never do the opposite.
- Never return unused chemicals to the original container.
- Never transfer substances by sucking on a pipette or straw; use a suction bulb.
- Follow instructions for proper disposal.

- Check the condition of glassware before and after using it. Inform your teacher of any broken, chipped, or cracked glassware, because it should not be used.
- Do not pick up broken glass with your bare hands. Place broken glass in a specially designated disposal container.

- Clean and decontaminate all work surfaces and personal protective equipment as directed by your instructor.
- Dispose of all sharps (broken glass and other contaminated sharp objects) and other contaminated materials (biological and chemical) in special containers as directed by your instructor.

Procedure

1. Translate each of the word equations shown below into chemical equations.

   hydrochloric acid + sodium hydroxide $\longrightarrow$
   sodium chloride + water

   sulfuric acid + sodium hydroxide $\longrightarrow$
   sodium sulfate + water

2. Copy Data Tables 1 and 2 in your lab notebook. Be sure that you have plenty of room for observations about each test.

3. Clean six test tubes, and rinse them with distilled water. They do not need to be dry.

4. Obtain approximately 10 mL of sodium hydroxide solution in a small beaker.

Data Table 1

HCl volume (mL)	NaOH (drops)
2.00	
2.00	
4.00	
4.00	

Data Table 2

H_2SO_4 volume (mL)	NaOH (drops)
2.00	
2.00	

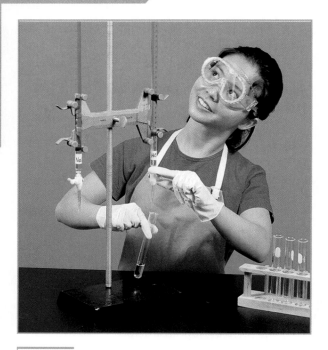

Figure A

Part I

5. Use a buret to put exactly 2.00 mL of hydrochloric acid directly into your test tube, as shown in **Figure A.**

6. Add two drops of phenolphthalein indicator solution to the test tube.

7. Use a pipet to add the sodium hydroxide solution dropwise to the test tube. Count the number of drops of sodium hydroxide as you add them. Gently shake the test tube from side to side after adding each drop. **Continue adding drops until the color just changes from colorless to pink.**

8. Record in your data table the total number of drops of sodium hydroxide needed to reach the color change. To obtain consistent results, repeat this trial.

Part II

9. Use a buret to add exactly 4.00 mL of hydrochloric acid directly into a clean test tube.

10. Add two drops of phenolphthalein indicator solution to the test tube.

11. Using a pipet, add one drop of sodium hydroxide solution at a time to the test tube. Count the number of drops of sodium hydroxide as you add them. Gently swirl the test tube after adding each drop. **Continue adding drops until the color just changes from colorless to a pink.**

12. Record in your data table the total number of drops of sodium hydroxide needed to reach the color change. Repeat this trial.

Part III Sulfuric Acid

13. Use a buret to add exactly 2.00 mL of sulfuric acid directly into your test tube.

14. Add two drops of phenolphthalein indicator solution to the test tube.

15. Using a pipet, add one drop of sodium hydroxide solution at a time to the test tube. Count the number of drops of sodium hydroxide as you add them. Gently swirl the test tube after adding each drop. **Continue adding drops until the color just changes from colorless to pink.**

16. Record in your data table the total number of drops of sodium hydroxide needed to reach the color change. Repeat this trial.

17. Clean all apparatus and your lab station. Return equipment to its proper place. Dispose of chemicals and solutions in the containers designated by your teacher. Do not pour any chemicals down the drain or in the trash unless your teacher directs you to do so. Wash your hands thoroughly after all work is finished and before you leave the lab.

Analysis

1. Examining Data

What was the average number of drops of sodium hydroxide required to consume 2.00 mL of HCl? Show your work. 4.00 mL of HCl? Show your work.

2. Examining Data

What was the average number of drops of sodium hydroxide required to consume 2.00 mL of H_2SO_4? Show your work.

3. Analyzing Results

Compare your responses to Analysis item 1. Is there a difference in the average number of drops? What is the ratio between these two numbers? Is it 1:1, 1:2, 2:1, or 1:3? Explain the "chemistry" behind this ratio.

4. Analyzing Results

Now compare your responses to Analysis and Interpretation items 1 and 2. Is there a difference in the average number of drops? What is the ratio between these two numbers? Is it 1:1, 1:2, 1:3, etc? Explain the "chemistry" behind this ratio.

Conclusions

5. Applying Conclusions

Based on your observed results, how many drops of sodium hydroxide would be needed to react completely with a 2.00 mL sample of HNO_3?

$$HNO_3 + NaOH \longrightarrow NaNO_3 + H_2O$$

15B Acid-Base Titration of an Eggshell

OBJECTIVES

◆ Determine the amount of calcium carbonate present in an eggshell.

◆ Relate experimental titration measurements to a balanced chemical equation.

◆ Infer a conclusion from experimental data.

◆ Apply reaction-stoichiometry concepts.

MATERIALS

◆ balance

◆ beaker, 100 mL

◆ bottle, 50 mL or small Erlenmeyer flask

◆ desiccator (optional)

◆ distilled water

◆ drying oven

◆ eggshell

◆ forceps

◆ graduated cylinder, 10 mL

◆ HCl, 1.00 M

◆ medicine droppers or thin-stemmed pipets (3)

◆ mortar and pestle

◆ NaOH, 1.00 M

◆ phenolphthalein solution

◆ weighing paper

◆ white paper or white background

Introduction

You are a scientist working with the Department of Agriculture. A farmer has brought a problem to you. In the past 10 years, his hens' eggs have become increasingly fragile. So many of them have been breaking that he is beginning to lose money. The farmer believes his problems are linked to a landfill upstream, which is being investigated for illegal dumping of PCBs and other hazardous chemicals. Your job is to find out if the PCBs are the cause of the hens' fragile eggs.

Birds have evolved a chemical process that allows them to rapidly produce the calcium carbonate, $CaCO_3$, required for eggshell formation. Research has shown that some chemicals, like DDT and PCBs, can decrease the amount of calcium carbonate in the eggshell, resulting in shells that are thin and fragile.

You need to determine how much calcium carbonate is in sample eggshells from chickens that were not exposed to PCBs. The farmer's eggshells contain about 78% calcium carbonate. The calcium carbonate content of eggshells can easily be determined by means of an acid-base back-titration. A carefully measured excess of a strong acid will react with the calcium carbonate. Because the acid is in excess, there will be some left over at the end of the reaction. The resulting solution will be titrated with a strong base to determine how much acid remained unreacted. Phenolphthalein will be used as an indicator to signal the endpoint of the titration. From this measurement, you can determine the following:

• the amount of excess acid that reacted with the eggshell

• the amount of calcium carbonate that was present to react with this acid

Safety Procedures

- Wear safety goggles when working around chemicals, acids, bases, flames, or heating devices. Contents under pressure may become projectiles and cause serious injury.
- Never look directly at the sun through any optical device or use direct sunlight to illuminate a microscope.
- Avoid wearing contact lenses in the lab.
- If any substance gets in your eyes, notify your instructor immediately and flush your eyes with running water for at least 15 minutes.

- Secure loose clothing, and remove dangling jewelry. Don't wear open-toed shoes or sandals in the lab.
- Wear an apron or lab coat to protect your clothing when working with chemicals.
- If a spill gets on your clothing, rinse it off immediately with water for at least 5 min while notifying your instructor.

- If a chemical gets on your skin or clothing or in your eyes, rinse it immediately, and alert your instructor.
- If a chemical is spilled on the floor or lab bench, alert your instructor, but do not clean it up yourself unless your teacher says it is OK to do so.

- Always use caution when working with chemicals.
- Never mix chemicals unless specifically directed to do so.
- Never taste, touch, or smell chemicals unless specifically directed to do so.
- Add acid or base to water; never do the opposite.
- Never return unused chemicals to the original container.
- Never transfer substances by sucking on a pipette or straw; use a suction bulb.
- Follow instructions for proper disposal.

- Whenever possible, use an electric hot plate as a heat source instead of an open flame.
- When heating materials in a test tube, always angle the test tube away from yourself and others.
- Glass containers used for heating should be made of heat-resistant glass.
- Know your school's fire-evacuation routes.

- Clean and decontaminate all work surfaces and personal protective equipment as directed by your instructor.
- Dispose of all sharps (broken glass and other contaminated sharp objects) and other contaminated materials (biological and chemical) in special containers as directed by your instructor.

Procedure

1. Make data and calculation tables like the following tables.

Data Table 1

Total volume of acid drops	
Average volume of each drop	
Total volume of base drops	
Average volume of each drop	

Data Table 2

Titration Steps	
Mass of entire eggshell	
Mass of ground eggshell sample	
Number of drops of 1.00 M HCl added	150
Volume of 1.00 M HCl added	
Number of drops of 1.00 M NaOH added	
Volume of 1.00 M NaOH added	
Volume of HCl reacting with NaOH	
Volume of HCl reacting with eggshell	
Number of moles of HCl reacting with eggshell	
Number of moles of $CaCO_3$ reacting with HCl	
Mass of $CaCO_3$	
Percentage of $CaCO_3$ in eggshell sample	

Data Table 3   Graduated Cylinder Readings (Pipet Calibration: Steps 3–5)

Trial	Initial acid pipet	Final acid pipet	Initial base pipet	Final base pipet
1				
2				
3				

Figure A

Figure B

Figure C

Figure D

2. Remove the white and the yolk from the egg, as shown in **Figure A.** Dispose of these according to your teacher's directions. Wash the shell with distilled water, and carefully peel all the membranes from the inside of the shell. Discard the membranes. Place ALL of the shell in a pre-massed beaker, and dry the shell in the drying oven at 110°C for about 15 min.

3. Put exactly 5.0 mL of water in the 10.0 mL graduated cylinder. Record this volume in the data table in your lab notebook. Fill the first dropper or pipet with water. This dropper should be labeled "Acid." **Do not use this dropper for the base solution.** Holding the dropper vertical, add 20 drops of water to the cylinder. For the best results, keep the sizes of the drops as even as possible throughout this investigation. Record the new volume of water in the first data table as Trial 1.

4. Without emptying the graduated cylinder, add an additional 20 drops from the dropper, as you did in step 3, and record the new volume as the final volume for Trial 2. Repeat this procedure once more for Trial 3.

5. Repeat steps 3 and 4 for the second thin-stemmed dropper. Label this dropper "Base." **Do not use this dropper for the acid solution.**

6. Make sure that the three trials produce data that are similar to each other. If one is greatly different from the others, perform steps 3–5 over again. If you're still waiting for the eggshell in the drying oven, calculate and record in the first data table the total volume of the drops and the average volume per drop.

7. Remove the eggshell and beaker from the oven. Cool them in a desiccator. Record the mass of the entire eggshell in the second data table. Place half the shell in a clean mortar, and grind it to a very fine powder, as shown in **Figure B.** This will save time when dissolving the eggshell. (If time permits, dry the crushed eggshell again, and cool it in the desiccator.)

8. Measure the mass of a piece of weighing paper. Transfer about 0.1 g of ground eggshell to a piece of weighing paper, and measure the eggshell's mass as accurately as possible. Record the mass in the second data table. Place this eggshell sample in a clean 50 mL bottle or Erlenmeyer flask. A flask will make it easier to swirl the mixture when needed.

9. Fill the acid dropper with the 1.00 M HCl acid solution, and then empty the dropper into an extra 100 mL beaker. Label the beaker "Waste." Fill the base dropper with the 1.00 M NaOH base solution, and then empty the dropper into the 100 mL beaker.

10. Fill the acid dropper once more with 1.00 M HCl. Using the acid dropper, add exactly 150 drops of 1.00 M HCl to the bottle (or flask) with the eggshell, as shown in **Figure C.** Swirl gently for 3–4 minutes. Rinse the sides of the flask with about 10 mL of distilled water. Using a third dropper, add two drops of phenolphthalein solution. Record the number of drops of HCl used in the second data table.

11. Fill the base dropper with the 1.00 M NaOH. Slowly add NaOH from the base dropper into the bottle or flask with the eggshell mixture until a faint pink color persists, even after it is swirled gently, as shown in **Figure D.** It may help to use a white piece of paper as a background so you will be able to see the color as soon as possible. **Be sure to add the base drop by drop, and be certain the drops end up in the reaction mixture and not on the side of the bottle or flask. Keep a careful count of the number of drops used.** Record the number of drops of base used in the second data table.

12. Clean all apparatus and your lab station. Return the equipment to its proper place. Dispose of chemicals and solutions in the containers designated by your teacher. Do not pour any chemicals down the drain or in the trash unless your teacher directs you to do so. Wash your hands thoroughly before you leave the lab and after all work is finished.

Analysis

1. Explaining Events
The calcium carbonate in the eggshell sample undergoes a double-replacement reaction with the hydrochloric acid in step 10. Then the carbonic acid that was formed decomposes. Write a balanced chemical equation for these reactions. (Hint: The gas observed was carbon dioxide.)

2. Explaining Events
Write the balanced chemical equation for the acid-base neutralization of the excess unreacted HCl with the NaOH.

3. Organizing Data
Make the necessary calculations from the first data table to find the number of milliliters in each drop. Using this milliliter/drop ratio, convert the number of drops of each solution in the second data table to volumes in milliliters.

4. Analyzing Results
Using the relationship between the molarity and volume of acid and the molarity and volume of base needed to neutralize the acid, calculate the volume of the HCl solution that was neutralized by the NaOH. Then subtract this amount from the initial volume of HCl to determine how much HCl reacted with $CaCO_3$.

Conclusions

5. Evaluating Data
Use the stoichiometry of the reaction in Analysis and Interpretation item 1 to calculate the number of moles of $CaCO_3$ that reacted with the HCl, and record this number in your table.

6. Evaluating Data
Workers in a lab in another city have also tested eggs, and they found that a normal eggshell is about 97% $CaCO_3$. Calculate the percent error for your measurement.

Extensions

1. Building Models
Calculate an estimate of the mass of $CaCO_3$ present in the entire eggshell, based on your results. (Hint: Apply the percent composition of your sample to the mass of the entire eggshell.)

2. Designing Experiments
What possible sources of error can you identify in this procedure? If you can think of ways to eliminate them, ask your teacher to approve your plan, and run the procedure again.

Inquiry LAB
Design Your Own Experiment

THE PROBLEM

DELIVER BY OVERNIGHT COURIER

Date: April 21, 2004
To: EPA National Headquarters
From: Anthony Wong, Plant Supervisor
Re: Vacaville Bleachex Corp. Plant Spill

As a result of last night's earthquake, the Bleachex plant in the industrial park south of Vacaville was severely damaged. The safety control measures failed because of the magnitude of the earthquake.

Bleachex manufactures a variety of products using concentrated acids and bases. Plant officials noticed a large quantity of liquid, which was believed to be either sodium hydroxide or hydrochloric acid solution, flowing through the loading bay doors. An Emergency Toxic Spill Response Team attempted to determine the source and identity of the unknown liquid. A series of explosions and the presence of chlorine gas forced the team to abandon its efforts. The unknown liquid continues to flow into the nearly full containment ponds.

We are sending a sample of the liquid to you by overnight courier, and we hope that you can quickly and accurately identify the liquid and notify us of the proper method for cleanup and disposal. We need your answer as soon as possible.

Sincerely,

Anthony Wong

Anthony Wong

THE PLAN

CheMystery Labs, Inc. 52 Fulton Street, Springfield, VA 22150

Memorandum

Date: April 22, 2004
To: Cicely Jackson
From: Marissa Bellinghausen

This project is a high priority. First we must determine the pH of the unknown so that we know whether it is an acid or a base. Then titrate the unknown using a standard solution to determine its concentration so that we can advise Bleachex on the amount of neutralizing agents that will be needed for the three containment ponds.

I need the following items:
- a detailed one-page plan for your procedure that includes all necessary data tables (include multiple trials)
- a detailed list of the equipment and materials you will need

When you have completed your experiment, prepare a report in the form of a two-page letter that we can fax to Anthony Wong. The letter must include the following:
- the identity of the unknown and its concentration
- the pH of the unknown and an explanation of how you determined the pH
- paragraph summarizing how you titrated the sample to determine its concentration
- a detailed and organized data table
- a detailed analysis section, including calculations, a discussion of the multiple trials, and a statistical analysis of your precision
- your proposed method for cleanup and disposal, including the amount of neutralizing agents that will be needed

Required Precautions

- Wear safety goggles when working around chemicals, acids, bases, flames, or heating devices. Contents under pressure may become projectiles and cause serious injury.
- Avoid wearing contact lenses in the lab.
- If any substance gets in your eyes, notify your instructor immediately and flush your eyes with running water for at least 15 min.
- Secure loose clothing, and remove dangling jewelry. Don't wear open-toed shoes or sandals in the lab.

- Wear an apron or lab coat to protect your clothing when working with chemicals.
- If a spill gets on your clothing, rinse it off immediately with water for at least 5 min while notifying your instructor.

- Always use caution when working with chemicals.
- Never taste, touch, or smell chemicals unless specifically directed to do so.
- Follow instructions for proper disposal.
- Whenever possible, use an electric hot plate as a heat source instead of an open flame.

- When heating materials in a test tube, always angle the test tube away from yourself and others.
- Know your school's fire-evacuation routes.
- Clean and decontaminate all work surfaces and personal protective equipment as directed by your instructor.
- Dispose of all sharps (broken glass and other contaminated sharp objects) and other contaminated materials (biological and chemical) in special containers as directed by your instructor.

Skills Practice Lab

OBJECTIVES

◆ Prepare and observe several different reaction mixtures.

◆ Demonstrate proficiency in measuring reaction rates.

◆ Analyze the results and relate experimental results to a rate law that you can use to predict the results of various combinations of reactants.

USING SCIENTIFIC **METHODS**

MATERIALS

◆ 8-well microscale reaction strips (2)

◆ distilled or deionized water

◆ fine-tipped dropper bulbs or small micro-tip pipets (3)

◆ solution A

◆ solution B

◆ stopwatch or clock with second hand

16 Reaction Rates

Introduction

Executive "toys" are a big business. Your company has been contacted by a toy company that wants technical assistance in designing a new executive desk gadget. The company wants to investigate a reaction that turns a distinctive color in a specific amount of time. Although it will not be easy to deter-

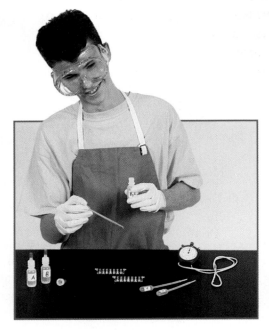

mine the precise combination of chemicals that will work, the profit the company stands to make would be worthwhile in the end.

In this experiment you will determine the rate of an *oxidation-reduction,* or *redox,* reaction. Reactions of this type are discussed in the chapter "Electrochemistry." The net equation for the reaction you will study is as follows:

$$3Na_2S_2O_5(aq) + 2KIO_3(aq) + 3H_2O(l) \xrightarrow{H^+}$$

$$2KI(aq) + 6NaHSO_4(aq)$$

One way to study the rate of this reaction is to observe how fast $Na_2S_2O_5$ is used up. After all the $Na_2S_2O_5$ solution has reacted, the concentration of iodine, I_2, an intermediate in the reaction, builds up. A starch indicator solution added to the reaction mixture will signal when this happens. The colorless starch will change to a blue-black color in the presence of I_2.

In the experiment, the concentrations of the reactants are given in terms of drops of Solution A and drops of Solution B. Solution A contains $Na_2S_2O_5$, the starch-indicator solution, and dilute sulfuric acid to supply the hydrogen ions needed to catalyze the reaction. Solution B contains KIO_3. You will run the reaction with several different concentrations of the reactants and record the time it takes for the blue-black color to appear.

To determine the best conditions and concentrations for the reaction, you will determine the following:

• how changes in reactant concentrations affect the reaction outcome

• how much time elapses for each reaction

• a rate law for the reaction that will allow you to predict the results with other combinations

Safety Procedures

- Wear safety goggles when working around chemicals, acids, bases, flames, or heating devices. Contents under pressure may become projectiles and cause serious injury.
- Never look directly at the sun through any optical device or use direct sunlight to illuminate a microscope.
- Avoid wearing contact lenses in the lab.
- If any substance gets in your eyes, notify your instructor immediately and flush your eyes with running water for at least 15 min.

- Secure loose clothing, and remove dangling jewelry. Don't wear open-toed shoes or sandals in the lab.
- Wear an apron or lab coat to protect your clothing when working with chemicals.
- If a spill gets on your clothing, rinse it off immediately with water for at least 5 min while notifying your instructor.

- If a chemical gets on your skin or clothing or in your eyes, rinse it immediately, and alert your instructor.
- If a chemical is spilled on the floor or lab bench, alert your instructor, but do not clean it up yourself unless your teacher says it is OK to do so.

- Always use caution when working with chemicals.
- Never mix chemicals unless specifically directed to do so.
- Never taste, touch, or smell chemicals unless specifically directed to do so.
- Add acid or base to water; never do the opposite.
- Never return unused chemicals to the original container.
- Never transfer substances by sucking on a pipette or straw; use a suction bulb.
- Follow instructions for proper disposal.

- Clean and decontaminate all work surfaces and personal protective equipment as directed by your instructor.
- Dispose of all sharps (broken glass and other contaminated sharp objects) and other contaminated materials (biological and chemical) in special containers as directed by your instructor.

Procedure

1. Copy the data table below in your lab notebook.

Data Table 1

	Well 1	Well 2	Well 3	Well 4	Well 5
Time reaction began					
Time reaction stopped					
Drops of A					
Drops of B					
Drops of H_2O					

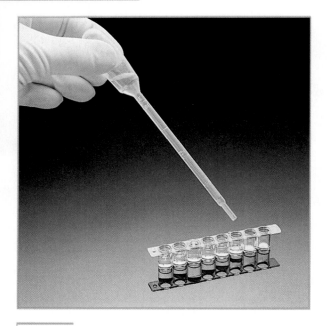

Figure A

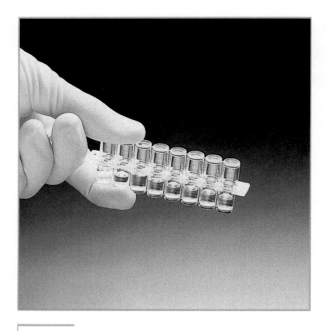

Figure B

2. Obtain three dropper bulbs or small microtip pipets, and label them "A," "B," and "H$_2$O."

3. Fill bulb or pipet A with solution A, fill bulb or pipet B with solution B, and fill the bulb or pipet for H$_2$O with distilled water.

4. Using the first 8-well strip, place five drops of Solution A into each of the first five wells, as shown in **Figure A.** (Disregard the remaining three wells.) Record the number of drops in the appropriate places in your data table. For best results, try to make all of the drops about the same size.

5. In the second 8-well reaction strip, place one drop of Solution B in the first well, two drops in the second well, three drops in the third well, four drops in the fourth well, and five drops in the fifth well. Record the number of drops in the appropriate places in your data table.

6. In the second 8-well strip that contains drops of Solution B, add four drops of water to the first well, three drops to the second well, two drops to the third well, and one drop to the fourth well. Do not add any water to the fifth well.

7. Carefully invert the second strip. The surface tension should keep the solutions from falling out of the wells. Place the second strip well-to-well on top of the first strip, as shown in **Figure B.**

8. Holding the strips tightly together, record the exact time or set the stopwatch as you shake the strips. This procedure should effectively mix the upper solutions with each of the corresponding lower ones.

9. Observe the lower wells. Note the sequence in which the solutions react, and record the number of seconds it takes for each solution to turn a blue-black color.

10. Dispose of the solutions in the container designated by your teacher. Wash your hands thoroughly after cleaning up the area and equipment.

Analysis

1. Organizing Data

Calculate the time elapsed for the complete reaction of each combination of Solution A and Solution B.

2. Constructing Graphs

Make a graph of your results. Label the x-axis "Number of drops of Solution B." Label the y-axis "Time elapsed." Make a similar graph for drops of Solution B versus rate (1/time elapsed).

3. Analyzing Data

Which mixture reacted the fastest? Which mixture reacted the slowest?

4. Explaining Events

Why was it important to add the drops of water to the wells that contained fewer than five drops of Solution B? (Hint: Figure out the total number of drops in each of the reaction wells.)

Conclusions

5. Evaluating Methods

How can you be sure that each of the chemical reactions began at about the same time? Why is this important?

6. Evaluating Results

Of the following variables that can affect the rate of a reaction, which is tested in this experiment: temperature, catalyst, concentration, surface area, or nature of reactants? Explain your answer.

7. Analyzing Graphs

Use your data and graphs to determine the relationship between the concentration of Solution B and the rate of the reaction. Describe this relationship in terms of a rate law.

8. Evaluating Data

Share your data with other lab groups, and calculate a class average for the rate of the reaction for each concentration of B. Compare the results from other groups with your results. Explain why there are differences in the results.

9. Evaluating Methods

What are some possible sources of error in this procedure? If you can think of ways to eliminate them, ask your teacher to approve your plan and run your procedure again.

10. Making Predictions

How would your data be different if the experiment were repeated but Solution A was diluted with one part solution for every seven parts distilled water?

Extensions

1. Designing Experiments

What combination of drops of Solutions A and B would you use if you wanted the reaction to last exactly 2.5 min? Design an experiment to test your answer. If your teacher approves your plan, perform the experiment, and record these results. Make another graph that includes both the old and new data.

2. Designing Experiments

How would you determine the smallest interval of time during which you could distinguish a clock reaction? Design an experiment to find out. If your teacher approves your plan, perform your experiment.

3. Designing Experiments

How would the results of this experiment be affected if the reaction took place in a cold environment? Design an experiment to test your answer using materials available. If your teacher approves your plan, perform your experiment and record the results. Make another graph, and compare it with your old data.

4. Designing Experiments

Devise a plan to determine the effect of Solution A on the rate law. If your teacher approves your plan, perform your experiment, and determine the rate law for this reaction.

5. Building Models

If Solution B contains 0.02 M KIO_3, calculate the value for the constant, k, in the expression below. (Hint: Remember that Solution B is diluted when it is added to Solution A.)

$$rate = k[KIO_3]$$

Skills Practice Lab

OBJECTIVES

- Demonstrate proficiency in performing redox titrations and recognizing the end point of a redox reaction.

- Determine the concentration of a solution using stoichiometry and volume data from a titration.

MATERIALS

- beaker 250 mL (2)
- beaker 400 mL
- burets (2)
- distilled water
- double buret clamp
- Erlenmeyer flask, 125 mL (4)
- $FeSO_4$ solution
- graduated cylinder, 100 mL
- H_2SO_4, 1.0 M
- $KMnO_4$, 0.0200 M
- ring stand
- wash bottle

17 Redox Titration

Introduction

You are a chemist working for a chemical analysis firm. A large pharmaceutical company has hired you to help salvage some products that were damaged by a small fire in their warehouse. Although there was only minimal smoke and fire damage to the warehouse and products, the sprinkler system ruined the labeling on many of the pharmaceuticals. The firm's best-selling products are iron tonics used to treat low-level anemia. The tonics are produced from hydrated iron(II) sulfate, $FeSO_4 \cdot 7H_2O$. The different types of tonics contain different concentrations of $FeSO_4$. You have been hired to help the pharmaceutical company figure out the proper label for each bottle of tonic.

In the chapter "Acids and Bases" you studied acid-base titrations in which an unknown amount of acid is titrated with a carefully measured amount of base. In this procedure a similar approach called a redox titration is used. In a redox titration, the reducing agent, Fe^{2+}, is oxidized to Fe^{3+} by the oxidizing agent, MnO_4^-. When this process occurs, the Mn in MnO_4^- changes from a +7 to a +2 oxidation state and has a noticeably different color. You can use this color change in the same way that you used the color change of phenolphthalein in acid-base titrations—to signify a redox reaction end point. When the reaction is complete, any excess MnO_4^- added to the reaction mixture will give the solution a pink or purple color. The volume data from the titration, the known molarity of the $KMnO_4$ solution, and the mole ratio from the following balanced redox equation will give you the information you need to calculate the molarity of the $FeSO_4$ solution.

$$5Fe^{2+}(aq) + MnO_4^-(aq) + 8H^+(aq) \longrightarrow$$
$$5Fe^{3+}(aq) + Mn^{2+}(aq) + 4H_2O(l)$$

To determine how to label the bottles, you must determine the concentration of iron(II) ions in the sample from an unlabeled bottle from the warehouse by answering the following questions:

- How can the volume data obtained from the titration and the mole ratios from the balanced redox reaction be used to determine the concentration of the sample?

- Which tonic is in the sample, given information about the concentration of each tonic?

Safety Procedures

- Wear safety goggles when working around chemicals, acids, bases, flames, or heating devices. Contents under pressure may become projectiles and cause serious injury.
- Never look directly at the sun through any optical device or use direct sunlight to illuminate a microscope.
- Avoid wearing contact lenses in the lab.
- If any substance gets in your eyes, notify your instructor immediately and flush your eyes with running water for at least 15 min.

- Secure loose clothing, and remove dangling jewelry. Don't wear open-toed shoes or sandals in the lab.
- Wear an apron or lab coat to protect your clothing when working with chemicals.
- If a spill gets on your clothing, rinse it off immediately with water for at least 5 min while notifying your instructor.

- If a chemical gets on your skin or clothing or in your eyes, rinse it immediately, and alert your instructor.
- If a chemical is spilled on the floor or lab bench, alert your instructor, but do not clean it up yourself unless your teacher says it is OK to do so.

- Always use caution when working with chemicals.
- Never mix chemicals unless specifically directed to do so.
- Never taste, touch, or smell chemicals unless specifically directed to do so.
- Add acid or base to water; never do the opposite.
- Never return unused chemicals to the original container.
- Never transfer substances by sucking on a pipette or straw; use a suction bulb.
- Follow instructions for proper disposal.

- Check the condition of glassware before and after using it. Inform your teacher of any broken, chipped, or cracked glassware, because it should not be used.
- Do not pick up broken glass with your bare hands. Place broken glass in a specially designated disposal container.

- Clean and decontaminate all work surfaces and personal protective equipment as directed by your instructor.
- Dispose of all sharps (broken glass and other contaminated sharp objects) and other contaminated materials (biological and chemical) in special containers as directed by your instructor.

Data Table 1

Trial	Initial KMnO$_4$ volume	Final KMnO$_4$ volume	Initial FeSO$_4$ volume	Final FeSO$_4$ volume
1				
2				
3				

Procedure

1. Organizing Data

Copy the data table above in your lab notebook.

2. Clean two 50 mL burets with a buret brush and distilled water. Rinse each buret at least three times with distilled water to remove any contaminants.

3. Label two 250 mL beakers "0.0200 M KMnO$_4$," and "FeSO$_4$ solution." Label three of the flasks

1, 2, and 3. Label the 400 mL beaker "Waste." Label one buret "KMnO$_4$" and the other "FeSO$_4$."

4. Measure approximately 75 mL of 0.0200 M KMnO$_4$, and pour it into the appropriately labeled beaker. Obtain approximately 75 mL of FeSO$_4$ solution, and pour it into the appropriately labeled beaker.

5. Rinse one buret three times with a few milliliters of 0.0200 M KMnO$_4$ from the appropriately labeled beaker. Collect these rinses in the

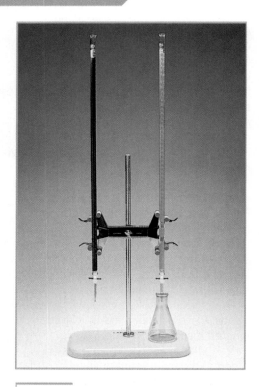

Figure A

Figure B

waste beaker. Rinse the other buret three times with small amounts of FeSO$_4$ solution from the appropriately labeled beaker. Collect these rinses in the waste beaker.

6. Set up the burets as shown in **Figure A.** Fill one buret with approximately 50 mL of the 0.0200 M KMnO$_4$ from the beaker, and fill the other buret with approximately 50 mL of the FeSO4 solution from the other beaker.

7. With the waste beaker underneath its tip, open the KMnO$_4$ buret long enough to be sure the buret tip is filled. Repeat for the FeSO$_4$ buret.

8. Add 50 mL of distilled water to one of the 125 mL Erlenmeyer flasks, and add one drop of 0.0200 M KMnO$_4$ to the flask. Set this aside to use as a color standard, as shown in **Figure B,** to compare with the titration and to determine the end point.

9. Record the initial buret readings for both solutions in your data table. Add 10.0 mL of the

hydrated iron(II) sulfate, FeSO$_4$·7H$_2$O, solution to flask 1. Add 5 mL of 1.0 M H$_2$SO$_4$ to the FeSO$_4$ solution in this flask. The acid will help keep the Fe^{2+} ions in the reduced state, allowing you time to titrate.

10. Slowly add KMnO$_4$ from the buret to the FeSO$_4$ in the flask while swirling the flask, as shown in **Figure C.** When the color of the solution matches the color standard you prepared in step 8, record the final readings of the burets in your data table.

11. Empty the titration flask into the waste beaker. Repeat the titration procedure in steps 9 and 10 with flasks 2 and 3.

12. Always clean up the lab and all equipment after use. Dispose of the contents of the waste beaker in the container designated by your teacher. Also pour the contents of the color-standard flask into this container. Wash your hands thoroughly after cleaning up the area and equipment.

Figure C

Analysis

1. **Analyzing Data**
   Calculate the number of moles of MnO_4^- reduced in each trial.

2. **Analyzing Data**
   Calculate the number of moles of Fe^{2+} oxidized in each trial.

3. **Analyzing Data**
   Calculate the average concentration (molarity) of the iron tonic.

4. **Explaining Events**
   Explain why it was important to rinse the burets with $KMnO_4$ or $FeSO_4$ before adding the solutions. (Hint: Consider what would happen to the concentration of each solution if it were added to a buret that had been rinsed only with distilled water.)

Conclusions

5. **Evaluating Data**
   The company makes three different types of iron tonics: Feravide A, with a concentration of 0.145 M $FeSO_4$; Feravide Extra-Strength, with 0.225 M $FeSO_4$; and Feravide Jr., with 0.120 M $FeSO_4$. Which tonic is your sample?

6. **Evaluating Methods**
   What possible sources of error can you identify with this procedure? If you can think of ways to eliminate them, ask your teacher to approve your plan, and run the procedure again.

Extensions

1. **Research and Communication**
   Blueprints are based on a photochemical reaction. The paper is treated with a solution of iron(III) ammonium citrate and potassium hexacyanoferrate(III) and dried in the dark. When a tracing-paper drawing is placed on the blueprint paper and exposed to light, Fe^{3+} ions are reduced to Fe^{2+} ions, which react with hexacyanoferrate(III) ions in the moist paper to form the blue color on the paper. The lines of the drawing block the light and prevent the reduction of Fe^{3+} ions, resulting in white lines. Find out how sepia prints are made, and report on this information.

2. **Building Models**
   Electrochemical cells are based on the process of electron flow in a system with varying potential differences. Batteries are composed of such systems and contain different chemicals for different purposes and price ranges. You can make simple experimental batteries using metal wires and items such as lemons, apples, and potatoes. What are some other "homemade" battery sources, and what is the role of these food items in producing electrical energy that can be measured as battery power? Explain your answers.

THE PROBLEM

Inquiry LAB
Design Your Own Experiment

May 11, 2004

George Taylor
Director of Analytical Services
CheMystery Labs, Inc.
52 Fulton Street
Springfield, VA 22150

Dear Mr. Taylor:

Because of the high quality of your firm's work in the past, Goldstake is again asking that you submit a bid for a mining feasibility study. A study site in New Mexico has yielded some promising iron ore deposits, and we are evaluating the potential yield.

Your bid should include the cost of evaluating the sample we are sending with this letter and the fees for 20 additional analyses to be completed over the next year. The sample is a slurry extracted from the mine using a special process that converts the iron ore into iron(II) sulfate, $FeSO_4$, dissolved in water. The mine could produce up to 1.0×10^5 L of this slurry daily, but we need to know how much iron is in that amount of slurry before we proceed.

The contract for the other analyses will be awarded based on the accuracy of this analysis and the quality of the report. Your report will be used for two purposes: to evaluate the site for quantity of iron and to determine who our analytical consultant will be if the site is developed into a mining operation. I look forward to reviewing your bid proposal.

Sincerely,

Lynn L. Brown

Lynn L. Brown
Director of Operations
Goldstake Mining Company

References

Review more information on redox reactions. Remember to add a small amount of sulfuric acid, H_2SO_4, so the iron will stay in the Fe^{2+} form. Calculate your disposal costs based on the mass of potassium permanganate, $KMnO_4$, and $FeSO_4$ in your solutions, as well as the mass of the H_2SO_4 solution.

THE PLAN

CheMystery Labs, Inc. 52 Fulton Street, Springfield, VA 22150

Memorandum

Date: May 12, 2004
To: Crystal Sievers
From: George Taylor

Good news! The quality of our work has earned us a repeat customer, Goldstake Mining Company. This analysis could turn into a long-term contract. Perform the analysis more than one time so that we can be confident of our accuracy. Before you begin your analysis, send Ms. Brown the following items:

- a detailed, one-page plan for the procedure and all necessary data tables
- a detailed sheet that lists all of the equipment and materials you plan to use

When you have completed the laboratory work, please prepare a report in the form of a two-page letter to Ms. Brown. Include the following information:

- moles and grams of $FeSO_4$ in 10 mL of sample
- moles, grams, and percentage of iron(II) in 10 mL of the sample
- the number of kilograms of iron that the company could extract from the mine each year, assuming that 1.0×10^5 L of slurry could be mined per day, year round
- a balanced equation for the redox equation
- a detailed and organized data and analysis section showing calculations of how you determined the moles, grams, and percentage of iron(II) in the sample (include calculations of the mean, or average, of the multiple trials)

Required Precautions

- Wear safety goggles when working around chemicals, acids, bases, flames, or heating devices. Contents under pressure may become projectiles and cause serious injury.
- Avoid wearing contact lenses in the lab.
- If any substance gets in your eyes, notify your instructor immediately and flush your eyes with running water for at least 15 min.
- Secure loose clothing, and remove dangling jewelry. Don't wear open-toed shoes or sandals in the lab.
- Wear an apron or lab coat to pro-

tect your clothing when working with chemicals.
- If a spill gets on your clothing, rinse it off immediately with water for at least 5 min while notifying your instructor.
- Always use caution when working with chemicals.
- Never taste, touch, or smell chemicals unless specifically directed to do so.
- Follow instructions for proper disposal.
- Whenever possible, use an electric hot plate as a heat source instead of an open flame.

- When heating materials in a test tube, always angle the test tube away from yourself and others.
- Know your school's fire-evacuation routes.
- Clean and decontaminate all work surfaces and personal protective equipment as directed by your instructor.
- Dispose of all sharps (broken glass and other contaminated sharp objects) and other contaminated materials (biological and chemical) in special containers as directed by your instructor.

Skills Practice Lab

OBJECTIVES

♦ Synthesize two different polymers.

♦ Prepare a small toy ball from each polymer.

♦ Observe the similarities and differences between the two types of balls.

♦ Measure the density of each polymer.

♦ Compare the bounce height of the two balls.

MATERIALS

♦ acetic acid solution (vinegar), 5% (10 mL)

♦ beaker, 2 L, or plastic bucket or tub

♦ distilled water

♦ ethanol solution, 50% (3 mL)

♦ gloves

♦ graduated cylinder, 10 mL

♦ graduated cylinder, 25 mL

♦ liquid latex (10 mL)

♦ meterstick

♦ paper cups, 5 oz (2)

♦ paper towels

♦ sodium silicate solution (12 mL)

♦ wooden stick

19 Polymers and Toy Balls

Introduction

Your company has been contacted by a toy company that specializes in toy balls made from vulcanized rubber. Recent legislation has increased the cost of disposing of the sulfur and other chemical byproducts of the manufacturing process for this type of rubber. The toy company wants you to research some other materials.

Rubber is a polymer of covalently bonded atoms. When rubber is vulcanized, it is heated with sulfur. The sulfur atoms form bonds between adjacent molecules of rubber, which increases its strength and making it more elastic.

Latex rubber is a colloidal suspension that can be made synthetically or found naturally in plants. Latex is composed of approximately 60% water, 35% hydrocarbon monomers, 2% proteins, and some sugars and inorganic salts.

The polymer formed from ethanol, C_2H_5OH, and a solution of sodium silicate, mostly in the form of $Na_2Si_3O_7$, also has covalent bonds. When the polymer is formed, water is also a product.

Latex rubber and the ethanol–sodium silicate polymer are the two materials you will become familiar with as you do the following:

• Synthesize each polymer.

• Make a ball 2–3 cm in diameter from each polymer.

• Make observations about the physical properties of each polymer.

• Measure how well each ball bounces.

Safety Procedures

- Wear safety goggles when working around chemicals, acids, bases, flames, or heating devices. Contents under pressure may become projectiles and cause serious injury.
- Never look directly at the sun through any optical device or use direct sunlight to illuminate a microscope.
- Avoid wearing contact lenses in the lab.
- If any substance gets in your eyes, notify your instructor immediately and flush your eyes with running water for at least 15 min.

- Secure loose clothing, and remove dangling jewelry. Don't wear open-toed shoes or sandals in the lab.
- Wear an apron or lab coat to protect your clothing when working with chemicals.
- If a spill gets on your clothing, rinse it off immediately with water for at least 5 min while notifying your instructor.

- If a chemical gets on your skin or clothing or in your eyes, rinse it immediately, and alert your instructor.

- If a chemical is spilled on the floor or lab bench, alert your instructor, but do not clean it up yourself unless your teacher says it is OK to do so.

- Always use caution when working with chemicals.
- Never mix chemicals unless specifically directed to do so.
- Never taste, touch, or smell chemicals unless specifically directed to do so.
- Add acid or base to water; never do the opposite.
- Never return unused chemicals to the original container.
- Never transfer substances by sucking on a pipette or straw; use a suction bulb.
- Follow instructions for proper disposal.

- Use flammable liquids only in small amounts.
- When working with flammable liquids, be sure that no one else in the lab is using a lit Bunsen burner or plans to use one. Make sure there are no other heat sources present.

Data Table 1

Trial	Height (cm)	Mass (g)	Diameter (cm)
1			
2			
3			

Procedure

1. Copy Data Table 1 above in your lab notebook. Be sure that you have plenty of room for observations about each test.

Organizing Data

2. Fill the 2 L beaker, bucket, or tub about half full with distilled water.

3. Using a clean 25 mL graduated cylinder, measure 10 mL of liquid latex and pour it into one of the paper cups.

4. Thoroughly clean the 25 mL graduated cylinder with soap and water, and then rinse it with distilled water.

5. Measure 10 mL of distilled water. Pour it into the paper cup with the latex.

6. Measure 10 mL of the 5% acetic acid solution, and pour it into the paper cup with the latex and water.

7. Immediately stir the mixture by using the wooden stick.

8. As you continue stirring, a polymer "lump" will form around the wooden stick. Pull the stick with the polymer lump from the paper cup, and immerse the lump in the 2 L beaker, bucket, or tub.

9. While wearing gloves, gently pull the lump from the wooden stick. Be sure to keep the lump immersed in the water, as shown in **Figure A.**

10. Keep the latex rubber underwater, and use your gloved hands to mold the lump into a ball, as shown in **Figure B.** Squeeze the lump several times to remove any unused chemicals. You may remove the latex rubber from the water as you roll it in your hands to smooth the ball.

11. Set aside the latex-rubber ball to dry. While it is drying, begin to make a ball from the ethanol and sodium silicate solutions.

12. In a clean 25 mL graduated cylinder, measure 12 mL of sodium silicate solution and pour it into the other paper cup.

13. In a clean 10 mL graduated cylinder, measure 3 mL of 50% ethanol. Pour the ethanol into the paper cup with the sodium silicate, and mix with the wooden stick until a solid substance is formed.

14. While wearing gloves, remove the polymer that forms and place it in the palm of one hand, as shown in **Figure C.** Gently press it with the palms of both your hands until a ball that does not crumble is formed. This process takes a little time and patience. The liquid that comes out of the ball is a combination of ethanol and water. Occasionally moisten the ball by letting a small amount of water from a faucet run over it. When the ball no longer crumbles, you are ready to go on to the next step.

15. Observe as many physical properties of the balls as possible, and record your observations in your lab notebook.

16. Drop each ball several times, and record your observations.

17. Drop each ball from a height of 1 m, and measure its bounce. Perform three trials for each ball, and record the values in your data table.

18. Measure the diameter and the mass of each ball, and record the values in your data table.

19. Clean all apparatus and your lab station. Dispose of any extra solutions in the containers indicated by your teacher. Clean up your lab area. Remember to wash your hands thoroughly when your lab work is finished.

Figure A

Figure B

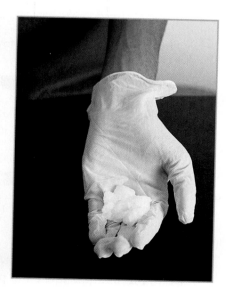

Figure C

Analysis

1. **Analyzing data** Give the chemical formula for the latex (isoprene) monomer and the ethanol–sodium silicate polymer.

2. **Analyzing data** List at least three observations you made of the properties of the two different balls.

3. **Explaining events** Explain how your observations in item 2 indicate that the polymers in each ball are not ionically bonded.

4. **Organizing data** Calculate the average height of the bounce for each type of ball.

5. **Organizing data** Calculate the volume for each ball. Even though the balls may not be perfectly spherical, assume that they are. (Hint: The volume of a sphere is equal to $\frac{4}{3} \times \pi \times r^3$, where r is the radius of the sphere, which is one-half of the diameter.)

6. **Organizing data** Using your measurements for the volumes from item 5 and the recorded mass, calculate the density of each ball.

Conclusions

7. **Evaluating data** Which polymer would you recommend for the toy company's new toy balls? Explain your reasoning.

8. **Evaluating results** Using the table shown below, calculate the unit cost, that is, the amount of money it costs to make a single ball. (Hint: Calculate how much of each reagent is needed to make a single ball.)

Data Table 2

Reagent	Price (dollars per liter)
Acetic acid solution	1.50
Ethanol solution	9.00
Latex solution	20.00
Sodium silicate solution	10.00

9. **Evaluating results** What are some other possible practical applications for each of the polymers you made?

10. **Making predictions** When a ball bounces up, kinetic energy of motion is converted into potential energy. With this in mind, explain which will bounce higher, a perfectly symmetrical, round sphere or an oblong shape that vibrates after it bounces.

11. **Evaluating methods** Explain why you didn't measure the volume of the balls by submerging them in water.

Extensions

1. **Research and communication** Polymers are used in our daily lives. Describe or list the polymers you come into contact with during a one-day period in your life.

2. **Designing experiments** Design a mold for a polymer ball that will make it symmetrical and smooth. If your teacher approves of your design, try the procedure again with the mold.

A CHEMICAL REFERENCE HANDBOOK

TABLE A-1   SI MEASUREMENT

Prefix	Symbol	Factor of Base Unit	Prefix	Symbol	Factor of Base Unit
giga	G	1 000 000 000	*centi*	c	0.01
mega	M	1 000 000	*milli*	m	0.001
kilo	k	1 000	*micro*	μ	0.000 001
hecto	h	100	*nano*	n	0.000 000 001
deka	da	10	*pico*	p	0.000 000 000 001
deci	d	0.1			

TABLE A-2   UNIT ABBREVIATIONS

amu	=	atomic mass unit (mass)	**mol**	=	mole (quantity)
atm	=	atmosphere (pressure, non-SI)	**M**	=	molarity (concentration)
Bq	=	becquerel (nuclear activity)	**N**	=	newton (force)
°C	=	degree Celsius (temperature)	**Pa**	=	pascal (pressure)
J	=	joule (energy)	**s**	=	second (time)
K	=	kelvin (temperature, thermo-dynamic)	**V**	=	volt (electric potential difference)

TABLE A-3   SYMBOLS

Symbol		Meaning	Symbol		Meaning
α	=	helium nucleus (also ^4_2He) emission from radioactive materials	ΔG^0	=	standard free energy of reaction
β	=	electron (also $^0_{-1}e$) emission from radioactive materials	ΔG^0_f	=	standard molar free energy of formation
γ	=	high-energy photon emission from radioactive materials	H	=	enthalpy
			ΔH^0	=	standard enthalpy of reaction
Δ	=	change in a given quantity (e.g., ΔH for change in enthalpy)	ΔH^0_f	=	standard molar enthalpy of formation
c	=	speed of light in vacuum	K_a	=	ionization constant (acid)
c_p	=	specific heat capacity (at constant pressure)	K_b	=	dissociation constant (base)
			K_{eq}	=	equilibrium constant
D	=	density	K_{sp}	=	solubility-product constant
E_a	=	activation energy	KE	=	kinetic energy
E^0	=	standard electrode potential	m	=	mass
E^0_{cell}	=	standard potential of an electrochemical cell	N_A	=	Avogadro's number
			n	=	number of moles
G	=	Gibbs free energy	P	=	pressure

TABLE A-3 CONTINUED

Symbol		Meaning	Symbol		Meaning
pH	=	measure of acidity $(-\log[H_3O^+])$	T	=	temperature (thermodynamic, in kelvins)
R	=	ideal gas law constant	t	=	temperature (in degrees Celsius)
S	=	entropy	V	=	volume
S^0	=	standard molar entropy	v	=	velocity or speed

TABLE A-4 PHYSICAL CONSTANTS

Quantity	Symbol	Value
Atomic mass unit	amu	$1.660\ 5402 \times 10^{-27}$ kg
Avogadro's number	N_A	$6.022\ 137 \times 10^{23}$/mol
Electron rest mass	m_e	$9.109\ 3897 \times 10^{-31}$ kg 5.4858×10^{-4} amu
Ideal gas law constant	R	8.314 L•kPa/mol•K 0.0821 L•atm/mol•K
Molar volume of ideal gas at STP	V_M	22.414 10 L/mol
Neutron rest mass	m_n	$1.674\ 9286 \times 10^{-27}$ kg 1.008 665 amu
Normal boiling point of water	T_b	373.15 K = 100.0°C
Normal freezing point of water	T_f	273.15 K = 0.00°C
Planck's constant	h	$6.626\ 076 \times 10^{-34}$ J•s
Proton rest mass	m_p	$1.672\ 6231 \times 10^{-27}$ kg 1.007 276 amu
Speed of light in a vacuum	c	$2.997\ 924\ 58 \times 10^8$ m/s
Temperature of triple point of water		273.16 K = 0.01°C

TABLE A-5 PROPERTIES OF COMMON ELEMENTS

Name	Form/color	Density (g/cm³)	Melting point (°C)	Boiling point (°C)	Common oxidation states
Aluminum	silver metal	2.702	660.37	2467	3+
Arsenic	gray metalloid	5.72714	817 (28 atm)	613 (sublimes)	3−, 3+, 5+
Barium	bluish white metal	3.51	725	1640	2+
Bromine	red-brown liquid	3.119	27.2	58.78	1−, 1+, 3+, 5+, 7+
Calcium	silver metal	1.54	839 ± 2	1484	2+
Carbon	diamond	3.51	3500 (63.5 atm)	3930	2+, 4+
	graphite	2.25	3652 (sublimes)	—	
Chlorine	green-yellow gas	3.214*	2100.98	234.6	1−, 1+, 3+, 5+, 7+
Chromium	gray metal	7.2028	1857 ± 20	2672	2+, 3+, 6+

continued on next page

TABLE A-5 CONTINUED

Name	Form/color	Density (g/cm³)	Melting point (°C)	Boiling point (°C)	Common oxidation states
Cobalt	gray metal	8.9	1495	2870	2+, 3+
Copper	red metal	8.92	1083.4 ± 0.2	2567	1+, 2+
Fluorine	yellow gas	1.69‡	2219.62	2188.14	1−
Germanium	gray metalloid	5.32325	937.4	2830	4+
Gold	yellow metal	19.31	1064.43	2808 ± 2	1+, 3+
Helium	colorless gas	0.1785*	2272.2 (26 atm)	2268.9	0
Hydrogen	colorless gas	0.0899*	2259.34	2252.8	1−, 1+
Iodine	blue-black solid	4.93	113.5	184.35	1−, 1+, 3+, 5+, 7+
Iron	silver metal	7.86	1535	2750	2+, 3+
Lead	bluish white metal	11.343716	327.502	1740	2+, 4+
Lithium	silver metal	0.534	180.54	1342	1+
Magnesium	silver metal	‑1.745	648.8	1107	2+
Manganese	gray-white metal	7.20	1244 ± 3	1962	2+, 3+, 4+, 6+, 7+
Mercury	silver liquid metal	13.5462	238.87	356.58	1+, 2+
Neon	colorless gas	0.9002*	2248.67	2245.9	0
Nickel	silver metal	8.90	1455	2730	2+, 3+
Nitrogen	colorless gas	1.2506*	2209.86	2195.8	3−, 3+, 5+
Oxygen	colorless gas	1.429*	2218.4	2182.962	2−
Phosphorus	yellow solid	1.82	44.1	280	3−, 3+, 5+
Platinum	silver metal	21.45	1772	3827 ± 100	2+, 4+
Potassium	silver metal	0.86	63.25	760	1+
Silicon	gray metalloid	2.33 ± 0.01	1410	2355	2+, 4+
Silver	white metal	10.5	961.93	2212	1+
Sodium	silver metal	0.97	97.8	882.9	1+
Strontium	silver metal	2.6	769	1384	2+
Sulfur	yellow solid	1.96	119.0	444.674	2−, 4+, 6+
Tin	white metal	7.28	231.88	2260	2+, 4+
Titanium	white metal	4.5	1660 ± 10	3287	2+, 3+, 4+
Uranium	silver metal	19.05 ± 0.0225	1132.3 ± 0.8	3818	3+, 4+, 6+
Zinc	blue-white metal	7.14	419.58	907	2+

* Densities of gases given in g/L at STP
† Densities obtained at 20°C unless otherwise noted (superscript)
‡ Density of fluorine given in g/L at 1 atm and 15°C

TABLE A-6 KEY OF ATOM COLORS USED IN MOLECULAR MODELS IN *HOLT CHEMISTRY*

Element	Color	Element	Color
Hydrogen, H		Silicon, Si	
Helium, He		Phosphorus, P	
Carbon, C		Sulfur, S	
Nitrogen, N		Chlorine, Cl	
Oxygen, O		Argon, Ar	
Fluorine, F		Iron, Fe	
Neon, Ne		Copper, Cu	
Sodium, Na (similar color used for all Group 1 metals)		Bromine, Br	
Magnesium, Mg (similar color used for all Group 2 metals)		Silver, Ag	
Aluminum, Al		Iodine, I	

TABLE A-7 COMMON IONS

Cation	Symbol	Cation	Symbol	Anion	Symbol	Anion	Symbol
Aluminum	Al^{3+}	Lead(II)	Pb^{2+}	Acetate	CH_3COO^-	Hydrogen sulfate	HSO_4^-
Ammonium	NH_4^+	Magnesium	Mg^{2+}	Bromide	Br^-	Hydroxide	OH^-
Arsenic(III)	As^{3+}	Mercury(I)	Hg_2^{2+}	Carbonate	CO_3^{2-}	Hypochlorite	ClO^-
Barium	Ba^{2+}	Mercury(II)	Hg^{2+}	Chlorate	ClO_3^-	Iodide	I^-
Calcium	Ca^{2+}	Nickel(II)	Ni^{2+}	Chloride	Cl^-	Nitrate	NO_3^-
Chromium(II)	Cr^{2+}	Potassium	K^+	Chlorite	ClO_2^-	Nitrite	NO_2^-
Chromium(III)	Cr^{3+}	Silver	Ag^+	Chromate	CrO_4^{2-}	Oxide	O_2^-
Cobalt(II)	Co^{2+}	Sodium	Na^+	Cyanide	CN^-	Perchlorate	ClO_4^-
Cobalt(III)	Co^{3+}	Strontium	Sr^{2+}	Dichromate	$Cr_2O_7^{2-}$	Permanganate	MnO_4^-
Copper(I)	Cu^+	Tin(II)	Sn^{2+}	Fluoride	F^-	Peroxide	O_2^{2-}
Copper(II)	Cu^{2+}	Tin(IV)	Sn^{4+}	Hexacyanoferrate(II)	$Fe(CN)_6^{4-}$	Phosphate	PO_4^{3-}
Hydronium	H_3O^+	Titanium(III)	Ti^{3+}	Hexacyanoferrate(III)	$Fe(CN)_6^{3-}$	Sulfate	SO_4^{2-}
Iron(II)	Fe^{2+}	Titanium(IV)	Ti^{4+}	Hydride	H^-	Sulfide	S^{2-}
Iron(III)	Fe^{3+}	Zinc	Zn^{2+}	Hydrogen carbonate	HCO_3^-	Sulfite	SO_3^{2-}

TABLE A-8 PREFIXES FOR NAMING COVALENT COMPOUNDS

Prefix	Number of Atoms	Example	Name		Prefix	Number of Atoms	Example	Name
mono-	1	CO	carbon monoxide		hexa-	6	CeB_6	cerium hexaboride
di-	2	SiO_2	silicon dioxide		hepta-	7	IF_7	iodine heptafluoride
tri-	3	SO_3	sulfur trioxide		octa-	8	Np_3O_8	trineptunium octoxide
tetra-	4	SCl_4	sulfur tetrachloride		nona-	9	I_4O_9	tetraiodine nonoxide
penta-	5	$SbCl_5$	antimony pentachloride		deca-	10	S_2F_{10}	disulfur decafluoride

TABLE A-9 ACTIVITY SERIES OF THE ELEMENTS

Activity of Metals

Li Rb K Ca Ba Sr Ca Na	react with cold H_2O and acids, replacing hydrogen; react with oxygen, forming oxides
Mg Al Mn Zn Cr Fe Cd	react with steam (but not cold water) and acids; replacing hydrogen; react with oxygen, forming oxides
Co Ni Sn Pb	do not react with water; react with acids, replacing hydrogen; react with oxygen, forming oxides
H_2 Sb Bi Cu Hg	react with oxygen, forming oxides
Ag Pt Au	fairly unreactive, forming oxides only indirectly.

Activity of Halogens

F_2

Cl_2

Br_2

I_2

TABLE A-10   STATE SYMBOLS AND REACTION CONDITIONS

Symbol	Meaning
(s), (l), (g)	substance in the solid, liquid, or gaseous state
(aq)	substance in aqueous solution (dissolved in water)
$\longrightarrow$	"produces" or "yields," indicating the result of a reaction
$\rightleftarrows$	reversible reaction in which products can reform into reactants; final result is a mixture of products and reactants
$\xrightarrow{\Delta}$ or $\xrightarrow{heat}$	reactants are heated; temperature is not specified
Pd	name or chemical formula of a catalyst, added to speed a reaction
(c), ↓	product is a solid precipitate
↑	product is a gas

TABLE A-11   THERMODYNAMIC DATA

Substance	ΔH_f^0 (kJ/mol)	S^0 (J/mol·K)	ΔG_f^0 (kJ/mol)	Substance	ΔH_f^0 (kJ/mol)	S^0 (J/mol·K)	ΔG_f^0 (kJ/mol)
$Ag(s)$	0.0	42.7	0.0	$C_6H_{14}(g, \textit{n-hexane})$	−167.1	388.4	0.0
$AgCl(s)$	−127.1	96.2	−109.8	$C_7H_{16}(g, \textit{n-heptane})$	−187.7	427.9	8.0
$AgNO_3(s)$	−124.4	140.9	−33.5	$C_8H_{18}(g, \textit{n-octane})$	−208.6	466.7	16.3
$Al(s)$	0.0	28.3	0.0	$C_8H_{18}(g, \textit{iso-octane})$	−224.0	423.2	12.6
$AlCl_3(s)$	−705.6	110.7	−628.9	$CaCO_3(s)$	−1206.9	92.9	−1128.8
$Al_2O_3(s, \textit{corundum})$	−1676.0	51.0	−1582.4	$CaCl_2(s)$	−795.8	108.4	−748.1
$Br_2(l)$	0.0	152.2	0.0	$Ca(OH)_2(s)$	−986.1	83.4	−898.6
$Br_2(g)$	30.9	245.5	30.9	$Ca(s)$	0.0	41.6	0.0
C (s, diamond)	1.9	2.4	2.90	$CaO(s)$	−634.9	38.2	−604.04
C (s, graphite)	0.0	5.7	0.0	$Cl_2(g)$	0.0	223.1	0.0
$CCl_4(l)$	−132.8	216.2	−65.3	$Cu(s)$	0.0	33.2	0.0
$CCl_4(g)$	−95.8	309.9	−60.2	$CuCl_2(s)$	−220.1	108.1	−175.7
$CH_3OH(l)$	−239.1	127.2	−166.4	$CuSO_4(s)$	−770.0	109.3	−661.9
$CH_4(g)$	−74.9	186.3	−50.8	$F_2(g)$	0.0	202.8	0.0
$CO(g)$	−110.5	197.6	−137.2	$Fe(s)$	0.0	27.3	0.0
$CO_2(g)$	−393.5	213.8	−394.4	$FeCl_3(s)$	−399.4	142.3	−334.05
$CS_2(g)$	117.1	237.8	67.2	$Fe_2O_3(s, \textit{hematite})$	−824.8	87.4	−742.2
$C_2H_6(g)$	−83.8	229.1	32.9	$Fe_3O_4(s, \textit{magnetite})$	−1120.9	145.3	−1015.5
$C_2H_4(g)$	52.5	219.3	68.1	$H_2(g)$	0.0	130.7	0.0
$C_2H_5OH(l)$	−277.0	161.0	−174.9	$HBr(g)$	−36.4	198.6	−53.4
$C_3H_8(g)$	−104.7	270.2	−24.3	$HCl(g)$	−92.3	186.8	−95.3
$C_4H_{10}(g, \textit{n-butane})$	−125.6	310.1	−16.7	$HCN(g)$	135.1	201.7	124.7
$C_4H_{10}(g, \textit{isobutane})$	−134.2	294.6	−20.9	$HCOOH(l)$	−425.1	129.0	−361.4

continued on next page

TABLE A-11 CONTINUED

Substance	ΔH_f^0 (kJ/mol)	S^0 (J/mol·K)	ΔG_f^0 (kJ/mol)	Substance	ΔH_f^0 (kJ/mol)	S^0 (J/mol·K)	ΔG_f^0 (kJ/mol)
HF(g)	−272.5	173.8	−273.2	NO₂(g)	33.1	240.0	51.3
HNO₃(g)	−134.3	266.4	−74.8	N₂O(g)	82.4	220.0	104.2
H₂O(g)	−241.8	188.7	−228.6	N₂O₄(g)	9.1	304.4	97.8
H₂O(l)	−285.8	70.0	−237.2	Na(s)	0.0	51.5	0.0
H₂O₂(l)	−187.8	109.6	−120.4	NaCl(s)	−411.2	72.1	−384.2
H₂S(g)	−20.5	205.7	−33.6	NaOH(s)	−425.9	64.4	−379.5
H₂SO₄(l)	−814.0	156.9	−690.1	O₂(g)	0.0	205.0	0.0
K(s)	0.0	64.7	0.0	O₃(g)	142.7	238.9	163.2
KCl(s)	−436.7	82.6	−409.2	Pb(s)	0.0	64.8	0.0
KNO₃(s)	−494.6	133.0	−394.9	PbCl₂(s)	−359.4	136.2	−317.9
KOH(s)	−424.7	78.9	−379.1	PbO(s, red)	−219.0	66.3	−188.95
Li(s)	0.0	29.1	0.0	S(s)	0.0	32.1	0.0
LiCl(s)	−408.6	59.3	−384.4	SO₂(g)	−296.8	248.1	−300.2
LiOH(s)	−484.9	42.8	−439.0	SO₃(g)	−395.8	256.8	−371.1
Mg(s)	0.0	32.7	0.0	Si(s)	0.0	18.8	0.0
MgCl₂(s)	−641.6	89.6	−591.8	SiCl₄(g)	−657.0	330.9	−617.0
Hg(l)	0.0	76.0	0.0	SiO₂(s, quartz)	−910.9	41.5	−856.7
Hg₂Cl₂(s)	−264.2	192.5	−210.8	Sn(s, white)	0.0	51.6	0.0
HgO(s, red)	−90.8	70.3	−55.6	Sn(s, gray)	−2.1	44.1	0.13
N₂(g)	0.0	191.6	0.0	SnCl₄(l)	−511.3	258.6	−440.2
NH₃(g)	−45.9	192.8	−16.5	Zn(s)	0.0	41.6	0.0
NH₄Cl(s)	−314.4	94.6	−203.0	ZnCl₂(s)	−415.0	111.5	−369.4
NO(g)	90.3	210.8	86.6	ZnO(s)	−348.3	43.6	−318.32

TABLE A-12   HEAT OF COMBUSTION

Formula	ΔH_c(kJ/mol)	Formula	ΔH_c(kJ/mol)	Formula	ΔH_c(kJ/mol)
H₂(g)	−285.8	C₆H₁₄(l)	−4163.2	C₁₀H₈(s)	−5156.3
C(s, graphite)	−393.5	C₇H₁₆(l)	−4817.0	C₁₄H₁₀(s)	−7076.5
CO(g)	−283.0	C₈H₁₈(l)	−5470.5	CH₃OH(l)	−726.1
CH₄(g)	−890.8	C₂H₄(g)	−1411.2	C₂H₅OH(l)	−1366.8
C₂H₆(g)	−1560.7	C₃H₆(g)	−2058.0	(C₂H₅)₂O(l)	−2751.1
C₃H₈(g)	−2219.2	C₂H₂(g)	−1301.1	CH₂O(g)	−570.7
C₄H₁₀(g)	−2877.6	C₆H₆(l)	−3267.6	C₆H₁₂O₆(s)	−2803.0
C₅H₁₂(g)	−3535.6	C₇H₈(l)	−3910.3	C₁₂H₂₂O₁₁(s)	−5640.9

TABLE A-13   WATER-VAPOR PRESSURE

Temperature (°C)	Pressure (mm Hg)	Pressure (kPa)	Temperature (°C)	Pressure (mm Hg)	Pressure (kPa)
0.0	4.6	0.61	23.0	21.1	2.81
5.0	6.5	0.87	23.5	21.7	2.90
10.0	9.2	1.23	24.0	22.4	2.98
15.0	12.8	1.71	24.5	23.1	3.10
15.5	13.2	1.76	25.0	23.8	3.17
16.0	13.6	1.82	26.0	25.2	3.36
16.5	14.1	1.88	27.0	26.7	3.57
17.0	14.5	1.94	28.0	28.3	3.78
17.5	15.0	2.00	29.0	30.0	4.01
18.0	15.5	2.06	30.0	31.8	4.25
18.5	16.0	2.13	35.0	42.2	5.63
19.0	16.5	2.19	40.0	55.3	7.38
19.5	17.0	2.27	50.0	92.5	12.34
20.0	17.5	2.34	60.0	149.4	19.93
20.5	18.1	2.41	70.0	233.7	31.18
21.0	18.6	2.49	80.0	355.1	47.37
21.5	19.2	2.57	90.0	525.8	70.12
22.0	19.8	2.64	95.0	633.9	84.53
22.5	20.4	2.72	100.0	760.0	101.32

TABLE A-14   DENSITIES OF GASES AT STP

Gas	Density (g/L)	Gas	Density (g/L)
Air, dry	1.293	Hydrogen	0.0899
Ammonia	0.771	Hydrogen chloride	1.639
Carbon dioxide	1.997	Hydrogen sulfide	1.539
Carbon monoxide	1.250	Methane	0.7168
Chlorine	3.214	Nitrogen	1.2506
Dinitrogen monoxide	1.977	Nitrogen monoxide (at 10°C)	1.340
Ethyne (acetylene)	1.165	Oxygen	1.429
Helium	0.1785	Sulfur dioxide	2.927

TABLE A-15   DENSITY OF LIQUID WATER

Temperature (°C)	Density (g/cm³)	Temperature (°C)	Density (g/cm³)
0	0.999 84	25	0.997 05
2	0.999 94	30	0.995 65
3.98 (maximum)	0.999 973	40	0.992 22
4	0.999 97	50	0.988 04
6	0.999 94	60	0.983 20
8	0.999 85	70	0.977 77
10	0.999 70	80	0.971 79
14	0.999 24	90	0.965 31
16	0.998 94	100	0.958 36
20	0.998 20		

TABLE A-16   MEASURES OF CONCENTRATION

Name	Symbol	Units	Areas of application
Molarity	M	$\dfrac{\text{mol solute}}{\text{L solution}}$	in solution stoichiometry calculations
Molality	m	$\dfrac{\text{mol solute}}{\text{kg solvent}}$	boiling-point elevation and freezing-point depression calculations
Mole fraction	X	$\dfrac{\text{mol solute}}{\text{total mol solution}}$	in solution thermodynamics
Volume percent	% V/V	$\dfrac{\text{volume solute}}{\text{volume solution}} \times 100$	with liquid-liquid mixtures
Mass or weight percent	% or %w/w	$\dfrac{\text{g solute}}{\text{g solution}} \times 100$	in biological research
Parts per million	ppm	$\dfrac{\text{g solute}}{1\ 000\ 000\ \text{g solution}}$	to express small concentrations
Parts per billion	ppb	$\dfrac{\text{g solute}}{1\ 000\ 000\ 000\ \text{g solution}}$	to express very small concentrations, as in pollutants or contaminants

TABLE A-17 SOLUBILITIES OF GASES IN WATER*

Gas	0°C	10°C	20°C	60°C
Air	0.029 18	0.022 84	0.018 68	0.012 16
Ammonia	1130	870	680	200
Carbon dioxide	1.713	1.194	0.878	0.359
Carbon monoxide	0.035 37	0.028 16	0.023 19	0.014 88
Chlorine	—	3.148	2.299	1.023
Hydrogen	0.021 48	0.019 55	0.018 19	0.016 00
Hydrogen chloride	512	475	442	339
Hydrogen sulfide	4.670	3.399	2.582	1.190
Methane	0.055 63	0.041 77	0.033 08	0.019 54
Nitrogen†	0.023 54	0.018 61	0.015 45	0.010 23
Nitrogen monoxide	0.073 81	0.057 09	0.047 06	0.029 54
Oxygen	0.048 89	0.038 02	0.031 02	0.019 46
Sulfur dioxide	79.789	56.647	39.374	—

* Volume of gas (in liters) at STP that can be dissolved in 1 L of water at the temperature (°C) indicated.
† Atmospheric nitrogen: 98.815% N_2, 1.185% inert gases

TABLE A-18 SOLUBILITY OF COMPOUNDS*

Formula	0°C	20°C	60°C	100°C
$Al_2(SO_4)_3$	31.2	36.4	59.2	89.0
NH_4Cl	29.4	37.2	55.3	77.3
NH_4NO_3	118	192	421	871
$(NH_4)_2SO_4$	70.6	75.4	88	103
$BaCl_2 \cdot 2H_2O$	31.2	35.8	46.2	59.4
$Ba(OH)_2$	1.67	3.89	20.94	$101.40^{80°}$
$Ba(NO_3)_2$	4.95	9.02	20.4	34.4
$Ca(HCO_3)_2$	16.15	16.60	17.50	18.40
$Ca(OH)_2$	0.189	0.173	0.121	0.076
$CuCl_2$	68.6	73.0	96.5	120
$CuSO_4 \cdot 5H_2O$	23.1	32.0	61.8	114
$PbCl_2$	0.67	1.00	1.94	3.20
$Pb(NO_3)_2$	37.5	54.3	91.6	133
$LiCl$	69.2	83.5	98.4	128
Li_2SO_4	36.1	34.8	32.6	$30.9^{90°}$
$MgSO_4$	22.0	33.7	54.6	68.3
$HgCl_2$	3.63	6.57	16.3	61.3

continued on next page

TABLE A-18 CONTINUED

Formula	0°C	20°C	60°C	100°C
KBr	53.6	65.3	85.5	104
$KClO_3$	3.3	7.3	23.8	56.3
KCl	28.0	34.2	45.8	56.3
K_2CrO_4	56.3	63.7	70.1	$74.5^{90°}$
KI	128	144	176	206
KNO_3	13.9	31.6	106	245
K_2SO_4	7.4	11.1	18.2	24.1
$AgC_2H_3O_2$	0.73	1.05	1.93	$2.59^{80°}$
$AgNO_3$	122	216	440	733
$NaC_2H_3O_2$	36.2	46.4	139	170
$NaClO_3$	79.6	95.9	137	204
NaCl	35.7	35.9	37.1	39.2
$NaNO_3$	73.0	87.6	122	180
$C_{12}H_{22}O_{11}$	179.2	203.9	287.3	487.2

* Solubilities are given in grams of solute that can be dissolved in 100 g of water at the temperature (°C) indicated.

TABLE A-19   EXAMPLES OF COMPLEX IONS

Complex cation	Color	Complex cation	Color	Complex anion	Color
$[Co(NH_3)_6]^{3+}$	yellow-orange	$[Fe(H_2O)_5SCN]^{2+}$	deep red	$[Co(CN)_6]^{3-}$	pale yellow
$[Co(NH_3)_5(H_2O)]^{3+}$	bright red	$[Ni(NH_3)_6]^{2+}$	blue-violet	$[CoCl_4]^{2-}$	blue
$[Co(NH_3)_5Cl]^{2+}$	violet	$[Ni(H_2O)_6]^{2+}$	green	$[Cu_2Cl_6]^{2-}$	red
$[Co(H_2O)_6]^{2+}$	pink	$[Zn(NH_3)_4]^{2+}$	colorless	$[Fe(CN)_6]^{3-}$	red
$[Cu(NH_3)_4]^{2+}$	blue-purple	$[Zn(NH_3)_6]^{2+}$	colorless	$[Fe(CN)_6]^{4-}$	yellow
$[Cu(H_2O)_4]^{2+}$	light blue			$[Fe(C_2O_4)_3]^{3-}$	green

TABLE A-20 EQUILIBRIUM CONSTANTS

Equation	K_{eq} expression	Values	
$N_2(g) + 3H_2(g) \rightleftarrows 2NH_3(g)$	$\dfrac{[NH_3]^2}{[N_2][H_2]^3}$	3.3×10^8 (25°C)	4.2×10^1 (327°C)
$NH_3(aq) + H_2O(l) \rightleftarrows NH_4^+(aq) + OH^-(aq)$	$\dfrac{[NH_4^+][OH^-]}{[NH_3]}$	1.8×10^{-5} (25°C)	
$2NO_2(g) \rightleftarrows N_2O_4(g)$	$\dfrac{[N_2O_4]}{[NO_2]^2}$	1.25×10^3 (0°C) 1.65×10^2 (25°C) 2.0×10^1 (100°C)	
$N_2(g) + O_2(g) \rightleftarrows 2NO(g)$	$\dfrac{[NO]^2}{[N_2][O_2]}$	4.5×10^{-31} (25°C)	6.7×10^{-10} (627°C)
$CO_2(g) + H_2(g) \rightleftarrows CO(g) + H_2O(g)$	$\dfrac{[CO][H_2O]}{[CO_2][H_2]}$	2.2 (1400°C)	4.6 (2000°C)
$H_2CO_3(aq) + H_2O(l) \rightleftarrows H_3O^+(aq) + HCO_3^-(aq)$	$\dfrac{[H_3O^+][HCO_3^-]}{[H_2CO_3]}$	4.3×10^{-7} (25°C)	
$HCO_3^-(aq) + H_2O(l) \rightleftarrows CO_3^{2-}(aq) + H_3O^+(aq)$	$\dfrac{[CO_3^{2-}][H_3O^+]}{[HCO_3^-]}$	4.7×10^{-11} (25°C)	
$H_2(g) + I_2(g) \rightleftarrows 2HI(g)$	$\dfrac{[HI]^2}{[H_2][I_2]}$	1.13×10^2 (250°C)	1.8×10^1 (1127°C)
$Hg^{2+}(aq) + Hg(l) \rightleftarrows Hg_2^{2+}(aq)$	$\dfrac{[Hg_2^{2+}]}{[Hg^{2+}]}$	8.1×10^1 (25°C)	

TABLE A-21 APPROXIMATE CONCENTRATION OF IONS IN OCEAN WATER

Ion	Concentration (mol/L)	Ion	Concentration (mol/L)
Cl^-	0.554	K^+	0.010
Na^+	0.470	Ca^{2+}	0.009
Mg^{2+}	0.047	CO_3^{2-}	0.002
SO_4^{2-}	0.015	Br^-	0.001

TABLE A-22   STANDARD ELECTRODE POTENTIALS

Electrode reaction	$E°$ (V)
$Li^+(aq) + e^- \rightleftarrows Li(s)$	−3.0401
$K^+(aq) + e^- \rightleftarrows K(s)$	−2.931
$Ca^{2+}(aq) + 2e^- \rightleftarrows Ca(s)$	−2.868
$Na^+(aq) + e^- \rightleftarrows Na(s)$	−2.71
$Mg^{2+}(aq) + 2e^- \rightleftarrows Mg(s)$	−2.372
$Al^{3+}(aq) + 3e^- \rightleftarrows Al(s)$	−1.662
$Zn(OH)_2(s) + 2e^- \rightleftarrows Zn(s) + 2OH^-(aq)$	−1.249
$2H_2O(l) + 2e^- \rightleftarrows H_2(g) + 2OH^-(aq)$	−0.828
$Zn^{2+}(aq) + 2e^- \rightleftarrows Zn(s)$	−0.7618
$Fe^{2+}(aq) + 2e^- \rightleftarrows Fe(s)$	−0.447
$PbSO_4(s) + H_3O^+(aq) + 2e^- \rightleftarrows Pb(s) + HSO_4^-(aq) + H_2O(l)$	−0.42
$Cd^{2+}(aq) + 2e^- \rightleftarrows Cd(s)$	−0.4030
$Pb^{2+}(aq) + 2e^- \rightleftarrows Pb(s)$	−0.1262
$Fe^{3+}(aq) + 3e^- \rightleftarrows Fe(s)$	−0.037
$2H_3O^+(aq) + 2e^- \rightleftarrows H_2(g) + 2H_2O(l)$	0.000
$AgCl(s) + e^- \rightleftarrows Ag(s) + Cl^-(aq)$	+0.222
$Cu^{2+}(aq) + 2e^- \rightleftarrows Cu(s)$	+0.3419
$O_2(g) + 2H_2O(l) + 4e^- \rightleftarrows 4OH^-(aq)$	+0.401
$I_2(s) + 2e^- \rightleftarrows 2I^-(aq)$	+0.5355
$Fe^{3+}(aq) + e^- \rightleftarrows Fe^{2+}(aq)$	+0.771
$Hg_2^{2+}(aq) + 2e^- \rightleftarrows 2Hg(l)$	+0.7973
$Ag^+(aq) + e^- \rightleftarrows Ag(s)$	+0.7996
$Br_2(l) + 2e^- \rightleftarrows 2Br^-(aq)$	+1.066
$MnO_2(s) + 4H_3O^+(aq) + 2e^- \rightleftarrows Mn^{2+}(aq) + 6H_2O(l)$	+1.224
$O_2(g) + 4H_3O^+(aq) + 4e^- \rightleftarrows 6H_2O$	+1.229
$Cl_2(g) + 2e^- \rightleftarrows 2Cl^-(aq)$	+1.358
$PbO_2(s) + 4H_3O^+(aq) + 2e^- \rightleftarrows Pb^{2+}(aq) + 6H_2O(l)$	+1.455
$MnO_4^-(aq) + 8H_3O^+(aq) + 5e^- \rightleftarrows Mn^{2+}(aq) + 12H_2O(l)$	+1.507
$PbO_2(s) + HSO_4^-(aq) + 3H_3O^+(aq) + 2e^- \rightleftarrows PbSO_4(s) + 5H_2O(l)$	+1.691
$Ce^{4+}(aq) + e^- \rightleftarrows Ce^{3+}(aq)$	+1.72
$Ag_2O_2(s) + 4H^+(aq) + e^- \rightleftarrows 2Ag(s) + 2H_2O(l)$	+1.802
$F_2(g) + 2e^- \rightleftarrows 2F^-(aq)$	+2.866

TABLE A-23 SOME CLASSES OF ORGANIC COMPOUNDS

Class	Functional group	Example	Use
Alcohol	—OH	2-propanol	disinfectant
Aldehyde	$-\overset{\overset{O}{\|\|}}{C}-H$	benzaldehyde	almond flavor
Halide	—F, Cl, Br, I	trichlorofluoromethane (Freon-11)	refrigerant
Amide	$-\overset{\overset{O}{\|\|}}{C}-NH_2$	niacinamide (nicotinamide)	nutrient
Amine	$-\overset{\|}{\underset{\|}{N}}$	caffeine	beverage ingredient
Carboxylic acid	$-\overset{\overset{O}{\|\|}}{C}-OH$	tetradecanoic acid (myristic acid)	soap-making ingredient
Ester	$-\overset{\overset{O}{\|\|}}{C}-O-$	ethyl butanoate	perfume ingredient
Ether	—O—	methyl phenyl ether (anisole)	perfume ingredient
Ketone	$-\overset{\overset{O}{\|\|}}{C}-$	propanone (acetone)	solvent in nail-polish remover

TABLE A-24   SOLUBILITY PRODUCT CONSTANTS AT 25°C

Salt	K_{sp}	Salt	K_{sp}
Ag_2CO_3	8.4×10^{-12}	$FeCO_3$	3.1×10^{-11}
$AgCl$	1.8×10^{-10}	$Fe(OH)_2$	4.9×10^{-17}
Ag_2CrO_4	1.1×10^{-12}	$Fe(OH)_3$	2.6×10^{-39}
Ag_2S	1.1×10^{-49}	FeS	1.6×10^{-19}
$AgBr$	5.4×10^{-13}	$MgCO_3$	6.8×10^{-6}
AgI	8.5×10^{-17}	$Mg(OH)_2$	5.6×10^{-12}
$AlPO_4$	9.8×10^{-21}	$Mg_3(PO_4)_2$	9.9×10^{-25}
$BaSO_4$	1.1×10^{-10}	$MnCO_3$	2.2×10^{-11}
$CaCO_3$	5.0×10^{-9}	$Pb(OH)_2$	1.4×10^{-20}
$Ca(OH)_2$	4.7×10^{-7}	PbS	9.0×10^{-29}
$Ca_3(PO_4)_2$	2.1×10^{-33}	$PbSO_4$	1.8×10^{-8}
$CaSO_4$	7.1×10^{-5}	$SrSO_4$	3.4×10^{-7}
CuS	1.3×10^{-36}	$ZnCO_3$	1.2×10^{-10}
		ZnS	2.9×10^{-25}

STUDY SKILLS FOR CHEMISTRY
TABLE OF CONTENTS

Succeeding in Your Chemistry Class

Your success in this course will depend on your ability to apply some basic study skills to learning the material. Studying chemistry can be difficult, but you can make it easier using simple strategies for dealing with the concepts and problems. Becoming skilled in using these strategies will be your keys to success in this and many other courses.

Reading the Text

- **Read the assigned material before class** so that the class lecture makes sense. Use a dictionary to help you interpret vocabulary. Remember while reading to figure out what information is important.

  Working together with others using Paired Reading and Discussion strategies can help you decide what is important and clarify the material. (For more discussion, see Other Reading Strategies on page 853.)

- **Select a quiet setting** away from distractions so that you can concentrate on what you are reading.

- **Have a pencil and paper nearby to jot down notes and questions** you may have. Be sure to get these questions answered in class. Power Notes (see page 849) can help you organize the notes you take and prepare you for class.

- **Use the Objectives in the beginning of each section as a list of what you need to know** from the section. Teachers generally make their tests based on the text objectives or their own objectives. Using the objectives to focus your reading can make your learning more efficient. Using the K/W/L strategy (see page 851) can help you relate new material to what you already know and what you need to learn.

Taking Notes in Class

- **Be prepared to take notes during class.** Have your materials organized in a notebook. Separate sheets of paper can be easily lost.

- **Don't write down everything your teacher says.** Try to tell which parts of the lecture are important and which are not. Reading the text before class will help in this. You will not be able to write down everything, so you must try to write down only the important things.

- **Recopying notes later is a waste of time** and does not help you learn material for a test. Do it right the first time. Organize your notes as you are writing them down so that you can make sense of your notes when you review them without needing to recopy them.

Reviewing Class Notes

- **Review your notes as soon as possible after class.** Write down any questions you may have about the material covered that day. Be sure to get these questions answered during the next class. You can work with friends to use strategies such as Paired Summarizing and L.I.N.K. (See page 853.)

- **Do not wait until the test to review.** By then you will have forgotten a good portion of the material.

- **Be selective about what you memorize.** You cannot memorize everything in a chapter. First of all, it is too time consuming. Second, memorizing and understanding are not the same thing. Memorizing topics as they appear in your notes or text does not guarantee that you will be able to correctly answer questions that require understanding of those topics. You should only memorize material that you understand. Concept Maps and other Reading Organizers, Sequencing/Pattern Puzzles, and Prediction Guides can help you understand key ideas and major concepts. (See pages 846, 852, and 854.)

Working Problems

In addition to understanding the concepts, the ability to solve problems will be a key to your success in chemistry. You will probably spend a lot of time working problems in class and at home. The ability to solve chemistry problems is a skill, and like any skill, it requires practice.

- **Always review the Sample Problems in the chapter.** The Sample Problems in the text provide road maps for solving certain types of problems. Cover the solution while trying to work the problem yourself.

- **The problems in the Chapter Review are similar to the Sample Problems.** If you can relate an assigned problem to one of the Sample Problems in the chapter, it shows that you understand the material.

- **The four steps: Gather information, Plan your work, Calculate, and Verify should be the steps you go through when working assigned problems.** These steps will allow you to organize your thoughts and help you develop your problem-solving skills.

- **Never spend more than 15 minutes trying to solve a problem.** If you have not been able to come up with a plan for the solution after 15 minutes, additional time spent will only cause you to become frustrated. What do you do? Get help! See your teacher or a classmate. Find out what it is that you do not understand.

- **Do not try to memorize the Sample Problems; spend your time trying to understand how the solution develops.** Memorizing a particular sample problem will not ensure that you understand it well enough to solve a similar problem.

- **Always look at your answer and ask yourself if it is reasonable and makes sense.** Check to be sure you have the correct units and numbers of significant figures.

Completing Homework

Your teacher will probably assign questions and problems from the Section Reviews and Chapter Reviews or assign Concept Review worksheets. The purpose of these assignments is to review what you have covered in class and to see if you can use the information to answer questions or solve problems. As in reviewing class notes, do your homework as soon after class as possible while the topics are still fresh in your mind. Do not wait until late at night, when you are more likely to be tired and to become frustrated.

Reviewing for an exam

- **Don't panic and don't cram!** It takes longer to learn if you are under pressure. If you have followed the strategies listed here and reviewed along the way, studying for the exam should be less stressful.

- When looking over your notes and concept maps, **recite ideas out loud.** There are two reasons for reciting:
  1. You are hearing the information, which is effective in helping you learn.
  2. If you cannot recite the ideas, it should be a clue that you do not understand the material, and you should begin rereading or reviewing the material again.

- **Studying with a friend provides a good opportunity for recitation.** If you can explain ideas to your study partner, you know the material.

Taking an exam

- **Get plenty of rest before the exam** so that you can think clearly. If you have been awake all night studying, you are less likely to succeed than if you had gotten a full night of rest.

- **Start with the questions you know.** If you get stuck on a question, save it for later. As time passes and you work through the exam, you may recall the information you need to answer a difficult question or solve a difficult problem.

Good luck!

Making Concept Maps

MAP Ⓐ

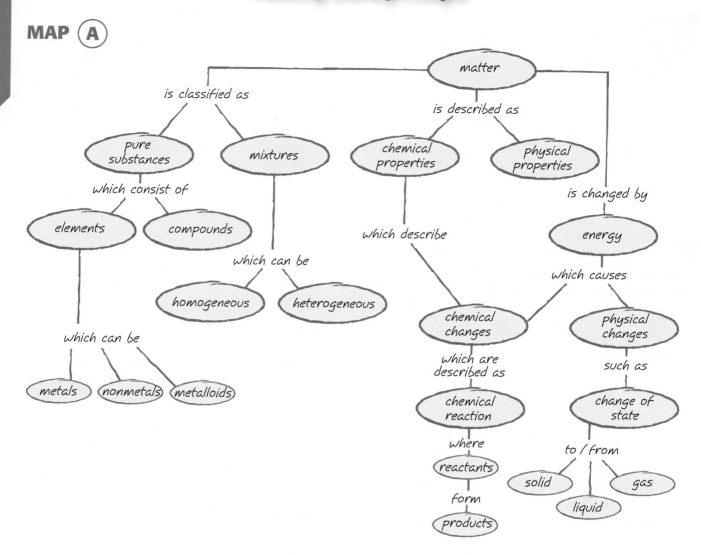

Making concept maps can help you decide what material in a chapter is important and how to efficiently learn that material. A concept map presents key ideas, meanings, and relationships for the concepts being studied. It can be thought of as a visual road map of the chapter. Learning happens efficiently when you use concept maps because you work with only the key ideas and how they fit together.

The concept map shown as **Map A** was made from vocabulary terms from the first few chapters of the book. Vocabulary terms are generally labels for concepts, and concepts are generally nouns. In a concept map, linking words are used to form propositions that connect concepts and give them meaning in context. For example, on the map above, "matter is described by physical properties" is a proposition.

Studies show that people are better able to remember materials presented visually. A concept map is better than an outline because you can see the relationships among many ideas. Because outlines are linear, there is no way of linking the ideas from various sections of the outline. Read through the map to become familiar with the information presented. Then look at the map in relation to all of the text pages in the first few chapters; which gives a better picture of the important concepts—the map or the full chapters?

To Make a Concept Map

1. List all the important concepts.

We'll use some of the boldfaced and italicized terms from the chapter "Matter and Energy."

matter	mixture
compound	pure substance
element	heterogeneous mixture
homogeneous mixture	

• From this list, group similar concepts together. For example, one way to group these concepts would be into two groups—one that is related to mixtures and one that is related to pure substances.

mixture	pure substance
heterogeneous mixture	compound
homogeneous mixture	element

2. Select a main concept for the map.

We will use matter as the main concept for this map.

3. Build the map by placing the concepts according to their importance under the main concept. For this map the main concept is *matter.*

One way of arranging the concepts is shown in **Map B.**

MAP B

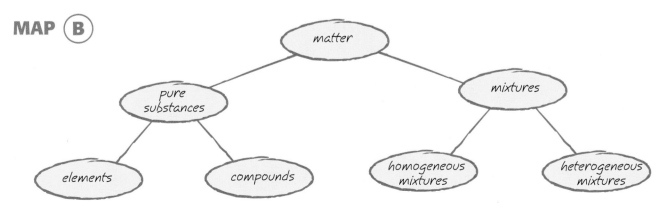

MAP C

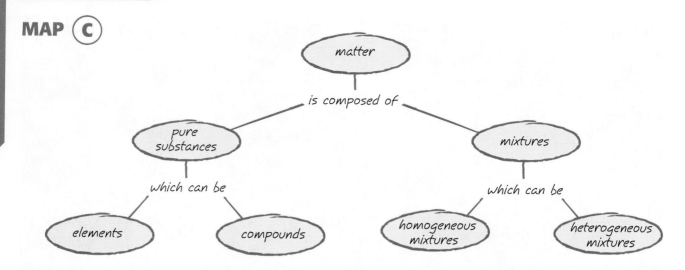

4. Add linking words to give meaning to the arrangement of concepts.

When adding the links, be sure that each proposition makes sense. To distinguish concepts from links, place your concepts in circles, ovals, or rectangles, as shown in the maps. Then make cross-links. Cross-links are made of propositions and lines connecting concepts across the map. Links that apply in only one direction are indicated with an arrowhead. **Map C** is a finished map covering the main ideas listed in Step 1.

Making maps might seem difficult at first, but the process forces you to think about the meanings and relationships among the concepts. If you do not understand those relationships, you can get help early on.

Practice mapping by making concept maps about topics you know. For example, if you know a lot about a particular sport, such as basketball, or if you have a particular hobby, such as playing a musical instrument, you can use that topic to make a practice map. By perfecting your skills with information that you know very well, you will begin to feel more confident about making maps from the information in a chapter.

Remember, the time you devote to mapping will pay off when it is time to review for an exam.

PRACTICE

1. Classify each of the following as either a concept or linking word(s).

   a. classification ———————————

   b. is classified as ———————————

   c. forms ———————————

   d. is described by ———————————

   e. reaction ———————————

   f. reacts with ———————————

   g. metal ———————————

   h. defines ———————————

2. Write three propositions from the information in **Map A.**

   ———————————————————

   ———————————————————

   ———————————————————

3. List two cross-links shown on **Map C.**

   ———————————————————

   ———————————————————

   ———————————————————

Making Power Notes

Power notes help you organize the chemical concepts you are studying by distinguishing main ideas from details. Similar to outlines, power notes are linear in form and provide you with a framework of important concepts. Power notes are easier to use than outlines because their structure is simpler. Using the power notes numbering system you assign a 1 to each main idea and a 2, 3, or 4 to each detail.

Power notes are an invaluable asset to the learning process, and they can be used frequently throughout your chemistry course. You can use power notes to organize ideas while reading your text or to restructure your class notes for studying purposes.

To learn to make power notes, practice first by using single-word concepts and a subject you are especially interested in, such as animals, sports, or movies. As you become comfortable with structuring power notes, integrate their use into your study of chemistry. For an easier transition, start with a few boldfaced or italicized terms. Later you can strengthen your notes by expanding these single-word concepts into more-detailed phrases and sentences. Use the following general format to help you structure your power notes.

> Power 1: Main idea
> > Power 2: Detail or support for power 1
> > > Power 3: Detail or support for power 2
> > > > Power 4: Detail or support for power 3

1. **Pick a Power 1 word from the text.**

   The text you choose does not have to come straight from your chemistry textbook. You may be making power notes from your lecture notes or from an outside source. We'll use the term atom found in the chapter entitled "Atoms and Moles" in your textbook.

   > Power 1: Atom

2. **Using the text, select some Power 2 words to support your Power 1 word.**

   We'll use the terms nucleus and electrons, which are two parts of an atom.

   > Power 1: Atom
   > > Power 2: Nucleus
   > > Power 2: Electrons

3. **Select some Power 3 words to support your Power 2 words.** We'll use the terms positively charged and negatively charged, two terms that describe the Power 2 words.

> Power 1: Atom
> > Power 2: Nucleus
> > > Power 3: Positively charged
> > Power 2: Electrons
> > > Power 3: Negatively charged

4. **Continue to add powers to support and detail the main idea as necessary.**

   There are no restrictions on how many power numbers you can use in your notes. If you have a main idea that requires a lot of support, add more powers to help you extend and organize your ideas. Be sure that words having the same power number have a similar relationship to the power above. Power 1 terms do not have to be related to each other. You can use power notes to organize the material in an entire section or chapter of your text. Doing so will provide you with an invaluable study guide for your classroom quizzes and tests.

> Power 1: Atom
> > Power 2: Nucleus
> > > Power 3: Positively charged
> > > Power 3: Protons
> > > > Power 4: Positively charged
> > > Power 3: Neutrons
> > > > Power 4: No charge
> > Power 2: Electrons
> > > Power 3: Negatively charged

Practice

1. **Use a periodic table and the power notes structure below to organize the following terms: alkaline-earth metals, nonmetals, calcium, sodium, halogens, metals, alkali metals, chlorine, barium, and iodine.**

   > 1 _____
   > > 2 _____
   > > > 3 _____
   > > 2 _____
   > > > 3 _____
   > > > 3 _____
   > 1 _____
   > > 2 _____
   > > > 3 _____
   > > > 3 _____

Making Two-Column Notes

Two-column notes can be used to learn and review definitions of vocabulary terms, examples of multiple-step processes, or details of specific concepts. The two-column-note strategy is simple: write the term, main idea, step-by-step process, or concept in the left-hand column, and the definition, example, or detail on the right.

One strategy for using two-column notes is to organize main ideas and their details. The main ideas from your reading are written in the left-hand column of your paper and can be written as questions, key words, or a combination of both. Details describing these main ideas are then written in the right-hand column of your paper.

1. **Identify the main ideas.** The main ideas for a chapter are listed in the section objectives. However, you decide which ideas to include in your notes. For example, here are some main ideas from the objectives in Section 4-2.

• Describe the locations in the periodic table and the general properties of the alkali metals, alkaline-earth metals, the halogens, and the noble gases.

2. **Divide a blank sheet of paper into two columns and write the main ideas in the left-hand column.** Summarize your ideas using quick phrases that are easy for you to understand and remember. Decide how many details you need for each main idea, and write that number in parentheses under the main idea.

3. **Write the detail notes in the right-hand column.** Be sure you list as many details as you designated in the main-idea column. Here are some main ideas and details about some of the groups in the Periodic Table.

The two-column method of review is perfect whether you use it to study for a short quiz or for a test on the material in an entire chapter. Just cover the information in the right-hand column with a sheet of paper, and after reciting what you know, uncover the notes to check your answers. Then ask yourself what else you know about that topic. Linking ideas in this way will help you to gain a more complete picture of chemistry.

Main Idea	Detail Notes
• Alkali metals (4 properties)	• Group 1 • highly reactive • ns^1 electron configuration • soft, silvery
• Alkaliine-earth metals (4 properties)	• Group 2 • reactive • ns^2 electron configuration • harder than alkali metal
• Hologens (3 properties)	• Group 17 • reactive • nonmetallic
• Noble gases	• Group 18 • low reactivity • stable ns^2-np^6 configuration

Using the K/W/L/ Strategy

The K/W/L strategy stands for "what I **K**now—what I **W**ant to know—what I **L**earned." You start by brainstorming about the subject matter before reading the assigned material. Relating new ideas and concepts to those you have learned previously will help you better understand and apply the new knowledge you obtain. The section objectives throughout your textbook are ideal for using the K/W/L strategy.

1. **Read the section objectives.** You may also want to scan headings, boldfaced terms, and illustrations before reading. Here are two of the objectives from Section 1-2 to use as an example.

   - Explain the gas, liquid, and solid states in terms of particles.
   - Distinguish between a mixture and a pure substance.

2. **Divide a sheet of paper into three columns, and label the columns "What I Know," "What I Want to Know," and "What I Learned."**

3. **Brainstorm about what you know about the information in the objectives, and write these ideas in the first column.** Because this chart is designed primarily to help you integrate your own knowledge with new information, it is not necessary to write complete sentences.

4. **Think about what you want to know about the information in the objectives, and write these ideas in the second column.** Include information from both the section objectives and any other objectives your teacher has given you.

5. **While reading the section or afterwards, use the third column to write down the information you learned.** While reading, pay close attention to any information about the topics you wrote in the "What I Want to Know" column. If you do not find all of the answers you are looking for, you may need to reread the section or reference a second source. Be sure to ask your teacher if you still cannot find the information after reading the section a second time.

It is also important to review your brainstormed ideas when you have completed reading the section. Compare your ideas in the first column with the information you wrote down in the third column. If you find that some of your brainstormed ideas are incorrect, cross them out. It is extremely important to identify and correct any misconceptions you had prior to reading before you begin studying for your test.

What I Know	What I want to Know	What I Learned
• gas has no definite shape or volume	• how gas, liquid, and solid states are related to particles	• molecules in solid and liquid states are close together, but are far apart in gas state
• liquid has no definite shape, but has definite volume	• how mixtures and pure substances are different	
• solid has definite shape and volume		• molecules in solid state have fixed positions, but molecules in liquid and gas states can flow
• mixture is a combination of substances		• mixtures are combinations of pure substances
• pure substance has only one component		• pure substances have fixed compositions and definite properties

Using Sequencing/Pattern Puzzles

You can use pattern puzzles to help you remember sequential information. Pattern puzzles are not just a tool for memorization. They also promote a greater understanding of a variety of chemical processes, from the steps in solving a mass-mass stoichiometry problem to the procedure for making a solution of specified molarity.

Here's a step-by-step example showing how to make a pattern puzzle, and how to use it to help you study. For other topics that require remembering information in a particular order, just follow the same steps.

1. Write down the steps of a process in your own words.

For an example, we will use the process for converting the amount of a substance in moles to mass in grams. (See Sample Problem D in the chapter on "Atoms and Moles.")

On a sheet of notebook paper, write down one step per line, and do not number the steps.

Do not copy the process straight from your textbook. Writing the steps in your own words promotes a more thorough understanding of the process.

You may want to divide longer steps into two or three shorter steps.

• Gather information about what you know and what you don't know.

• Write a set-up that shows what is given and what is desired.

• Look at the periodic table to determine the molar mass of the substance.

• Write the correct conversion factor that has mass in the numerator and amount in moles in the denominator.

• Multiply the amount of substance by the conversion factor.

• Solve the equation and verify that your answer is reasonable.

2. Cut the sheet of paper into strips with only one step per strip of paper. Shuffle the strips of paper so that they are out of sequence.

• Look at the periodic table to determine the molar mass of the substance.

• Solve the equation and verify that your answer is reasonable.

• Write the correct conversion factor that has mass in the numerator and amount in moles in the denominator.

• Write a set-up that shows what is given and what is desired.

• Gather information about what you know and what you don't know.

• Multiply the amount of substance by the conversion factor.

3. Place the strips in their proper sequence. Confirm the order of the process by checking your text or your class notes.

Pattern puzzles are especially helpful when you are studying for your chemistry tests. Before tests, use your puzzles to practice sequencing and to review the steps of chemistry processes. You and a classmate can also take turns creating your own pattern puzzles of different chemical processes and putting each other's puzzles in the correct sequence. Studying with a classmate in this manner will help make studying fun and will enable you to help each other.

• Gather information about what you know and what you don't know.

• Write a set-up that shows what is given and what is desired.

• Look at the periodic table to determine the molar mass of the substance.

• Write the correct conversion factor that has mass in the numerator and amount in moles in the denominator.

• Multiply the amount of substance by the conversion factor.

• Solve the equation and verify that your answer is reasonable.

Other Reading Strategies

Brainstorming

Brainstorming is a strategy that helps you recognize and evaluate the knowledge you already have before you start reading. It works well individually or in groups. When you brainstorm, you start with a central term or idea, then quickly list all the words, phrases, and other ideas that you think are related to it.

Because there are no "right" or "wrong" answers, you can use the list as a basis for classifying terms, developing a general explanation, or speculating about new relationships. For example, you might brainstorm a list of terms related to the word element before you read about elements early in the textbook. The list might include gold, metals, chemicals, silver, carbon, oxygen, and water. As you read the textbook, you might decide that some of the terms you listed are not elements. Later, you might use that information to help you distinguish between elements and compounds.

Building/Interpreting Vocabulary

Using a dictionary to look up the meanings of prefixes and suffixes as well as word origins and meanings helps you build your vocabulary and interpret what you read. If you know the meaning of prefixes like kilo- (one thousand) and milli- (one thousandth), you have a good idea what kilograms, kilometers, milligrams, and millimeters are and how they are different. (See Appendix A for a list of SI Prefixes.)

Knowledge of prefixes, suffixes, and word origins can help you understand the meaning of new words. For example, if you know the prefix –poly comes from the word meaning *many,* it will help you understand what polysaccharides and polymers are.

Reading Hints

Reading hints help you identify and bookmark important charts, tables, and illustrations for easy reference. For example, you may want to use a self-adhesive note to bookmark the periodic table in the chapter describing it or on the inside back cover of your book so you can easily locate it and use it for reference as you study different aspects of chemistry and solve problems involving elements and compounds.

Interpreting Graphic Sources of Information

Charts, tables, photographs, diagrams, and other illustrations are graphic, or visual, sources of information. The labels and captions, together with the illustrations help you make connections between the words and the ideas presented in the text.

Reading Response Logs

Keeping a reading response log helps you interpret what you read and gives you a chance to express your reactions and opinions about what you have read. Draw a vertical line down the center of a piece of paper. In the left-hand column, write down or make notes about passages you read to which you have reactions, thoughts, feelings, questions, or associations. In the right-hand column, write what those reactions, thoughts, feelings, questions, or associations are. For example, you might keep a reading response log when studying about Nuclear Energy.

Graphing Skills

Line Graphs

In laboratory experiments, you will usually be controlling one variable and seeing how it affects another variable. Line graphs can show these relations clearly. For example, you might perform an experiment in which you measure the growth of a plant over time to determine the rate of the plant's growth. In this experiment, you are controlling the time intervals at which the plant height is measured. Therefore, time is called the *independent variable*. The height of the plant is the *dependent variable*. **Table 1** gives some sample data for an experiment to measure the rate of plant growth.

The independent variable is plotted on the *x*-axis. This axis will be labeled *Time (days),* and will have a range from 0 days to 35 days. Be sure to properly label your axis including the units on the values.

The dependent variable is plotted on the *y*-axis. This axis will be labeled *Plant Height (cm)* and will have a range from 0 cm to 5 cm.

Think of your graph as a grid with lines running horizontally from the y-axis, and vertically from the *x*-axis. To plot a point, find the *x* (in this example time) value on the *x*-axis. Follow the vertical line from the *x*-axis until it intersects

the horizontal line from the *y*-axis at the corresponding *y* (in this case height) value. At the intersection of these two lines, place your point. **Figure 1** shows what a line graph of the data in **Table 1** might look like.

Figure 1

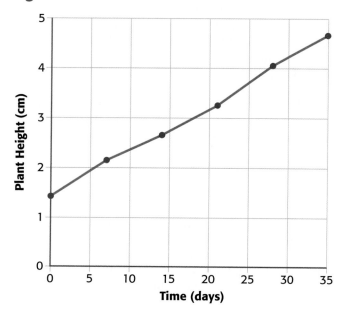

Table 1   Experimental Data for Plant Growth versus Time

Time (days)	Plant height (cm)
0	1.43
7	2.16
14	2.67
21	3.25
28	4.04
35	4.67

Practice

1. What does the line in **Figure 1** show, and what can you conclude about the plants used in the experiment?

2. Create a line graph of the following data.

Number of Days	Plant height (cm)
0	1.46
7	2.67
14	3.89
21	4.82

3. Compare the graph you made with **Figure 1**. What can you conclude about the two different groups of plants?

Scatter Plots

Some experiments or groups of data are best represented in a graph that is similar to a line graph and that is called a scatter plot. As in a line graph, the data points are plotted on the graph by using values on an *x*-axis and a *y*-axis. Scatter plots are often used to find trends in data. Instead of connecting the data points with a line, a trend can be represented by a best-fit line. A best-fit line is a line that represents all of the data points without necessarily going through all of them. To find a best-fit line, pick a line that is equidistant from as many data points as possible. Examine the graph below.

Figure 2

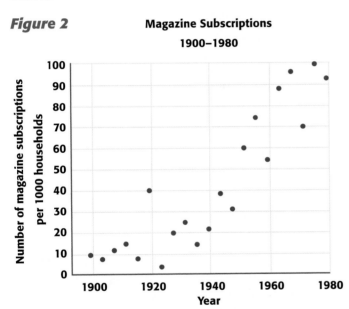

Magazine Subscriptions
1900–1980

If we connected all of the data points with lines, the lines would create a zigzag pattern that would not tell us much about our data. But if we find a best-fit line, we can see a trend more clearly. Furthermore, if we pick two points on the best-fit line, we can estimate its slope. Examine the dotted lines on *Figure 3.*

The points can be estimated as 18 magazine subscriptions per 100 households in 1920, and 42 magazine subscriptions per 100 households in 1940. If we subtract 1920 from 1940, and 18 sub-

scriptions from 42 subscriptions (using the point slope formula), we see that the line shows a trend of an increase of 24 subscriptions per 1000 households acres every 20 years. Scatter plots can also be used when there are two or more trends within one group of data or when there is no distinct trend at all.

Figure 3

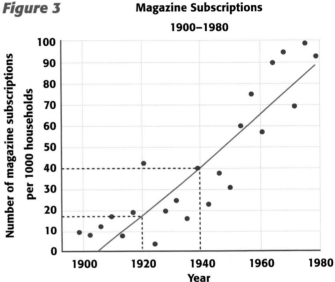

Magazine Subscriptions
1900–1980

Practice

1. Copy the graph below, and draw a best-fit line.

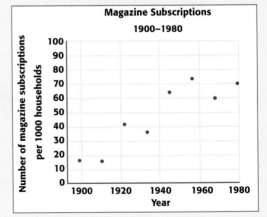

Magazine Subscriptions
1900–1980

2. What does that line represent?
3. If these were the data from a different city than the data in *Figure 3,* what conclusions could you draw about the two cities?

GRAPHING CALCULATOR TECHNOLOGY

Downloading

To solve the Graphing Calculator problems in Technology & Learning sections of the Chapter Reviews, you will need to download the programs and the datasets. These files can be found at the website go.hrw.com.

internet connect

www.scilinks.org
Topic: go.hrw.com
SciLinks code: HW4 TI83

SCILINKS Maintained by the
National Science
Teachers Association

This Web site contains links for downloading programs and applications you will need for Technology and Learning exercises.

Note: In order to transfer programs and applications from a computer to your calculator, you will need a TI-Graph Link cable. Programs can also be transferred directly between calculators using a unit-to-unit cable. Refer to the TI Web site or to your calculator's user's manual for instructions.

If your computer does not already have TI-Graph Link software installed, click Step 1: TI-Graph Link Software and follow the links for downloading and installing TI-Graph Link from the TI Web site.

Click Step 2: HChmProg. This will load the file HCHMPROG.ZIP onto your computer. Once the file is downloaded, double-click the icon and the file will be extracted into a file called hchmprog.8xg.

Click Step 3: Getting Started and follow the instructions for your TI-Graph Link to load hchmprog.8xg onto your TI calculator. When the file is sent to the calculator, it should expand into 17 programs. These programs should appear in the PRGM menu.

Download the CHEMAPPS application from go.hrw.com.

Using the Data Sets

Once you have loaded the application, you can use it via the "apps" function on the calculator. Press the blue [APPS] button in the upper left portion of the keypad. A menu will be displayed of the applications in your calculator's memory. Look for an application titled **CHEMAPPS**. Select it either by using the arrow keys and pressing **ENTER** or by using the number keys. Do this each time you choose to open the application. A title screen will appear for a few seconds, and then a menu will be displayed, listing all of the available data sets. The sets are listed by chapter and can be selected with either the arrow keys and [ENTER] or with the number keys.

The instructions are essentially the same for every question that uses the datasets. You can select the lists to be loaded by placing the cursor beside them and pressing the [ENTER] key or, if you would like to load all of the lists from that chapter at one time, you can use the arrow keys to go to the **All** menu and choose the first option **All+** and then press [ENTER]. When you have chosen all of the data sets that you need, use the arrow keys to go to the **Load** menu. The **SetUpEditor** allows you to decide where you would like the lists to be stored. 1: Add to Editor adds the lists to the end of the List editor. By choosing [STAT] and **1: Edit…,** you can see these lists behind L_6.
2: Exchange Lists replaces the data values in those with the sets from the application. They will be replaced in order, so if you have only one list, L_1 will be replaced, and so on. **3: No change** will not add the list to the List editor; however, by going to the List menu via [2nd] [LIST], you will see the names of the data sets listed under L_6. After you have finished, choose **Load,** and the sets will be loaded into the desired location.

Troubleshooting

- Calculator instructions in the Holt Chemistry program are written for the TI-83 Plus. You may use other graphing calculators, but some of the programs and instructions may not work exactly as described.

- If you have problems loading programs or applications onto your calculator, you may need to clear programs or other data from your calculator's memory.

- Always make sure that you are downloading correct versions of the software. TI-Graph Link has different versions for Windows and for Macintosh as well as different versions for different calculators.

- If you need additional help, Texas Instruments can provide technical support at 1-800 TI-CARES.

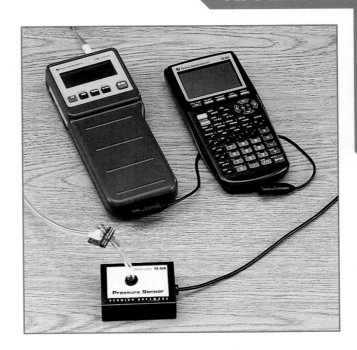

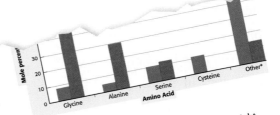

which ...presents the pro... found in wool?

56. According to the graph, what is significant about spider fibroin protein?

57. What are the mole percentages of alanine in α-keratin and fibroin?

58. Why do you think the mole percentages of all of the amino acids are not shown?

59. Spider fibroin protein is a much stronger material than α-keratin in wool. Violet would like to create a strong protein for

manufacturing fishing line. What amino acids might she decide to use to build the protein? Use the graph to support your answer.

TECHNOLOGY AND LEARNING

60. Graphing Calculator

Polypeptides and Amino Acids

Go to Appendix C. If you are using a TI-83 Plus, you can download the program PEPTIDE and run the application as directed. If you are using another calculator, your teacher will provide you with keystrokes to use. There are 20 amino acids that occur in proteins found in nature. The program will prompt you to input a number of amino acids. After you do, press **ENTER**. The program will respond with the number of different straight-chain polypeptides possible given that number of amino acid units.

a. Aspartame is an artificial sweetener that is a dipeptide, a protein made of two amino acids. How many possible dipeptides are there?

b. Enkephalins produced in the brain serve to help the body deal with pain. Several of them are pentapeptides. That is, they are polypeptides made of five amino acids. How many different pentapeptides are there?

c. The calculator uses the following equation:

number of polypeptides $= 20^{(\text{number of amino acids})}$

This equation can also be expressed as:

number of polypeptides $=$
$(2^{(\text{number of amino acids})})(10^{(\text{number of amino acids})})$

Given this equation, estimate how many possible polypeptides there are that are made of 100 amino acids? (Hint: The answer is too large for your calculator. However, you can use the graphing calculator to find the value of $2^{(\text{number of amino acids})}$.)

PROBLEM BANK

The Science of Chemistry

Section: Describing Matter

1. What is the density of a block of marble that occupies 310 cm^3 and has a mass of 853 g?

2. Diamond has a density of 3.26 g/cm^3. What is the mass of a diamond that has a volume of 0.350 cm^3?

3. What is the volume of a sample of liquid mercury that has a mass of 76.2 g, given that the density of mercury is 13.6 g/mL?

4. What is the density of a sample of ore that has a mass of 74.0 g and occupies 20.3 cm^3?

5. Find the volume of a sample of wood that has a mass of 95.1 g and a density of 0.857 g/cm^3?

6. Express a length of 16.45 m in centimeters.

7. Express a length of 16.45 m in kilometers.

8. Express a mass of 0.014 mg in grams.

9. Complete the following conversion:
   10.5 g = _____ kg

10. Complete the following conversion:
    1.57 km = _____ m

11. Complete the following conversion:
    3.54 μg = _____ g

12. Complete the following conversion:
    3.5 mol = _____ μmol

13. Complete the following conversion:
    1.2 L = _____ mL

14. Complete the following conversion:
    358 cm^3 = _____ m^3

15. Complete the following conversion:
    548.6 mL = _____ cm^3

16. What is the density of an 84.7 g sample of an unknown substance if the sample occupies 49.6 cm^3?

17. What volume would be occupied by 7.75 g of a substance with a density of 1.70766 g/cm^3?

18. Express a time period of exactly 1 day in terms of seconds. Try to write out all the equalities needed to solve this problem.

19. How many centigrams are there in 6.25 kg?

20. Polycarbonate plastic has a density of 1.2 g/cm^3. A photo frame is constructed from two 3.0 mm sheets of polycarbonate. Each sheet measures 28 cm by 22 cm. What is the mass of the photo frame?

21. Find the volume of a cube that is 3.23 cm on each edge.

22. Calculate the density of a 17.982 g object that occupies 4.13 cm^3.

Matter and Energy

Section: Measurements and Calculations in Chemistry

1. Determine the specific heat of a material if a 35.0 g sample absorbed 48.0 J as it was heated from 293 K to 313 K.

2. A piece of copper alloy with a mass of 85.0 g is heated from 30.0°C to 45.0°C. In the process, it absorbs 523 J of energy as heat.
   a. What is the specific heat of this copper alloy?
   b. How much energy will the same sample lose if it is cooled to 25.0°C?

3. The temperature of a 74.0 g sample of material increases from 15.0°C to 45.0°C when it absorbs 2.00 kJ of energy as heat. What is the specific heat of this material?

4. How much energy is needed to raise the temperature of 5.00 g of gold ($c_p = 0.129$ J/(g•K)) by 25.0°C?

5. Energy in the amount of 420 J is added to a 35.0 g sample of water ($c_p = 4.18$ J/(g•K)) at a temperature of 10.0°C. What will be the final temperature of the water?

6. What mass of liquid water ($c_p = 4.18$ J/(g•K)) at room temperature (25°C) can be raised to its boiling point with the addition of 24.0 kJ of energy?

7. How much energy would be absorbed as heat by 75 g of iron (c_p = 0.449 J/(g•K)) when heated from 295 K to 301 K?

Atoms and Moles

Section: Structure of Atoms

1. How many protons, electrons, and neutrons are in an atom of bromine-80?

2. Write the nuclear symbol for carbon-13.

3. Write the hyphen notation for the element that contains 15 electrons and 15 neutrons.

4. How many protons, electrons, and neutrons are in an atom of carbon-13?

5. Write the nuclear symbol for oxygen-16.

6. Write the hyphen notation for the element whose atoms contains 7 electrons and 9 neutrons.

Section: Counting Atoms

7. What is the mass in grams of 2.25 mol of the element iron, Fe?

8. What is the mass in grams of 0.375 mol of the element potassium, K?

9. What is the mass in grams of 0.0135 mol of the element sodium, Na?

10. What is the mass in grams of 16.3 mol of the element nickel, Ni?

11. What is the mass in grams of 3.6 mol of the element carbon, C?

12. What is the mass in grams of 0.733 mol of the element chlorine, Cl?

13. How many moles of calcium, Ca, are in 5 g of calcium?

14. How many moles of gold, Au, are in 3.6 × 10^{-10} g of gold?

15. How many moles of copper, Cu, are in 3.22 g of copper?

16. How many moles of lithium, Li, are in 2.72 × 10^{-4} g of lithium?

17. How many moles of lead, Pb, are in 1.5 × 10^{12} atoms of lead?

18. How many moles of tin, Sn, are in 2500 atoms of tin?

19. How many atoms of aluminum, Al, are in 2.75 mol of aluminum?

20. How many moles of carbon, C, are in 2.25 × 10^{22} atoms of carbon?

21. How many moles of oxygen, O, are in 2 × 10^6 atoms of oxygen?

22. How many atoms of sodium, Na, are in 3.8 mol of sodium?

23. What is the mass in grams of 7.5 × 10^{15} atoms of nickel, Ni?

24. How many atoms of sulfur, S, are in 4 g of sulfur?

25. What mass of gold, Au, contains the same number of atoms as 9.00 g of aluminum, Al?

26. What is the mass in grams of 5 × 10^9 atoms of neon, Ne?

27. How many atoms of carbon, C, are in 0.02 g of carbon?

28. What mass of silver, Ag, contains the same number of atoms as 10 g of boron, B?

29. How many moles of atoms are there in 3.25 × 10^5 g Pb?

30. How many moles of atoms are there in 150 g S?

The Mole and Chemical Composition

Section: Avogadro's Number and Mole Conversions

1. What is the mass in grams of 3.25 mol $Fe_2(SO_4)_3$?

2. How many moles of molecules are there in 250 g of hydrogen nitrate, HNO_3?

3. How many molecules of aspirin, $C_9H_8O_4$, are there in a 100 mg tablet of aspirin?

4. What is the mass in grams of 3.04 mol of ammonia vapor, NH_3?

5. Calculate the mass of 0.257 mol of calcium nitrate, $Ca(NO_3)_2$.

6. How many moles are there in 6.60 g of $(NH_4)_2SO_4$?

7. How many moles are there in 4.50 kg of $Ca(OH)_2$?

8. How many molecules are there in 25.0 g of H_2SO_4?

9. How many molecules are there in 125 g of sugar, $C_{12}H_{22}O_{11}$?

10. What is the mass in grams of 6.25 mol of copper(II) nitrate?

11. How many moles are there in 3.82 g of SO_2?

12. How many moles are there in 4.15×10^{-3} g of $C_6H_{12}O_6$?

13. How many moles are there in 77.1 g of Cl_2?

14. How many molecules are there in 3.82 g of SO_2?

15. How many molecules are there in 4.15×10^{-3} g of $C_6H_{12}O_6$?

16. Determine the number of moles in 4.50 g of H_2O?

17. Determine the number of moles in 471.6 g of $Ba(OH)_2$?

18. Determine the number of moles in 129.68 g of $Fe_3(PO_4)_2$?

19. What is the mass in grams of 1.00 mol NaCl?

20. What is the mass in grams of 2.00 mol H_2O?

21. What is the mass in grams of 3.500 mol $Ca(OH)_2$?

22. What is the mass in grams of 0.6250 mol $Ba(NO_3)_2$?

Section: Relative Atomic Mass and Chemical Formulas

23. Find the formula mass for the following: H_2SO_4.

24. Find the formula mass for the following: $Ca(NO_3)_2$.

25. Find the formula mass for the following: PO_4^{3-}.

26. Find the formula mass for the following: $MgCl_2$.

27. Find the formula mass for the following: Na_2SO_3.

28. Find the formula mass for the following: $HClO_3$.

29. Find the formula mass for the following: MnO_4^-.

30. Find the formula mass for the following: C_2H_6O.

31. Find the molar mass for the following: Al_2S_3.

32. Find the molar mass for the following: $NaNO_3$.

33. Find the molar mass for the following: $Ba(OH)_2$.

34. Find the molar mass for the following: K_2SO_4.

35. Find the molar mass for the following: $(NH_4)_2CrO_4$.

36. Determine the formula mass of the following: calcium acetate, $Ca(CH_3COO)_2$.

37. Determine the molar mass of the following: glucose, $C_6H_{12}O_6$.

38. Determine the molar mass of the following: calcium acetate, $Ca(CH_3COO)_2$.

39. Determine the formula mass of the following: the ammonium ion, NH_4^+.

40. Determine the molar mass of the following: the chlorate ion, ClO_3^-.

41. Determine the molar mass of XeF_4.

42. Determine the molar mass of $C_{12}H_{42}O_6$.

43. Determine the molar mass of Hg_2I_2.

44. Determine the molar mass of CuCN.

45. Determine the formula mass of the following: the ammonium ion, NH_4^+.

46. Determine the formula mass of the following: the chlorate ion, ClO_3^-.

Section: Formulas and Percentage Composition

47. Calculate the percentage composition of lead (II) chloride, $PbCl_2$.

48. Calculate the percentage composition of barium nitrate, $Ba(NO_3)_2$.

49. Find the mass percentage of water in $ZnSO_4 \cdot 7H_2O$.

50. a. Magnesium hydroxide is 54.87% oxygen by mass. How many grams of oxygen are in 175 g of the compound?

 b. How many moles of oxygen is this?

51. Calculate the percent composition of sodium nitrate, $NaNO_3$.

52. Calculate the percent composition of silver sulfate, Ag_2SO_4.

53. What is the mass percentage of water in the hydrate $CuSO_4 \cdot 5H_2O$?

54. a. Zinc chloride, $ZnCl_2$ is 52.02% chlorine by mass. What mass of chlorine is contained in 80.3 g of $ZnCl_2$?

 b. How many moles of Cl is this?

55. A compound is found to contain 63.52% iron and 36.48% sulfur. Find its empirical formula.

56. Determine the formula mass and molar mass of ammonium carbonate, $(NH_4)_2CO_3$.

 a. The formula mass is _____ ?

 b. The molar mass is _____ ?

57. Calculate the percent composition of $(NH_4)_2CO_3$.

58. A compound is found to contain 36.48% Na, 25.41% S, and 38.11% O. Find its empirical formula.

59. Calculate the percent composition of sodium chloride, NaCl.

60. Calculate the percent composition of silver nitrate, $AgNO_3$.

61. Calculate the percent composition of magnesium hydroxide, $Mg(OH)_2$.

62. What is the mass percentage of water in the hydrate $CuSO_4 \cdot 5H_2O$?

63. Determine the percent composition for Na-ClO.

64. Determine the percent composition for H_2SO_3.

65. Determine the percent composition for C_2H_5COOH.

66. Determine the percent composition for $BeCl_2$.

67. a. A compound is found to contain 54.5% carbon, 9.1% hydrogen, and 36.4% oxygen Determine the simplest formula.

 b. The molar mass of a compound is 88.1 g. What is the molecular formula if the simplest formula is C_2H_4O?

68. Find the empirical formula of a compound found to contain 26.56% potassium, 35.41% chromium, and the remainder oxygen.

69. Analysis of 20.0 g of a compound containing only calcium and bromine indicates that 4.00 g of calcium are present. What is the empirical formula of the compound formed?

70. A compound is analyzed and found to contain 36.70% potassium, 33.27% chlorine, and 30.03% oxygen. What is the empirical formula of the compound?

71. Determine the empirical formula of the compound that contains 17.15% carbon, 1.44% hydrogen, and 81.41% fluorine.

72. A 60.00 g sample of tetraethyl-lead, a gasoline additive, is found to contain 38.43 g lead, 17.83 g carbon, and 3.74 g hydrogen. Find its empirical formula.

73. A 100.00 g sample of an unidentified compound contains 29.84 g sodium, 67.49 g chromium, and 72.67 g oxygen. What is the compound's empirical formula?

74. A compound is found to contain 53.70% iron and 46.30% sulfur. Find its empirical formula.

75. Analysis of a compound indicates that it contains 1.04 g K, 0.70 g Cr, and 0.86 g O. Find its empirical formula.

76. If 4.04 g of N combine with 11.46 g of O to produce a compound with the formula mass of 108.0 amu, what is the molecular formula of this compound?

77. Determine the empirical formula of a compound containing 63.50% silver, 8.25% nitrogen, and the remainder oxygen.

78. Determine the empirical formula of a compound found to contain 52.11% carbon, 13.14% hydrogen, and 34.75% oxygen.

79. Chemical analysis of citric acid shows that it contains 37.51% C, 4.20% H, 58.29% O. What is its empirical formula?

80. A 175.0 g sample of a compound contains 56.15 g C, 9.43 g H, 74.81 g O, 13.11 g N, and 21.49 g Na. What is its empirical formula?

81. In the laboratory, a sample of pure nickel was placed in a clean, dry, weighed crucible. The crucible was heated so that the nickel would react with the oxygen in the air. After the reaction appeared complete, the crucible was allowed to cool and the mass was determined. The crucible was reheated and allowed to cool. Its mass was then determined again to be certain that the reaction was complete. The following data were collected:
Mass of crucible = 30.02 g
Mass of nickel and crucible = 31.07 g
Mass of nickel oxide and crucible = 31.36 g
Determine the following information based on the data given above:
Mass of nickel =
Mass of nickel oxide =
Mass of oxygen =
Based on your calculations, what is the empirical formula for the nickel oxide?

82. Determine the molecular formula of the compound with an empirical formula of CH and a formula mass of 78.110 amu.

83. A sample compound with a formula mass of 34.00 amu is found to consist of 0.44 g H and 6.92 g O. Find its molecular formula.

84. The empirical formula for trichloroisocyanuric acid, the active ingredient in many household bleaches, is OCNCl. The molar mass of this compound is 232.41 g/mol. What is the molecular formula of trichloroisocyanuric acid?

85. Determine the molecular formula of a compound with an empirical formula of NH_2 and a formula mass of 32.06 amu.

86. The molar mass of a compound is 92 g/mol. Analysis of a sample of the compound indicates that it contains 0.606 g N and 1.390 g O. Find its molecular formula.

87. What is the molecular formula of the molecule that has an empirical formula of CH_2O and a molar mass of 120.12 g/mol?

88. A compound with a formula mass of 42.08 amu is found to be 85.64% carbon and 14.36% hydrogen by mass. Find its molecular formula.

Stoichiometry

Section: Calculating Quantities in Reactions

1. A seashell, composed largely of calcium carbonate, is placed in a solution of HCl. As a result, 1500 mL of dry CO_2 gas at STP is produced. The other products are $CaCl_2$ and H_2O.

   Based on this information, how many grams of $CaCO_3$ are consumed in the reaction?

2. *Acid precipitation* is the term generally used to describe rain or snow that is more acidic than normal. One cause of acid precipitation is the formation of sulfuric and nitric acids from various sulfur and nitrogen oxides produced in volcanic eruptions, forest fires, and thunderstorms. In a typical volcanic eruption, for example, 3.50×10^8 kg of SO_2 may be produced. If this amount of SO_2 were converted to H_2SO_4 according to the two-step process given below, how many kilograms of H_2SO_4 would be produced from such an eruption?

$$SO_2 + \frac{1}{2}O_2 \longrightarrow SO_3$$

$$SO_3 + H_2O \longrightarrow H_2SO_4$$

3. Solid iron(III) hydroxide decomposes to produce iron(III) oxide and water vapor. If 0.750 L of water vapor is produced at STP,

   a. How many grams of iron(III) hydroxide were used?

   b. How many grams of iron(III) oxide are produced?

4. Balance the following chemical equation.
   $$Mg(s) + O_2(g) \longrightarrow MgO(s)$$
   Then, based on the quantity of reactant or product given, determine the corresponding quantities of the specified reactants or products, assuming that the system is at STP.

a. How many moles of MgO are produced from 22.4 L O_2?

b. How many moles of MgO are produced from 11.2 L O_2?

c. How many moles of MgO are produced from 1.40 L O_2?

5. Assume that 8.50 L of I_2 are produced using the following reaction that takes place at STP:
$KI(aq) + Cl_2(g) \longrightarrow KCl(aq) + I_2(g)$
Balance the equation before beginning your calculations.

a. How many moles of I_2 are produced?

b. How many moles of KI were used?

c. How many grams of KI were used?

6. Suppose that 650 mL of hydrogen gas are produced through a replacement reaction involving solid iron and sulfuric acid, H_2SO_4, at STP. How many grams of iron(II) sulfate are also produced?

Section: Limiting Reactants and Percentage Yield

7. Methanol, CH_3OH, is made by causing carbon monoxide and hydrogen gases to react at high temperature and pressure. If 450 mL of CO and 825 mL of H_2 are mixed,

a. Which reactant is present in excess?

b. How much of that reactant remains after the reaction?

c. What volume of CH_3OH is produced, assuming the same pressure?

Causes of Change

Section: Using Enthalpy

1. When 1 mol of methane is burned at constant pressure, −890 kJ/mol of energy is released as heat. If a 3.2 g sample of methane is burned at constant pressure, what will be the value of ΔH? (Hint: Convert the grams of methane to moles. Also make sure your answer has the correct sign for an exothermic process.)

2. How much energy is needed to raise the temperature of a 55 g sample of aluminum from 22.4°C to 94.6°C? The specific heat of aluminum is 0.897 J/(g•K).

3. 3500 J of energy are added to a 28.2 g sample of iron at 20°C. What is the final temperature of the iron in kelvins? The specific heat of iron is 0.449 J/(g•K).

Section: Changes in Enthalpy During Reactions

4. The combustion of methane gas, CH_4, forms $CO_2(g) + H_2O(l)$. Calculate the energy as heat produced by burning 1 mol of the methane gas.

5. Calculate ΔH for the following reaction:
$$2N_2(g) + 5O_2(g) \longrightarrow 2N_2O_5(g)$$
Use the following data in your calculation:

$$H_2(g) + \frac{1}{2}O_2(g) \longrightarrow H_2O(l)$$
$$\Delta H_f^0 = -285.8 \text{ kJ/mol}$$

$$N_2O_5(g) + H_2O(l) \longrightarrow 2NHO_3(l)$$
$$\Delta H = -76.6 \text{ kJ/mol}$$

$$\frac{1}{2}N_2(g) + \frac{3}{2}O_2(g) + \frac{1}{2}H_2(g) \longrightarrow HNO_3(l)$$
$$\Delta H_f^0 = -174.1 \text{ kJ/mol}$$

6. Calculate the enthalpy of formation of butane, C_4H_{10}, using the following balanced chemical equation and information. Write out the solution according to Hess's law.
$$C(s) + O_2(g) \longrightarrow CO_2(g)$$
$$\Delta H_f^0 = -393.5 \text{ kJ/mol}$$

$$H_2(g) + \frac{1}{2}O_2(g) \longrightarrow H_2O(l)$$
$$\Delta H_f^0 = -285.8 \text{ kJ/mol}$$

$$4CO_2(g) + 5H_2O(l) \longrightarrow C_4H_{10}(g) + \frac{13}{2}O_2(g)$$
$$\Delta H = 2877.6 \text{ kJ/mol}$$

7. Calculate the enthalpy of combustion of 1 mol of nitrogen, N_2, to form NO_2.

8. Calculate the enthalpy of formation for 1 mol sulfur dioxide, SO_2, from its elements, sulfur and oxygen. Use the balanced chemical equation and the following information.

$$S(s) + \frac{3}{2}O_2(g) \longrightarrow SO_3(g)$$
$$\Delta H = -395.2 \text{ kJ/mol}$$

$$2SO_2(g) + O_2(g) \longrightarrow 2SO_3(g)$$
$$\Delta H = -198.2 \text{ kJ/mol}$$

9. Use enthalpy data given after the question to calculate the enthalpy of reaction for each of the following. Solve each by combining the known thermochemical equations.

a. $CaCO_3(s) \rightarrow CaO(s) + CO_2(g)$

   Given:

   $2CaCO_3(s) \rightarrow 2Ca(s) + 2C(s) + 3O_2(g)$
   $\Delta H = 2413.8$ kJ/mol
   $2Ca(s) + O_2(g) \rightarrow 2CaO(s)$
   $\Delta H = -1269.8$ kJ/mol
   $C(s) + O_2(g) \rightarrow CO_2(g)$
   $\Delta H = -393.51$ kJ/mol

b. $Ca(OH)_2(s) \rightarrow CaO(s) + H_2O(g)$

   Given:

   $Ca(OH)_2(s) \rightarrow Ca(s) + O_2(g) + H_2(g)$
   $\Delta H = 983.2$ kJ/mol
   $2Ca(s) + O_2(g) \rightarrow 2CaO(s)$
   $\Delta H = -1269.8$ kJ/mol
   $2H_2(g) + O_2(g) \rightarrow 2H_2O(g)$
   $\Delta H = -483.6$ kJ/mol

c. $Fe_2O_3(s) + 3CO(g) \rightarrow 2Fe(s) + 3CO_2(g)$

   Given:

   $2Fe_2O_3(s) \longrightarrow 4Fe(s) + 3O_2(g)$
   $\Delta H = 1651$ kJ/mol

   $CO(g) \longrightarrow C(s) + \frac{1}{2}O_2(g)$

   $\Delta H = 110.5$ kJ/mol

   $C(s) + O_2(g) \longrightarrow CO_2(g)$

   $\Delta H = -393.5$ kJ/mol

10. Calculate the enthalpies for reactions in which ethane, C_6H_6, are the respective reactants and $CO_2(g)$ and $H_2O(l)$ are the products in each. Solve each by combining the known thermochemical equations using the ΔH values given below. Verify the result by using the general equation for finding enthalpies of reaction from enthalpies of formation.

a. $C_2H_6(g) + O_2(g) \rightarrow$

   Given:

   $C_2H_6(g) \rightarrow 2C(s) + 3H_2(g)$
   $\Delta H = 83.8$ kJ/mol
   $C(s) + O_2(g) \rightarrow CO_2(g)$
   $\Delta H = -393.5$ kJ/mol

   $H_2(g) + \frac{1}{2}O_2(g) \rightarrow H_2O(l)$

   $\Delta H = -285.8$ kJ/mol

b. $C_6H_6(g) + O_2(g) \rightarrow$

   Given:

   $C_6H_6(l) \rightarrow 6C(s) + 3H_2(g)$
   $\Delta H = -49.08$ kJ/mol
   $C(s) + O_2(g) \rightarrow CO_2(g)$
   $\Delta H = -393.5$ kJ/mol

   $H_2(g) + \frac{1}{2}O_2(g) \rightarrow H_2O(l)$

   $\Delta H = -285.8$ kJ/mol

11. The enthalpy of formation of ethanol, C_2H_5OH, is -277 kJ/mol at 298.15 K. Calculate the enthalpy of combustion of one mole of ethanol from the information given below, assuming that the products are $CO_2(g)$ and $H_2O(l)$.

   $C_2H_5OH(l) \rightarrow 2C(s) + 3H_2(g) + \frac{1}{2}O_2(g)$

   $\Delta H = -(-277$ kJ/mol$)$
   $C(s) + O_2(g) \rightarrow CO_2$
   $\Delta H = -393.5$ kJ/mol

   $H_2(g) + \frac{1}{2}O_2(g) \rightarrow H_2O(l)$

   $\Delta H = -285.8$ kJ/mol

12. The enthalpy of formation for sulfur dioxide gas is -0.2968 kJ/(mol•K). Calculate the amount of energy given off in kJ when 30 g of $SO_2(g)$ is formed from its elements.

Section: Order and Spontaneity

13. Predict the sign of ΔS for each of the following reactions:

a. the thermal decomposition of solid calcium carbonate

   $$CaCO_3(s) \rightarrow CaO(s) + CO_2(g)$$

b. the oxidation of SO_2 in air

   $$2SO_2(g) + O_2(g) \rightarrow 2SO_3(g)$$

14. Calculate the value of ΔG for the reaction below, given the values of ΔH and ΔS. The temperature is 298 K.
   $Cu_2S(s) + S(s) \rightarrow 2CuS(s)$
   $\Delta H = -26.7$ kJ/mol
   $\Delta S = -0.0197$ kJ/(mol•K)

15. Will the reaction in item 14 be spontaneous at 298 K?

16. Predict whether the value of ΔS for each of the following reactions will be greater than, less than, or equal to zero.

   a. $3H_2(g) + N_2(g) \longrightarrow 2NH_3(g)$

   b. $2Mg(s) + O_2(g) \longrightarrow 2MgO(s)$

   c. $C_6H_{12}O_6(s) + 6O_2(g) \longrightarrow$
$$6CO_2(g) + 6H_2O(g)$$

   d. $KNO_3(s) \longrightarrow K^+(aq) + NO_3^-(aq)$

17. Based on the following values, compute ΔG values for each reaction and predict whether the reaction will occur spontaneously.

   a. $\Delta H = 125$ kJ/mol, $T = 293$ K,
$\Delta S = 0.035$ kJ/(mol·K)

   b. $\Delta H = -85.2$ kJ/mol, $T = 400$ K,
$\Delta S = 0.125$ kJ/(mol·K)

   c. $\Delta H = -275$ kJ/mol, $T = 773$ K,
$\Delta S = 0.45$ kJ/(mol·K)

18. The ΔS for the reaction shown, at 298.15 K, is 0.003 kJ/(mol·K). Calculate the ΔG for this reaction, and determine whether it will occur spontaneously at 298.15 K.
$C(s) + O_2(g) \longrightarrow CO_2(g) + 393.51$ kJ/mol

19. When graphite reacts with hydrogen at 300K, ΔH is -74.8 kJ/mol and ΔS is -0.0809 kJ/(mol·K). Will this reaction occur spontaneously?

20. The thermite reaction used in some welding applications has the following enthalpy and entropy changes at 298.15 K. Assuming ΔS and ΔH are constant, calculate ΔG at 448 K.
$Fe_2O_3(s) + 2Al(s) \longrightarrow 2Fe(s) + Al_2O_3(s)$
$\Delta H = -851.5$ kJ/mol
$\Delta S = -0.0385$ kJ/(mol·K)

21. Calculate the change in enthalpy for the following reaction.
$4FeO(s) + O_2(g) \longrightarrow 2Fe_2O_3(s)$
Use the following enthalpy data.
$FeO(s) \longrightarrow Fe(s) + \frac{1}{2}O_2(g)$

   $\Delta H = 272$ kJ/mol

$2Fe(s) + \frac{3}{2}O_2(g) \longrightarrow Fe_2O_3(g)$

   $\Delta H = -824.2$ kJ/mol

States of Matter and Intermolecular Forces

Section: Intermolecular Forces

1. a. Which contains more molecules of water, 5.00 cm^3 of ice at 0.0°C or 5.00 cm^3 of liquid water at 0.0°C?

   b. How many more molecules?

GAS

Section: Characteristics of Gases

1. Convert a pressure of 1.75 atm to kPa.

2. Convert the pressure of 1.75 atm to mm Hg.

3. Convert a pressure of 570 torr to atmospheres.

4. Convert a pressure of 570 torr to kPa.

5. A weather report gives a current atmospheric pressure reading of 745.8 mm Hg. Express this reading in atmospheres.

6. Convert the pressure of 745.8 mm Hg to torrs.

7. Convert the pressure of 745.8 mm Hg to kilopascals.

8. Convert a pressure of 151.98 kPa to pressure in standard atmospheres.

9. Convert a pressure of 456 torr to pressure in standard atmospheres.

10. Convert a pressure of 912 mm Hg to pressure in standard atmospheres.

Section: The Gas Laws

11. A balloon filled with helium gas has a volume of 500 mL at a pressure of 1 atm. The balloon is released and reaches an altitude of 6.5 km, where the pressure is 0.5 atm. Assuming that the temperature has remained the same, what volume does the gas occupy at this height?

12. A gas has a pressure of 1.26 atm and occupies a volume of 7.4 L. If the gas is compressed to a volume of 2.93 L, what will its pressure be, assuming constant temperature?

13. Divers know that the pressure exerted by the water increases about 100 kPa with every 10.2 m of depth. This means that at 10.2 m below

the surface, the pressure is 201 kPa; at 20.4 m below the surface, the pressure is 301 kPa; and so forth. Given the volume of a balloon is 3.5 L at STP and that the temperature of the water remains the same, what is the volume 51 m below the water's surface?

14. The piston of an internal combustion engine compresses 450 mL of gas. The final pressure is 15 times greater than the initial pressure. What is the final volume of the gas, assuming constant temperature?

15. A helium-filled balloon contains 125 mL of gas at a pressure of 0.974 atm. What volume will the gas occupy at standard pressure?

16. A weather balloon with a volume of 1.375 L is released from Earth's surface at sea level. What volume will the balloon occupy at an altitude of 20.0 km, where the air pressure is 10 kPa?

17. A sample of helium gas has a volume of 200 mL at 0.96 atm. What pressure, in atm, is needed to reduce the volume at constant temperature to 50 mL?

18. A certain mass of oxygen was collected over water when potassium chlorate was decomposed by heating. The volume of the oxygen sample collected was 720 mL at 25°C and a barometric pressure of 755 torr. What would the volume of the oxygen be at STP? (Hint: First calculate the partial pressure of the oxygen. Then use the combined gas law.)

19. Use Boyle's law to solve for the missing value in the following:
$P_1 = 350$ torr, $V_1 = 200$ mL, $P_2 = 700$ torr, $V_2 = ?$

20. Use Boyle's law to solve for the missing value in the following:
$P_1 = 0.75$ atm, $V_2 = 435$ mL, $P_2 = 0.48$ atm, $V_1 = ?$

21. Use Boyle's law to solve for the missing value in the following:
$V_1 = 2.4 \times 10^5$ mL, $P_2 = 180$ mm Hg, $V_2 = 1.8 \times 10^3$ mL, $P_1 = ?$

22. The pressure exerted on a 240 mL sample of hydrogen gas at constant temperature is increased from 0.428 atm to 0.724 atm. What will the final volume of the sample be?

23. A flask containing 155 cm³ of hydrogen was collected under a pressure of 22.5 kPa. What pressure would have been required for the volume of the gas to have been 90 cm³, assuming the same temperature?

24. A gas has a volume of $V_1 = 450$ mL. If the temperature is held constant, what volume would the gas occupy if the pressure $P_2 = 2P_1$?

25. What volume would the gas occupy if the pressure $P_2 = 0.25P_1$?

26. A sample of oxygen that occupies 1.00×10^6 mL at 575 mm Hg is subjected to a pressure of 1.25 atm. What will the final volume of the sample be if the temperature is held constant?

27. A helium-filled balloon has a volume of 2.75 L at 20°C. The volume of the balloon decreases to 2.46 L after it is placed outside on a cold day.

   a. What is the outside temperature in K?

   b. What is the outside temperature in °C?

28. A gas at 65°C occupies 4.22 L. At what Celsius temperature will the volume be 3.87 L, assuming the same pressure?

29. A certain quantity of gas has a volume of 0.75 L at 298 K. At what temperature, in degrees Celsius, would this quantity of gas be reduced to 0.50 L, assuming constant pressure?

30. A balloon filled with oxygen gas occupies a volume of 5.5 L at 25°C. What volume will the gas occupy at 100°C?

31. A sample of nitrogen gas is contained in a piston with a freely moving cylinder. At 0°C, the volume of the gas is 375 mL. To what temperature must the gas be heated to occupy a volume of 500 mL?

32. Use Charles's law to solve for the missing value in the following:
$V_1 = 80$ mL, $T_1 = 27$°C, $T_2 = 77$°C, $V_2 = ?$

33. Use Charles's law to solve for the missing value in the following:
$V_1 = 125$ L, $V_2 = 85$ L, $T_2 = 127$°C, $T_1 = ?$

34. Use Charles's law to solve for the missing value in the following:
$T_1 = -33°C$, $V_2 = 54$ mL, $T_2 = 160°C$, $V_1 = ?$

35. A sample of air has a volume of 140 mL at 67°C. At what temperature will its volume be 50 mL at constant pressure?

36. At standard temperature, a gas has a volume of 275 mL. The temperature is then increased to 130°C, and the pressure is held constant. What is the new volume?

37. An aerosol can contains gases under a pressure of 4.5 atm at 20°C. If the can is left on a hot sandy beach, the pressure of the gases increases to 4.8 atm. What is the Celsius temperature on the beach?

38. Before a trip from New York to Boston, the pressure in an automobile tire is 1.8 atm at 293 K. At the end of the trip, the pressure gauge reads 1.9 atm. What is the new Celsius temperature of the air inside the tire? (Assume tires with constant volume.)

39. At 120°C, the pressure of a sample of nitrogen is 1.07 atm. What will the pressure be at 205°C, assuming constant volume?

40. A sample of helium gas has a pressure of 1.2 atm at 22°C. At what Celsius temperature will the helium reach a pressure of 2 atm?

41. An empty aerosol-spray can at room temperature (20°C) is thrown into an incinerator where the temperature reaches 500°C. If the gas inside the empty container was initially at a pressure of 1.0 atm, what pressure did it reach inside the incinerator? Assume the gas was at constant volume and the can did not explode.

42. The temperature within an automobile tire at the beginning of a long trip is 25°C. At the conclusion of the trip, the tire has a pressure of 1.8 atm. What is the final Celsius temperature within the tire if its original pressure was 1.75 atm?

43. A sample of gas in a closed container at a temperature of 100°C and a pressure of 3.0 atm is heated to 300°C. What pressure does the gas exert at the higher temperature?

44. A sample of hydrogen at 47°C exerts a pressure of 0.329 atm. The gas is heated to 77°C at constant volume. What will its new pressure be?

45. To what temperature must a sample of nitrogen at 27°C and 0.625 atm be taken so that its pressure becomes 1.125 atm at constant volume?

46. The pressure on a gas at −73°C is doubled, but its volume is held constant. What will the final temperature be in degrees Celsius?

Section: Molecular Composition of Gases

47. Quantitatively compare the rates of effusion for the following pairs of gases at the same temperature and pressure.
a. Hydrogen and nitrogen
b. Fluorine and chlorine

48. Some hydrogen gas is collected over water at 20°C. The levels of water inside and outside the gas-collection bottle are the same. The partial pressure of hydrogen is 742.5 torr. What is the barometric pressure at the time the gas is collected?

49. What is the volume, in liters, of 0.100 g of $C_2H_2F_4$ vapor at 0.928 atm and 22.3°C?

50. What is the molar mass of a 1.25 g sample of gas that occupies a volume of 1.00 L at a pressure of 0.961 atm and a temperature of 27.0°C?

51. What pressure, in atmospheres, is exerted by 0.325 mol of hydrogen gas in a 4.08 L container at 35°C?

52. A gas sample occupies 8.77 L at 20.0°C. What is the pressure, in atmospheres, given that there are 1.45 mol of gas in the sample?

53. A 2.07 L cylinder contains 2.88 mol of helium gas at 22°C. What is the pressure in atmospheres of the gas in the cylinder?

54. A tank of hydrogen gas has a volume of 22.9 L and holds 14.0 mol of the gas at 12°C. What is the reading on the pressure gauge in atmospheres?

55. A sample that contains 4.38 mol of a gas at 250.0 K has a pressure of 0.857 atm. What is the volume?

56. How many liters are occupied by 0.909 mol of nitrogen at 125°C and 0.901 atm pressure?

57. A reaction yields 0.00856 mol of O_2 gas. What volume in mL will the gas occupy if it is collected at 43.0°C and 0.926 atm pressure?

58. A researcher collects 0.00909 mol of an unknown gas by water displacement at a temperature of 16.0°C and 0.873 atm pressure (after the partial pressure of water vapor has been subtracted). What volume of gas in mL does the researcher have?

59. How many grams of carbon dioxide gas are there in a 45.1 L container at 34.0°C and 1.04 atm?

60. What is the mass, in grams, of oxygen gas in a 12.5 L container at 45.0°C and 7.22 atm?

61. A sample of carbon dioxide with a mass of 0.30 g was placed in a 250 mL container at 400.0 K. What is the pressure exerted by the gas?

62. What mass of ethene gas, C_2H_4, is contained in a 15.0 L tank that has a pressure of 4.40 atm at a temperature of 305 K?

63. NH_3 gas is pumped into the reservoir of a refrigeration unit at a pressure of 4.45 atm. The capacity of the reservoir is 19.4 L. The temperature is 24.0°C. What is the mass of the gas in g?

64. What is the molar mass of a gas if 0.427 g of the gas occupies a volume of 125 mL at 20.0°C and 0.980 atm?

65. What is the density of a sample of ammonia gas, NH_3, if the pressure is 0.928 atm and the temperature is 63.0°C?

66. The density of a gas was found to be 2.0 g/L at 1.50 atm and 27.0°C. What is the molar mass of the gas?

67. What is the density of argon gas, Ar, at a pressure of 551 torr and a temperature of 25.0°C?

68. A chemist determines the mass of a sample of gas to be 3.17 g. Its volume is 942 mL at a temperature of 14.0°C and a pressure of 1.09 atm. What is the molar mass of the gas?

69. The density of dry air at sea level (1 atm) is 1.225 g/L at 15.0°C. What is the average molar mass of the air?

70. How many liters of gaseous carbon monoxide at 27.0°C and 0.247 atm can be produced from the burning of 65.5 g of carbon according to the following equation?
$$2C(s) + O_2(g) \longrightarrow 2CO(g)$$

71. Calculate the pressure, in atmospheres, exerted by each of the following.
    a. 2.50 L of HF containing 1.35 mol at 320 K
    b. 4.75 L of NO_2 containing 0.860 mol at 300 K
    c. 750 mL of CO_2 containing 2.15 mol at 57.0°C

72. Calculate the volume, in liters, occupied by each of the following.
    a. 2.00 mol of H_2 at 300.0 K and 1.25 atm
    b. 0.425 mol of NH_3 at 37.0°C and 0.724 atm
    c. 4.00 g of O_2 at 57.0°C and 0.888 atm

73. Determine the number of moles of gas contained in each of the following.
    a. 1.25 L at 250 K and 1.06 atm
    b. 0.800 L at 27.0°C and 0.925 atm
    c. 750 mL at –50.0°C and 0.921 atm

74. Find the mass of each of the following.
    a. 5.60 L of O_2 at 1.75 atm and 250 K
    b. 3.50 L of NH_3 at 0.921 atm and 27.0°C
    c. 0.125 mL of SO_2 at 0.822 atm and –53.0°C

75. Find the molar mass of each gas measured at the specified conditions.
    a. 0.650 g occupying 1.12 L at 280 K and 1.14 atm
    b. 1.05 g occupying 2.35 L at 37.0°C and 0.840 atm
    c. 0.432 g occupying 750 mL at –23.0°C and 1.03 atm

76. If the density of an unknown gas is 3.20 g/L at –18.0°C and 2.17 atm, what is the molar mass of this gas?

77. One method of estimation the temperature of the center of the sun is based on the assumption that the center consists of gases that have an average molar mass of 2.00 g/mol. If the density of the center of the sun is 1.40 g/cm^3 at a pressure of 1.30×10^9 atm, calculate the temperature in degrees Celsius.

78. Three of the primary components of air are carbon dioxide, nitrogen, and oxygen. In a sample containing a mixture of only these gases at exactly one atmosphere pressure, the partial pressures of carbon dioxide and nitrogen are given as $P_{CO_2} = 0.285$ torr and $P_{N_2} = 593.525$ torr. What is the partial pressure of oxygen?

79. Determine the partial pressure of oxygen collected by water displacement if the water temperature is 20.0°C and the total pressure of the gases in the collection bottle is 730 torr. P_{H_2O} at 20.0°C is equal to 17.5 torr.

80. A sample of hydrogen effuses through a porous container about 9.00 times faster than an unknown gas. Estimate the molar mass of the unknown gas.

81. Compare the rate of effusion of carbon dioxide with that if hydrogen chloride at the same temperature and pressure.

82. If a molecule of neon gas travels at an average of 400 m/s at a given temperature, estimate the average speed of a molecule of butane gas, C_4H_{10}, at the same temperature.

83. Nitrogen effused through a pinhole 1.7 times as fast as another gaseous element at the same conditions. Estimate the other element's molar mass.

84. Determine the molecular mass ratio of two gases whose rates of diffusion have a ratio of 16.0:1.

85. Estimate the molar mass of a gas that effuses at 1.60 times the effusion rate of carbon dioxide.

86. List the following gases in order of increasing average molecular velocity at 25°C: H_2O, He, HCl, BrF, and NO_2.

87. What is the ratio of the average velocity of hydrogen molecules to that of neon atoms at the same temperature and pressure?

88. At a certain temperature and pressure, chlorine molecules have an average velocity of 0.0380 m/s. What is the average velocity of sulfur dioxide molecules under the same conditions?

89. A sample of helium effuses through a porous container 6.50 times faster than does unknown gas X. What is the molar mass of the unknown gas?

90. How many liters of H_2 gas at STP can be produced by the reaction of 4.60 g of Na and excess water, according to the following equation?
$$2Na(s) + 2H_2O(l) \longrightarrow H_2(g) + 2NaOH(aq)$$

91. How many grams of Na are needed to react with H_2O to liberate 400 mL H_2 gas at STP?

92. What volume of oxygen gas in liters can be collected at 0.987 atm pressure and 25.0°C when 30.6 g of $KClO_3$ decompose by heating, according to the following equation?
$$2KClO_3(s) \xrightarrow[MnO_2]{\Delta} 2KCl(s) + 3O_2(g)$$

93. What mass of sulfur must be used to produce 12.6 L of gaseous sulfur dioxide at STP according to the following equation?
$$S_8(s) + 8O_2(g) \longrightarrow 8SO_2(g)$$

94. How many grams of water can be produced from the complete reaction of 3.44 L of oxygen gas, at STP, with hydrogen gas?

95. Aluminum granules are a component of some drain cleaners because they react with sodium hydroxide to release both energy and gas bubbles, which help clear the drain clog. The reaction is
$$2NaOH(aq) + 2Al(s) + 6H_2O(l) \longrightarrow$$
$$2NaAl(OH)_4(aq) + 3H_2(g)$$
What mass of aluminum would be needed to produce 4.00 L of hydrogen gas at STP?

96. What volume of chlorine gas at 38.0°C and 1.63 atm is needed to react completely with 10.4 g of sodium to form NaCl?

97. Air bags in cars are inflated by the sudden decomposition of sodium azide, NaN_3, by the following reaction.
$$2NaN_3(s) \longrightarrow 3N_2(g) + 2Na(s)$$
What volume of N_2 gas, measured at 1.30 atm and 87.0°C, would be produced by the reaction of 70.0 g of NaN_3?

98. Assume that 5.60 L of H_2 at STP react with CuO according to the following equation:
$$CuO(s) + H_2(g) \longrightarrow Cu(s) + H_2O(g)$$
How many moles of H_2 react?

99. A modified Haber process for making ammonia is conducted at 550°C and 250 atm. If 10.0 kg of nitrogen (the limiting reactant) is used and the process goes to completion, what volume of ammonia is produced?

100. When liquid nitroglycerin, $C_3H_5(NO_3)_3$, explodes, the products are carbon dioxide, nitrogen, oxygen, and water vapor. If 500.0 g of nitroglycerin explode at STP, what is the total volume, at STP, for all the gases produced?

101. The principal source of sulfur on Earth is deposits of free sulfur occurring mainly in volcanically active regions. The sulfur was initially formed by the reaction between the two volcanic vapors SO_2 and H_2S to form $H_2O(l)$ and $S_8(s)$. What volume of SO_2, at 0.961 atm and 22.0°C, was needed to form a sulfur deposit of 4.50×10^5 kg on the slopes of a volcano in Hawaii?

102. What volume of H_2S, at 0.961 atm and 22.0°C, was needed to form a sulfur deposit of 4.50×10^5 kg on the slopes of a volcano in Hawaii?

103. A 3.25 g sample of solid calcium carbide, CaC_2, reacted with water to produce acetylene gas, C_2H_2, and aqueous calcium hydroxide. If the acetylene was collected over water at 17.0°C and 0.974 atm, how many milliliters of acetylene were produced?

104. Assume that 13.5 g of Al react with HCl according to the following equation, at STP:
$Al(s) + HCl(aq) \longrightarrow AlCl_3(aq) + H_2(g)$
Remember to balance the equation first.
a. How many moles of Al react?
b. How many moles of H_2 are produced?
c. How many liters of H_2 at STP are produced?

Solutions

Section: Concentration and Molarity

1. What is the molarity of a 2.000 L solution that is made from 14.60 g of NaCl?

2. What is the molarity of a HCl solution that contains 10.0 g of HCl in 250 mL of solution?

3. How many moles of NaCl are in 1.25 L of 0.330 M NaCl?

4. What is the molarity of a solution composed of 6.250 g of HCl in 0.3000 L of solution?

5. 5.00 grams of sugar, $C_{12}H_{22}O_{11}$, are dissolved in water to make 1.00 L of solution. What is the concentration of this solution expressed as molarity?

6. Suppppose you wanted to dissolve 40.0 g NaOH in enough H_2O to make 6.00 L of solution. You want to calculate the molarity, M, of the resulting solution.
a. What is the molar mass of NaOH?
b. What is the molarity of this solution?

7. What is the molarity of a solution of 14.0 g of NH_4Br in enough H_2O to make 150 mL of solution?

8. Suppose you wanted to produce 1.00 L of a 3.50 M solution of H_2SO_4. How many grams of solute are needed to make this solution?

9. How many grams of solute are needed to make 2.50 L of a 1.75 M solution of $Ba(NO_3)_2$?

10. How many moles of NaOH are contained in 65.0 mL of a 2.20 M solution of NaOH in H_2O?

11. A solution is made by dissolving 26.42 g of $(NH_4)_2SO_4$ in enough H_2O to make 50.00 mL of solution.
a. What is the molar mass of $(NH_4)_2SO_4$?
b. What is the molarity of this solution?

12. Suppose you wanted to find out how many milliliters of 1.0 M $AgNO_3$ are needed to provide 168.88 of pure $AgNO_3$.
a. What is the molar mass of $AgNO_3$?
b. How many mL of solution are needed?

13. Na_2SO_4 is dissolved in water to make 450 mL of a 0.250 M solution.
a. What is the molar mass of Na_2SO_4?
b. How many moles of Na_2SO_4 are needed?

14. Citric acid is one component of some soft drinks. Suppose that a 2.00 L solution is made from 150 mg of citric acid, $C_6H_8O_7$.
a. What is the molar mass of $C_6H_8O_7$?
b. What is the molarity of citric acid in the solution?

15. Suppose you wanted to know how many grams of KCl would be left if 350 mL of a 6.0 M KCl solution were evaporated to dryness.
    a. What is the molar mass of KCl?
    b. How many grams of KCl would remain?

16. Sodium metal reacts violently with water to form NaOH and release hydrogen gas. Suppose that 10.0 g of Na reacts completely with 1.00 L of water, and the final volume of the system is 1.00 L.

    $$2Na(s) + 2H_2O(l) \longrightarrow 2NaOH(aq) + H_2(g)$$

    a. What is the molar mass of NaOH?
    b. What is the molarity, M, of the NaOH solution formed by the reaction?

Section: Physical Properties of Solutions

17. Given 0.01 m aqueous solutions of each of the following, arrange the solutions in order of increasing change in the freezing point of the solution.
    a. NaI
    b. $CaCl_2$
    c. K_3PO_4
    d. $C_6H_{12}O_6$ (glucose)

Chemical Equilibrium

Section: Systems at Equilibrium

1. At equilibrium a mixture of N_2, H_2, NH_3 gas at 500°C is determined to consist of 0.602 mol/L of N_2, 0.420 mol/L of H_2, and 0.113 mol/L of NH_3. What is the equilibrium constant for the reaction $N_2(g) + 3H_2(g) \rightleftharpoons 2NH_3(g)$ at this temperature?

2. The reaction $AB_2C(g) \rightleftharpoons B_2(g) + AC(g)$ reached equilibrium at 900 K in a 5.00 L vessel. At equilibrium 0.0840 mol of AB_2C, 0.0350 mol of B_2, and 0.0590 mol of AC were detected. What is the equilibrium constant at this temperature for this system?

3. At equilibrium at 1.0 L vessel contains 20.00 mol of H_2, 18.00 mol of CO_2, 12.00 mol of H_2O, and 5.900 mol of CO at 427°C. What is the value of K_{eq} at this temperature for the following reaction?

    $$CO_2(g) + H_2(g) \rightleftharpoons CO(g) + H_2O(g)$$

4. A reaction between gaseous sulfur dioxide and oxygen gas to produce gaseous sulfur trioxide takes place at 600°C. At that temperature, the concentration of SO_2 is found to be 1.50 mol/L, the concentration of O_2 is 1.25 mol/L, and the concentration of SO_3 is 3.50 mol/L. Using the balanced chemical equation, calculate the equilibrium constant for this system.

5. At equilibrium at 2500 K, [HCl] = 0.0625 and $[H_2] = [Cl_2] = 0.00450$ for the reaction

    $$H_2(g) + Cl_2(g) \rightleftharpoons 2HCl(g).$$

    Find the value of K_{eq}.

6. An equilibrium mixture at 435°C is found to consist of 0.00183 mol/L of H_2, 0.00313 mol/L of I_2, and 0.0177 mol/L of HI. Calculate the equilibrium constant, K_{eq}, for the reaction $H_2(g) + I_2(g) \rightleftharpoons 2HI(g)$.

7. For the reaction

    $$H_2(g) + I_2(g) \rightleftharpoons 2HI(g)$$

    at 425°C, calculate [HI], given $[H_2] = [I_2] = 0.000479$ and $K_{eq} = 54.3$.

8. At 25°C, an equilibrium mixture of gases contains 0.00640 mol/L PCl_3, 0.0250 mol/L Cl_2, and 0.00400 mol/L PCl_5. What is the equilibrium constant for the following reaction?

    $$PCl_5(g) \rightleftharpoons PCl_3(g) + Cl_2(g)$$

9. At equilibrium a 2 L vessel contains 0.360 mol of H_2, 0.110 mol of Br, and 37.0 mol of HBr. What is the equilibrium constant for the reaction at this temperature?

    $$H_2(g) + Br_2(g) \rightleftharpoons 2HBr(g)$$

10. Calculate the solubility-product constant, K_{sp}, of lead(II) chloride, $PbCl_2$, which has a solubility of 1.00 g/100.0 g H_2O at a temperature other than 25°C.

11. 5.00 g of Ag_2SO_4 will dissolve in 1.00 L of water. Calculate the solubility product constant for this salt.

12. What is the value of K_{sp} for tin(II) sulfide, given that its solubility is 5.2×10^{-12} g/100.0 g water?

13. Calculate the solubility product constant for calcium carbonate, given that it has a solubility of 5.3×10^{-5} g/L of water.

14. Calculate the solubility of cadmium sulfide, CdS, in mol/L, given the K_{sp} value as 8.0×10^{-27}.

15. Determine the concentration of strontium ions in saturated solution of strontium sulfate, $SrSO_4$, if the K_{sp} for K_{sp} is 3.2×10^{-7}.

16. What is the solubility in mol/L of manganese(II) sulfide, MnS, given that its K_{sp} value is 2.5×10^{-13}?

17. Calculate the concentration of Zn^{2+} in saturated solution of zinc sulfide, ZnS, given that K_{sp} of zinc sulfide equals 1.6×10^{-24}.

18. What is the value of K_{sp} for Ag_2SO_4 if 5.40 g is soluble in 1.00 L of water?

19. Calculate the concentration of Hg^{2+} ions in a saturated solution of HgS(s). K_{sp} is 1.6×10^{-52}.

20. At 25°C, the value of K_{eq} is 1.7×10^{-13} for the following reaction.

$$N_2O(g) + \frac{1}{2}O_2(g) \rightleftarrows 2NO(g)$$

It is determined that $[N_2O] = 0.0035$ mol/L and $[O_2] = 0.0027$ mol/L. Using this information, what is the concentration of NO(g) at equilibrium?

21. Tooth enamel is composed of the mineral hydroxyapatite, $Ca_5(PO_4)_3OH$, which has a K_{sp} of 6.8×10^{-37}. The molar solubility of hydroxyapatite is 2.7×10^{-5} mol/L. When hydroxyapatite is reacted with fluoride, the OH^- is replaced with the F^- ion on the mineral, forming fluorapatite, $Ca_5(PO_4)_3F$. (The latter is harder and less susceptible to caries.) The K_{sp} of fluorapatite is 1×10^{-60}. Calculate the solubility of fluorapatite in water. Given your calculations, can you support the fluoridation of drinking water? Your answer must be within ± 0.5%.

Acids and Bases

Section: Acidity, Basicity and pH

1. Identify the following as being true of acidic or basic solutions at 25°C:
$[H_3O^+] = 1.0 \times 10^{-3}$ M

2. Identify the following as being true of acidic or basic solutions at 25°C:
$[OH^-] = 1.0 \times 10^{-4}$ M

3. Identify the following as being true of acidic or basic solutions at 25°C:
pH = 5

4. Identify the following as being true of acidic or basic solutions at 25°C:
pH = 8

5. The pH of a hydrochloric acid solution for cleaning tile is 0.45. What is the $[H_3O^+]$ in the solution?

6. A $Ca(OH)_2$ solution has a pH of 8.
a. Determine $[H_3O^+]$ for the solution.
b. Determine $[OH^-]$.
c. Determine $[Ca(OH)_2]$.

7. Determine the pH of the following solution:
1×10^{-3} M HCl.

8. Determine the pH of the following solution:
1×10^{-5} M HNO_3.

9. Determine the pH of the following solution:
1×10^{-4} M NaOH.

10. Determine the pH of the following solution:
1×10^{-2} M KOH.

11. The pH of a solution is 10.
a. What is the concentration of hydroxide ions in the solution?
b. If the solution is $Sr(OH)_2(aq)$, what is its molarity?

12. Determine the hydronium ion concentration in a solution that is 1×10^{-4} M HCl.

13. Determine the hydroxide ion concentration in a solution that is 1×10^{-4} M HCl.

14. Determine the hydronium ion concentration in a solution that is 1×10^{-3} M HNO_3.

15. Determine the hydroxide ion concentration in a solution that is 1×10^{-3} M HNO_3.

16. Determine the hydroxide ion concentration in a solution that is 3×10^{-2} M NaOH.

17. Determine the hydronium ion concentration in a solution that is 3.00×10^{-2} M NaOH.

18. a. Determine the hydroxide ion concentration in a solution that is 1×10^{-4} M $Ca(OH)_2$.
    b. Determine the hydronium ion concentration in a solution that is 1×10^{-4} M $Ca(OH)_2$.

19. a. Determine the $[H_3O^+]$ in a 0.01 M solution of $HClO_4$.
    b. Determine the $[OH^-]$ in a 0.01 M solution of $HClO_4$.

20. An aqueous solution of $Ba(OH)_2$ has a $[H_3O^+]$ of 1×10^{-11} M.
    a. What is the $[OH^-]$?
    b. What is the molarity of $Ba(OH)_2$ in the solution?

Section: Neutralization and Titration

21. If 20 mL of 0.01 M aqueous HCl is required to neutralize 30 mL of an aqueous solution of NaOH, determine the molarity of the NaOH solution.

Reaction Rates

Section: How Can Reaction Rates Be Explained?

1. A reaction involving reactants A and B is found to occur in the one-step mechanism: $2A + B \longrightarrow A_2B$. Write the rate law for this reaction, and predict the effect of doubling the concentration of either reactant on the overall reaction rate.

2. A chemical reaction is expressed by the balanced chemical equation $A + 2B \longrightarrow C$. Using the data below, determine the rate law for the reaction.
   Experiment # 1. initial $[A] = 0.2$ M
   initial $[B] = 0.2$ M
   initial rate of formation of C = 0.0002 M/min

   Experiment # 2. initial $[A] = 0.2$ M
   initial $[B] = 0.4$ M
   initial rate of formation of C = 0.0008 M/min

   Experiment # 3. initial $[A] = 0.4$ M
   initial $[B] = 0.4$ M/min
   initial rate of formation of C = 0.0016 M

3. A particular reaction is found to have the following rate law.
   $R = k[A][B]^2$
   How is the rate affected if
   a. the initial concentration of A is cut in half?
   b. the initial concentration of B is tripled?
   c. the initial concentration of A is doubled, but the concentration of B is cut in half?

Oxidation, Reduction, and Electrochemistry

Section: Oxidation-Reduction Reactions

1. Name the following acid: HNO_2

2. Assign oxidation numbers to each atom in H_2SO_3.

3. Assign oxidation numbers to each atom in H_2CO_3.

4. Assign oxidation numbers to each atom in HI.

5. Assign oxidation numbers to each atom in CO_2.

6. Assign oxidation numbers to each atom in NH_4^+.

7. Assign oxidation numbers to each atom in MnO_4^-.

8. Assign oxidation numbers to each atom in $S_2O_3^{2-}$.

9. Assign oxidation numbers to each atom in H_2O_2.

10. Assign oxidation numbers to each atom in P_4O_{10}.

11. Assign oxidation numbers to each atom in OF_2.

12. Assign oxidation numbers to each atom in SO_3.

13. Determine the oxidation state of the metal in CdS.

14. Determine the oxidation state of the metal in ZnS.

15. Determine the oxidation state of the metals in $PbCrO_4$.

16. Determine the oxidation state of the metal in $Fe(SCN)^{2+}$.

17. Determine the oxidation state of the metal in MnO_4^-.

18. Determine the oxidation state of the metals in $CoCl_2$.

19. Determine the oxidation state of the metal in $[Cu(NH_3)_4](OH)_2$.

20. Determine the oxidation state of the nitrogen in N_2O_3.

21. Determine the oxidation state of the nitrogen in N_2O_5.

22. Which of the following equations represent redox reactions?

    a. $2KNO_3(s) \longrightarrow 2KNO_2(s) + O_2(g)$

    b. $H_2(g) + CuO(s) \longrightarrow Cu(s) + H_2O(l)$

    c. $NaOH(aq) + HCl(aq) \longrightarrow NaCl(aq) + H_2O(l)$

    d. $H_2(g) + Cl_2(g) \longrightarrow 2HCl(g)$

    e. $SO_3(g) + H_2O(l) \longrightarrow H_2SO_4(aq)$

23. Identify if the following reactions are redox or nonredox:

    a. $2NH_4Cl(aq) + Ca(OH)_2(aq) \longrightarrow$
    $2NH_3(aq) + 2H_2O(l) + CaCl_2(aq)$

    b. $2HNO_3(aq) + 3H_2S(g) \longrightarrow 2NO(g) + 4H_2O(l) + 3S(s)$

    c. $[Be(H_2O)_4]^{2+}(aq) + H_2O(l) \longrightarrow$
    $H_3O^+(aq) + [Be(H_2O)_3OH]^+(aq)$

Nuclear Chemistry

Section: Atomic Nuclei and Nuclear Stability

1. The mass of a $^{20}_{10}Ne$ atom is 19.992 44 amu. Calculate the mass defect.

2. The mass of a $^{7}_{3}Li$ atom is 7.016 00 amu. Calculate its mass defect.

3. Calculate the nuclear binding energy of one lithium-6 atom. The measured atomic mass of lithium-6 is 6.015 amu.

4. Calculate the nuclear binding energy of the nucleus $^{35}_{19}K$. The measured atomic mass of $^{35}_{19}K$ is 34.988011 amu.

5. Calculate the nuclear binding energy of the nucleus $^{23}_{11}Na$. The measured atomic mass of $^{23}_{11}Na$ is 22.989 767 amu.

6. The nuclear binding energy of $^{35}_{19}K$ is 4.47×10^{-11} J. Calculate the binding energy per nucleon for $^{35}_{19}K$.

7. The nuclear binding energy of $^{23}_{11}Na$ is 2.99×10^{-11} J. Calculate the binding energy per nucleon for $^{23}_{11}Na$.

8. Calculate the binding energy per nucleon of $^{238}_{92}U$ in joules. The atomic mass of $^{238}_{92}U$ is 238.050 784 amu.

9. The energy released by the formation of a nucleus of $^{56}_{26}Fe$ is 7.89×10^{-11} J. Use Einstein's equation, $E = mc^2$, to determine how much mass is lost (in kilograms) in this process.

10. Calculate the nuclear binding energy of one mole of deuterium atoms. The measured mass of deuterium is 2.0140 amu.

Section: Nuclear Charge

11. Balance the nuclear equation:
    $^{43}_{19}K \longrightarrow {}^{43}_{20}Ca + \underline{\quad ? \quad}$

12. Balance the nuclear equation:
    $^{233}_{92}U \longrightarrow {}^{229}_{90}Th + \underline{\quad ? \quad}$

13. Balance the nuclear equation:
    $^{11}_{6}C + \underline{\quad ? \quad} \longrightarrow {}^{11}_{5}B$

14. Balance the nuclear equation:
    $^{13}_{7}N \longrightarrow {}^{0}_{-1}e + \underline{\quad ? \quad}$

15. Write the nuclear equation for the release of an alpha particle by $^{210}_{84}Po$.

16. Write the nuclear equation for the release of an alpha particle by $^{210}_{82}Pb$.

17. Balance the nuclear equation:
    $^{239}_{93}Np \longrightarrow {}^{0}_{-1}e + \underline{\quad ? \quad}$

18. Balance the nuclear equation:
    $^{9}_{4}Be + {}^{4}_{2}He \longrightarrow \underline{\quad ? \quad}$

19. Balance the nuclear equation:
$$^{32}_{15}P + \underline{\quad ? \quad} \longrightarrow \, ^{33}_{15}P$$

20. Balance the nuclear equation:
$$^{236}_{92}U \longrightarrow \, ^{94}_{36}Kr + \underline{\quad ? \quad} + 3\,^{1}_{0}n$$

Section: Uses of Nuclear Chemistry

21. The Environmental Protection Agency and health officials nationwide are concerned about the levels of radon gas in homes. The half-life of the radon-222 isotope is 3.8 days. If a sample of gas taken from a basement contains 4.38 μg of radon-222, how much radon will remain in the sample after 15.2 days?

22. Uranium-238 decays through alpha decay with a half-life of 4.46×10^9 years. How long would it take for 7/8 of a sample of uranium-238 to decay?

23. The half-life of carbon-14 is 5715 years. How long will it be until only half of the carbon-14 in a sample remains?

24. The half-life of iodine-131 is 8.040 days. What percentage of an iodine-131 sample will remain after 40.2 days?

25. The half-life of plutonium-239 is 241 10 years. Of an original mass of 100 g, how much remains after 96 440 years?

26. The half-life of thorium-227 is 18.72 days. How many days are required for three-fourths of a given amount to decay?

27. The half-life of protactinium-234 is 6.69 hours. What fraction of a given amount remains after 26.76 hours?

28. How many milligrams remain of a 15 mg sample of radium-226 after 6396 years? The half life of radium-226 is 1599 years.

29. After 4797 years, how much of an original 0.25 g of radium-226 remains? Its half-life is 1599 years.

30. The half-life of radium-224 is 3.66 days. What was the original mass of radium-224 if 0.05 g remains after 7.32 days?

SELECTED ANSWERS

The Science of Chemistry

Practice Problems A

1. a. 0.000 765 kg
   b. 1340 mg
   c. 0.0342 g
   d. 23 745 000 000 mg
3. shortest: 0.0128 km; longest: 17 931 mm

Section 2 Review

9. a. 17 300 ms
   b. 0.000 002 56 km
   c. 5.67 g
   d. 0.005 13 km
11. about 1081 beans

Chapter Review

31. a. 0.357 L
    b. 2.5×10^7 mg
    c. 35 L
    d. 2460 cm^3
    e. 2.5×10^{-4} g
    f. 2.5×10^{-7} kg
    g. 1500 ms
    h. 10 500 mmol
37. 151 g

Matter and Energy

Section 1 Review

7. a. 373 K
   b. 1058 K
   c. 273 K
   d. 236 K

Practice Problems A

1. a. 1273 mL
   b. 98.5 cm^2
   c. 8.2 g
3. 4593 kJ/min

Practice Problems B

1. 0.069 J/g•K
3. 329 K

Section 3 Review

7. 0.30 J/g•K
9. 5.2×10^3 s

Chapter Review

25. 6.411 g
27. 2.79 m^2
29. 8.82×10^{-4} g
31. a. 9.225×10^{-2} km
    b. 9.225×10^3 cm
33. 3.6×10^3 J
35. 13°C
37. a. 6.730×10^{-4}
    b. $5.000\ 00 \times 10^4$
    c. 3.010×10^{-6}
39. 8.57×10^8 m^2
41. 4

Atoms and Moles

Practice Problems A

1. 11 protons and 11 electrons
3. 80

Practice Problems B

1. Both isotopes have 17 protons and 17 electrons. Cl-35 has 18 neutrons and Cl-37 has 20 neutrons.

Section 2 Review

5. a. 35 electrons, 35 protons, 45 neutrons
   b. 46 electrons, 46 protons, 60 neutrons
   c. 55 electrons, 55 protons, 78 neutrons

Practice Problems C

1. $1s^2 2s^2 2p^4$ or $[He]2s^2 2p^4$

Section 3 Review

5. $1s^2 2s^2 2p^6 3s^2 3p^1$ or $[Ne]3s^2 3p^1$
7. 5

Practice Problems D

1. 238 g
3. 0.84 mol; 0.86 g

Practice Problems E

1. 4.2×10^{23} atoms
3. 0.58 mol

Section 4 Review

7. 2.4×10^{24} atoms
9. 1.3 mol

Chapter Review

39. 54
41. 9
43. a. $^{234}_{92}U$
    b. $^{235}_{92}U$
    c. $^{238}_{92}U$
45. $^{12}_6C$ and $^{13}_6C$
47. $1s^2 2s^2 2p^6 3s^2 3p^6 3d^8 4s^2$
49. 6
51. a. 0.500 mol
    b. 4.7 mol
    c. 0.100 mol
53. 0.005 g
55. a. 1.2×10^{24} atoms
    b. 6.02×10^{23} atoms
    c. 2.7×10^{24} atoms
59. 99.8g
61. $1s^2 2s^2 2p^6 3s^2 3p^6 3d^{10} 4s^2$
63. a. 42 g
    b. 1.7×10^{-13} g
    c. 0.449 g
69. 0.307 kg
71. 0.39 mol; 2.3×10^{23} atoms
73. 5
79. $1s^2 2s^2 2p^6 3s^2 3p^3$

Ionic Compounds

Practice Problems A

1. a. $Ca(CN)_2$
   b. $Rb_2S_2O_3$
   c. $Ca(CH_3COO)_2$
   d. $(NH_4)_2SO_4$

Chapter Review

27. a. Li_2SO_4
    b. $Sr(NO_3)_2$
    c. NH_4CH_3COO
    d. $Ti_2(SO_4)_3$

29. barium; Cl^-; chromium(III); fluoride; Mn^{2+}; O^{2-}; **a.** $MnCl_2$; **b.** CrF_3; **c.** BaO

31. a. peroxide
  b. chromate
  c. ammonium
  d. carbonate

33. a. CN^-
  b. SO_4^{2-}
  c. NO_3^-
  d. MnO_4^-

35. a. 3
  b. 1
  c. 4
  d. 7

Covalent Compounds

Practice Problems A

1.

Practice Problems B

1.

Practice Problems C

1.

Practice Problems D

1. trigonal pyramidal
3. trigonal planar

Section 3 Review

7. a. bent
  b. trigonal pyramidal
  c. trigonal pyramidal
  d. tetrahedral

Chapter Review

33. a.

b.

c.

d.

e.

35. a. tetrahedral
  b. bent

36. a.

b.

41. a. silicon tetrachloride, tetrahedral

  b. boron trichloride, trigonal planar

  c. nitrogen tribromide, trigonal pyramidal

The Mole and Chemical Composition

Practice Problems A

1. 1.13×10^{23} ions Na^+
3. 2.544×10^{24} molecules $C_2H_4O_2$

Practice Problems B

1. 0.940 mol Xe
3. 4.5×10^{-7} mol termites
5. a. 1.050×10^{-2} mol O
  b. 5.249×10^{-3} mol C
  c. 3.690 mol O
  d. 8.841×10^{-8} mol K^+
  e. 3.321×10^{-10} mol Cl^-
  f. 6.64×10^{-10} mol N
  g. 6.63×10^2 mol Cl^-

Practice Problems C

1. 223 g Cu
3. 1063 g CH_4

Practice Problems D

1. 2.25×10^{24} atoms Cu
3. 9.33×10^{25} atoms As

Section 1 Review

7. a. 3.61×10^{24} Na^+ ions
  b. 7.23×10^{24} Na^+ ions
  c. 3.08×10^{24} Na^+ ions
9. a. 2.86×10^{-7} g He
  b. 15.22 g CH_4
  c. 200.5 g Ca^{2+}
11. 206.3 g ibuprofen
13. a. 26.7 g Ca
  b. 50. g boron-11
  c. 7.032×10^{-4} g Na^+

Practice Problems E

1. 69.73 amu

Practice Problems F

1. a. 259.80 g/mol
  b. 136.06 g/mol
  c. 342.34 g/mol
  d. 253.80 g/mol
  e. 60.06 g/mol
  f. 262.84 g/mol
3. a. 92.15 g/mol
  b. 0.0815 mol $C_6H_5CH_3$

Section 2 Review

9. 10.80 amu; boron
11. a. SrS, 119.69 g/mol, 1.76×10^{-2} mol SrS

b. PF_3, 87.97 g/mol, 2.40×10^{-2} mol PF_3

c. $Zn(C_2H_3O_2)_2$, 183.49 g/mol, 1.15×10^{-2} mol $Zn(C_2H_3O_2)_2$

d. $Hg(BrO_3)_2$, 456.39 g/mol, 4.62×10^{-3} mol $Hg(BrO_3)_2$

e. $Ca(NO_3)_2$, 164.10 g/mol, 1.29×10^{-2} mol $Ca(NO_3)_2$

Practice Problems G

1. Mn_2O_3

3. Fe_3O_4

Practice Problems H

1. C_6H_6

3. NO_2

Practice Problems I

1. 93.311% Fe, 6.689% C

3. 35.00% N, 5.05% H, 59.96% O

5. a. Both are 39.99% C, 6.73% H, and 53.28% O because, if you combine the hydrogen atoms in acetic acid, the empirical formulas are the same.

b. The percentage composition of the empirical formula is the same as the percentage compositions of the molecular formulas.

Section 3 Review

5. SO_2

7. a. 64.62% Ag, 14.39% C, 1.82% H, and 19.17% O

b. 55.39% Pb, 18.95% Cl, and 25.66% O

c. 27.93% Fe, 24.06% S, and 48.01% O

d. 39.81% Cu, 20.09% S, and 40.10% O

Chapter Review

23. 1.20×10^{24} molecules $C_{12}H_{22}O_{11}$

25. 7.53×10^{21} atoms Hg

27. 5.815×10^{24} ions Ni^{2+}

29. 41.5 mol MgO

31. 12.5 mol C_6H_6

33. 6.82×10^{-2} mol Na^+

35. 0.006192 g P

37. 1.46×10^{-8} g CO_2

39. 81.6 g O_2

41. 160 g NaCl

43. a. 58.44 g NaCl

b. 36.04 g H_2O

c. 260 g $Ca(OH)_2$

d. 163 g $Ba(NO_3)_2$

45. 1.337×10^{24} formula units $ZnCl_2$

47. 2.79×10^{24} atoms Al

49. a. 4.99×10^{-2} mol $(NH_4)_2SO_4$

b. 61 mol $Ca(OH)_2$

c. 7.49×10^{-2} mol H_2SO_4

51. 7.25 mol $NaNO_2$

53. 57.8 mol C_3H_8

55. 79.90 amu

57. 55.84 amu

59. 94.97 g/mol

61. 75.08 g/mol

63. P_2O_3

65. $C_9H_{18}N_6$

67. $Co_2C_8O_8$

69. $C_4H_4N_2$

71. 37.56% NH_4^+

73. 22.57% N, 6.51% H, 19.35% C, 51.56% O

75. a. 0.00152 mol Na^+

b. 0.0072 mol Ca^{2+}

77. 40 mol H

79. 28.09 amu

81. a. 180.18 g/mol

b. 180.18 g/mol

83. $AlPO_4$ is 22.12% Al, while $AlCl_3$ is 20.23% Al, so aluminum phosphate has more aluminum per gram.

Stoichiometry

Practice Problems A

1. a. 0.670 mol O_2

b. 1.34 mol H_2O

Practice Problems B

1. 45.6 g Al

3. 679 g Fe_2O_3

Practice Problems C

1. 315 mL C_5H_8

3. 113 mL C_5H_{12}

Practice Problems D

1. 2.89×10^{24} molecules BrF_5

Section 1 Review

5. a. 1.42 mol CO_2

b. 47.2 mL CO_2

Practice Problems E

1. PCl_3 is excess, H_2O is limiting, theoretical yield is 109 g HCl

3. PCl_3 is excess, H_2O is limiting, theoretical yield is 101 g HCl

Practice Problems F

1. N_2 is limiting, 85.3%

3. Br_2 is limiting, 90.9%

Practice Problems G

1. 1.04×10^3 g NH_3

3. 439 g BrCl

Section 2 Review

7. a. $P_4O_{10} + 6H_2O \longrightarrow 4H_3PO_4$

b. 138.1 g H_3PO_4

c. 91.4%

9. 3.70 g Cu

11. a. $Mg + 2H_2O \longrightarrow Mg(OH)_2 + H_2$

b. 86.8%

c. 55 g $Mg(OH)_2$

Practice Problems H

1. 33 g Na

3. 121 g $NaHCO_3$

Practice Problems I

1. 2.17 cycles, so after 3 full cycles all of the 1.00 mL of isooctane will have reacted

3. $2CH_3OH + 3O_2 \longrightarrow 2CO_2 + 4H_2O$; 2.2×10^2 L air

Practice Problems J

1. 4.01 g CO_2

Section 3 Review

5. 10.7 g Na_2O

Chapter Review

27. a. 6.6 mol H_2

b. 3.36 mol O_2

c. 8.12 mol H_2

29. a. 1.08 mol O_2

b. 2.62 mol Al_2O_3

c. 1.99 mol Al_2O_3

31. a. 49.0 g O_2

b. 748 g $KClO_3$

c. 12.7 g KCl

33. 107 g O_2
35. a. 113 L O_2
   b. 25 L O_2
   c. 7.41 L O_2
37. a. 4.61×10^{23} molecules H_2
   b. 4.31×10^{23} atoms Na
   c. 3.30×10^{20} molecules H_2
39. a. excess, H_2O; limiting, CaC_2
   b. 26 g C_2H_2
   c. 74 g $Ca(OH)_2$
41. 75.6%
43. 72.5%
45. 4.7 g Al
47. 46.6 L CO_2
49. $\dfrac{25 \text{ mol } O_2}{2 \text{ mol } C_8H_{18}}$, or 25:2
51. 1.71×10^3 L O_2
53. 2.16×10^3 g CO_2; 88.0%
55. a. $\dfrac{2 \text{ mol NaN}_3}{2 \text{ mol Na}}$, $\dfrac{2 \text{ mol NaN}_3}{3 \text{ mol N}_2}$,

   $\dfrac{2 \text{ mol Na}}{2 \text{ mol NaN}_3}$,

   $\dfrac{2 \text{ mol Na}}{3 \text{ mol N}_2}$, $\dfrac{3 \text{ mol N}_2}{2 \text{ mol NaN}_3}$,

   $\dfrac{3 \text{ mol N}_2}{2 \text{ mol Na}}$

   b. 4.0 mol N_2
   c. 11.0 g Na
   d. 4.86×10^{23} molecules N_2
57. 1.44 L N_2
59. 1.74×10^3 g CO_2

Causes of Change

Practice Problems A
1. 97 J
3. 220 J

Section 1 Review
9. 29.2 kJ
11. 0.52 mol
13. 301 K
15. 0.864 J/K•mol

Practice Problems B
1. 2.60×10^3 J/mol
3. 3.6×10^2 J/mol

Practice Problems C
1. -2.56×10^3 J/mol
3. -2.8×10^2 J/mol

Section 2 Review
5. 120 J/mol
7. 3690 J/mol
9. 42.8 J/K•mol

Practice Problems D
1. -57.2 kJ

Practice Problems E
1. -1428.6 kJ; exothermic

Section 3 Review
5. -818.6 kJ

Practice Problems F
1. -332.2 J/K
3. -95 J/K

Practice Problems G
1. -41 kJ, spontaneous
3. -1.2 kJ, spontaneous

Practice Problems H
1. -394.4 kJ, spontaneous

Section 4 Review
7. -146.5 J/K
9. 182.1 kJ, no
11. 60 kJ; The result is half the result of problem 10.

Chapter Review
27. 35.2 J/K•mol
29. 7040 J
31. -5030 J
33. -510 kJ
35. 117.6 J
37. -663 J/K
39. -345 kJ   The reaction is spontaneous.
41. -227.9 kJ   The reaction is spontaneous.
45. -57 kJ
51. -184.614 kJ; yes
53. Reaction 1: $\Delta G = 115$ kJ, not spontaneous; Reaction 2: $\Delta G = -101$ kJ, spontaneous; Reaction 3: $\Delta G = -310$ kJ, spontaneous
55. 108.7 J/K

States of Matter and Intermolecular Forces

Practice Problems A
1. $T_{mp} = 159$ K, $T_{bp} = 351$ K
3. $T_{mp} = 195$ K, $T_{bp} = 240$ K

Section 3 Review
7. 266 K
9. 233 K

Practice Problems B
1. a. Phase diagram for sulfur dioxide, SO_2

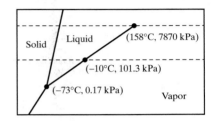

   b. solid
   c. vapor
   d. The sulfur dioxide changes from a liquid to a vapor.
   e. The sulfur dioxide remains a liquid.

Section 4 Review
7. a. Phase diagram for benzene, C_6H_6

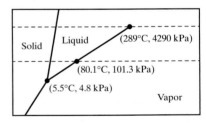

   b. liquid
   c. vapor
   d. The benzene changes from a solid to a liquid.
   e. The benzene changes from a liquid to a vapor.

Chapter Review
53. 290 K
55. 273 K
57. 391 K
59. 334 K

61. Phase diagram for ammonia, NH_3

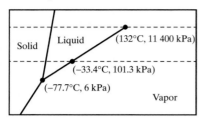

63. Phase diagram for iodine, I_2

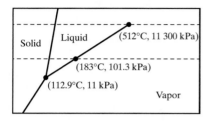

65. 3130 K

Gases

Practice Problems A

1. 7.37×10^6 Pa

3. 0.9869 atm

Section 1 Review

9. 610.5 Pa

Practice Problems B

1. 142 mL

3. 7.9×10^5 L

Practice Problems C

1. 0.67 L

3. −11.0°C

Practice Problems D

1. 1.29 atm

3. 491 K, or 218°C

Section 2 Review

5. 31.0 mL

7. 114 kPa

9. 5.00 L

Practice Problems E

1. 7.97×10^{-2} mol

3. 1500 kPa

Practice Problems F

1. N_2 has a higher speed; 1.069 times faster

3. 48.6 g/mol

Practice Problems G

1. 11.4 L

3. 3.87 g Na

Section 3 Review

7. 0.781 mol

9. 5.3×10^{-3} mol SO_2

11. 15.0 L

Chapter Review

31. 101 325 newtons

33. 290 kPa

35. 113 mL

37. 1100 mL

39. 66.3 mL

41. 93.3 mL

43. 0.540 L

45. 3.1 L

47. 152 kPa

49. 26 kPa

51. 8.4 atm

53. 0.781 mol

55. 266 kPa

57. 4.0×10^3 L

59. $M = 64$ g/mol. It is SO_2.

61. 1.91×10^3 m/s

63. 10.4 L

65. 3.56×10^{-2} g C_8H_{18}

67. 2.28×10^3 L

69. 1.80 g CO_2; 1.80 g/L CO_2; 1.31 g/L O_2

71. 8.1×10^3 g LiOH

73. 5.71 g/L

75. 171 kPa

77. 12.5 g O_2

79. 2.64 L H_2

81. 171°C

83. 44 g He

91. 0.65 g O_2

Solutions

Practice Problems A

1. 15 ppm

3. 4250 ppm

5. 63 ppm

7. 26 ppm

Practice Problems B

1. 0.83 M acetic acid

3. 0.816 M sulfuric acid

5. 0.2501 M $Ba(OH)_2$

7. 11 g NaCl

Practice Problems C

1. 109 g HCl

3. 451 g CdS

Section 2 Review

5. 438 ppm Cd

7. 4.00 g NaOH

9. 0.838 M NaOCl

11. 5.8×10^3 g $Ca_3(PO_4)_2$ and 2.0×10^3 g H_2O

Chapter Review

43. 1.7 ppm

45. 5×10^{-2} g Cl_2

47. 0.7776 M NaOH

49. 2.0 mol $AgNO_3$

51. 1.61 M $CuCl_2$

53. 0.123 M H_3PO_4

55. 5.85 g NaOH

57. 5.02 M HCl

59. 0.259 g LiF

61. 0.74 M $C_6H_{12}O_6$

63. 0.50 M $Ba_3(PO_4)_2$

65. 0.0309 M $AgNO_3$

67. 440 g H_2SO_4

69. 23 g K_2CrO_4

71. 0.111 M $C_6H_{12}O_6$

73. 0.0230 M $PbBr_2$

75. 1.9 M NaCl

77. 1.93 M NaCl

Chemical Equilibrium

Practice Problems A

1. 2.0

Practice Problems B

1. 3.5×10^{-5}

Practice Problems C

1. 6.2×10^{-9}

3. 1.80×10^{-10}

Practice Problems D

1. 2.6×10^{-4}

3. 1.43×10^{-2}

Section 2 Review

  5. 3.53×10^{-3}

  7. $C(s) + CO_2(g) \rightleftharpoons 2CO(g)$;
    2.5×10^{-2}

Chapter Review

  33. a. 0.67

    b. 0.52

    c. 311

  35. 0.048 mol/ L

  37. 0.046 mol/ L

  39. 2.9×10^{-26}

  41. 8.22×10^{-96}

  43. $1.7 \times 10^{-14}, 1.3 \times 10^{-7},$
    $2.7 \times 10^{-7}, 1.6 \times 10^{-4}$

  45. 4.0×10^{-5} M

  57. 2.8×10^{-3} mol/ L

Acids and Bases

Practice Problems A

  1. $[H_3O^+] = 1.38 \times 10^{-11}$ M

  3. $[H_3O^+] = 2.67 \times 10^{-13}$ M

  5. $[OH^-] = 2.4 \times 10^{-4}$ M; $[H_3O^+]$
    $= 4.2 \times 10^{-11}$ M

Practice Problems B

  1. 2.3

  3. 11.3

Practice Problems C

  1. $[H_3O^+] = 5.0 \times 10^{-4}$ M

  3. $[H_3O^+] = 7.9 \times 10^{-9}$ M; $[OH^-] =$
    1.3×10^{-6} M

Section 2 Review

  7. $[OH^-] = 3.16 \times 10^{-12}$ M; pH =
    2.50

  9. $[OH^-] = 0.088$ M; $[H_3O^+] = 1.1$
    $\times 10^{-13}$ M; pH = 12.95

  11. 1.00×10^{-2} mol HBr

Practice Problems D

  1. 6.9×10^{-3} M

  3. 4.674×10^{-3} moles

Section 3 Review

  9. 13.5 mL

  11. 0.18 M

Practice Problems E

  1. $[H_3O^+] = 1.62 \times 10^{-3}$ M

  3. $K_a = 6.4 \times 10^{-5}$

Section 4 Review

  7. $K_a = 3.97 \times 10^{-8}$

  9. $K_a = 6.8 \times 10^{-4}$

  13. 1.0×10^{-4}

  15. $K_{eq} = 9.19 \times 10^3$

Chapter Review

  41. 2.74×10^{-14} M

  43. 5.35×10^{-12} M

  45. 1.41×10^{-7} M

  47. 12.13

  49. 12.91

  51. a. 2.3

    b. 1.3

    c. 0.3

    d. −0.7

  53. 13.48

  55. 0.17

  57. 5.72

  59. $[H_3O^+] = 3.2 \times 10^{-10}$ M;
    $[OH^-] = 3.2 \times 10^{-5}$ M

  61. 3×10^{-5} mol

  63. $[OH^-] = 5.25 \times 10^{-6}$ M

  65. $[H_3O^+] = 7.9 \times 10^{-11}$ M;
    $[OH^-] = 1.3 \times 10^{-4}$ M

  67. $[OH^-] = 2.0 \times 10^{-10}$ M

  69. $[H_3O^+] = 1.0 \times 10^{-3}$ M

  71. $[H_3O^+] = 5.0 \times 10^{-14}$ M

  75. 55.0 mL

  77. 0.5260 M

  79. 0.798 M

  81. 0.1544 M

  83. 1.8×10^{-4}

  85. 1.6×10^{-5}

  87. 2.5×10^{-3}

  89. 1.3

  91. $[H_3O^+] = 0.400$ M; pH = 0.40;
    $[OH^-] = 2.5 \times 10^{-14}$ M

  93. 1.7×10^{-7}

  95. $[H_3O^+] = 1.1 \times 10^{-3}$ M;
    $[C_6H_5COOH] = 0.020$ M;
    $[C_6H_5COO^-] = 1.2 \times 10^{-3}$ M

  99. 1.24 g

  101. 2.42

Reaction Rates

Practice Problems A

  1. $rate = \dfrac{-\Delta[Br^-]}{2\Delta t} = \dfrac{-\Delta[H_2O_2]}{\Delta t}$

  3. 9.0×10^{-5} M/s

Practice Problems B

  1. 1

  3. a factor of 3.2

Section 2 Review

  7. 8.1×10^{-5} M/s

Chapter Review

  25. 1.4×10^{-6} M/s

  27. 1.8×10^{-6} M/s

  29. $rate = k[NO]^2[Br_2]$

  31. double

Oxidation, Reduction, and Electrochemistry

Practice Problems A

  1. a. N = −3; H = +1

    c. H = +1; O = −2

    e. H = 0

    g. K = +1; Cl = +5; O = −2

    i. Ca = +2; O = −2; H = +1

    k. H = +1; P = +5; O = −2

Practice Problems B

  1. half-reactions: $4Fe(s) \longrightarrow$
    $4Fe^{3+}(aq) + 12e^-$ and $3O_2(aq) +$
    $12H_3O^+(aq) + 12e^- \longrightarrow$
    $18H_2O(l)$; overall: $4Fe(s) +$
    $3O_2(aq) + 12H_3O^+(aq) \longrightarrow$
    $4Fe^{3+}(aq) + 18H_2O(l)$

  3. half-reactions: $2Br^-(aq) \longrightarrow$
    $Br_2(aq) + 2e^-$ and $H_2O_2(aq) +$
    $2H_3O^+(aq) + 2e^- \longrightarrow 4H_2O(l)$;
    overall: $2Br^-(aq) + H_2O_2(aq) +$
    $2H_3O^+(aq) \longrightarrow Br_2(aq) +$
    $4H_2O(l)$

Section 1 Review

  7. a. C = −4; H = +1

    b. H = +1; S = +4; O = −2

    c. Na = +1; H = +1; C =
                         +4; O = −2

    d. Na = +1; Bi = +5; O = −2

Practice Problems C

  1. (+0.401 V) − (−0.037 V) =
                      +0.438 V

  3. (+0.7996 V) − (+0.3419 V) =
                      +0.4577 V

Section 3 Review

  5. The cell voltage is (+1.358 V) −
    (+0.222 V) = +1.136 V. The sil-
    ver electrode is the anode.

Chapter Review

39. C = +4; O = −2
41. Ba = +2; Cl = −1
43. S = −2
45. C = −4; H = +1
47. Ca = +2; C = +4; O = −2
49. C = +4; O = −2; Cl = −1
51. $Fe(s) \longrightarrow Fe^{2+}(aq) + 2e^-$
53. $Fe(s) + Cl_2(g) \longrightarrow$
$ Fe^{2+}(aq) + 2Cl^-(aq)$
55. $6H_2O(l) \longrightarrow O_2(g) +$
$ 4H_3O^+(aq) + 4e^-$
57. $O_2(g) + 4H_3O^+(aq) + 4e^- \longrightarrow$
$ 6H_2O(l)$
59. $4H_2O(l) + 2SO_2(aq) +$
$O_2(g) \longrightarrow 2HSO_4^-(aq) + 2H_3O^+(aq)$
61. (+1.30 V) − (+0.45 V) = +0.85 V
63. (0.0000 V) − (+0.771 V) =
$ -0.771$ V
65. (+0.401 V) − (−0.828 V) =
$ +1.229$ V
67. (+1.691 V) − (−0.42 V) =
$ +2.11$ V
75. 0
77. $Fe^{3+}(aq) + e^- \longrightarrow Fe^{2+}(aq)$;
reduction
79. (−0.4030 V) − (+1.358 V) =
$ -1.761$ V

Nuclear Chemistry

Practice Problems A

1. $^{214}_{82}Pb$
3. $^{256}_{101}Md$

Section 2 Review

5. a. $^{233}_{92}U \longrightarrow ^{229}_{90}Th + ^{4}_{2}He$
b. $^{66}_{29}Cu \longrightarrow ^{66}_{30}Zn + ^{0}_{-1}e$
c. $^{9}_{4}Be \longrightarrow ^{4}_{2}He \longrightarrow ^{13}_{6}C$
$ ^{13}_{6}C \longrightarrow ^{12}_{6}C + ^{1}_{0}n$
d. $^{238}_{92}U + ^{1}_{0}n \longrightarrow ^{239}_{92}U$
$ ^{239}_{92}U \longrightarrow ^{239}_{93}Np + ^{0}_{-1}e$
$ ^{239}_{93}Np \longrightarrow ^{239}_{94}Pu + ^{0}_{-1}e$

Practice Problems B

1. 6396 y
3. 12 min

Practice Problem C

1. 0.25 mg
3. 0.32 g

Section 3 Review

5. 1/8
7. 1/16

Chapter Review

33. $^{235}_{92}U \longrightarrow ^{231}_{90}Th + ^{4}_{2}He$
$^{231}_{90}Th \longrightarrow ^{231}_{91}Pa + ^{0}_{-1}e$
$^{231}_{91}Pa \longrightarrow ^{227}_{89}Ac + ^{4}_{2}He$
$^{227}_{89}Ac \longrightarrow ^{227}_{90}Th + ^{0}_{-1}e$
$^{227}_{90}Th \longrightarrow ^{223}_{88}Ra + ^{4}_{2}He$
$^{223}_{88}Ra \longrightarrow ^{219}_{86}Rn + ^{4}_{2}He$
$^{219}_{86}Rn \longrightarrow ^{215}_{84}Po + ^{4}_{2}He$
$^{215}_{84}Po \longrightarrow ^{211}_{82}Pb + ^{4}_{2}He$
$^{211}_{82}Pb \longrightarrow ^{211}_{83}Bi + ^{0}_{-1}e$
$^{211}_{83}Bi \longrightarrow ^{211}_{84}Po + ^{0}_{-1}e$
$^{211}_{84}Po \longrightarrow ^{207}_{82}Pb + ^{4}_{2}He$
35. a. $^{2}_{1}H$
b. $^{12}_{6}C$
c. $^{0}_{-1}e$
37. 12.8 h
39. 5715 y
41. 0.274 g
43. 0.056 g
45. a. 1.55:1; outside
b. 1:1; within
c. 1.15:1; within
d. 1.6:1; outside
47. 8.77×10^{-28} kg
51. for example,
$^{238}_{92}U + ^{1}_{0}n \longrightarrow ^{239}_{92}U$
$^{239}_{92}U \longrightarrow ^{239}_{93}Np + ^{0}_{-1}e$
$^{239}_{93}Np \longrightarrow ^{239}_{94}Pu + ^{0}_{-1}e$
53. 0.200 g
55. 0.04049 amu per atom
57. 1.34×10^{-10} y
59. 3%
61. 4.9×10^{-12} J
63. 10.5 days
65. $^{217}_{89}Ac \longrightarrow ^{213}_{87}Fr + ^{4}_{2}He$

67. The total mass of this nucleus—56 amu—will be greater than 55.847 amu. The mass is not equal to 55.847 amu because that value is an average of the masses of several different isotopes.
69. 29.1 y
71. 64.4 min
73. 22 860 y
75. $^{130}_{52}Te \longrightarrow ^{130}_{53}I + ^{0}_{-1}e$

Carbon and Organic Compounds

Practice Problems A

1. a. 2,2,4-trimethylpentane
b. 1-pentyne
c. 2,3,4-trimethylnonane
d. 2-methyl-3-hexene

Practice Problems B

1. a. ethanol
b. 2-pentanone
c. butanoic acid
d. 3-hexanol

Section 2 Review

7. methylbenzene

Chapter Review

41. a. iodomethane
b. 1, 3-dichloroethane
47. a. 3-hexanone
b. 3-bromo-1,1-dichlorobutane
c. cyclohexanol

GLOSSARY

A

accuracy
a description of how close a measurement is to the true value of the quantity measured (p. 55)

acid-ionization constant, K_a
the equilibrium constant for a reaction in which an acid donates a proton to water (p. 559)

actinide
any of the elements of the actinide series, which have atomic numbers from 89 (actinium, Ac) through 103 (lawrencium, Lr) (p. 130)

activated complex
a molecule in an unstable state intermediate to the reactants and the products in the chemical reaction (p. 590)

activation energy
the minimum amount of energy required to start a chemical reaction (p. 590)

activity series
a series of elements that have similar properties and that are arranged in descending order of chemical activity; examples of activity series include metals and halogens (p. 280)

actual yield
the measured amount of a product of a reaction (p. 316)

addition reaction
a reaction in which an atom or molecule is added to an unsaturated molecule (p. 694)

alkali metal
one of the elements of Group 1 of the periodic table (lithium, sodium, potassium, rubidium, cesium, and francium) (p. 125)

alkaline-earth metal
one of the elements of Group 2 of the periodic table (beryllium, magnesium, calcium, strontium, barium, and radium) (p. 126)

alkane
a hydrocarbon characterized by a straight or branched carbon chain that contains only single bonds (p. 681)

alkene
a hydrocarbon that contains one or more double bonds (p. 681)

alkyne
a hydrocarbon that contains one or more triple bonds (p. 681)

alloy
a solid or liquid mixture of two or more metals (p. 130)

amino acid
any of 20 different organic molecules that contain a carboxyl and an amino group and that combine to form proteins (p. 717)

amphoteric
describes a substance, such as water, that has the properties of an acid and the properties of a base (p. 538)

anion
an ion that has a negative charge (p. 161)

anode
the electrode on whose surface oxidation takes place; anions migrate toward the anode, and electrons leave the system from the anode (p. 614)

aromatic hydrocarbon
a hydrocarbon that contains six-carbon rings and is usually very reactive (p. 682)

atom
the smallest unit of an element that maintains the properties of that element (p. 21)

atomic mass
the mass of an atom expressed in atomic mass units (p. 100)

atomic number
the number of protons in the nucleus of an atom; the atomic number is the same for all atoms of an element (p. 84)

ATP
adenosine triphosphate, an organic molecule that acts as the main energy source for cell processes; composed of a nitrogenous base, a sugar, and three phosphate groups (p. 737)

Aufbau principle
the principle that states that the structure of each successive element is obtained by adding one proton to the nucleus of the atom and one electron to the lowest-energy orbital that is available (p. 97)

average atomic mass
the weighted average of the masses of all naturally occurring isotopes of an element (p. 235)

Avogadro's law
the law that states that equal volumes of gases at the same temperature and pressure contain equal numbers of molecules (p. 432)

Avogadro's number
6.02×10^{23}, the number of atoms or molecules in 1 mol (p. 101, p. 224)

B

beta particle
a charged electron emitted during certain types of radioactive decay, such as beta decay (p. 649)

boiling point
the temperature and pressure at which a liquid and a gas are in equilibrium (p. 382)

bond energy
the energy required to break the bonds in 1 mol of a chemical compound (p. 192)

bond length
the distance between two bonded atoms

at their minimum potential energy; the average distance between the nuclei of two bonded atoms (p. 192)

bond radius
half the distance from center to center of two like atoms that are bonded together (p. 135)

Boyle's law
the law that states that for a fixed amount of gas at a constant temperature, the volume of the gas increases as the pressure of the gas decreases and the volume of the gas decreases as the pressure of the gas increases (p. 424)

Brønsted-Lowry acid
a substance that donates a proton to another substance (p. 535)

Brønsted-Lowry base
a substance that accepts a proton (p. 536)

buffer
a solution made from a weak acid and its conjugate base that neutralizes small amounts of acids or bases added to it (p. 561)

C

calorimeter
a device used to measure the heat absorbed or released in a chemical or physical change (p. 351)

calorimetry
the measurement of heat-related constants, such as specific heat or latent heat (p. 351)

carbohydrate
any organic compound that is made of carbon, hydrogen, and oxygen and that provides nutrients to the cells of living things (p. 712)

catalysis
the acceleration of a chemical reaction by a catalyst (p. 593)

catalyst
a substance that changes the rate of a chemical reaction without being consumed or changed significantly (p. 593)

cathode
the electrode on whose surface reduction takes place (p. 613)

cation
an ion that has a positive charge (p. 161)

chain reaction
a reaction in which a change in a single molecule makes many molecules change until a stable compound forms (p. 654)

Charles's law
the law that states that for a fixed amount of gas at a constant pressure, the volume of the gas increases as the temperature of the gas increases and the volume of the gas decreases as the temperature of the gas decreases (p. 426)

chemical
any substance that has a defined composition (p. 4)

chemical change
a change that occurs when one or more substances change into entirely new substances with different properties (p. 39)

chemical equation
a representation of a chemical reaction that uses symbols to show the relationship between the reactants and the products (p. 263)

chemical equilibrium
a state of balance in which the rate of a forward reaction equals the rate of the reverse reaction and the concentrations of products and reactants remain unchanged (p. 497)

chemical kinetics
the area of chemistry that is the study of reaction rates and reaction mechanisms (p. 576)

chemical property
a property of matter that describes a substance's ability to participate in chemical reactions (p. 18)

chemical reaction
the process by which one or more substances change to produce one or more different substances (p. 5, p. 260)

clone
an organism that is produced by asexual reproduction and that is genetically identical to its parent; to make a genetic duplicate (p. 731)

coefficient
a small whole number that appears as a factor in front of a formula in a chemical equation (p. 268)

colligative property
a property that is determined by the number of particles present in a system but that is independent of the properties of the particles themselves (p. 482)

colloid
a mixture consisting of tiny particles that are intermediate in size between those in solutions and those in suspensions and that are suspended in a liquid, solid, or gas (p. 456)

combustion reaction
the oxidation reaction of an organic compound, in which heat is released (p. 276)

common-ion effect
the phenomenon in which the addition of an ion common to two solutes brings about precipitation or reduces ionization (p. 517)

compound
a substance made up of atoms of two or more different elements joined by chemical bonds (p. 24)

concentration
the amount of a particular substance in a given quantity of a mixture, solution, or ore (p. 460)

condensation
the change of state from a gas to a liquid (p. 382)

condensation reaction
a chemical reaction in which two or more molecules combine to produce water or another simple molecule (p. 699, p. 715)

conductivity
the ability to conduct an electric current (p. 478)

conjugate acid
an acid that forms when a base gains a proton (p. 537)

conjugate base
a base that forms when an acid loses a proton (p. 537)

conversion factor
a ratio that is derived from the equality of two different units and that can be used to convert from one unit to the other (p. 13)

corrosion
the gradual destruction of a metal or alloy as a result of chemical processes such as oxidation or the action of a chemical agent (p. 620)

covalent bond
a bond formed when atoms share one or more pairs of electrons (p. 191)

critical mass
the minimum mass of a fissionable isotope that provides the number of neutrons needed to sustain a chain reaction (p. 654)

critical point
the temperature and pressure at which the gas and liquid states of a substance become identical and form one phase (p. 402)

crystal lattice
the regular pattern in which a crystal is arranged (p. 174)

D

Dalton's law of partial pressures
the law that states that the total pressure of a mixture of gases is equal to the sum of the partial pressures of the component gases (p. 439)

decomposition reaction
a reaction in which a single compound breaks down to form two or more simpler substances (p. 278)

density
the ratio of the mass of a substance to the volume of the substance; often expressed as grams per cubic centimeter for solids and liquids and as grams per liter for gases (p. 16)

denature
to change irreversibly the structure or shape—and thus the solubility and other properties—of a protein by heating, shaking, or treating the protein with acid, alkali, or other species (p. 723)

detergent
a water-soluble cleaner that can emulsify dirt and oil (p. 484)

diffusion
the movement of particles from regions of higher density to regions of lower density (p. 436)

dipole
a molecule or a part of a molecule that contains both positively and negatively charged regions (p. 195)

dipole-dipole forces
interactions between polar molecules (p. 386)

disaccharide
a sugar formed from two monosaccharides (p. 712)

dissociation
the separating of a molecule into simpler molecules, atoms, radicals, or ions (p. 472)

DNA
deoxyribonucleic acid, the material that contains the information that determines inherited characteristics (p. 726)

DNA fingerprint
the pattern of bands that results when an individual's DNA sample is fragmented, replicated, and separated (p. 730)

double bond
a covalent bond in which two atoms share two pairs of electrons (p. 204)

double-displacement reaction
a reaction in which a gas, a solid precipitate, or a molecular compound forms from the apparent exchange of atoms or ions between two compounds (p. 283)

E

effusion
the passage of a gas under pressure through a tiny opening (p. 437)

electrochemical cell
a system that contains two electrodes separated by an electrolyte phase (p. 613)

electrochemistry
the branch of chemistry that is the study of the relationship between electric forces and chemical reactions (p. 612)

electrode
a conductor used to establish electrical contact with a nonmetallic part of a circuit, such as an electrolyte (p. 613)

electrolysis
the process in which an electric current is used to produce a chemical reaction, such as the decomposition of water (p. 627)

electrolyte
a substance that dissolves in water to give a solution that conducts an electric current (p. 478)

electrolytic cell
an electrochemical device in which electrolysis takes place when an electric current is in the device (p. 627)

electromagnetic spectrum
all of the frequencies or wavelengths of electromagnetic radiation (p. 92)

electron
a subatomic particle that has a negative charge (p. 80)

electron shielding
the reduction of the attractive force between a positively charged nucleus and its outermost electrons due to the cancellation of some of the positive charge by the negative charges of the inner electrons (p. 133)

electron configuration
the arrangement of electrons in an atom (p. 96)

electronegativity
a measure of the ability of an atom in a chemical compound to attract electrons (p. 137)

electroplating
the electrolytic process of plating or coating an object with a metal (p. 630)

element
a substance that cannot be separated or broken down into simpler substances by chemical means; all atoms of an element have the same atomic number (p. 22)

elimination reaction
a reaction in which a simple molecule, such as water or ammonia, is removed and a new compound is produced (p. 699)

empirical formula
a chemical formula that shows the composition of a compound in terms of the relative numbers and kinds of atoms in the simplest ratio (p. 242)

emulsion
any mixture of two or more immiscible liquids in which one liquid is dispersed in the other (p. 484)

endothermic
describes a process in which heat is absorbed from the environment (p. 40)

end point
the point in a titration at which a marked color change takes place (p. 554)

energy
the capacity to do work (p. 38)

enthalpy
the sum of the internal energy of a system plus the product of the system's volume multiplied by the pressure that the system exerts on its surroundings (p. 340)

entropy
a measure of the randomness or disorder of a system (p. 358)

enzyme
a type of protein that speeds up metabolic reactions in plant and animals without being permanently changed or destroyed (p. 595, p. 722)

equilibrium
in chemistry, the state in which a chemical process and the reverse chemical process occur at the same rate such that the concentrations of reactants and products do not change (p. 400)

equilibrium constant
a number that relates the concentrations of starting materials and products of a reversible chemical reaction to one another at a given temperature (p. 503)

evaporation
the change of a substance from a liquid to a gas (p. 39, p. 382)

excess reactant
the substance that is not used up completely in a reaction (p. 313)

excited state
a state in which an atom has more energy than it does at its ground state (p. 94)

exothermic
describes a process in which a system releases heat into the environment (p. 40)

freezing
the change of state in which a liquid becomes a solid as heat is removed (p. 383)

freezing point
the temperature at which a solid and liquid are in equilibrium at 1 atm pressure; the temperature at which a liquid substance freezes (p. 383)

functional group
the portion of a molecule that is active in a chemical reaction and that determines the properties of many organic compounds (p. 683)

gamma ray
the high-energy photon emitted by a nucleus during fission and radioactive decay (p. 649)

Gay-Lussac's law
the law that states that the pressure of a gas at a constant volume is directly proportional to the absolute temperature (p. 430)

Gay-Lussac's law of combining volumes of gases
the law that states that the volumes of gases involved in a chemical change can be represented by the ratio of small whole numbers (p. 439)

gene
a segment of DNA that is located in a chromosome and that codes for a specific hereditary trait (p. 728)

Gibbs energy
the energy in a system that is available for work (p. 362)

Graham's law of diffusion
the law that states that the rate of diffusion of a gas is inversely proportional to the square root of the gas's density (p. 437)

ground state
the lowest energy state of a quantized system (p. 94)

group
a vertical column of elements in the periodic table; elements in a group share chemical properties (p. 119)

half-life
the time required for half of a sample of a radioactive substance to disintegrate by radioactive decay or by natural processes (p. 658)

half-reaction
the part of a reaction that involves only oxidation or reduction (p. 608)

halogen
one of the elements of Group 17 (fluorine, chlorine, bromine, iodine, and astatine); halogens combine with most metals to form salts (p. 126)

heat
the energy transferred between objects that are at different temperatures; energy is always transferred from higher-temperature objects to lower-temperature objects until thermal equilibrium is reached (p. 41, p. 338)

Henry's law
the law that states that at constant temperature, the solubility of a gas in a liquid is directly proportional to the partial pressure of the gas on the surface of the liquid (p. 477)

Hess's law
the law that states that the amount of heat released or absorbed in a chemical reaction does not depend on the number of steps in the reaction (p. 353)

heterogeneous
composed of dissimilar components (p. 26)

homogeneous
describes something that has a uniform structure or composition throughout (p. 26)

Hund's rule
the rule that states that for an atom in the ground state, the number of unpaired electrons is the maximum possible and these unpaired electrons have the same spin (p. 98)

hydration
the strong affinity of water molecules for particles of dissolved or suspended substances that causes electrolytic dissociation (p. 472)

hydrocarbon
an organic compound composed only of carbon and hydrogen (p. 680)

hydrogen bond
the intermolecular force occurring when a hydrogen atom that is bonded to a highly electronegative atom of one molecule is attracted to two unshared electrons of another molecule (p. 387)

GLOSSARY

hydrolysis
a chemical reaction between water and another substance to form two or more new substances; a reaction between water and a salt to create an acid or a base (p. 716)

hydronium ion
an ion consisting of a proton combined with a molecule of water; H_3O^+ (p. 480)

hypothesis
a theory or explanation that is based on observations and that can be tested (p. 50)

ideal gas
an imaginary gas whose particles are infinitely small and do not interact with each other (p. 433)

ideal gas law
the law that states the mathematical relationship of pressure (P), volume (V), temperature (T), the gas constant (R), and the number of moles of a gas (n); $PV = nRT$ (p. 434)

immiscible
describes two or more liquids that do not mix with each other (p. 470)

indicator
a compound that can reversibly change color depending on the pH of the solution or other chemical change (p. 546)

intermediate
a substance that forms in a middle stage of a chemical reaction and is considered a stepping stone between the parent substance and the final product (p. 589)

intermolecular forces
the forces of attraction between molecules (p. 386)

ion
an atom, radical, or molecule that has gained or lost one or more electrons and has a negative or positive charge (p. 161)

isomer
one of two or more compounds that have the same chemical composition but different structures (p. 686)

isotope
an atom that has the same number of protons (atomic number) as other atoms of the same element do but that has a different number of neutrons (atomic mass) (p. 88)

kinetic energy
the energy of an object that is due to the object's motion (p. 42)

kinetic-molecular theory
a theory that explains that the behavior of physical systems depends on the combined actions of the molecules constituting the system (p. 421)

lanthanide
a member of the rare-earth series of elements, whose atomic numbers range from 58 (cerium) to 71 (lutetium) (p. 130)

lattice energy
the energy associated with constructing a crystal lattice relative to the energy of all constituent atoms separated by infinite distances (p. 168)

law
a summary of many experimental results and observations; a law tells how things work (p. 52)

law of conservation of energy
the law that states that energy cannot be created or destroyed but can be changed from one form to another (p. 40)

law of conservation of mass
the law that states that mass cannot be created or destroyed in ordinary chemical and physical changes (p. 52, p. 76)

law of definite proportions
the law that states that a chemical compound always contains the same elements in exactly the same proportions by weight or mass (p. 75)

law of multiple proportions
the law that states that when two elements combine to form two or more compounds, the mass of one element that combines with a given mass of the other is in the ratio of small whole numbers (p. 77)

Le Châtelier's principle
the principle that states that a system in equilibrium will oppose a change in a way that helps eliminate the change (p. 512)

Lewis structure
a structural formula in which electrons are represented by dots; dot pairs or dashes between two atomic symbols represent pairs in covalent bonds (p. 199)

limiting reactant
the substance that controls the quantity of product that can form in a chemical reaction (p. 313)

London dispersion force
the intermolecular attraction resulting from the uneven distribution of electrons and the creation of temporary dipoles (p. 390)

main-group element
an element in the *s*-block or *p*-block of the periodic table (p. 124)

mass
a measure of the amount of matter in an object; a fundamental property of an object that is not affected by the forces that act on the object, such as the gravitational force (p. 10)

mass defect
the difference between the mass of an atom and the sum of the masses of the atom's protons, neutrons, and electrons (p. 644)

mass number
the sum of the numbers of protons and neutrons in the nucleus of an atom (p. 85)

matter
anything that has mass and takes up space (p. 10)

melting
the change of state in which a solid becomes a liquid by adding heat or changing pressure (p. 383)

melting point
the temperature and pressure at which a solid becomes a liquid (p. 383)

miscible
describes two or more liquids that can dissolve into each other in various proportions (p. 470)

mixture
a combination of two or more substances that are not chemically combined (p. 25)

molarity
a concentration unit of a solution expressed as moles of solute dissolved per liter of solution (p. 462)

molar mass
the mass in grams of 1 mol of a substance (p. 101, p. 230)

mole
the SI base unit used to measure the amount of a substance whose number of particles is the same as the number of atoms of carbon in exactly 12 g of carbon-12 (p. 101, p. 224)

molecular formula
a chemical formula that shows the number and kinds of atoms in a molecule, but not the arrangement of the atoms (p. 244)

molecular orbital
the region of high probability that is occupied by an individual electron as it travels with a wavelike motion in the three-dimensional space around one of two or more associated nuclei (p. 191)

molecule
the smallest unit of a substance that keeps all of the physical and chemical properties of that substance; it can consist of one atom or two or more atoms bonded together (p. 23)

monosaccharide
a simple sugar that is the basic subunit of a carbohydrate (p. 712)

neutral
describes an aqueous solution that contains equal concentrations of hydronium ions and hydroxide ions (p. 542)

neutralization reaction
the reaction of the ions that characterize acids (hydronium ions) and the ions that characterize bases (hydroxide ions) to form water molecules and a salt (p. 548)

neutron
a subatomic particle that has no charge and that is found in the nucleus of an atom (p. 82)

newton
the SI unit for force; the force that will increase the speed of a 1 kg mass by 1 m/s each second that the force is applied (abbreviation, N) (p. 419)

noble gas
an unreactive element of Group 18 of the periodic table; the nobles gases are helium, neon, argon, krypton, xenon, or radon (p. 127)

nonelectrolyte
a liquid or solid substance or mixture that does not allow an electric current (p. 479)

nonpolar covalent bond
a covalent bond in which the bonding electrons are equally attracted to both bonded atoms (p. 194)

nuclear fission
the splitting of the nucleus of a large atom into two or more fragments; releases additional neutrons and energy (p. 654)

nuclear fusion
the combination of the nuclei of small atoms to form a larger nucleus; releases energy (p. 656)

nuclear reaction
a reaction that affects the nucleus of an atom (p. 143)

nucleic acid
an organic compound, either RNA or DNA, whose molecules are made up of one or two chains of nucleotides and carry genetic information (p. 725)

nucleon
a proton or neutron (p. 642)

nucleus
in physical science, an atom's central region, which is made up of protons and neutrons (p. 81)

nuclide
an atom that is identified by the number of protons and neutrons in its nucleus (p. 642)

octet rule
a concept of chemical bonding theory that is based on the assumption that atoms tend to have either empty valence shells or full valence shells of eight electrons (p. 159)

orbital
a region in an atom where there is a high probability of finding electrons (p. 91)

order
in chemistry, a classification of chemical reactions that depends on the number of molecules that appear to enter into the reaction (p. 586)

oxidation
a reaction that removes one or more elec-

trons from a substance such that the substance's valence or oxidation state increases (p. 604)

oxidation number
the number of electrons that must be added to or removed from an atom in a combined state to convert the atom into the elemental form (p. 606)

oxidation-reduction reaction
any chemical change in which one species is oxidized (loses electrons) and another species is reduced (gains electrons); also called redox reaction (p. 605)

oxidizing agent
the substance that gains electrons in an oxidation-reduction reaction and that is reduced (p. 611)

partial pressure
the pressure of each gas in a mixture (p. 439)

pascal
the SI unit of pressure; equal to the force of 1 N exerted over an area of 1 m^2 (abbreviation, Pa) (p. 419)

Pauli exclusion principle
the principle that states that two particles of a certain class cannot be in exactly the same energy state (p. 96)

peptide bond
the chemical bond that forms between the carboxyl group of one amino acid and the amino group of another amino acid (p. 718)

percentage composition
the percentage by mass of each element in a compound (p. 241)

period
in chemistry, a horizontal row of elements in the periodic table (p. 122)

periodic law
the law that states that the repeating chemical and physical properties of elements change periodically with the atomic numbers of the elements (p. 119)

pH
a value that is used to express the acidity or alkalinity (basicity) of a system; each whole number on the scale indicates a tenfold change in acidity; a pH of 7 is neutral, a pH of less than 7 is acidic, and a pH of greater than 7 is basic (p. 542)

phase
in chemistry, a part of matter that is uniform (p. 399)

phase diagram
a graph of the relationship between the physical state of a substance and the temperature and pressure of the substance (p. 402)

photosynthesis
the process by which plants, algae, and some bacteria use sunlight, carbon dioxide, and water to produce carbohydrates and oxygen (p. 734)

physical change
a change of matter from one form to another without a change in chemical properties (p. 39)

physical property
a characteristic of a substance that does not involve a chemical change, such as density, color, or hardness (p. 15)

polar covalent bond
a covalent bond in which a pair of electrons shared by two atoms is held more closely by one atom (p. 194)

polyatomic ion
an ion made of two or more atoms (p. 178)

polymer
a large molecule that is formed by more than five monomers, or small units (p. 696)

polypeptide
a long chain of several amino acids (p. 718)

polysaccharide
one of the carbohydrates made up of long chains of simple sugars; polysaccharides include starch, cellulose, and glycogen (p. 712)

precision
the exactness of a measurement (p. 55)

pressure
the amount of force exerted per unit area of a surface (p. 419)

product
a substance that forms in a chemical reaction (p. 8)

protein
an organic compound that is made of one or more chains of amino acids and that is a principal component of all cells (p. 717)

proton
a subatomic particle that has a positive charge and that is found in the nucleus of an atom; the number of protons of the nucleus is the atomic number, which determines the identity of an element (p. 82)

pure substance
a sample of matter, either a single element or a single compound, that has definite chemical and physical properties (p. 22)

quantity
something that has magnitude, size, or amount (p. 12)

quantum number
a number that specifies the properties of electrons (p. 95)

radioactivity
the process by which an unstable nucleus emits one or more particles or energy in the form of electromagnetic radiation (p. 648)

rate-determining step
in a multistep chemical reaction, the step that has the lowest velocity, which determines the rate of the overall reaction (p. 589)

rate law
the expression that shows how the rate of formation of product depends on the concentration of all species other than the solvent that take part in a reaction (p. 586)

reactant
a substance or molecule that participates in a chemical reaction (p. 8)

reaction mechanism
the way in which a chemical reaction takes place; expressed in a series of chemical equations (p. 586)

reaction rate
the rate at which a chemical reaction takes place; measured by the rate of formation of the product or the rate of disappearance of the reactants (p. 578)

recombinant DNA
DNA molecules that are artificially created by combining DNA from different sources (p. 732)

reducing agent
a substance that has the potential to reduce another substance (p. 611)

reduction
a chemical change in which electrons are gained, either by the removal of oxygen, the addition of hydrogen, or the addition of electrons (p. 605)

resonance structure
in chemistry, any one of two or more possible configurations of the same compound that have identical geometry but different arrangements of electrons (p. 206)

respiration
in chemistry, the process by which cells produce energy from carbohydrates; atmospheric oxygen combines with glucose to form water and carbon dioxide (p. 736)

reversible reaction
a chemical reaction in which the products re-form the original reactants (p. 497)

S

salt
an ionic compound that forms when a metal atom or a positive radical replaces the hydrogen of an acid (p. 167)

saturated hydrocarbon
an organic compound formed only by carbon and hydrogen linked by single bonds (p. 688)

saturated solution
a solution that cannot dissolve any more solute under the given conditions (p. 474)

scientific method
a series of steps followed to solve problems, including collecting data, formulating a hypothesis, testing the hypothesis, and stating conclusions (p. 46)

self-ionization constant of water, K_w
the product of the concentrations of the two ions that are in equilibrium with water; $[H_3O^+][OH^-]$ (p. 540)

significant figure
a prescribed decimal place that determines the amount of rounding off to be done based on the precision of the measurement (p. 56)

single bond
a covalent bond in which two atoms share one pair of electrons (p. 200)

soap
a substance that is used as a cleaner and that dissolves in water (p. 484)

solubility
the ability of one substance to dissolve in another at a given temperature and pressure; expressed in terms of the amount of solute that will dissolve in a given amount of solvent to produce a saturated solution (p. 468)

solubility equilibrium
the physical state in which the opposing processes of dissolution and crystallization of a solute occur at equal rates (p. 476)

solubility product constant
the equilibrium constant for a solid that is in equilibrium with the solid's dissolved ions (p. 507)

solute
in a solution, the substance that dissolves in the solvent (p. 455)

solution
a homogeneous mixture of two or more substances uniformly dispersed throughout a single phase (p. 454)

solvent
in a solution, the substance in which the solute dissolves (p. 455)

specific heat
the quantity of heat required to raise a unit mass of homogeneous material 1 K or 1°C in a specified way given constant pressure and volume (p. 45)

spectator ions
ions that are present in a solution in which a reaction is taking place but that do not participate in the reaction (p. 286)

standard electrode potential
the potential developed by a metal or other material immersed in an electrolyte solution relative to the potential of the hydrogen electrode, which is set at zero (p. 622)

standard solution
a solution of known concentration (p. 550)

standard temperature and pressure
for a gas, the temperature of 0°C and the pressure 1.00 atm (p. 420)

states of matter
the physical forms of matter, which are solid, liquid, gas, and plasma (p. 6)

stoichiometry
the proportional relationships between two or more substances during a chemical reaction (p. 303)

strong acid
an acid that ionizes completely in a solvent (p. 532)

strong base
a base that ionizes completely in a solvent (p. 534)

strong force
the interaction that binds nucleons together in a nucleus (p. 643)

sublimation
the process in which a solid changes directly into a gas (The term is sometimes also used for the reverse process.) (p. 383)

substitution reaction
a reaction in which one or more atoms replace another atom or group of atoms in a molecule (p. 696)

superheavy element
an element whose atomic number is greater than 106 (p. 147)

supersaturated solution
a solution that holds more dissolved solute than is required to reach equilibrium at a given temperature (p. 475)

surface tension
the force that acts on the surface of a liquid and that tends to minimize the area of the surface (p. 380)

surfactant
a compound that concentrates at the boundary surface between two immiscible phases, solid-liquid, liquid-liquid, or liquid-gas (p. 484)

suspension
a mixture in which particles of a material are more or less evenly dispersed throughout a liquid or gas (p. 454)

synthesis reaction
a reaction in which two or more substances combine to form a new compound (p. 277)

T

temperature
a measure of how hot (or cold) something is; specifically, a measure of the average kinetic energy of the particles in an object (p. 43, p. 339)

thermodynamics
the branch of science concerned with the energy changes that accompany chemical and physical changes (p. 348)

titrant
a solution of known concentration that is used to titrate a solution of unknown concentration (p. 550)

titration
a method to determine the concentration of a substance in solution by adding a

solution of known volume and concentration until the reaction is completed, which is usually indicated by a change in color (p. 550)

transition range
the pH range over which a variation in a chemical indicator can be observed (p. 554)

transition metal
one of the metals that can use the inner shell before using the outer shell to bond (p. 129)

triple bond
a covalent bond in which two atoms share three pairs of electrons (p. 205)

triple point
the temperature and pressure conditions at which the solid, liquid, and gaseous phases of a substance coexist at equilibrium (p. 402)

unit
a quantity adopted as a standard of measurement (p. 12)

unit cell
the smallest portion of a crystal lattice that shows the three-dimensional pattern of the entire lattice (p. 175)

unsaturated hydrocarbon
a hydrocarbon that has available valence bonds, usually from double or triple bonds with carbon (p. 688)

unsaturated solution
a solution that contains less solute than a saturated solution does and that is able to dissolve additional solute (p. 474)

unshared pair
a nonbonding pair of electrons in the valence shell of an atom; also called lone pair (p. 200)

valence electron
an electron that is found in the outermost shell of an atom and that determines the atom's chemical properties (p. 119, p. 199)

vapor pressure
the partial pressure exerted by a vapor that is in equilibrium with its liquid state at a given temperature (p. 400)

voltage
the potential difference or electromotive force, measured in volts; it represents the amount of work that moving an electric charge between two points would take (p. 613)

volume
a measure of the size of a body or region in three-dimensional space (p. 10)

VSEPR theory
a theory that predicts some molecular shapes based on the idea that pairs of valence electrons surrounding an atom repel each other (p. 209)

weak acid
an acid that releases few hydrogen ions in aqueous solution (p. 532)

weak base
a base that releases few hydroxide ions in aqueous solution (p. 534)

weight
a measure of the gravitational force exerted on an object; its value can change with the location of the object in the universe (p. 10)

GLOSARIO

A

accuracy/exactitud
término que describe qué tanto se aproxima una medida al valor verdadero de la cantidad medida (p. 55)

acid-ionization constant/constante de ionización ácida, el término K_a
la constante de equilibrio para una reacción en la cual un ácido dona un protón al agua (p. 559)

actinide/actínido
cualquiera de los elementos de la serie de los actínidos, los cuales tienen números atómicos del 89 (actinio, Ac) al 103 (laurencio, Lr) (p. 130)

activated complex/complejo activado
una molécula que está en un estado inestable, intermedio entre los reactivos y los productos en una reacción química (p. 590)

activation energy/energía de activación
la cantidad mínima de energía que se requiere para iniciar una reacción química (p. 590)

activity series/serie de actividad
una serie de elementos que tienen propiedades similares y que están ordenados en orden descendiente respecto a su actividad química; algunos ejemplos de series de actividad incluyen a los metales y los halógenos (p. 280)

actual yield/rendimiento real
la cantidad medida del producto de una reacción (p. 316)

addition reaction/reacción de adición
una reacción en la que se añade un átomo o una molécula a una molécula insaturada (p. 696)

alkali metal/metal alcalino
uno de los elementos del Grupo 1 de la tabla periódica (litio, sodio, potasio, rubidio, cesio y francio) (p. 125)

alkaline-earth metal/metal alcalinotérreo
uno de los elementos del Grupo 2 de la tabla periódica (berilio, magnesio, calcio, estroncio, bario y radio) (p. 126)

alkane/alcano
un hidrocarburo formado por una cadena simple o ramificada de carbonos que únicamente contiene enlaces sencillos (p. 681)

alkene/alqueno
un hidrocarburo que contiene uno o más enlaces dobles (p. 681)

alkyne/alquino
un hidrocarburo que contiene uno o más enlaces triples (p. 681)

alloy/aleación
una mezcla sólida o líquida de dos o más metales (p. 130)

amino acid/aminoácido
cualquiera de las 20 distintas moléculas orgánicas que contienen un grupo carboxilo y un grupo amino y que se combinan para formar proteínas (p. 717)

amphoteric/anfotérico
término que describe una substancia, como el agua, que tiene propiedades tanto de ácido como de base (p. 538)

anion/anión
un ion que tiene carga negativa (p. 161)

anode/ánodo
el electrodo en cuya superficie ocurre la oxidación; los aniones migran hacia el ánodo y los electrones se alejan del sistema por el ánodo (p. 614)

aromatic hydrocarbon/hidrocarburo aromático
un hidrocarburo que tiene anillos de seis carbonos y que normalmente es muy reactivo (p. 682)

atom/átomo
la unidad más pequeña de un elemento que conserva las propiedades de ese elemento (p. 21)

atomic mass/masa atómica
la masa de un átomo, expresada en unidades de masa atómica (p. 100)

atomic number/número atómico
el número de protones en el núcleo de un átomo; el número atómico es el mismo para todos los átomos de un elemento (p. 84)

ATP/ATP
adenosín trifosfato; molécula orgánica que funciona como la fuente principal de energía para los procesos celulares; formada por una base nitrogenada, un azúcar y tres grupos fosfato (p. 737)

Aufbau principle/principio de Aufbau
el principio que establece que la estructura de cada elemento sucesivo se obtiene añadiendo un protón al núcleo del átomo y un electrón a un orbital de menor energía que se encuentre disponible (p. 245)

average atomic mass/masa atómica promedio
el promedio ponderado de las masas de todos los isótopos de un elemento que se encuentran en la naturaleza (p. 235)

Avogadro's law/ley de Avogadro
la ley que establece que volúmenes iguales de gases a la misma temperatura y presión contienen el mismo número de moléculas (p. 432)

Avogadro's number/número de Avogadro
6.02×10^{23}, el número de átomos o moléculas que hay en 1 mol (p. 101, p. 224)

B

beta particle/partícula beta
un electrón con carga, emitido durante ciertos tipos de desintegración radiactiva, como por ejemplo, durante la desintegración beta (p. 649)

boiling point/punto de ebullición
la temperatura y presión a la que un líquido y un gas están en equilibrio (p. 382)

bond energy/energía de enlace
la energía que se requiere para romper los enlaces de 1 mol de un compuesto químico (p. 192)

bond length/longitud de enlace
la distancia entre dos átomos que están enlazados en el punto en que su energía potencial es mínima; la distancia promedio entre los núcleos de dos átomos enlazados (p. 192)

bond radius/radio de enlace
la distancia mitad del centro al centro de dos como los átomos que se pegan juntos (p. 135)

Boyle's law/ley de Boyle
la ley que establece que para una cantidad fija de gas a una temperatura constante, el volumen del gas aumenta a medida que su presión disminuye y el volumen del gas disminuye a medida que su presión aumenta (p. 424)

Brønsted-Lowry acid/ácido de Brønsted-Lowry
una substancia que le dona un protón a otra substancia (p. 535)

Brønsted-Lowry base/base de Brønsted-Lowry
una substancia que acepta un protón (p. 536)

buffer/búfer
una solución que contiene un ácido débil y su base conjugada y que neutraliza pequeñas cantidades de ácidos y bases que se le añaden (p. 561)

C

calorimeter/calorímetro
un aparato que se usa para medir la cantidad de calor absorbido o liberado en un cambio físico o químico (p. 351)

calorimetry/calorimetría
la medida de las constantes relacionadas con el calor, tales como el calor específico o el calor latente (p. 351)

carbohydrate/carbohidrato
cualquier compuesto orgánico que está hecho de carbono, hidrógeno y oxígeno y que proporciona nutrientes a las células de los seres vivos (p. 712)

catalysis/catálisis
la aceleración de una reacción química por un catalizador (p. 593)

catalyst/catalizador
una substancia que cambia la tasa de una reacción química sin ser consumida ni cambiar significativamente (p. 593)

cathode/cátodo
el electrodo en cuya superficie ocurre la reducción (p. 613)

cation/catión
un ion que tiene carga positiva (p. 161)

chain reaction/reacción en cadena
una reacción en la que un cambio en una sola molécula hace que muchas moléculas cambien, hasta que se forma un compuesto estable (p. 654)

Charles's law/ley de Charles
la ley que establece que para una cantidad fija de gas a una presión constante, el volumen del gas aumenta a medida que su temperatura aumenta y el volumen del gas disminuye a medida que su temperatura disminuye (p. 426)

chemical/substancia química
cualquier substancia que tiene una composición definida (p. 4)

chemical change/cambio químico
un cambio que ocurre cuando una o más substancias se transforman en substancias totalmente nuevas con propiedades diferentes (p. 39)

chemical equation/ecuación química
una representación de una reacción química que usa símbolos para mostrar la relación entre los reactivos y los productos (p. 263)

chemical equilibrium/equilibrio químico
un estado de equilibrio en el que la tasa de la reacción directa es igual a la tasa de la reacción inversa y las concentraciones de los productos y reactivos no sufren cambios (p. 497)

chemical kinetics/cinética química
el área de la química que se ocupa del estudio de las tasas de reacción y de los mecanismos de reacción (p. 576)

chemical property/propiedad química
una propiedad de la materia que describe la capacidad de una substancia de participar en reacciones químicas (p. 18)

chemical reaction/reacción química
el proceso por medio del cual una o más substancias cambian para producir una o más substancias distintas (p. 5, p. 260)

clone/clon
un organismo producido por reproducción asexual que es genéticamente idéntico a su progenitor; clonar significa hacer un duplicado genético (p. 731)

coefficient/coeficiente
un número entero pequeño que aparece como un factor frente a una fórmula en una ecuación química (p. 268)

colligative property/propiedad coligativa
una propiedad que se determina por el número de partículas presentes en un sistema, pero que es independiente de las propiedades de las partículas mismas (p. 482)

colloid/coloide
una mezcla formada por partículas diminutas que son de tamaño intermedio entre las partículas de las soluciones y las de las suspensiones y que se encuentran suspendidas en un líquido, sólido o gas (p. 456)

combustion reaction/reacción de combustión
la reacción de oxidación de un compuesto orgánico, durante la cual se libera calor (p. 276)

common-ion effect/efecto del ion común
el fenómeno en el que la adición de un ion común a dos solutos produce precipitación o reduce la ionización (p. 517)

compound/compuesto
una substancia formada por átomos de dos o más elementos diferentes unidos por enlaces químicos (p. 24)

concentration/concentración
la cantidad de una cierta substancia en una cantidad determinada de mezcla, solución o mena (p. 460)

condensation/condensación
el cambio de estado de gas a líquido (p. 382)

condensation reaction/reacción de condensación
una reacción química en la que dos o más moléculas se combinan para producir agua u otra molécula simple (p. 699, p. 715)

conductivity/conductividad
la capacidad de conducir una corriente eléctrica (p. 478)

conjugate acid/ácido conjugado
un ácido que se forma cuando una base gana un protón (p. 537)

conjugate base/base conjugada
una base que se forma cuando un ácido pierde un protón (p. 537)

conversion factor/factor de conversió
una razón que se deriva de la igualdad entre dos unidades diferentes y que se puede usar para convertir una unidad en otra (p. 13)

corrosion/corrosión
la destrucción gradual de un metal o de una aleación como resultado de procesos químicos tales como la oxidación o la acción de un agente químico (p. 620)

covalent bond/enlace covalente
un enlace formado cuando los átomos comparten uno más pares de electrones (p. 191)

critical mass/masa crítica
la cantidad mínima de masa de un isótopo fisionable que proporciona el número de neutrones que se requieren para sostener una reacción en cadena (p. 654)

critical point/punto crítico
la temperatura y presión a la que los estados líquido y gaseoso de una substancia se vuelven idénticos para formar una fase (p. 402)

crystal lattice/red cristalina
el patrón regular en el que un cristal está ordenado (p. 174)

D

Dalton's Law of Partial Pressures/ley de Dalton de las presiones parciales
la ley que establece que la presión total de una mezcla de gases es igual a la suma de las presiones parciales de los gases componentes (p. 439)

decomposition reaction/reacción de descomposición
una reacción en la que un solo compuesto se descompone para formar dos o más substancias más simples (p. 278)

density/densidad
la relación entre la masa de una substancia y su volumen; comúnmente se expresa en gramos por centímetro cúbico para los sólidos y líquidos, y como gramos por litro para los gases (p. 16)

denature/desnaturalice
para hacer una proteína perder sus estructuras terciarias y cuaternarios (p. 723)

detergent/detergente
un limpiador no jabonoso, soluble en agua, que emulsiona la suciedad y el aceite (p. 484)

diffusion/difusión
el movimiento de partículas de regiones de mayor densidad a regiones de menor densidad (p. 436)

dipole/dipolo
una molécula o parte de una molécula que contiene regiones cargadas tanto positiva como negativamente (p. 195)

dipole-dipole forces/fuerzas dipolo-dipolo
interacciones entre moléculas polares (p. 386)

disaccharide/disacárido
un azúcar formada a partir de dos monosacáridos (p. 712)

dissociation/disociación
la separación de una molécula en moléculas más simples, átomos, radicales o iones (p. 472)

DNA/ADN
ácido desoxirribonucleico, el material que contiene la información que determina las características que se heredan (p. 726)

DNA fingerprint/huella de ADN
el patrón de bandas que se obtiene cuando los fragmentos de ADN de un individuo se separan (p. 730)

double bond/doble enlace
un enlace covalente en el que dos átomos comparten dos pares de electrones (p. 204)

double-displacement reaction/reacción de doble desplazamiento
una reacción en la que un gas, un precipitado sólido o un compuesto molecular se forma a partir del intercambio aparente de iones entre dos compuestos (p. 283)

E

effusion/efusión
el paso de un gas bajo presión a través de una abertura diminuta (p. 437)

electrochemical cell/celda electroquímica
un sistema que contiene dos electrodos separados por una fase electrolítica (p. 615)

electrochemistry/electroquímica
la rama de la química que se ocupa del estudio de la relación entre las fuerzas eléctricas y las reacciones químicas (p. 614)

electrode/electrodo
un conductor que se usa para establecer contacto eléctrico con una parte no metálica de un circuito, tal como un electrolito (p. 615)

electrolysis/electrólisis
el proceso por medio del cual se utiliza una corriente eléctrica para producir una reacción química, como por ejemplo, la descomposición del agua (p. 629)

electrolyte/electrolito
una substancia que se disuelve en agua y crea una solución que conduce la corriente eléctrica (p. 478)

electrolytic cell/celda electrolítica
un aparato electroquímico en el que se da lugar la electrólisis cuando hay una corriente eléctrica en el aparato (p. 629)

electromagnetic spectrum/espectro electromagnético
todas las frecuencias o longitudes de onda de la radiación electromagnética (p. 92)

electron/electrón
una partícula subatómica que tiene carga negativa (p. 80)

electron shielding/blindaje de los electrónes
la reducción de la fuerza atractiva entre un núcleo positivamente cargado y sus electrones exteriores debido a la cancelación de algo de la carga positiva por las cargas negativas de los electrones internos (p. 133)

electron configuration/configuración electrónica
el ordenamiento de los electrones en un átomo (p. 96)

electronegativity/electronegatividad
una medida de la capacidad de un átomo de un compuesto químico de atraer electrones (p. 137)

electroplating/electrochapado
el proceso de recubrir o aplicar una capa de un metal a un objeto (p. 630)

element/elemento
una substancia que no se puede separar o descomponer en substancias más simples por medio de métodos químicos; todos los átomos de un elemento tienen el mismo número atómico (p. 22)

elimination reaction/reacción de eliminación
una reacción en la que se remueve una molécula simple, como el agua o el amoníaco, y se produce un nuevo compuesto (p. 699)

empirical formula/fórmula empírica
la composición de un compuesto en función del número relativo y el tipo de átomos que hay en la proporción más simple (p. 242)

emulsion/emulsión
cualquier mezcla de dos o más líquidos inmiscibles en la que un líquido se encuentra disperso en el otro (p. 484)

endothermic/endotérmico
término que describe un proceso en que se absorbe calor del ambiente (p. 40)

end point/punto de equivalencia
el punto en una titulación en el que ocurre un cambio marcado de color (p. 554)

energy/energía
la capacidad de realizar un trabajo (p. 38)

enthalpy/entalpía
la suma de la energía interna de un sistema más el producto del volumen del sistema multiplicado por la presión que el sistema ejerce en su ambiente (p. 340)

entropy/entropía
una medida del grado de aleatoriedad o desorden de un sistema (p. 358)

enzyme/enzima
un tipo de proteína que acelera las reacciones metabólicas en las plantas y animales, sin ser modificada permanentemente ni ser destruida (p. 595, p. 722)

equilibrium/equilibrio
en química, el estado en el que un proceso químico y el proceso químico inverso ocurren a la misma tasa, de modo que las concentraciones de los reactivos y los productos no cambian (p. 400)

equilibrium constant/constante de equilibrio
un número que relaciona las concentraciones de los materiales de inicio y los productos de una reacción química reversible a una temperatura dada (p. 503)

evaporation/evaporación
el cambio de una substancia de líquido a gas (p. 39, p. 382)

excess reactant/reactivo en exceso
la substancia que no se usa por completo en una reacción (p. 313)

excited state/estado de excitación
un estado en el que un átomo tiene más energía que en su estado fundamental (p. 94)

exothermic/exotérmico
término que describe un proceso en el que un sistema libera calor al ambiente (p. 40)

F

freezing/congelamiento
el cambio de estado de líquido a sólido al eliminar calor del líquido (p. 383)

freezing point/punto de congelación
la temperatura a la que un sólido y un líquido están en equilibrio a 1 atm de presión; la temperatura a la que una substancia en estado líquido se congela (p. 383)

functional group/grupo funcional
la porción de una molécula que está involucrada en una reacción química y que determina las propiedades de muchos compuestos orgánicos (p. 683)

G

gamma ray/rayo gamma
el fotón de alta energía emitido por un núcleo durante la fisión y la desintegración radiactiva (p. 649)

Gay-Lussac's law/ley de Gay-Lussac
la ley que establece que la presión por un gas a volumen constante es directamente proporcional a la temperatura absoluta (p. 430)

Gay-Lussac's law of combining volumes of gases/ley de combinación de los volúmenes de los gases de Gay-Lussac
la ley que establece que los volúmenes de los gases que participan en un cambio químico se pueden representar por razones de números pequeños enteros (p. 439)

gene/gene
un segmento de ADN ubicado en un cromosoma, que codifica un carácter hereditario específico (p. 728)

Gibbs energy/energia Gibbs
la energía de un sistema disponible para realizar un trabajo (p. 362)

Graham's law of diffusion/ley de efusión de Graham
la ley que establece que la tasa de difusión de un gas es inversamente proporcional a la raíz cuadrada de su densidad (p. 437)

ground state/estado fundamental
el estado de energía más bajo de un sistema cuantificado (p. 94)

group/grupo
una columna vertical de elementos de la tabla periódica; los elementos de un grupo comparten propiedades químicas (p. 119)

H

half-life/vida media
el tiempo que tarda la mitad de una muestra de una substancia radiactiva en desintegrarse por desintegración radiactiva o por procesos naturales (p. 658)

half-reaction/media reacción
la parte de una reacción que sólo involucra oxidación o reducción (p. 608)

halogen/halógeno
uno de los elementos del Grupo 17 (flúor, cloro, bromo, yodo y ástato); se combinan con la mayoría de los metales para formar sales (p. 126)

heat/calor
la transferencia de energía entre objetos que están a temperaturas diferentes; la energía siempre se transfiere de los objetos que están a la temperatura más alta a los objetos que están a una temperatura más baja, hasta que se llega a un equilibrio térmico (p. 41, p. 338)

Henry's law/ley de Henry
la ley que establece que a una temperatura constante, la solubilidad de un gas en un líquido es directamente proporcional a la presión parcial de un gas en la superficie del líquido (p. 477)

Hess's law/ley de Hess
la ley que establece que la cantidad de calor liberada o absorbida en una reacción química no depende del número de pasos que tenga la reacción (p. 353)

heterogeneous/heterogéneo
compuesto de componentes que no son iguales (p. 26)

homogeneous/homogéneo
término que describe a algo que tiene una estructura o composición global uniforme (p. 26)

Hund's rule/regla de Hund
la regla que establece que para un átomo en estado fundamental, el número de electrones no apareados es el máximo posible y que estos electrones no apareados tienen el mismo espín (p. 98)

hydration/hidratación
la fuerte afinidad de las moléculas del agua a partículas de substancias disueltas o suspendidas que causan disociación electrolítica (p. 472)

hydrocarbon/hidrocarburo
un compuesto orgánico compuesto únicamente por carbono e hidrogeno (p. 680)

hydrogen bond/enlace de hidrógeno
la fuerza intermolecular producida por un átomo de hidrógeno que está unido a un átomo muy electronegativo de una molécula y que experimenta atracción a dos electrones no compartidos de otra molécula (p. 387)

hydrolysis/hidrólisis
una reacción química entre el agua y otras substancias para formar dos o más substancias nuevas; una reacción entre el agua y una sal para crear un ácido o una base (p. 716)

hydronium ion/ion hidronio
un ion formado por un protón combinado con una molécula de agua; H_3O^+ (p. 480)

hypothesis/hipótesis
una teoría o explicación basada en observaciones y que se puede probar (p. 50)

ideal gas/gas ideal
un gas imaginario con partículas que son infinitamente pequeñas y que no interactúan unas con otras (p. 433)

ideal gas law/ley de los gases ideales
la ley que establece la relación matemática entre la presión (P), volumen (V), temperatura (T), la constante de los gases (R) y el número de moles de un gas (n); $PV = nRT$ (p. 434)

immiscible/inmiscible
término que describe dos o más líquidos que no se mezclan uno con otro (p. 470)

indicator/indicador
un compuesto que puede cambiar de color de forma reversible dependiendo del pH de la solución o de otro cambio químico (p. 546)

intermediate/intermediario
una substancia que se forma en un estado medio de una reacción química y que se considera un paso importante entre la substancia original y el producto final (p. 589)

intermolecular forces/fuerzas intermoleculares
las fuerzas de atracción entre moléculas (p. 386)

ion/ion
un átomo, radical o molécula que ha ganado o perdido uno o más electrones y que tiene una carga negativa o positiva (p. 161)

isomer/isómero
uno de dos o más compuestos que tienen la misma composición química pero diferentes estructuras (p. 686)

isotope/isótopo
un átomo que tiene el mismo número de protones (número atómico) que otros átomos del mismo elemento, pero que tiene un número diferente de neutrones (masa atómica) (p. 88)

kinetic energy/energía cinética
la energía de un objeto debido al movimiento del objeto (p. 42)

kinetic-molecular theory/teoría cinética molecular
una teoría que explica que el comportamiento de los sistemas físicos depende de las acciones combinadas de las moléculas que constituyen el sistema (p. 421)

lanthanide/lantánido
la serie de elementos de tierras raras, cuyos números atómicos van del 58 (cerio) al 71 (lutecio) (p. 130)

lattice energy/energía de la red cristalina
la energía asociada con la construcción de una red cristalina en relación con la energía de todos los átomos que la constituyen cuando éstos están separados por distancias infinitas (p. 168)

law/ley
un resumen de muchos resultados y observaciones experimentales; una ley dice cómo funcionan las cosas (p. 52)

law of conservation of energy/ley de la conservación de la energía
la ley que establece que la energía ni se crea ni se destruye, sólo se transforma de una forma a otra (p. 40)

law of conservation of mass/ley de la conservación de la masa
la ley que establece que la masa no se crea ni se destruye por cambios químicos o físicos comunes (p. 52, p. 76)

law of definite proportions/ley de las proporciones definidas
la ley que establece que un compuesto químico siempre contiene los mismos elementos en exactamente las mismas proporciones de peso o masa (p. 75)

law of multiple proportions/ley de las proporciones múltiples
la ley que establece que cuando dos elementos se combinan para formar dos o más compuestos, la masa de un elemento que se combina con una cantidad determinada de masa de otro elemento es en la proporción de número enteros pequeños (p. 77)

Le Chatelier's principle/principio de Le Chatelier
el principio que establece que un sistema en equilibrio se opondrá a un cambio de modo tal que ayude a eliminar el cambio (p. 512)

Lewis structure/estructura de Lewis
una fórmula estructural en la que los electrones se representan por medio de puntos; pares de puntos o líneas entre dos símbolos atómicos representan pares en los enlaces covalentes (p. 199)

limiting reactant/reactivo limitante
la substancia que controla la cantidad de producto que se puede formar en una reacción química (p. 313)

London dispersion force/fuerza de dispersión de London
la atracción intermolecular que se produce como resultado de la distribución desigual de los electrones y la creación de dipolos temporales (p. 390)

main-group element/elemento de grupo principal
un elemento que está en el bloque s- o p- de la tabla periódica (p. 124)

mass/masa
una medida de la cantidad de materia que tiene un objeto; una propiedad fundamental de un objeto que no está afectada por las fuerzas que actúan sobre el objeto, como por ejemplo, la fuerza gravitacional (p. 10)

GLOSARIO

mass defect/defecto de masa
la diferencia entre la masa de un átomo y la suma de la masa de los protones, neutrones y electrones del átomo (p. 644)

mass number/número de masa
la suma de los números de protones y neutrones que hay en el núcleo de un átomo (p. 85)

matter/materia
cualquier cosa que tiene masa y ocupa un lugar en el espacio (p. 10)

melting/fusión
el cambio de estado en el que un sólido se convierte en líquido al añadir calor o al cambiar la presión (p. 383)

melting point/punto de fusión
la temperatura y presión a la cual un sólido se convierte en líquido (p. 383)

miscible/miscible
término que describe a dos o más líquidos que son capaces de disolverse uno en el otro en varias proporciones (p. 470)

mixture/mezcla
una combinación de dos o más substancias que no están combinadas químicamente (p. 25)

molarity/molaridad
una unidad de concentración de una solución, expresada en moles de soluto disuelto por litro de solución (p. 462)

molar mass/masa molar
la masa en gramos de 1 mol de una substancia (p. 101, p. 230)

mole/mol
la unidad fundamental del sistema internacional de unidades que se usa para medir la cantidad de una substancia cuyo número de partículas es el mismo que el número de átomos de carbono en exactamente 12 g de carbono-12 (p. 101, p. 224)

molecular formula/fórmula molecular
una fórmula química que muestra el número y los tipos de átomos que hay en una molécula, pero que no muestra cómo están distribuidos (p. 244)

molecular orbital/orbital molecular
una región entre dos núcleos donde hay grande probabilidad de tener un electron que mueve como una onda (p. 191)

molecule/molécula
la unidad más pequeña de una substancia que conserva todas las propiedades físicas y químicas de esa substancia; puede estar formada por un átomo o por dos o más átomos enlazados uno con el otro (p. 23)

monosaccharide/monosacárido
un azúcar simple que es una subunidad fundamental de los carbohidratos (p. 712)

neutral/neutro
describe una solución acuosa que contenga concentraciones iguales de los iones del hydronium y de los iones del hidróxido (p. 542)

neutralization reaction/reacción de neutralización
la reacción de los iones que caracterizan a los ácidos (iones hidronio) y de los iones que caracterizan a las bases (iones hidróxido) para formar moléculas de agua y una sal (p. 548)

neutron/neutrón
una partícula subatómica que no tiene carga y que se encuentra en el núcleo de un átomo (p. 82)

newton/newton
la unidad de fuerza del sistema internacional de unidades; la fuerza que aumentará la rapidez de un kg de masa en 1 m/s cada segundo que se aplique la fuerza (abreviatura: N) (p. 419)

noble gas/gas noble
un elemento no reactivo del Grupo 18 de la tabla periódica; los gases nobles son: helio, neón, argón, criptón, xenón o radón (p. 127)

nonelectrolyte/no-electrolito
una substancia o una mezcla líquida o sólida que no permite el flujo de una corriente eléctrica (p. 479)

nonpolar covalent bond/enlace covalente no polar
un enlace covalente en el que los electrones de enlace tienen la misma atracción por los dos átomos enlazados (p. 194)

nuclear fission/fisión nuclear
la partición del núcleo de un átomo grande en dos o más fragmentos; libera neutrones y energía adicionales (p. 654)

nuclear fusion/fusión nuclear
combinación de los núcleos de átomos pequeños para formar un núcleo más grande; libera energía (p. 656)

nuclear reaction/reacción nuclear
una reacción que afecta el núcleo de un átomo (p. 143)

nucleic acid/ácido nucleico
un compuesto orgánico, ya sea ARN o ADN, cuyas moléculas están formadas por una o más cadenas de nucleótidos y que contiene información genética (p. 725)

nucleon/nucleón
un protón o neutrón (p. 642)

nucleus/núcleo
en ciencias físicas, la región central de un átomo, la cual está constituida por protones y neutrones (p. 81)

nuclide/nucleido
un átomo que se identifica por el número de protones y neutrones que hay en su núcleo (p. 642)

octet rule/regla del octeto
un concepto de la teoría de formación de enlaces químicos que se basa en la suposición de que los átomos tienden a tener orbitales de valencia vacíos u orbitales de valencia llenos de ocho electrones (p. 159)

orbital/orbital
una región en un átomo donde hay una alta probabilidad de encontrar electrones (p. 91)

order/orden
en la química, una clasificación de reacciones químicas que depende del número de las moléculas que aparecen entrar en la reacción (p. 586)

oxidation/oxidación
una reacción en la que uno o más electrones son removidos de una substancia, aumentado su valencia o estado de oxidación (p. 604)

oxidation number/número de oxidación
el número de electrones que se deben añadir o remover de un átomo en estado de combinación para convertirlo a su forma elemental (p. 606)

oxidation-reduction reaction/reacción de óxido-reducción
cualquier cambio químico en el que una especie se oxida (pierde electrones) y otra especie se reduce (gana electrones); también se denomina reacción redox (p. 605)

oxidizing agent/agente oxidante
la substancia que gana electrones en una reacción de óxido-reducción y que es reducida (p. 611)

partial pressure/presión parcial
la presión de cada gas en una mezcla (p. 439)

pascal/pascal
la unidad de presión del sistema internacional de unidades; es igual a la fuerza de 1 N ejercida sobre un área de 1 m^2 (abreviatura: Pa) (p. 419)

Pauli exclusion principle/principio de exclusión de Pauli
el principio que establece que dos partículas de una cierta clase no pueden estar en exactamente el mismo estado de energía (p. 96)

peptide bond/enlace peptídico
el enlace químico que se forma entre el grupo carboxilo de un aminoácido y el grupo amino de otro aminoácido (p. 718)

percentage composition/composición porcentual
el porcentaje en masa de cada elemento que forma un compuesto (p. 241)

period/período
en química, una hilera horizontal de elementos en la tabla periódica (p. 122)

periodic law/ley periódica
la ley que establece que las propiedades químicas y físicas repetitivas de un elemento cambian periódicamente en función del número atómico de los elementos (p. 119)

pH/pH
un valor que expresa la acidez o la alcalinidad (basicidad) de un sistema;

cada número entero de la escala indica un cambio de 10 veces en la acidez; un pH de 7 es neutro, un pH de menos de 7 es ácido y un pH de más de 7 es básico (p. 542)

phase/fase
en química, una parte de la materia que es uniforme (p. 399)

phase diagram/diagrama de fases
una gráfica de la relación entre el estado físico de una substancia y la temperatura y presión de la substancia (p. 402)

photosynthesis/fotosíntesis
el proceso por medio del cual las plantas, algas y algunas bacterias utilizan la luz solar, dióxido de carbono y agua para producir carbohidratos y oxígeno (p. 734)

physical change/cambio físico
un cambio de materia de una forma a otra sin que ocurra un cambio en sus propiedades químicas (p. 39)

physical property/propiedad física
una característica de una substancia que no implica un cambio químico, tal como la densidad, el color o la dureza (p. 15)

polar covalent bond/enlace covalente polar
un enlace en el que un par de electrones que está siendo compartido por dos átomos se mantiene más unido a uno de los átomos (p. 194)

polyatomic ion/ion poliatómico
un ion formado por dos o más átomos (p. 178)

polymer/polímero
una molécula grande que está formada por más de cinco monómeros, o unidades pequeñas (p. 698)

polypeptide/polipéptido
una cadena larga de varios aminoácidos (p. 718)

polysaccharide/polisacárido
uno de los carbohidratos formados por cadenas largas de azúcares simples; algunos ejemplos de polisacáridos incluyen al almidón, celulosa y glucógeno (p. 712)

precision/precisión
la exactitud de una medición (p. 55)

pressure/presión
la cantidad de fuerza ejercida en una superficie por unidad de área (p. 419)

product/producto
una substancia que se forma en una reacción química (p. 8)

protein/proteína
un compuesto orgánico que está hecho de una o más cadenas de aminoácidos y que es el principal componente de todas las células (p. 717)

proton/protón
una partícula subatómica que tiene una carga positiva y que se encuentra en el núcleo de un átomo; el número de protones que hay en el núcleo es el número atómico, y éste determina la identidad del elemento (p. 82)

pure substance/substancia pura
una muestra de materia, ya sea un solo elemento o un solo compuesto, que tiene propiedades químicas y físicas definidas (p. 22)

Q

quantity/cantidad
algo que tiene magnitud o tamaño (p. 12)

quantum number/número cuántico
un número que especifica las propiedades de los electrones (p. 95)

R

radioactivity/radiactividad
el proceso por medio del cual un núcleo inestable emite una o más partículas o energía en forma de radiación electromagnética (p. 648)

rate-determining step/paso determinante de la tasa
en una reacción química de varios pasos, el paso que tiene la velocidad más baja, el cual determina la tasa global de la reacción (p. 589)

rate law/ley de la tasa
la expresión que muestra la manera en que la tasa de formación de producto depende de la concentración de todas las especies que participan en una reacción, excepto del solvente (p. 586)

reactant/reactivo
una substancia o molécula que participa en una reacción química (p. 8)

reaction mechanism/mecanismo de reacción
la manera en la que ocurre una reacción química; se expresa por medio de una serie de ecuaciones químicas (p. 586)

reaction rate/tasa de reacción
la tasa a la que ocurre una reacción química; se mide por la tasa de formación del producto o por la tasa de desaparición de los reactivos (p. 578)

recombinant DNA/ADN recombinante
moléculas de ADN que son creadas artificialmente al combinar ADN de diferentes fuentes (p. 732)

reducing agent/agente reductor
una substancia que tiene el potencial de reducir otra substancia (p. 611)

reduction/reducción
un cambio químico en el que se ganan electrones, ya sea por la remoción de oxígeno, la adición de hidrógeno o la adición de electrones (p. 605)

resonance structure/estructura de resonancia
en la química, una de dos o más configuraciones posibles del mismo compuesto que tienen geometría idéntica pero diversos arreglos de electrones (p. 206)

respiration/respiración
en química, el proceso por medio del cual las células producen energía a partir de los carbohidratos; el oxígeno atmosférico se combina con la glucosa para formar agua y dióxido de carbono (p. 736)

reversible reaction/reacción inversa
una reacción química en la que los productos vuelven a formar los reactivos originales (p. 497)

S

salt/sal
un compuesto iónico que se forma cuando el átomo de un metal o un radical positivo reemplaza el hidrógeno de un ácido (p. 167)

saturated hydrocarbon/hidrocarburo saturado
un compuesto orgánico formado sólo por carbono e hidrógeno unidos por enlaces simples (p. 688)

saturated solution/solución saturada
una solución que no puede disolver más soluto bajo las condiciones dadas (p. 474)

scientific method/método científico
una serie de pasos que se siguen para solucionar problemas, los cuales incluyen recopilar información, formular una hipótesis, comprobar la hipótesis y sacar conclusiones (p. 46)

self-ionization constant of water, K_w constante de la auto-ionización de agua
el producto de las concentraciones de los dos iones que están en equilibrio con agua (p. 540)

significant figure/cifra significativa
un lugar decimal prescrito que determina la cantidad de redondeo que se hará con base en la precisión de la medición (p. 56)

single bond/enlace simple
un enlace covalente en el que dos átomos comparten un par de electrones (p. 200)

soap/jabón
una sustancia que se usa como limpiador y que se disuelve en el agua (p. 484)

solubility/solubilidad
la capacidad de una sustancia de disolverse en otra a una temperatura y presión dadas; se expresa en términos de la cantidad de soluto que se disolverá en una cantidad determinada de solvente para producir una solución saturada (p. 468)

solubility equilibrium/equilibrio de solubilidad
el estado físico en el que los procesos opuestos de disolución y cristalización de un soluto ocurren a la misma tasa (p. 476)

solubility product constant/constante del producto de solubilidad
la constante de equilibrio de un sólido que está en equilibrio con los iones disueltos del sólido (p. 507)

solute/soluto
en una solución, la sustancia que se disuelve en el solvente (p. 455)

solution/solución
una mezcla homogénea de dos o más sustancias dispersas de manera uniforme en una sola fase (p. 454)

GLOSARIO

solvent/solvente
en una solución, la sustancia en la que se disuelve el soluto (p. 455)

specific heat/calor específico
la cantidad de calor que se requiere para aumentar una unidad de masa de un material homogéneo 1 K ó 1°C de una manera especificada, dados un volumen y una presión constantes (p. 45)

spectator ions/iones espectadores
iones que están presentes en una solución en la que está ocurriendo una reacción, pero que no participan en la reacción (p. 286)

standard electrode potential/potencial estándar del electrodo
el potencial que desarrolla un metal u otro material que se encuentre sumergido en una solución de electrolitos respecto al potencial del electrodo de hidrógeno, al cual se le da un valor de cero (p. 622)

standard solution/solución estándar
una solución de concentración conocida (p. 550)

standard temperature and pressure/temperatura y presión estándar
para un gas, la temperatura de 0°C y la presión de 1.00 atm (p. 420)

states of matter/estados de la materia
las formas físicas de la materia, que son sólida, líquida, gaseosa y plasma (p. 6)

stoichiometry/estequiometría
las relaciones proporcionales entre dos o más substancias durante una reacción química (p. 303)

strong acid/ácido fuerte
un ácido que se ioniza completamente en un solvente (p. 532)

strong base/base fuerte
un base que se ioniza completamente en un solvente (p. 534)

strong force/fuerza fuerte
la interacción que mantiene unidos a los nucleones en un núcleo (p. 643)

sublimation/sublimación
el proceso por medio del cual un sólido se transforma directamente en un gas o un gas se transforma directamente en un sólido (p. 383)

substitution reaction/reacción de sustitución
una reacción en la cual uno o más átomos reemplazan otro átomo o grupo de átomos en una molécula (p. 696)

superheavy element/elemento superheavy
un elemento que número atómico es mayor de 106 (p. 147)

supersaturated solution/solución sobresaturada
una solución que contiene más soluto disuelto que el que se requiere para llegar al equilibro a una temperatura dada (p. 475)

surface tension/tensión superficial
la fuerza de atracción entre las moléculas que están debajo de la superficie de un líquido, la cual crea una fuerza hacia adentro que tiende a evitar que el líquido fluya (p. 380)

surfactant/surfactant
un compuesto que se concentra en la superficie del límite entre dos fases, solid-liquid, líquido-líquidos, o líquido-gases inmiscibles (p. 484)

suspension/suspensión
una mezcla en la que las partículas de un material se encuentran dispersas de manera más o menos uniforme a través de un líquido o de un gas (p. 454)

synthesis reaction/reacción de síntesis
una reacción en la que dos o más sustancias se combinan para formar un compuesto nuevo (p. 277)

temperature/temperatura
una medida de qué tan caliente (o frío) está algo; específicamente, una medida de la energía cinética promedio de las partículas de un objeto (p. 43, p. 339)

thermodynamics/termodinámica
la ramificación de la ciencia referida a los cambios de la energía que acompañan cambios del producto químico y de la comprobación (p. 348)

titrant/titrant
una solución de la concentración sabida que se utiliza para titular una solución de la concentración desconocida (p. 550)

titration/titulación
un método para determinar la concentración de una sustancia en una solución al añadir una solución de volumen y concentración conocidos hasta que se completa la reacción, lo cual normalmente es indicado por un cambio de color (p. 550)

transition range/intervalo de transición
el rango de pH en el cual se puede observar una variación en un indicador químico (p. 554)

transition metal/metal de transición
uno de los metales que tienen la capacidad de usar su orbital interno antes de usar su orbital externo para formar un enlace (p. 129)

triple bond/enlace triple
un enlace covalente en el que dos átomos comparten tres pares de electrones (p. 205)

triple point/punto triple
las condiciones de temperatura y presión en las que las fases sólida, líquida y gaseosa de una sustancia coexisten en equilibrio (p. 402)

unit/unidad
una cantidad adoptada como un estándar de medición (p. 12)

unit cell/celda unitaria
la porción más pequeña de una red cristalina, la cual muestra el patrón tridimensional de la red completa (p. 175)

unsaturated hydrocarbon/hidrocarburo no saturado
un hidrocarburo que tiene enlaces de valencia disponibles, normalmente de enlaces dobles o triples con carbono (p. 688)

unsaturated solution/solución no saturada
una solución que contiene menos soluto que una solución saturada, y que tiene la capacidad de disolver más soluto (p. 474)

unshared pair/par compartido
un par de electrones que no están enlazados en el orbital de valencia de un átomo; también se llama par solitario (p. 200)

valence electron/electrón de valencia
un electrón que se encuentra en el orbital más externo de un átomo y que determina las propiedades químicas del átomo (p. 119, p. 199)

vapor pressure/presión de vapor
la presión parcial ejercida por el vapor, la cual está en equilibrio con su estado líquido a una temperatura dada (p. 400)

voltage/voltaje
la diferencia de potencial o fuerza electromotriz, medida en voltios; representa la cantidad de trabajo que tomaría mover una carga eléctrica entre dos puntos (p. 613)

volume/volumen
una medida del tamaño de un cuerpo o región en un espacio de tres dimensiones (p. 10)

VSEPR theory/teoría VSEPR
una teoría que predice algunas formas moleculares con base en la idea de que los pares de electrones de valencia que rodean un átomo se repelen unos a otros (p. 209)

weak acid/ácido débil
un ácido que libera pocos iones de hidrógeno en una solución acuosa (p. 532)

weak base/base débil
un base que libera pocos iones de hidroxido en una solución acuosa (p. 534)

weight/peso
una medida de la fuerza gravitacional ejercida sobre un objeto; su valor puede cambiar en función de la ubicación del objeto en el universo (p. 10)

chlorofluorocarbons (CFCs), 52

chromatography, 458, 458*f*

cloning, 731

coefficients, 268, 303

cohesion, 379

colligative properties, 481–483, 481*f*, 482*f*, 483*f*

colloids, 456, 456*f*

color, 15

color change, 8, 9*f*. *See also* **indicators**

combining volumes, Gay-Lussac's law of, 439, 439*f*

combustion, 276, 276*f*
calorimetry of, 351–352, 351*f*
fire extinguishers and, 290
of gasoline, 323–324, 323*f*, 325
rate of, 582

common ion effect, 517, 517*f*, 561–562

completion reactions, 302, 496, 496*f*

complex ions, 500, 500*f*, 500*t*
equilibria with, 500–501, 513, 513*f*, 515, 515*f*

composition, percentage, 241–243, 241*f*, 246–248, 246*f*

compounds, 24–25, 24*f*. *See also* **ionic compounds; molecular compounds**
atomic theory and, 74–78, 75*f*, 76*f*, 77*f*, 77*t*
versus mixtures, 27

compressibility, of gases, 417, 417*f*, 421

computer chips, 214

concentration, 460–461, 460*t*. *See also* **solutions**
defined, 460
from equilibrium constant, 506
Le Châtelier's principle and, 512, 513, 513*f*, 517, 561–562
molarity, 460*t*, 462–467, 462*f*
parts per million, 460, 460*t*, 461
reaction rate and, 582, 582*f*, 586–588, 588*f*
from solubility product constant, 510
in stoichiometry problems, 308
from titration data, 555–556

condensation, 381, 381*f*, 382, 382*f*
energy and, 396, 396*f*

condensation reactions, 699–700, 699*f*, 700*f*, 715, 718

conductivity, 478. *See also* **electrical conductivity**

conjugate acid, 537, 537*t*, 558, 559, 559*t*

conjugate base, 537, 537*t*, 558, 559, 559*t*
in buffer, 561–563

conservation of energy, 40–41, 41*f*

conservation of mass, 52, **76**, 76*f*, 77
balanced equation and, 263, 267, 267*t*, 270

controlled experiment, 51

conversion factors, 13–14, 226–227, 230
significant figures and, 58

coordination compounds, 500–501, 500*f*, 500*t*

corrosion, 620, 620*f*
prevention of, 131, 621, 621*f*, 631

Coulomb's law, 83

counting units, 101, 225, 225*t*

count value, 58

covalent bonds, 190–198. *See also* **Lewis structures**
defined, 191
electronegativity and, 194–197, 194*f*, 195*f*, 196*t*
energy and, 192–193, 192*f*, 193*t*
polar, 194–196, 195*f*, 197, 212–213, 212*f*
shared electrons in, 190–191, 191*f*, 194, 198, 200

covalent compounds. *See also* **molecular compounds**
as electrolytes, 479
formulas of, 236
naming of, 206–207, 207*t*
properties of, 197*t*, 198

critical mass, 654

critical point, 402, 402*f*

CRT (cathode-ray tube), 79–80, 79*f*, 80*f*

crystal lattice, 173, **174–175**, 174*f*
energy of, 168–169, 169*f*

crystallization, 476, 476*f*

cyclotron, 145

D

Dalton (unit of atomic mass), 100

Dalton, John, 77–78, 439, 643

Dalton's law of partial pressures, 439

Daniell cell, 616, 616*f*, 622*f*

dating objects, 658–662, 658*f*, 661*f*

decanting, 457*f*

decay series, 651, 651*f*

decomposition reaction, 278–279, 278*f*

definite proportions, law of, 75, 75*f*, 77

denaturing, 723

density, 16–17, 16*f*, 17*f*, 17*t*
calculation of, 17, 57
defined, 16
of gases, 417, 417*f*
QuickLAB experiment on, 18
in stoichiometry problems, 308, 309

deoxyribonucleic acid (DNA), 726–728, 726*f*, 727*f*, 728*f*
HIV treatment and, 738
hydrogen bonding in, 388, 388*f*, 726*f*, 727, 728
QuickLAB isolation of, 727
recombinant, 731

deposition, 381, 381*f*, 384

derived units, 15, 15*f*

detergent, 484, 484*f*, 485–486, 486*f*

diamond, 679, 679*f*

diatomic elements, 23, 23*f*
molar masses of, 237

diffraction
of electrons, 91
of X rays, 175

diffusion
of gases, 436–438, 436*f*
in solution, 358*f*, 359

dipole, 195

dipole-dipole forces, 386–389, 387*t*, 388*f*, 389*f*. *See also* **hydrogen bonds**
dispersion forces and, 391, 392

disaccharides, 712, 713, 714, 714*f*, 715. *See also* **sucrose**

disorder, 358, 359, 363

displacement reactions, 280–282, 280*f*, 281*t*
double, 283, 283*f*
net ionic equation for, 287

dissociation, 472

distillation, 459

disulfide bridge, 719, 719*f*

DNA (deoxyribonucleic acid), 726–728, 726*f*, 727*f*, 728*f*
HIV treatment and, 738
hydrogen bonding in, 388, 388*f*, 726*f*, 727, 728
QuickLAB isolation of, 727
recombinant, 731

DNA fingerprinting, 730–731

***d* orbitals,** 95, 96, 96*f*, 97, 97*f*
periodic table and, 122, 129

double bonds
defined, 204
in Lewis structures, 204, 205, 206, 206*f*
in organic compounds, 681, 682, 688, 688*f*, 689–691

double-displacement reaction, 283, 283*f*

Downs cell, 628, 628*f*

dry cells, 617, 617*f*

ductility, 129

dyes, synthetic, 49

dynamic equilibrium, 499, 499*f*

E

effective nuclear charge, 135, 136, 137, 138, 139

effusion, 437–438, 437*f*

Einstein, Albert, 62, 91, 143

electrical conductivity
of acid solutions, 480, 531, 531*f*
bond type and, 197, 197*t*
of metals, 129, 197
of salts, 172, 172*f*
of solutions, 478–481, 480*f*
of superconductors, 148

electric current, 479, 481, 531, 613

INDEX

STAFF CREDITS

**Executive Editor,
High School Physical Sciences**
Mark Grayson

Managing Editor
Debbie Starr

Editorial Development Team
John A. Benner
Molly Frohlich
Michael Mazza
Micah Newman
Kharissia Pettus

Copyeditors
Dawn Marie Spinozza, Copyediting
  Manager
Anne-Marie De Witt
Jane A. Kirschman
Kira J. Watkins

Editorial Support Staff
Jeanne Graham
Mary Helbling
Shannon Oehler
Stephanie S. Sanchez
Tanu'e White

Editorial Interns
Kristina Bigelow
Erica Garza
Sarah Ray
Kenneth G. Raymond
Kyle Stock
Audra Teinert

Online Products
Bob Tucek, Executive Editor
Wesley M. Bain
Catherine Gallagher
Douglas P. Rutley

Production
Eddie Dawson, Production Manager
Suzanne Brooks, Production Coordinator

Design
Book Design
Joe Melomo, Design Director
Lori Male, Senior Designer
Ed Diaz, Design Associate
Alicia Sullivan, Designer, Teacher Edition
Sally Bess, Designer, Teacher Edition
Charlie Taliaferro, Design Associate,
Teacher Edition

Image Acquistions
Curtis Riker, Director
Jeannie Taylor, Photo Research Supervisor
Elaine Tate, Art Buyer Supervisor

Design New Media
Ed Blake, Design Director
Kimberly Cammerata, Design Manager
Michael Rinella, Senior Designer

Media Design
Richard Metzger, Design Director
Chris Smith, Senior Designer

Graphic Services
Kristen Darby, Director
Jeff Robinson, Senior Ancillary Designer
Jane Dixon, Image Designer

Cover Design
Joe Melomo, Design Director

**Design Implementation and
Page Production**
AARTPACK, Inc.

Electronic Publishing
EP Manager
Robert Franklin

EP Team Leaders
Juan Baquera
Sally Dewhirst
Christopher Lucas
Nanda Patel
JoAnn Stringer

Senior Production Artists
Katrina Gnader
Lana Kaupp
Kim Orne
Production Artists
Sara Buller
Ellen Kennedy
Patty Zepeda
Quality Control
Barry Bishop
Becky Golden-Harrell
Angela Priddy
Ellen Rees

New Media
Armin Gutzmer, Director of Development
Melanie Baccus, New Media Coordinator
Lydia Doty, Senior Project Manager
Marsh Flournoy, Quality Assurance Analyst
Cathy Kuhles, Technical Assistant
Tara F. Ross, Senior Project Manager

Ancillary Development and Production
GTS, Inc., Boston, Massachusetts

PHOTO & ART CREDITS

Art

Abbreviated as follows: (t) top; (b) bottom; (l) left; (r) right; (c) center.

All art unless otherwise noted by Holt, Rinehart, and Winston.

Chapter 1: Page 6-8, Kristy Sprott; 16 (br), David Chapman; 22-26, Kristy Sprott; 28 (flow chart), David Chapman, 28 (molecules), Kristy Sprott; 32, Kristy Sprott; 34, David Chapman. **Chapter 2:** Page 39,40, Kristy Sprott; 41-46, David Chapman; 48-53, Kristy Sprott; 70, David Chapman. **Chapter 3:** Page 74 (br), Kristy Sprott; 74 (bl), David Chapman; 75 (cr), Kristy Sprott; 75 (br), David Chapman; 76-80, Kristy Sprott; 81 (bl), Uhl Studios, Inc.; 83-91, Kristy Sprott; 92, Uhl Studios, Inc.; 93, David Chapman; 96, J/B Woolsey Associates; 97, David Chapman. **Chapter 4:** Page 117, Jack Scott; 119-130, David Chapman; 133, Kristy Sprott; 134, David Chapman; 135 (t), Kristy Sprott; 135 (b), David Chapman; 136-140 David Chapman; 143-144, Kristy Sprott; 146, 151, David Chapman; 153, Kristy Sprott; 154, David Chapman.**Chapter 5:** Page 160, David Chapman; 161, Kristy Sprott; 162-163, David Chapman; 169 (flow chart), David Chapman, (molecules), Kristy Sprott; 172, 174 (zooms), 183, Kristy Sprott; 186, David Chapman. **Chapter 6:** Page 190-192, Kristy Sprott; 194, David Chapman; 195-212, Kristy Sprott. **Chapter 7:** Page 236-238, Kristy Sprott; 241, David Chapman; 242, Kristy Sprott; 246-256, David Chapman. **Chapter 8:** Page 261-297, Kristy Sprott. **Chapter 9:** Page 315, (zooms) Kristy Sprott, (realia) Stephen Durke/Washington Artists; 321, 323, Stephen Durke/Washington Artists 326, Kristy Sprott; 334, David Chapman. **Chapter 10:** Page 339, Leslie Kell; 341, David Chapman; 348, Kristy Sprott; 351, Uhl Studios, Inc.; 353, Stephen Durke/Washington Artists; 363-366, Kristy Sprott; 373,374, David Chapman; **Chapter 11:** Page 379, 380, Kristy Sprott; 381, Stephen Durke/Washington Artists; 384-390, Kristy Sprott; 394, David Chapman; 399, 400 (tc), Kristy Sprott; 400 (bl), Stephen Durke/ Washington Artists; 401-412, David Chapman. **Chapter 12:** Page 417 Kristy Sprott; 418 (t,b),419, Stephen Durke/Washington Artists; 421, Kristy Sprott; 422, David Chapman; 424 (t), Stephen Durke/Washington Artists; 424 (br), 427, David Chapman; 429 (t), Stephen Durke/Washington Artists; 429 (b), David Chapman; 431, Kristy Sprott; 434, David Chapman; 437, Kristy Sprott; 439-450, Stephen Durke/Washington Artists. **Chapter 13:** Page 455, 456, 462, 464, 471, Kristy Sprott; 472, David Chapman; 473, Kristy Sprott; 474, David Chapman; 476, 479, 480, 482, Kristy Sprott; 483, David Chapman; 484, David Chapman/Kristy Sprott; 485, Kristy Sprott. **Chapter 14:** Page 496, 497, Kristy Sprott; 498, 505, David Chapman; 507, 513, Kristy Sprott; 514, David Chapman; 515-524, Kristy Sprott. **Chapter 15:** Page 531-539, Kristy Sprott; 540, David Chapman; 549, Kristy Sprott; 551, David Chapman; 557-572, Kristy Sprott.

Chapter 16: Page 578, Kristy Sprott; 580, David Chapman; 588 (clocks) David Chapman, (zooms) Kristy Sprott; 591 (chart) David Chapman, (molecules) Kristy Sprott; 592, Kristy Sprott; 594, David Chapman. **Chapter 17:** Page 605, 608, Kristy Sprott; 613 (bl), Stephen Durke/Washington Artists; 613 (br), Kristy Sprott; 614 (art), Stephen Durke/ Washington Artists; 614 (zoom), Kristy Sprott; 615 (tr), Stephen Durke/Washington Artists; 616 (art), Stephen Durke/Washington Artists; 616 (zooms), Kristy Sprott; 616, Kristy Sprott; 617-620 (art), Stephen Durke/Washington Artists; 620 (zooms), Kristy Sprott; 626 (art), Stephen Durke/Washington Artists; 626 (zooms), Kristy Sprott; 628-635, Stephen Durke/Washington Artists; 638, David Chapman. **Chapter 18:** Page 643-644, Kristy Sprott; 645-648, David Chapman; 649,650, Kristy Sprott; 651, David Chapman; 655, Uhl Studios, Inc.; 661, 663, David Chapman. **Chapter 19:** Page 680-700 Kristy Sprott. **Chapter 20:** Page 717-720, Stephen Durke/Washington Artists; 722, Christy Krames; 726-728, 733, Morgan-Cain & Assoc.; 743, John White/The Neis Group; 744, David Chapman. **Labs:** Page 758, Uhl Studios, Inc.; 763, David Chapman; 764, 788, Uhl Studios, Inc. **Appendix:** Page 831, Kristy Sprott.

Photo

Abbreviated as follows: (t) top; (b) bottom; (l) left; (r) right; (c) center.

All photos unless otherwise noted by Holt, Rinehart, and Winston.

Cover and page i: Charlie Winters; **Table of Contents:** Page vi, © Russell Johnson (t); © Kraft/Explorer/Photo Researchers, Inc. (b); vii, IBM Research, Almaden Research Center (t); © Science Pictures Limited/CORBIS (b); viii, © CORBIS (t); © Adastra/Getty Images (b); ix, © Ted Mahieu/Corbis Stock Market (t); © Roland Birke/Peter Arnold (b); x, © David Nardini /Getty Images (t); Peter Van Steen/HRW (b); xi, © Charles O'Rear/ CORBIS (t); Brand X (b); xii, © H. Abernathy/H. Armstrong Roberts Stock Photography (t); © Jerry Shulman/ SuperStock (b); **Chapter 1:** Page 2, © Russell Johnson; 4, Victoria Smith/HRW; 6, Peter Van Steen/HRW; 7, Victoria Smith/HRW; 8, Charlie Winter/HRW; 9 (all), Sergio Purtell/Foca/HRW; 10, Victoria Smith/HRW; 10 (inset), Victoria Smith/HRW; 11, Charlie Winters/HRW; 12, Andrew Brookes Creative/Corbis Stock Market; 15, Victoria Smith/HRW; 17, Denis Fagan/HRW; 19, Charlie Winter/HRW; 20, © S. Feld/H. Armstrong Roberts Stock Photography; 21, © Larry Lefever/Grant Heilman Photography; 22 (tl), © Leonard Lessin/Peter Arnold; 22 (tm), © Tom Pantages Photography; 22 (tr) © Charles D. Winters/Photo Researchers, Inc.; 23 (l), Victoria Smith/HRW; 23 (r), © Steve Vidler/SuperStock; 24, © Francisco Cruz/SuperStock; 25 (l), © Neal Mishler/FPG International/Getty Images; 25 (r), Sergio Purtell/Foca/HRW; 26 (both), Charlie Winter/HRW; 29, © A. Ramey/Photo Edit. **Chapter 2:** Page 36, © Kraft/Explorer/Photo Researchers, Inc.; 38, Charlie Winters/HRW; 39, Ice Sculpture by Duncan Hamilton. Photograph by Mike Venables; 40, Charlie Winters/HRW; 42, © Bettmann/CORBIS; 43, © Ana/The Image Works; 48, © Andre Jenny/South Stock/PictureQuest; 49, Photo Researchers, Inc.; 50, HRW Photo; 51, © Photo by Joe Raedle/Newsmakers/Getty Images ; 52, © Newsmakers/Getty Images; 54, © Kristen Brochmann/Fundamental Photography; 56, Charlie Winters/HRW; 57, Victoria Smith/HRW; 64, Scoones/SIPA Press. **Chapter 3:** Page 72, IBM Research, Almaden Research Center; 75, Peter Van Steen/HRW; 79, © Kelvin Murray/Stone/ Getty Images; 80, Charlie Winters/HRW; 82 (l), Sergio Purtell/Foca; 82 (r), © Steve Bronstein/Getty Images; 100, Sam Dudgeon/HRW; 104, John Macfie/ImageState; 105 (t), E.R. Degginger/ Color-Pic, Inc.; 105 (b), © Carl Frank/ Photo Researchers, Inc. **Chapter 4:** Page 114, © Phil Schermeister/CORBIS; 116 (all), Charlie Winters/HRW; 123, © Linda S. Nye/Phototake; 125, 126, Charlie Winters/HRW; 127 (t), Charlie Winters/ HRW; 127 (b), © Bob Burch/Index Stock Imagery; 128, © Tom Pantages; 129, © Leonard Lessin/Peter Arnold; 130, © Roger Ressmeyer/CORBIS; 131, © Fritz Henle/Photo Researchers, Inc.; 132 (all), Charlie Winters; 142, NASA; 143, © Roger Russmeyer/ CORBIS; 148 (l), NASA; 148 (r), © Gabe Palmer/Corbis Stock Market. **Chapter 5:** Page 156, © Science Pictures Limited/CORBIS; 158 (both), 159, 164 (both), Charlie Winters/ HRW; 166, © Eric Simmons/Stock Boston/PictureQuest; 170, © David Young-Wolff/Photo Edit; 172 (all), Charlie Winters/HRW;174 (l), © Edward R. Degginger/ Bruce Coleman Inc./ PictureQuest; 174 (r), © Anthony Mercieca/ SuperStock; 181, Rick Lance/ Phototake; 181 (inset), Sam Dudgeon/HRW. **Chapter 6:** Page 188, Peter Van Steen/HRW; 190, © T. McCarthy/Custom Medical Stock Photo; 191, 193 (both), 195, Victoria Smith/HRW; 197 (all), Charlie Winters/HRW; 199, Courtesy of Edgar Fahs Smith Collection/University of Pennsylvania Library; 200, Sam Dudgeon/HRW; 203, 204 (both), Sergio Purtell/Foca; 206, © IFA/Peter Arnold, Inc.; 208, Victoria Smith/ HRW; 214, © Jim Karageorge/FPG International/ Getty Images. **Chapter 7:** Page 222, © CORBIS; 224, Charlie Winters/HRW; 225, Victoria Smith/HRW; 234, 236 (all), 241 (both), Charlie Winters/HRW; 244, Sam Dudgeon/HRW; 249, © Jim Karageorge/ FPG International/Getty Images. **Chapter 8:** Page 258, © Adastra/Getty Images; 260, © Phil Schermeister/CORBIS; 261 (both), ©1988 Richard Megna Fundamental Photographs; 262 (l), Sam Dudgeon/HRW; 262 (r), Charlie Winters/HRW; 263, 264 (both), Sam Dudgeon/HRW; 272 (both), 275,

Periodic Table of the Elements

Key:

	6	— Atomic number
	C	— Symbol
	Carbon	— Name
	12.0107	— Average atomic mass
	$[He]2s^2 2p^2$	— Electron configuration

Period

Group 1

1
H
Hydrogen
1.007 94
$1s^1$

Group 2

3
Li
Lithium
6.941
$[He]2s^1$

4
Be
Beryllium
9.012 182
$[He]2s^2$

11
Na
Sodium
22.989 770
$[Ne]3s^1$

12
Mg
Magnesium
24.3050
$[Ne]3s^2$

Group 3

21
Sc
Scandium
44.955 910
$[Ar]3d^1 4s^2$

39
Y
Yttrium
88.905 85
$[Kr]4d^1 5s^2$

57
La
Lanthanum
138.9055
$[Xe]5d^1 6s^2$

89
Ac
Actinium
(227)
$[Rn]6d^1 7s^2$

Group 4

22
Ti
Titanium
47.867
$[Ar]3d^2 4s^2$

40
Zr
Zirconium
91.224
$[Kr]4d^2 5s^2$

72
Hf
Hafnium
178.49
$[Xe]4f^{14} 5d^2 6s^2$

104
Rf
Rutherfordium
(261)
$[Rn]5f^{14} 6d^2 7s^2$

Group 5

23
V
Vanadium
50.9415
$[Ar]3d^3 4s^2$

41
Nb
Niobium
92.906 38
$[Kr]4d^4 5s^1$

73
Ta
Tantalum
180.9479
$[Xe]4f^{14} 5d^3 6s^2$

105
Db
Dubnium
(262)
$[Rn]5f^{14} 6d^3 7s^2$

Group 6

24
Cr
Chromium
51.9961
$[Ar]3d^5 4s^1$

42
Mo
Molybdenum
95.94
$[Kr]4d^5 5s^1$

74
W
Tungsten
183.84
$[Xe]4f^{14} 5d^4 6s^2$

106
Sg
Seaborgium
(263)
$[Rn]5f^{14} 6d^4 7s^2$

Group 7

25
Mn
Manganese
54.938 049
$[Ar]3d^5 4s^2$

43
Tc
Technetium
(98)
$[Kr]4d^6 5s^1$

75
Re
Rhenium
186.207
$[Xe]4f^{14} 5d^5 6s^2$

107
Bh
Bohrium
(264)
$[Rn]5f^{14} 6d^5 7s^2$

Group 8

26
Fe
Iron
55.845
$[Ar]3d^6 4s^2$

44
Ru
Ruthenium
101.07
$[Kr]4d^7 5s^1$

76
Os
Osmium
190.23
$[Xe]4f^{14} 5d^6 6s^2$

108
Hs
Hassium
(265)
$[Rn]5f^{14} 6d^6 7s^2$

Group 9

27
Co
Cobalt
58.933 200
$[Ar]3d^7 4s^2$

45
Rh
Rhodium
102.905 50
$[Kr]4d^8 5s^1$

77
Ir
Iridium
192.217
$[Xe]4f^{14} 5d^7 6s^2$

109
Mt
Meitnerium
(268)†
$[Rn]5f^{14} 6d^7 7s^2$

19
K ·
Potassium
39.0983
$[Ar]4s^1$

20
Ca
Calcium
40.078
$[Ar]4s^2$

37
Rb
Rubidium
85.4678
$[Kr]5s^1$

38
Sr
Strontium
87.62
$[Kr]5s^2$

55
Cs
Cesium
132.905 45
$[Xe]6s^1$

56
Ba
Barium
137.327
$[Xe]6s^2$

87
Fr
Francium
(223)
$[Rn]7s^1$

88
Ra
Radium
(226)
$[Rn]7s^2$

† Estimated from currently available IUPAC data.

* The systematic names and symbols for elements greater than 109 will be used until the approval of trivial names by IUPAC.

58
Ce
Cerium
140.116
$[Xe]4f^1 5d^1 6s^2$

59
Pr
Praseodymium
140.907 65
$[Xe]4f^3 6s^2$

60
Nd
Neodymium
144.24
$[Xe]4f^4 6s^2$

61
Pm
Promethium
(145)
$[Xe]4f^5 6s^2$

62
Sm
Samarium
150.36
$[Xe]4f^6 6s^2$

90
Th
Thorium
232.0381
$[Rn]6d^2 7s^2$

91
Pa
Protactinium
231.035 88
$[Rn]5f^2 6d^1 7s^2$

92
U
Uranium
238.0289
$[Rn]5f^3 6d^1 7s^2$

93
Np
Neptunium
(237)
$[Rn]5f^4 6d^1 7s^2$

94
Pu
Plutonium
(244)
$[Rn]5f^6 7s^2$

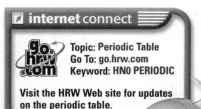